International Conference on Biomedical Engineering 2019

Journal of Physics: Conference Series Volume 1372

Tanjung Bungah, Malaysia
26 - 27 August 2019

Editors:

Chong Yen Fook
Haniza Binti Yazid
Saufiah Binti Abdul Rahim
Nazahah Binti Mustafa
Siti Khadijah Zaaba
Ahmad Nasrul Bin Norali
Anas Bin Mohd Noor

ISBN: 978-1-7138-1129-9
ISSN: 1742-6588

Printed from e-media with permission by:

Curran Associates, Inc.
57 Morehouse Lane
Red Hook, NY 12571

CURRAN ASSOCIATES INC. proceedings .com

Some format issues inherent in the e-media version may also appear in this print version.

Institute of Physics
Dirac House, Temple Back
Bristol BS1 6BE UK

Phone: 44 1 17 929 7481
Fax: 44 1 17 920 0979

techtracking@iop.org

Additional copies of this publication are available from:

Curran Associates, Inc.
57 Morehouse Lane
Red Hook, NY 12571 USA
Phone: 845-758-0400
Fax: 845-758-2633
Email: curran@proceedings.com
Web: www.proceedings.com

TABLE OF CONTENTS

Journal of Physics: Conference Series **1372** (2019) 011001 doi:10.1088/1742-6596/1372/1/011001

Preface

With great pleasure, we would like to welcome all delegates to our 2019 International Conference on Biomedical Engineering (ICoBE). It is a great pleasure and honor for the Biomedical Eletronic Engineering Programme, under the purview of School of Mechatronic Engineering, Universiti Malaysia Perlis (UniMAP) to organize this conference.

ICoBE conference covers a wide range of current research areas which covers important elements in the related engineering fields. This also reflects the diverse research interests conducted under the School. This conference gathers experts from different countries to discuss the state of the art and to present new research findings and perspectives. We are sure all participants will use this conference as a platform to present their work and address current problems in this challenging and promising field and subsequently create new avenues for the commercialization of new products.

We want to express our gratitude to all reviewers, both local and international, for their time and effort in reviewing the papers. Their feedbacks to the authors have been excellent and useful to ensure the high quality of papers for the conference. Our sincere thank also goes to all sponsors for their generosity. Last but not least, we very much appreciate and thank UniMAP and all members of the organizing committee for their hard work, cooperation and suggestions in ensuring the success of this conference.

Thank you and see you again in our next conference.

Dr. Chong Yen Fook
Dr. Haniza Binti Yazid
Dr. Saufiah Binti Abdul Rahim
Dr. Nazahah Binti Mustafa
Dr. Siti Khadijah Zaaba
Mr. Ahmad Nasrul Bin Norali
Mr. Anas Bin Mohd Noor

Editors
2019 International Conference on Biomedical Engineering (ICoBE)

Sponsor or funding acknowledgements

We would like to thank our sponsors, who has supported the 2019 Internationa Conference on Biomedical Engineering (ICoBE).

Main Sponsor: STAR LABS BIOSCIENCE

aBy res ABYRES
Absolute Binary Resources

CENFOTEC SDN BHD

NORTHERN RUNNERS

TMS PASSION IN SCIENCE

NEXT LEVEL Technologies Sdn Bhd
Redefining Engineering for Healthcare

NATURAL INTENTION SDN. BHD.
1057227-M
No. 11592, Kampung Tersusun, Jalan Kampung Dew, 34700 Simpang, Taiping, Perak.

Conference Photographs

icomms'19
ICADME
5TH JOINT INTERNATIONAL CONFERENCE
ON
MECHATRONIC, MECHANICAL AND BIOMEDICAL ENGINEERING
(JICOMMBE 2019)
26TH- 27TH AUGUST 2019
RAINBOW PARADISE BEACH RESORT, PENANG
ORGANIZER:
SPONSORSHIP: STAR LABS
PARADISE

icomms'19
ICADME
ICoBE

2019
(ICoBE 2019)

Feature Extraction using Discriminant Graph
Laplacian PCA with Application to
Biomedical Datasets

Peer review statement

All papers published in this volume of *Journal of Physics: Conference Series* have been peer reviewed through processes administered by the proceedings Editors. Reviews were conducted by expert referees to the professional and scientific standards expected of a proceedings journal published by IOP Publishing.

Stress Detection using Galvanic Skin Response: An Android Application

Roy Francis Navea*, Paul John Buenvenida, Christian Dave Cruz

De La Salle University, Manila, Philippines

*roy.navea@dlsu.edu.ph

Abstract. Stress is one of the factors that affect human health in many aspects. It is considered as one of the culprits in increasing the risk of getting sick that could probably lead to critical physical or mental illnesses. Stress can be experienced everywhere and in different circumstances. Hence, stress should be controlled and managed by monitoring its progress or regress. Physiological information can be used to determine stress levels. One of these is the Galvanic Skin Response (GSR) that utilizes skin conductance which is known to be directly involved in the emotional behavioural regulation in humans. In this study, a method on how to determine stress when a person is engaged in mobile communication is proposed. An Android application was developed that is capable of determining the stress level of a person while doing SMS composition. GSR data were utilized and the performance of the proposed method was found of no significant difference with a commercially available device. Factors like phone size and period of texting was investigated and were found out that these only contribute an extremely low level of stress. The developed App could be used to determine stress levels especially if emotional conversations are considered.

1. Introduction

Modern communication systems transformed the way people talk and perceive information. This is brought about by the advanced telecommunication devices which are easily accessible by hand, the smartphones. A recent survey shows that 66.72% of the world owns a cell phone with mobile connections close to 9 billion consisting of unique mobile users approaching 5.2 billion out of the approximate 7.7 billion population of the world. People ages between 18-29 are all found to have mobile phones while 94% own a smartphone only [1]. The demand for more voice and data services is ever increasing parallel to the increasing contents that the Internet can provide.

The most common method of communication is via text messaging or technically known as SMS (short message service) which is available on-line or off-line. This is viewed by many people as a key part of daily social life. Receiving a text message somehow provides an instant urge to read and send a reply to the sender regardless of whatever the recipient is doing [2]. Nonetheless, it is anticipated to experience stress during this instance in communication. A direct link exists between stress levels and texting using mobile phones [3]. Texting induces stress to a person that probably varies with the period of texting time and the size of the phone aside from the thoughts of the message received. Long term exposure to mental stress could lead to some health problems as well as disruptions of the different systems of the human body [4]. Aside from mental issues, the physical and social well-being of a person can also be affected.

Physiological information can be used to identify if a person is stressed or not. They are used to possibly predict increasing or decreasing levels of stress once a person is engaged in a certain activity. In a more practical and economical way, the electrical conductance of the skin can be utilized through what is known as galvanic skin response (GSR) or electro dermal activation (EDA) [4]–[6]. GSR or EDA is the measure of the continuous variations in the electrical characteristics (conductance) of the skin caused by the variation of the sweating activity of the human body. This concept is based on the assumption that skin resistance varies with the state of the sweat glands in the skin. The resistance was just reciprocated to determine the conductance. This concept is based on the activity of the Autonomic Nervous System (ANS) with which if its sympathetic branch is highly aroused, the activity of the sweat gland follows. This results to decrease in skin resistance thus, increasing skin conductance. This instance is used to measure human sympathetic nervous system (SNS) responses which are directly involved in the emotional behavioural regulation in humans. Hence, GSR data were used in relation to mental state, such as stress, drowsiness and engagement [7].

This study would like to look at the effect of the ergonomic design of the mobile phone in terms of its length, height and width as well as the length of time that a person is engaged in a texting task. Two types of mobile phones were used; conventional phones and smartphones. Phone size matters and has been a constantly changing design parameter since mobile phones were made available in the market. Mobile phones in the 1990's are big and bulky. Their size was reduced around the year 2000's. But when the smartphones were invented with touchscreens, phone sizes varied proportional to the area of its screen [8]. Keypads are now made available on screens and other characters are available for additional means of conveying information. The physical appearance of mobile phones has evolved to further improve user-experience and ergonomics.

The utilization of conventional phones that uses T9 keypads versus the modern touchscreen smartphones with "qwerty" keypads poses an issue of which among the two induces lesser or greater stress levels to the users especially when they are engaged in emotional conversations. Factors like the size or dimensions of the phone, and the time of composing messages were investigated in relation to the possible stress level induced. GSR was used as the physiological information which was then processed to assess stress levels. Moreover, a stress level detection application was developed in the Android platform which was benchmarked from the performance of a commercial grade eSense GSR skin response application [9].

2. Methods

Stress level detection using GSR is guided by the block diagram shown in Figure 1. An initial skin conductance is required to establish baseline readings when the person is in a relaxed state. A specialized sensor attached to the fingers of the participant and connected to the headphone socket of the mobile phone was used to gather the GSR. Readings should be obtained when the person is idle or basically at rest. GSR is taken again when the person is engaged in a texting task or when composing a pre-defined text message. These readings are then pre-processed to determine the upper and lower thresholds or cut-off values.

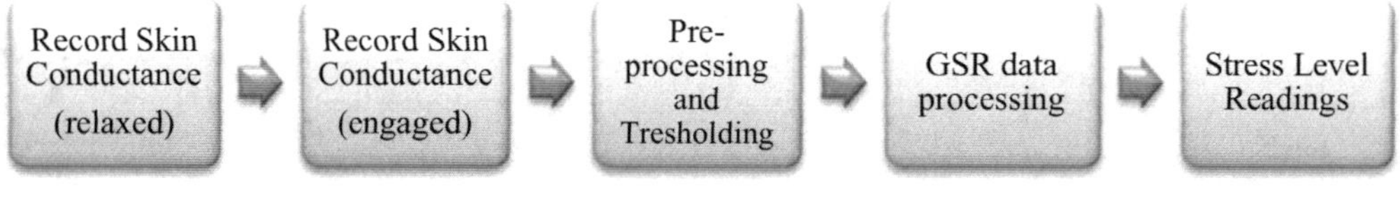

Figure 1. Stress level detection block diagram

A total of 25 tests were conducted from five different participants doing a texting task five times. The baseline data was recorded when the participants are in a relaxed position for a minute and the actual test data follows with a variable period of time. Two apps were used: the eSense App and the GSR App. Initial data acquisition was performed in a quiet and controlled environment. This was performed to calibrate the conductance readings of the GSR App as compared to the commercial grade eSense App. Six mobile phones were used. Three mobile phones are of the conventional type while the other three are of the smartphone type.

2.1 GSR data processing

The next part is to convert the obtained readings into a set of values that can be graphed similar to what the eSense App does. This process took several iterations before coming close to the commercial grade output of the eSense App. The results were repeatedly analyzed and the processing was recalibrated in order to arrive at values similar to how the eSense outputs were obtained.

Five participants were invited to test the system by texting the predefined message 5 times. After all the GSR data were gathered, the values were converted into its corresponding stress levels. This was done by calculating the mean maximum and mean minimum GSR's of the participants which were obtained from both apps. Given the minimum and maximum bounds, the interval was equally divided into seven to provide stress level points. The average GSR values of the participants during text composition were then marked against their corresponding stress levels in the scale. All stress levels data for each app were then statistically compared.

2.2 GSR application program logic

The GSR sensor sends data to the mobile phone through the audio jack and socket terminals. The recording can be started and can be stopped by the user at any given time. The readings are saved in a *csv* file format which can be accessed using the smartphone itself or a computer for further processing and analysis. They also serve as inputs of the real-time graph displayed on screen. The graph shows a skin conductance vs. time plot with a 0.2-second interval.

The program logic flowchart of the application is in Figure 2. The start / stop button is used to trigger the start and stop of the data recording task of the application. The GSR App will continue to read skin conductance and calculate the stress levels until the start / stop button is pressed. Results are saved in a csv file and the user can have the option to view or share this data. Initial sharing feature is via electronic mail only. In addition, a new session can be started for new set of data or for another user. The threshold values can be altered for new users. This was made possible because of the understanding that people have difference skin conductance thresholds. This feature allows the application to be customized according to who is using it. Results may also vary depending on the person's skin conductance. The stress level of one person may or will be really different from the other. Hence, the recorded files can be stored for future reference mapping the history of stress level tests done using the application.

2.3 Hardware overview

There are six phones used in this study. Three phones (Phones 1-3) are conventional and the other three are smartphones (Phones 4-6). Conventional phones use T9 keypads while smartphones uses "QWERTY" touchscreen keypads. These phones were used to gather data using the eSense App and the GSR App. Figure 3 shows the mobile phones used in the study. Among the conventional phones, Phone 1 is the biggest and Phone 3 being the smallest. Among the smartphones, Phone 6 is the biggest while Phone 4 is the smallest. The brand and models are different in consideration of design variability.

2.4 Experimental procedures

Thirty participants were randomly selected to test the developed application. The participants were asked to compose a text message using conventional and smart phones. The GSR App and the eSense App were both used to gather skin conductance data and later be processed to determine the stress level. The stress levels of the 30 participants during the texting task were averaged per phone and were compared to the average baseline of the participant to determine if there is a significant difference between the reading obtained between the GSR App and the eSense App.

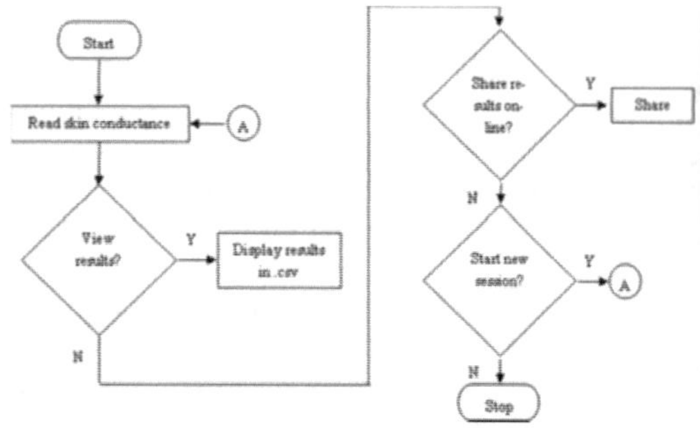

Figure 2. Stress level detection block diagram

Figure 3. Mobile Phones used

3. Results and discussion

The GSR App was developed in Android Studio. The graphical user interface (GUI) is shown in Figure 4. The parts of the App are labelled accordingly and their functions are literally understandable.

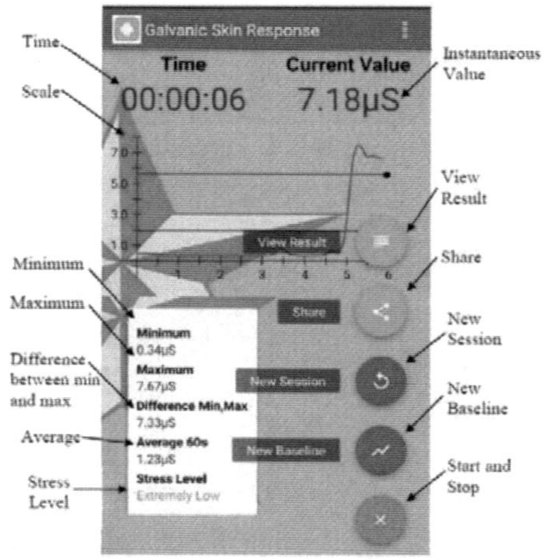

Figure 4. GSR App Graphical User Interface

3.1 eSense App and GSR App result validation

Figure 5 shows the graphical representation of some of the average skin conductance of the participants as recorded by the eSense and GSR Apps using Phones 1-6, respectively. The left bar of each pair is the result of the eSense App while the other of from the GSR App.

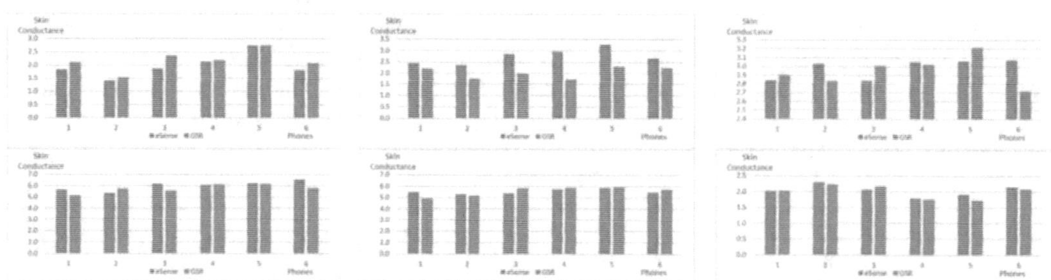

Figure 5. Average skin conductance of the participants

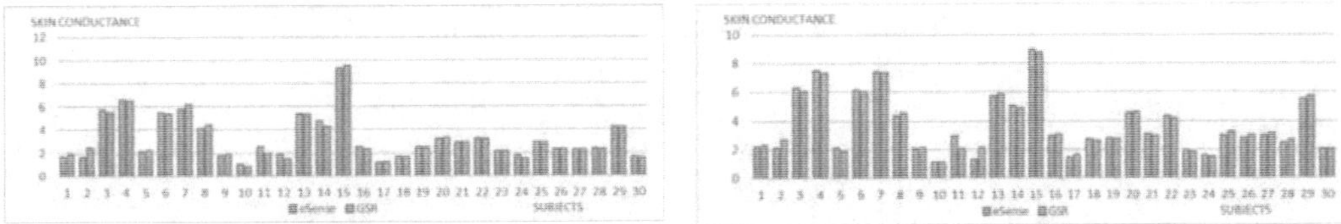

(a) (b)

Figure 6. Average skin conductance using (a) conventional phone, (b) smartphone

Figure 6 shows the average skin conductance of the participants when using conventional phones and smartphones, respectively. Results were obtained using the eSense and GSR apps. Again, the left bar of each pair is the result of the eSense App while the other of from the GSR App.

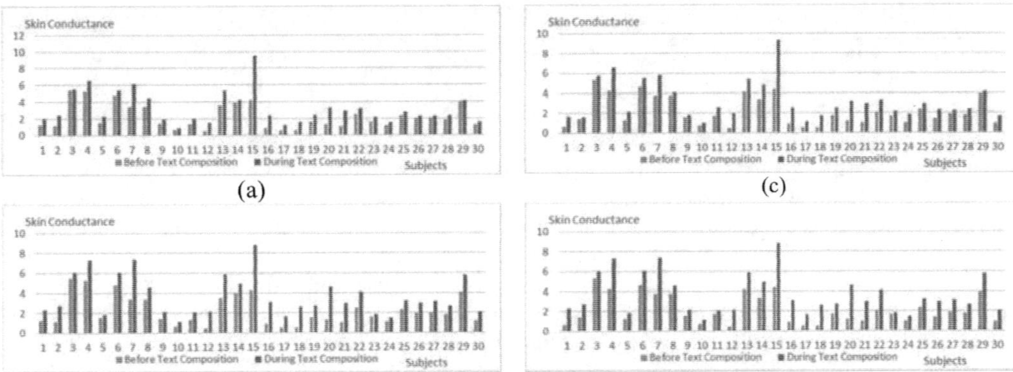

(a) (c)

(b) (d)

Figure 7. Skin conductance in the (a), (b) GSR App and (c),(d) eSense App before and during texting using conventional and smart phones, respectively

Figure 7 shows the increased level of skin conductance when the participants are engaged in a texting task using conventional phones and smart phones, respectively. Results were obtained using the GSR and eSense Apps. The increase in skin conductance is evident in the graphs.

3.2 Test for significant difference

To validate the results of the GSR App, the skin conductance measurements were compared to the skin conductance measurements of the eSense App. A total of 180 tests were conducted involving 30 participants who used 6 different mobile phones to perform a texting task.

The t-test was used in order to statistically determine and test the null hypothesis (Ho) that there is no significant difference between the skin conductance measured using the eSense App and GSR App before and during text composition. The t-values obtained ranges from -0.0036 to 0.1278. Results fall below the t-critical which is 1.6716. This is indicative that there is no significant difference between the skin conductance measured by eSense App and the GSR App during a texting task for all the phones considered. The developed GSR App can perform at par with the commercial grade eSense App.

3.3 Stress level analysis

Stress levels were determined using of a 7-point equally divided Likert scale. Each average skin conductance value was converted and gauged according to its corresponding stress levels as defined by the range of values in Table 1. Extremely low stress levels starts with 1.0 going up to 7.0 to indicate extremely high stress levels. In Table 2, the average stress levels of the participants were averaged per phone used. Results show an extremely low level of stress when the participants are using the phones doing a texting task using the mobile phones as presented.

Table 1. Stress level assessment scale

Description	Extremely Low	Very Low	Slightly low	Moderate	Slightly high	Very high	Extremely high
Range	1.0 – 1.8	1.9 – 2.7	2.8 – 3.6	3.7 – 4.5	4.6 – 5.4	5.5 – 6.3	6.4 - 7.0

Table 2. Average stress levels

	Phone 1	Phone 2	Phone 3	Phone 4	Phone 5	Phone 6
eSense App	1.0333	1.0333	1.0667	1.0667	1.1000	1.1333
GSR App	1.0333	1.0667	1.0333	1.0667	1.0667	1.1000

4. Conclusion

The developed Android application (GSR App) can simulate the performance of the commercial Galvanic Skin Response (eSense) application. The GSR App can record skin conductance and systematically determine the stress level of a person during a texting task. The GSR App was able to display the stress levels of an individual on the smartphone. The stress level was based on the value of skin conductance of the person while doing a texting task. The skin conductance recorded by the eSense and GSR App was compared and results show that there is no significant difference between the results given by the two apps. The texting task provides an extremely low level of stress though there are evidences that task increased some of the participants' skin conductance values. However, it is not enough to elevate the low stress levels to higher level.

References

[1] Worldometers 2019 Global Smartphone Penetration Data https://www.bankmycell.com/blog/how-many-phones-are-in-the-world.
[2] Vitelli R 2019 Stress, Texting, and Being Social *Psychology Today* https://www.psychologytoday.com/us/blog/media-spotlight/201312/stress-texting-and-being-social.
[3] Vahedi Z and Saiphoo A 2018 The association between smartphone use, stress, and anxiety: A meta-analytic review, *Stress Heath Vol. 34*, **3** 347–358
[4] Tang T B, Yeo L W and Lau D J H 2014 Activity awareness can improve continuous stress detection in galvanic skin response *Proceedings of IEEE Sensors* 1980–1983
[5] Subramanya K, Bhat V V, and Kamath S 2013 A wearable device for monitoring galvanic skin response to accurately predict changes in blood pressure indexes and cardiovascular dynamics *Annual IEEE India Conference* **2** 1–4.
[6] Fernandes A, Helawar R, Lokesh R, Tari T, and Shahapurkar A V 2015 Determination of stress using Blood Pressure and Galvanic Skin Response *International Conference on Communication and Network Technologies* 165–168
[7] BrainSigns 2018 Galvanic Skin response (GSR) https://www.brainsigns.com/en/science/s2/technologies/gsr.
[8] Mancha A 2015 Mobile Phone Size Evolution https://www.dailyinfographic.com/mobile-phone-size-evolution
[9] WorldWorks 1996 eSense GSR Skin Response http://www.lifematters.com/esense-gsr.asp.

Journal of Physics: Conference Series **1372** (2019) 012002 doi:10.1088/1742-6596/1372/1/012002

Feature Extraction Using Discriminant Graph Laplacian Principal Component Analysis with Application to Biomedical Datasets

Muhammad Aminu[1] and Noor Atinah Ahmad[2*]

[1,2]School of Mathematical Sciences, Universiti Sains Malaysia, 11800, Penang, Malaysia

*nooratinah@usm.my

Abstract. In this paper, we propose a manifold learning method called discriminant graph Laplacian principal component analysis (DGLPCA) for feature extraction. The proposed method projects high dimensional data into a lower dimensional subspace while preserving much of the intrinsic structure of the data. Moreover, DGLPCA integrates maximum margin criterion into its objection function to improve class separability in the lower dimensional space. The effectiveness of the proposed method is demonstrated on two publicly available biomedical datasets taken from UCI machine learning repository. The results show that our proposed method provides more discriminative power compared to other similar approaches.

1. Introduction

Biomedical datasets are rapidly growing and becoming more and more complex [1]. These datasets are generally associated with high-dimensional features and multiple classes. The ill-structured nature of the biomedical data requires intelligent machine learning and data mining algorithms for automated analysis in order to make logical inferences from the stored raw data [2]. A diverse set of machine learning and data mining algorithms have been previously used to extract useful information from biomedical data [3]. The research in this area continues to grow partly because machine learning makes it possible to model data extremely well, without using strong assumptions about the modeled system.

Feature extraction is a machine learning technique that maps the original high dimensional data to a lower dimensional subspace while maximizing the variability of the new feature space. Several feature extraction methods have been used in the literature [4-6]. The most widely used feature extraction method is the principal component analysis (PCA). It is well-known that PCA preserves only global Euclidean structure of data. The method is not designed to detect the underlying manifold structure hidden in the high dimensional data. The efforts to include manifold learning techniques into PCA give rise to several variants of graph-Laplacian PCA [7-9]. However, this transformation matrix does not include class information and therefore may not be optimal for classification. When data comes with class labels, we would like to include this information into our feature extraction method so that the classification ability of the method is improved. Although it has been shown that the PgLPCA method in [9] is superior to the graph-Laplacian PCA (GLPCA) in [8] in terms of both clustering performance and noise reduction, neither method actually take classification explicitly into consideration. PgLPCA and GLPCA are completely unsupervised with regards to the class label information, and this might sometimes lead to unsatisfactory performances of the methods in classification problems. Also, both PgLPCA and GLPCA use the k -nearest neighbor method to construct the graph. However, this approach is non-parameter free and sensitive to noise.

In this paper, we propose a novel algorithm for feature extraction and discriminant analysis called discriminant graph Laplacian PCA (DGLPCA). DGLPCA is essentially developed from PCA but has significant advantages over the ordinary PCA method. Benefiting from the objective of maximum margin criterion (MMC) [10], we integrate the objective function of MMC into GLPCA to improve its classification ability. We show how the solutions to this method can be obtained by solving a generalized eigenvalue problem.

The following points highlight the contributions of this paper:

- A new approach for PCA based on graph embedding is developed, which is called discriminant graph Laplacian PCA (DGLPCA). DGLPCA encodes discriminant information into the objective of graph Laplacian PCA to improve its classification ability.
- Instead of using k -nearest neighbour method for graph construction, in our proposed DGLPCA method, we construct a graph based on the class information of samples that capture as much of the discriminative information of data as possible. Our approach for constructing the graph is parameter free and can capture the discriminative information of data to a higher extent.
- We use the same approach as in maximum margin criterion [10] to maximize class separation in our proposed methods.
- The performance of DGLPCA is evaluated using two biomedical datasets publicly available from the UCI Machine Learning repository [11].

The remainder of the paper is organized as follows: In Section 2, we describe some related works from [8] and [9] followed a description of our newly proposed discriminant graph Laplacian PCA (DGLPCA) in Section 3. The extensive experimental results are presented in Section 4. Finally, we give some concluding remarks in Section 5.

2. Related Works

Assuming we have a data matrix $X = \left[x_1, x_2, \ldots, x_n\right]^T$, where each row represents a sample (or a patient) and each column represent a feature. Each patient x_i belongs to one of the C classes $\{X_1, X_2, \ldots, X_C\}$ where $X_c = \left[x_1^c, x_2^c, \ldots, x_{n_c}^c\right]$, ($c = 1,2,\ldots,C$) and n_c denotes the number of patients in class c . PCA aims to find a projection matrix U that minimizes the reconstruction error. The projection matrix U is obtained by solving the following minimization problem:

$$\min_U \left\| X - UU^T X \right\|_F^2, \tag{1}$$

subject to $U^T U = I$, where I is an identity matrix.

To incorporate manifold learning into PCA, the graph Laplacian PCA (GLPCA) [8] is introduced and the modified minimization problem is given by:

$$\min_U (-tr(U^T XX^T U) + \alpha tr(U^T XLX^T U)), \tag{2}$$

subject to $U^T XX^T U = I$, where $\alpha \geq 0$ controls the smoothness of the new representation. The matrix $L = D - S$ is called the graph Laplacian [12] where S is an $n \times n$ weight matrix defined by

$$S_{ij} = \begin{cases} \exp\left(-\dfrac{x_i - x_j^{\,2}}{\sigma} \right); & if\ x_i \in N_k\left(x_j\right) or\ x_j \in N_k\left(x_i\right) \\ 0; & otherwise. \end{cases} \tag{3}$$

such that $N_k\left(x_j\right)$ denotes the k nearest neighbour set of x_j, σ is a user specified parameter, and D is a diagonal matrix whose entries are equal to the column (or row) sum of S .The optimal transformation matrix U of GLPCA can be computed by applying an eigen decomposition on $XX^T - \alpha XLX^T$ and XX^T.

3. Proposed Methodology

3.1. Maximum Margin Criterion

In order to enhance the classification ability of GLPCA, we seek to capture the discriminatory information using the Maximum Margin Criterion (MMC) as proposed in [10]. Let $L = D - S$ be the graph Laplacian in (2) where S is an $n \times n$ weight matrix defined as

$$S = diag\left(S^{(1)}, S^{(2)}, \ldots, S^{(C)}\right), \tag{4}$$

such that $S^{(c)}$ is an $n_c \times n_c$ weight matrix having entries all equal to $1/n_c$ (i.e., $S_{ij}^{(c)} = 1/n_c$ for $i, j = 1, 2, \ldots, n_c$) and D is a diagonal matrix whose entries are equal to the column (or row) sum of S. With definitions $S_b = XSX^T$ and $S_w = XLX^T$, the MMC finds the maximizing subspace of the following objective function,

$$\max_U tr\left(U^T\left(S_b - S_w\right)U\right); \quad s.t\ U^T XX^T U = I \tag{5}$$

It can be shown that the matrices S_b and S_w are respectively the between class scatter matrix and the within class scatter matrix which are used to define the Linear Discriminant Analysis (LDA). The LDA method has long been a benchmark method for discrimination and classification. But it is well known that the objective function of LDA can easily lead to singularities in small sample-size problem. This problem is affectively avoided in MMC.

3.2. The proposed DGLPCA method

In PgLPCA and GLPCA, the graph Laplacian is constructed using the k -nearest neighbor method. The method defines S as in (3) which requires user-defined parameters k and σ which are the number of nearest neighbors and the width of the heat kernel respectively. Although the performance of GLPCA (and PgLPCA) depends on the choices of k and σ, it is not obvious how these parameters should be chosen for a particular problem. In MMC, S is defined based on the information of the class labels and it is parameter free. Hence, by combining the objective function of GLPCA (2) and the objective function of MMC (5), we can improve discriminative ability of GLPCA while eliminating the need to adjust k and σ.

Similar to [13-15], we propose combining the objective functions in a multi-objective optimization problem of the form

$$\begin{cases} \min_U tr\left(U^T\left(-XX^T + \alpha XLX^T\right)U\right); \\ \max_U tr\left(U^T\left(S_b - S_w\right)U\right); \end{cases} \quad s.t\ U^T XX^T U = I. \tag{6}$$

Solving the multi-objective optimization problem (6) is an attempt to find a lower dimensional subspace that preserves the manifold structure of the data and maximizes the margin between different classes at the same time. These two properties are very important for clustering and classification tasks. The objective function (6) can be rewritten in the following equivalent form:

$$\max_U tr\left(U^T X\left(I - \alpha L + \eta\left(S - L\right)\right)X^T U\right), \tag{7}$$

$$s.t\ U^T XX^T U = I,$$

where the regularization parameter $\eta > 0$ balances the contribution from the two parts. Using Lagrange multiplier method, the constraint optimization problem (7) can be transformed to a generalized eigenvalue equation of the form

$$(X(I - \alpha L + \eta(S - L))X^T)U = \Lambda XX^T U. \quad (8)$$

where the solution to (8), namely U_{opt}, is the principle eigenspace of U. Let the columns of U_{opt} be $u_0, u_2, \ldots, u_{d-1}$ ordered according to their eigenvalues $\lambda_0 \geq \lambda_1 \geq \cdots \geq \lambda_{d-1}$. For an arbitrary sample $x_i \in R^m$, the embedding is represented by:

$$x_i \rightarrow y_i = U_{opt}^T x_i, \quad U_{opt} = [u_0, u_2, \ldots, u_{d-1}],$$

where y_i is a representation of x_i in a d-dimensional space ($d \ll m$) and U_{opt} is a $m \times d$ matrix. Note that, in biomedical datasets, the number of features is very high compared to the number of the samples, i.e. the dimensionality of the data is usually higher than the number of samples. In such cases, the matrix XX^T in (9) can become singular. We apply the Tikhonov regularization technique [16] to solve the singularity problem. The generalized eigenvalue problem (8) becomes:

$$\left(X\left(I - \alpha L + \eta\left(S - L\right)\right)X^T\right)U = \Lambda\left(XX^T + \sigma I\right)U, \quad (9)$$

where $\sigma > 0$ is the regularization parameter.

4. Experimental Results

In this section, we investigate the performance of our proposed discriminant graph Laplacian principal component analysis (DGLPCA) in extracting features capable of separating classes in data. All of our experiments have been performed on an intel core i7, 3.20GHz windows PC with 8GB memory.

4.1. Data and evaluation criteria

Two publicly available datasets taken from UCI machine learning repository are used in our experiments. They are Thyroid disease and Wisconsin diagnostic breast cancer data sets. The Wisconsin (diagnostic) data contains features computed from a digitized image of a fine needle aspirant (FNA) of a breast mass. This dataset consists of 569 patterns with 30 features categorized into two classes: benign and malignant. The original data was obtained by Wolberg and Mangasarian [17]. The Thyroid disease dataset contains observations from 7200 patients characterized by 15 binary and 6 continuous patients attributes. Each of the patient is assigned to one of the three following classes: 1) Normal, not hyperthyroid, 2) Hyperfunction and 3) Subnormal functioning.

We used the K-NN classifier to identify the observations after feature extraction where K refers to the number of nearest neighbors. The classification accuracies are measured based on $K = 5$ for the breast cancer dataset, while for the Thyroid disease dataset, we set $K = 20$. The performance of our proposed DGLPCA is evaluated in comparison with the ordinary PCA and GLPCA methods.

4.2. Discussion

To obtain reliable experimental results, we reshuffle all the datasets randomly in the experiments. For each of the datasets, $p = (33, 50, 67)$ percent of samples are randomly selected as training sets and the rest are used as test sets. The training samples are used to determine the projection vectors. The test samples are then mapped into lower dimensional subspace using the obtained projection vectors. Classification is then performed in the lower dimensional subspace using the nearest neighbor classifier.

Table 1 presents the classification accuracies obtained on the breast cancer data using the PCA, GLPCA, DGLPCA and the baseline methods. For the baseline method, classification is performed in the original 30-dimensional space without any dimensionality reduction. The number of nearest neighbors for constructing the nearest neighbor graph of GLPCA was set to $k = 3$ and the heat kernel width was set to $\sigma = 70$ for optimum performance (empirically determined). For DGLPCA, the regularization parameters α and η are set to 1×10^4. As can be seen from Table 1, the performance of PCA, GLPCA and DGLPCA varies with different value of p. However, our proposed DGLPCA obtained the least classification error in all of the three cases.

Similar to the experiment on breast cancer data, we used the nearest neighbor classifier to classify the test samples of the Thyroid data using features extracted by the PCA, GLPCA, DGLPCA and the baseline method. The two parameters k and σ in GLPCA were set to 10 and 0.02 respectively (empirically optimized). While the regularization parameters α and η in our proposed DGLPCA method were both set to 1×10^4. We extracted only 3 features using the different techniques and report the results in Table 2. As can be seen, our algorithm outperforms all the other methods in all cases. We observed that, even when the training data is small, the performance of DGLPCA is significantly higher than that of GLPCA and all other methods with large training data. This indicates that our method is able to learn from small training sets.

In summary, our proposed DGLPCA method obtained the best performance in all experiments. The experimental results further illustrate that, based on the datasets we used, the newly proposed DGLPCA method demonstrates more discriminating power than the ordinary GLPCA method. This results highlight the effectiveness of combining ideas from GLPCA and MMC in feature extraction.

Table 1. Classification results on the Wisconsin diagnostic breast cancer data.

Method	33% Training		50% Training		67% Training	
	error (%)	# of features	error (%)	# of features	error (%)	# of features
Baseline	8.16	30	7.72	30	7.89	30
PCA	8.42	2	8.77	2	7.89	2
GLPCA	13.68	2	9.47	2	6.84	2
DGLPCA	3.68	2	4.56	2	2.11	2

Table 2. Classification results on the Thyroid disease data.

Method	33% Training		50% Training		67% Training	
	error (%)	# of features	error (%)	# of features	error (%)	# of features
Baseline	7.04	21	6.80	21	6.75	21
PCA	7.40	3	7.33	3	7.58	3
GLPCA	7.38	3	6.92	3	7.63	3
DGLPCA	5.00	3	3.75	3	3.75	3

5. Conclusions

In this paper, we have introduced an efficient feature extraction technique called discriminant graph Laplacian PCA. Our approach combines ideas from spectral graph analysis and maximum margin criterion to provide an efficient and effective approach for feature extraction. We showed empirically that the newly proposed approach performed better than the previous approaches in extracting features capable of discriminating the classes in a data. Other than improving the discriminative power of graph-Laplacian PCA, the proposed DGLPCA also provides a parameter-free graph construction method so that its performance is independent on the graph parameters.

However, more work needs to be done in evaluating the performance of the proposed method on other datasets. Also, the performance of the proposed method using different classification techniques such as support vector machine, random forest and neural networks should further be investigated.

Acknowledgements
The authors would like to acknowledge UNIVERSITI SAINS MALAYSIA for the financial support through Research University Grant (RUI) (acc. No. 1001/PMATHS/8011040) and the USM Bridging Grant (acc. No. 304/PMATHS/6316049).

References

[1] Wasan S, Bhatnagar V and Kaur H 2006 The impact of data mining techniques on medical diagnostics, *Data Science Journal* 5, pp. 119–126.

[2] Huang S, Zhou J, Wang Z, Ling Q and Shen Y 2017 Biomedical informatics with optimization and machine learning, *EURASIP J Bioinform Syst Biology* 2017:**4**.

[3] Foster K R, Koprowski R and Skufca J D 2014 Machine learning, medical diagnosis, and biomedical engineering research: Commentary, *BioMedical Engineering OnLine*. 13–94.

[4] Boulesteix A -L 2004 PLS dimension reduction for classification with microarray data, *Statistical applications in genetics and molecular biology,* **3**(1), 1–30.

[5] Liu J -X, Wang D, Gao Y -L, Zheng C -H, Shang J -L, Liu F and Xu Y 2017 A joint-$L_{2,1}$-norm-constraint-based semi-supervised feature extraction for RNA-Seq data analysis, *Neurocomputing,* **228**, 263–269.

[6] Liu H, Zhang Z, Zhang X, Yang Y, Zhang Z, Liu X, Wang F, Han Y and Zhang C 2018 Dimensionality Reduction for Identification of Hepatic Tumor Samples Based on Terahertz Time-Domain Spectroscopy, *IEEE Transactions on Terahertz Science and Technology,* **8**(3), 271-277.

[7] Shahid N, Kalofolias V, Bresson X, Bronstein M, and Vandergheynst P 2015 Robust principal component analysis on graphs, in *Proceedings of the IEEE International Conference on Computer Vision*, 2812-2820.

[8] Jiang B, Ding C, Luo B, and Tang J 2013 Graph-Laplacian PCA: Closed-form solution and robustness, in *Proceedings of the IEEE Conference on Computer Vision and Pattern Recognition*, 3492-3498.

[9] Feng C -M, Gao Y -L, Liu J -X, Zheng C -H, and Yu J 2017 PCA based on graph laplacian regularization and P-norm for gene selection and clustering, *IEEE transactions on nanobioscience,* **16**(4), 257-265.

[10] Li H, Jiang T, and Zhang K 2004 Efficient and robust feature extraction by maximum margin criterion, in *Advances in neural information processing systems*, 97-104.

[11] UCI repository of machine learning databases, University of California-Irvine, Department of Information and Computer Science.

[12] He X and Niyogi P 2004 Locality preserving projections, in *Advances in neural information processing systems*, 153-160.

[13] Gui J, Jia W, Zhu L, Wang S -L, and Huang D -S 2010 Locality preserving discriminant projections for face and palmprint recognition, *Neurocomputing,* **73**(13-15), 2696-2707.

[14] Gui J, Sun Z, Jia W, Hu R, Lei Y, and Ji S 2012 Discriminant sparse neighborhood preserving embedding for face recognition, *Pattern Recognition,* **45**, (8), 2884-2893.

[15] Zhang J, Wang J, and Cai X 2017 Sparse locality preserving discriminative projections for face recognition, *Neurocomputing,* **260**, 321-330.

[16] Wang Q, Meng Z, and Li X 2017 Locality adaptive discriminant analysis for spectral–spatial classification of hyperspectral images, *IEEE Geoscience and Remote Sensing Letters,* **14**(11), 2077-2081.

[17] Wolberg W H and Mangasarian O L 1990 Multisurface method of pattern separation for medical diagnosis applied to breast cytology, *Proceedings of the national academy of sciences,* **87**(23) 9193-9196.

Q-switching Zirconia-Erbium-doped Pulsed Fiber Laser with MWCNTs-PEO as Saturable Absorber

Arni Munira Markom[1*], Muhammad Taufiq Ahmad[2], Hazlihan Haris[3], Ahmad Razif Muhammad[3], Zakiah Mohd Yusoff[1], Norhafizah Burham[4], Mukul Chandra Paul[5], Anirban Dhar[5], Shyamal Das[5], Mrinmay Pal[5], Shyamal Kumar Bhadra[5] and Sulaiman Wadi Harun[3]

[1] Faculty of Electrical Engineering, Universiti Teknologi MARA (UiTM) Cawangan Johor, Kampus Pasir Gudang, 81750 Johor, Malaysia.
[2] Faculty of Electronics & Computer Engineering, Universiti Teknikal Malaysia Melaka (UTeM), 76100 Melaka, Malaysia
[3] Department of Electrical Engineering, University of Malaya, 50603 Kuala Lumpur, Malaysia.
[4] Faculty of Electrical Engineering, Universiti Teknologi Mara (UiTM) Selangor, 40450 Shah Alam, Selangor, Malaysia.
[5] Fiber Optics and Photonics Division, CSIR-Central Glass and Ceramic Research Institute, Kolkata 700 032, India

*arnimunira@uitm.edu.my

Abstract. Q-switched pulses provide a significant contribution to recent biomedical applications such as laser treatment, medical imaging, biomedical diagnoses and spectroscopy due to high efficiency, compact device, less footprint, flexibility and cost effective. A successful generation of Q-switching pulsed fiber laser by using a homemade fiber of zirconia-based erbium-doped fiber (Zr-EDF) with the utilization multi-walled carbon nanotubes-polyethylene oxide (MWCNTs-PEO) as saturable absorber is reported. The active medium is 1 meter length of Zr-EDF with -0.6 ps^2 group delay dispersion (GDD) for overall setup arrangement. At maximum pump power 126 mW, the repetition rate, pulse duration, output power and pulse energy are 17.3 kHz, 7.57 μs, 1.13 mW and 65.03 nJ, respectively.

1. Introduction

Q-switching pulsed fiber lasers are frequently used for developing applications with high energy pulses at low repetition rate. It have been applied in many fields especially in biomedical technologies such as laser treatment, medical imaging, biomedical diagnoses and spectroscopy due to efficiency, compactness, simple, flexible to design, high safety and low cost [1-2] . Recently, Q-switching laser treatment is gained significant attention including removal hair, unwanted brown spots, sun freckles, spider veins, and tattoos as well as skin resurfacing for anti-aging faces [3-6]. The energy pulsed laser releases the pigment into the skin, thus naturally it can reabsorbed and disposed to the body without harm. Medical imaging provides high resolution and deepest in vivo imaging with broadest tuning using laser ultrasound Q-switching together with baby development, tumour progression and electrography that maps elastic properties and stiffness of soft tissue [7-9]. Biomedical diagnosis help so much people especially for early detection such as tumour, cancer and fungal infection [10-11]. Finally spectroscopy where radiation intensity measurement such as toxic compound in blood samples and optical fiber cure monitoring [12-13].

The most popular rare earth material that used as active medium is Erbium for more than 70 years. However, this material is saturated technology and leads to unfavourable effect to collect optical gain, yet the material has outstanding optical properties. It is also stuck at C-band transmission and need help with various complicated structure of experiment to enhance the transmission to the L-band to open wide opportunity for many applications, nowadays. Therefore, a homemade optical fiber of zirconia-yttria-aluminium co-doped erbium-doped (Zr-EDF) with a massive erbium concentration of 4000 ppm is develop to create a simple experiment structure of pulsed generation. Moreover, our aim is to achieve pulsed creation with using a minimum length of active medium to realize a compact devices compared to 2 m to 5 m for normal Erbium fiber.

The development of Q-switching pulsed fiber laser is preferable by using passive method where saturable absorber (SA) as the key element. Commonly, SA is a piece of thin film material whose absorptions decline as increasing irradiances. When useful light go through the SA, the process of molecules absorption become stronger and push significant number of atoms excite to the upper levels. The absorption becomes very small when the ground and excited level from population inversion are almost equivalent, and thus the material of SA is understood to be saturated [14]. An increasing irradiance creates to the further saturating until the material is changed from absorbing to transmitting, and cause the generation of intense and short (in range micro to nanoseconds) pulse duration. In this experiment, the SA used is multi-walled carbon nanotubes (MWCNTs) due to excellent throwbacks that proven in many researches in optical fiber lasers [15-17]. This MWCNTs is then mixed with polyethylene oxide (PEO) polymer compound because it easily dispersed in water and make it simple and easier for the fabrication process.

2. Methodology

The saturable absorber (SA) was named as multi-walled carbon nanotubes - polyethylene oxide (MWCNTs-PEO). About 10 to 20 nm mean diameter and 1to 2 μm length distribution were used for SA preparation. To improve liquid solubility, 400 ml pre-dissolved Sodium Dodecyl Sulphate (SDS) solution (0.01g/ml) is homogenized with 250 mg MWCNTs which were stirred in 50 W ultrasonic cleaner for one hour. Then, the undispersed MWCNTs were detached by using centrifugation (1000 rpm). Next, to form MWCNTs-PEO compound mixture, the dispersed MWCNTs solution was sonicated at about one hour after poured with 10.1 wt % concentration of PEO solution. At last, in room temperature for seven days were used to dry the MWCNTs-PEO compound and successfully to fabricate 50 μm thickness of thin film. The Raman spectrum of the MWCNTs-PEO thin film as shown in Figure 1. The D-band at 1356 cm^{-1} defines for multi-layer of walls whereas G^+ and G' bands were formed at 1579 cm^{-1} and 2719 cm^{-1} due to extending tangential carbon-carbon bonds within graphene sheet.

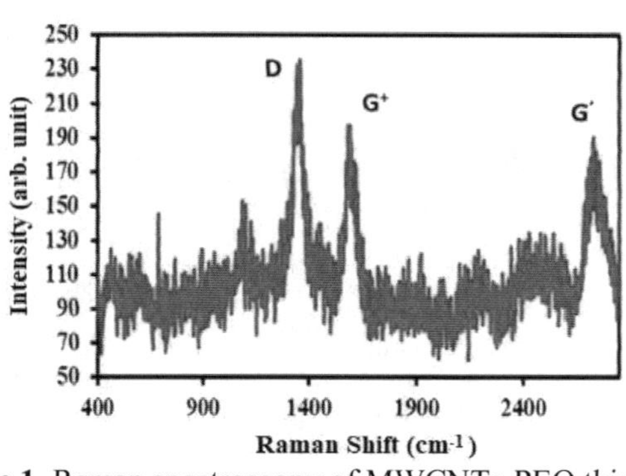

Figure 1. Raman spectroscopy of MWCNTs-PEO thin film.

Figure 2 demonstrates the arrangement setup of the proposed Zr-EDFL where it consists a 1 m Zr-EDF as active medium, 980/1550 nm wavelength division multiplexer to multiplex the optical signal carrier to the two different wavelengths of 980 nm and 1550 nm in the optical fiber, polarization controller (PC) used to alter the polarization of light, isolator to enable only one direction of light transmission and a coupler 95/5 to divide 5 % to observe the output while the balance will be back to the setup arrangement to produce pulsed fiber laser. While fabricated MWCNTs-PEO SA was cut to a small piece of 2 x 2 mm and inserted between two fiber ferrules with the aid of index-matching gel. Finally, a 3 dB coupler was used to divide each 50 % to observe the output of pulsed fiber laser and optical spectrum by using Oscilloscope (OSC) and optical spectrum analyzer (OSA), respectively. Besides, the 2 m length of Zr-EDF had -56 ps^2/km for group velocity dispersion (GVD), the 4 m length of WDM had -38 ps^2/km GVD and remaining single mode fiber (SMF) from PC, isolator and coupler had -21 ps^2/km GVD. Thus, the total length arrangement setup was 21.7 m and obtained -0.6 ps^2 group delay dispersion (GDD). This small amount of GDD contributes in normal Q-switching pulsed fiber laser which defines in anomalous dispersion.

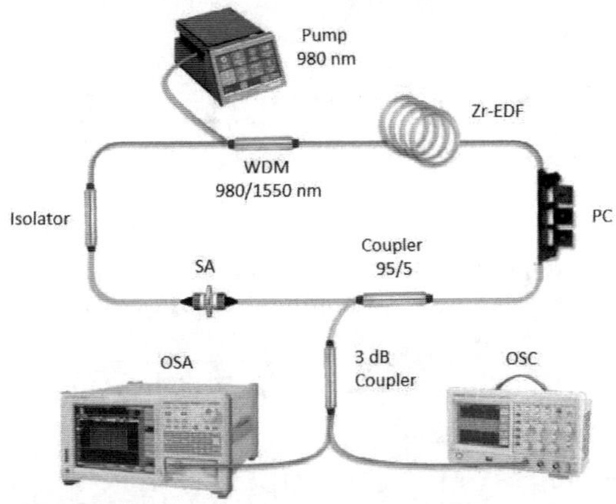

Figure 2. Arrangement setup for Zr-EDFL.

3. Results and Discussion

Optical spectrum with and without SA were illustrated in Figure 3 (a). The threshold laser without SA was 57 mW as it expanded from 1540 nm to 1640 nm with the peak power of – 43.4 dBm at 1568.8 nm. Meanwhile the threshold laser with SA was 87.2 mW as the wideband spectrum from 1520 nm to 1640 nm with the peak power of – 30.9 dBm at 1560.8 nm wavelength. The result shows with the aid of SA, the spectrum of fiber laser was able to develop a wideband transmission and sharp lasing due to intra-cavity losses by the added the SA. Moreover, with MWCNTs-PEO SA, a stable and self-stating pulsed train Q-switching was developed at the threshold pump power. The fundamental behind Q-switching theory was the initial loss cavity at the top whereas Q factor was sat at the ground level. As the pump power was slowly increased, the optical gain also increased due to the population inversion in the active medium until the loss is surpassed by the accumulated optical gain. This phenomena activates a high Q factor and develop pulsed fiber laser of Q-Switching.

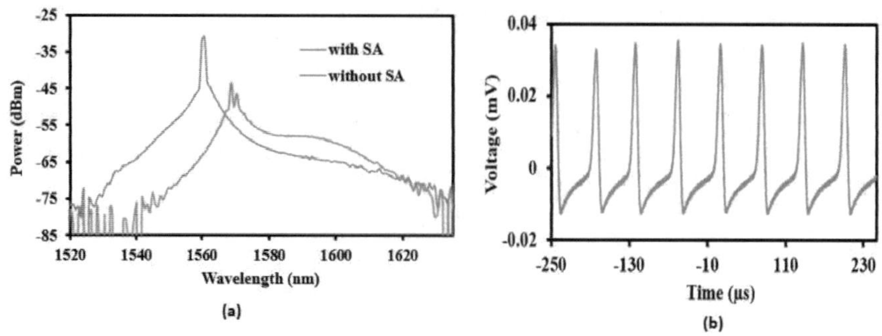

Figure 3. (a) Optical spectrum. (b) Q-Switching pulse train at pump power of 121 mW.

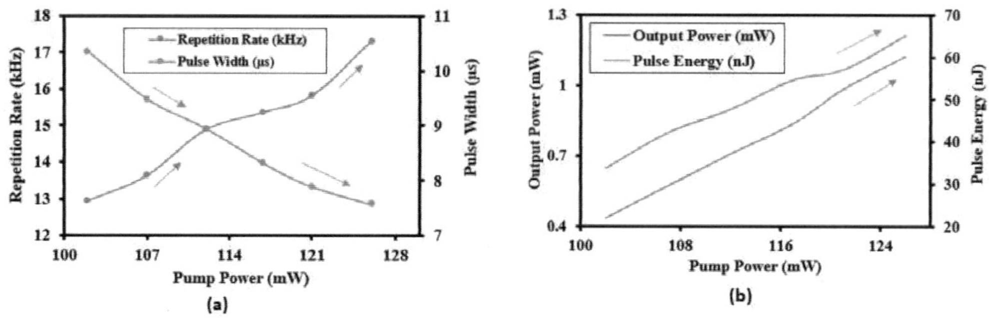

Figure 4. (a) Repetition rate and pulse width against pump power. (b) Output power and pulse energy against pump power.

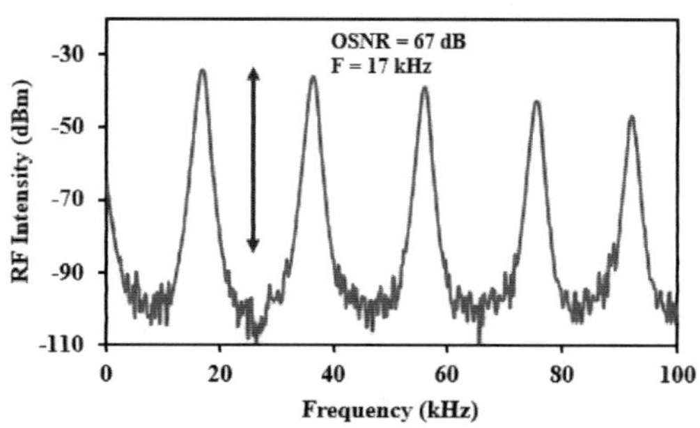

Figure 5. Optical signal to noise ratio (OSNR) at repetition rate of 17 kHz.

Figure 3 (b) shows the pulse train of normal Q-switching at pump power of 121 mW which was obtained 15.82 kHz and 7.89 µm for repetition rate and pulse duration. The repetition rate and the pulse duration generation were caused by the active medium of Zr-EDF with high erbium concentration about 4000 ppm. The Zr-EDF is allows to activate population inversion and thus generate optical gain with the increasing pump power. Therefore, with the increasing pump power that varied from 102 mW to 126 mW, the Q-switching repetition rate was consistently increased from 12.94 kHz until 17.30 kHz as illustrated in Figure 4 (a). Meanwhile the pulse duration was inversely proportional with the pump powers started by 10.34 µm until 7.57 µm. The trends of Q-switching were increasing repetition rate and decreasing pulse duration against increment pump power.

Figure 4 (b) shows output power and pulse energy against pump power. The pulse energy was increases as increasing pump power from 33.77 nJ to the highest of 65.03 nJ. This pulse energy was defined by multiplication of pulse duration with peak power for each pump power increment. Then, the output power increases from 0.44 mW until 1.13 mW at maximum pump power and calculated by multiply the pulse energy with the repetition rate. Both of the calculation was actually affected by the length of Zr-EDF and the thicker layer of MWCNTs-PEO where more atoms stored at energy level were stimulated to create coherent photons of useful light fiber laser. Besides, Figure 5 shows the OSNR at repetition rate of 17 kHz with 67 dB, which translates the high stability of the Q-switching pulsed fiber laser.

Table 1 shows the optical gain characteristics for three different length of Zr-EDF using simple double pass arrangement of experiment. 1 m length shows the highest gain of 40.3 dB compared to 0.5 m and 2 m with only 32.7 dB and 34.0 dB, respectively. Again, 1 m shows the highest and consistent uniform gain of 38.6 dB with a small fluctuation of 1.1 dB from 1525 nm to 1560 nm. Meanwhile 0.5 m and 2 m show lower and almost similar results for uniform gain around 27 dB with specific wavelength as detailed in the table. The fluctuations are also large with 7.2 dB and 32.5 dB for 0.5 m and 2 m that leads to huge loss and noise for real device applications.

Table 1. Zr-EDF characteristics of optical gain

Length of Zr-EDF	**0.5 m**	**1 m**	**2 m**
Highest Gain	32.7 dB	40.3 dB	34.0 dB
Uniform Gain	27.5 ± 7.2 dB	38.6 ± 1.1 dB	27.7 ± 32.5 dB
Uniform Gain Wavelength	1530 nm – 1565 nm	1525 nm – 1560 nm	1540 nm – 1575 nm

Table 2 is the comparison between this Zr-EDF with a commercial Erbium-doped fiber (EDF) which is done by F. Ahmad et. al by using the same length of 1 m active medium and the same SA of multiwall CNT-PEO [16]. Both results are Q-switching pulsed fiber lasers. Zr-EDF shows excellent candidate as alternative optical fiber with a better results for pulse duration, pulse energy and output power. This is with the help of zirconia-yttria-aluminium doping with erbium element that successfully to increase high and uniform optical gain to create a better Q-switched pulsed fiber laser. However, the repetition rate has a slight low due to loss in experiment arrangement.

Table 2. Q-switched pulsed fiber laser comparison between EDF and Zr-EDF.

Characteristics	**EDF**	**Zr-EDF**
Repetition Rate	4.5 kHz – 20.0 kHz	12.9 kHz – 17.3 KHz
Pulse Duration	8.8 μs	7.57 μs
Pulse Energy	15.3 nJ	65.3 nJ
Output Power	0.03 mW – 0.31 mW	0.44 mW – 1.13 mW

4. Conclusion

The development Q-switching pulsed fiber laser was successfully using zirconia-yttria-aluminium co-doped erbium-doped fiber with 4000 ppm erbium concentration without any detrimental effect. Together with the use of MWCNTs-PEO saturable absorber, the majority population inversion was achieved and thus provides 17.3 kHz repetition rate, 7.57 μs pulse duration and 65.03 nJ pulse energy at maximum pump power of 126 mW.

Acknowledgements

The authors acknowledge Geran Lestari, 600 IRMI 5/3/LESTARI (038/2019) UiTM, Department of Electrical Engineering, Faculty of Engineering,Universiti Malaya of Malaysia and partially, Fiber Optics and Photonics Division, CSIR-Central Glass and Ceramic Research Institute of India for supporting this research.

References

[1] Busser B, Moncayo S, Coll JL, Sancey L, Motto-Ros V. 2018 Elemental imaging using laser-induced breakdown spectroscopy: A new and promising approach for biological and medical applications *Coordination Chemistry Reviews* **358** 70-9.

[2] Goldman L. 2018 Applications of the Laser. *Crc Press.*

[3] Reiter O, Atzmony L, Akerman L, Levi A, Kershenovich R, Lapidoth M, Mimouni D. 2016 Picosecond lasers for tattoo removal: a systematic review *Lasers in Medical Science* **7** 1397-405.

[4] Molnár E, Molnár B, Lohinai Z, Tóth Z, Benyó Z, Hricisák L, Windisch P, Vág J. 2017 Evaluation of laser speckle contrast imaging for the assessment of oral mucosal blood flow following periodontal plastic surgery: an exploratory study. *BioMed Research International.* Hruza GJ, Tanzi EL. 2017 Lasers and Lights E-Book: Procedures in Cosmetic Dermatology Series *Elsevier Health Sciences.*

[5] Rahman HA, Harun SW, Batumalay M, Muttalib FA, Ahmad H. 2012 Fiber optic displacement sensor using multimode plastic fiber probe and tooth surface *IEEE Sensors Journal* **13(1)** 294-8.

[6] Liu D, Liang Y, Jin L, Sun H, Cheng L, Guan BO. 2016 Highly sensitive fiber laser ultrasound hydrophones for sensing and imaging applications *Optics Letters* **41(19)** 4530-3.

[7] Wang S, Larin KV. 2015 Optical coherence elastography for tissue characterization: a review. *Journal of Biophotonics* **8(4)** 279-302.

[8] Sartori S, Mauri G, Tombesi P, Di Vece F, Bianchi L, Pacella CM. 2018 Ultrasound-guided percutaneous laser ablation is safe and effective in the treatment of small renal tumors in patients at increased bleeding risk *International Journal of Hyperthermia* **11** 1-7.

[9] Vo-Dinh T, editor 2014 Biomedical photonics handbook: biomedical diagnostics. *CRC press.*

[10] Galvan Garcia HR. 2014 Onychomycosis: 1064-nm Nd: YAG q-switch laser treatment *Journal of Cosmetic Dermatology* **13(3)** 232-5.

[11] Haris H, Harun SW, Jusoh Z. 2019 Generation of bound state of solitons pulses with graphene in Erbium-doped fiber laser cavity *In Journal of Physics: Conference Series* **1151** No. 1, p. 012017

[12] Kinet D, Mégret P, Goossen K, Qiu L, Heider D, Caucheteur C. 2014 Fiber Bragg grating sensors toward structural health monitoring in composite materials: Challenges and solutions *Sensors* **14(4)** 7394-419.

[13] Harun SW, Ismail MA, Ahmad F, Ismail MF, Nor RM, Zulkepely NR, Ahmad H. 2012 A Q-switched erbium-doped fiber laser with a carbon nanotube based saturable absorber *Chinese Physics Letters* **29(11)** 114202.

[14] Taib NA, Bidin N, Haris H, Adnan NN, Ahmad MF, Harun SW. 2016 Multi-walled carbon nanotubes saturable absorber in Q-switching flashlamp pumped Nd: YAG laser *Optics & Laser Technology* **79** 193-7.

[15] Ahmad F, Haris H, Nor RM, Zulkepely NR, Ahmad H, Harun SW. 2014 Passively Q-switched EDFL using a multi-walled carbon nanotube polymer composite based on a saturable absorber *Chinese Physics Letters* **31(3)** 034204.

[16] Haris H, Anyi CL, Muhammad AR, Ahmad F, Nor RM, Zulkepely NR, Ali NM, Harun SW, Arof H. 2015 Performance of passively Q-switched ring erbium-doped fiber laser using a multiwalled carbon nanotubes polyethylene oxide (PEO) polymer composite-based saturable absorber *Microwave and Optical Technology Letters* **57(8)** 1897-901.

A Review of Frequency Selection and Antenna Arrangement System for Microwave Imaging in Early Breast Cancer Detection

Hadi Jumaat[1*], Mohd Tarmizi Ali[1], Kismet Hong Ping[2], Najwa Mohd Faudzi[1], Nurfarahin Miswadi[1], Hamizan Yon[1]

[1]Faculty of Electrical Engineering, Universiti Teknologi MARA, Malaysia.
[2]Department of Electrical & Electronic Engineering, Faculty of Electrical Engineering, Universiti Malaysia Sarawak, Sarawak, Malaysia

*jumaathadi@gmail.com

Abstract. Microwave imaging technique is one of the developing techniques that having attention by researcher around the world in early breast cancer monitoring due it is non-ionizing and the potential for cost saving and simpler system as compared to the current imaging techniques. The frequency selection is crucial in this application due to no bandwidth proven best. However, the amount of antenna is said affected the image resolution. This paper is an attempt to review the frequency selection and the antenna arrangement system in the past decades works for future antenna development in early breast cancer detection.

1. Introduction

According to the breast cancer mortality statistic [1], it is in the stressing situation to the world and moves toward becoming test to the specialists in exploring the fix, particularly in early breast cancer detection method. As result, in the past two decades the techniques in early malignant growth discovery have evolve; mammography, tomography, thermography, ultrasound and microwave imaging [2]–[6]. Microwave imaging is potential to become alternative technique besides X-ray mammography and magnetic resonance imaging (MRI) in early breast cancer detection due to the non-ionizing radiation, safer, potential for cost saving and simpler system [7].

The microwave imaging adopted the techniques on dielectric contrast by reconstructed the data obtained from transmitter and receiver whereby object is located between the transceiver. Then, the reconstructed data produce the contrast of malignant tumour and benign tissue. The tissues identified clearly since water density in malignant tumour is more than the normal breast tissue. At present, no bandwidth proven best for microwave imaging [7]. The frequency selection is crucial in antenna design since all equations involve in antenna shaping is depending on the operating frequencies. Thus, literature on frequency selection has been held with considering the works by signal processing field to justify the suitable frequency range for next work on designing antenna for microwave imaging on early breast cancer detection application. Some of the works basically using a point radiator representing antennas suggested at 1 GHz to 3 GHz in detecting the malignant tumour [8]. Besides that, a few researchers in antenna design field in [9]–[12] have been highlighted the potential of UWB frequency range in microwave imaging for breast cancer monitoring range from 1.3 GHz up to 20 GHz. Furthermore, a work highlighted in [7] is using 3.2 GHz to 5.7 GHz operating frequency with consideration of details tissue parameters. Overall from the previous researches, numerous frequency range have been suggested to cover the tissues.

In microwave imaging a single pair of antennas is insufficient for effective classification of tumour as per mentioned in [13]. More antenna will have better image resolution. Image resolution in microwave imaging is important to display better image of scanning object. Anyhow, not much literature presented to date regarding the arrangement of antenna for breast cancer monitoring. The optimization work on arrangement of antenna elements (antenna arrangement system) need to be done to have the best fundamental theory towards the arrangement system. Since the antenna arrangement system should be at the circumference of human breast, this limitation will lead to limited space between antenna. Without compromising the necessity of having much antenna in producing better image resolution, miniaturization of antenna should be proposed so it can be arranged with the maximum number of antennas on a level. However, miniaturizing antennas in contact with biological tissues is challenging due to having different propagation behaviour and return loss compared to those in free space [15].

The importance of frequency selection and good antenna arrangement system for microwave imaging in monitoring breast cancer has been highlighted. As per mentioned, in microwave imaging, multiple identical single antenna design is arranged around the object. Thus, the works by previous researchers is investigated including the size of the single antenna design involve in the multiple antenna arrangement system. These will help in precise frequency range selection in developing antenna arrangement system for microwave imaging in early breast cancer detection application in future. In this work, the reviews are focusing on the frequency range, number of antennas adopted, and the level in the antenna arrangement.

2. Antenna Arrangement for Microwave Imaging Breast Cancer Detection

The antenna arrangement in microwave imaging is developed at the area of around the human breast. However, the arrangement could be on one level, two level or more as at the height of the human breast. This is depending on the overall size of the single antenna to fit in the arrangement system without performance degradation. In this section, the review works is classified into two categories of antenna arrangement; one level and multiple level.

2.1. One Level Antenna Arrangement

An annular antenna system for microwave imaging for breast cancer detection has been developed by Amit S. Narkhede [14] in 2016. A circular patch antenna in conjunction with circular slot as shown in **Figure 1** (a) is fabricated on FR4 substrate ($\varepsilon r = 4.4$) with the thickness of 1.6 mm. The overall single antenna size is 47mm x 37 mm (L x W). The antenna design operates from 2.2 GHz to 12 GHz with gain of 6 dBi and having an omnidirectional radiation pattern. Next, the antenna arrangement has been simulated by using HFSS simulation tool whereby 12 identical antennas are assembled in one level arrangement as depicted in **Figure 1** (b) along a cylinder circumference with radius of 200 mm to form an antenna system.

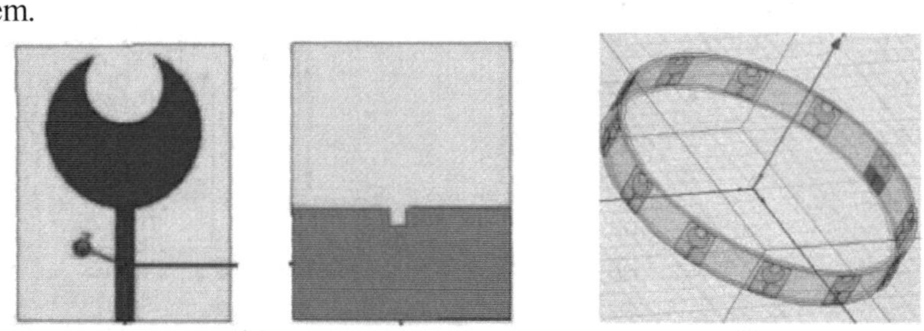

(a) (b)

Figure 1. The annular antenna system for microwave imaging for breast cancer detection.
(a) The geometry of the single antenna and (b) simulated one level antenna arrangement. [14]

Journal of Physics: Conference Series **1372** (2019) 012004 doi:10.1088/1742-6596/1372/1/012004

Another simulation work on one-layer antenna array arrangement have been proposed in 2016 by using 8-element microstrip patch antenna array [36]. The antenna has been designed purposely for microwave imaging application, but the measurement of the antenna system is using representative method such as plexiglass and ethanol samples to be detected instead of breast phantom. The authors develop the antenna system to be operated at 3 GHz to detect the representative object. The microstrip patch antenna is designed on FR4 substrate (εr = 4.3, tanδ = 0.025) with 8 antennas are assembled in octagon shape around the object with the overall system footprint exhibit 136 mm x 136 mm (L x W) as shown in **Figure 2**.

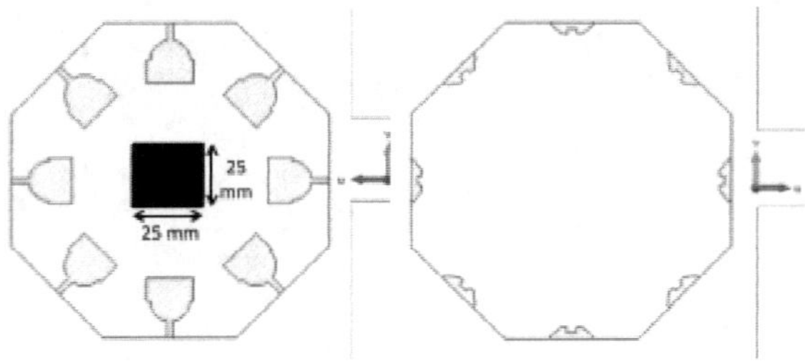

Figure 2. Microwave imaging using 8-element microstrip patch antenna array on one level arrangement. [15]

An experimental work on early breast cancer detection using circular antenna array have been reported by S. S Tiang in [16]. The overall work including antenna development and image reconstruction. At first, a p-shaped wide slot antenna configuration has been proposed by the author with 16 mm x 16 mm (L x W) overall size of the single antenna to be placed later in circular antenna arrangement circumference the breast phantom. The antenna is designed to operates from 4.5 GHz to10.9 GHz on Rogers RT6010 microstrip substrate with dielectric constant of 10.2. The author proposed using 88 mm diameter phantom with contain of fat, skin, glandular and tumour layers. The distance between two opposite antennas is set to 90 mm. Next, 16 identical single antennas are arranged on one layer with circular arrangement as depicted in **Figure 3**.

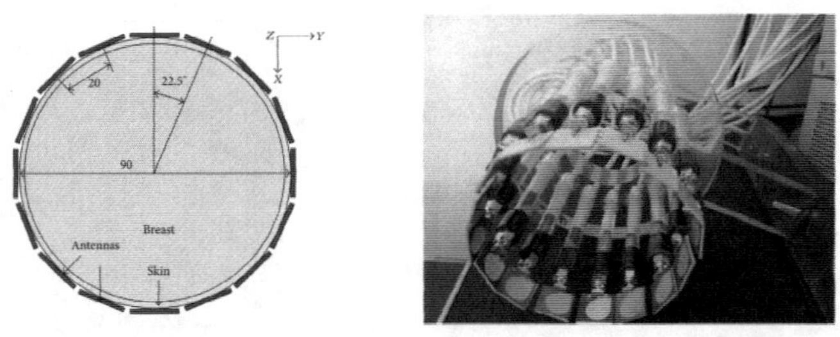

(a) (b)

Figure 3. Development of compact wide-slot antenna for early stage breast cancer detection with (a) circular antenna arrangement and (b) experiment set up.[16]

2.2. Multiple Level Antenna Arrangement

In 2016, a work done by Yunpeng Li [13] classified in the comparison of microwave breast cancer detection results with breast phantom data and clinical trial data by varying the number of antennas. The experimented data has been collected and illustrates the microwave imaging technique (using breast phantom experiment) contribute to easier task as compared to clinical trial data collection. It is also summarized in this work the insufficient for effective classification of tumor by using a single antenna pair. The variations of the point antenna arrangement are from 2, 4, 8 and 16 have been investigated. The simulation is working on points radiating element representing as antennas instead of using real antenna structure as exhibit in **Figure 4** simulated between 2 GHz to 4 GHz. This experiment shows that at least four antennas are needed for good classification performance, and the performance becomes more reliable as the increasing to 8 or 16 points of radiating elements. Hence, this condition showing that the numbers of antenna will enhance the image output by producing more data towards the object. Furthermore, in this research work do mention the importance in adopting reliable breast phantom specification to get more robust result of the breast tumor detection.

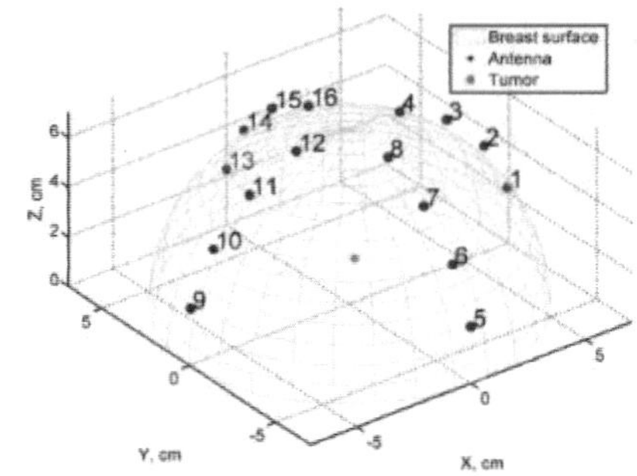

Figure 4. The simulation of 1 to 16 points of antenna arrangement. [13]

In 2017, an assembly conformal antenna array for wearable microwave breast imaging application has been conducted by F. Wang [17] in simulation using CST Studio Suite. The author designing a single UWB antenna with triangle shape of fractal slot to reduce the overall size of the antenna with the final footprint is 32.7 mm x 26 mm (L x W) in size as shown in **Figure 5** (a). The designated antenna operating frequency is from 3.2 GHz to 5.7 GHz with 2.5 GHz of bandwidth on DuPont Kapton polyimide substrate (εr = 3.1) and 0.075 um thickness. Next, simulation works is done for two-level antenna array mounting with 16 reassemble antennas involve on realistic Voxel-based breast model as depicted in **Figure 5** (b). The proposed work could provide the estimate location and size of breast tumor less than 2.5 mm radium.

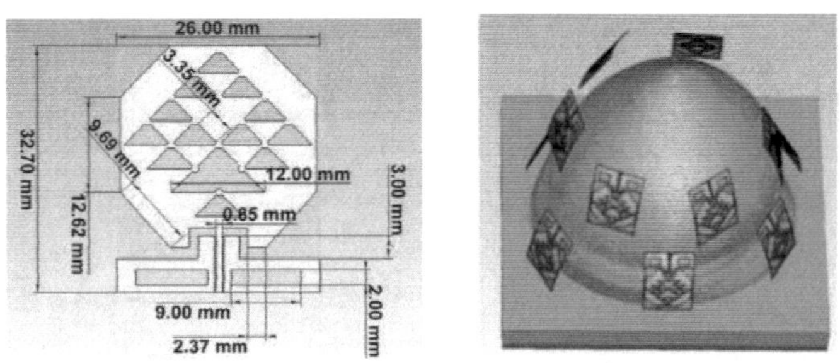

(a) (b)

Figure 5. The assembly conformal antenna array for wearable microwave breast imaging application. (a) The geometry of UWB antenna and (b) Two-level antenna array mounting on breast model. [17]

Another interesting work by Hadi Bahramiabarghouei [18] developing a flexible 16 antenna array for microwave breast cancer detection. Two antenna designs with arrangement system produced by the author. A single and dual polarization antenna has been investigated and fabricated on Kapton polyimide (εr = 3.5) with 0.05 mm thickness. The geometry of the antennas is shown in **Figure 6** (a). The simulated result analyzed in this work is using HFSS simulation tool. The antennas have been designed to be operated from 2 GHz to 4 GHz. The single polarized antenna design adopted the patch tapping technique to resonate from the lowest to higher desired frequency. On the other hand, the dual polarized antenna design adopted spiral shape to produce dual-polarization and attaining the desired frequency. In addition, the dual-polarization is supposed by the author due to it may add information to the collected signal by enabling recording of backscattered signal of two polarizations. Both antenna design achieved 20 mm x 20 mm (L x W) size. Moreover, the antennas were arranged by using 4 x 4 array arrangement system as exhibited in **Figure 6** (b) that creates 4 levels of antenna arrangement.

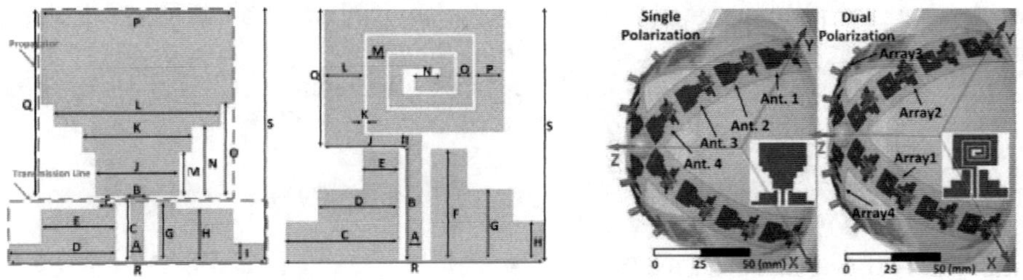

(a) (b)

Figure 6. The flexible 16 antenna array for microwave breast cancer detection. (a) The single antenna geometry for single and dual-polarization and (b) 16 antenna array assembly system. [18]

3. Conclusion and Future Research

A several works related to the development of antenna arrangement system for microwave imaging in early breast cancer detection has been done. In this work, several arrangement techniques that researchers have undertaken within this decade have been reported and tabulated in Table 1 for guidance and enhancement of future antenna arrangement. Frequency selection has been observed and found the wideband frequency suitable for use in this application. The reviews on antenna arrangement system has been presented on one level and multiple level. The arrangement of a single antenna at each stage is aimed at achieving the maximum number of antennas which will be benefited in the array system to be used. This is due to the theory that the number of antennas will affect the imaging resolution. In addition, the breast phantom model adopted in the investigated work were also identified and tabulated for consideration in future related to design work on breast phantom model. In summary, the measurement result of each investigated work shows agreement with their hypothesis by exhibited imaging result. However, improvement of antenna arrangement in the system probably will enhance the data to be reconstructed to produce better image resolution.

Table 1. Comparison of Developed Antenna Arrangement System

Name of antenna arrangement system	Frequency (GHz)	Level	No. of antenna	Breast phantom model
The annular antenna system [14]	2.2 - 12	One	12	No phantom
8-element microstrip patch antenna array [15]	3	One	8	Plexiglass and ethanol samples
Wide-slot antenna for early stage breast cancer detection with circular antenna arrangement [16]	4.5 – 10.9	One	16	Chemical-based heterogeneous phantom (oil-gelatin mixture)
Assembly conformal antenna array [17]	3.2 - 5.7	Two	16	Voxel-based breast model
1 to 16 points of antenna arrangement [13]	2 - 4	Four	16	Heterogeneous phantom (mimic dielectric properties)
The flexible 16 antenna array [18]	2 - 4	Four	16	Three layers homogeneous medium

References

[1] R. L. Siegel, K. D. Miller, and A. Jemal 2018 Cancer statistics *CA. Cancer J. Clin.* **68** no. 1, pp. 7–30

[2] J. Dengler, S. Behrens, and J. F. Desaga 1993 Segmentation of Microcalcifications in Mammograms *IEEE Trans. Med. Imaging* **12** no. 4, pp. 634–642

[3] S. I. Parker *et al.*1994 Breast Cancer Calcification Measurements Using Direct X-ray Detection in a Monolithic Silicon Pixel Detector *IEEE Trans. Nucl. Sci.* **41**, no. 6, pp. 2862–2873

[4] C. K. Abbey, R. J. Zemp, L. Jie, K. K. Lindfors, and M. F. Insana 2006 Observer efficiency in discrimination tasks Simulating Malignant and benign breast lesions imaged with ultrasound *Med. Imaging, IEEE Trans.* **25**, no. 2, pp. 198–209

[5] W. Liu, H. M. Jafari, S. Hranilovic, and M. J. Deen 2006 Time domain analysis of UWB breast cancer detection *23rd Biennial Symposium on Communications* no. 3, pp. 336–339.

[6] Z. . Zheng and T. . Zhang 2012 Microwave Imaging for Early Breast Cancer Detection *Sch. Enviromental Sci.*

[7] Z. Wang, E. G. Lim, Y. Tang, and M. Leach 2014 Medical applications of microwave imaging *Sci. World J.*

[8] K. A. H. Ping 2008 Inverse Scattering Analysis Applied to Breast Cancer Detection, Nagasaki University

[9] Z. N. Y. Chen, S. Yang, S. He 2009 Design and Analysis of Wideband Planar *Electromagnetics* **15**, pp. 95–112

[10] M. A. Sulaiman, M. T. Ali, I. Pasya, N. Ramli, H. Alias, and N. Ya'Acob 2012 UWB microstrip antenna based on circular patch topology with stepped blocks (wing) *IEEE Symp. Wirel. Technol. Appl. ISWTA* no. 600, pp. 262–265

[11] B. Biswas, R. Ghatak, and D. R. Poddar 2017 A Fern Fractal Leaf Inspired Wideband Antipodal Vivaldi Antenna for Microwave Imaging System *IEEE Trans. Antennas Propag* **65** no. 11, pp. 6126–6129

[12] D. Backstrom 2017 A Survey on Antennas used for Early Detection of Breast Cancer *Australian family physician* **35** no. 4, pp. 1736–1739.

[13] Y. Li, A. Santorelli, and M. Coates 2016 Comparison of microwave breast cancer detection results with breast phantom data and clinical trial data: Varying the number of antennas *2016 10th Eur. Conf. Antennas Propagation, EuCAP*

[14] A. S. Narkhede, A. Gandhi, S. Bhaskara Naik, and J. Prajapati 2016 Annular antenna system for Microwave Imaging for Breast Cancer Detection *Proc. - 7th IEEE Indian Antenna Week 2016 A Work. Adv. Antenna Technol. IAW 2016*, no. Iaw, pp. 21–24

[15] G. Govind, E. Johari, and M. J. Akhtar 2016 Microwave imaging using 8-element microstrip patch antenna array *2016 IEEE Antennas Propag. Soc. Int. Symp. APSURSI 2016 - Proc.* pp. 857–858

[16] S. S. Tiang, M. S. Hathal, N. S. N. Anwar, M. F. Ain, and M. Z. Abdullah 2014 Development of a compact wide-slot antenna for early stage breast cancer detection featuring circular array full-view geometry *Int. J. Antennas Propag.*

[17] F. Wang 2017 Assembly Conformal Antenna Array for Wearable Microwave Breast Imaging Application *Loughbrgh. Antennas Propag. Conf. (LAPC 2017)* pp. 1–5

[18] H. Bahramiabarghouei, E. Porter, A. Santorelli, B. Gosselin, M. Popovic, and L. A. Rusch 2015 Flexible 16 Antenna Array for Microwave Breast Cancer Detection *IEEE Trans. Biomed. Eng.*, **62** no. 10, pp. 2516–2525

Simulation Study of a Modified Rectangular Microstrip for the Hyperthermia Breast Cancer Procedure with SEMCAD X Solver

Kasumawati Lias[1*], Nabilah Abdul Jaffar[2], Norlida Buniyamin[2], Hazrul Mohamed Basri[1], Wan Azlan Wan Zainal Abidin[1], Kuryati Kipli[1], Nur Alia Athirah Mohtadzar[1]

[1]Faculty of Engineering, Universiti Malaysia Sarawak (UNIMAS), 94300, Kota Samarahan, Sarawak
[2]Faculty of Electrical Engineering, Universiti Teknologi MARA (UiTM), 40450, Shah Alam, Selangor

*lkasumawati@unimas.my

Abstract. A modified rectangular microstrip is introduced and presented as an applicator for hyperthermia cancer procedure. SEMCAD X solver is used as the simulation tool to investigate the rectangular microstrip, which is applied and radiated to the breast phantom. The research is carried out in order to enhance the penetration depth, which is up to 100mm and improve the focusing capability, which means that to reduce the unwanted hot spots at the vicinity area of the treated tissue. The specific absorption rate (SAR) distributions of 1 and 10 g weights of breast tissue are obtained as simulation results. The operating frequency utilized for the study is 915 MHz, with three operating powers of 10, 100, and 200 W. Beside rectangular, a circular microstrip applicator with microstrip feed line is also simulated in order to determine which shape of the microstrip applicator presented the greatest results in term of the depth penetration and focusing capability. As expected, the greatest SAR is provided by the microstrip applicator with rectangular shape. By that, the rectangular microstrip applicator is further explored and modified. From the outcomes, the rectangular microstrip with structure 4 presents the utmost specific absorption rate deposition; the required penetration depth is achieved, and focusing is enhanced with minimal unwanted hot spots at the vicinity of the treated area.

1. Introduction

Microstrip antenna introduces many inventive works and is one of the most dynamic fields in communication applications. Microstrip patch is generally designed and developed with various shapes, such as square, rectangular, circular, triangular, and elliptical or some other common shape. The most widely used among all are rectangular and circular shapes.

The fundamental structure of microstrip antenna comprises a metallic radiating patch element, which is integrated into a grounded dielectric substrate. Rectangular and circular patches are extensively used because they easily provide feed line flexibility, multiple frequency operation, linear and circular polarizations, frequency agility, and good bandwidth [1]. The basic configurations of both patches are illustrated in Figures 1 (a) and (b).

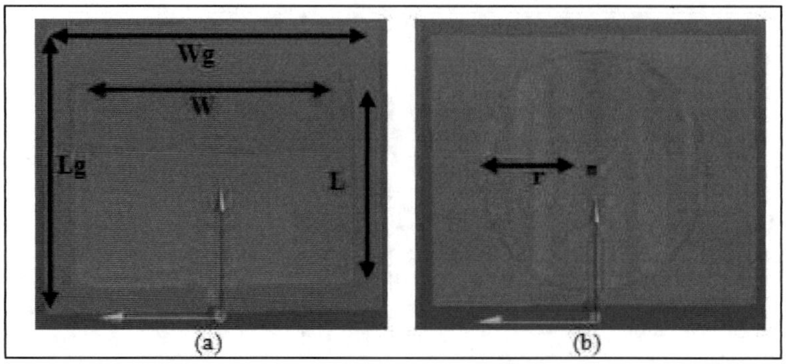

Figure 1. Microstrip patch antenna configuration
(a) rectangular (b) circular

Microstrip patch antennas are currently highly demanded in a broad range of microwave system applications, such as satellite communications, missile systems, and global positioning system for remote sensing. They have been introduced for biomedical use in this recent year as well. For instance, they are employed in hyperthermia cancer procedure [2], [3]. In articles, single and array arrangement of microstrip antenna is proposed, with different shapes, sizes, and structures. In [2], a circular patch with an operating frequency of 434 MHz was designed and developed to observe the radiation deposition into the area to be radiated. The effective field size (EFS) resulted from the simulation study was approximately 4 cm in diameter and 2 cm in penetration depth. Depth penetration was achieved with a wide range of radiation deposition. John Stang proposed an array arrangement of microstrip patch antennas in [3]. This arrangement aimed to enhance the focusing capability toward the targeted area to be radiated. Hence, the unwanted hot spots, which might lead to damaging of other healthy surrounding tissue, could be minimized.

Depth penetration and focusing capability are two crucial factors in the hyperthermia cancer procedure, especially when it involves a non-invasive microwave or radio wave applicator. In such an applicator, penetration depth and focusing capability may be needed to enhance to ensure the accessibility toward deep targeted areas and to minimize the effect of the unwanted hot spots resulted from the radiation absorption distribution by the applicator.
In this study, rectangular and circular patch microstrip antennas, which are also called "applicators for hyperthermia procedure," are introduced. The specific absorption rate (SAR) is simulated to determine and observe the radiation deposition of the applicators into the targeted area to be treated, which is the breast phantom in this research.

2. Design of simulation (DoS)

SEMCAD X is used as a simulation tool to design, develop, and investigate the rectangular and circular patches toward the hyperthermia breast cancer procedure. Before simulating SAR, the applicators and breast phantom are required to be constructed. The length and width of the rectangular patch are calculated with Equations (1)–(7) [1], [4]–[6]. For the rectangular patch, $w/L >> \lambda/2$, and the dielectric constant is usually in the range of $2.2 < \varepsilon r < 12$ [4]. The radius of the circular patch is presented Equations (8) and (9) [1], [6], [7]. The operating frequency to be used is 915 MHz, which is included under the industry, scientific, and medical (ISM) frequency. The substrate used for the investigation is silicon with $\varepsilon r = 11.9$.

$$w = \frac{c}{2f} x \sqrt{\frac{2}{\varepsilon r + 1}} \tag{1}$$

$$\varepsilon eff = \frac{\varepsilon r + 1}{2} + \frac{\varepsilon r - 1}{2}\left[1 + \frac{10h}{w}\right]^{-0.5} \tag{2}$$

$$Leff = \frac{c}{2f\sqrt{\varepsilon eff}} \tag{3}$$

$$\Delta L = 0.412XhX\frac{\varepsilon eff + 0.3}{\varepsilon eff - 0.258}X\frac{\frac{w}{h} + 0.264}{\frac{w}{h} + 0.8} \tag{4}$$

$$L = Leff - 2\Delta L \tag{5}$$

$$Lg = 6h + L \tag{6}$$

$$Wg = 6h + w \tag{7}$$

$$a = \frac{F}{\left(\sqrt{1 + \frac{2h}{\pi\varepsilon rF}\left[ln(\frac{\pi F}{2h}) + 1.7726\right]}\right)} \tag{8}$$

Where;

$$F = \frac{8.791X10^{9}}{f\sqrt{\varepsilon r}} \tag{9}$$

In the above equations, w is the width of the patch; L is the length of the patch; εeff and $Leff$ are the effective dielectric constant and the effective length of the patch, respectively. c is the free-space velocity of light, h is the substrate thickness, ΔL is the extension of length caused by the fringing effect, f is the frequency, and a is the radius of the circular patch. According to[8], Equations (8) and (9) for circular patch do not consider the fringing effect. The fringing effect makes the patch electrically large. Hence, the effective radius, ae, as written in Equation (10), is used.

$$ae = a\left(1 + \frac{2h}{\pi\varepsilon ra}\left[ln\left(\frac{\pi a}{2h}\right) + 1.7726\right]\right)^{1/2} \tag{10}$$

After constructing the applicators, the breast phantom, which consists of the breast fat and breast cancer/tumor, is developed. The breast cancer/tumor phantom is positioned 100 mm deep from the outer side of the breast skin. The radius for the breast fat is 100 mm, whereas the breast tumor phantom has a radius of 50 mm. The electrical and thermal properties are provided in Table 1. The properties are retrieved from the Gabriel database of the SEMCAD X solver. The simulation is conducted with the applicator 2 mm from the breast phantom.

Further modification on either one of the patches, which provides the utmost SAR deposition with required penetration depth and focusing capability, is conducted to further enhance the outcomes and to observe the effect of the various shapes and structures on the SAR deposition of the breast phantom.

The simulation study also includes the investigations on the different operating powers and applicator distances from the breast phantom. Finally, water bolus is integrated into the hyperthermia procedure. Water bolus is used to provide a cooling environment to minimize skin burn and to provide a comfortable situation to the patient during the hyperthermia procedure. Distilled water is utilized as a coolant fluid, with εr, σ, and ρ of 76.7, 5e-005, and 1000, respectively. The thickness of the water bolus is stipulated to 15 mm.

Table 1. Breast phantom electrical and thermal properties

	915 MHz				
	Relative Permittivity, εr	**Electrical Conductivity, σ (S/m)**	**Density, ρ(kg/m3)**	**Specific Heat Capacity, C (J/kg/K)**	**Thermal Conductivity, K (W/m/K)**
Breast Fat	4.6988	0.125134	1000	2348.33	0.209
Breast Cancer/ Breast Tumor	48.3624	2.65306	1000	2352.55	0.789

The summary of the DoS is illustrated in Figure 2.

3. Result and discussion

Based on Equations (1)–(7), the parameter specifications of the rectangular and circular microstrip applicators are provided in Table 2.

Table 2. Rectangular microstrip applicator specifications

	915 MHz	
Dimension		**Unit**
	Silicon	
Dielectric Constant (εr)	11.9	
Substrate Thickness (h)	2	mm
Length (L)	48	mm
Width (W)	65	mm
Ground Length (Lg)	60	mm
Ground Width (Wg)	77	mm
Radius (a)	1	mm
Effective Radius (ae)	2	mm

Table 3 shows the SAR deposition for the rectangular and circular patches with an operating frequency of 915 MHz and an operating power of 10 W. Based on this table, the rectangular patch provides the utmost result of SAR deposition compared with the circular patch. The rectangular patch is then modified to enhance the required penetration depth and focusing capability toward the breast phantom. The SAR distributions of different rectangular structures and operating powers are provided in Table 4. The rectangular patch with structure 4 provides the utmost results. Meanwhile, in terms of different powers, the SAR distribution is wider, when higher operating power is utilized. In addition, the SAR value increases when the operating power increases, either for 1 or 10 g weight of the breast tissue. The brightest color of the SAR deposition represents the highest radiation absorption. SAR refers to the transferred power divided by the mass of the object, which is provided in Equation (11). According to[9], SAR takes a value proportional to the square of the electric field around the antenna and is equivalent to the heating source generated by the electric field in the tissue.

$$SAR = \frac{\sigma}{2\rho}|E|^2 \text{ (W/kg) or (mW/g)} \tag{11}$$

where σ is the conductivity of tissue (S/m), E is the electric field (V/m), and ρ is the density of tissue (kg/m^3).

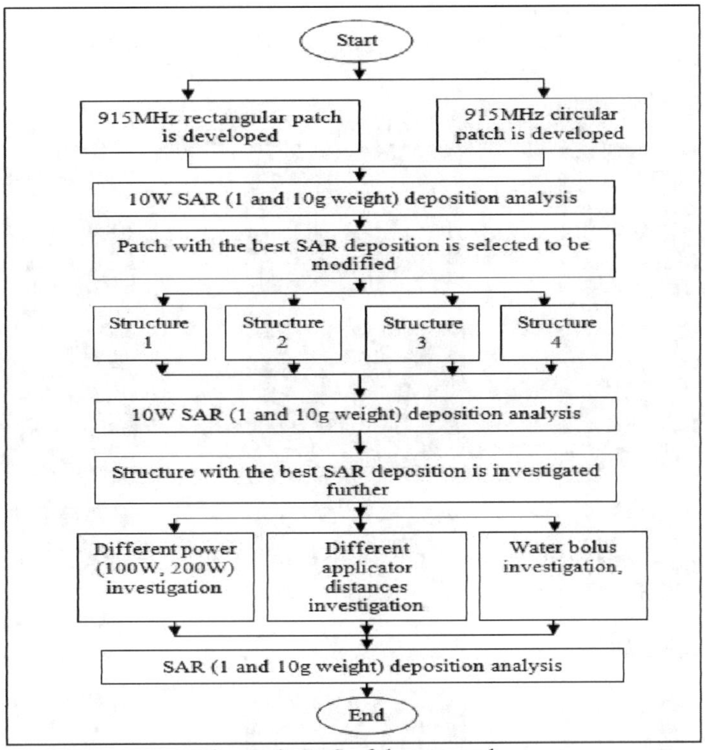

Figure 2. DoS of the research

Table 3. SAR deposition for rectangular and circular patches

Patch/Mass of Breast Tissue	Rectangular Patch	Circular Patch
1 g	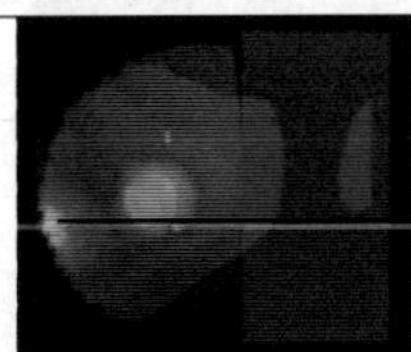	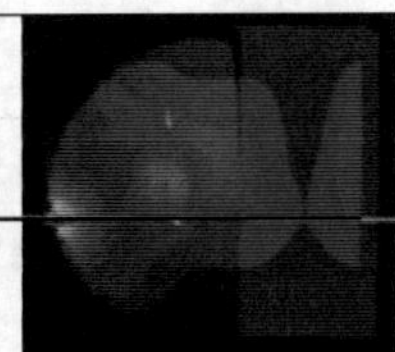
10 g	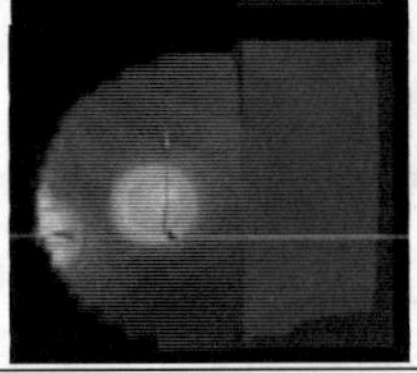	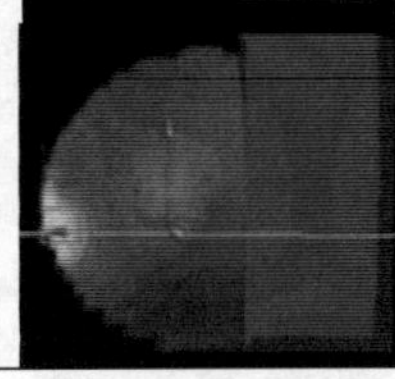

The SAR deposition of different distances of the rectangular patch with structure 4 applicators from the breast phantom is indicated in Table 5, either with the addition of water bolus to the treatment procedure or without water bolus. From the simulated SAR distribution in Table 5, 2mm water bolus thickness and 2mm distances without water bolus have presented the greatest results of penetration depth and focusing capability of the applicator.

Table 4. SAR for different rectangular structures and operating powers

915 MHz		
Operating Power / Rectangular Structures	**10 W**	
	SAR 1 g	**SAR 10 g**
	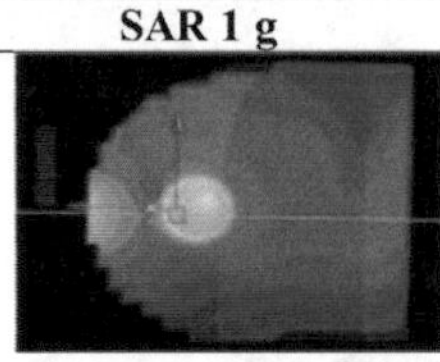	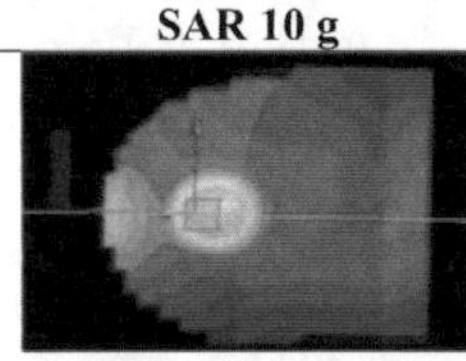
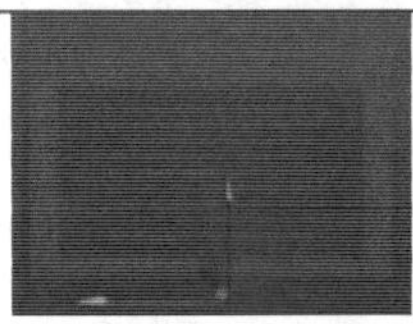	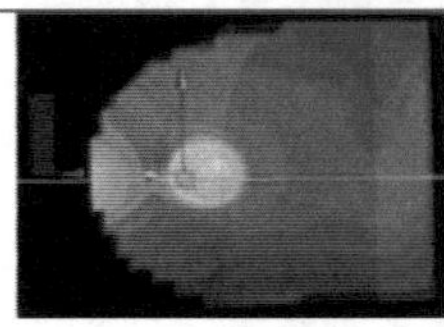	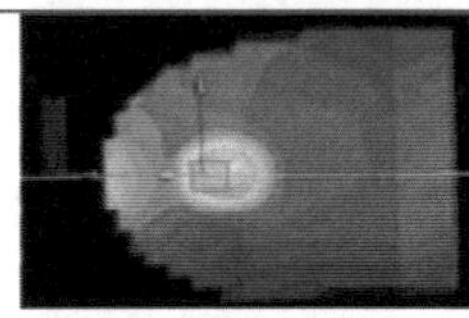
	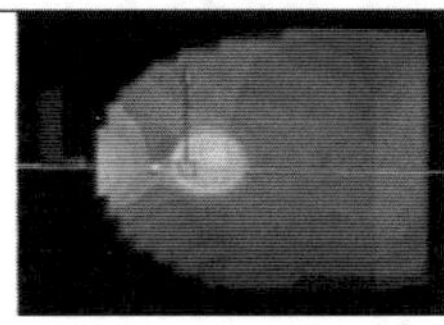	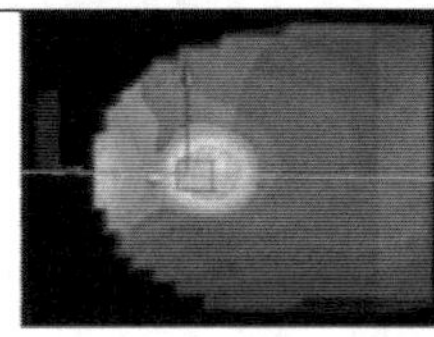
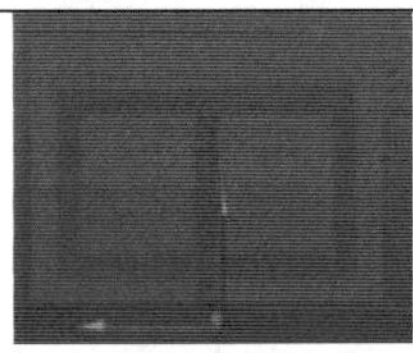	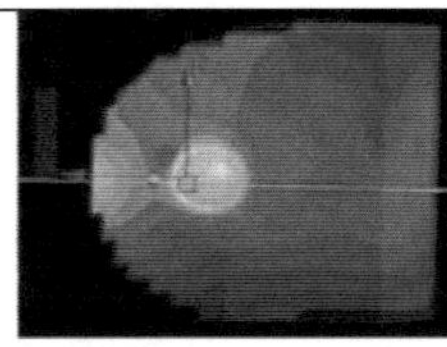	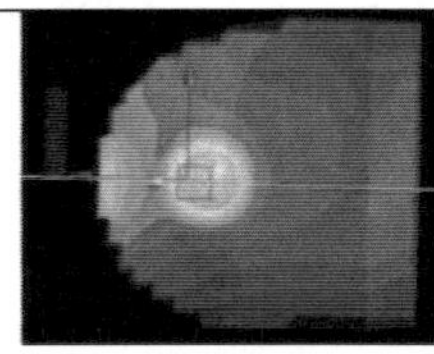
	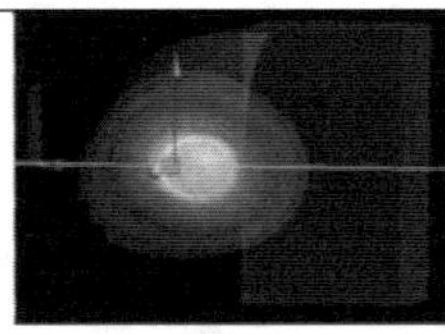	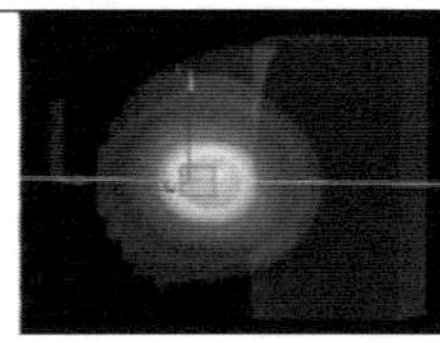

Table 5. SAR for different rectangular patches with structure 4 applicator distance and different operating powers

915 MHz			
Without Water Bolus			
Operating Power /Distances	**10 W**	**100 W**	**200 W**
SAR 1 g			
2 mm	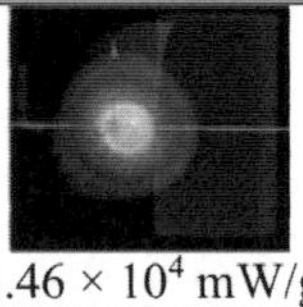 1.46×10^4 mW/g	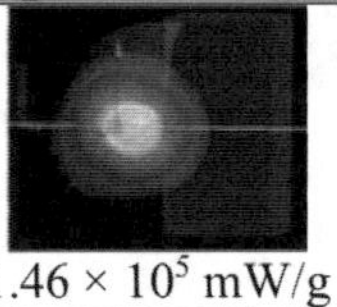 1.46×10^5 mW/g	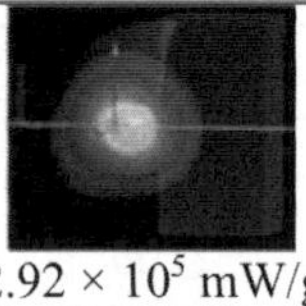 2.92×10^5 mW/g

10 mm	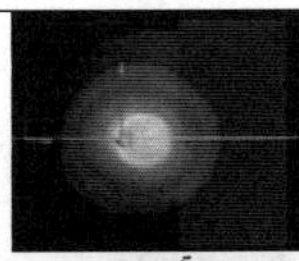 3.97×10^5 mW/g	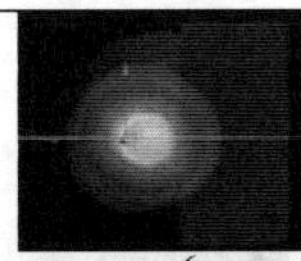 3.97×10^6 mW/g	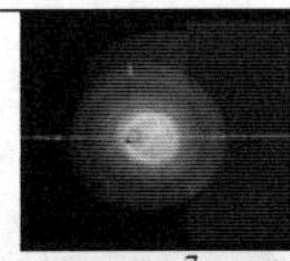 1.79×10^7 mW/g
100 mm	 1.39×10^4 mW/g	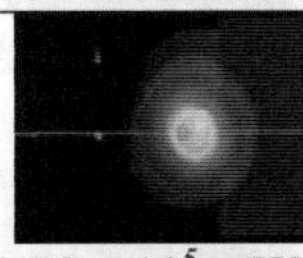 1.39×10^5 mW/g	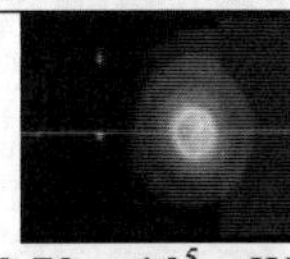 2.79×10^5 mW/g
SAR 10 g			
2 mm	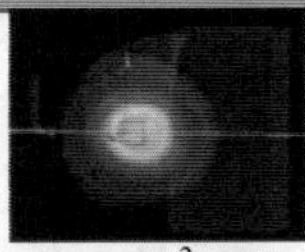 8.93×10^3 mW/g	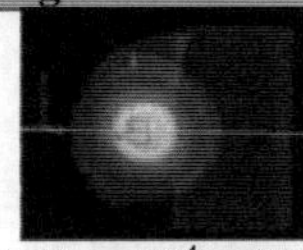 8.93×10^4 mW/g	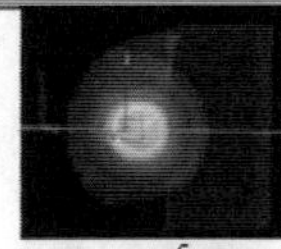 1.79×10^5 mW/g
10 mm	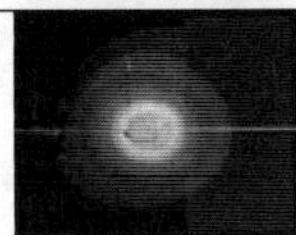 6.16×10^5 mW/g	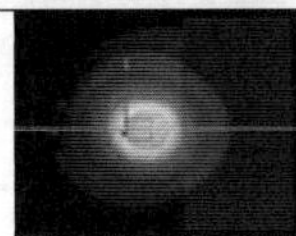 6.16×10^6 mW/g	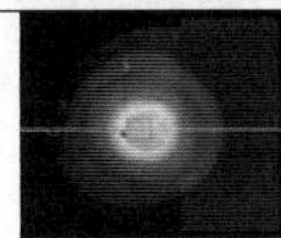 1.23×10^7 mW/g
100 mm	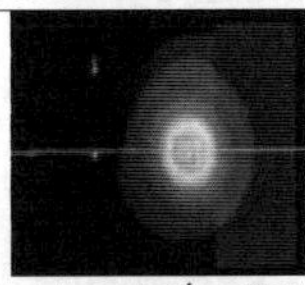 1.11×10^4 mW/g	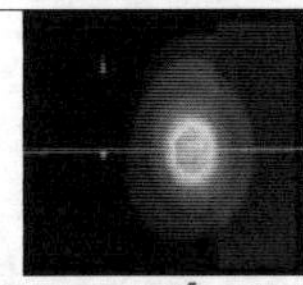 1.11×10^5 mW/g	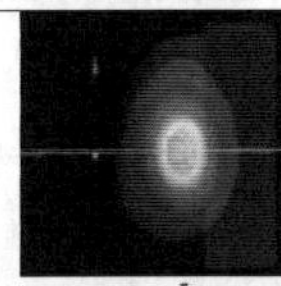 2.22×10^5 mW/g

With Water Bolus

Operating Power /Distances	**10 W**	**100 W**	**200 W**
SAR 1 g			
2 mm	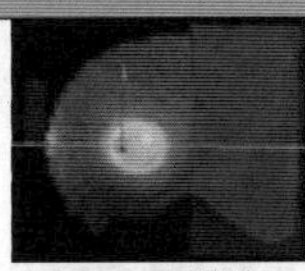 29.7 mW/g	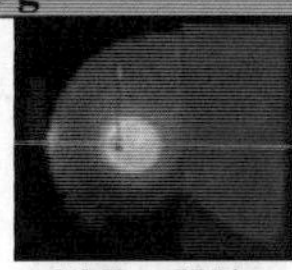 297 mW/g	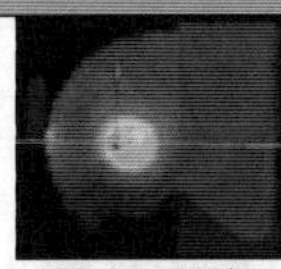 594 mW/g
10 mm	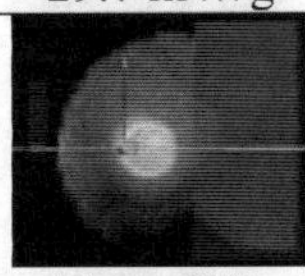 11.4 mW/g	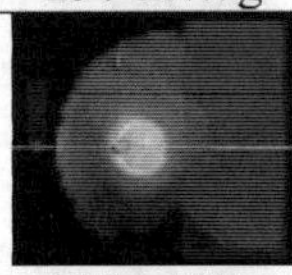 114 mW/g	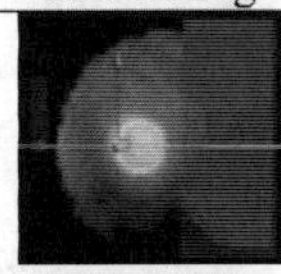 228 mW/g
100 mm	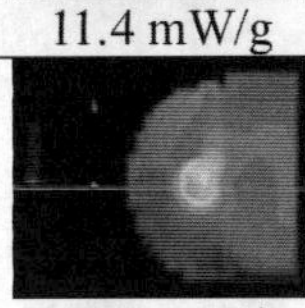 0.162 mW/g	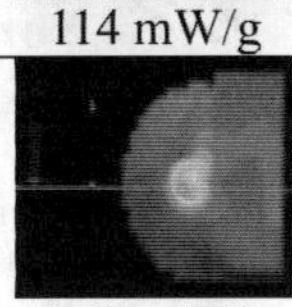 1.62 mW/g	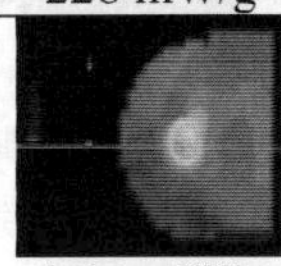 3.24 mW/g
SAR 10 g			
2 mm	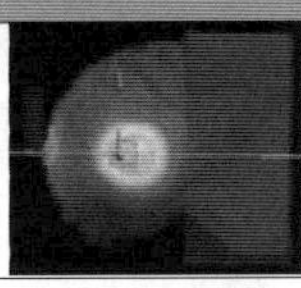	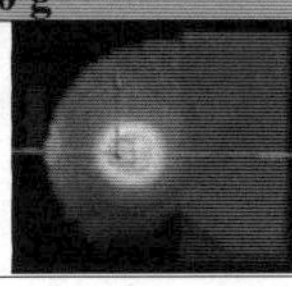	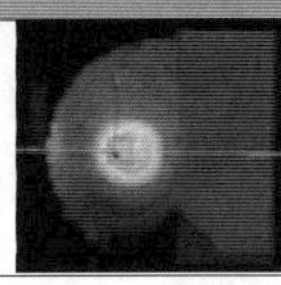

	18.1 mW/g	181 mW/g	362 mW/g
10 mm	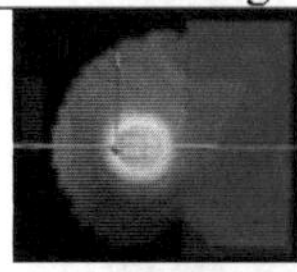	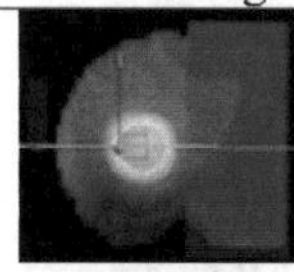	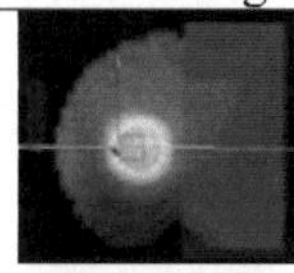
	7.59 mW/g	75.9 mW/g	152 mW/g
100 mm	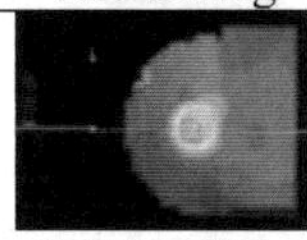	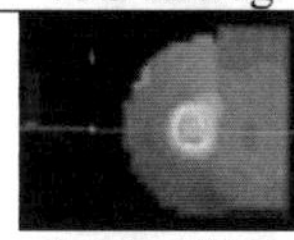	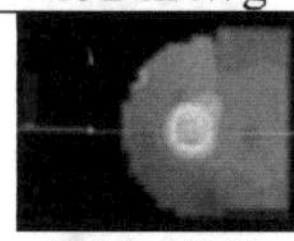
	0.119 mW/g	1.19 mW/g	2.37 mW/g

4. Conclusion

Rectangular patch applicator provides better SAR deposition than circular patch applicator. When the rectangular patch is modified, the rectangular patch with structure 4 presents the utmost SAR deposition, in which the penetration depth and focusing capability are at the greatest agreement.

Acknowledgments

The authors gratefully acknowledge and thank the University Malaysia Sarawak (UNIMAS) and Universiti Teknologi MARA (UiTM) for providing facilities for this research. The appreciation also goes to *Geran Initiatif Penyeliaan* (600-IRMI/GIP 5/3 (088/2018) for financial supports of this research.

References

[1] T. Ferdous, A. Nayna, and F. Ahmed 2014 Comparative Study of Rectangular and Circular Microstrip Patch Antennas in X Band *International Conference on Electrical Engineering and Information & Communication Technology (ICEEICT)* pp. 1–4

[2] T. Drizdal, P. Togni, and J. Vrba 2007 Microstrip Applicator for Local Hyperthermia *2007 International Conference on Electromagnetics in Advanced Applications*pp. 1047–1049.

[3] J. Stang, M. Haynes, P. Carson, and M. Moghaddam 2012 A preclinical system prototype for focused microwave thermal therapy of the breast *IEEE Trans. Biomed. Eng.* **59**, no. 9, pp. 2431–2438

[4] P. Subbulakshmi and R. Rajkumar 2013 Rectangular Microstrip Patch Array Antenna *IEEE Int. Conf. Emerg. Trends Comput. Commun. Nanotechnol. (ICECCN 2013)* pp. 547–552

[5] M. Ahamed, K. Bhowmik, and A. Al Suman 2012 Analysis And Design of Rectangular Microstrip Patch Antenna On Different Resonant Frequencies For Pervasive Wireless Communication *Int. J. Sci. Technol. Res* **1**, no. 5, pp. 108–111

[6] A. Arya, S. Sharma, P. Yadav, and R. Dindha 2012 Performance Comparison between Rectangular & Circular Patch Antenna Array with EBG Structure Abstract *Int. J. Appl. Eng. Res.* **7**, no. 11

[7] A. Dalli, L. Zenkouar, and S. Bri 2012 Comparison of Circular Sector and Rectangular Patch Antenna Arrays in C-Band *J. Electromagn. Anal. Appl.* pp. 457–467

[8] B. Kwaha, O. Inyang, and P. Amalu 2011 The Circular Microstrip Patch Antenna-Design And Implementation. *IJRRAS* **8**, pp. 86–95

[9] D. U. Yong-xing, 2009 The Analysis and Simulation of microstrip Spiral Antenna for Microwave Hyperthermia *2009 3rd Int. Conf. Bioinforma. Biomed. Eng.*, pp. 1–4

Segmentation of Cortical and Cancellous Bone with Osteogenesis Imperfecta using Thresholding-based Method

Tan Tze Voon[1], Mohd Hanafi Mat Som[1*], Haniza Yazid[1], Khairul Salleh Basaruddin[1] and Abdul Razak Sulaiman[2]

[1]School of Mechatronic Engineering, Universiti Malaysia Perlis, Pauh Putra Campus, 02600, Arau, Perlis, Malaysia.
[2]Department of Orthopaedics, School of Medical Science, Universiti Sains Malaysia, 16150 Kubang Kerian, Kelantan, Malaysia

*mhanafi@unimap.edu.my

Abstract. Osteogenesis Imperfecta (OI) is a genetic bone disorders that mainly affect the bones which commonly leads to the multiple fractures. The purpose of this study is to segment the cortical and cancellous bone of tibia affected with the OI from the CT images. This project consists of two sections, segmentation of the cortical and cancellous bone, and the evaluation of the image performance. Contrast adjustment was implemented to enhance the contrast of images. K-Means and multi-threshold were implemented to segment the cortical bone at the proximal, diaphysis, and distal tibia. Post-processing was applied to further refine the segmented images. All segmented images were then evaluated by using the ground truth images. From the results, the process with contrast adjustment and multi-threshold obtained the highest accuracy and Dice-coefficient of 99.91% and 81.42% respectively at the proximal region while 99.93% and 73.88% at the distal region. However, pathway that make use of K-Means performed best in diaphysis region. This method has obtained the highest accuracy and Dice-coefficient of 99.98% and 95.59% respectively.

1. Introduction

Osteogenesis Imperfecta (OI) is a type of genetic bone disorder where the bones break or fracture easily by means of fragile bone [1]. It is estimated to happen in one per 30,000 of live births. OI is caused by the mutation of type 1 collagen genes (COL1A1 and COL1A2) [2]. Multiple fracture is the most common symptom for this disease. Prediction of the bone fracture become the main issue for this disease to determine the maximum force that the bone can withstand. Hence, segmentation of cortical and cancellous bone of tibia from CT images have become an important step for the purpose of reconstruct the bone which can be used in the prediction of the fracture risk [3–5].

Sahama et al. and Mansor et al. [6,7] implemented multi-threshold method, Canny edge detection, and single threshold in the segmentation of long bone with different threshold levels for proximal, diaphysis, and distal region. Threshold-based method is considered as the easiest way in the perform the image segmentation. Multi- threshold obtained the highest accuracy compared to Canny edge detection and single threshold old. Clustering method is one of the unsupervised process that grouping a set of similar objects in the same cluster or by means of separated into uniform groups. A comparison between the K-means and Fuzzy C-Means based on clustering accuracy for different shaped of clustered structures where there are in regular or irregularly scattered has been proposed by Cebeci et al. [8]. In this study, image processing techniques were applied on the CT images of the tibia to segment the cortical and cancellous bone. The segmentation of the image was challenging due to the abnormal shape of type III OI bone. The approach includes pre-processing of the image to enhance the image, segmentation of the images using different techniques, and post-processing of the image for further refinement.

2. Methodology

The 388 CT images of type III OI tibia were obtained from Hospital Universiti Sains Malaysia (HUSM) with patient consent. Out of 388 tibia images, 31 are proximal, 330 are diaphysis, and 27 are distal tibia. Each slice of image has a dimension of 512 × 512.

2.1. Segmentation of Cortical Bone at Proximal, Diaphysis and Distal Region

There were two pathways implemented to segment the cortical bone. One includes with the pre-processing stage before segmentation while another path excludes the pre-processing stage. Figure 1 and Figure 2 show the summary of the pathways.

2.1.1. Overview of Pathway 1

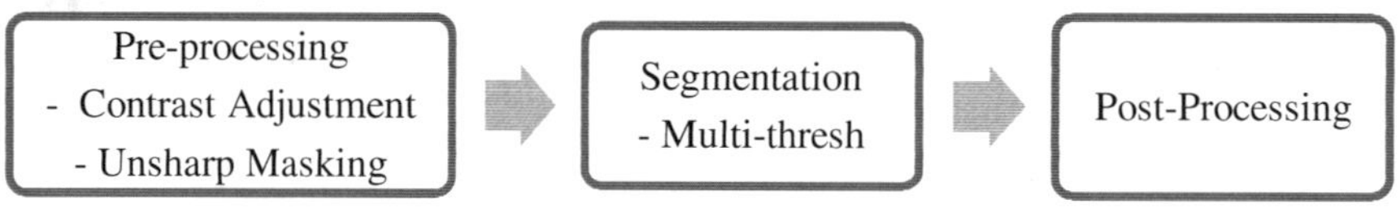

Figure 1. Block diagram of Pathway 1

The intensity values of the image are adjusted to increase the contrast of the image especially between the cortical bone and the soft tissues. The bright areas appear brighter and conversely the dark areas appear darker. Unsharp masking was applied to enhance the edge details. Multi-threshold method was applied in the segmentation process to separate the interest region into foreground while unwanted image become the background. The output of the multi-threshold was set into binary.

2.1.2. Overview of Pathway 2

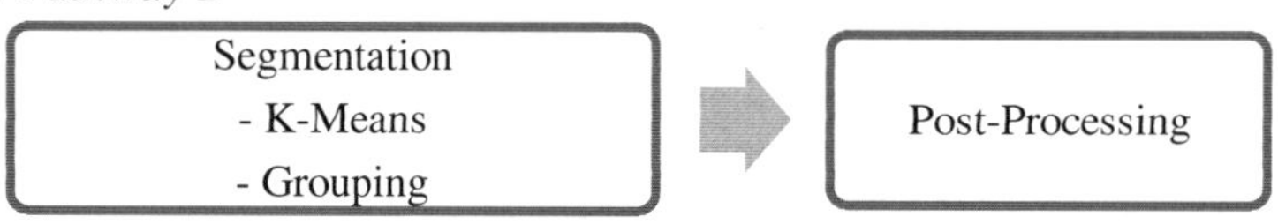

Figure 2. Block diagram of Pathway 2

The image was clustered into few groups according to the selected centroids of the clusters. Euclidean distance was used to assign the object to the closest cluster center and recalculated until there is no changes for the centroids. j is the objective function, k is the number of clusters, x is the number of cases, c_j is the centroid for the cluster j. Equation (1) shows the function of K-Means.

$$j = \sum_{j=1}^{k} \sum_{i=1}^{x} \left\| x_i^{(j)} - c_j \right\|^2 \tag{1}$$

The output image of K-Means is in grayscale form. The images were then grouped based on the intensity value of the region of interest into foreground (one) while the unwanted regions are grouped together into background (zero).

2.2. Post-processing in Pathway 1 and 2

There were two methods of post-processing, the first method was used for the cortical bone at proximal and distal regions while the second method was applied for the diaphysis region of tibia. This approach was taken address the different of bone thickness where the cortical bone at proximal and distal regions is thinner than the cortical bone at the diaphysis region. The illustration of the methods is shown in Figure 3 and Figure 4.

2.2.1. Post-processing for Proximal and Distal Region of Tibia

Figure 3. Summary of post-processing in proximal and distal region

After segmentation process, there were still some unwanted small pixel regions around of inside the cortical bone. It was necessary to remove the small pixel regions and extract the cortical where the cortical had the largest area compared to other small areas. The cortical at proximal and distal region had almost similar intensity values as the soft tissue, thus, some of the cortical that had been extracted often face issues of incomplete boundary. Hence, edge linking process was performed to connect the endpoint of the broken boundary. Since the cortical bone at the proximal and distal region were thinner compared to the diaphysis region, therefore, morphological operation was applied to obtain the thickness that almost similar with the original grayscale image. There were few steps used in morphological operation:

i) The completed boundary of the cortical is filled, then the boundary of the filled region is extracted.
ii) The filled region of the cortical bone undergo erosion, then the boundary after erosion is extracted.
iii) The boundary extracted in step (i) combined with the boundary extracted in step (ii).

2.2.2. Post-processing for Diaphysis Region of Tibia

Figure 4. Summary of post-processing in diaphysis region

The output of the segmentation for diaphysis region consists of the tibia with fibula. In order to remove the unwanted region, fibula was labelled. Output of the labelled fibula was matched with the output of segmentation. The pixels with the same intensity value of 1 between labelled fibula with the output of segmentation was set as background (0) while the unmatched pixels were set as foreground (1).

2.3. Whole Tibia and Segmentation of Cancellous Bone

Since the cancellous bone is bounded by cortical bone, this indicates that the center region of the cortical segmented was considered as the cancellous bone. In order to segment the cancellous bone as foreground while other parts include cortical as the background, region filling was first applied to the segmented cortical. This results in the formation of whole bone region of tibia. The output of the region filling is matched with the cortical segmented. Both images with the same intensity value of 1 was set to 0 while the unmatched pixel was set to 1.

2.4. Image Validation

The performance of image segmented was then evaluated by comparing with the ground truth image. In this study, ground truth images were created manually by using GIMP software. Evaluation techniques includes sensitivity, specificity, accuracy, and Dice-Coefficient which based on the confusion matrix. The terms that associated with the confusion matrix are True Positive (TP), True Negative (TN), False Positive (FP) and False Negative (FN).

Dice-Coefficient is one of the similarities measures used to calculate the similarities and variation between the output image segmented and ground truth image. Sensitivity is used to measure the positive pixels in the ground truth that are also identified as positive by the result of segmented image. Specificity used to measure the negative pixels in the ground truth that are also identified as the negative in the segmented image. Accuracy is the number of correct predictions made by overall of the predictions made. Equation (2), (3), (4) and (5) shows the calculation of Dice-Coefficient, Sensitivity, Specificity and Accuracy.

$$Dice = 2TP / (2TP + FP + FN) \quad (2)$$

$$Sensitivity = TPR = TP / (TP + FN) \quad (3)$$

$$Specificity = TNR = TN / (TN + FP) \quad (4)$$

$$Accuracy = (TP + TN) / (TP + TN + FP + FN) \quad (5)$$

3. Results and Discussion

3.1. Segmentation of Cortical Bone

Both examples of process outputs for first and second pathway are shown in Figure 5 and Figure 6. Two pathways were used to segment the cortical bone at the proximal, diaphysis, and distal region of tibia.

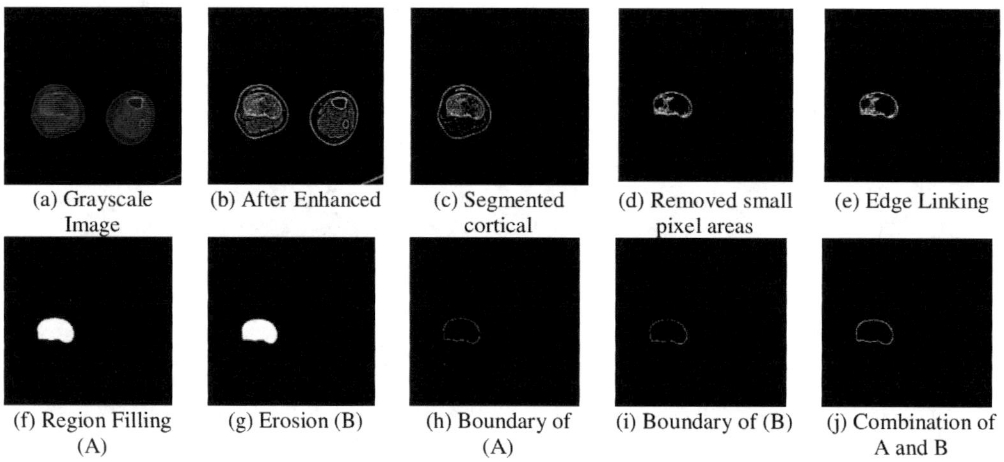

Figure 5. Process of Pathway 1 at Proximal Region

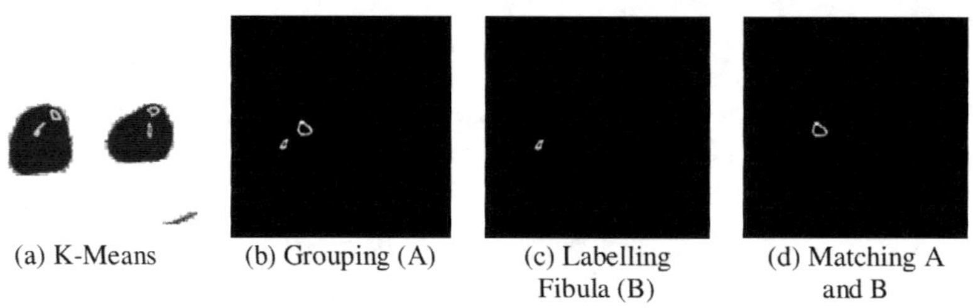

Figure 6. Process of Pathway 2 at Diaphysis Region

The results of segmented cortical for each proximal, diaphysis, and distal region were almost similar in shape and thickness. There were only differences in some small region of white pixels. In order to compare the results, evaluation was carried out by using the ground truth images. Table 1 shows the average results for both pathways at different regions of tibia.

First pathway performed the best in proximal and distal regions while second pathway provide a better result than first pathway at the diaphysis region. Both accuracy and specificity obtained results with 99% or almost perfect. However, the similarity that obtained at the proximal and distal region were lower compared to the diaphysis region. Since the cortical bone at the proximal and distal regions are much thinner compared to the diaphysis region and has almost same intensity as the soft tissue that located around the cortical bone, it was easily to cause incomplete boundary upon segmentation. Hence, pre-processing was applied to the image to enhance the contrast between the cortical bone and surrounding soft tissues. The second pathway without the pre-processing stage obtained the highest similarity at the diaphysis region compared to the proximal and distal regions. This is because the contrast between the cortical bone and the soft tissue is very good compared to other regions of tibia. Hence, the first pathway with pre-processing stage that applied in diaphysis region caused some over-enhanced which would affect the similarity of the cortical segmented. The segment size directly affects the Dice-coefficient. The smaller the segment size, the higher the sensibility of the image. It would be easily affected if one of the positive pixels of output image is unmatched to the ground truth image [9].

Table 1. Performance of segmented image of cortical bone

	Proximal		**Diaphysis**		**Distal**	
	Path 1	Path 2	Path 1	Path 2	Path 1	Path 2
Average Sensitivity	0.8510	0.8412	0.8889	0.9840	0.7053	0.6290
Average Specificity	0.9994	0.9994	1.0000	0.9999	0.9997	0.9996
Average Accuracy	0.9991	0.9990	0.9998	0.9998	0.9993	0.9991
Average Dice-Coefficient	0.8142	0.8051	0.9320	0.9559	0.7388	0.6575

3.2. Whole Tibia and Segmentation of Cancellous Bone

First pathway performed the best in segmentation of cortical bone in proximal and distal regions. Thus, the output of this pathway was used to segment the cancellous bone. On the contrary, second pathway that obtained the highest accuracy in diaphysis region was implemented to segment the cancellous bone. It was challenging to segment the cancellous bone prior to cortical bone since the intensity of the cancellous bone was almost similar with the background intensity. Hence, the cortical was segmented first and then followed by the segmentation of cancellous bone. However, there was a limitation where the cortical bone must be successfully segmented with a complete boundary. Only then, one can apply the method of segmentation of cancellous bone that has been proposed earlier. The results of the performance were shown in Table 2.

Table 2. Performance of segmented image of whole tibia and cancellous bone

	Whole Tibia			**Cancellous Bone**		
	Proximal	Diaphysis	Distal	Proximal	Diaphysis	Distal
Average Sensitivity	0.9994	0.9950	0.9990	0.9953	0.9255	0.9996
Average Specificity	0.9995	0.9999	0.9997	0.9996	1.0000	0.9996
Average Accuracy	0.9995	0.9999	0.9997	0.9995	0.9999	0.9996
Average Dice-Coefficient	0.9905	0.9836	0.9769	0.9901	0.9571	0.9588

In term of similarity, the segmented cancellous bone was better compared to the cortical bone. In term of sensitivity, the results for both cancellous and whole tibia were 90% and above. In term of specificity and accuracy, both segmented cancellous and whole tibia obtained 99%. This indicates that the segmentation algorithms were able to identify the region of interest correctly. From the results, it can be concluded that the methods of segmentation of cortical and cancellous bone performed well in this study.

3.3. 3D Reconstruction of Tibia

Figure 7 shows the 3D reconstruction of tibia by using the VOXELCON software. This model was reconstructed from a total of 388 cortical and cancellous images.

(a) (b) (c)

Figure 7. (a) Cortical bone model, (b) cancellous bone model and (c) Combination model

4. Conclusion

This study employs a total of 388 CT images of type III OI tibia. The cortical and cancellous bone that composed in the CT images were segmented successfully using the image processing techniques that proposed in this paper. Results that were obtained has proved the ability of the segmentation algorithms to separate the cortical and cancellous bone where it is important in reconstruction of the 3D bone model for the future analysis such as prediction of fracture risk or in the study of the mechanical stress distribution on the tibia that affected with OI.

Acknowledgement

The authors would like to acknowledge the support from the Fundamental Research Grant Scheme (FRGS) under a grant number of FRGS/1/2016/TK03/UNIMAP/02/6 from the Ministry of Education Malaysia.

References

[1] Bishop N 2010 Characterising and treating osteogenesis imperfecta *Early Hum. Dev.* **86** 743–6

[2] Van Dijk F S, Pals G, Van Rijn R R, Nikkels P G J and Cobben J M 2010 Classification of Osteogenesis Imperfecta revisited *Eur. J. Med. Genet.* **53** 1–5

[3] Wanna S B C, Basaruddin K S, Mat Som M H, Sulaiman A R, Shukrimi A, Khan S F, Abdul Majid M S and Ridzuan M J M 2018 Fracture risk prediction on children with Osteogenesis Imperfecta subjected to loads under activity of daily living *IOP Conference Series: Materials Science and Engineering* vol 429 p 012004

[4] Wanna S B C, Basaruddin K S, Mat Som M H, Mohamad Hashim M S, Daud R, Abdul Majid M S and Sulaiman A R 2017 Prediction on fracture risk of femur with Osteogenesis Imperfecta using finite element models: Preliminary study *Journal of Physics: Conference Series* vol 908 p 012022

[5] Wanna S B C, Basaruddin K S, Som M H M, Salleh A F and Sulaiman A R 2018 Effect of loading direction on fracture of bone with osteogenesis imperfecta (OI) during standing *AIP Conference Proceedings* vol 2030 p 020094

[6] Rathnayaka K, Sahama T, Schuetz M A and Schmutz B 2010 Effects of CT image segmentation methods on the accuracy of long bone 3D reconstructions *Med. Eng. Phys.* **33** 226–33

[7] Mansor N, Beh Y M, Mat Som M H, Basaruddin K S, Mustafa N, Yazid H and Salleh A F 2018 Reconstruction of cortical and cancellous bone in tibia with osteogenesis imperfecta *J. Telecommun. Electron. Comput. Eng.* **10** 115–9

[8] Cebeci Z and Yildiz F 2015 Comparison of K-Means and Fuzzy C-Means Algorithms on Different Cluster Structures *J. Agric. Informatics* **6** 13–23

[9] Taha A A and Hanbury A 2015 Metrics for evaluating 3D medical image segmentation: analysis, selection, and tool *BMC Med. Imaging* **15** 29

Automated Intelligent Diagnostic Procedure for Chronic Leukemia

Chan Lok Pooi[1], Mohd Yusoff Mashor[1*], Robiyanti Adollah[1] and Rosline Hassan[2]

[1] School of Mechatronics Engineering, Universiti Malaysia Perlis, 02600 Arau, Perlis, Malaysia.

[2] Department of Haemotology, School of Medical Science, Univeristi Sains Malaysia, Kubang Kerian, Kelantan, Malaysia.

*yusoff@unimap.edu.my

Abstract. Due to growing statistics and the important role of early diagnosis for chronic Leukemia, an automated intelligent diagnostic procedure for chronic Leukemia is needed. This paper presents an automated procedure for this diagnostic system which consisted of four main stages; namely Image Segmentation, Feature Extraction, Feature Selection and Classification. Colour Thresholding and Gradient Edge Detection were applied to segment the nucleus of white blood cells from the image. A total of 548 cells nuclei were extracted from 100 images and used to develop and validate the proposed procedure. Three feature selection algorithms and three classifiers were analyzed and evaluated in order to obtain a reliable and robust automated diagnostic procedure. Based on the results, the ReliefF algorithm was selected and applied for feature selection, the method yields the best accuracy of 98.1% for testing accuracy. Besides, the selected classifier was Multilayer Perceptron (MLP) network trained by Levenberg Marquardt (LM) algorithm.

1. Introduction

Leukocytes (or White Blood Cells) is the cell of the immune system that protects the human body by fighting infection or foreign invaders. However, Leukemia disrupts the function of blood. Generally, there are four types of Leukemia, which are Chronic Lymphocytic Leukemia (CLL), Acute Lymphocytic Leukemia (ALL), Chronic Myelogenous Leukemia (CML), and Acute Myelogenous Leukemia (AML). Generally, the family history and physical checking are the starting point in the diagnosis of Leukemia, but this is not enough to determine the presence of Leukemia for a person. Tissue biopsies taken from the bone marrow (bone marrow biopsy) or lymph nodes (lymph node biopsy) is done to look for the evidence of Leukemia.

According to statistics of the Leukemia & Lymphoma Society (LLS) [1] and MediMedia Asia (MIMS) Malaysia [2], the problem of Leukemia in Malaysia is growing, but the survivor rate for chronic leukemia is relatively high, provided it can be diagnosed and treated at the early stage [3]. In addition, World Life Expectancy [4] claimed that the death rate of leukemia in Malaysia is 4.18 that corresponding to 75 ranks over 183 countries in the world. Due to growing statistics and the important role of early diagnosis for chronic leukemia, an intelligent screening system is proposed to fulfill the demand.

2. Methodology

In general, there are four main stages involved in this study, namely Image Segmentation, Feature Extraction, Feature Selection and Classification. The general methodology was shown in Figure 1. In this section, the development process of the intelligent screening system was presented.

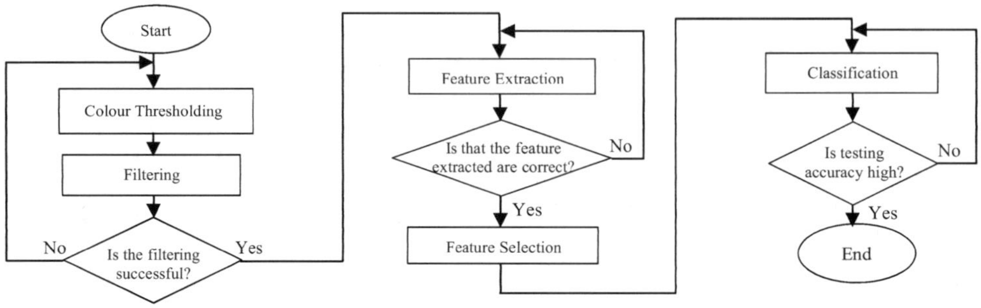

Figure 1. General methodology flowchart of an intelligent screening system

2.1. Image Segmentation

Image segmentation technique is important to segment the irrelevance and unwanted information with object of interest. The purpose of image segmentation is to provide the meaningful representation of the image to the human [5]. The techniques applied in this study were Colour Thresholding [6] and Gradient Edge Detection [7].

In Colour Thresholding, determination of the threshold value was the main task in this study. The threshold value was determined by the pixel value of RGB components for different regions, red blood cells (RBCs), nucleus of white blood cells (N) and the background (BG) with 50 arbitrarily selected points, thus a total of 2700 points will be analyzed in this section (50 points of pixel value were selected for each regions for six samples in which the pixel value consisted of RBG information).

After segmenting the nucleus from an image, Gradient Edge Detection was applied to separate the cell in the image as individual object. Selection of the Gradient Edge Detection technique was done by feeding the segmented image into different operators, included Sobel, Prewitt and Roberts. The operator which is able to provide the clearer boundary of nucleus between each other would be selected and applied in this study.

Besides, Filtering technique [8] was applied to remove the artefact and prepared the image for the next stage. In this study, the filtering range was determined by the convex area of each nucleus (for both CML and CLL) which only focused on the maximum and minimum of convex area.

2.2. Feature Extraction

Feature extraction is a stage that transform the large input data into a reduced representation set of features [9]. Different type of input will result in different type of feature. In this study, geometrical, colour and textural features were used to distinguish the CML and CLL. A total of 28 features were extracted from these categories, included area, major axis length, minor axis length, eccentricity, convex area, solidity, extent, perimeter, compactness, mean intensity, minimum intensity, mean of R, G and B components, variance of R, G and B components, standard deviation of R, G and B components, Hue (H), Saturation (S), Intensity (I), entropy, contrast, correlation, energy and homogeneity.

2.3. Feature Selection

Feature selection algorithm was used to reduce the dimension of the feature space and improve the performance of classification [10]. The feature selection algorithms that have been applied in this study were Genetic Algorithm (GA), ReliefF algorithm (RfF) and Neighbourhood Component Analysis (NCA) algorithm.

Optimization of feature selection algorithm was conducted to ensure the algorithm was fit with the data pattern. After adjusting the optimal arguments and parameters, a total of 28 features were fed into the related algorithm and recorded the results accordingly. However, with this selected features alone will never showed any effect toward the performance of the system, therefore evaluation of these selected features with a classifier was needed in which the effect can be seen by examining the accuracy of the classifier, given that the classifiers used for the evaluation are same for all.

2.4. Classification

Classification is a process that classifier is implemented to divide the feature vector into various classes based on feature similarity. In order to ensure the performance of the classifier in this system, three types of classifiers were chosen to be evaluated, which are k-Nearest Neighbour (kNN), Support Vector Support (SVM) and Multilayer Perceptron (MLP) network trained by Levenberg Marquardt (LM) algorithm.

Before the selection of a classifier, each classifier was optimized by the related optimization procedure. Since *k*-value is the factor that affect the performance of classifier, hence, *k*-value analysis was the optimization of *k*NN, the *k*-value involved included 3, 5, 7, 9, 11, 13 and 15. For SVM, the optimization was more challenge than *k*NN as the optimization procedure is longer than *k*NN. Kernel function, box constraint was the basic parameters to be optimize in SVM, if the selected SVM was under the polynomial family, the polynomial order analysis needed to organize. On the other hand, the optimization that was applied in MLP included number of hidden nodes analysis, learning rate analysis and number of epochs. The range of these optimization parameters for MLP was [1, 30], [0.1, 0.5] and [1, 20], respectively.

After optimizing the classifiers, evaluation the classifier was conducted to select the best classifier in which the evaluation parameter involved testing accuracy for both CML and CLL, the difference of testing accuracy between CML and CLL, and the performance error. Each classifier would run the relevant algorithm with the optimal setting and the evaluation parameters, as mentioned above, were recorded for analysis. A weightage scoring method was used for the evaluation of classifier in order to overcome the uneven contribution of parameter in which the mark was given by the ranking method, for instance, 3 marks was given to the classifier with the best performance and vice versa. Then, the classifier with the highest final score would be selected and implemented in this study.

3. Result and discussion

To develop an accurate and reliable diagnostic procedure for chronic Leukemia, several parameters needed to be investigated and analyzed. Hence, this chapter was discussed and presented the results of the development process. The results were presented based on the development stage in which the first section was about the determination of threshold values for Colour Thresholding and Convex Area Filtering, and the selection of Gradient Edge operators. The second section was discussed about the evaluation and selection of feature selection algorithm. The last section of this chapter presented about the optimization of each classifier, and selection of the best classifier for the screening system.

3.1. Image Segmentation

From the pixel analysis, the threshold value can be set from the G component as it showed a gap between nucleus of WBCs, RBCs and background regions, while other component values (R and B components) were overlapped between each other. After the analysis, it was found that the nucleus of WBCs candidate threshold value for CML was 115 and 117 while CLL was within the range of 118 and 120 in which the common point of threshold value did not exist. Although the threshold value of Colour Thresholding for CML and CLL was different, the average between the ranges was able to be determined. Hence, 117 was selected for Colour Thresholding.

Another threshold value was determined by the mean of total value of RGB components for a pixel. This was used to improve the performance of segmentation. Based on the result, the threshold value will be set between 300 and 450 as the mean value for total RGB components was always maintain in this range.

Selection of Gradient Edge operator was conducted after the determination of these threshold values. Based on result, the Roberts operator was selected and implemented in this system as it able to produce low noise image.

After that, the determination of filtering range for Convex Area Filtering was conducted which only focused on the minimum and maximum value of convex area for both CML and CLL. Based on the collected data, the minimum and maximum value was 2520 and 8803, respectively, hence, the filtering range was set between 2500 and 9000.

3.2. Feature Extraction

A total of 28 features was extracted successfully that included three categories: geometrical, colour and also textural. The geometrical category were area, major axis length, minor axis length, eccentricity, convex area, solidity, extent, perimeter and compactness, while the colour feature were mean intensity, minimum intensity, mean of R, G, B component, variance of R, G, B component, standard deviation of R, G, B components and HSI. Entropy, contrast, correlation, energy and homogeneity were extracted for the textural category.

3.3. Feature Selection

Although only three feature selection algorithms were involved, five set of features have been analyzed as RfF produced three different set of features at three different cut off weights, namely RfF group 1, RfF group 2 and RfF group 3 with the cut off weights of 0.080, 0.035 and 0.028, respectively.

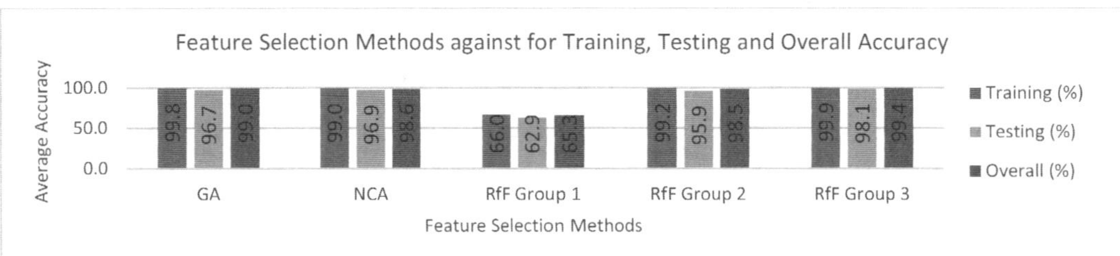

Figure 2. Graph of average accuracies against feature selection algorithm

Based on Figure 2, RfF group 1 exhibited the lowest average accuracies, this is because the number of features involved in RfF group 1 was only two features that caused the underfitting phenomena to occur. On the other hand, the RfF group 3 exhibited the highest average accuracies, therefore, the feature set of RfF group 3 was selected and applied for the following analysis, included area, minor axis length, convex area, perimeter, compactness, mean intensity, minimum intensity, mean of R component, variance of R, G, B components, standard deviation of R, G, B components, energy and correlation

3.4. Classification

In *k*NN, the optimization parameter was only focused on the *k*-value which vary from 1 to 15 with the increment of 2 and testing accuracy was the optimization target. Based on the result, the *k*-value of 1 was selected as it achieved the highest testing accuracy that was 91.5%.

For SVM, the optimization involved were kernel function and box constraint. The result showed that the quadratic SVM, which was under the polynomial family of SVM, was selected as it exhibited the highest accuracy (98.2%). Based on the selected kernel function, box constraint analysis was conducted. The result (as shown in Figure 3) showed that the box constraint of 0.001 achieved the highest testing accuracy (100%), thus 0.001 was applied in SVM. Since the selected SVM was under the SVM family of polynomial, hence optimization of polynomial order was also conducted. From the Figure 4, the polynomial order of 3 was selected in this study as it achieved the highest testing accuracy (100%) and less computational complexity.

Number of hidden node analysis, learning rate analysis and number of epochs analysis were the optimization procedure for MLP network classifier. The evaluation target for these three analysis were testing accuracy and Mean Square Error (MSE) that expressed in 10^{-6}. The optimization was started with number of hidden nodes analysis and followed by the learning rate analysis and then number of epochs analysis. Figure 5, Figure 6 and Figure 7 were the results for the optimization, respectively.

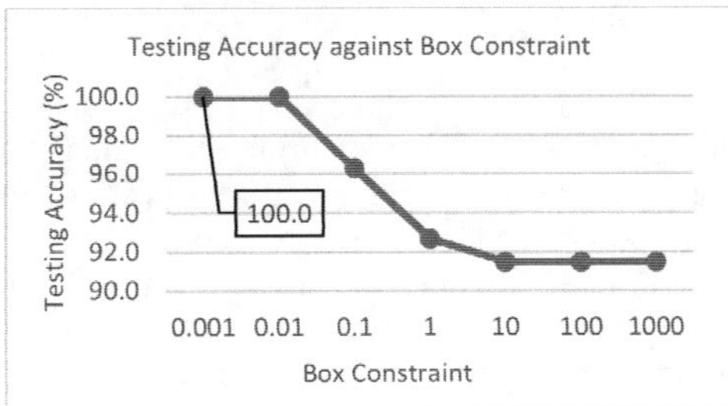

Figure 3. Graph of testing accuracy against box constraint

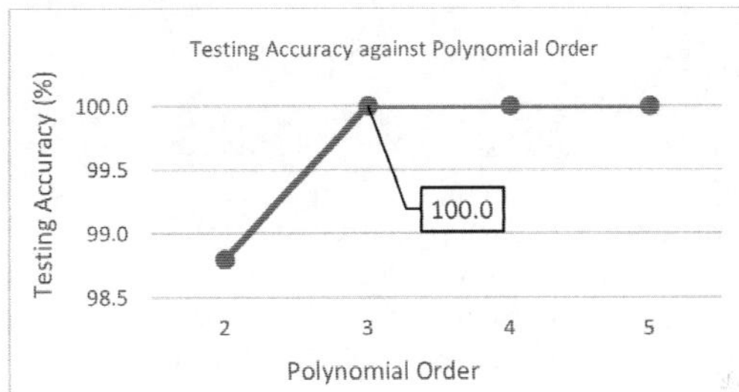

Figure 4. Graph of testing accuracy against polynomial order

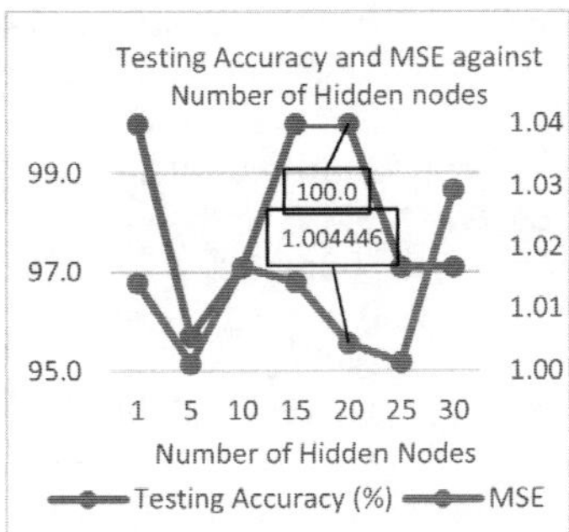

Figure 5. Graph of testing accuracy and MSE against number of hidden nodes

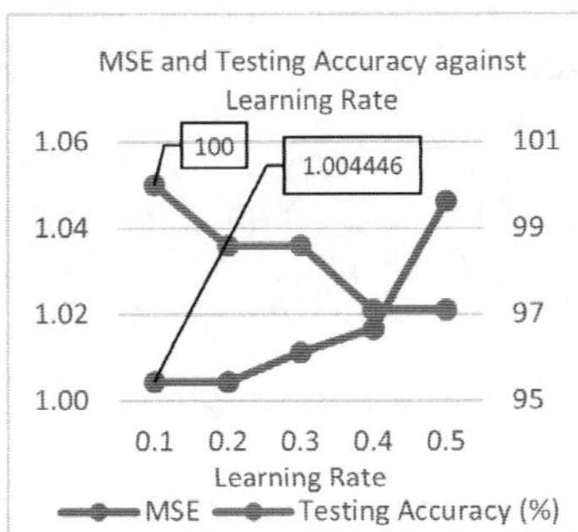

Figure 6. Graph of MSE and testing accuracy against learning rate

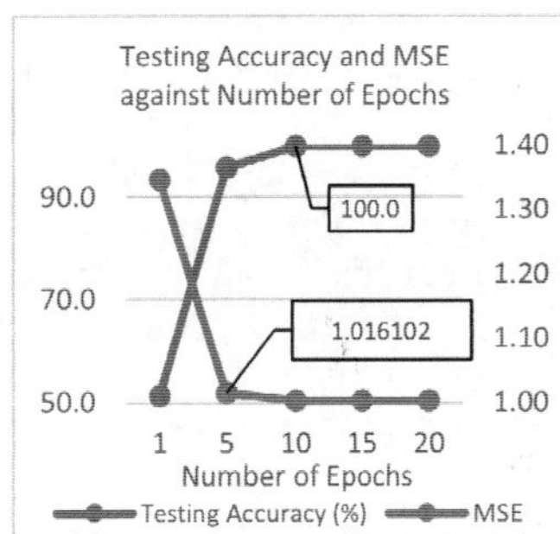

Figure 7. Graph of testing accuracy and MSE against the number of epochs

Based on Figure 5, the highest testing accuracy were achieved by 1, 15 and 20 hidden nodes, however 20 hidden nodes was selected as the MSE was the lowest. According to the results in Figure 6, it can be concluded that the learning rate of 0.1 was selected since it yields the highest accuracy and the lowest MSE. As this analysis was used to avoid the overfitting of a classifier, hence the starting point of the constant trend (as shown in Figure 7) would be selected. Based on this criteria, the 10 epochs was selected and implemented in this system.

Evaluation and selection of classifier was conducted after the optimization of classifier. This weightage scoring method was used to overcome the problem of uneven contribution of evaluation parameter. The result of scores and the weightage distribution were shown in Table 1. Table 1 presented the highest score in this analysis was MLP with score of 3.0 while *k*NN only score 1.0 and SVM was score of 2.0. Hence, MLP trained by LM algorithm was the selected classifier.

Table 1. The summary of scores and weightage distribution

Evaluation Parameter	Weightage	*k*NN		SVM		MLP	
		Value	Score	Value	Score	Value	Score
Testing Accuracy (both CML and CLL)	0.5	91.5%	1	98.8%	2	100.0%	3
Difference Testing Accuracy (between CML and CLL)	0.3	18.9%	1	2.7%	2	0%	3
Error (10^{-6})	0.2	1.0891	1	1.0123	2	1.0000	3
Total	1.0		1.0		2.0		3.0

4. Conclusion

In conclusion, the study was able to propose a diagnostic system to classify the chronic leukemia slide images by analyzing the stage of image segmentation, feature extraction, feature selection and classification by LM algorithm. The proposed method was able to produce a testing accuracy of 100% with 1.0 x 10^{-6} error which able to help the hematologist eliminate the step of screening of microscopic slide in the conventional diagnosis system. However, this diagnostic system is only software part of the overall system, therefore, a further study could focus on the development a fully automated diagnostic system that enable the system to be interfaced and controlled the microscope and digital camera.

References

[1] LaRussaA, "Facts and Statistics," *Leukemia and Lymphoma Society*, 03-Mar-2015. [Online]. Available: http://www.lls.org/http:/llsorg.prod.acquia-sites.com/facts-and-statistics/facts-and-statistics-overview/facts-and-statistics. [Accessed: 06-Oct-2018].

[2] V. J. Khoo, "Health DG: Malaysian National Cancer Registry Report 2012 – 2014 in progress," *MIMS News*. [Online]. Available: https://today.mims.com/health-dg--malaysian-national-cancer-registry-report-2012---2014-in-progress. [Accessed: 06-Oct-2018].

[3] Leukemia: Risk Factors, Symptoms, Types, and Treatment," *Healthline*. [Online]. Available: https://www.healthline.com/health/leukemia#outlook. [Accessed: 06-Oct-2018].

[4] "Leukemia death rate by country," *World Life Expectancy*. [Online]. Available: https://www.worldlifeexpectancy.com/cause-of-death/leukemia/by-country/. [Accessed: 06-Oct-2018].

[5] A. Garg 2016 A Review on Image Segmentation Techniques *International Journal of Recent Research Aspects (Conscientious and Unimpeachable Technologies 2016)* 53–55

[6] R. Adollah, M. Y. Mashor, H. Rosline, and N. H. Harun 2012 Multilevel Thresholding as a Simple Segmentation Technique in Acute Leukemia Images *Journal of Medical Imaging and Health Informatics* **2** 285–288

[7] S. Shubhashree 2014 A Review on Edge Detection Techniques for Image Segmentation *International Journal of Computer Science and Information Technologies* **5** 5898–5900

[8] S. Eddins, "Binary image area filtering," *Steve on Image Processing and MATLAB*, 22-Aug-2016. [Online]. Available: https://blogs.mathworks.com/steve/2016/08/22/binary-image-area-filtering/. [Accessed: 10-Oct-2018].

[9] K. Chinnathambi, A. Ramasamy, and P. Rajendran 2014 Modified segmentation algorithm and its feature extraction of cancer affected white blood cells *2014 International Conference on Circuits, Power and Computing Technologies [ICCPCT-2014]* 1202–1210

[10] Z. Wang, Y. Zhang, Z. Chen, H. Yang, Y. Sun, J. Kang, Y. Yang, and X. Liang 2016 Application of ReliefF algorithm to selecting feature sets for classification of high resolution remote sensing image *IEEE International Geoscience and Remote Sensing Symposium* 755–758

A Study of Informative EEG Channel and Brain Region for Typing Activity

Ng Wei Bin[1], Saidatul Ardeenawatie Awang[1*], Chong Yen Fook[1], Lim Chee Chin[1], Ong Zhi Ying[1]

[1]Biosignal Processing Research Group (BioSIM), School of Mechatronic Engineering, Universiti Malaysia Perlis, Arau, Perlis, Malaysia

*saidatul@unimap.edu.my

Abstract. Electroencephalography (EEG) is an electrophysiological monitoring method to record electrical activity of the brain. Brain consists of four lobes which is frontal, parietal, temporal and occipital lobe. Each lobe has their own respective function and it release different rhythmic wave when carry out different actions. However, when performing a specific activity, not all the EEG channels tend to be informative to the particular activity. The need to optimize the number of channels is crucial to reduce computational complexity. The aim of this paper is to determine the informative EEG channel/s and brain region for typing activity. 20 healthy with right-handed subjects from Universiti Malaysia Perlis (UniMAP) were enrolled in this study. Typing task was performed for 3 trials and 5 minutes per trial. In EEG signal processing, Notch filter and Butterworth bandpass filter were used to remove powerline artefact and to filter the signal into alpha (8-13Hz) and beta waves (13-30Hz). Welch method was applied to extract features from typing task. The obtained results were then undergoing the statistical analysis before load into the K-Nearest Neighbour (KNN) and Linear Discriminant Analysis (LDA) classifier. Based on this study, it is found that channel P3 in parietal region and channel T6 in temporal region give highest accuracy which is 99.44% for typing task activity.

1. Introduction

Brain is a gifted structure for human being as it controls and coordinates the actions and reactions, moreover enable us to feel, memories, learn and think for that every single particular [1]. The brain can be divided into four lobes that respect to different function which is frontal, parietal, temporal, and occipital lobe. The frontal lobe related to personality, emotion, focus, planning, voluntary action, and problem solving [2]. Whereas orientation, proprioception, recognition, and perception of stimuli are more related to parietal lobe [2]. The smallest lobe which is occipital lobe take care of visual processing [2]. Temporal lobe is related to perception and recognition of auditory stimuli, speech and memory [2].

The brain consists billions of nerve cells that communicate with each other through electrical impulses [3]. These electrical activity can be measure by the electrodes that place on human scalp and it is known as Electroencephalogram (EEG) [2], [4]. The cost of EEG recording device can be reduce if one can identified the relevant and effective EEG channels for a specific activity. This is because not all the EEG channels are informative for an activity. As stated by [5], besides the informative channels, the other channels will be either impair or do not influence the identification results. Hence, the channels of EEG and brain regions during typing activity are evaluated to identify the degree of informative and identification result in this paper. The EEG signals are recorded when the subjects were asked to perform the typing tasks. In this research, the parietal lobe from the brain were assumed to have the most informative channel and region compare to the other in this typing activity as it is correlated to forming motor intention task [6].

2. Experiment Protocol and Methodology

2.1. Data Set and Protocol

Twenty undergraduate students from Universiti Malaysia Perlis (UniMAP) who are right handed and mentally healthy are recruited to join the study. The EEG recording device used in this research is "Truscan32" (Deymed Diagnostic, Czech Republic) consists of 19 channels and using 10/20 International System for electrode configuration. The EEG sampling rate is 256 Hz. The subjects were asked to perform typing task for 5 minutes and it were repeated for 3 times. Rest condition with closing eyes for 30 seconds was given between the trials. Figure 1 and Figure 2 show the photo during the data collection and the flow of the experiment.

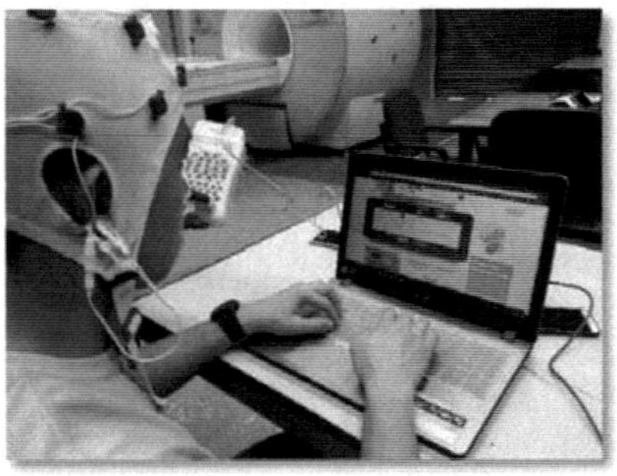

Figure 1. Subject performing typing task.

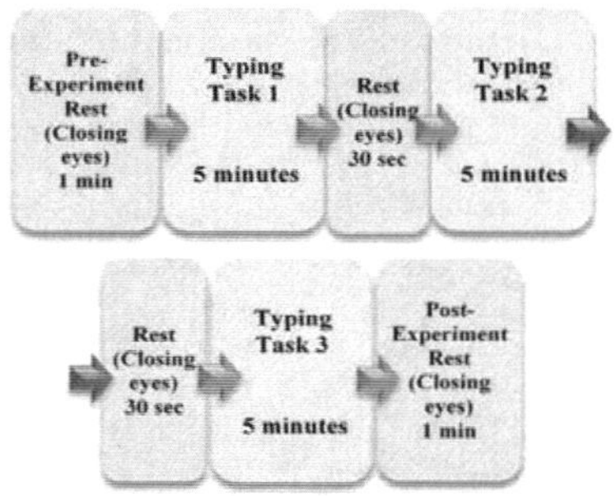

Figure 2. The flowchart of experiment protocol.

2.2. EEG Signal Processing

The recorded EEG signals were analysed through the MATLAB software. Three main stages were carried out in the EEG signals processing which is pre-processing, feature extraction and classification.

2.2.1. Pre-processing

The raw EEG signals require denoising as it contain noise such as eye blinking, powerline interference and etc. In the pre-processing, the Notch filter and second order Butterworth bandpass filter was implemented. Notch filter is used to remove the 50Hz powerline interference while the Butterworth filter provides a good transient characteristics response that is maximally flat in the passband and stopband and almost no ripple [7]. Most of the previous study were done on Alpha and Beta therefore this two frequencies is selected in this research [8]. The frequency band of the Alpha and Beta are 8 Hz to 13 Hz and 13 Hz to 30 Hz respectively [9]. The raw EEG signals of 3 typing trials which is 5 minutes long for each trial were framed into 300 columns (represent 300 seconds) and filtered for 20 subjects in each channel.

2.2.2. Feature Extraction

Feature extraction is a process that reduces the attribute of a data therefore making the data become a smaller and richer set of attributes which will ease the result of classification later. In this study, the Power Spectral Density (PSD) was used as the feature extraction method as it is a common technique that distributes the signal power over frequency and show the strength of the energy as function of frequency [9]. The Welch method were adopted under PSD in this research. In this study, Welch method was adopted to extract the features. The filtered EEG signals of 3 typing trials from 20 subjects were combined and undergo the Welch method feature extraction separately in 19 channels. As a result, there are 19 channels' EEG signals which represent the average value from 20 subjects over 3 typing trials are readily to use in the classification stage. Welch method is usually used for estimate the power spectrum over time by dividing the signals into modified segments and obtain the average spectra periodogram after computes the magnitude of it [10]. It is a non-parametric spectral methods that commonly used for tackling the EEG signals [11].

2.3 Statistical Features

The location parameters such as computation of means or medians and the dispersion parameters like standard errors or interquartile range are often used as statistical features analysis of EEG data [12]. Once the feature extraction is done to the framed data, it will be having 129 rows of value in each column. The featured data is then undergoing calculation of statistical features which is mean, median, standard deviation, and variance. The obtained data will be combined as input for EEG classification.

2.4 Classifier

2.4.1 K-Nearest Neighbour (K-NN)

K-Nearest Neighbour is known as non-parametric method and instant-learner because it does not make any assumptions on the input data and directly take the available data as training. The rule of the algorithm of K-NN is training the samples themselves without require any supplementary data [13]. "K" means the number of the referred neighbour and this value plays an important factor in the classification since it described the distance metric. Larger the value of K, the noise effect on the classification of noise can be minimize [13]. The value of K used in this classification is 5. This mean that, 5 closest training samples are selected based on a distance metric and then a voting for most number of samples per class is done. So if 3 samples belong to Class-1 and 2 belong to Class-5, then that test sample is classified as Class-1. So the value of k indicates the number of training samples that are needed to classify the test sample.

2.4.2 Linear Discriminant Analysis (LDA)

It is a method to explain the distinctive nature of more than two classes of object by finding a linear combination of the extracted features [14]. In the Linear Discriminant Analysis, it assumes that the conditional probability density functions of the classes are normally distributed with same covariance matrix [4]. This classifier will compute the mean of all sample for each class. Then it is subtracting from each observation of that class with their sample mean to obtain the covariance. The situation where the class-frequencies unequal can be tackled by LDA handy [14].

3. Result and Discussion

To evaluate the most informative EEG channels and brain region for typing activity, the raw EEG signals of 20 subjects from all 19 channels has been processed. In this part, the brain region will be evaluated by few groups according to the channel location. For example, first group is the frontal region that represent by FP1, FP2, F7, F3, FZ, F4, and F8. Second group is temporal region that consists of T3, T4, T5 and T6. While central region only consists 3 channels which is C3, CZ, and C4. Next, it is parietal region that indicate the channels of P3, PZ and P4. Lastly the group is occipital region which mean O1 and O2 channels. Table 1 shows the classification result for evaluating the most informative EEG channels.

Based on Table 1, it can be seen that the most informative individual channel is P3 which it is able to provide 99.44% result in the Alpha frequency through KNN classification. Based on the result, the most informative brain region is around the parietal lobe of the brain, which is P3, Pz and P4. Its informative allow the classification achieved 98.33% in Alpha frequency band and 95.56% in Beta frequency band in average. The performance from temporal region also not much significant difference and similar to the parietal region which it able to achieve 98.20% in Alpha frequency band and 92.92% in Beta frequency band in average. The T6 channel in the temporal region can obtained same accuracy result with P3 channel from parietal region which is 99.44%.

Table 1. The classification results of EEG channels from different brain regions.

		Classification Accuracy (%)			
Classifier		KNN		LDA	
Frequency Band		Alpha	Beta	Alpha	Beta
Frontal	Fp1	86.03	82.68	64.25	57.06
	Fp2	86.11	83.89	73.89	56.67
	F7	96.67	90.00	93.33	81.11
	F3	97.22	90.45	90.56	70.22
	Fz	83.80	73.33	64.25	66.86
	F4	94.38	77.78	76.11	71.67
	F8	96.67	88.33	88.33	80.56
	Average	**91.55**	**83.78**	**78.67**	**69.16**
Temporal	T3	97.78	89.44	93.89	84.44
	T4	96.67	94.44	83.33	86.67
	T5	98.89	92.78	95.00	93.33
	T6	99.44	95.00	96.11	87.22
	Average	**98.20**	**92.92**	**92.08**	**87.92**
Central	C3	98.89	95.56	86.11	85.56
	Cz	81.67	78.33	73.74	78.89
	C4	97.22	95.00	89.44	90.00
	Average	**92.59**	**89.63**	**83.10**	**84.82**
Parietal	P3	99.44	93.89	94.44	86.11
	Pz	98.33	97.22	92.22	85.00
	P4	97.22	95.56	92.78	91.11
	Average	**98.33**	**95.56**	**93.15**	**87.41**
Occipital	O1	97.78	90.00	95.56	80.56
	O2	96.67	94.44	91.11	86.67
	Average	**97.23**	**92.22**	**93.34**	**83.62**

Keyboard typing is a complex perceptual, cognitive, and motor process which involves the generation of motor commands that are not directly related to letter shape [15]. The parietal lobe contains the primary sensory cortex which is responsible for forming of motor intention and interpreting the somatosensory signals like touch, movement coordination, visual perception [6], [16]. Whereas in the medial surface of the temporal lobe, which is responsible for the recognition memory [16]. During the 3 typing trials, the subjects tend to recognize an incoming object such as the words or letter which is fixed content in this experiment through recollection and familiarity through repeating the typing trials for 3 times [16].

In the research done by [17], it is reported that the anterior part of the left superior parietal lobule (SPL) is the crucial area for the neural process for typing. Their study also stated that the left posteromedial portion of the left Intra-Parietal Sulcus (IPS) was dominant in typewriting [17]. This can be explained that channel P3 has higher accuracy because it located at left hemisphere of brain.

Typing activity required positioning the finger to touch the keyboard so therefore the parietal lobe more active during the given typing task. Hence, it can be concluded that typing activity is closely related and dominant in the parietal lobe. The results were led by parietal then follow by the temporal, occipital, central and lastly frontal as shown in Figure 3 and Figure 4. In general, the KNN classifier is performed better than LDA classifier to determine the most informative EEG channels in typing task study.

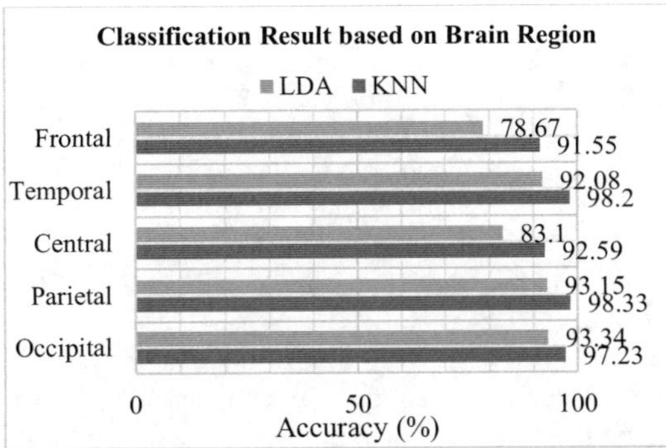

Figure 3. The Classification Accuracy for Alpha waves.

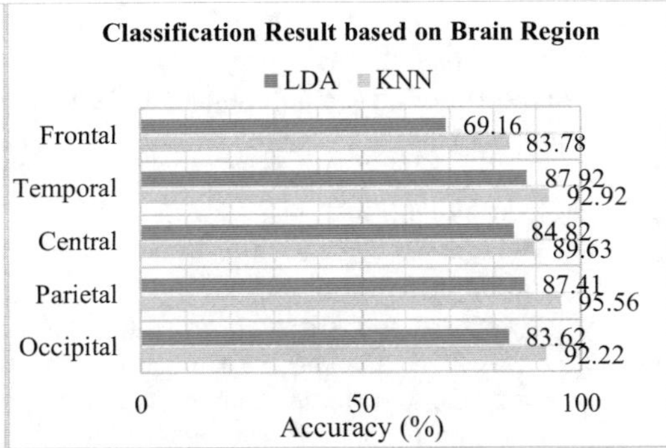

Figure 4. The Classification Accuracy for Beta waves.

4. Conclusion

In this study, the Welch method under PSD is used as feature extraction where KNN and LDA are used as classifier. The goal in this research is to determine the most informative EEG channels and brain region for typing trial activities. The highest performance of individual EEG channel is P3 and T6 that able to give result of 99.44% in the Alpha frequency band through KNN classification. When the channels are grouped according to their brain region, the parietal lobe has seemed to be the most informative brain region for because it achieve the 98.33% in Alpha frequency band and 95.56% Beta frequency band in average for KNN classification. It is followed by the temporal region which able to achieve 98.20% in Alpha frequency band and 92.92% in Beta frequency band which is very close result to the parietal region. In general, the result obtained in this study suggested that the parietal region is the most informative region for typing activity, then followed by the temporal region.

Acknowledgement

This research was funded by the Fundamental Research Grant Scheme of Ministry of Higher Education (FRGS Grant) under supervision of Research Management Institute Centre, Universiti Malaysia Perlis (FRGS/1/2018/TK04/UNIMAP/02/11).

References

[1] M. S. Gaffrey, J. L. Luby, and D. M. Barch 2013 Towards the study of functional brain development in depression: An Interactive Specialization approach *Neurobiol. Dis.*, **52** pp. 38–48

[2] L.-E. Notevarp Bjørge and T. Hübertz Emaus 2017 Identification of EEG-based signature produced by visual exposure to the primary colors RGB

[3] S. Kodagoda, E. A. S. M. Hemachandra, P. A. A. R. Pannipitiya, L. S. Bartholomeuz, and A. A. Pasqual 2006 Minimal Invasive Headband for Brain Computer Interfacing and Analysis *2006 International Conference on Information and Automation* pp. 109–114.

[4] S. Yang and F. Deravi 2017 On the Usability of Electroencephalographic Signals for Biometric Recognition: A Survey *IEEE Trans. Human-Machine Syst* **47** no. 6, pp. 958–969

[5] K. V. R. Ravi and R. Palaniappan 2007 A Minimal Channel Set for Individual Identification with EEG Biometric Using Genetic Algorithm *International Conference on Computational Intelligence and Multimedia Applications (ICCIMA 2007)* pp. 328–332

[6] Y. Kamikawa and T. Tanaka 2015 Responses in posterior parietal cortex to movement intention task with visual and tactile cues *Proceedings of the Annual International Conference of the IEEE Engineering in Medicine and Biology Society, EMBS*,

[7] M. Lewandowski and J. Walczak 2015 Comparison of single-level and multi-level filtering systems of EEG signal *2015 Signal Processing: Algorithms, Architectures, Arrangements, and*

Applications (SPA) pp. 34–37.

[8] K. Yaomanee, S. Pan-ngum, and P. I. N. Ayuthaya 2012 Brain signal detection methodology for attention training using minimal EEG channels *2012 Tenth International Conference on ICT and Knowledge Engineering* pp. 84–89.

[9] Z. Y. Ong and Z. Ibrahim 2018 Power Spectral Density Analysis for Human EEG- based Biometric Identification *2018 Int. Conf. Comput. Approach Smart Syst. Des. Appl.* pp. 1–6

[10] A. Saidatul, M. P. Paulraj, S. Yaacob, and M. A. Yusnita 2011 Analysis of EEG signals during relaxation and mental stress condition using AR modeling techniques *Proc. - 2011 IEEE Int. Conf. Control Syst. Comput. Eng. ICCSCE 2011* pp. 477–481

[11] P. F. Diez, E. Laciar, V. Mut, E. Avila, and A. Torres 2009 A comparative study of the performance of different spectral estimation methods for classification of mental tasks *2008 30th Annual International Conference of the IEEE Engineering in Medicine and Biology Society* pp. 1155–1158.

[12] P. Abrahantes, José & SERROYEN, Jan & Geys, Helena & Molenberghs, Geert & Drinkenburg, 2007 "Statistical methods for EEG data," Limburgs Universitair Centrum

[13] H. N. Oon, A. Saidatul, and Z. Ibrahim 2018 Analysis on Non-Linear Features of Electroencephalogram (EEG) Signal for Neuromarketing Application *2018 Int. Conf. Comput. Approach Smart Syst. Des. Appl.* pp. 1–8

[14] A. Bhardwaj, A. Gupta, P. Jain, A. Rani, and J. Yadav 2015 Classification of human emotions from EEG signals using SVM and LDA Classifiers in *2015 2nd International Conference on Signal Processing and Integrated Networks (SPIN)* pp. 180–185.

[15] T. A. Salthouse 1986 Perceptual, Cognitive, and Motoric Aspects of Transcription Typing *Psychological Bulletin*

[16] K. H. Jawabri and S. Sharma 2019 *Physiology, Cerebral Cortex Functions*. StatPearls Publishing

[17] Y. Higashiyama, K. Takeda, Y. Someya, Y. Kuroiwa, and F. Tanaka 2015 The Neural Basis of Typewriting: A Functional MRI Study *PLoS One* **10** no. 7, p. e0134131

Design and Development of Artificial Spinal Ligaments for Paediatric Synthetic Spine

Nor Amalina Muhayudin[1,2*], Khairul Salleh Basaruddin[2], Fiona McEvoy[3], Anthony Tansey[3]

[1] School of Manufacturing Engineering, Universiti Malaysia Perlis, 02600, Arau Perlis, Malaysia
[2] School of Mechatronic Engineering, Universiti Malaysia Perlis, 02600, Arau Perlis, Malaysia
[3] Mechanical Engineering Department, Institute of Technology Tallaght, Dublin, D 24, Ireland

*amalmuhayudin85@gmail.com

Abstract. A synthetic spine is a model fabricated from artificial materials consisting of the vertebrae, intervertebral discs and ligaments for spinal testing. The synthetic spine overcomes many difficulties associated with biological specimens such as handling, biohazard concerns, high costs, and limited availability of specimens, quality and large inter-specimen variability. This paper presents the design and development of spinal ligaments to mimic the stiffness of the paediatric ligaments for use in the synthetic spine. Spinal ligaments are uniaxial structures in the spine that carry tensile loads along the direction of the fibres. Early in the research, silicone materials were used to cover the whole spinal unit, but it became apparent that the material responses were inadequate. The synthetic spine was revised to use fibreglass tape to more closely simulate the natural spinal structures. The composite design applied in this paper consisted of soft silicone rubber and fibreglass tape to obtain the natural stiffness that normally occurred in the spinal ligaments.

1. Introduction

Spinal ligaments are naturally designed so that when it was subjected to different forces and moments it will only resist tensile forces along the directions of the fibres. The main functions of the ligaments are to hold the vertebrae together, stabilise the spine and protect the spinal cord including the disc by allowing limited physiologic motion. There are two primary ligament systems in the spine, the intra-segmental and inter-segmental systems. The intra-segmental system attached individual vertebra together, which includes the ligamentum flavum (LF), interspinous (ISL) and intertransverse (TL) ligaments. The inter-segmental system that holds the vertebrae together includes the anterior (ALL) and posterior (PLL) longitudinal ligaments, and the supraspinous (SSL) ligaments. The spinal ligaments are primarily collagenous except for the ligamentum flavum, which is primarily comprised of elastin. Facet capsulary ligaments (CL) attach the articular processes of the vertebra and encapsulate the facet joint.

Adams et al. using a stepwise reduction method to characterise the resistance of flexion by posterior ligaments on 27 human cadaveric lumbar spines [1]. The specimens were tested with all ligaments intact, and then each ligament was dissected accordingly, starting with ISL/SSL, followed by LF and finally CL. The study suggested that CL provided about 40% of the joint resistance in full flexion, while others had only a small role to play. Interestingly the intervertebral disc contributed 30% resistance [1]. The findings were in agreement with Gillespie et al. on porcine spine [2]. Heuer et al. investigated the function of anatomical components using eight human lumbar spinal segments under multiple loadings [3]. Similar to Adams et al. and Gillispie et al., the ligaments were intact for the first test and then tested after dissection of different components [1–3]. The results were in agreement with them where the dissection of posterior ligaments increased flexion movement and ALL contributed the most in resisting extension [1–3].

The main challenge in investigating the effects of individual ligament is to keep the ligaments intact during testing as once one ligament began to detach, it was impossible to reattach it back. Most of the studies to date used Finite Element Analysis (FEA) models as it provided more control of all the ligaments. Among FEA studies were Zander et al. , Lee et al. and Brolin et al., who investigated the effects of individual ligaments on spinal movements [4–6]. Zander et al. conducted nonlinear studies using L3-L4 to investigate the influence of ligament stiffness on rotation and forces in ligaments [4]. They found ALL was completely unloaded during flexion and was the only ligament loaded during extension. This was in agreement with Heuer et al [3,4]. Lee et al. conducted a material sensitivity study on L2-L3 under sagittal plane loading and took into consideration the variation in humans by using probabilistic method [5]. The model was then compared against experimental data with and without posterior elements. The results differed from experiments under high extension loadings when the posterior elements were dissected. They also found the ALL significantly affect the extension spinal movements [5].

Recently, Lapirre et al. in collaboration with Sawbones to develop synthetic spinal ligaments using polyester fibres and a polyurethane matrix for their lumbar analogue model [7]. Based on their preliminary test, all the posterior spinal ligaments including ISL, SSL, and TL were fabricated mainly for aesthetic appeal since these ligaments do not have a large effect on the spinal motions as compared with ALL and PLL.

This paper aims to develop spinal ligaments for paediatric synthetic spine to mimic the movement of the human spine suitable for spinal testing and no interest in clinical purpose. The objectives of this paper are to determine the design and materials for the paediatric spinal ligaments and to replicate the stiffness that normally occurred in human spinal ligaments. Since most of the studies found that ALL significantly affected the spinal movements, this paper focused on developing ALL and PLL individually as an improvement from the authors previous developed synthetic spine [8].

2. Materials and Methods

The difficulties in developing the spinal ligaments lay in finding appropriate materials to mimic the behaviour of the ligaments and attaching the fibres within the material. The materials selected for spinal ligaments consisted of fibre tape and two different types of silicone rubber (MoldMax 40 and SortaClear 40). A pre-made fibre tape was embedded within the silicone materials. The fibreglass tape was used because it is extremely strong and lightweight. Fibreglass is a fibre reinforced polymer made of a plastic matrix reinforced by fine fibres of glass. The weave pattern of the fibre tape was measured based on the 45 degree bias direction of a woven fabrics as shown in Figure 1. The fibre tape provided the stiffness when embedded within the silicone rubber material and controlled the non-linear force displacement properties of the ligaments.

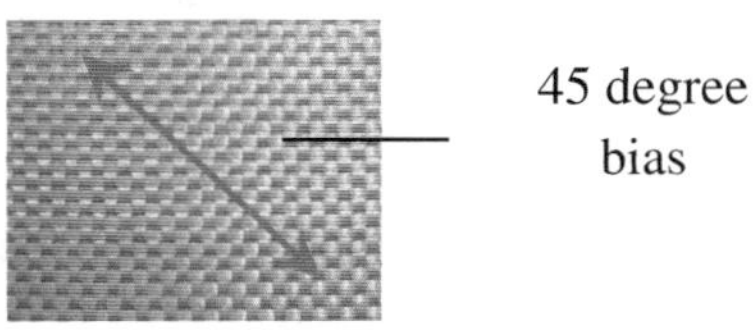

Figure 1. The 45 degree bias direction of the fibre tape

Journal of Physics: Conference Series **1372** (2019) 012009 doi:10.1088/1742-6596/1372/1/012009

The mould forming the spinal ligaments was designed to embed the fibre into the middle of the silicone rubber (Sorta Clear 40). Two clamping holders were fixed each end to ensure the fibre tape did not deform during the remaining curing process (24 hours at room temperature) as shown in Figure 2. It was essential to ensure that the fibre tape remained within the silicone rubber because the elastic zone (EZ) stiffness of the synthetic ligaments is controlled by the volume fraction of the fibre while the neutral zone (NZ) stiffness is based on the silicone rubber material itself. Tensile tests were performed on all five specimens until failure at 0.0423 mm/sec at a constant rate with 4 N preload based on the experimental set up used by Lapirre et al. on synthetic spinal ligaments for adult lumbar model [7].

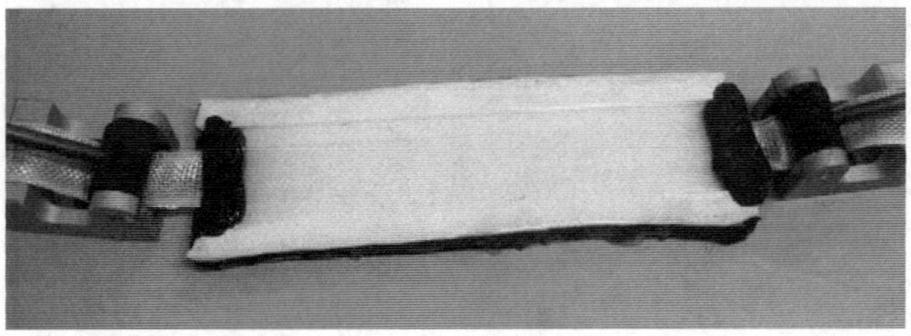

Figure 2. The fibre tape was clamped at both ends of the mould.

Synthetic ligaments fabricated in this present study used a combination of two different type of materials; fibre tape and silicone rubber in order to replicate the mechanical properties of human spinal ligaments. The synthetic ligaments can be considered as a composite material; therefore, the rule of mixtures could be applied to predict the synthetic ligament stiffness. In order to apply the rule of mixtures, these assumptions had to be taken into considerations; fibres were uniformly distributed throughout the matrix (silicone rubber), perfect bonding between fibres and matrix, matrix is free of voids, applied loads are either parallel or normal to the fibre direction, no residual stress and fibre and matrix materials behave linearly like elastic materials.

The first step to verify the synthetic ligament was to ensure that the completed synthetic ligament was a mixture of both materials as shown in Figure 3. An Equation 1 was used.

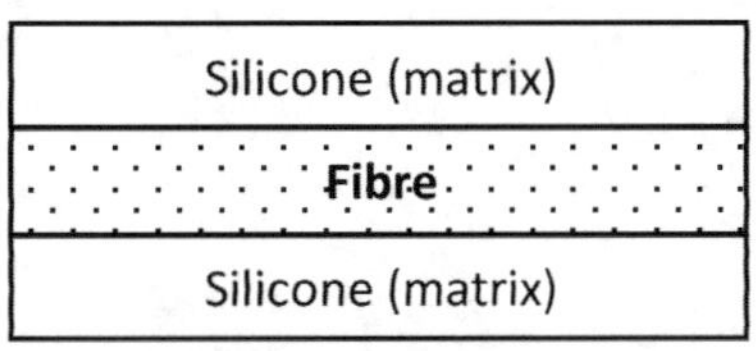

Figure 3. The schematic diagram of synthetic ligament.

$$\boldsymbol{\rho_c A_c = \rho_f A_f + \rho_m A_m} \tag{1}$$

Where A_c, A_f and A_m are composite, fibre and matrix cross-sections andρ_c, ρ_f and ρ_m are composite, fibre and matrix density. Assuming that the strains of all elements along one direction are equal therefore based on the parallel combination of rule of mixtures; the total density of the composite material is given using Equation 2.

$$\boldsymbol{\rho_c = \rho_f V_f + \rho_m V_m \; with \; parallel \; combination, \rho_c = \rho_f V_f + \rho_m (1 - V_f)} \tag{2}$$

The percentage difference of the expected and the achieved composite material was only 0.1%, thus suggesting that the composite material fabricated in this present study followed the static equilibrium principle. The next step was to predict the upper and lower limit of the elastic modulus of the synthetic ligaments. The general rule of mixtures used to predict various properties of composite materials made up of continuous and unidirectional fibres. In this present study, the upper limit of the modulus corresponded to the loading parallel to fibres could be simplified as Equation 3. The lower limit of the modulus corresponded with the transverse loading and was simplified in Equation 4.

$$E_u = E_f V_f + E_m V_m \tag{3}$$

$$E_L = \frac{E_f E_m}{E_f V_m + E_m V_f} \tag{4}$$

Where E_f and E_m were the elastic modulus of fibre and matrix and V_f and V_m were the volume of fibre and matrix.

3. Results and Discussion

The elastic modulus of the fibre tape was measured from stress-strain curves from the tensile tests of five specimens, (Refer to Figure 4). The results for fibre tape were consistent for all five specimens with an average elastic modulus of 1.87 GPa. The elastic modulus was calculated at the slope of linear region.

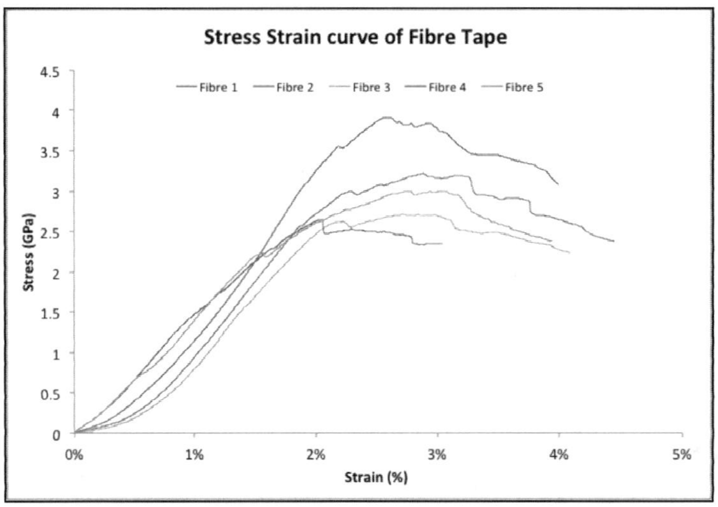

Figure 4. Stress strain curve of fibre tape.

The elastic modulus of the Sorta Clear 40 (matrix) used in this present study was calculated from Mechanical Data Sheet (MDS) which approximately around 0.62 MPa. Based on these two moduli, the lower and upper limit moduli for the synthetic ligaments could be predicted. By using Equation 3 and 4, the upper limit (E_u) the lower limit (E_L) of the synthetic ligament were 112.78 MPa and 0.66 MPa. Two variables were measured to compare with human spinal ligaments they were the NZ stiffness and EZ stiffness. Figure 5 showed the typical load displacement curve obtained from synthetic ligament.

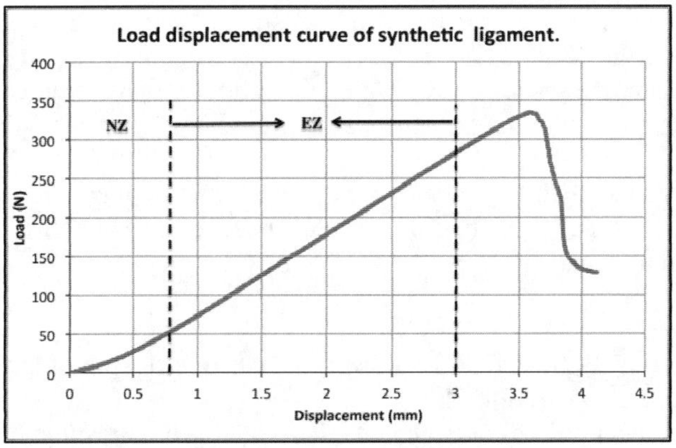

NZ-Neutral Zone stiffness, EZ –Elastic zone stiffness

Figure 5. Typical load displacement curve of synthetic spinal ligament.

The average measured NZ was 37.8 N/mm and the EZ stiffness was 109.6 MPa. A comparison between the findings of present study and literatures are summarised in Table 1 [9–12] . The synthetic ligament in this present study has been simplified to one composite material for both ALL and PLL. The synthetic ligament was dissected to two parts to represent ALL (2/3) and PLL (1/3) and since it followed the static equilibrium principle, both parts carried similar mechanical loads. As presented in Table 1, the stiffness for both ALL and PLL of human adult spinal ligament had a large range of data. Since no data was found for paediatric spinal ligaments, the NZ stiffness and EZ stiffness of synthetic ligament measured in this study were compared with adult data and were within the ranges provided. The paediatric spinal ligament was expected to follow the pattern of intervertebral disc where the stiffness of paediatric ligament was lower than adult ligaments because both are soft tissues. The value of both stiffness were closer to the lower limit of the adult ligaments provided by literature, it was assumed that the synthetic ligament had more flexibility compared to adult ligament [9-12] . The average elastic modulus measured for synthetic ligament was approximately 22.5 MPa, which was within the upper and lower limit of the modulus calculated using rule of mixture. Based on the above argument, the synthetic ligament fabricated in the present study was used as spinal ligaments in the synthetic paediatric spine.

Table 1. Comparison of NZ and EZ stiffness of ALL and PLL between the present study and literature [9-12]

Ligament	NZ stiffness (N/mm)	EZ stiffness (N/mm)
ALL	**19.64 - 78.64** (literature) **37.8** (present study)	**17.3-632.36** (literature) **109.6** (present study)
PLL	**9.52 - 296.57** (literature) **37.8** (present study)	**8.5-573.33** (literature) **109.6** (present study)

4. Conclusion

The objective of this paper is to design synthetic spinal ligaments to replicate the stiffness in paediatric synthetic spine. This study focused only on ALL and PLL since the rest of spinal ligaments do not have an essential effect to the spinal movements. The design of synthetic spinal ligaments presented in this paper was simplified for ease of fabrication. Since there is no experimental data of paediatric spinal ligaments and range of motion to date, the results then were compared with experimental and FEA spinal ligaments of adult data. The results indicated that the stiffness of fabricated synthetic spinal ligaments in this study was closer to the lower limit of the adult range and it was assumed that it may provide more flexibility. Future works is to determine if the synthetic spinal ligament provided the expected movement on the spinal unit by testing the range of motion of the paediatric synthetic spine using MTS Bionix ServoHydraulic spine simulator.

References

[1] Adams MA, Hutton WC, Stott JRR 1980 The resistance to flexion of the lumbar intervertebral joint. *Spine (Phila Pa 1976)* **5** 245–53

[2] Gillespie KA and Dickey JP 2004 Biomechanical role of lumbar spine ligaments in flexion and extension: determination using a parallel linkage robot and a porcine model. *Spine (Phila Pa 1976)* **29** 1208–16

[3] Heuer F, Schmidt H, Klezl Z, Claes L, Wilke H-J 2007 Stepwise reduction of functional spinal structures increase range of motion and change lordosis angle *J Biomech [Internet]* **40** 271–80

[4] Zander T, Rohlmann A, Bergmann G 2004 Influence of ligament stiffness on the mechanical behavior of a functional spinal unit *J Biomech [Internet]* **37** 1107–11

[5] Lee K.K and Teo EC 2005 Material sensitivity study on lumbar motion segment (L2-L3) under sagittal plane loadings using probabilistic method *J Spinal Disord Tech* **18** 163–70

[6] Brolin, K and Halldin P 2004 Development of a finite element model of the upper cervical spine and a parameter study of ligament characteristics *Spine (Phila Pa 1976)* **29** 376–85.

[7] Lapierre LJ. Control of the Mechanical Properties of the Synthetic Anterior Longitudinal Ligament and its Effect on the Mechanical Analogue Lumbar Spine Model By Control of the Mechanical Properties of the Synthetic Anterior Longitudinal Ligament and its Effect on the. *Mechanical Engineering. University of Kansas*

[8] Muhayudin NA, Tansey A, McEvoy F, Kiely P. 2010 Characterisation of Correction Forces of Spinal Fusion Surgery *IFMBE Proceedings vol 31.* 573–576

[9] Goel VK, Ebraheim NA, Biyani A, Rengachary S, Faizan A 2005 Role of mechanical factors in the evaluation of pedicle screw type spinal fixation devices. *Neurol India* **53** 399–407.

[10] Augustus A.White III and Punjabi MM.1990 Clinical Biomechanics of the spine 2nd edition. *Clinical biomechanics.*

[11] Chazal J, Tanguy A, Bourges M, Gaurel G, Escande G, Guillot M Vanneuville G. 1985 Biomechanical properties of spinal ligaments and a histological study of the supraspinal ligament in traction. *J Biomech.* **18** 167–76.

[12] A. Shirazi-Adl. 2006 *Analysis of large compression loads on lumbar spine in flexion and in torsion using a novel wrapping element.* J Biomech.**39** 267–75.

Investigation on gas plasma and liquid plasma treatment for polyether ether ketone (PEEK) surface for wettability enhancement

Theng Saw Vira[1], Siti Khadijah Zaaba[1,2*], Mohammad Taufiq Mustaffa[3]

[1]School of Mechatronic Engineering, Universiti Malaysia Perlis, Arau, Perlis, Malaysia
[2]Centre of Excellence for Advanced Sensor Technology (CEASTech), Universiti Malaysia Perlis
[3]School of Manufacturing Engineering, Universiti Malaysia Perlis, Arau, Perlis, Malaysia

*khadijah@unimap.edu.my

Abstract. The results of liquid and gas plasma treatment on polyether ether ketone PEEK surfaces are reported. The experiments were conducted by using variation of treatment time (5 s to 60 s) using Helium as its working gas. The results were obtained by measuring the water contact angle (WCA) and the studying the surface morphology of the PEEK surface. Comparisons of WCA and Scanning Electron Microscopy (SEM) analysis were conducted for samples which undergone gas plasma treatment and liquid plasma treatment. After gas plasma treatment, WCA of the PEEK surface decreased from 65° to 21°. Results of WCA for liquid plasma treatment showed decrement from 65° to 45°. Images from SEM showed PEEK surface area after gas and liquid plasma showed surface area increment of 40% for gas plasma and 20% for liquid plasma. The decrement of WCA and increment of surface area showed that PEEK surface properties changed from hydrophobic to hydrophilic via wettability enhancement.

1. Introduction

PEEK has great physical and chemical properties, including outstanding mechanical properties, chemical resistance and thermal performance. It can withstand long term continuous temperatures of 248 °C as well as exposure to hot water or steam with low moisture absorption. PEEK also have good compatibility with all normal cleansing strategies, protection from erosion, and simplicity of framing by trim or machining and a thickness equivalent to human tissue.

In general, plasma is well known as the fourth state of matter. Plasma is also known as ionized gas. The plasma treatment conducted is atmospheric pressure low temperature process which could modify surface. To modify the surface properties of materials such as solid polymer, plasma treatment (low and atmospheric pressure) is extensively used [1]. Our previous work has shown that microorganism can be deactivated when plasma discharges is in water compared to discharge using gas. [2]. Thus, it is interesting to investigate its surface properties using WCA and SEM analysis. Another example of plasma treatment using air as its working gas also has shown promising results [3].

Plasma technique allows the surface to be more hydrophilic with the formation of new groups which can interact specifically with other groups when the reactive gases in the plasma such as Helium or Argon are used. Plasma surface modification is getting more attention in biomedical engineering because it is an inexpensive and effective surface treatment technique for polymers. Plasma treatment is another simple technique used to modify polymers to the surface. Plasma treatment on PEEK has shown an increment of surface wettability at a fast rate and to create hydroxyl groups on the surface [4]. However, this depends on the type of material, the suitable choice of plasma, and time taken enable to increasing in hydrophobicity or hydrophilicity.

The surface property can be changed from hydrophobic to hydrophilic by using plasma treatment method [5]. In order to increase the wettability, while has little effect on its bulk properties, plasma is a relatively low cost tool, highly-efficient and easy to handle.

2. Plasma Technology and Surface Treatment

The transformations of non-thermal plasma at the atmospheric temperature plasma are significant. The economic and operational advantages compared to low pressure have led to the establishment of a variety of non-thermal atmospheric pressure plasma as a latest trend in research [6]. Atmospheric pressure plasma also produces reactive species to improve the material's surface characteristics and have an equivalent result in low pressure plasma treatment [7].

Atmospheric pressure plasma system was developed using a high voltage ignition coil power supply. This system can support plasma discharge by use of helium and argon as operating gas. The discharge temperature was found to be around 35 °C even in long term operation and the surface of thermal-sensitive materials such as polyethylene (PE) without altering its bulk properties can be modified using this type of non-thermal plasma. In an attempt to investigate the use of the plasma system to modify the PE surface, the results showed that the PE surface can be altered by 60 seconds of plasma treatment to a hydrophilic state. The reduction of water contact angle confirmed this [8].

Engineering of thermoplastic mostly show low energy surface and bad adhesion. Thus, it does not have the properties of surface that is needed in some application. In order to make the surface active, different methods are used. Non-thermal plasma is extensively used to modify the surface of various polymers to produce and improve the adhesion and wettability properties without giving an effect to the bulk properties [9].

3. Methodology

To investigate the effect of treating PEEK surface by using atmospheric gas and liquid plasma, experiments on WCA and SEM were conducted as follows.

3.1. Preparation of PEEK before Plasma Treatment

A 1 cm thick PEEK was cut into 3 cm x 3 cm. The PEEK was placed on a glass plate before exposure to plasma. The center of each samples were exposed to atmospheric plasma. For liquid plasma, the PEEKs were submerged into the liquid. The samples were exposed to plasma for 5 to 60 seconds.

3.2. Plasma Surface Treatment

For gas plasma, the parameters of treatment such as helium gas flow rate, the working distance of the plasma discharge and treatment time were optimized. Helium gas flowed from a cylinder gas tank through the gas tube into the gas regulator at 1 L/min. Then, the gas travelled through the gas tube into the flow meter system then through quartz glass tube. Treatment time was varied, which are between 5 and 60 seconds. Plasma plumes on the surface of the PEEK sample is shown in Figure 1.

For liquid plasma, the quartz glass with the high voltage electrode and ground electrode was submerged into distilled water of 30 ml to activate the liquid. Then, the PEEK sample was inserted into the beaker as shown in Figure 2. Treatment time was varied, which are between 5 and 60 seconds.

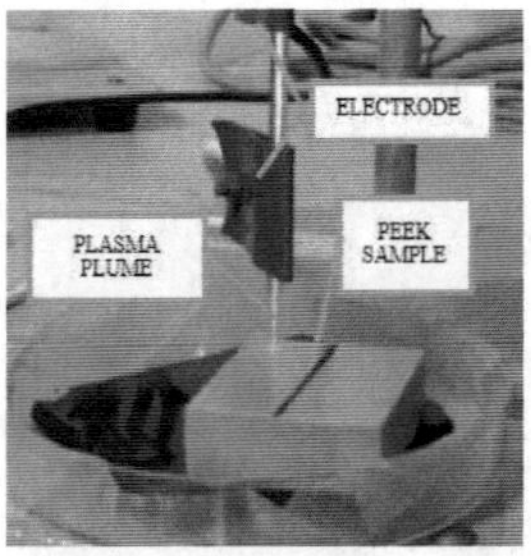

Figure 1. PEEK sample exposed to gas plasma

Figure 2. PEEK sample submerged inside liquid plasma

3.3. Water Contact Angle (WCA) Measurements

The WCA measurements were conducted to determine the wettability of the surface. Untreated PEEK that acts as a control sample were placed on flat and a smooth surface, then 1 drop of distilled water were dropped by using a syringe on the surface of the sample. An example of image of the water droplet on the PEEK surface is shown in Figure 3. To determine the WCA the angle between the baseline of the droplet and the tangent of the drop boundary were determined using ImageJ1.52a software. Measurements were conducted 3 times and results are averaged values.

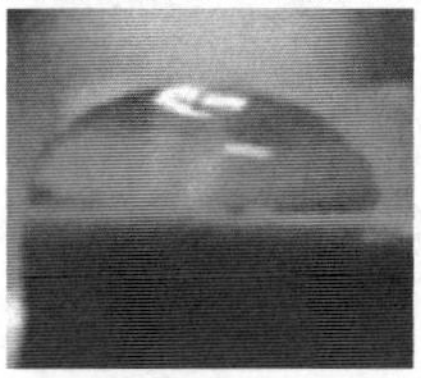

Figure 3. Water droplet on Untreated PEEK Surface

3.4. Scanning Electron Microscopy (SEM)

PEEK samples were coated with platinum before SEM analysis. This is because a direct exposure to electron will cause surface charge on PEEK surface. Images captured will be distorted due to this charging effect. By coating with platinum a conductive layer will be created on the PEEK enabling SEM image to be captured without distortion.

4. Results and Discussions

Each sample was treated by using Helium as working gas at flow rate of 1 L/min. The treatment time for each sample is 5, 15, 25, 35 and 60 seconds by using gas and liquid plasma. SEM images were taken only for treatment time 5s and 60s for gas and liquid experiments.

4.1. Comparison of Water Contact Angle (WCA) between Gas and Liquid Plasma of PEEK Surface

Figure 4 presents the results comparing WCA of PEEK after gas and liquid plasma treatment. It is shown that gas plasma treatment produces higher wettability compared to the liquid plasma treatment. Analyzing the treatment time, it is shown that gas plasma produces smaller WCA to the PEEK surface starting at 5 seconds compared to liquid plasma. Table 1 presents the results of WCA percentage decrement comparison between gas plasma and liquid plasma treatment. Surface processing using gas plasma produces smaller WCA faster.

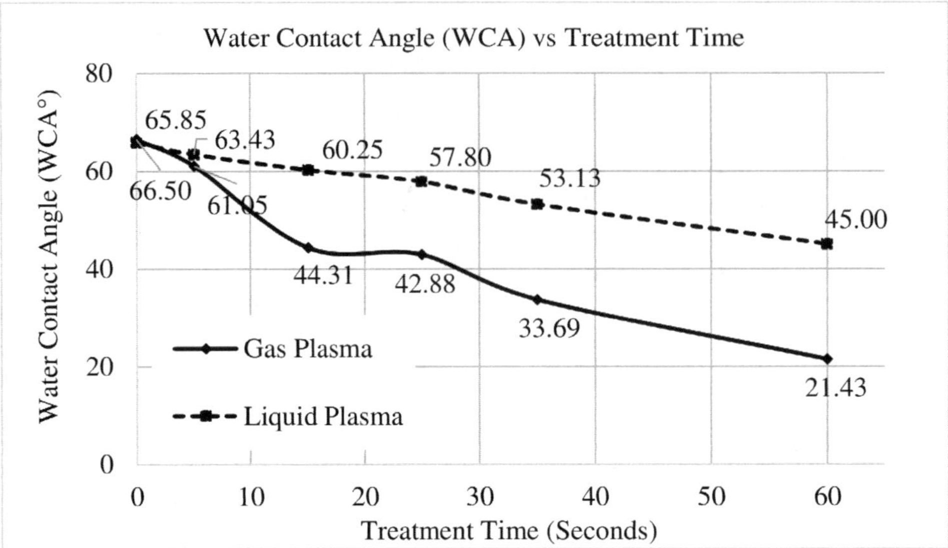

Figure 4. Comparison of WCA and treatment time between gas plasma and liquid plasma treatment

Table 1. Percentage decrement of Water Contact Angle (WCA) of PEEK Surface after Gas and Liquid Plasma Treatment

	GAS PLASMA			LIQUID PLASMA		
Treatment Time (s)	Water Contact Angle (°)	WCA decrement (°)	**WCA decrement (%)**	Water Contact Angle (°)	WCA decrement (°)	**WCA decrement (%)**
0	66.501	-	-	65.854	-	-
5	61.049	5.45	8.02	63.435	2.42	3.67
15	44.307	22.19	33.37	60.255	5.60	8.61
25	42.879	23.62	35.52	57.804	8.05	12.22
35	33.690	32.81	49.34	53.130	12.72	19.32
60	21.434	45.07	67.77	45	20.85	31.67

4.2. Comparison of Scanning Electron Microscopy (SEM)

Figure 5 and 6 shows the SEM images for PEEK samples after 5 s and 60 s of gas plasma treatment respectively. Figure 7 and 8 shows the SEM images for PEEK samples after 5 s and 60 s of liquid plasma treatment respectively. By referring to the SEM images in Figure 5 - Figure 8, the PEEK surface area after gas plasma and liquid plasma treatment were analyzed and the changes on the surface area were measured. The measurement results in Figure 9 showed that the PEEK surface area after gas plasma treatment is larger than liquid plasma treatment by approximately 40%. The results also showed the increasing value of surface area when the treatment time increased.

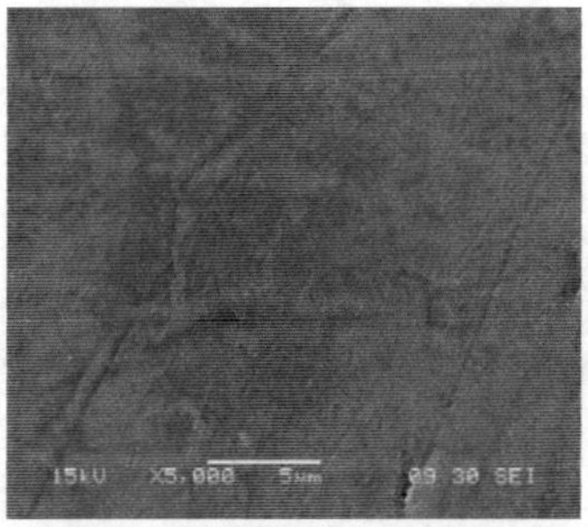

Figure 5. SEM image for 5 s of Gas Plasma Treatment

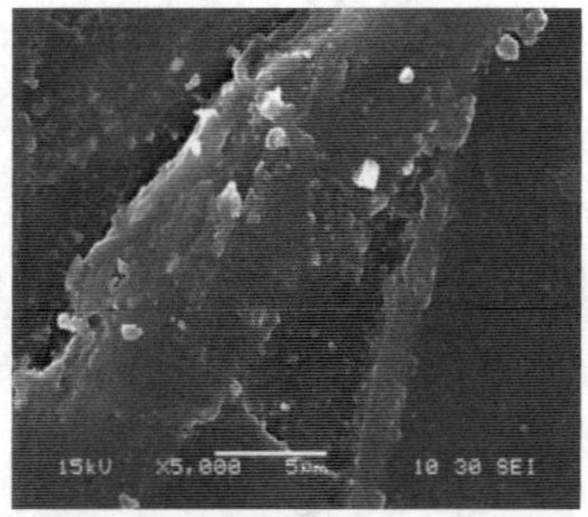

Figure 6. SEM image for 5 s of Liquid Plasma Treatment

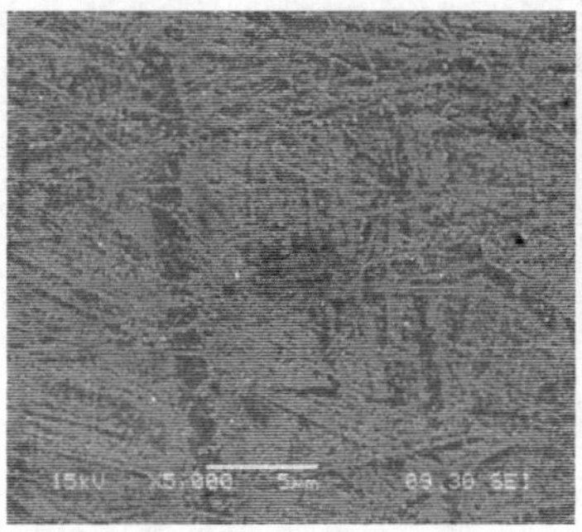

Figure 7. SEM image for 60 s of Gas Plasma Treatment.

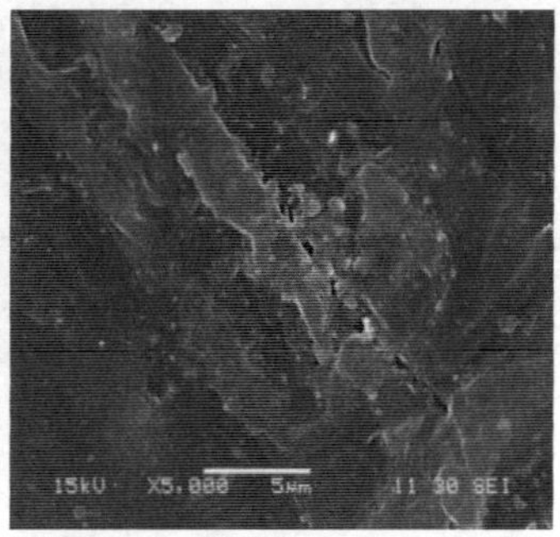

Figure 8. SEM image for 60 s of Liquid Plasma Treatment

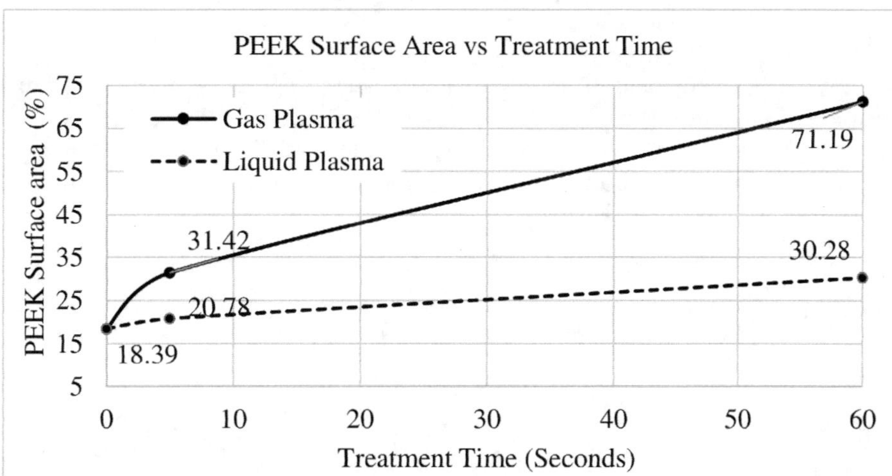

Figure 9. Comparison of PEEK surface area and treatment between gas and liquid plasma treatment

5. Conclusion

This paper presents a comparative study of surface treatment between gas plasma and liquid plasma. WCA and surface area measurements were conducted to evaluate the effect of both plasma treatments. When the water contact angle (WCA) decreases, the wettability of the surface increase. Subsequently, when the wettability of the surface increases, surface hydrophilic properties will also increases [9]. Our results showed that gas plasma treatment produces higher wettability compared to the liquid plasma treatment. It is because of the gas plasma can modify the surface of PEEK in short period of treatment time [3]. From SEM image analysis, the surface area of the treated PEEK showed the 40%

increment when treated with gas plasma. The longer treatment time, the larger the area of treated PEEK surface is produced. Gas plasma treatment is more effective plasma treatment because of the hydrophilic surface properties.

Acknowledgement

The author would like to acknowledge the support from the Fundamental Research Grant Scheme (FRGS) under grant number of FRGS/1/2015/SG02/UNIMAP/02/5 from the Ministry of Education Malaysia.

References

[1] Lu P, Cullen P J and Ostrikov K 2016 Atmospheric Pressure Nonthermal Plasma Sources, *Cold Plasma in Food and Agriculture: Fundamentals and Applications* 83–116.
[2] Ramli N A H, Zaaba S K, Zakaria A, Mamduh S M and Nordin N A 2005 Development of atmospheric plasma gas discharge in liquid, *Proceedings 2015 2nd International Conference on Biomedical Engineering (ICoBE)* 30–31.
[3] Rymuszka D, Terpiłowski K, Borowski P and Holysz L 2016 Time-dependent changes of surface properties of polyether ether ketone caused by air plasma treatment, *Polym. Int.*, **65**(7) 827–834.
[4] Fukuda A, Tsuchiya N, Sunarso and Toita R 2019 Surface plasma treatment and phosphorylation enhance the biological performance of poly(ether ether ketone), *Colloids Surfaces B Biointerfaces* **173**(4) 36–42.
[5] Daniela M.Correia *et al.* 2019 Surface wettability modification of poly(vinylidene fluoride) and copolymer films and membranes by plasma treatment, *Polymer (Guildf).* **169** 138–147.
[6] Fridman A, Gutsol A and Cho Y L 2007 Non-thermal atmospheric pressure plasma, *Adv. Heat Transf.* **40**.
[7] Setsuhara Y 2016 Low-temperature atmospheric-pressure plasma sources for plasma medicine *Arch. Biochem. Biophys.* **605** 3–10.
[8] Kasih T P 2016 Development of non-thermal atmospheric pressure plasma system for surface modification of polymeric materials, *WSEAS Trans. Syst. Control* **9**(2) 2–7.
[9] Kaczorowski W, Batory D, Szymanski W and Niedzielski P 2015 Evaluation of the surface properties of PEEK substrate after two-step plasma modification: Etching and deposition of DLC coatings, *Surf. Coatings Technol.* **265** 92–98.

Gender Prediction by Speech Analysis

Nur Adlin Nazifa [1], Chong Yen Fook[1*], Lim Chee Chin[1], Vikneswaran Vijean[1], Eng Swee Kheng[2]

[1] Biosignal Processing Research Group (BioSIM), School of Mechatronic Engineering, Universiti Malaysia Perlis, Arau, Perlis, Malaysia
[2] School of Mechatronic Engineering, Universiti Malaysia Perlis, 02600 Arau, Perlis, Malaysia.

*yenfook@unimap.edu.my

Abstract. Speech is one of the important methods of communication for humans. The speech signal itself contain linguistic information that can be used to identify the speaker information such as gender, emotions and many more. There are some problems that involve in detection gender of the speaker. In forensic analysis, the police need to detect criminal profile from any evidence such as voice from any calls and while in healthcare aspect, some vocal fold pathologies can be bias to a particular gender such as vocal fold cyst can be seen particularly in female patients and the patient will have problem with their voices. Three features are extract from the speech signal which are Mel Frequency Cepstrum Coefficient (MFCC), Linear Prediction Coding (LPC) and Linear Prediction Coding Coefficient (LPCC). While for the classification, two classifier are used which are Support Vector Machine (SVM) and k-Nearest Neighbour (KNN). The recognition rate is higher for the combination of MFCC and LPCC compared to other features. SVM classifier had outperformed KNN classifier and obtained highest recognition rate of 97.45%. Lastly a graphical user interface system is develop that will record the voice of the speakers, pre-process the signal, extract MFCC and LPCC and classify it using SVM.

1. Introduction

Humans can deliver message and information between each other through speech. Human voice is particularly a piece of human sound creation in which the human vocal chords is the essential sound source, which plays an important role in the conversation [1]. Linguistic information such as gender, emotions and others can be recognize through speech signal. Speech processing is the study of speech signals, and various of methods used to process them. Many applications such as speech coding, speech synthesis, speech recognition and speaker recognition technologies is employed using speech processing. One of important step in speech and speaker recognition is gender recognition [2]. Physiological differences such as vocal fold thickness or vocal tract length and differences in speaking style of humans can be identify as partly the reason gender based differences in human speech [3]. Since these differences are reflected in the speech signal, acoustic measures related to those properties can be helpful for gender classification. Female voice usually has higher resonance frequencies and higher pitch than male voice [4]. The main objective of this paper is to investigate the relationship between the speech signals towards the gender of the speakers. In order to predict the gender of the speaker, this research has proposed a method to extract several features from the speaker voice and class it according to its

Journal of Physics: Conference Series **1372** (2019) 012011 doi:10.1088/1742-6596/1372/1/012011

gender. This study involved the methods of pre-processing of speech signal, extracting features from the filtered signal and classified it to two classes which are male or female.

2. Experiment

2.1. Data Acquisition

There are twenty four subjects which included twelve male and twelve female had volunteered to participate in this experiment. Microphone used for data acquisition in this experiment is Philips Speechmike Premium LFH3500. It is USB dictation microphone with push-button operation that can connected with the laptop. A Graphical User Interface (GUI) in MATLAB is developed to record the voice for each subject and stored the data. Sampling frequency was set to 16 000 Hz and 16 bit. User can record the voice and also play back the recorded voice.

2.2. Data Acquisition

First, a brief is given to the subjects on how the workflow of the experiment and what they need to do. Then, demonstration is given to them on how to handle the microphone and what button they need to push. The subjects need to sit in front of the laptop and sit with relax. The device is placed 3 to 4 cm from the mouth. The recorder will start and the subjects will pronounce a sequence of number from '0 to 9'. The sequence of number is repeat for ten trials with random sequence. Lastly, the recorder will stop and end of data acquisition process.

3. Speech Signal Processing

The recorded speech signal was analysed in signal processing which comprises of three stages. It started with pre-processing of the signal, feature extraction from the signal and lastly classification into two classes which are male and female.

3.1. Pre-processing

The signal needs to be processed before undergoes feature extraction. If there are some noises, it needs to be eliminate and filter to enhance the speech signal features on the waveform. Most of the gender information in the speech signal lies in the voiced region of the speech signal. The signal needs to be segmented for each number from '0' to '9'. The silence and unvoiced region are removed by set up a threshold value of energy for the voiced region. The energy value of signal above the threshold value is considered as voiced region and the energy value below the threshold value will considered as unvoiced region. Next, pre-emphasis is used to neutralize pronunciation effect of speech signal from the lip. It boosts input frequency range that most susceptible to noise and makes the position of dominant energy in frequency more crucial. In this case, the value of pre-emphasis used is 0.98.

3.2. Feature Extraction

Some useful information needs to extract from the signal to distinguish whether the speaker is male or female. The feature extraction that will be use in this project is Mel Frequency Cepstrum Coefficient (MFCC), Linear Prediction Coefficient (LPC) and Linear Predictive Cepstral Coefficient (LPCC).

3.2.1. Mel Frequency Cepstrum Coefficient (MFCC)

Mel Frequency Cepstrum Coefficient (MFCC) are based on the known variation of the human ear's critical bandwidth frequencies with filters spaced linearly at low frequencies and logarithmically at high frequencies used to capture the important characteristics of speech [5]. The signals were divided into small frame size with the length 30 ms. Hamming window is used in this research to eliminate discontinuities at the edges and integrates all the closest frequency lines. The third step is to apply Fast Fourier Transform (FFT) which convert each frame of N samples from time domain to frequency domain and to extract the frequency components of the signal. After that, the signal needs to go through

Mel Filter Bank to filter the signal so it approximates to Mel scale. Each filter output is the sum of its filtered spectral components. Lastly, Discrete Cosine Transform (DCT) is computed by converting log Mel spectrum into time domain. Final result of the conversion is what it called Mel Frequency Cepstrum Coefficient.

3.2.2. Linear Predictive Coding (LPC)

Linear predictive coding (LPC) is a tool used mostly in audio signal processing and speech processing for representing the spectral envelope of a digital signal of speech in compressed form, using the information of a linear predictive model [6]. Pre-processing techniques of this method is including pre-emphasis, frame blocking and Hamming window. The autocorrelation analysis is used to identify fundamental frequency or pitch from the speech signal. Lastly, the signals need to undergo LP analysis.

3.2.3. Linear Predictive Coding Coefficient (LPCC)

Linear Predictive Cepstral Coefficient (LPCC) can generates vocal tract coefficients and it can represent the spectral magnitude of speech signal. It behaves in the Cepstrum domain which is also used in Linear Predictive Coefficient (LPC). In this method, speech signal also need to undergo pre-processing technique such as pre-emphasis, frame blocking and Hamming window and auto correlation analysis. Auto correlation is used as the method to obtain the parameters by assuming that the output of the system is predicted as linear combination of past output samples while input is unknown [7]. Next, in LPC analysis, it will convert auto correlation coefficient into LPC parameters. The Levinson-Durbin recursive algorithm can be used to identify the coefficients [8].

3.3. Classifier

Classifier that had been chosen is Support Vector Machine (SVM) and K-Nearest Neighbour (KNN).

3.3.1. Support Vector Machine (SVM)

Firstly, kernel function is applied that performs a low to high dimensional feature transformation and in this case linear kernel function, Gaussian kernel function and polynomial kernel function is used. Next, this classifier constructs a maximum margin hyperplane to draw the decision boundary between classes. SVM search for an optimal hyperplane H that separates this data space into two distinct sub-spaces and each corresponding to a class. This hyperplane can be described as the following equation [9-10]:

$$H: x^T w + b = 0 \tag{1}$$

Where w is the perpendicular vector to the hyperplane H and b is the bias of this function.

3.3.2. K-Nearest Neighbour (KNN)

K-Nearest Neighbour is base of an unknown sample on the “votes” of K of its nearest neighbour rather than only it’s single neighbour. The input of KNN classifier in the feature space consists of the k closest training example [11-13]. However, the output of the KNN are depends on the usage of the KNN either it is used for classification or regression. For the KNN classification, the class membership is the output of the classifiers. The data is classified using majority vote of the neighbours and the data is assigned to its classes according to its most common its k nearest neighbour [14-15]. The nearest neighbour are found by minimising a distance function, usually the Euclidean distance. K-value used is varies from 1 to 10 folds with 10-fold cross validation.

4. Result and Discussion

4.1. Comparison of SVM and KNN

Table 1 shows the recognition rate of SVM based on features extracted using several kernel functions.

Table 1. Recognition rate of SVM based on features using several kernel functions

Kernel function	Features extracted			
	MFCC	LPC	LPCC	MFCC+LPCC
Linear	89.43%	89.70%	89.08%	92.82%
Gaussian	92.90%	91.25%	93.31%	95.06%
Polynomial	95.54%	95.59%	95.96%	97.45%

From Table 1, it can be seen that polynomial kernel function has highest recognition rate of gender prediction compared to the linear and Gaussian kernel functions. The accuracy achieved is 95.54% for MFCC, 95.59% for LPC, 95.96% for LPCC and the highest is 97.45% for MFCC+LPCC. Table 2 shows the recognition rate of KNN based on features extracted using different k-values.

Table 2. Recognition rate of KNN based on features using different k-values

k-value	Features extracted			
	MFCC	LPC	LPCC	MFCC+LPCC
1	88.87	87.45	89.76	90.89
2	87.63	85.77	88.67	89.37
3	86.31	84.48	87.25	87.98
4	85.69	83.96	86.69	87.44
5	85.55	83.83	86.51	87.40
6	85.50	83.65	86.36	87.36
7	85.36	83.36	86.17	87.26
8	85.25	83.13	85.99	87.19
9	85.15	83.00	85.89	87.11
10	85.07	82.91	85.81	87.05

From Table 2, combination of MFCC and LPCC has the highest recognition rate using KNN in all k-value compared to the others feature which the highest value is 90.89%. Followed by MFCC and LPCC that have higher accuracy than LPC. Then, the lowest rate is achieved by LPC which most of its recognition rate is below than 90%. Furthermore, the result shows that combination of MFCC and LPCC is the most suitable features in gender prediction and able to distinguish the gender since MFCC and LPCC is more sensitive and nearest feature to human voice in both classifiers.

4.2. Graphical User Interface (GUI) Development

GUI was developed to build a system for prediction of gender of the speaker by using MATLAB software. This GUI can be used to record the voice and display the speech signal, segment word by word and display the signal, extract MFCC and LPCC features and classify the gender of the speaker. The layout of the GUI for gender prediction system is shown in Figure 1

.

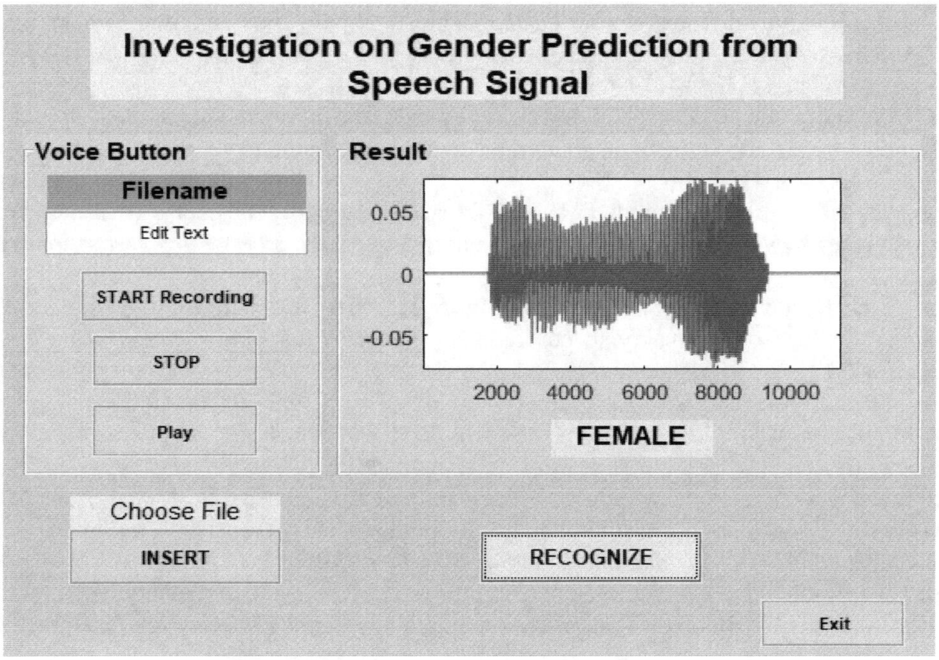

Figure 1. GUI for gender prediction system

5. Conclusion

From the analysis of results, SVM classifier had outperformed KNN classifier and obtained highest recognition rate of 97.45%. This could be because KNN is a lazy algorithm that depends barely on statistics and comparison, and it must keep track of large amount of features whereas SVM uses offline learning to find the optimal hyperplane so it can distinguish gender well and very effective.

6. References

[1] Kumar P, Baheti P, Jha R K, Sarmah P and Sathish K 2018 Voice Gender Detection Using Gaussian Mixture Model *Journal of Network Communications and Emerging Technologies (JNCET)* **8**(4) 132-136.

[2] Deiv D S 2011 Automatic Gender Identification for Hindi Speech Recognition *International Journal of Computer Applications* **31**(5) 1–8.

[3] Upadhyaya P, Farooq O, Abidi M R and Varshney Y V 2018 Continuous Hindi speech recognition model based on Kaldi ASR toolkit *Proc. 2017 Int. Conf. Wirel. Commun. Signal Process. Networking, WiSPNET* **2018**(2) 786–789.

[4] Rashmi C R 2014 Review of Algorithms and Applications in Speech Recognition System *International Journal of Computer Science and Information Technologies* **5**(4) 5258–5262.

[5] Jayasankar T, Vinothkumar K and Vijayaselvi A 2017 Automatic gender identification in speech recognition by genetic algorithm *Appl. Math. Inf. Sci.* **11**(3) 907–913.

[6] Carol Y Espy-Wilson 1994 A feature-based semivowel recognition system *The Journal of the Acoustical Society of America* **96**(1) 65-72.

[7] Kaur G, Srivastava M and Kumar A 2018 Genetic Algorithm for Combined Speaker and Speech Recognition using Deep Neural Networks *Journal of Telecommunications and Information Technology* **2018**(2) 23–31.

[8] Yusnita M A, Hafiz A M, Fadzilah M N, Zulhanip A Z and Idris M 2017 Automatic gender recognition using linear prediction coefficients and artificial neural network on speech signal *2017 7th IEEE International Conference on Control System, Computing and Engineering (ICCSCE), Penang* 372-377.

[9] Ahmad J, Muhammad K and Kwon S 2016 Dempster-Shafer Fusion based Gender Recognition for Speech Analysis Applications *2016 International Conference on Platform Technology and Service (PlatCon), Jeju,* 1-4.

[10] Fook C Y, Hariharan M, Yaacob S, and Adom A H 2012 Malay speech recognition in normal and noise condition *IEEE 8th International Colloquium on Signal Processing and its Applications (CSPA 2012)* 409–412.

[11] Qawaqneh Z, Mallouh A A and Barkana B D 2017 Age and gender classification from speech and face images by jointly fine-tuned deep neural networks *Expert Systems with Applications* **85** 76–86.

[12] Fook C Y, Hariharan M, Yaacob S, Adom A H 2016 Subtractive clustering based feature enhancement for isolated malay speech recognition *ARPN Journal of Engineering and Applied Sciences* **11**(13) 8517-8524.

[13] Jonsson P and Wohlin C 2004 An evaluation of k-nearest neighbour imputation using Likert data *10th International Symposium on Software Metrics, 2004. Proceedings., Chicago, Illinois, USA* 108-118.

[14] Chin L C, Fook C Y, Vikneswaran V, Abdul Rahim M A H, Hariharan M 2018 Body Mass Index (BMI) Of Normal and Overweight/Obese Individuals Based on Speech Signals *Journal of Telecommunication, Electronic and Computer Engineering* **10** 1-16.

[15] Chin L C, Basah S N, Yaacob S, Juan Y E and Kadir A K A 2015 Camera Systems in Human Motion Analysis for Biomedical Applications *AIP Conference Proceedings* **1660(1)** 090006,.

Pearson's Correlation Coefficient Analysis of non-invasive Jaundice Detection based on Colour Card Technique

Asyraf Hakimi Abu Bakar[1], Mohd Najmuddin Mohd Hassan[2], Ammar Zakaria[1], Ahmad Ashraf Abdul Halim[2]

[1]School of Mechatronic Engineering, Universiti Malaysia Perlis (UniMAP), Pauh Putra Campus, 02600, Arau, Perlis.
[2]School of Computer and Communication Engineering, Universiti Malaysia Perlis (UniMAP), Pauh Putra Campus, 02600, Arau, Perlis.

*asyrafhakimi@unimap.edu.my

Abstract. Jaundice is describes as yellow discoloration of the skin and other tissues of a newborn infant. It happens when the bilirubin pigment rises up to 5 mg/dL or 85 µmol/L due the organs just started to developed. 75% of newborns in Malaysia had jaundice in the first week of life. Conventional assessment method is not convenient since it unable to detect all type of human skin colours, caused traumatization and costly. This research work proposed jaundice detection system based on the colour card technique that represent all types of human skin colours. This research does not involve any subject due to parent's cooperation and ethics limitation. The input data is represented by the colour card shades. It represents random bilirubin colour in patient's body, all types of human skin colour, and standard reference bilirubin concentration colour. It acquires input data by capturing different type of colour card shades using OPT101 photodiode sensor and USB4000-XR1-ES spectrometer. The input data collected were verified through Anderson-Darling normality test and analysed using Pearson's correlation coefficient to measure the strength of the association between the two variables for each input sample tested. Based on the correlation results, it shows high correlation value (r=0.997). Percentage error analysis is used to validate the experimental device and shows value of 9.90% which prove this technique is reliable. The significant contributions of this work is the improvement of the accuracy in detecting jaundice level for all type of human skin colours through this system.

1. Introduction

Jaundice in newborn or neonatal hyperbilirubinemia, is a yellow discoloration and shows common sign of liver not functioning appropriately caused by the pigment of bilirubin that rises above 85µmol/l or 5mg/dl [1, 2, 3]. A survey conducted by Ministry of Health Malaysia found that about 75% of newborns experience jaundiced in the first week of life, based on data collected from government hospitals and health centers [4]. The statistic shows how crucial this medical condition among Malaysians. Jaundice disorder happens due to the organs and metabolism of a newborn is just starting to develop [5]. It typically results from the deposition of unconjugated bilirubin pigment in the skin and mucous membranes [6]. For infants, their liver is immature and potentially cannot fully remove the bilirubin from body which causes the yellowing of the skin [7]. Babies may develop kernicterus when the excess bilirubin traverse from the bloodstream to the brain tissue, which can cause death, while survivors will suffer cerebral palsy and high frequency hearing loss [9, 10]. Treatments use to cure jaundice are phototherapy (exposing baby to fluorescent light) and blood exchange transfusion [10].

2. Jaundice Assessment Method

Jaundice assessment method is a marker used to identify those infants who may be at risk for developing severe hyperbilirubinemia which can potentially lead to brain damage. Clinical examination usually reveals the degree of hyperbilirubinemia by referring to the TSB (Total Serum Bilirubin) levels [11]. Jaundice assessment is divided into two types that are invasive and non-invasive method.

2.1. Invasive Method: Blood Test

Blood test is considered the best way to detect any disease in human body which bring into accurate results, including jaundice [12]. Most hospitals in Malaysia uses this invasive method, by taking a specific amount of infant's blood and test it in the laboratory. The blood samples are taken from the baby's heels using capillary stick as shown in Figure 1 [13]. Unfortunately, invasive jaundice measurement has some drawbacks which are painful for the infants, caused delay in obtaining results which can lead to serious problem to the infant with high TSB.

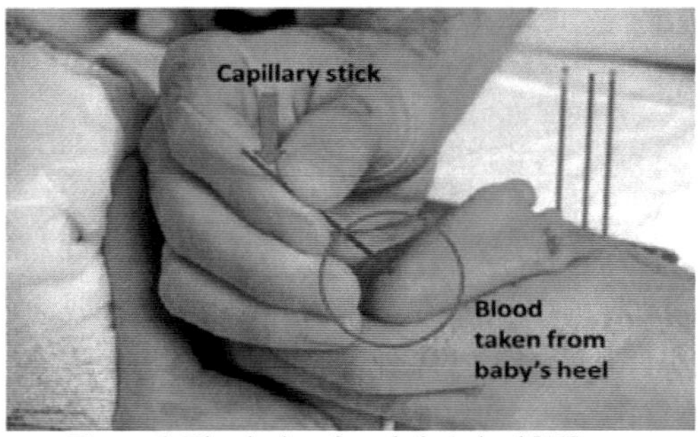

Figure 1. Blood taken from baby's heel [13].

2.2. Non-invasive Method: Kramer's Rule (visual inspection)

Kramer's Rule is a method that fully relies on human manual visual assessment. Kramer divide the body into five zones; starts from head and neck, upper trunk, lower trunk and thigh, arms and lower legs and lastly the palms and soles as shown in Figure 2 [14]. Unfortunately, Kramer's Rule is not beneficial if the baby has dark skin. Also, this manual visual assessment may lead to human error and unreliability issues [15].

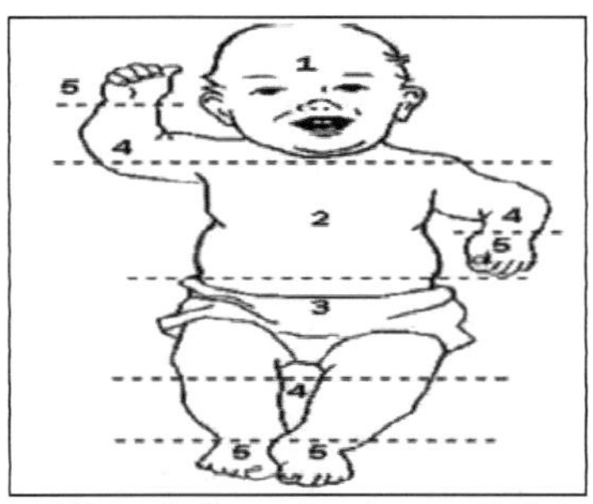

Figure 2. Kramer's rule zones[14]

2.3. Non-invasive Method: Jaundice Meter

Jaundice meter or mostly known as transcutaneous bilirubinometer is a non-invasive technique that is painless, eliminates traumatization and offer fast result by using the concept of optical reflection and absorption [14]. It works by directing light into the skin of the neonate and measuring the intensity of a specific wavelength that is returned to the photocell [16]. Indeed, jaundice meter available in the market offers painless effect and fast results, but not all jaundice meter concern on the human skin factor based on the human skin colour variations [17].

3. Data Collection Method

This system consists of a sensing module, microcontroller for data processing and GUI for display and monitoring. Figure 3 shows the block diagram of the system with OPT101 silicon photodiode sensor is used as the sensing module, white LED (Light Emitting Diode) functioned as light source, and spectral reflectance method is applied as a standard optical method. The sensor senses the change of intensity of the colour, then the input data is processed by a microcontroller that translates the colour intensity to an arbitrary displayed voltage value. The data are then transmitted wirelessly using XBee transmitter to the receiver, processed and displayed through GUI (Graphical User Interface) on the computer in real-time manner. All data are collected using experimental device (OPT101 silicon photodiode sensor) and actual device (Near Infrared (NIR) spectrometer USB4000-XR1-ES). Each sample measurement is taken for 10 seconds per reading and repeated for five times for each sample. The reason of this process is to make sure the device sensor is fully stabilize, and ready to read the data. Figure 4 shows the data collection method for both experimental and actual device.

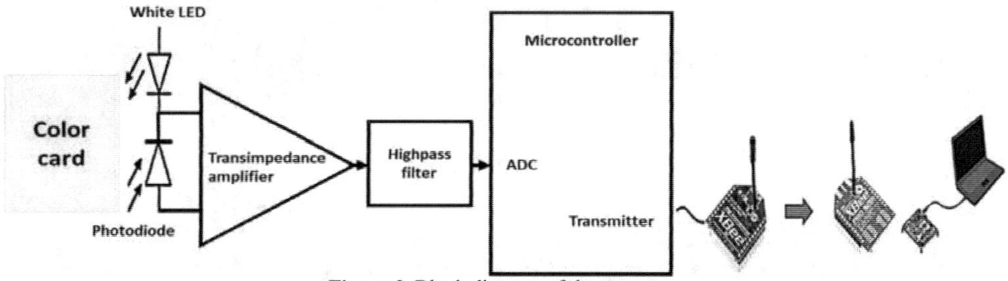

Figure 3. Block diagram of the system

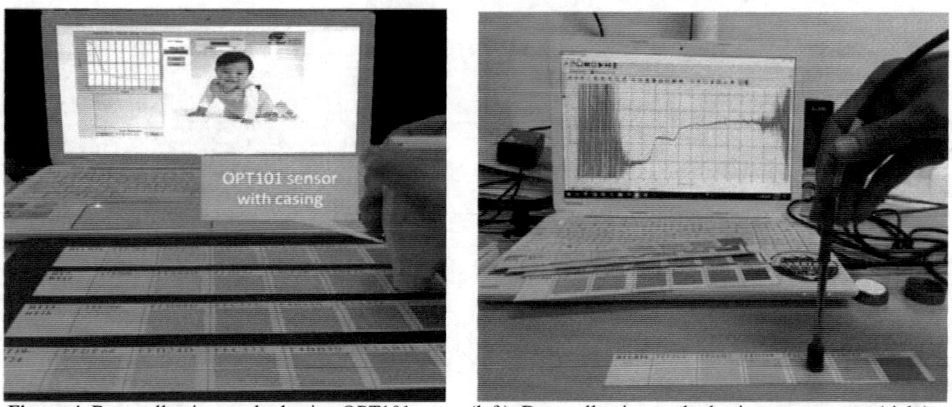

Figure 4. Data collection method using OPT101sensor (left), Data collection method using spectrometer (right)

3.1. Colour Card Prototype

Three types of colour card are used as the input and standard reference data for this research. Yellow colour card shades shows the arrangement of seven yellow colour represents random bilirubin colour tissue of jaundice patient because bilirubin appears to be in yellow colour [18]. Fitzpatrick human skin colour card is used to represents a classification scheme for human skin colour standardization [19]. This is to highlight the concern usage of human skin colour element in detecting jaundice as it is one of the main disadvantage in the current jaundice detection system. The third colour card is the bilirubin reference card which represents the standard reference colour of bilirubin and concentration for each colour as well as the treatment required [20]. This bilirubin colour card is used as reference chart in this research.

3.1.1. Colour Card Combination.

This section describes the combination and mixture of yellow shade, Fitzpatrick human skin colour shade, and bilirubin reference card. Each type of yellow shade (Y1 to Y7) and Fitzpatrick shade (T1 to T6) are combined together using colour mixer generator from http://kotoritone.com/color/en/ website, resulting a sample name S1 to S42 [21]. Also, the colour combination involves the mixture of yellow shade and bilirubin reference card (B1 to B4), resulting new reference standard name BT1 to BT24 [21]. Some example of this input data are shown in Figure 5 and Figure 6.

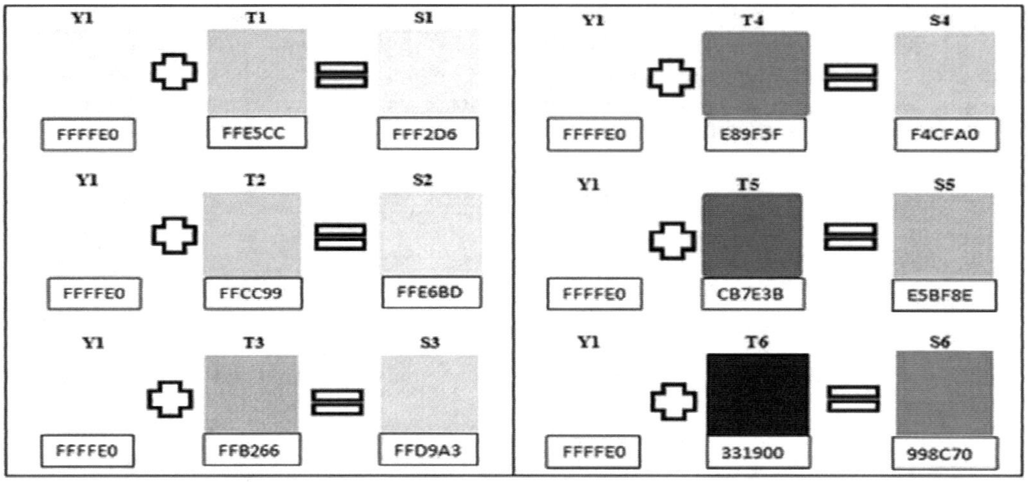

Figure 5. Sample from S1-S6

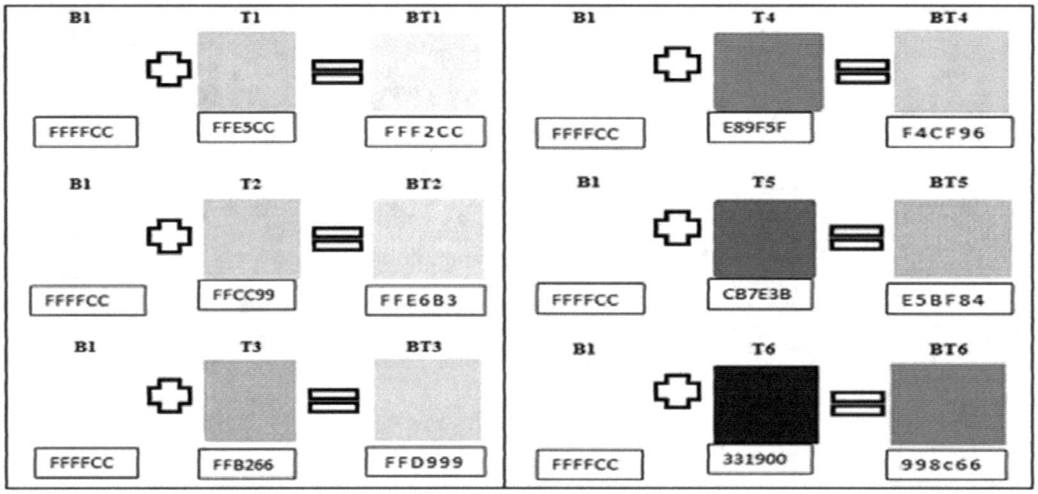

Figure 6. BT reference standard shade from BT1-BT6 (right).

4. Results and Discussion

A total of 42 S samples (S1 to S42) of input data representing random patient bilirubin colour combined with human skin colour, and 24 shades (BT1 to BT24) reference chart are collected and verified through Anderson-Darling Normality Test to measures how well the data follow and fits a specified distribution.

4.1. Pearson's Correlation Coefficient

The verified data then being analyzed using Pearson's correlation coefficient. It is a measure of the strength of the association between the two variables. Table 1 shows the correlation result for data collected using experimental device (OPT101). There are significant positive correlation between S1-S6 and BT1-BT6 with (r =0.988, p =0.000). From the correlation result of sample S7-S12, S13-S18, and S19-S24, it is apparent that the sample have high correlation with BT7-BT12 at (r =0.991, p =0.000); (r =0.997, p =0.000); (r =0.997, p =0.000). There is a clear statement saying that sample S25-S30 have high correlation with BT13-BT18 at (r =0.995, p =0.000). The sample S31-S36 and S37-S42 have strong correlation of (r =0.997, p =0.000); (r =0.995, p =0.000) with BT19-BT24.

Table 1. Correlation results for experimental device (OPT101 sensor)

		Pearson's correlation coefficient result (r)						
BT reference data	Correlation	S1-S6	S7-S12	S13-S18	S19-S24	S25-S30	S31-S36	S37-S42
BT1-BT6 (0-4 mg/dL)	Pearson Correlation	0.988	0.98	0.948	0.942	0.947	0.901	0.908
	p value	0.000	0.004	0.004	0.005	0.004	0.014	0.012
BT7-BT12 (5-14 mg/dL)	Pearson Correlation	0.895	0.991	0.997	0.997	0.994	0.955	0.961
	p value	0.016	0.000	0.000	0.000	0.000	0.003	0.002
BT13-BT18 (15-19 mg/dL)	Pearson Correlation	0.862	0.967	0.978	0.992	0.995	0.991	0.993
	p value	0.027	0.002	0.001	0.000	0.000	0.000	0.000
BT19-BT24 (> 20mg/dL)	Pearson Correlation	0.827	0.904	0.924	0.952	0.963	0.997	0.995
	p value	0.043	0.013	0.008	0.003	0.002	0.000	0.000

Similar method of analysis was carried out for USB4000-XR1-ES spectrometer. This is to support the result obtained from the OPT101 sensor and to prove whether the device sensor is reliable or otherwise. The Pearson's correlation coefficient results for S sample vs BT data using USB4000-XR1-ES spectrometer are presented in Table 2 with highlighted area shows the highest correlation between S samples tested vs BT reference data.

Table 2. Correlation result for actual device (USB4000-XR1-ES spectrometer)

		Pearson's correlation coefficient result (r)						
BT reference data	Correlation	S1-S6	S7-S12	S13-S18	S19-S24	S25-S30	S31-S36	S37-S42
BT1-BT6 (0-4 mg/dL)	Pearson Correlation	0.976	0.858	0.905	0.854	0.955	0.961	0.977
	p value	0.001	0.029	0.013	0.030	0.003	0.002	0.001
BT7-BT12 (5-14 mg/dL)	Pearson Correlation	0.918	0.931	0.966	0.918	0.949	0.863	0.885
	p value	0.010	0.007	0.002	0.010	0.004	0.027	0.019
BT13-BT18 (15-19 mg/dL)	Pearson Correlation	0.935	0.905	0.920	0.886	0.991	0.898	0.930
	p value	0.006	0.013	0.009	0.019	0.000	0.015	0.007
BT19-BT24 (> 20mg/dL)	Pearson Correlation	0.890	0.669	0.809	0.808	0.858	0.980	0.986
	p value	0.017	0.047	0.051	0.052	0.029	0.001	0.000

Both correlation results between experimental and actual device are compared and it shows that the experimental device OPT101 follows the same output from USB4000-XR1-ES spectrometer. Each S sample found its own matched BT reference data based on the highest r value which indicate strong correlation between both variables. It can be summarizes that the experimental device OPT101 sensor have the similar outcome results as actual device USB4000-XR1-ES spectrometer. The highlighted area in Table 3 shows the correlation between both devices.

Table 3. Pearson's correlation result for OPT101 sensor and USB4000-XR1-ES spectrometer

		Pearson's correlation coefficient result (r)						
BT reference data	**Device**	**S1-S6**	**S7-S12**	**S13-S18**	**S19-S24**	**S25-S30**	**S31-S36**	**S37-S42**
BT1-BT6	**OPT101**	0.988	0.98	0.948	0.942	0.947	0.901	0.908
(0-4 mg/dL)	**USB4000**	0.976	0.858	0.905	0.854	0.955	0.961	0.977
BT7-BT12	**OPT101**	0.895	0.991	0.997	0.997	0.994	0.955	0.961
(5-14 mg/dL)	**USB4000**	0.918	0.931	0.966	0.918	0.949	0.863	0.885
BT13-BT18	**OPT101**	0.862	0.967	0.978	0.992	0.995	0.991	0.993
(15-19 mg/dL)	**USB4000**	0.935	0.905	0.920	0.886	0.991	0.898	0.930
BT19-BT24	**OPT101**	0.827	0.904	0.924	0.952	0.963	0.997	0.995
(> 20mg/dL)	**USB4000**	0.890	0.669	0.809	0.808	0.858	0.980	0.986

4.2. Percentage Error

Percentage error analysis is used to validate the experimental device compared to actual device. The percentage error results for BT reference data shows the highest value of % error obtained is 9.62%, whereas S sample show the highest % error of 9.90%. Based on statement by Najmuddin [22], the output obtained from experimental reading must not exceeds 10% of the actual reading output to prove the reliability of the experimental device. Both results for BT reference and S sample shows % error below than 10%, thus, proves that the experimental device is valid and reliable to be used.

5. Results and Discussion

The significant contributions of this work has been to improve the accuracy of non-invasive jaundice assessment for various type of human skin colour by the implementation of colour card technique. Additional to the previous work, this research introduces colour card technique addressing the concern of human skin factor as it is the major disadvantages in most of the jaundice assessment method. Thus, human skin colours represented by Fitzpatrick shades are considered. Since lack of research been made for detecting jaundice in dark skin colour that leads to the disadvantages to previous device, this research combines the variations on human skin colour ranging from type one to six with the random bilirubin colour in human body. In conclusion, it has been shown that the jaundice assessment using colour card technique is a promising enabling technique for non-invasive and non-traumatize. This technique demonstrate reliability, precision in its results and enhance the safety for user.

References

[1] Z. Zakaria *et al.* 2015 A Short Review on Kramel's Rule and Magnetic Induction Spectroscopy **6** 103–106.

[2] C. M. Lewandowski 2015 Management of Neonatal Jaundice *Eff. Br. mindfulness Interv. acute pain Exp. An Exam. Individ. Differ.* **1** 1689–1699.

[3] S. Ellairaja, K. Shenbagavalli, S. Ponmariappan, and V. S. Vasantha 2017 A green and facile approach for synthesizing imine to develop optical biosensor for wide range detection of bilirubin in human biofluids *Biosens. Bioelectron.* **91** 82–88.

[4] N. K. Ives 2015 Management of neonatal jaundice *Paediatr. Child Health (Oxford).* **25** no. 6, pp. 276–281.
[5] A. M. Osman, Zulfadhli, Afandi Ahmad 2014 Rapid Prototyping of Neonatal Jaundice Detector Using Skin Optics Theory *IEEE* pp. 8–10.
[6] Mohamed Abd El-Halim Abd El-Aal 2012 Rebound hyperbilirubinemia after phototherapy *Cairo University*.
[7] S. Asmah and B. Daud 2012 Automatic Phototherapy Vest (APV) *Universiti Teknologi Malaysia*.
[8] L. Holland and K. Blick 2009 Implementing and validating transcutaneous bilirubinometry for neonates *Am. J. Clin. Pathol.* **132** no. 4, pp. 555–561.
[9] M. N. Mansor, M. Hariharan, S. N. Basah, and S. Yaacob 2013 New newborn jaundice monitoring scheme based on combination of pre-processing and color detection method *Neurocomputing* **120** 258–261.
[10] M. A. Erlandsen, T. Willy, and R. Hansen 2010 Treatment of neonatal jaundice - more than phototherapy and exchange transfusions *East. J. Med.* **15** 175–185.
[11] C. M. da Conceição, M. F. P. da S. Dornaus, M. A. Portella, A. D. A. Deutsch, and C. M. Rebello 2014 Influence of assessment site in measuring transcutaneous bilirubin. *Einstein (Sao Paulo).* **12** no. 1, pp. 11–5.
[12] W. D. Engle, G. L. Jackson, E. K. Stehel, D. M. Sendelbach, and M. D. Manning 2005 Evaluation of a Transcutaneous Jaundice Meter Following Hospital Discharge in Term and Near-Term Neonates *J. Perinatol.* **25** 486–490.
[13] National screening Unit of the Ministry of health 2010 The Newborn Metabolic Screening Programme *Newborn Screening Free health checks for your baby*, pp. 1–8.
[14] N. Ali, S. Z. M. Muji, A. Joret, R. Amirulah, N. Podari, and N. F. Dol Risep 2015 Optical technique for jaundice detection *ARPN J. Eng. Appl. Sci.* **10** no. 20, pp. 9929–9933.
[15] C. L. Davis, B. Rappaport, and T. I. Insti- 2008 Is Visual Assessment of Jaundice Reliable as a Screening Tool to Detect Significant Neonatal Hyperbilirubinemia?.
[16] T. S. K. Lam, K. L. Tsui, and C. W. Kam 2008 Evaluation of a point-of-care transcutaneous bilirubinometer in Chinese neonates at an accident and emergency department *Hong Kong Med. J.* **14**, no. 5, pp. 356–360.
[17] T. M. Slusher, A. Zipursky, and V. K. Bhutani 2011 A Global Need for Affordable Neonatal Jaundice Technologies *Semin. Perinatol. Elsevier Inc* **35**, no. 3, pp. 185–191.
[18] L. Mundt 2010 Chemical Analysis of Urine *Graff's Textb. Urin. Body Fluids*, pp. 35–45.
[19] K. R. L. Duteil and P. Bahadoran 2016 Measuring the Skin *Agache's Measuring the Skin*, P. A. Philippe Humbert, Ferial Fanian, Howard Maibach, Ed. Springer pp. 35–48.
[20] J. Kriangsak 2010 The Bilirubin Color Scale: a low-cost device for measuring plasma bilirubin.
[21] A. H. A. Bakar, N. M. Hassan, A. Zakaria, K. A. A. Halim, and A. A. A. Halim 2017 Jaundice (Hyperbilirubinemia) detection and prediction system using color card technique *Proceedings - 2017 IEEE 13th International Colloquium on Signal Processing and its Applications* pp. 208–213.
[22] A. F. Najmuddin and I. M. Ibrahim 2010 A Simulation Approach : Improving Patient Waiting Time for Multiphase Patient Flow of Obstetrics and Gynecology Department (O & G Department) *Local Specialist Centre* **9** no. 10, pp. 778–790.

Fabrication and Biomechanical Evaluation of Polyurethane Material for Synthetic Bone

Siti Adawiyah Zulkefli[1,3], Maryam Hanim Shaikhani[1,3], Ahmad Kafrawi Nasution[4], Asnida Abdul Wahab[1], Mohammed Rafiq Abdul Kadir[2,3] and Muhammad Hanif Ramlee[1,3*]

[1] Medical Device and Technology Centre (MEDITEC), Institute of Human Centered Engineering (i-Humen), Universiti Teknologi Malaysia, 81310, Johor, Malaysia.
[2] Sport Innovation & Technology Centre (SITC), Institute of Human Centered Engineering (i-Humen), Universiti Teknologi Malaysia, 81310, Johor, Malaysia.
[3] Bioinspired Innovation Group (BIOINSPIRA), School of Biomedical Engineering and Health Sciences, Faculty of Engineering, Universiti Teknologi Malaysia, 81310, Johor, Malaysia.
[4] Faculty of Engineering, Universiti Muhammadiyah Riau, Pekan Baru, Indonesia.

*muhammad.hanif.ramlee@biomedical.utm.my

Abstract. Synthetic bone has become an important tool for education in the orthopaedics field and biomechanical research. The use of the conventional tool such as cadaver has become a challenge to the researcher and orthopaedic surgeon, as it was highly cost, ethical issues and storage. In addition, the available synthetic bones in the market were expensive and only in European size. This research was focusing on developing the synthetic bone using polyurethane material, evaluate the mechanical strength of the developed synthetic bone and characterize the morphological structure of the developed synthetic bone. The development of the synthetic bone was using Polyurethane (PU) as its material. Two type composition of PU was used for this research. The material mixed and poured into the mould that has been designed according to American Society Testing and Material (ASTM) standard which was ASTM-F1839. Then, composition 1 and composition 2 undergo mechanical testing and morphology analysis by using Instron and Scanning Electron Microscopy (SEM) respectively. The average Young's Modulus of composition 1 and composition 2 were 208.6963 MPa and 384.6304 MPa respectively. In short, the developed synthetic bone has greater Young's modulus compared to the synthetic bone that available in the market and which are 132.3254 MPa (Sawbone) and 68.1483 MPa (Synbone). The morphological analysis has shown the fabricated synthetic bone has compact structure compare to the one in the market that mimic the cancellous bone. In conclusion, composition 2 of PU have the highest Young's modulus but from the morphological analysis it does not mimic the morphology of cancellous bone.

1. Introduction

Bones were very essential in daily life activities as it gives support to the locomotion such as walking, standing, running and many more [1]. Moreover, in the body there was a bone, which was tibia that plays an essential role in our life. The study was focus on optimizing the Young' modulus of the synthetic bone to meet Young's modulus of cancellous bone. Young's modulus for cancellous bone was range between 350 MPa to 700 MPa. The value of cancellous bone of 350 MPa was for the bone that encounters metabolic bone disorder while 700 MPa was for the normal bone [2]. Nonetheless, the tibia was exposed to the fractures that might occur due to the impact while doing various type of activities. There are several types of fracture that might occur such as transverse, oblique, comminuted and compound fractures [3,4].

In the orthopaedic field, it was very essential for the surgeon to prescribe the best treatment to the patient. In addition, the orthopaedic surgeon also needs skills in order to perform the surgery to prevent any unwanted event. Hence, the use of the synthetic bone will provide the surgeon and medical students to familiarize with the surgical tools and simulate the surgery in the operation theatre [5,6]. For the biomechanics researcher, to simulate implant would need cadaver in order to have accurate data. This cadaver would cause several issues due to the limited storage and life span, cost and the most crucial is ethical approval [7–9]. Even though several companies such as Synbone and Sawbone are providing synthetic bone, the size of the bone was in European size. Thus, the biomechanics researcher has limitation in their study where the bones that can be available were only in European size.

In order to fabricate the synthetic tibia bone, PU was used as the material. PU was first discovered by Otto Bayer and his coworkers at the laboratories of I.G. Farben in Leverkusen, Germany on 1937. In recent year, PU was widely used in automotive, building and home construction, medical devices and implants [10,11]. As reported by Nuno V. Gama et al, the consumption of PU in 2017 was estimated at 60.5 billion USD in 2017. However, it was predicted to increase which is over 79 billion USD in 2021. Furthermore, PU was formed by the reaction of exothermic of isocyanates and polyol. Isocyanates acted as a hard segment which has NCO functional group while polyol acted as a soft segment that contains OH group [10]. N. Sidek et al mentioned that the ideal ratio of isocyanates and polyol was 1:1. In other words, the ratio of 1:1 will give the best result for the formation of Polyurethane. In this study, the ratio to produce Polyurethane is 1:1 [11]. To summarize, this study was intended on developing the synthetic bone by using polyurethane material, evaluate the mechanical strength of the developed synthetic bone and characterize the morphological structure.

2. Methodology

The experimental work of the study started by preparing the mould based on ASTM-F1839. Consequently, the preparation mixture of Polyurethane was made for composition 1 and composition 2. Then, composition 1, composition 2 and tibia synthetic bone from Sawbone and Synbone undergo mechanical testing and morphological analysis.

2.1. Preparation of Mould

The mould was designed to meet the requirement from ASTM-F1839 as shown in Fig. 1. The standard specification is for Rigid Polyurethane Foam for Use as a Standard Material for Testing Orthopedic Devices and Instruments. Figure 1 illustrates the flow for preparation of mould. Firstly, the mould was sketched on 3D Computer Aided (CAD) software, Solidworks (Dassault System Solidworks Corp., USA). The dimension of the mould was based ASTM F1839 which were 50.8mm x 50.8mm x 25.4mm. After sketching, the file was saved in the STL file. In order to 3D print the mould, the STL file was converted into the x3g file by using Markerbot software. Next, the mould was 3D printed by using the Flashforge 3D printer. The time consumed for the mould to be completely printed was approximately 5 hours. Once the 3D printed mould was ready, the mixture of the Moldmaker Silicone and hardener was mix together to make the silicone rubber mould. After the mixture was ready, the mixture was poured into the 3D printed mould. The mixture was left to dry for 24 hours. After 24 hours, the silicone mould was pulled out from the 3D printed mould.

Figure 1. Show the flow to produce mould based on ASTM F1939.

2.2. Preparation of PU

There were two type composition of PU were used in this study. Basically, for composition 1 and composition 2 were the fabricated PU which given in two parts which were Part A (Polyol) and Part B (isocyanate). As stated in the introduction the ideal ratio for the polyurethane was 1:1. Nonetheless, the ideal ratio for composition 1 was 2:1 based on the manufacturer's guideline. In contrast, composition 2 was using ratio 1:1 to get the optimum mechanical strength based on manufacturer's guideline. Hence, to prepare the specimen the mixture of Part A and Part B from both composition were mixed according manufacturer's guideline as shown in Fig 2. The mixture then was poured into silicone rubber mould. Composition 1 take longer time to cure which is 2 to 3 days while for composition 2 take only 2 minutes to cure.

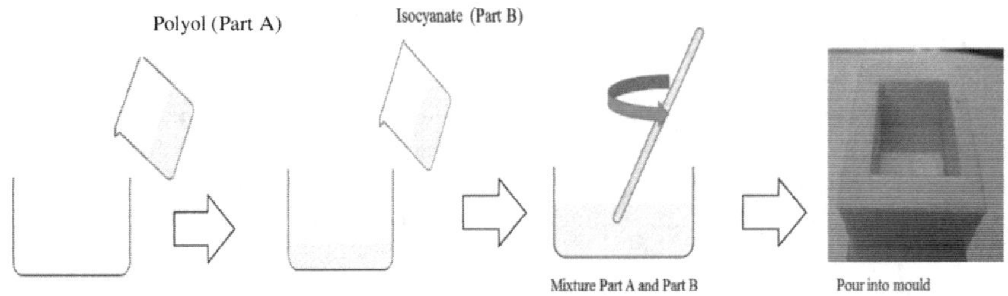

Figure 2. Show the procedure to mix isocyanate (Part A) and polyol (Part B) for both composition.

2.3. Mechanical Testing

The specimens will undergo mechanical testing by using a universal testing machine (8874 Model, Instron). Particularly, the specimen will undergo compression testing to evaluate its mechanical strength. The specimen was placed under the 25 kN load where the load was adjusted until the gap between the load and the specimen is approximately 0.5 cm in distance (Fig 2). During the compression test, the crosshead speed was set to 2.5 ± 0.25 mm/min based on ASTM-F1839. Furthermore, after the specimens have undergone the compression testing Bluehill software will provide data of the sample such as Young's modulus, compression stress and strain and many more. Then, the data was plotted by using Microsoft Excel.

2.4. Morphological analysis

The specimen undergo the morphological analysis using Scanning Electron Microscopy (SEM). SEM was used to visualize the morphology of the cell structure of the PU and existing synthetic bone. Before entering the SEM, several steps need to be conducted for the best visualization. The specimens of the different composition and existing synthetic bone (Sawbone and Synbone) were cut by using the band saw into 1cm x 0.5cm x 1cm. Following that, the specimens were coated with gold by Leica EN ACE to prevent charging on the surface. For this study, the magnification t use to visualize the morphology of the specimens are 500X and 1000X.

3. Results and Discussion.

This study focused on mechanical strength and microstructure of the specimen. The results are shown at the following part.

3.1. Mechanical Testing using Instron

Fig. 3 depicts the Young's modulus for Sample A, Sample B, Sawbone and Synbone after undergone compression testing by using Instron. This finding has significant impact as it show the Young's modulus from the available product in the market. Comparatively, based on the results of Young's modulus of the Synbone and Sawbone, it showed that Sawbone has higher Young's modulus compared to Synbone. The value of Young's modulus for Sawbone is 132.3254 MPa while Synbone is 68.1483 MPa. In contrast, the Young's modulus for the sample A and Sample B are 208.6963 MPa and 384.6304 MPa respectively. This evident prove that Sample B has higher Young's modulus compared to Sample A. As states earlier, cancellous bone for tibia range from 350 MPa to 700MPa. Based on the experiment it proved that sample B has Young's Modulus which in the range of tibia cancellous bone. Together, the present findings confirm that sample A, bone from Synbone and Sawbone does not meet the Young's Modulus for tibia cancellous bone.

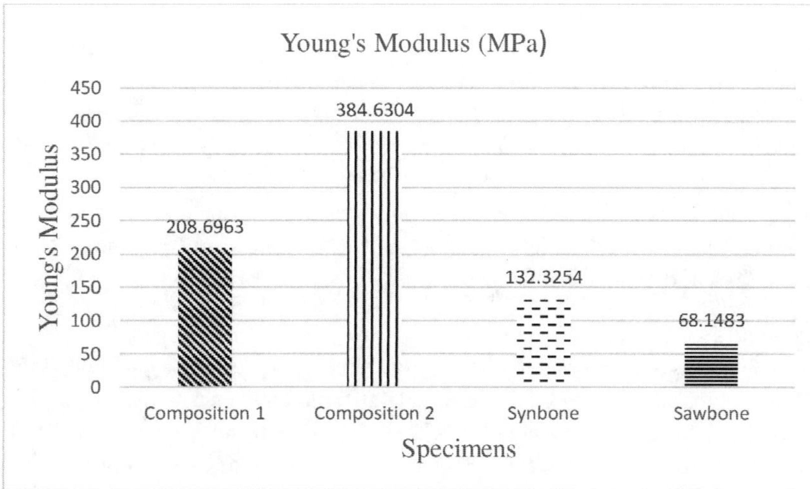

Figure 3. Depict the bar graph for average of Young's Modulus for the specimen

3.2. Microstructure testing by using SEM

Overall, the visualization of morphology of specimens done under Table Top SEM with 500X. Hence, this technique was used to facilitate the third objective of this study which to enable a comparative analysis in regards to microstructural properties of synthetic bone. The results of the experiment found that composition 1 possess some inhomogeneous microstructural properties as compared to commercial products, Sawbone and Synbone (Fig. 3). In addition, composition 2 does not possess microstructure of the bone (Fig. 4) as it does not any pore to mimic cancellous bone. Hence, composition 2 need alternation in regard of microstructure by using additives such as blowing agent and surfactant (Fig. 5). Nonetheless, the microstructure of the Synbone and Sawbone in Fig. 6 and Fig. 7 respectively illustrate slightly different in term of the pore size. Yet, the microstructure of the commercial synthetic bone does not mimic the microstructure of cancellous bone. In addition, Fig. 6 and Fig. 7 appeared with slight contamination as compared to composition 1 and composition 2.

Figure 4. Depicts composition 1 of PU.

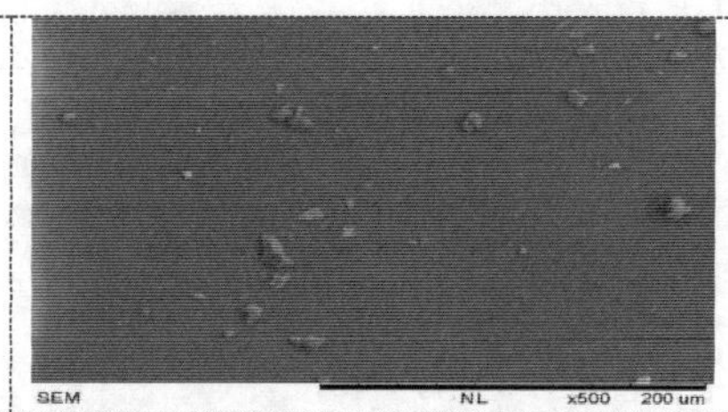

Figure 5. Illustrate composition 2 of PU.

Figure 6. Show sample for Sawbone.

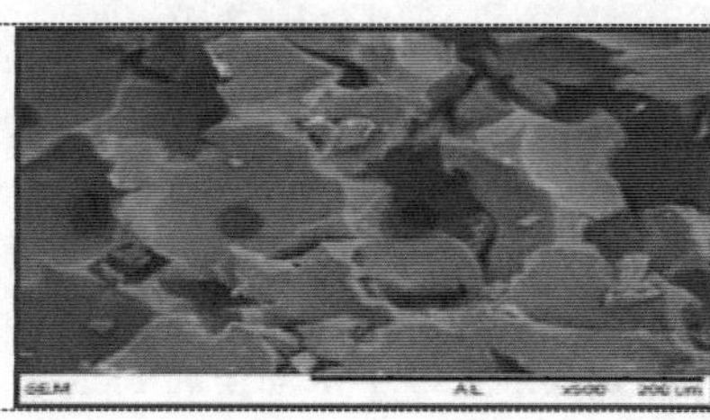

Figure 7. Depict sample for Synbone.

4. Conclusion

From this study, it proved that the average Young's modulus for composition 2 was higher compared to composition 1 and commercial product from Sawbone and Synbone that was 384.6304 MPa. This value of Young's modulus is in the range of cancellous bone. In addition, the commercial product does not meet the value of Young's Modulus of the cancellous bone that ranges 350 MPa (metabolic bone disorder) to 700 MPa (healthy bone). However, composition 1 and composition 2 does not mimic the microstructure of cancellous bone if compared to commercial product and need the addition of additives to form the porous structure. Nevertheless, it can be showed that the synthetic bone can be used for the biomechanical studies as demonstrated in the previous studies [12-13].

Acknowledgement.

This work was financially supported by the Universiti Teknologi Malaysia under the Research University Grant Grant (20H20, 4J358, 15J84 and 20H26), Fundamental Research Grant Scheme (FRGS) Ministry of Education Malaysia (278784-294883), Ministry of Energy, Science, Technology, Environment and Climate Change (EF0618I1166 and 4S144) and The Ministries of Research, Technology, and Higher Education Republic of Indonesia through Director of Research and Community Service with contract number: 009/L10/AK.04/KONTRAK-PENELITIAN/2019.

Reference

[1] Whiteman A, Dam A. Skeletal System Skeletal System [Internet]. Most. Elsevier Inc.; 2018. 53-70 p. Available from: http://dx.doi.org/10.1016/B978-0-12-381361-9.00005-6

[2] Thompson SM, Yohuno D, Bradley WN, Crocombe AD. Finite element analysis: a comparison of an all-polyethylene tibial implant and its metal-backed equivalent. *Knee Surgery, Sport Traumatol Arthrosc*. 2016;24(8):2560–6.

[3] Kaman MO, Celik N, Karakuzu S. Numerical Stress Analysis of the Plates Used to Treat the Tibia Bone Fracture. J *Appl Math Phys*. 2014;2(6):304–9.

[4] Gardner MP, Chong ACM, Pollock AG, Wooley PH. Mechanical evaluation of large-size fourth-generation composite femur and tibia models. *Ann Biomed Eng*. 2010;38(3):613–20.

[5] Möhlhenrich SC, Kniha K, Heussen N, Hölzle F, Modabber A. Effects on primary stability of three different techniques for implant site preparation in synthetic bone models of different densities. *Br J Oral Maxillofac Surg* [Internet]. 2016;54(9):980–6. 6. Hetaimish BM. Sawbones laboratory in orthopedic surgical training. Saudi Med J. 2016;37(4):348–53.

[6] Jaber FT, Al-Jayyousi AB. A phantom for cadaverless evaluation of targeting systems for distal locking of intramedullary nails. *Middle East Conf Biomed Eng MECBME*. 2014;115–8.

[7] Martin R, Menorca G, Reed JD. NIH Public Access. 2014;22(2):111–20.

[8] Leboulanger N. First cadaver dissection: Stress, preparation, and emotional experience. *Eur Ann Otorhinolaryngol Head Neck Dis* [Internet]. 2011;128(4):175–83.

[9] Gama N V., Ferreira A, Barros-Timmons A. Polyurethane foams: Past, present, and future. Materials (Basel). 2018;11(10).

[10] Solanki A, Das M, Thakore S. A review on carbohydrate embedded polyurethanes : An emerging area in the scope of biomedical applications. *Carbohydr Polym* [Internet]. 2018;181(August 2017):1003–16.

[11] Sidek N, Bakar IAA, Azman AA, Rahman ASA, Austin WA. Strength characteristic of polyurethane with variation of polyol to isocyanate mix ratio : A numerical analysis. *Proc - 2017 IEEE 2nd Int Conf Autom Control Intell Syst I2CACIS 2017*. 2017;2017– Decem(October):31–4.

[12] Ramlee MH, Sulong MA, Garcia-Nieto E, Penaranda DA, Felip AR, Abdul Kadir MR. Biomechanical features of six design of the delta external fixator for treating Pilon fracture: a finite element study. Medical & biological engineering & computing, 2018. **56**(10): p. 1925-1938.

[13] Ramlee MH, Derus A. The use of external fixator ankle and foot injuries management- a review on biomechanical perspectice. Medical Devices and Diagnosis Engineering, 2016. **1**(1): p. 5-10.

Journal of Physics: Conference Series **1372** (2019) 012014 doi:10.1088/1742-6596/1372/1/012014

Development and Verification of Three-Dimensional Model of Femoral Bone: Finite Element Analysis

Aishah Umairah Abd Aziz[1,4], Hong Seng Gan[2], Ahmad Kafrawi Nasution[5], Mohammed Rafiq Abdul Kadir[3,4] and Muhammad Hanif Ramlee[1,4*]

[1] Medical Devices and Technology Centre (MEDITEC), Institute of Human Centered Engineering (i-Humen), Universiti Teknologi Malaysia, 81310, Johor, Malaysia.
[2] Universiti Kuala Lumpur, British Malaysian Institute, Batu 8 Jalan Sg. Pusu, 53100 Gombak, Selangor, Malaysia
[3] Sport Innovation & Technology Centre (SITC), Institute of Human Centered Engineering (i-Humen), Universiti Teknologi Malaysia, 81310, Johor, Malaysia.
[4] Bioinspired Innovation Group (BIOINSPIRA), School of Biomedical Engineering and Health Sciences, Faculty of Engineering, Universiti Teknologi Malaysia, 81310, Johor, Malaysia.
[5] Faculty of Engineering, Universiti Muhammadiyah Riau, Pekan Baru, Indonesia.

[*] muhammad.hanif.ramlee@biomedical.utm.my

Abstract. Development of reconstructed three dimensional (3D) model of bones has successfully emerged from time to time. It is used in many engineering applications especially by researchers to help them in the simulation process through finite element analysis (FEA). Besides, the 3D model can also be used by surgeons with the help of engineers, in improving their pre-surgical planning and designing of patient-specific implants or surgical instruments. Focusing on the femur, this paper proposes a method in developing a reconstructed three-dimensional femoral bone model and verifying it through analysis by using the Finite Element Method. Load of 600N was applied on the most proximal point of the femoral head under axial loading. The strain value observed in the model were 0.35 mε, 0.242 mε, and 0.146 mε. These values were slightly higher than the value measured in cadaveric specimen in the previous study, but it showed the same pattern. Hence, it can be concluded that the development method is reliable.

1. Introduction

Femoral bone is the longest and heaviest bone in the human body. It helps to bear the body weight and hence, it plays significant roles in most of the daily activities. On that account, vast studies have been continuously done to better understand the femur. Researchers need to simulate the bone in various circumstances that sometimes, are not feasible in a normal test setup but, is possible through a numerical study by using Finite Element Method (FEM) [1]. FEM helps to observe the mechanical effects (stress, strain, displacement and etc.) at any desired point through its finite elements. For that, various studies have successfully employed Finite Element Analysis (FEA) as their research method [2-7]. A close up to one of the cases, researchers are studying the healing of a broken bone. There are many variables that need to be considered and observed to ensure the study is in the correct direction. To begin the analysis, the bone needs to be remodeled appropriately. An accurate and reliable three-dimensional model (3D) will help in producing reliable results [8]. Previously, researchers opted for a simple rod or tube to represent the bone. Henderson D. J. et. al., claimed that the model was symmetric and uniform thus, any variability throughout the testing could be minimized [9]. This method had raised several issues. In most of the cases where the rod or tube was chosen, the actual bone anatomy and mechanical properties were almost totally neglected, in which it will affect the reliability and accuracy of the result [10]. On the result and analysis part, strain is always observed in evaluating bone mechanical behavior [11]. For example, in the bone healing process, the callus formation is dependent on the strain value [12-16]. Thus, it is important for researchers to observe the strain of the bone and fundamentally, the strain is measured following equation (1) [15].

$$\text{strain}, \varepsilon = \frac{\Delta L}{L} \quad (1)$$

The accuracy and reliability of the modeling could be improved by using a correct anatomical model of bone through the reconstruction method. The stacks of two dimensional (2D) ct data of a real bone will be reconstructed into a 3D model, which will be used in the analysis, superseded the tube. To ensure that the developed model is correct and reliable, it needs to be validated and verified. Validated means a direct comparison with the physical occurrence, whereas, for verification, it relates to the actual solving of the equations, which is influenced by modeling representation [1]. Thus, in this paper, it is aiming to propose a method of developing a reconstructed 3D model of femoral bone and also to provide a verified analysis on the developed model by comparing the results with other literature. The model was tested under axial loading through FEM.

2. Methodology

Since this is basically a preliminary study, the modeling was made to mimic a healthy environment experienced by everyone most of the time. Hence, CT dataset of a healthy 27 years old man with 169 cm height and 75kg weight was chosen as the subject of the study. The data was obtained from Hospital Tengku Ampuan Afzan, Kuantan, Pahang, Malaysia. CT datasets, stored in DICOM format file, is a collection of grey scale images resulted from the attenuation of x-ray beams. The darker the image, the less attenuate the x-ray beam is. The ct images were captured for every 1.5 mm along the femur and since the attenuation rate is higher in bone, the image of the femur appears to be white.

2.1. Development of the reconstructed 3D femoral bone model

An image processing software Mimics (Materialise, Leuven, Belgium) was used in reconstructing and modeling of the femur. The ct data was loaded into the software and the bone was segmented based on the Hounsfield Unit (HU) set in 'thresholding' tool. The HU for bone ranges from 226 to 3071 HU. 2D image of the bone was then automatically calculated as shown in figure 1(a). To segment the region of interest (ROI), which is the femur, both 'edit mask' and 'region growing' tools were utilized (figure 1(b)). Then, the segmented femur was converted to a 3D form as in figure 1(c). All files are then saved in the stereolithographic (STL) format. The process can be best summarized in figure 1.

2.2. Mesh generation

As in figure 1(d), the next step done was mesh generation by using 3-MATIC software (Materialise, Leuven, Belgium). The tetrahedral element was used as it had been discussed to have higher accuracy in contrast to its equivalent lesser node element [17]. From the mesh convergence analysis, it was found that the results converge at 4.5mm. To highlight, mesh convergence analysis was done to determine the optimum mesh size to be used so that the analysis result does not change with any changes in mesh size. Smaller size results in more accurate and reliable results [18] but, it requires a longer running time and bigger computer resources. Hence, optimized mesh size is needed in order to balance them. Reducing the element size or *h*-refinement method was used throughout the modeling domain.

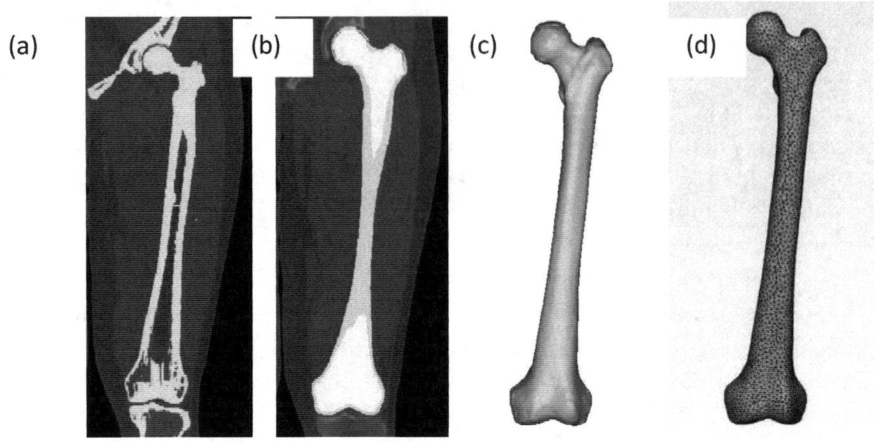

Figure 1. 3D femoral bone model development process from (a) 2D scanned image into a (b) segmented image based on the ROI, following by a (c) 3D reconstructed image and finally the (d) meshed model of the developed model.

2.3. Finite Element Analysis

The 3D model was then analyzed through Finite Element Method in Marc Mentat (MSC.Software, Santa Ana, CA). For simplification purpose, the material properties for bone were set as isotropic, homogenous and linearly elastic. The Young's modulus and Poisson's ratio were 7GPa and 0.3 respectively [7, 19]. The boundary condition applied in this study were as follows: distal end of the femur was set to fix in all cases, and to mimic the stance phase, as well as to verify the analysis with study made by Hensley et al [11], 600N point load was assigned at the femoral head in axial direction. Point load was chosen to represent a hip joint force in which it simplifies the hip muscle reaction as widely used in prior studies [3, 20, 21]. Figure 2 shows the developed 3D model of the femur with an illustrated boundary condition used in the analysis. The predicted linear strain was observed and compared in the analysis to verify the 3D femoral bone model. For that reason, three points for strain measurement were chosen following the cadaveric study made by Hensley S. et al [11]. They were located medially, starting with increment 25mm, starting 25mm below the lesser trochanter of the femur.

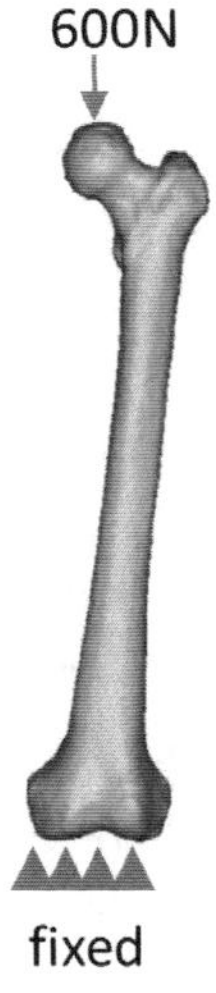

Figure 2. Boundary condition used in the Finite Element Analysis of developed femoral bone model.

3. Results and discussion

3.1. 3D model of the femur

The development steps in modeling the bone were successfully adopted. The model was visually compared to the standard anatomy of a femur to ensure the morphology of the model is correct. Through the proposed method, researchers are allowed to construct the bone in a faster way besides improving the reliability of their results as they are using actual bone anatomy. However, there are also some limitations and drawbacks in modeling the femur. For example, the model was assumed to be homogenous, linearly isotropic material. It contradicted the real bone composition which is non-linearly, orthotropic structure. Nonetheless, previous studies had successfully shown a good agreement between the cadaveric specimen and finite element model in spite of their simplified finite element material properties [22]. Unlike the real bone structure, the developed model was set as one material. It supposedly consists of both cortical and cancellous layer as done in many works of literature [3, 10, 23]. However, there is also literature that had been using a correct anatomical shape of the femoral model with a single material property [7, 17, 24, 25]. These mean that the simplifications of the bone material properties do give reliable results. As it is merely a preliminary study and considering the time restriction, only one subject was used throughout the analysis [26]. Thus, in the future, more subjects could be considered to further verified the proposed method.

3.2. Finite element analysis

From the analysis of the developed bone model, the maximum strain found at the three points were 0.35mε, 0.242mε, and 0.146mε respectively. These measurements were obtained at the white circle region as shown in figure 3. Low strain ranges from 0.082 mε and below was observed about half of the lower part of the model. This is due to the fix boundary condition at the distal condyle. Although the values obtained were higher than the cadaveric study by Hensley S. et al., it showed the same pattern with the DIC measured data. As the interested points went downwards, the strain values decreased. However, it was not in the case of strain gage measurement (table 1). Hensley S. et al., in their study, they were comparing the performance of DIC strain measurement techniques with strain gage for the use in biomechanical test materials [11]. They found that DIC measurement is more reliable compared to strain gage in measuring a material with more variability. Since the results from this study are in agreement to the DIC data, the 3D model developed is concluded to be verified and can be reliably used for future work.

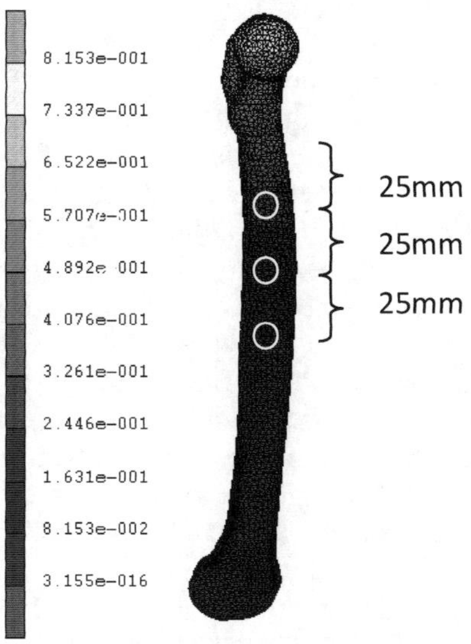

Figure 3. Maximum strain value for the analysis of the developed model.

Table 1. The maximum strain on the reconstructed model of the femur as compared to the model used by Hensley S. et al. [11].

	Point 1 (mε)	Point 2 (mε)	Point 3 (mε)
Hensley S. et al. (Strain gage)	0.125	0.120	0.125
Hensley S. et al. (DIC)	0.130	0.122	0.060
New method	0.350	0.242	0.146

4. Conclusion

The 3D femoral bone model was developed following the sequence of development steps. It was also analyzed through FEM, and despite some simplifications used in this study, the results obtained were still in agreement with the prior study. However, in the future, authors shall improve the modelling process by including more realistic material properties of human bone in an effort to obtain higher accuracy.

Acknowledgment

This work was financially supported by the Universiti Teknologi Malaysia under the Research University Grant (20H20, 4J358, 15J84 and 20H26), Fundamental Research Grant Scheme (FRGS) Ministry of Education Malaysia (278784-294883), Ministry of Energy, Science, Technology, Environment and Climate Change (EF0618I1166 and 4S144) and The Ministries of Research, Technology, and Higher Education Republic of Indonesia through Director of Research and Community Service with contract number: 009/L10/AK.04/KONTRAK-PENELITIAN/2019.

References

[1] MacLeod, A.R., Rose, H., Gill, H.S.: A validated open-source multisolver fourth-generation composite femur model. Journal of biomechanical engineering. **138**(12) (2016) 124501-124509

[2] Mehboob, A., Chang, S.H.: Biomechanical simulation of healing process of fractured femoral shaft applied by composite intramedullary nails according to fracture configuration. Composite structures. **185** (2018) 81-93

[3] Zhou, J.J., Zhao, M., Liu, D., Liu, H.Y., Du, C.F.: Biomechanical property of a newly designed assembly locking compression plate: Three-dimensional finite element analysis. Journal of healthcare engineering. (2017) 1-10

[4] Lopes, V.M.M., Neto, M.A., Amaro, A.M., Roseiro, L.M., Paulino, M.F.: FE and experimental study on how the cortex material properties of synthetic femurs affect strain levels. Medical engineering & physics. **46** (2017) 96-109

[5] Konya, M.N., Verim, O.: Numerical optimization of the position in femoral head of proximal locking screws of proximal femoral nail system; biomechanical study. Balkan medical journal. **34**(5) (2017) 425-431

[6] Giordano, V., Godoy-Santos, A.L., Belangero, W.D., Pires, R.E.S., Labronici, P.J., Koch, H.A.: Finite element analysis of the equivalent stress distribution in schanz screws during the use of a femoral fracture distractor. Revista brasiliera de ortopedia. **52**(4) (2017) 396-401

[7] Kumar, P.B., Parhi, D.R.: Vibrational characteristics and stress analysis in a human femur bone. Materials today: Proceedings. **4**(9) (2017) 10084-10087

[8] Fengjiao, G., Guanjun, Z., Jie, L., Shujing, W., Xu, L.: Study on validation method for femur finite element model under multiple loading conditions. IOP conference series: Materials science and engineering. **322**(2) (2018) 022061

[9] Henderson, D.J., Rushbrook, J.L., Harwood, P.J., Stewart, T.D.: What are the biomechanical properties of the taylor spatial frame™? Clinical orthopaedics and related research®. **475**(5) (2017) 1472-1482

[10] Mohd Aspar, M.A.S., Abd Razak, N.A., Abdul Kadir, M.R., Manap, H.: Development of 3-dimensional model of femur bone considering cortical and cancellous structures. International journal of engineering technology and sciences. **7** (1) (2017)

[11] Hensley, S., Christensen, M., Small, S., Archer, D., Lakes, E., Rogge, R.: Digital image correlation techniques for strain measurement in a variety of biomechanical test models. Acta of bioengineering and biomechanics. **19**(3) (2017)

[12] Epari, D.R., Taylor, W.R., Heller, M.O., Duda, G.N.: Mechanical conditions in the initial phase of bone healing. Clinical biomechanics. **21**(6) (2006) 646-655

[13] Boccaccio, A., Pappalettere, C. Mechanobiology of fracture healing: Basic principles and applications in orthodontics and orthopaedics. Theoretical biomechanics. Rijeka. (2011) 21-48.

[14] Pauwels, F.: "Grundriß einer biomechanik der frakturheilung. Verh dtsch orthop ges 34. Kongres 62-108," ges abh. Springer, Berlin. (1980)

[15] Hak, D.J., Toker, S., Yi, C., Toreson, J., Stahel, P.F.: The influence of fracture fixation biomechanics on fracture healing. Orthopedics. **33**(10) (2010) 752-755

[16] Tan, B.B., Shanmugam, R., Gunalan, R., Chua, Y.P., Hossain, G., Saw, A.: A biomechanical comparison between taylor's spatial frame and ilizarov external fixator. Malaysian orthopedic journal. **8**(2) (2014) 35-39

[17] N. Mughal, U., Khawaja, H., Moatamedi, M.: Finite element analysis of human femur bone. **9** (2015) 101-108

[18] Miura, M., Nakamura, J., Matsuura, Y., Wako, Y., Suzuki, T., Hagiwara, S., et al.: Prediction of fracture load and stiffness of the proximal femur by ct-based specimen specific finite element analysis: Cadaveric validation study. BMC musculoskeletal disorders. **18**(1) (2017) 536

[19] Vijayakumar, R., Madheswaran, M.: Modal analysis of femur bone using finite element method for healthcare system. 2017 conference on emerging devices and smart systems (2017) 224-228

[20] Reimeringer, M., Nuno, N.: The influence of contact ratio and its location on the primary stability of cementless total hip arthroplasty: A finite element analysis. Journal of biomechanics. **49**(7) (2016) 1064-1070

[21] Montanini, R., Filardi, V.: In vitro biomechanical evaluation of antegrade femoral nailing at early and late postoperative stages. Medical engineering & physics. **32**(8) (2010) 889-897

[22] Bettamer, A., Hambli, R., Allaoui, S., Almhdie-Imjabber, A.: Using visual image measurements to validate a novel finite element model of crack propagation and fracture patterns of proximal femur. Computer methods in biomechanics and biomedical engineering: imaging & visualization. **5**(4) (2017) 251-262

[23] Bayoglu, R., Okyar, A.F.: Implementation of boundary conditions in modeling the femur is critical for the evaluation of distal intramedullary nailing. Medical engineering & physics. **37**(11) (2015) 1053-1060

[24] Belaid, D., Bouchoucha, A.: Modeling of the mechanical behavior of the human femur: Stress analysis and strain 1702 (2015) 190023

[25] Dhanopia, A., Bhargava, M.: Finite element analysis of human fractured femur bone implantation with pmma thermoplastic prosthetic plate. Procedia engineering. **173** (2017) 1658-1665

[26] Ramlee MH, Sulong MA, Garcia-Nieto E, Penaranda DA, Felip AR, Abdul Kadir MR. Biomechanical features of six design of the delta external fixator for treating Pilon fracture: a finite element study. Medical & biological engineering & computing. **56**(10) 2018 1925-1938.

Low cost real-time finger flexion strength measurement system

Yin Qing Tan, Shyh Jeh Hwan and Siow Cheng Chan

Department of Mechatronics and Biomedical Engineering, Universiti Tunku Abdul Rahman, Malaysia
*tanyq@utar.edu.my

Abstract. Human grip strength is important in carrying out daily activities. However, the grip strength may be decreases or affected due to aging or diseases. Thus, hand exercising ball is invented as a rehabilitation device in order improve the physical grip strength. The commercial hand exercising balls comes with different stiffness which are soft, medium and hard without quantitative data. In the rehabilitation, diagnosis and intervention should be done quickly and accurately. The data are crucial for therapist to evaluate and determine the suitable treatment for the patients. A low cost real-time finger flexion strength measurement system is developed to measure the forces required to squeeze the hand exercising balls with different stiffness. The prototype is able to measure the subjects' grip strength from four fingers without the thumb. Data collected from 10 subjects (6 healthy subjects and 4 patients) with different gender, age, height, weight and health status. The force required to grip the hand exercising ball range from 7.71N to 79.11N for soft ball, 12.80N to 104.08N for medium ball and 19.35N to 165.82N for hard ball

1. Introduction

Hand is the most active part of human body in daily life, and finger flexion is so important to allow someone to hold the handset, carrying bags and baskets, hold the mug etc.

Studies had shown that grip strength has a high correlation with age where the grip strength will decrease gradually due to aging [1]. The reducing of grip strength will begin from approximately 30 years old based on the cross-sectional and longitudinal studies [2]. Study showed that the grip strength have an average decrement 0.5 % to 1 % annually after age of 30 [3]. In certain cases, people may loss their hand movement ability due to accident such as stroke. In such cases, grip strength can be used as a measurement of the degree of impairment in respect to neuromuscular or neurological disabilities suffered after trauma or surgery [4]. In addition, grip strength also served as a good indicator of an individual physical state, especially in an elderly population [5]. Grip-strength measurement has also correlated with various clinical tests [6, 7].

Exercise had been implemented into the treatment of large number of medical disorders[8]. A combination of strength training, aerobic exercise and flexibility exercises can help to reduce medication dependence while maintaining functional independence of the patients [9]. Therapeutic hand exercising ball has been invented as a device for the patients to retain or improve their affected grip strength by constantly gripping it in the palm which can increase their muscle performance [10]. The flexible and spongy therapeutic hand exercising ball is used to exercise the hand muscles and joints by repetitive gripping.

There are few types of therapeutic hand exercising balls can be found in the market. The most common shapes of the ball are the spherical shape and egg shape (Figure 1). Egg shape is chosen in this study because it conforms to the human hand.

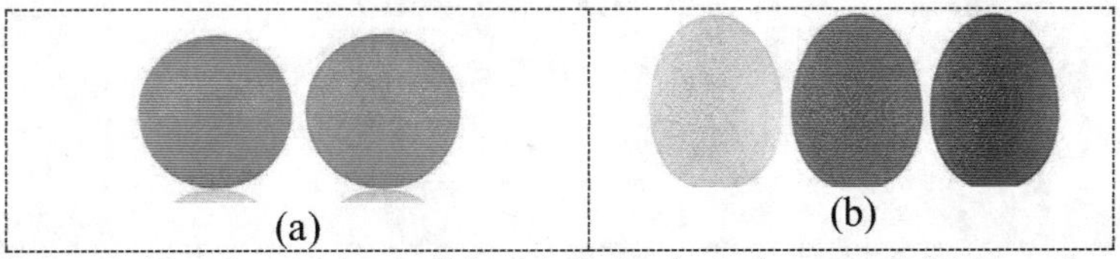

Figure 1. Hand exercise ball in (a) Spherical shape and (b) Egg shape

Manufacturers had produced hand exercising balls with a variety of stiffness to accommodate various treatment levels. However, the stiffness of the balls is normally categorized as strong, medium and soft, this may lead to controversy because different manufacturers have different standard of strong, medium and soft. The physical therapists had to decide which stiffness should be used for the patients based on their experience. This may have negative impacts for inexperience physical therapists if they use the wrong stiffness. This not only will affect the effectiveness of the treatment, but also may further injure the muscle of the patients.

According to Guyatt, the measurements of the effectiveness of a physic therapy intervention should be valid, responsive and interpretable [11]. In the rehabilitation and physiotherapy center, any diagnosis and intervention should be done quickly and accurately [12]. The accurate data are crucial for physical therapist to evaluate their patients and provide suitable treatments or interventions.

Thus, quick and accurate measurement of hand grip strength is crucial for physical therapist to evaluate their patients' performance and then suggest a suitable treatment or interventions for them.

2. Prototype

Flexible filament type force sensors, Interlink FSR 402 are selected due to its high accuracy, fast response time, high consistency, long lifespan and low cost.

The size of the hand exercising ball is small to be gripped firmly using four fingers only; hence the thumb force is not been included this study. Figure 2 shows the standard gripping pattern used to hold the hand exercising ball, it can be seen that the thumb is not pressing or touching any part of the ball. Thus, the sensors are placed at the tip of index, middle, ring and little fingers which contributed most of the gripping force.

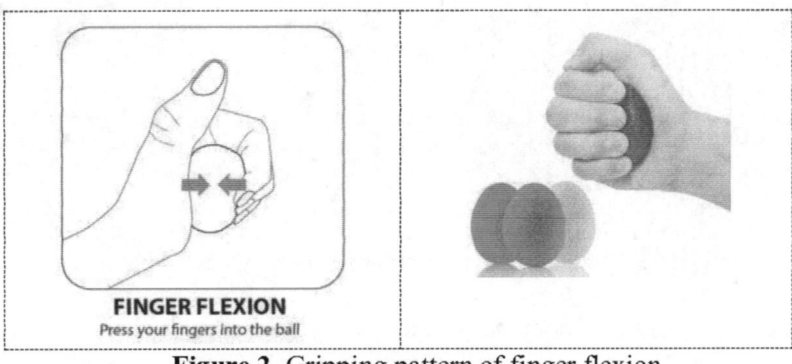

Figure 2. Gripping pattern of finger flexion

The sensor is fixed on the elastic rubber which can be secured on finger with Velcro tape. The elasticity of the rubber allows the sensor to be fixed on any size of finger (Figure 3).

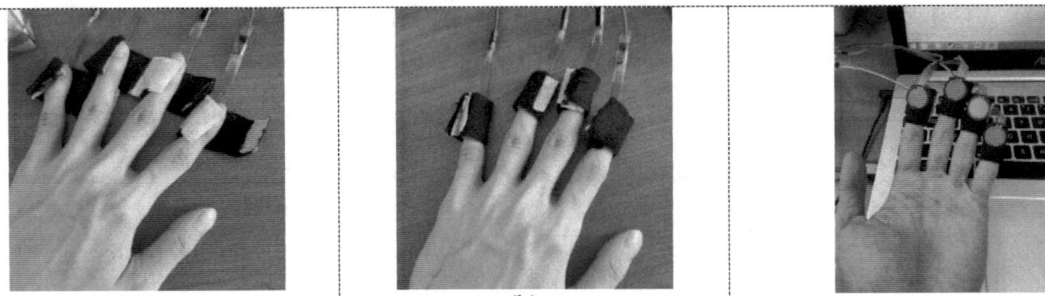

(a) (b) (c)

Figure 3. Sensor placement. (a) Fingers place on the elastic rubber with sensors. (b) The elastic rubbers are wrapped around the finger and secured with Velcro tape. (c) Sensors secured on the finger tips

The circuit design of the force sensor is shown in Figure 4. The system is power up with a 9V battery, and connects to computer via Arduino Uno. Two LED feedback is added to attract the attention of children, the first LED will light up when first threshold value been achieved, then both LED will light up when second threshold been achieved (Figure 5).

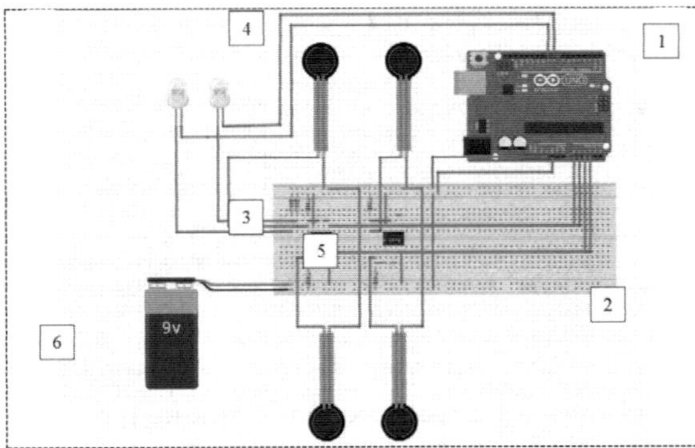

Figure 4. Circuit Design of the system

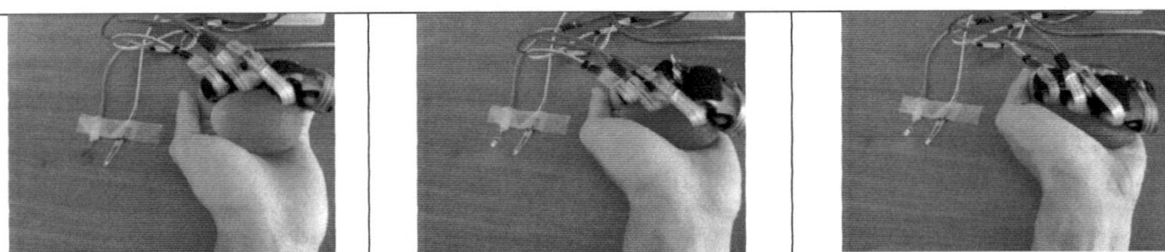

(a) (b) (c)

Figure 5. LED feedback. (a) No LED light without force. (b) Red LED light up when first threshold achieved. (c) Both LED light up when second threshold achieved

Study of Haidar showed that the dominant hand of the subjects has higher maximum grip strength than the non-dominant hand [4]. Grip strength was found to be closely correlated with the position of elbow and shoulder [13]. As recommended by American Society of Hand Therapists, the subject will flex their elbow at 90 degrees and rest their hand on table / chair arm / on the lap when performing the hand gripping exercise. Soft ball is put at horizontally on the palm and subject will fully flex their 4 fingers to apply maximum force on the ball repeatedly in 30 seconds. 5 minutes rest time is provided before the subject continues with middle stiffness and hard ball. Average of three maximum forces of each trial will be calculated.

A real-time display on monitor shows the force result and the maximum force will display at the end of each trial, as shown in Figure 6.

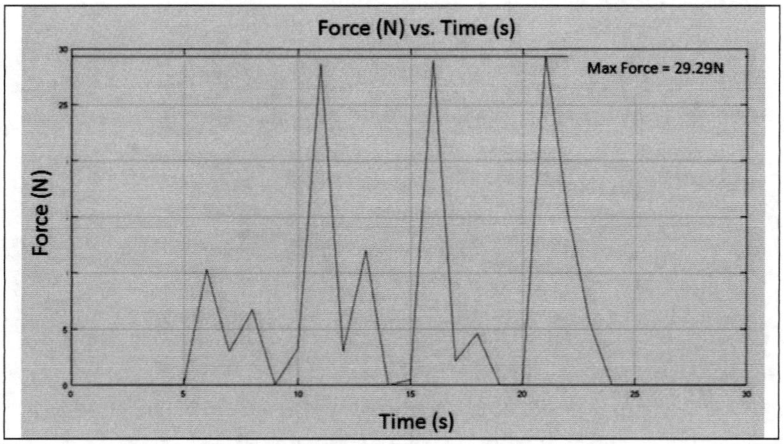

Figure 6. Real-time result for 1 trial

3. Data collection

The system has tested on 6 healthy subjects and 4 patients with different gender and age. The sensors are placed onto the fingers of dominant hand of the subject and the forces exerted onto the therapeutic hand balls with different stiffness are recorded. The gender, age, height and weight of the subjects are shown in Table 1 and Table 2.

Table 1. Normal subject information

Subject	Gender	Age	Height	Weight
1	Female	23	165cm	44kg
2	Female	58	160cm	55kg
3	Male	23	171cm	50kg
4	Male	25	170cm	75kg
5	Male	60	171cm	80kg
6	Male	7	115cm	18kg

Table 2. Patient subject information

Subject	Gender	Age	Height	Weight	Disability
7	Female	19	163cm	55kg	Learning disability
8	Male	11	145cm	38kg	Learning disability
9	Male	14	155cm	44kg	Hydrocephalus
10	Female	14	157cm	40kg	Apert Syndrome

The 4 fingers full flexion gripping force for each subject is summarized in Table 3.

The system success to capture the gripping force created by 4 fingers full flexion, except subject 10 whose suffer with Apert syndrome unable to bend her hand, therefore no gripping force been recorded.

Table 3. Result

Subject	Force(N)		
	Soft Ball	Medium Ball	Hard Ball
1	28.97	29.29	39.21
2	36.69	46.97	45.08
3	37.33	61.91	64.13
4	67.40	104.08	165.82
5	79.11	94.91	102.32
6	16.15	19.45	20.35
7	17.88	20.58	35.65
8	7.71	12.80	20.00
9	12.36	17.27	19.35
10	-	-	-

4. Discussion Conclusion

The developed low cost real-time finger flexion gripping strength measurement system able to collect data from various age group, gender, healthy subjects and patients.

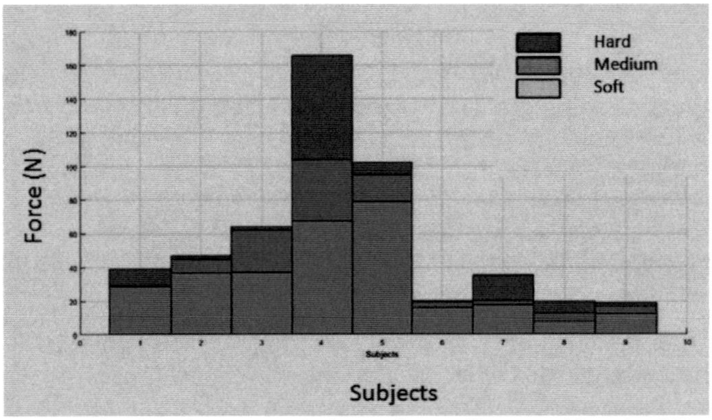

Figure 7. Gripping force result for soft, medium and hard stiffness ball.

The results showed that the grip force increases as the stiffness of the hand exercising balls increases from soft to hard. The force required to grip the hand exercising ball range from 7.71N to 79.11N for soft ball, 12.80N to 104.08N for medium ball and 19.35N to 165.82N for hard ball. Generally, adult male has higher average grip strength than adult female. Patients have relatively lower grip force compared to healthy subjects. Subject 4 (male, bodybuilder) has the highest average grip force which is 165.82N and subject 9 (patient with hydrocephalus) has the lowest average grip force which is 19.35N. On the other hand, the elderly male and female has higher grip force than the adult male and female.

Overall, the result obtained from this study is much lower than standard grip force recorded from healthy subject. This may due to the lack of thumb force in measurement. According to the fingers force sharing analysis done by Quaine [14], thumb force has a contribution of approximately 100N. Hence, this explained why the experimental results show much lower reading than other studies.

5. Conclusion

This study success to develop a low cost real-time finger flexion strength measurement system, which is light weight and portable, it is also a user friendly system which required minimum setup time. This system able to measure the gripping force regardless of age, gender, size of finger and health status of subject.

6. Limitation and future work

The system only able to measure 4 fingers reading without thumb, it is recommended to top up another sensor at thumb in the future. The system may modify to include the finger extension strength for further rehabilitation purpose. Current system had been provided with visual feedback, however, sound feedback like buzzer or alarm may introduce to improve the attention of subject, and this may also benefit to the subject with vision impairment.

References

[1] Kallman D A, Plato C C and Tobin J D 1990 The role of muscle loss in the age-related decline of grip strength: cross-sectional and longitudinal perspectives *Journal of Gerontology* **45**(3) pp 82-88

[2] Frederiksen H, Hjelmborg J, Mortensen J, Mcgue M, Vaupel J W and Christensen K 2006 Age trajectories of grip strength: cross-sectional and longitudinal data among 8,342 Danes aged 46 to 102 *Annals of Epidemiology* **16**(7) pp 554-562

[3] Sehl M E and Yates F E 2001 Kinetics of human aging: I. Rates of senescence between ages 30 and 70 years in healthy people *The Journals of Gerontology Series A: Biological Sciences and Medical Sciences* **56**(5) pp 198-208

[4] Haidar S, Kumar D, Bassi R and Deshmukh S 2004 Average versus maximum grip strength: which is more consistent? *Journal of Hand Surgery* **29**(1) pp 82-84

[5] Fried L P, Tangen C M, Walston J, Newman A B, Hirsch C, Gottdiener J, Seeman T, Tracy R, Kop W J, Burke G and McBurnie M A 2001 Frailty in older adults: evidence for a phenotype *The Journals of Gerontology Series A: Biological Sciences and Medical Sciences* **56**(3) pp 146-157

[6] Nybo H, Gaist D, Jeune B, McGue M, Vaupel J W and Christensen K 2001 Functional status and self-rated health in 2,262 nonagenarians: the Danish 1905 Cohort Survey *Journal of the American Geriatrics Society* **49**(5) pp 601-609

[7] Schaubert K L and Bohannon R W 2005 Reliability and validity of three strength measures obtained from community-dwelling elderly persons *Journal of Strength and Conditioning Research* **19**(3) pp717

[8] Roberts C K and Barnard R J 2005 Effects of exercise and diet on chronic disease *Journal of Applied Physiology* **98**(1) pp 3-30

[9] McDermott A Y and Mernitz H 2006 Exercise and older patients: prescribing guidelines *American Family Physician* **74**(3)

[10] Chow A E 2001 Therapeutic squeeze ball *Google Patents*

[11] Guyatt G H, Osoba D, Wu A W, Wyrwich K W, Norman G R and Clinical Significance Consensus Meeting Group 2002 Methods to explain the clinical significance of health status measures *Mayo Clinic Proceedings* **77**(4) pp 371-383

[12] Jensen G M, Gwyer J, Shepard K F and Hack L M 2000 Expert practice in physical therapy *Physical Therapy* **80**(1) pp28-43

[13] Kaushik A, Prakash N and Jagga V 2016 Effects of hand position with relation to elbow and shoulder position on maximum grip strength *Journal of Exercise Science and Physiotherapy* **12**(1) pp76

[14] Quaine F, Vigouroux L, Paclet F and Colloud F 2011 The thumb during the crimp grip *International Journal of Sports Medicine* **32**(01) pp 49-53

Finite Element Approach Using Electrical Tomography for Bioassay Application

Wan Aizatul Nadihah Binti Ropandi[1], Nur Amira Zulkiflli[1], Nina Diana Nawi[2], Fatin Aliah Phang Abdullah[3], Jaysuman Pusppanathan[4*], Zulkarnay Zakaria[5]

[1] School of Biomedical Engineering & Health Sciences, Faculty of Engineering, Universiti Teknologi Malaysia, 81310 Johor, Malaysia
[2] School of Education, Faculty of Social Sciences and Humanities, Universiti Teknologi Malaysia, 81310 Skudai Johor, Malaysia
[3] Centre of Engineering Education (CEE), Universiti Teknologi Malaysia, 81310 Skudai Johor, Malaysia
[4] Sports Innovation & Technology Centre, Institute of Human Centered Engineering, Universiti Teknologi Malaysia, 81310 Johor, Malaysia
[5] School of Mechatronic Engineering, universiti Malaysia 01000 Kangar, Perlis, Malaysia

*Corresponding author: jaysuman@utm.my

Abstract. Single-cell imaging is a part bioassay application that is highly used in the studies of microbiology. This bioassay application is used in order to determine the cell activity, behaviour and reaction as well as electrical properties and characteristic. Most of this can be achieved by using mass spectrometry, impedance analyser and flow cytometry. Currently, there is an emerging trend of on-chip integrated device that allows bioassay application to be performed real time. However, in terms of cell monitoring, which allows us to view the cross section image of cell, there is much to be explored. As each cell is composed of different material with specific relative permittivity, Electrical Capacitance Tomography is chosen to determine the electrical properties and spatial distribution of cell. Therefore, the purpose of this research is to simulate Electrical Capacitance Tomography for yeast cell growth in order to determine the most suitable configuration needed for ECT system.

1. Introduction

Tomography refer to the cross-sectional imaging, either 2-dimensional or 3-dimensional, of an object obtained by illuminating the object from multiple viewpoint [1] using the technique of transmission or reflection . Aside from medicinal field, tomography technique is also used in biochemical field as they enabled the researchers to view the reaction of certain substances towards living organism, called bioassay. Bioassay is a successful tool used in estimating and discovering of biologically active substances specifically for the microbiology applications.

One of the method that is use to study the reaction of bioassay is single-cell imaging. Single cell imaging is use to provide cross section image of cell activities, cell pathological as well as its reaction when foreign substances is added. Noted that a single cell imaging assay can also be considered as a technique that allow quantitative analysis of cell signalling pathway. Normally, imaging of single-cell can be acquired through microscopy techniques [3] for example mass spectroscopy and impedance flow cytometry. Nowadays, there is an emergence of Lab-On-Chip (LoC) technology [4], [5] which combines the function of laboratory operations of single-cell analysis into a miniaturized handheld device.

A research on cell impedance measurement done by [6] using the principle of Electrical Impedance Tomography (EIT) for single cell imaging of P. Polycephalum cell culture. The aim of the simulation is to determine the electrode configuration of the hardware device by introducing two regions with separate conductivity. From the simulation obtain, S. Tsuda created an on-chip EIT device and experiment was done by transferring a section of the cell culture into the sensing region. The result obtain shows that there is difference in cell electrical properties. J Hong on the other hand, develop a microfluidic device using the concept of single cell impedance measurement [7]. The device developed was used to analysed four different kind of cancer cell by supplying several ranges of impedance spectra towards the cells. supplied and based on the impedance measured, the results show that every cell has different tolerances to an electric field.

It can be seen majority of the application of LoC device is used to study the electrical properties of cell. However, there is only few application of cell monitoring provided. Taking the cell properties into consideration, single cell is made up of several components that has different dielectric constants. For example, range of relative permittivity for cell cytoplasm can be from 50-120 while for the cell membrane, the relative permittivity is 2-10 [8]. Thus, Electrical Capacitance Tomography (ECT) is used for cell monitoring in this study by calculating the dielectric constant spatial distribution inside a cell by measuring the capacitance values between electrodes placed on the insulating surface [9].

1.1. Fundamental of Electrical Capacitance Tomography

Electrical Capacitance Tomography (ECT) is a method that belongs to the soft-field tomography. ECT is a part of Electrical Tomography (ET) that uses the principle of electrostatic charge distribution [10] to reconstruct a tomographic image. ECT generally is used to reconstruct a 2-dimensional cross sectional image by measuring the variations of capacitance that exist between electrode pair which represented the spatial distribution [11] of the electrical permittivity of a material.

1.1.1. Electrostatic Behaviour in Electrical Capacitance Tomography. The ECT applied the principle of electrostatics and the system uses electrodes as a measurement tools where the electrode is considered in pair [12, 16]. In order to find the distribution of the electric potential, V, an electrode is excited which causes the electrode to become transmitter (Tx) while the grounded electrode will become receiver (Rx) and an electrical field distribution will exist between those transmitter and receiver. For homogenous condition of capacitance at a given electrode between Tx and Rx, the equation are as follows,

$$C_{ij} = -\frac{\left(\int_s \varepsilon \frac{\delta\varphi}{\delta n} \delta S\right)_i}{V_j} \tag{1}$$

When a dielectric material is presence is taken into consideration for the heterogeneous condition, the equation is as shown below

$$\Delta C_{ij} = \frac{\left(\varepsilon_0 \int_s \left(\frac{\delta\varphi\varepsilon_0}{\delta n} - \varepsilon_r \frac{\delta\varphi\varepsilon_r}{\delta n}\right) dS\right)_i}{V_j} \tag{2}$$

where ΔC_{ij} represent change in capacitance due to the presence of dielectric material given ε_0 is the permittivity of free space while ε_r is related to the permittivity of the sample material [13]. Therefore, based on the equation shown above, we can conclude that the presence of foreign matter will cause changes to electrostatic potential distribution inside the ECT system.

2. Methodology

This research will be using a Finite Element Method (FEM) to perform Finite Element Analysis that is use in solving capacitance calculation between electrode pairs with condition of known sensor structure and given permittivity distribution. Simulation is done by using software COMSOL Multiphysics focusing on single cell imaging of yeast. In this FEM simulation, excited electrode will generate line projection, that represent electrical field distribution, towards the grounded electrode that act as sensing port.

2.1. Experimental Set-Up

The growth of yeast cell was illustrated by using different radius at different coordinate, in order to study the reaction of sensing electrode and the projection line of electrical field. The coordinate chosen in this experiment are (0,0) and (2,1) while the radius is set to three ranges which were 1.5 mm, 2.5 mm and 3.5 mm. The sensing region is set to radius of 7.5 mm. The size of electrode on the other hand, is set to 2.08 mm width and length is adjusted to 5.0 mm to allow the sensing electrode to fall into the sensing region.

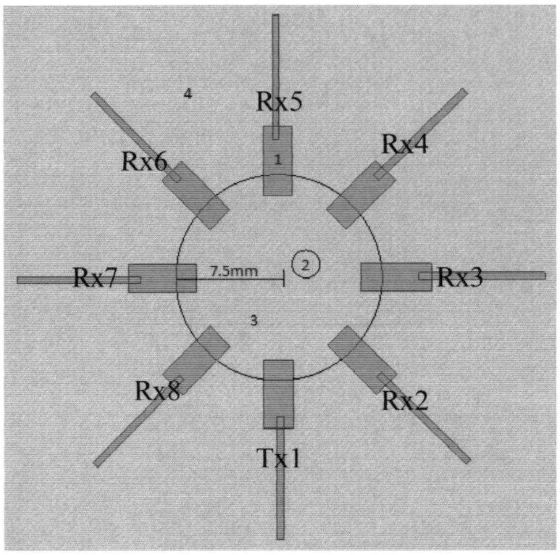

Figure 1. Arrangement of ECT system where (1) copper electrode with measurement 5mm x 2.08mm, (2) yeast cell with radius 1mm, (3) water, and (4) air region

The number of electrode chosen is 8 which produces 28 independent capacitance measurement [4], [14], [15], based on the formula given below

$$M = \frac{N\,(N-1)}{2} \tag{3}$$

where,

M = capacitance measurement

N = number of electrode

3. Results and Discussion

The following Fig. 2 shows the simulation result of cell growth.

Radius of cell (mm)	Coordinate (0, 0)	Coordinate (2, 1)
1.5	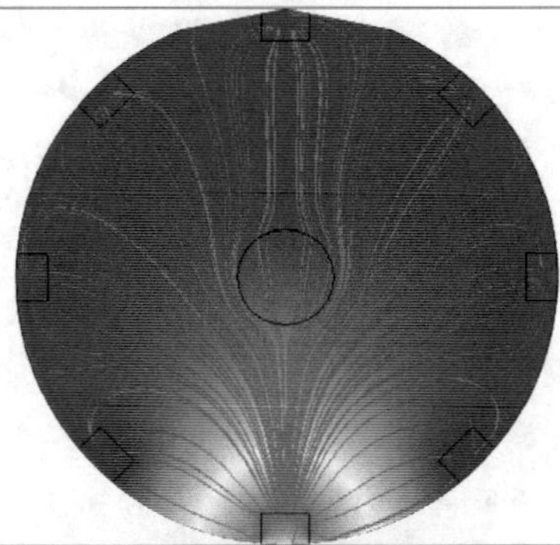	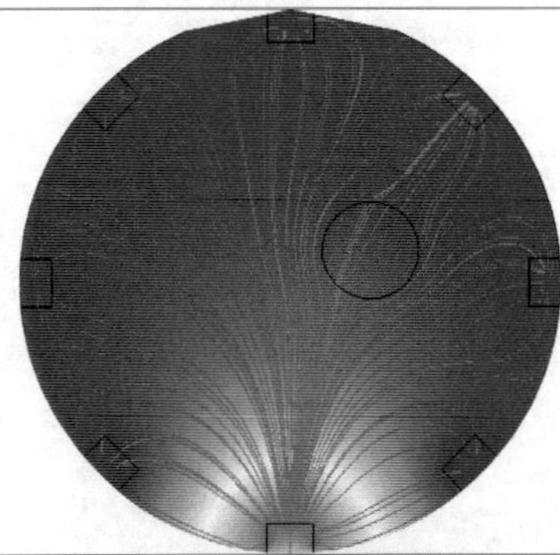
2.5	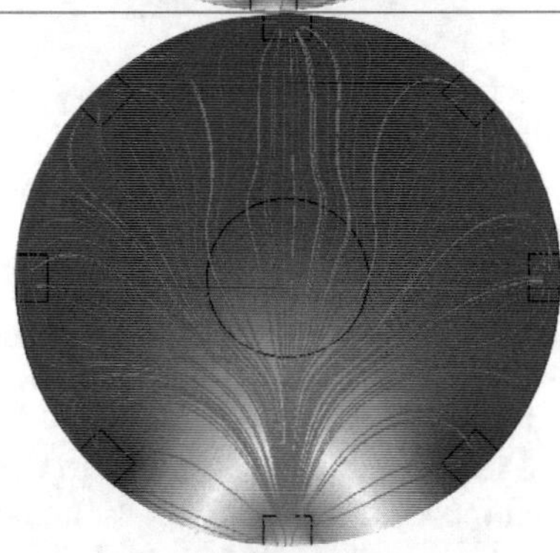	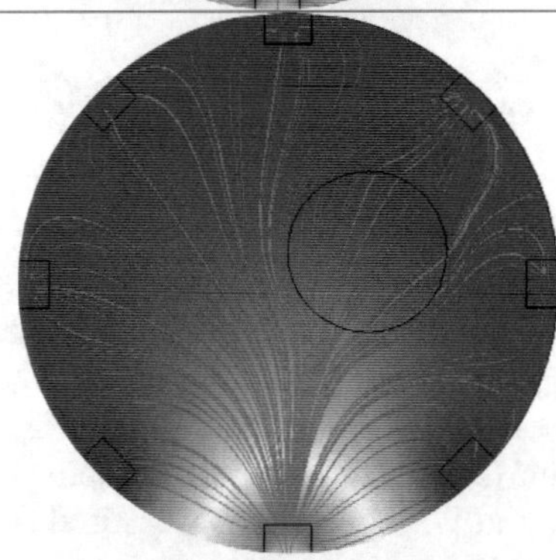
3.5	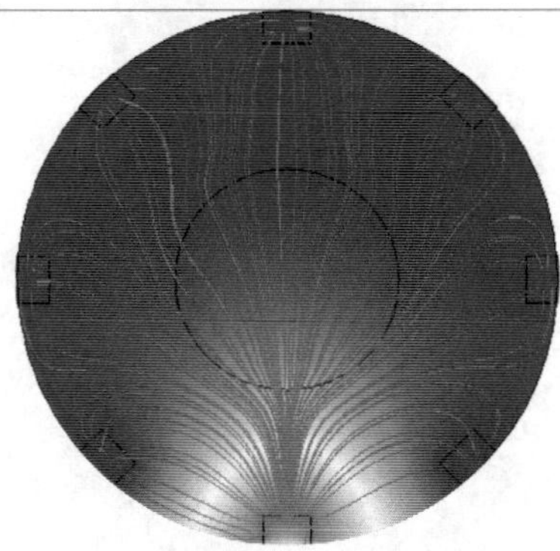	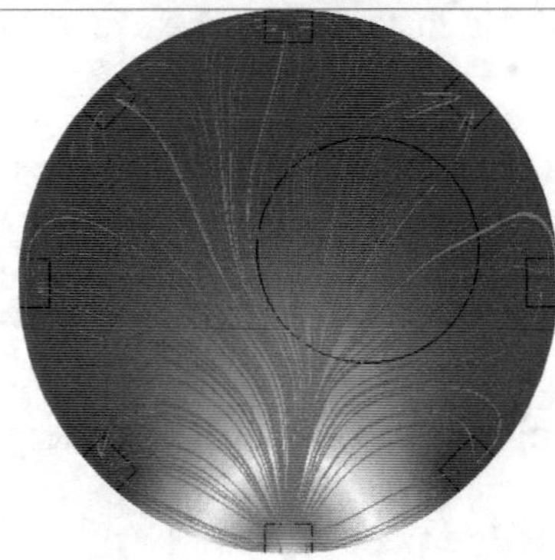

Figure 2 Projection of the electrical field lines from the excitation electrode to the receiving electrode

Based on the simulation, the value at the receiver electrode for yeast at both coordinate (0,0) and (2,1) was captured and the graph of measured capacitance when excitation done on electrode 1 (Tx1) is taken. The acquired capacitance value at each receiving electrode (RX 2 to RX 8 as in Fig. 1) is depicted in Fig.3 follows.

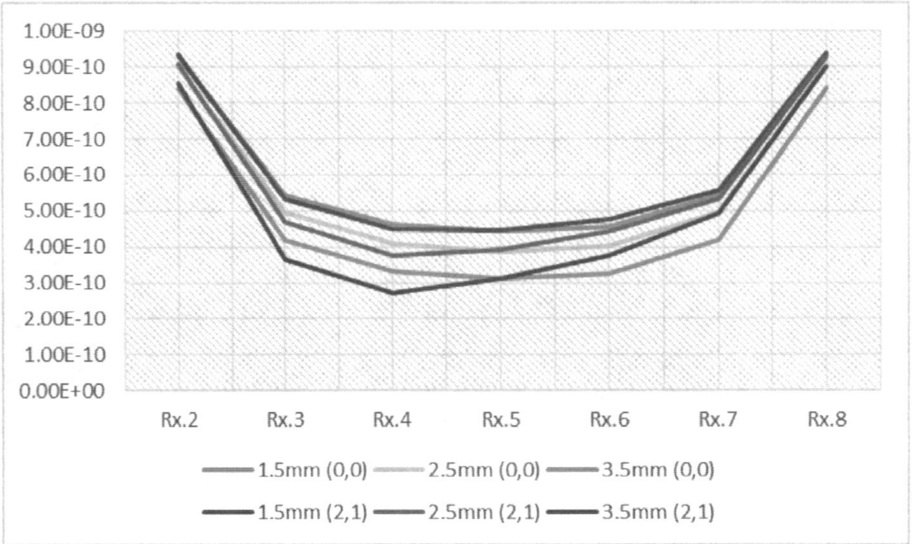

Figure 3 Line graph pattern for each receiving electrode response

The line projection of electrical field diverges between two different dielectric materials within the region of interest. Thus, by observing the projection line for electrical field distribution and the capacitance measurement captured at all grounded electrode, Rx1, Rx2, Rx3, Rx4, Rx5, Rx6 and Rx7, it shows that the value obtains at region adjacent to the excitation electrode, which is Rx2 and Rx8, is the highest. On the other hand, the measured capacitance indicates that the voltage measured for electrical field at sensor electrode Rx3 and Rx4 is the weaker, especially for yeast cell placed at coordinate (2,1). The different placement of yeast cell also plays a part where it will determine the diverging path of streamline. Measured capacitance for yeast cell positioned at coordinate (0,0) is more symmetrically arranged compared to the capacitance measured for coordinate (2,1). This is because, the yeast placed at that position causes the streamline to diverges more towards the other electrode.

4. Conclusion

Based on this research, ECT sensor configuration plays a major part in determining the functionality of ECT device. In this case, the size, angle between electrode and the number of electrode need to be taken into consideration so that the electrode will fall into the sensing region and at the same time, the placement of electrode must not affect the reading taken. The higher the number of electrode will generate more data measurement; however, it can also cause the streamline for adjacent electrode to coincide with one another.

Acknowledgement

The author would like to acknowledge the research and funding support from Universiti Teknologi Malaysia GUP grant 06G17 and 05G04.

References

[1] N. A. Mohd Razali et al.: Miniaturized planar tomography for multiphase stagnant sample detection. J. Teknol. (2015) 99–102

[2] D. Mullassery, C. A. Horton, C. D. Wood, and M. R. H. White.: Europe PMC Funders Group Single Live Cell Imaging for Systems Biology. Essays Biochem. (2009) 121–133

[3] H. Tang et. al.: High throughput spinning-disk imaging bioassay : from single-cell imaging to whole slide histopathology. The University of Hong Kong (Pokfulam, Hong Kong). (2017)

[4] Y. G. Zhou, H. Yan, and Z. Fang.: Design of an Electrical Capacitance Tomography System based on LabVIEW. International Journal of Signal Processing, Image Processing and Pattern Recognition. (2015) 165–178

[5] C. Parent, F. Boizot, M. Cubizolles, N. Verplanck, J.-L. Achard, and Y. Fouillet.: Quantitative biological assays with on-chip calibration using versatile architecture and collapsible chambers. Sensors and Actuators B: Chemical (2018) 106–114

[6] T. Sun, S. Tsuda, K. P. Zauner, and H. Morgan.: Single cell imaging using electrical impedance tomography. 4th IEEE International Conference Nano/Micro Engineering. (2009) 858–863

[7] J. L. Hong, K. C. Lan, and L. S. Jang.: Electrical characteristics analysis of various cancer cells using a microfluidic device based on single-cell impedance measurement. Sensors and Actuators B:Chemical. (2012) 927–934

[8] G. H. Markx and C. L. Davey.: The dielectric properties of biological cells at radiofrequencies : Applications in biotechnology. Enzyme Microb. Technol. (1999) 161–171

[9] Y. Kang, S. Liu, and J. Liu.: Image reconstruction algorithm for electrical capacitance tomography based on data correlation analysis. Flow Measurement Instrument. (2018) 113–122

[10] Z. Ren, S. Member, and W. Yang.: A Miniature Two-Plate Electrical Capacitance Tomography Sensor. IEEE Sensor Journal. (2015) 3037–3049

[11] C. Gunes, Q. M. Marashdeh, and F. L. Teixeira.: A Comparison Between Electrical Capacitance Tomography and Displacement-Current Phase Tomography. IEEE Sensor Journal. (2017) 8037–8046

[12] B. B. Abraham, and G. Anitha.: Designing of Lab View Based Electrical Capacitance Tomography System for the Imaging. Bonfring International Journal of Power System and Integrated Circuits. (2012) 1–6

[13] M. A. Nurge.: Electrical Capacitance Volume Tomography of High Contrast Dielectrics Using a Cuboid Geometry. Measurement Science and Technology. (2007) 1511-1520.

[14] J. Pusppanathan.: Hardware Development of Electrical Capacitance Tomography (ECT) System with Capacitance Sensor for Liquid Measurements. Jurnal Teknologi (2015) 13–22

[15] C. Deyun, L. Zhiqiang, G. Ming, W. Lili, and Y. Xiaoyang.: A Superresolution Image Reconstruction Algorithm Based on Landweber in Electrical Capacitance Tomography. Mathematical Problem in Engineering (2013) 1-8

[16] J. Pusppanathan. e*t. al* Single-plane dual-modality tomography for multiphase flow imaging by integrating electrical capacitance and ultrasonic sensors, IEEE Sensors Journal (2017) 17 (19), 6368-6377

Development of an Affordable, Accessible, and Available (3As) IoT-Enabled Motion Analysis System for Stroke Recovery Monitoring – An Introductory Approach

Yu Zheng CHONG[1*], Peng Yee NG[1]

[1] Department of Mechatronics and Biomedical Engineering, Lee Kong Chian Faculty of Engineering and Science, Universiti Tunku Abdul Rahman, Sungai Long Campus, MALAYSIA.

*chongyz@utar.edu.my

Abstract. According to the United Nations Sustainable Development Goals (SDGs) formulated in 2015, there are seventeen (17) identified goals aimed to be achieved by 2030. This project was focussed on SDG 3, SDG 10, and SDG 11 [1]. The developed motion analysis system adopted principles of Affordability, Accessibility, and Availability of this system for monitoring of stroke patients who are undergoing physical therapy (PT). Such system will enable stakeholders involved in rehabilitation of these communities to monitor the progress of PT regime that will be vital in determining efficacy of the regime. Furthermore, the proposed system has the capabilities of remote monitoring, and data acquisition via the IoT platform. The cost of developing this modular system was less than RM 1,000 (approximately USD 243) [2]. Hence, it is anticipated that this project will be a stepping stone towards 3As of motion analysis for stroke patients.

1. Introduction

Stroke is a disease that well defined in numerous literatures, and it will happen when blood delivered to brain tissue is disrupted [4]. According to statistics obtained from United Nations, there are approximately 15 million peoples suffer from stroke globally [3]. In the local Malaysian context, stroke is one of the leading causes of disability, and the third highest cause of mortality. National Stroke Association of Malaysia has estimated some 40,000 Malaysians suffer from stroke annually [4]. These figures grow annually at the rate of 29.5 % and 18.7 % for ischemic and haemorrhagic stroke yearly [4]. Hence, it is vital to have proper post-stroke treatments to enable recovery, and enhancing the quality of life of this community. The Ministry of Health Malaysia has proposed various treatment plans [4].

1.1. Clinical Gait Analysis and Applications of Depth Sensors

Clinical Gait Analysis (CGA) have been widely used in analysing gait for various purposes, from gait monitoring to intervention studies [5]. The adoption of such technologies in Malaysia is still in its infancies due to various factors, and one of the identified factors are the cost of adopting these technologies. Based on 2016 statistics, median monthly household income stands at RM 5,228 [6], hence, under these circumstances the affordability of household to own such monitoring devices at home will be a great financial challenge. In line with this, the project aims to develop a gait monitoring system that costs around RM1,000 which fulfil requirements of 3As.

Three-dimensional gait analysis (3DGA) has been widely used in analysis of human motion joint kinematics [7]. This quantitative assessment of human motion has been utilised specifically in clinical setting whereby various joint kinematics were measured and further analysed to obtain gait parameters that are useful in determination of various clinical considerations from diagnosing to analysing the effectiveness of rehabilitation regimes [7]. These technologies are considerably high cost, hence the 3As of such system is still not widely utilised [7].

With rapid advancement in technologies, there are various readily-available products in the market could be adopted to datalog human joint kinematics. The availabilities of depth sensors technology have enabled motion tracking possible. This adoption may be considered as the supplementary applications of the systems due to the primary functions are for electronic-gaming purposes [8]. Various studies have been conducted on different depth sensors available in the market, namely the Microsoft™ Kinect XBox 360, Asus XTion sensors. Functional assessment of human gait have been investigated in various studies [5][8]. One study noted that the Kinect XBox 360 sensor is affordable, highly portable, and user-friendly to enable clinicians to integrate the system into their clinical practice in assisting human motion monitoring and assessments, particularly in the field of rehabilitation [8][9].

1.2. Motivation
The primary aim of this project is to develop an Internet-of-Things (IoT)-enabled markerless human motion analysis system based on the Kinect XBox 360 sensor to monitor movement of patients with hemiplegia during physical rehabilitation sessions. The system also incorporated joint force reaction feedbacks to act as feedback aiming to provide motivation to the patient during rehabilitation sessions. In the nutshell there are three (3) objectives identified for this study as the following:-

- Development of testing protocol for marker-less motion analysis system for stroke rehabilitation monitoring
- Integration of IoT features to the Motion Analysis System
- Interactive feedback system as motivation to stroke patient

2. Methodology
This section describes methodology design and implementation for the system.

2.1. SWOT Analysis for the Proposed System
The strengths, weaknesses, opportunities, and threats (SWOT) analysis were conducted to evaluate integration of Kinect XBox 360 in stroke rehabilitation.

2.2 System Architecture
The proposed system architecture is shown in Figure 2. Firstly, the patient is required to perform sit-to-stand exercise. The manoeuvre will be recorded by camera system facing the frontal plane of the patient. Skeleton modelling was directly generated by via the Kinect System Development Kit (SDK) which is interfaced with a dedicated program installed in the computer. Joint landmarks were recorded in real-time for all movement performed by the patient. All information recorded will be stored in an encrypted secured database. This is essential due to the lack of standard database in public-domain utilising the Kinect System. The database is stored in a remote database which will enable retrieval of data from remote location which serves as part of the IoT technology.

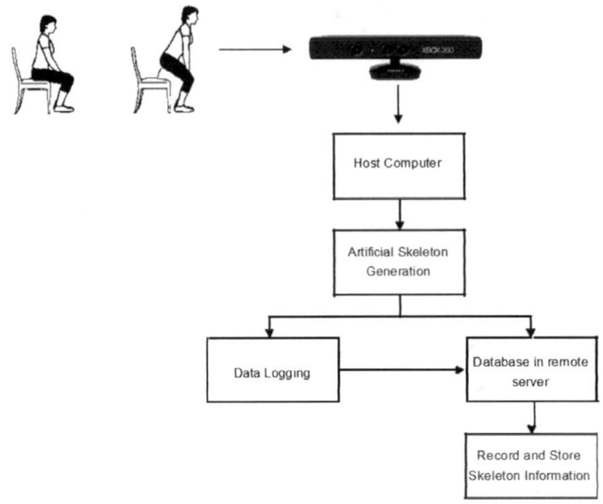

Figure 1. The System Architecture

2.2. Subjects Selection

Total of thirty (30) healthy subjects (as control) were recruited for the studies, and one (1) stroke patient volunteered for testing the prototype developed. The study protocol has been approved by the University Ethical Committee prior to data collection. All participants were briefed thoroughly on the study protocol before obtaining informed consent. Selected participants will need to fulfil some requirements prior selection to participate. The participant should be able to stand up from sitting position on their own, ability to understand instructions and understand commands. Exclusion criteria will be tetraplegic or non-independent participants that requiring assistance in accomplishing sit-to-stand manoeuvre.

2.3. Experimental Setup

Figure 2 shows the experimental setup for the sit-to-stand protocol.

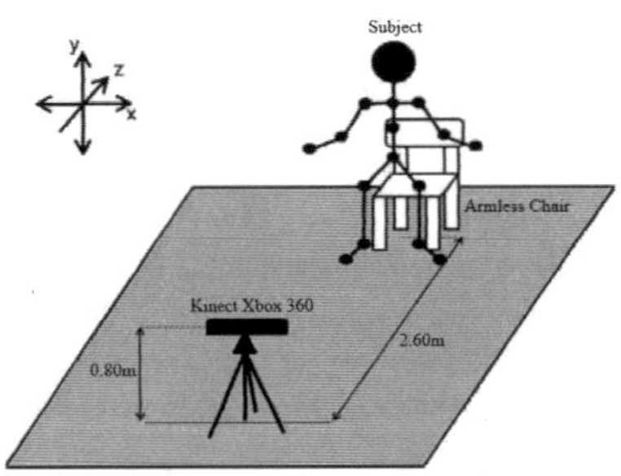

Figure 2. Experimental Setup for the sit-to-stand Motion Analysis

The Kinect Xbox 360 sensor was placed 2.60 meter away from the chair. The camera is affixed to a tripod at the height of 0.8 m from the ground as shown in Figure 2. The armless chair selected is of 0.4 cm high. According to [10], majority of movement involved in the sit-to-stand manoeuvre involves sagittal plane movement such as flexion and extension of both hip and knee joint; plantar-and-dorsi-flexion of the ankle joint. Furthermore, the orientation of the camera is placed in front of the subject to detect differences of both sides of the body.

Detailed protocol is summarised in Table 1.

Table 1. Detailed Protocol

No.	Details
1	The Kinect Xbox 360 is placed at 0.8m from the ground level;

2	Distance between The Kinect Xbox 360 is 2.6m away from the frontal plane of the subject;
3	Subject is seated on an armless chair with backrest. Fifty percent of the thigh length is sustained by the chair;
4	The original feet position and the height of chair are adjusted to ensure the position of knee and hip angle is 90 degrees.;
5	Subject is requested to rise from the chair in his/ her usual manner with self-paced and comfortable speed and sit down. This is to be repeated for three times.

2.4. Skeletal Tracking

Based on the capabilities of the camera system, total of twenty (20) anatomical landmarks coordinates were collected by the system. Three-dimensional coordinates were tracked in real-time to determine respective movement parameters of the movement. Dedicated datalogging system were developed utilising the Microsoft™ Visual Basic Studio 2012. Estimation of body position was performed by depth map computation, which is a proprietary machine learning capability embedded in the camera system.

Mathematical modelling of the lower limb has been performed utilising the link-segment model to model the moment of forces of respective lower extremities joints. In addition, the joint reaction forces were simulated utilising anthropometric data sourced from [11]. Figure 3 shows the link-segment model and anthropometric data adopted in the mathematical modelling.

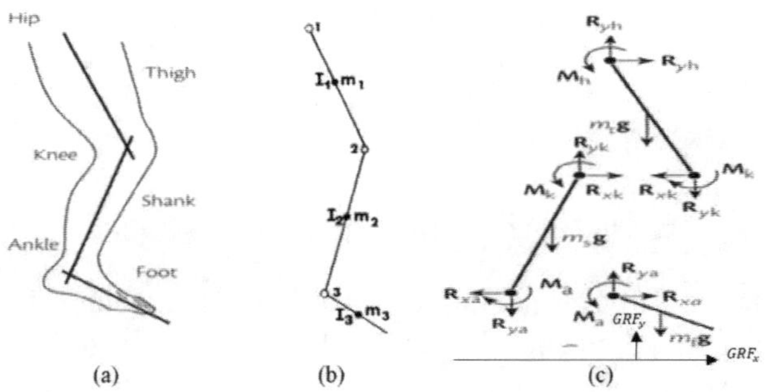

Segment	Segment Mass / Total Body Mass	Center of Mass / Segment Length Proximal	Distal	Density (kg/l)
Hand	0.006	0.506	0.494	1.16
Forearm	0.016	0.430	0.570	1.13
Upper Arm	0.028	0.436	0.564	1.07
Forearm and Hand	0.022	0.682	0.318	1.14
Total Arm	0.050	0.530	0.470	1.11
Foot	0.0145	0.500	0.500	1.10
Lower Leg (calf)	0.0465	0.433	0.567	1.09
Foot and Lower Leg	0.061	0.606	0.394	1.09
Upper Leg (thigh)	0.100	0.433	0.567	1.05
Total Leg	0.161	0.447	0.553	1.06
Head and Neck	0.081	1.000	—	1.11
Trunk	0.497	0.500	0.500	1.03

(d)

Figure 3. (a) Anatomical Model of the Lower Limb; (b) Link Segment Model of Lower Limbs; (c) Free Body Diagram of Lower Limbs; (d) The Normalized Mass and Length of Body Segments of Standard Human [11]

2.6 IoT Platform

Data acquired from camera system is filtered via the dedicated interface with the Joint Reaction Forces were tabulated and exported in Microsoft Excel (.xlsx) format before being stored in Microsoft OneDrive cloud storage. Data update and retrieval were handled by the Microsoft PowerApps as the back-end interface of the IoT platform. Dedicated mobile application known as 'TeleK" was developed utilising PowerApps Studio which enables distributions on mobile devices. Figure 4 shows the development process in TeleK:-

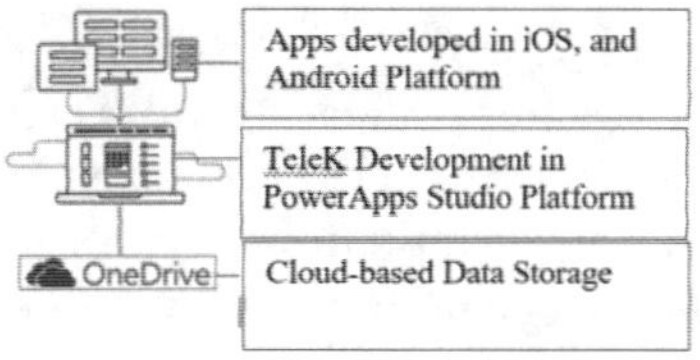

Figure 4. IoT Platform Development

3. Results and Discussions

Lower extremities joints kinematics and kinetics were obtained from the system which agree with [10][12]. As the data collection is still on-going, the results reported in this paper will form the preliminary results obtained from the system.

During the sitting phase of testing, lower extremities of the subject remained in neutral original position whereby the ankle, knee, and hip joints reaction forces are near to zero. When the subject started to change position from sitting to standing, it was noted that large forces were exerted on both knee and hip joints. This were due to the transfer of loading due to weight and balancing mechanism of the lower extremities in reaction to the dynamical changes in the upper extremities. The next event will be transition between sitting to full standing posture. Whilst the subject is standing in the upright position, both extremities will be transiting from dynamic situation to static situation, hence, the joint reaction forces will be back to equilibrium i.e. in static situation. Dynamical changes were recorded once again when the subject transited from standing position back to sitting position. During these changes, the knee joint will exert higher force to support the upper body weight of the subject, and at the same instances, hip will be lowered down until the subject fully resting on the chair. Figure 5 shows an example of the transition of joint reaction forces in the testing cycle.

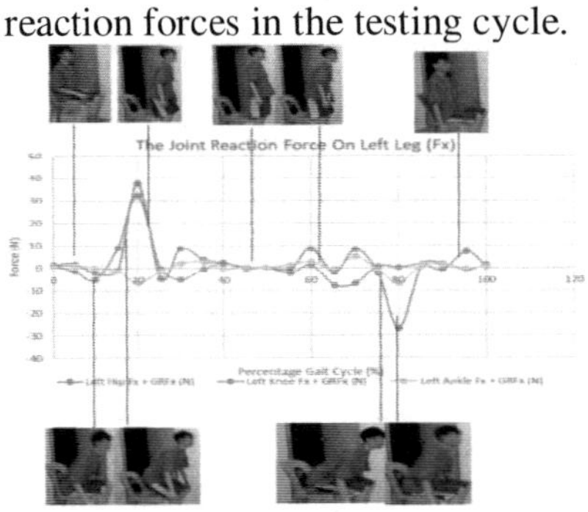

Figure 5. The joint reaction forces of Hip, Knee, and Ankle during sit-to-stand testing

It was found that the kinematics and kinetics of the test subjects are in accordance to results obtained in [10][12]. Figure 6 shows samples of joint kinematics collected by the system. In addition, it should be noted that these results are still in its preliminary stages, therefore, more testing will be performed to further validate the efficacy of the developed system.

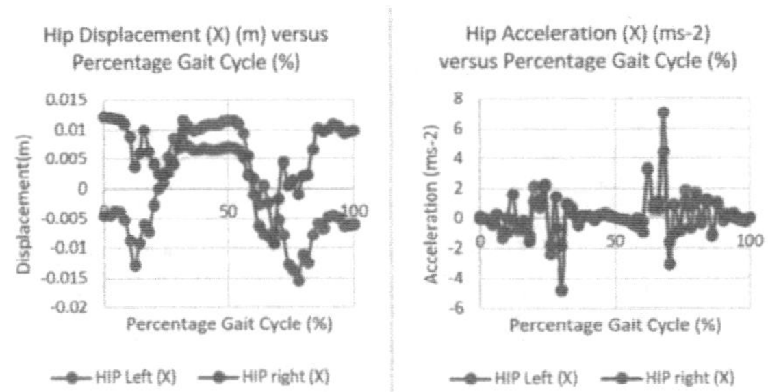

Figure 6. Sample Kinematics Parameters Collected by the System

As for the IoT-enabled platform, apps named as TeleK has been developed. The feature of the system enables secured remote access of gait parameters of patients. This will enable remote access to this vital data for the purposes of monitoring, and even diagnosing by healthcare professionals in determining the progress, modification of rehabilitation regime, and home monitoring by the patient caretakers. User profiles, training records, and comments are also made available in the apps. Figure 7 shows the screenshot of the developed apps.

Figure 7. The TeleK apps interface

4. Conclusions and Future Recommendations

This paper demonstrates that with the currently available depth camera system, it will be possible to develop a motion analysis system that may assist in improving the quality of life of stroke patients especially in rehabilitation. More importantly, this will be a stepping stone in the development of 3As in the broad area of clinical motion analysis. With the forthcoming technologies in wireless communications, such as 5G, the future will be very exciting especially in making biomechanical assessment technologies to be 3As and in-turn would be able to fulfil the SDGs [1].

References

[1] United Nations. *About the Sustainable Development Goals* *https://www.un.org/sustainabledevelopment/sustainable-development-goals/ (accessed 10 June 2019).*

[2] Bank Negara Malaysia. *Exchange rates.* *http://www.bnm.gov.my/index.php?ch=statistic&pg=stats_exchangerates (accessed 10 June 2019).*

[3] Aziz ZA, Lee YYL, Ngah BA, Sidek NN, Looi I, Hanip MR, et al. . Acute Stroke Registry Malaysia, 2010-2014: Results from the National Neurology Registry. . *J Stroke Cerebrovasc Dis* 2015; 24(12): . https://linkinghub.elsevier.com/retrieve/pii/S1052305715004280 (accessed 10 June 2019).

[4] Islam MR. . Rights of the People with Disabilities and Social Exclusion in Malaysia. . *Int J Soc Sci Humanit* 2015; 15(2): . http://www.ijssh.org/index.php?m=content&c=index&a=show&catid=52&id=763 (accessed 16 April 2019).

[5] Taborri J, Palermo E, Rossi S, Cappa P. . Gait partitioning methods: A systematic review. . *Sensors* 2016; 16(1):40-2 .

[6] Statistics Department Malaysia. . *Findings of the Household Income Survey* *https://www.dosm.gov.my/v1/index.php?r=column/cthemeByCat&cat=120&bul_id=RUZ5REwveU1ra1hGL21JWVlPRmU2Zz09&menu_id=amVoWU54UTl0a21NWmdhMjFMMWcyZz09 (accessed 10 June 2019).*

[7] Tucker CA, Wren TAL, Gorton GE, O S. . Efficacy of clinical gait analysis : A systematic review. . *Gait & Posture* 2011; 34():149-53 .

[8] Clark RA, Pua Y-H, Oliveira CC, Bower KJ, Thilarajah S, McGaw R, et al. . Reliability and concurrent validity of the Microsoft Kinect V2 for assessment of standing balance and postural control.. *Gait & Posture* 2015; 42(2): . http://linkinghub.elsevier.com/retrieve/pii/S0966636215000740 (accessed 16 April 2019).

[9] Ruff J, Wang TL, Quatman-Yates CC, Phieffer LS, Quatman CE. . Commercially available gaming systems as clinical assessment tools to improve value in the orthopaedic setting: A systematic review. Injury . *Gait & Posture* 2015; 46(2):179-83 . http://linkinghub.elsevier.com/retrieve/pii/S0966636215000740 (accessed 16 April 2019).

[10] Frykberg GE, Häger CK. . Movement analysis of sit-to-stand – research informing clinical practice. *Phys Ther Rev* 2015; 20(3): . http://www.tandfonline.com/doi/full/10.1179/1743288X15Y.0000000005 (accessed 16 April 2019).

[11] David A. Winter. *Biomechanics and Motor Control of Human Movement*, 4th ed. New York: John Wiley & Sons Inc; 2009.

[12] Crockett K, Ardell K, Hermanson M, Penner A, Lanovaz J, Farthing J, et al. . The Relationship of Knee-Extensor Strength and Rate of Torque Development to Sit-to-Stand Performance in Older Adults.. *Physiother Canada* 2013; 65(3): . https://utpjournals.press/doi/10.3138/ptc.2012-04 (accessed 16 April 2019).

Effect of swelling agent on pore properties of mesoporous carbonated hydroxyapatite

Nur Farahiyah Mohammad[1*], Nadia Liyana Amiruddin[1], Siti Shuhadah Md Saleh[2], Mustafa Ali Azhar Taib[3] and Nashrul Fazli Mohd Nasir[1]

[1] Biomedical Engineering Research Group (BERG), School of Mechatronic Engineering, Pauh Putra Campus, Universiti Malaysia Perlis, 02600 Arau, Perlis, Malaysia
[2] School of Materials Engineering, Universiti Malaysia Perlis, Jejawi, 02600 Arau, Perlis, Malaysia
[3] Advanced Technology Training Centre (ADTEC) Taiping, PT 15643, Kamunting Raya, Mukim Asam Kumbang, 34600 Kamunting, Perak, Malaysia

* farahiyah@unimap.edu.my

Abstract. Mesoporous carbonated hydroxyapatite (CHA) has been seen to be suitable as an adsorbent material due to the proven biocompatibility, bioactivity and chemical stability. However, obtaining the high surface area and desired pore size for mesoporous materials become the main challenge in producing good quality adsorbent materials. Thus, this study aims to investigate the effect of addition of swelling agents on pore properties of mesoporous CHA. Mesoporous CHA was synthesised by precipitation method. Triblock co-polymer, P123 surfactant was used as a soft template to introduce pores within the CHA particles and 1-dodecanethiol as a swelling agent. Two different mixing temperature (60°C and 80°C) were used during the mixing process of P123-calcium solution and 1-dodecanethiol. The effect of different mixing temperature (60°C and 80°C) and various concentration of swelling agent on the phase, morphology and pore characteristics of mesoporous CHA were investigated using various material characterization techniques. Mixing temperature 60°C, produced mesoporous CHA with much higher surface area (92.8 m^2/g), larger pore size (22.9 nm) and higher pore volume (0.502 cm^2/g) than those produce at 80°C (surface area = 62.7 m^2/g, pore size = 2.6 nm and pore volume = 0.105 cm^2/g). The highest surface area (146.9 m^2/g) of mesoporous CHA was obtained when there was no swelling agent added. However, the pore size of the sample was very small. Therefore, the optimum pore characteristics (surface area = 104.5 m^2/g, pore size = 24.4 nm and pore volume = 0.530 cm^2/g) of mesoporous CHA was obtained when the swelling agent concentration is 0.6 M and the mixing temperature is 60°C. Both mixing temperature and swelling agent concentration have profound effect on the pore characteristics of mesoporous carbonated hydroxyapatite (CHA).

1. Introduction

Conventional haemodialysis machine mostly used cellulose membranes for filtration. However, the filtration process can only remove small molecular weight uremic toxins [1]. New materials such as activated carbon [2-4], zeolites [5] and graphene oxide have been utilized to enhance the dialysis treatment. However, these material have some limitations such as the inability to filter out medium and large molecular weight [6]. Some of these materials even producing another uremic toxin such as methyl guanidine as reported when activated carbon is utilized as filter material [4]. Medium and large molecular weight uremic toxin for example *p-cresol* and *β2-microglobulin* could not pass through the membrane. As for the *p-cresol* assembled, the patients might face immune defect, inhibition of blood clot formation and increase cardiovascular mortality [7]. Basically, the inability to remove *p-cresol* will lead to intoxication. Meanwhile, the accumulation of *β2-microglobulin* inside human body will promote cancer progression. Thus, it is essential to ensure that a haemodialysis system is able to filter out these uremic toxins.

So, there is a need for a new material that able to adsorb large molecular weight uremic toxin. The material should have high surface area that will promote high adsorption capacity. Mesoporous CHA is defined as a nanoporous bioceramic which indicated the size of pore within 2 nm to 50 nm in diameter [1]. Mesoporous CHA previously has been proven to have high surface area up to 78 m^2/g [8]. Mesoporous CHA microspheres also demonstrated higher drug loading efficiency of 70-75% with gentamicin [9]. Consequently, this has proved that mesoporous CHA is a capable adsorbent material. However, obtaining the high surface area and desired pore size for mesoporous materials become the main challenge in producing good quality adsorbent materials. Therefore, this study aimed to synthesize mesoporous CHA by precipitation method using swelling agent as pore characteristics controller and to investigate the effect of mixing temperature and concentration of swelling agent on pore characteristics of mesoporous CHA.

2. Methodology

2.1 Materials

The chemicals used as the calcium and phosphate precursor were calcium nitrate tetrahydrate ($Ca(NO_3)_2.4H_2O$) and diammonium hydrogen phosphate ($(NH_4)_2.HPO_4$) respectively. Ammonium hydrogen carbonate (NH_4HCO_3) was used as a carbonate precursor. Non-ionic triblock co-polymer Pluronics® P123, ($PEO_{19}PPO_{69}$ PEO_{19}) donated by BASF, USA was used as the structure directing agent for the pore template. 1-Dodecanethiol ($C_{12}H_{26}S$) was used as swelling agent. Besides that, sodium hydroxide (NaOH) was used to maintained the pH of the mixtures at pH 11. Commercial hydroxyapatite (HA), obtained from Taihei Chemical Industries, Osaka, Japan was used as the control sample.

2.2 Sample preparation

P123, a non-ionic surfactant was dissolved in 100 ml of distilled water, followed by the addition of 9.446 g Ca $(NO_3)_2.4H_2O$ to obtain the clear micellar solution. Then, 6.3 g of 1-dodecanethiol was added into the mixtures. The surfactant-calcium-swelling mixed solution was stirred at 60°C. In the meantime, 3.17 g of diammonium hydrogen phosphate and 1.897 g of ammonium hydrogen carbonate was dissolved in 60 ml distilled water in another beaker. The phosphate-carbonate mixture was dripped slowly into surfactant-calcium-swelling mixed solution under continuous stirring motion until a milky suspension was yielded. During the mixing process, pH of the mixture was maintained at pH 11 throughout the mixing process using sodium hydroxide (NaOH). The mixture was then aged for 24 hours at room temperature. Then, the resulted white precipitate was centrifuged and washed for five times. Next, the white precipitate was dried in an oven at 100 ^{0}C for 24 hours and ground into fine powders using mortar and pestle. Finally, the as-synthesised white powders were calcined at 550°C for 6 hours. The final product was name as CHA-60-6.3. The whole process was then repeated for 80°C mixing temperature, which donated as CHA-80-6.3. Similarly, the process repeated by using different swelling agent concentrations and control sample (without swelling agent) which designated in this article as listed in Table 1.

Table 1. Samples denotation

Sample	Concentration of swelling agent (Molarity)	Synthesize temperature (°C)
HA (control)	-	Room temperature
CHA	-	Room temperature
CHA-80-6.3	0.3 M	80°C
CHA-60-6.3	0.3 M	60°C
CHA-60-12.6	0.6 M	60°C
CHA-60-18.9	0.9 M	60°C

2.3 Characterization of sample

Rikagu Ultima IV, X-ray Diffraction (XRD) diffractometer was used to identify the phase present in the sample. Meanwhile, Fourier Transform Infrared (FTIR) analysis was performed for the identification of functional group that represent the CHA sample in the range of wave numbers of 400 cm^{-1} to 4000 cm^{-1}. Morphological surface of the sample was studied by Scanning Electron Microscope (SEM). Nitrogen adsorption-desorption isotherms were measured with a Micromeritrics TriStar. Brunauer-Emmett- Teller (BET) equation was used to measured specific surface area. Barrett-Joyner-Halenda (BJH) equations was used to calculate pore size distribution (PSD) of the samples.

3. Result and discussion

3.1 Effect of swelling agent mixing temperature on mesoporous CHA

The XRD patterns of the synthesized sample of mesoporous CHA with different swelling agent mixing temperature are shown in Figure 1. The XRD results show that the phase of synthesis powder is a pure phase of CHA with reference to PDF pattern 01-089-7834 (CHA). No characteristic peaks of intermediate phase are observed in the XRD patterns, indicating that the main inorganic phase of the sample is CHA. Three main characteristics peaks of CHA $2\theta = 25.8°$, $31.8°$ and $33.0°$ which are indexed at (002), (211) and (300), are clearly seen in all XRD patterns of CHA and HA samples. However, the characteristic diffraction peak (112) is present only in XRD spectrum of HA at $2\theta = 32.2°$. The absent of this peak on XRD patterns of CHA most probably due to the broadening of the peaks indicate that CHA has lower crystallinity than HA. Besides that, diffraction of (102) located at $2\theta = 28.1°$ is disappear from the XRD analysis on sample of CHA-80-6.3 compared to the others CHA sample.

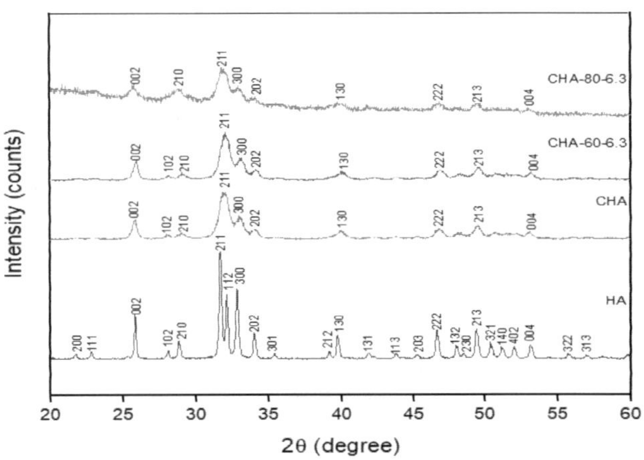

Figure 1. XRD patterns of samples synthesised using different mixing temperature

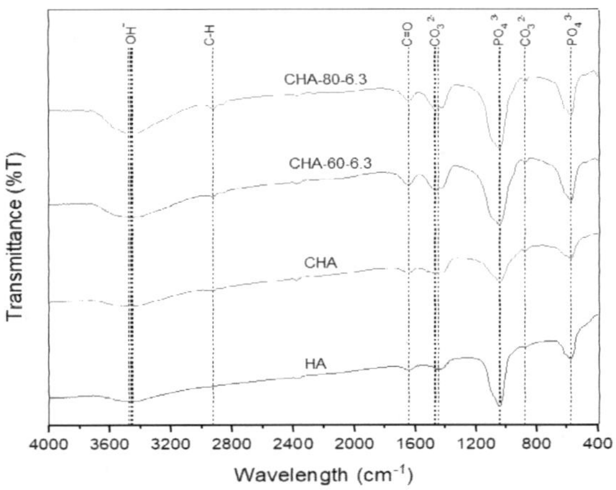

Figure 2. FTIR spectra band of the samples synthesised using different mixing temperature

The FTIR spectra of all samples are consistent (Figure 2) whereby the characteristic bands for mesoporous CHA are present in the samples. There are five functional groups presents in all mesoporous CHA samples. The bands appeared at wavelength = 3478.6 cm^{-1}, 3461.21 cm^{-1}, 3452.5 cm^{-1} and 3465.18 cm^{-1} are the characteristics for the OH^- functional group. Meanwhile, the absorption peaks for CO_3^{2-} functional group presents in two wavelength areas which are at 1450 cm^{-1} to 1410 cm^{-1} and 800 cm^{-1} to 880 cm^{-1}. The absorption peaks at 1017 cm^{-1}, 962 cm^{-1} and 567 cm^{-1}, 603 cm^{-1} contributed to the stretching vibration (V_3) and vibration (V_4) for PO_4^{3-} functional group respectively. Furthermore, C=O functional group appeared in the sample at wavelength ranging from 1640 cm^{-1} to 1690 cm^{-1}. Lastly, at wavelength of 2850 cm^{-1} to 3000 cm^{-1} which is for C-H stretching band presents in the sample.

The morphology of mesoporous CHA and the HA was observed by SEM as shown in Figure 3. The SEM images revealed that the CHA powders are consists of very fine agglomerated particles, regardless of the mixing temperatures used. All the CHA particles are in irregular shape that formed fine agglomerates with some hollow spaces between them. Meanwhile, the HA particles demonstrated a rod-like structure that look like rice grains.

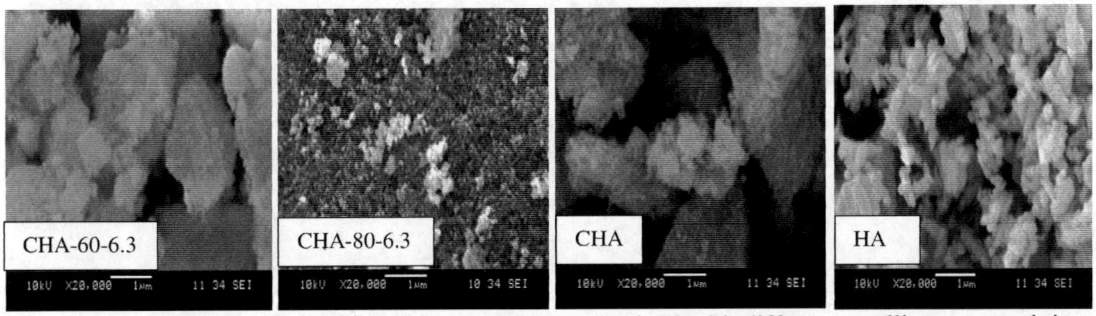

Figure 3. Morphological images of CHA samples synthesised with different swelling agent mixing temperature.

Figure 4 shows the nitrogen adsorption–desorption isotherms and the corresponding Barrett–Joyner–Halenda (BJH) pore size distribution. All mesoporous CHA samples exhibited type IV character with H1 hysteresis loop. An obvious step could be seen at P/Po= 0.8 – 1.0. The result obtained indicates that CHA-60-6.3 sample has the highest sorption properties among the samples followed by CHA, CHA-80-6.3 and HA. The isotherm curve of HA is identified as Type II isotherm, which is the common form isotherms of non-porous materials.

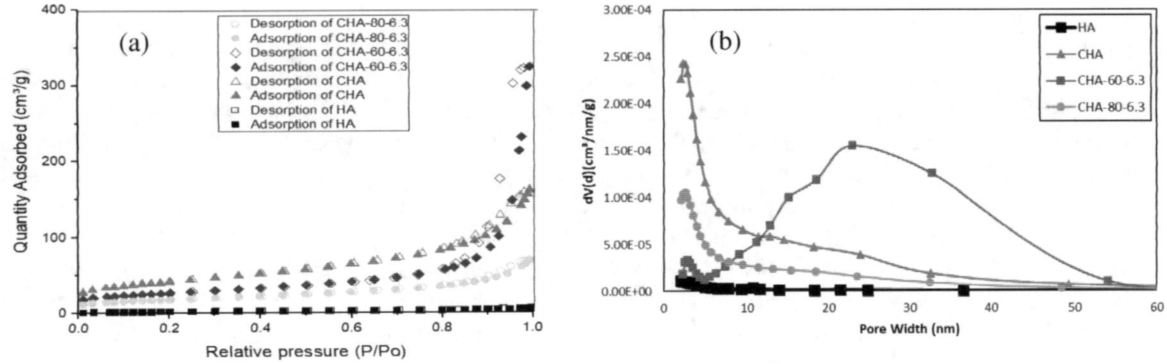

Figure 4. (a) Nitrogen adsorption–desorption isotherms and (b) pore size distribution of CHA-80-6.3, CHA-60-6.3, CHA and HA.

Table 2 presented the pore properties of mesoporous CHA samples with different synthesis temperature. The mesoporous CHA that synthesised without swelling agent have the highest BET surface area which is 146.9 m^2/g compared to the other samples. However, the surface area decreased as the 1-dodecanethiol was added during the synthesis process as shown by the BET surface area of CHA-80-6.3 and CHA-60-6.3 (62.7 m^2/g and 92.8 m^2/g). Lower mixing temperature produce mesoporous CHA with higher surface area, pore size and pore volume compared to higher mixing temperature. Increasing the synthesis temperature is expected to increase the pore properties of the sample since the hydrophobic molecule are expected to interact with the surfactant. However, in this study the increasing of mixing temperature could not promote bigger pore properties of the samples. As stated in previous studies, the interaction between these molecules could be enhanced by having other controllable parameters such as basic pH conditions and fixed amount of NaOH solution [10-12].

Table 2. The surface area, pore volume, pore size of mesoporous CHA samples

Sample	BET surface area (m^2/g)	BJH pore size (nm)	BJH pore volume (cm^3/g)
HA	7.8	2.0	0.008
CHA	146.9	2.3	0.244
CHA-80-6.3	62.7	2.6	0.105
CHA-60-6.3	92.8	22.9	0.502
CHA-60-12.6	104.5	24.4	0.530
CHA-60-18.9	108.6	22.9	0.525

3.2 Effect of swelling agent concentration on mesoporous CHA

The samples prepared with different concentration of swelling agent were characterized using XRD. As shown in Figure 5, the XRD of CHA-60-6.3, CHA-60-12.6 and CHA-60-18.9 show diffraction peaks similar to CHA (PDF#01-089-7834), indicating that pure phase had successfully synthesised. No characteristic peaks of intermediate phase observed diffraction pattern indicating that the main inorganic phase of the sample is CHA. XRD pattern of CHA samples are compared to the HA in order to further confirmed the formation of pure CHA. Three main characteristic peaks of CHA at $2\theta = 25.8^o$, 31.8^o and 33.0^o which can be indexes at (002), (211) and (300), are clearly seen in all XRD patterns of CHA and HA samples. However, the characteristic diffraction peak (112) is present only in XRD spectrum of HA at $2\theta = 32.2^o$. The absent of the peak characteristic on XRD pattern of CHA samples most probably due to the broadening of the sample. Besides, the present of P123 surfactant during synthesis process might disturbed the long-range order of the crystal structure.

Furthermore, the FTIR spectroscopy of the CHA-60-6.3, CHA-60-12.6 and CHA-60-18.9 all was investigated (Figure 6). As shown in Figure 6, the bands appeared at wavelength = 3465.18 cm^{-1}, 3478.6 cm^{-1}, 3461.21 cm^{-1} and 3457.33 cm^{-1} are the characteristic band for the OH^- functional group where OH^- stretch presents at 3200 cm^{-1} to 3500 cm^{-1}. Meanwhile, the absorption peaks for CO_3^{2-} functional group presents at 1450 cm^{-1} to 1410 cm^{-1} and 800 cm^{-1} to 880 cm^{-1} where all samples have CO_3^{2-} functional group as in Figure 6. The absorption peaks at 1017 cm^{-1}, 962 cm^{-1} and 567 cm^{-1}, 603 cm^{-1} are the stretching vibration (V_3) and vibration (V_4) for PO_4^{3-} functional group respectively. Furthermore, there was also C=O functional group appeared in the sample at wavelength ranging from 1640 cm^{-1} to 1690 cm^{-1}. Lastly, at wavelength of 2850 cm^{-1} to 3000 cm^{-1} which is for C-H stretching band presents in the sample. All functional groups appeared in the samples are compared from several reported works [10-16].

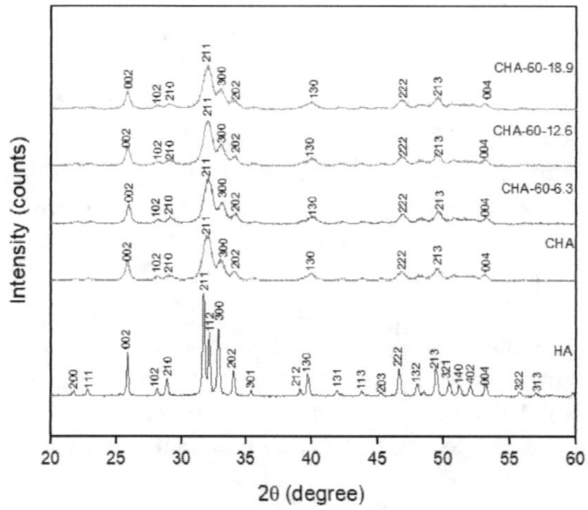

Figure 5. XRD patterns of samples synthesised using different swelling agent concentration

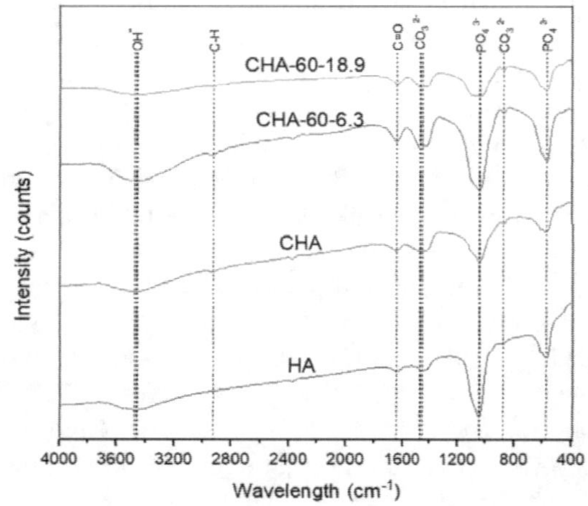

Figure 6. FTIR spectra band of the samples synthesised using swelling agent concentration

Figure 7 shows the morphological structure of HA, CHA, CHA-60-6.3 and CHA-60-18.9. Cylindrical particles are observed in the control HA sample while control CHA sample has an irregular shape. Close examination on the CHA sample showed the shapes were in transition process to a sphere shape micelle. Meanwhile, by adding swelling agent to the CHA samples, there is significant changes in the morphology of the samples. Samples synthesised with swelling agent has higher agglomeration compared to the control CHA sample. CHA particles become more agglomerated when concentration of swelling agent increase. The increment of the swelling agent concentration lead to the further diffusion of hydrophobic element of the swelling agent into hydrophobic section of the surfactant as stated in previous study[10]. Thus, the swelling process is occurred in the samples and foamy structured micelle could be obtained with further increment of swelling agent concentration.

Figure 7. Morphological images of CHA samples synthesised with different swelling agent concentration

The nitrogen adsorption-desorption isotherm of CHA samples with different concentration of swelling agent are shown in Figure 8 (a). All CHA samples exhibited a Type IV character with H1 hysteresis loop where H1 hysteresis loop is associated with porous materials consists of agglomerates or compacts of approximately uniform spheres in a fairly regular array. It can obviously be seen that all CHA samples synthesized with swelling agent has higher distribution of adsorption-desorption properties when compared to CHA and HA sample. The result obtained indicates that CHA samples with swelling agent has the high sorption properties among the other sample. An obvious step could be seen at $P/P_0= 0.8 – 1.0$. The isotherm curve of HA is identified as Type II isotherm, which is the common form isotherms of non-porous materials. Figure 8(b) presented the corresponding BJH pore size distribution (PSD) graph for mesoporous CHA samples with different concentration of swelling agent. Basically CHA samples synthesised with swelling agent have similar pore size distribution pattern between 2.0 to 50 nm . The pore sizes for CHA-60-6.3 are mainly distributed at 22.9 nm, CHA-60-12.6 at 24.4 nm and CHA-60-18.9 at 22.9 nm. CHA are mainly distributed at 2.3 nm and HA are at 2.0 nm. CHA-60-12.6 has the highest peak, while HA sample has the lowest pore size distribution since it is non-porous material.

As shown in Table 2, BET surface area of the samples increase with the increase of concentration of swelling agent. BJH pore size and BJH pore volume increased when concentration of swelling agent is twice (0.6 M) of the initial concentration (0.3 M). Meawhile, the pore size and volume slightly decrease while the swelling agent concentration is three times (0.9 M) of the initial concentration. The introduction of swelling agent is expected to promote higher degree of swelling in the micelle structures. The hydrophobic molecules of the swelling agent will interact with surfactant molecules to initiate swelling process. Higher degree of pore expansion could be obtained by introducing more concentrated swelling agent. However, with further increase of the swelling agent from 0.6 M to 0.9 M, the pore size and pore volume suddenly decrease. This reduction might be due to the excessive amount of swelling agent introduced [10] into the mixture of the solution.

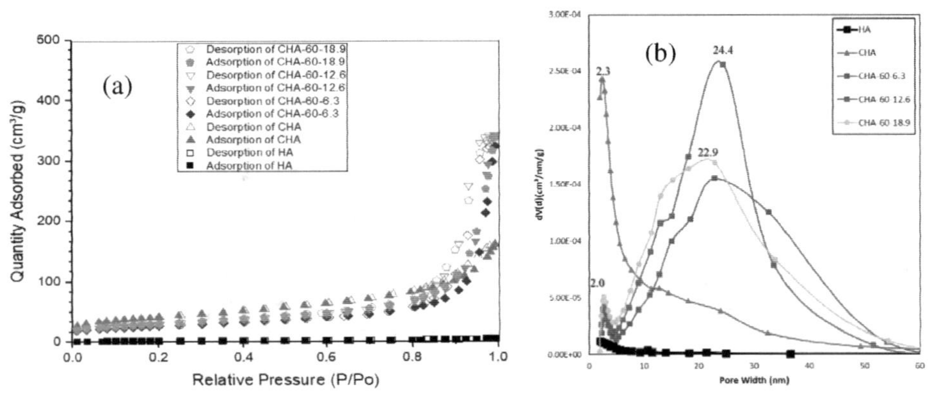

Figure 8. (a) Nitrogen adsorption–desorption isotherms and (b) pore size distribution of CHA-60-6.3, CHA-60-12.6, CHA-60-18.9, CHA and HA.

4. Conclusion

Mesoporous CHA samples have been successfully synthesized by chemical precipitation method. 1-dodecanethiol is utilized as swelling agent in order to study its effect on pore characteristics of mesoporous CHA sample. From the result obtained, swelling agent do affect the surface area and pore size of the mesoporous CHA sample. Introduction of swelling agent drastically reduced the surface area of mesoporous CHA. Therefore, to counter this drawback, 0.6 M of swelling agent should be mixed at 60°C to obtain much higher surface area (104 m^2/g) and pore size about 20-24 nm.

5. References

[1] Ooi C H 2014 Urea adsorption by activated carbon derived from oil palm kernel shell and empty fruit bunch fiber. (Penang: Universiti Sains Malaysia) p 157

[2] Krawczyk H 2009 Production of uremic toxin methylguanidine from creatinine via creatol on activated carbon *Journal of Pharmaceutical and Biomedical Analysis* **49** 945-9

[3] Mitome T, Uchida Y, Egashira Y, Hayashi K, Nishiura A and Nishiyama N 2013 Adsorption of indole on KOH-activated mesoporous carbon *Colloids and Surfaces A: Physicochemical and Engineering Aspects* **424** 89-95

[4] Howell C A, Sandeman S R, Zheng Y, Mikhalovsky S V, Nikolaev V G, Sakhno L A and Snezhkova E A 2016 New dextran coated activated carbons for medical use *CARBON* **97** 134-46

[5] Raharjo Y, Ismail A F, Othman M H D, Malek N A N N and Santoso D 2019 Preparation and characterization of imprinted zeolite-Y for p-cresol removal in haemodialysis *Materials Science and Engineering: C* **103** 109722

[6] Cheng Y-c, Fu C-c, Hsiao Y-s, Chien C-c and Juang R-s 2018 Clearance of low molecular-weight uremic toxins p -cresol , creatinine , and urea from simulated serum by adsorption *Journal of Molecular Liquids* **252** 203-10

[7] De Smet R, David F, Sandra P, Van Kaer J, Lesaffer G, Dhondt A, Lameire N and Vanholder R 1998 A sensitive HPLC method for the quantification of free and total p-cresol in patients with chronic renal failure *Clinica Chimica Acta* **278** 1-21

[8] Mohammad N F, Othman R, Abdullah A A and Yeoh F Y 2018 Effect of surfactant concentrations on pore characteristics of mesoporous carbonated hydroxyapatite prepared by soft-templating hydrothermal method *International Journal of Nanoelectronics and Materials* **11** 59-70

[9] Guo Y J, Long T, Chen W, Ning C Q, Zhu Z A and Guo Y P 2013 Bactericidal property and biocompatibility of gentamicin-loaded mesoporous carbonated hydroxyapatite microspheres *Materials Science and Engineering C* **33** 3583-91

[10] Bakhtiari L, Javadpour J, Rezaie H R, Erfan M and Shokrgozar M A 2015 The effect of swelling agent on the pore characteristics of mesoporous hydroxyapatite nanoparticles *Progress in Natural Science: Materials International* **25** 185-90

[11] Bakhtiari L, Rezaie H R, Javadpour J, Erfan M and Shokrgozar M A 2015 The effect of synthesis parameters on the geometry and dimensions of mesoporous hydroxyapatite nanoparticles in the presence of 1-dodecanethiol as a pore expander *Materials Science & Engineering C* **53** 1-6

[12] Bakhtiari L, Javadpour J, Rezaie H R, Erfan M, Mazinani B and Aminian A 2016 Pore size control in the synthesis of hydroxyapatite nanoparticles: The effect of pore expander content and the synthesis temperature *Ceramics International 42 (2016) 11259–11264*

[13] Ooi C H, Ling Y P, Pung S Y and Yeoh F Y 2019 Mesoporous hydroxyapatite derived from surfactant-templating system for p-Cresol adsorption: Physicochemical properties, formation process and adsorption performance *Powder Technology* **342** 725-34

[14] Mohammad N F, Othman R, Asma Abdullah N and Yee Yeoh F 2016 In vitro Evaluation of Mesoporous Carbonated Hydroxyapatite in MC3T3-E1 Osteoblast Cells *Procedia Chemistry* **19** 259-66

[15] Mohammad N F, Yeoh F-y and Othman R 2012 Synthesis & Characterization of Mesoporous Hydroxyapatite by Co- Synthesis & Characterization of Mesoporous Hydroxyapatite by Co-Precipitation Method

[16] Mohammad N F, Othman R and Yee Yeoh F 2015 Controlling the pore characteristics of mesoporous apatite materials_ Hydroxyapatite and carbonate apatite *Ceramics International 41 (2015) 10624–10633*

The Characterization Of Bario Rice Starch-nanoHA Scaffolds using SEM and Dielectric Measurement

M.Riza Roslan[2], N.F. Mohd Nasir[1,2], M.F. Abdul Khalid [1], E.M Cheng[1,2], N.A.M. Amin[3] , N. Mohammad[1], C.Y. Beh[1] and M.S.A. Majid[3]
[1]Microwave Research Institute (MRI), Universiti Teknologi Mara, Shah Alam, Malaysia
[2]Biomedical Electronic Engineering Program, School of Mechatronic Engineering, Universiti Malaysia Perlis, Perlis, Malaysia
[3]Mechanical Engineering Program, School of Mechatronic Engineering, Universiti Malaysia Perlis, Perlis, Malaysia

[a]Corresponding author: nashrul@unimap.edu.my

Abstract. Bone tissue scaffold had been ventured for over the decades as a solution for bone failure and trauma. Starch is the most common natural polymer that has been used as a biomaterials and a good candidate for scaffold fabrication as there are many resources of starch in Malaysia. Native starches may contribute differently in their structure in terms of the amylose content, interactions between granules, swelling ability and solubility in which those differences can be mainly due to the botanical origin. In Malaysia, there are various resources of rice starches. Here, the Bario rice starch from Sarawak was used to fabricate the scaffold via solvent casting and salt leaching technique. Bario rice is usually collected from various sources and variety which may contribute to different physical affect in comparison of using other type of rice. From the study, the morphologies and microstructures of the scaffold were evaluated using Scanning Electron Microscopic (SEM) and this showed no trend of pore size increment when starch percentage varied. Then, the dielectric properties were obtained via the values of dielectric constant and dielectric loss which are another type of indirect measurement to study the porosity of the scaffolds. The porosity has the value of the dielectric constant and the loss air matrix whereby this air matrix is represented by the pore in the scaffold. Here, the results were consistent with the observation made through SEM. Thus, the relationship between the porosity and the dielectric properties of the Bario Rice Starch-nHA tissue scaffolds had been established and there were no particular pattern could be concluded by using Bario rice for this scaffold.

1. Introduction

Hydroxyapatite is known to have similar chemical composition, mineral phase, and crystallinity of natural bone [1] which majorly exhibit high osteoconduction and osteoinduction when implanted in the human body [2]. Unfortunately, the nature of the hydroxyapatite is brittle and low in mechanical strengths which limit its application [3]. Besides that, hydroxyapatite alone is not mechanically competent to support various applications in biomedical engineering [3], thus various types of materials were used to fabricate tissue engineering scaffolds [4-7].

The starch is recognized to have favourable criteria such as biocompatible, low in cost, good biodegradable term, and non-toxic [8, 9]. Biopolymers formulated from starch have been widely applied in biomedical application such as for bone replacement, bone cement, and bone tissue engineering [9, 10]. Charoenrein et al. stated that different starches would exhibit different criteria and in order to improve the textural qualities of starch, the incorporation of starch with different botanical sources had been performed [11]. The differences in starch behaviour is mainly due to the concentration, mixing ratio, amylose content, interactions between granules, swelling power and solubility, and granule characteristics [12]. Considering the Sarawakian rice has numerous varieties and differs in term of botanical resource, it had been used as reinforcement to hydroxyapatite for bone tissue scaffold. Bario rice which is well known originated from Sarawak was chosen to be used in this study. The properties of the newly fabricated scaffolds were characterized in detail including their mechanical and electrical properties. Few dielectric properties studies were conducted on various types of starches including tapioca, corn, wheat, rice, waxymaize, amylomaize and also basmathi rice but none are performed especially on the scaffold fabricated from starch [13]. Thus, it is hope that by venturing the dielectric properties of scaffold to analyze its porosity may contribute to the new method in examining the porosity in scaffolds.

2. Materials and Methods

The Bario rice used was purchased from a market in Sarawak. Hydroxyapatite and 50% of glutaraldehyde was supplied by Sigma Aldrich and sodium chloride used was also purchased from a local market.

2.1. Scaffold Fabrications

Figure 1 shows the step to fabricate the Bario rice starch-hydroxyapatite (HA) scaffold. Five different ratio of Bario rice starch-nHA scaffolds were prepared which are 50:50(wt%), 60:40(wt%), 70:30(wt%), 80:20(wt%) and 90:10(wt%) were fabricated respectively. At first, the starch powder was mixed with distilled water to form the starch solution. Then, the porogen of NaCl was added into the starch solution followed by the addition of hydroxyapatite solution and stirred for 3 minutes. The slurry solution obtained was then casted onto the Teflon mold with dimension of 25mm x 15mm x 15mm. The next process was drying the slurry in the mold for 48 hours at 60oC. After the drying process, the scaffolds were soaked in 25% glutaraldehyde for 5 hours to enhance the structure of the scaffold and followed by immersion in distilled water to remove the glutaraldehyde. Lastly, the scaffolds were dried again at room temperature before characterizations proceed.

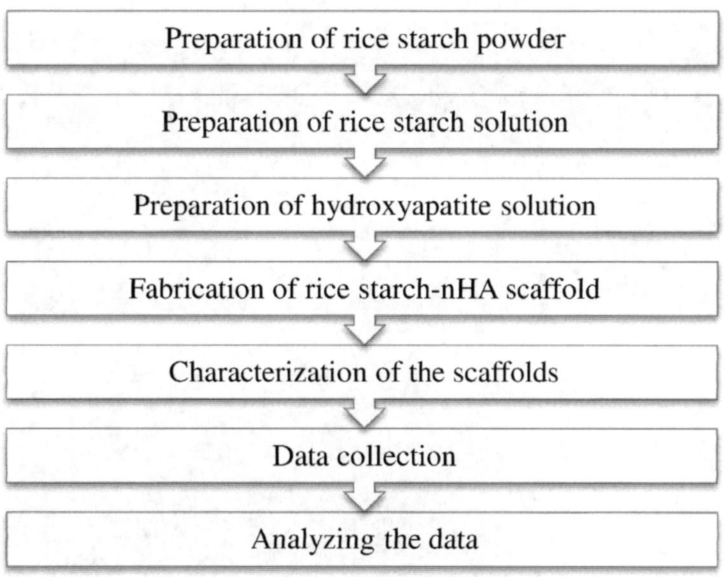

Figure 1: Overall methodology

2.2. Scanning Electron Microscopy (SEM)

The samples for Scanning Electron Microscopy (SEM) were cut symmetrically to observe the scaffold's porosity at the inner part and coated with platinum. The SEM model used was the Table Top TM3000, Hitachi with the excitation voltage of 15kV which was bombarded upon the samples. The magnifications used were about 100X to 300X.

2.3. Dielectric Measurement

The measurement of the dielectric properties was based on the transmission line method by using Agilent E8362B Performance Network Analyzer (PNA) to obtained the dielectric constant ε' and dielectric loss ε". The samples were prepared with the dimension of 8 mm x 16 mm x 5 mm which was basically based on the dimension of WR62 waveguide dielectric holder. After setting up the measurement, the frequency was set in the range of 12.4GHz to 18.0GHz. Then, the open air condition measurement was taken without the sample. Only then, the sample was placed in the WR62 waveguide holder with the wave guide adapter and the measurements were taken in triplicate.

3. Results and Discussion

3.1. Scanning Electron Microscopy (SEM)

For the SEM analysis, the pore size obtained from the 50wt%, 60wt%, and 70wt% of Bario rice starch-nHA composite scaffolds are 246μm, 282μm, and 163μm respectively as shown in Figure 2. This trend was observed for all the Bario rice starch-nHA composite scaffolds. Usually for other starch based scaffolds, the pore size would increase by increasing the starch percentages [7]. However, the Bario rice starch-nHA composite scaffolds did not show this particular trend. The high amount of starch contributes to the ruptures in scaffolds' structure where Ahmed et al., claimed that such high starch percentages would lead to scaffold's cracking and rupture [13].

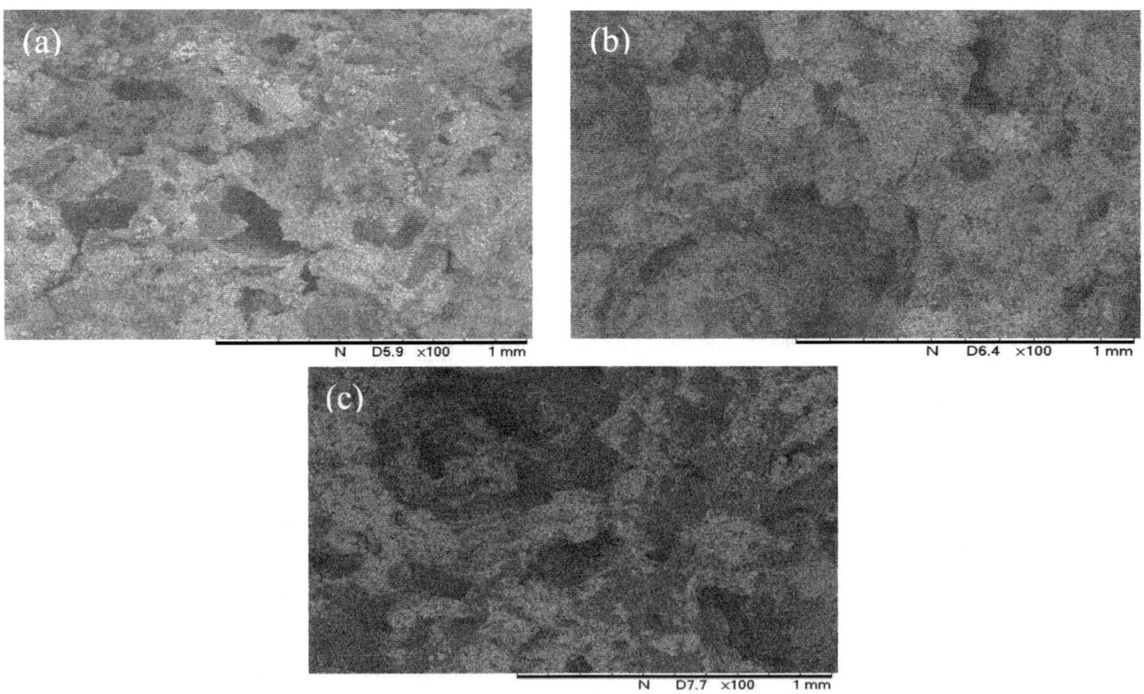

Figure 2: a) 50wt% Bario rice starch, b) 60wt% Bario rice starch, c) 70wt% Bario rice starch scaffolds

Collectively, the pores' size from 50wt% to 70wt% of starch percentage for Bario rice starch-nHA scaffolds would impose the range of pore size in between 10μm to 400μm which suitable to furnish the nutrient and osteoblast cellular infusion while preserving the structure of the scaffolds [14]. Besides, the mean pores size between 20μm to 1500μm is desirable for tissue engineering purpose [15]. But, most of the studies recommend that a pore size larger than 300 μm is suitable for the formation of new bone and vascularization within the scaffold since this size could facilitate capillary formation and able to guide direct osteogenesis [16]. At the same time, stimulation of osteochondral ossification might occur at the pore size smaller than 300μm. Cell seeding occur at average pore size with diameter of 20-125μm [17].

3.2. Dielectric Measurement

The measurement of the dielectric properties was based on transmission line method by using Performance Network Analyzer (PNA) to obtained the dielectric constant ε' and dielectric loss ε". Figure 3 shows the graphs of dielectric constant for 50wt%, 60wt%, and 70wt% Bario rice starch percentages.

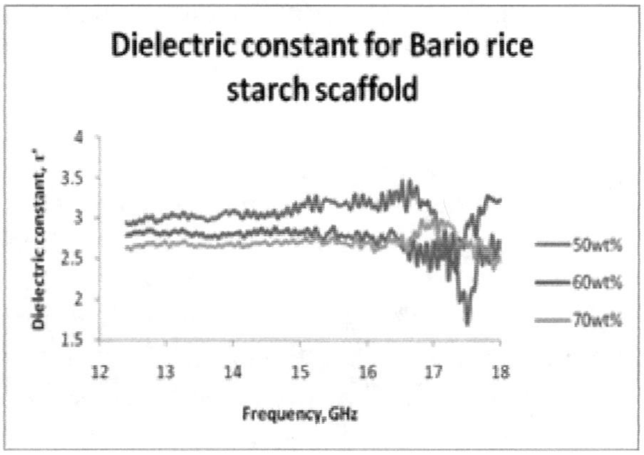

Figure 3: Dielectric constant

Figure 4 shows the graphs of dielectric loss for 50wt%, 60wt%, and 70wt% Bario rice starch percentages. The value of dielectric constant and dielectric loss remained constant but at high frequency the dielectric constant and the dielectric loss is not stable and drastically dropped. This phenomenon is due to the energy absorption caused by phase lag between the dipole rotation and the field [18].

From the Figure 3, it is noticed that at the rice starch percentage 50wt% and 60wt% of Bario rice starch-nHA scaffolds has the highest dielectric constant the dielectric constant are reduced as increasing the rice starch percentage 50wt% to 70wt%. Ahmed et al. also reported that the dielectric constant and the dielectric loss will be reduced while increasing the concentration of the Indian Basmati rice flour slurry [19]. Besides, Ndife *et al.* agreed that by increasing the starch concentration, the dielectric constant and the dielectric loss value will be diminished [20]. However, this trend is not observed with the Bario rice starch-nHA scaffolds. The dielectric constant is highly affected by the water content or the moisture content [21] as water is a main polar component liable for any changes in the dielectric properties of a material [21].

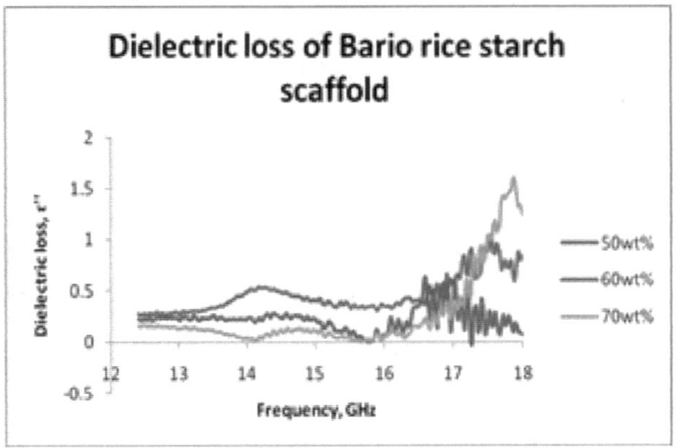

Figure 4: Dielectric loss

Higher water content or moisture content may increase the mobility of the water molecule or the polar molecule that couple the electromagnetic energy portrayed by the larger value of the dielectric properties. Consequently, the 50wt% has a higher dielectric constant and dielectric loss compared to the 60wt% and the 70wt% as increasing starch percentages may reduce the water content [19], thereby contributing to lesser mobility of water molecules in the system. It is suggested that the higher starch percentages may induced higher porosity and this should be indicated by the dielectric properties of the air matrix [22] where the dielectric constant for air is 1 and the dielectric loss is 0 [23] but this is not shown here in the results.

Nevertheless, there is inconsistency trend in the dielectric loss value for the Bario rice starch-nHA scaffolds in correspondent to the porosity as the starch percentages increased. This might due to the structural shrinkage of the scaffolds. It is suggested that this could affect the fixation of the scaffold into the waveguide holder during the experimental set up, therefore influencing the dielectric loss value. As mentioned earlier, the resource of Bario rice starch which might come from several type of rice which might affect the results. Hence, the porosity of the Bario rice starch scaffold cannot be controlled and thus, affected the dielectric properties as well.

4. Conclusion

Native rice starch is a potential material to fabricate the bone tissue scaffold. Here, dielectric properties characterizations were used as indirect measurement techniques to determine the porosity of the scaffolds. The characterization of dielectric for porosity analysis shows that the porosity behave as dielectric properties of air matrix whereby, initially the porosity is claimed as the air matrix. Thereby, dielectric measurement for porosity analysis may lay new preliminary study for porosity in bone tissue scaffold. For the SEM analysis, it is showed that by increasing the rice starch percentage, the porosity distributed on the scaffold will increase increase but this did not occur for the 70wt% starch percentage. It is realised that the Bario rice starch-nHA scaffolds, there is an increment from 50wt% to 60wt% starch percentages but decrement started to occur thereafter. Thus, it is interesting to discover this peculiarity which could be explained further, perhaps by understanding the biochemistry interactions happened at the molecular level of these scaffolds composition.

5. Acknowledgement

The author would like to acknowledge the support from Geran Bestari Perdana under a grant number of 600-IRMI/PERDANA 5/3 BESTARI (092/2018) from the Universiti Teknologi Mara.

6. References

[1] Mcnamara S L, Rnjak-kovacina J, Schmidt D F, Lo T J and Kaplan D L 2014 Biomaterials **35**, 6941
[2] Elayaraja K, Arul K T, Ferraris S, Spriano S, Ferraris M, Asokan K, Kalkura S N 2015 Ceram. Int. **41,** 13153
[3] Cipreste M F and Sousa E M B 2014 J Biomater Nanobiotechnol **January,** 24
[4] Nasir N F M, Khan S F, Baling J S, Roslan M R, Cheng E M, Amin N A M, Abdullah M N 2018 J. Eng. Res.& Edu. **10**, 83
[5] Razali K R, Nasir N F M, Cheng E M, Mamat N, Mazalan M, Wahab Y and Roslan M M 2014 2014 IEEE Conference on Biomedical Engineering and Sciences (IECBES) (Miri:IEEE) p 256
[6] Normahira M, Raimi, R K, Nasir F M N, Norazian A R and Adilah H 2014 Advanced Materials Research 970 (Trans Tech Publications) p 3
[7] Roslan M R, Nasir N F M, Cheng E M and Mamat N 2016 International Conference on Electrical, Electronics, and Optimization Techniques (ICEEOT) (Chennai: IEEE) p 1560
[8] Xie F, Pollet E, Halley P J and L. Avérous L 2013 Prog. Polym. Sci. **38,** 1590
[9] Roslan M R, Nasir N F M, Cheng E M and Amin N A M 2016 International Conference on Electrical, Electronics, and Optimization Techniques (ICEEOT) (Chennai: IEEE) p 1857
[10] Sundaram J, Durance T D and Wang R 2008 Acta Biomater. **4,** 932
[11] Charoenrein S and Preechathammawong N 2012 Carbohydr. Polym. **90,** 1032
[12] Seetapan N, Limparyoon N, Gamonpilas C, Methacanon P and Fuongfuchat A 2015 J. Food Eng, **151**, 51
[13] Ahmed Y M Z, Ewais E M M and El-sheikh S M 2014 Integr. Med. Res.
[14] Woodard J R et al. 2007 Biomaterials **28,** 45
[15] Xin H M, Max J, Peng T M CX and Turng L 2014 J. Mater. Sci. **49,** 2324
[16] Zeng F, Ma F, Kong F, Gao Q and Yu S 2015 Food Chem.**172,** 92
[17] Wani A A, Singh P, Shah M A, Schweiggert-weisz U, Gul K, and Wani I A 2016 Compr Rev Food Sci F. **11**, 417
[18] Piyasena P, Ramaswamy H S, Awuah G B and Defelice C 2002 J. Food Process Eng. **26**, 93
[19] Ahmed J, Ramaswamy H S and Raghavan V G S 2007 J. Food Eng. **80**,1125
[20] Ndife L, Sumnu M K , and Bayindirli 1998 Food Res. Int. **31**, 143
[21] Motwani T, Seetharaman K, and Anantheswaran R C 2007 Carbohydr Polym. **67**, 73
[22] H I and Agag T 2011 Handbook of benzoxazine resins (Amsterdam: Elsevier)
[23] Sun D W 2012 Thermal food processing: new technologies and quality issues (CRC Press)

Implementing a Brain-Computer Interface Wheelchair in Home-Care Setting: Preliminary Result

Kamala Krishnan[1*], Tan Lee Fan[1], Danny Ng Wee Kiat[1], Goh Sing Yau[1]

[1]Faculty of Engineering and Science, Universiti Tunku Abdul Rahman, Kajang, Malaysia.

*kamalakrishnan497@gmail.com

Abstract. This main aim of this study is to implement the brain-computer interface wheelchair in a home-care setting for the occupant's benefit. Two patients suffered from motor neuron diseases (MND) and were recruited from local elderly home care. The wheelchair was set-up and test trials were conducted prior to the data collection. All the challenges faced during the process of implementation were recorded. Solutions were suggested accordingly.

1. Introduction

Brain-Computer Interface (BCI) is a system which acquires and analyzes the brain signals and translates them into commands which later will be transferred into the desired action by an output device [8]. BCI wheelchair research records a rapid growth in the last two decades, however, studies with actual patients are limited compared with healthy subjects towards a new modality-independent interface for a robotic wheelchair [1].

Researchers studied applying the BCI system on patients suffering from chronic diseases [2]. Motor Neuron Disease (MND) is a collective term for a progressive neurological disorder that affects efferent neurons which control the voluntary muscles [4]. These MND sufferers will experience gradual atrophy and weakness of muscles which cause body stiffness leaving optic muscle movement control as last to be damaged [5]. Few types of MND include Amyotrophic Lateral Sclerosis (ALS), spinal cord injury, multiple sclerosis, and stroke. Approximately, there are greater than hundred million motor neuron diseased patients around the globe could be benefited from the BCI technology, with many affected with stroke [2].

Statistics showed that the incidence of MND worldwide is found 2/100,000 cases annually while there are about 600 new cases reported each year in Malaysia [6]. Almost 50% of MND patients are unable to control an electric wheelchair through a common method which is wheeling by hands and walking with the wheelchair by feet. Particularly, these patients are only able to exercise their willpower in the presence of vision and brain signals [3].

BCIs has been developed as neurorehabilitation tools for paralysis patients with an aim to improve their quality-of-life (QoL) and to allow them to live in a normal and comfortable life [7,8]. However, there is a lack of BCI study in the home-care setting. Thus, the present study intends to implement the BCI-wheelchair in the home-care setting for the occupant's benefit. However, there were many challenges encountered during the process of implementation. All the challenges were identified and addressed. The potential technical solutions found in this study may contribute to the future problem-solving in the BCI research.

2. Materials and Methods

2.1 Participants

This case study obtained an ethical approval both from Universiti Tunku Abdul Rahman (UTAR) Scientific and Ethical Review Committee and Jabatan Kebajikan Masyarakat Malaysia (JKMM) respectively. Two subjects were recruited from JKMM and informed consent is obtained prior to data collection. The inclusion criteria were the mentally stable patients who suffer from motor neuron disease and wheelchair bound with voluntary participation.

2.2 Equipment

2.2.1 Steady-State Visual Evoked Potential (SSVEP) and Electroencephalogram (EEG) Electrode. SSVEP is used to record a person's visual frequency with the aid of the EEG electrode. Three standard 10mm gold-plated EEG electrodes are placed on the participant's scalp: Oz as the signal channel, A1 (left mastoid) and A2 (right mastoid) as reference and ground channels, respectively. Initially, pre-test practice is conducted and followed by the selection practice. The stimulus frequency tested were: 7, 8, 11, 12, 13, 14, and 15 Hz.

2.2.2 SSVEP Test. SSVEP test is conducted using 2 types of selection modes which are (1) SSVEP PreTest and (2) SSVEP Selection. First, a brief on SSVEP PreTest, it starts with 5 seconds of rest/ black screen and 5 seconds of flashing and was repeated for 7 times. The test lasts for more than 1 minute. Next, is the SSVEP Selection mode. In this mode, there will be 6 flickering boxes appear at the same time with an arrow pointing to one of the boxes. The end user is required to focus at the pointed box.

2.2.3 Wheelchair Navigation. An autonomous wheelchair is used in this study. Each of the stimuli or flashing box represent a pre-set location which aids in wheelchair mobility. The navigation software will drive the wheelchair to the destination once it is selected. The top laser, SICK NAV350 laser navigator is used for localization and navigation purposes. NAV350 is capable of localizing its current location through the use of reflectors which serves as fixed landmarks [13]. In this study, a total of 32 reflectors (Figure 1) were placed along the navigation pathway. 32 reflectors were pasted on the pillars along the navigation pathway. The reflectors were pasted at the height of 158cm from the base or ground. Length and width of each reflector is 37cm and 8.5cm respectively. The path planning and navigation can be conducted with a predefined map and its known current location. A second laser which is the front laser is used for more accurate measurement of the surrounding obstacle(s). The obstacles can be living or non-living things. Once an obstacle is sensed, the wheelchair will stop

Journal of Physics: Conference Series **1372** (2019) 012020 doi:10.1088/1742-6596/1372/1/012020

immediately and only moves after the obstacles have been removed. The wheelchair preset navigation route (Figure 2) is set for 46.0m (one-way) and a total of 92.0m (two-way).

Figure 1. Reflectors pasted on the pillars along the wheelchair route.

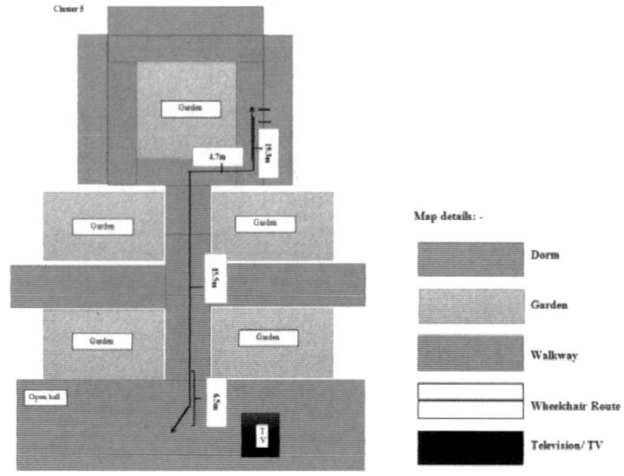

Figure 2. Wheelchair Route Map.

2.3 Experiment Setup

The experiment was conducted with the participant sat on their wheelchair (Figure3). The laptop screen was closely aligned in a range of 30-50cm from the subject's eye level. The trial session was allowed for the subjects to familiarize with the system. Before each trial session, instructions were given in the verbal form.

To be specific, the actual study was intended to be conducted as: The subject will be instructed to select their preset destination just by looking at the flickering box displayed on the selection menu (Figure 4). Each of the flickering boxes represents a preset location which aids in the BCI wheelchair mobility. The selection confirmation will appear as in a binary choice of flickering boxes which are "Yes" and "No". Once the selected destination is confirmed, the wheelchair will drive them. There is no exact preset duration for steady-state visual evoked potential (SSVEP) selection using an EEG electrode as it solely depends on the users' will to move around to the preset destination. Furthermore, once the subject makes a selection, the wheelchair will start to move and the subject is free to rest, sightseeing or do any other activities of their interest until they reach their destination.

The BCI wheelchair was tested without the end user prior several times in order to check the safety, control, and its smooth running.

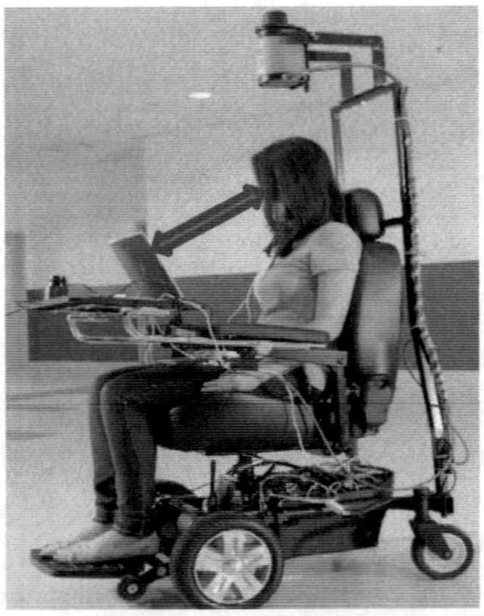

Figure 3. Model sitting on a BCI wheelchair. The arrow signifies the distance between the model's eye and the laptop screen.

2.4 Data Collection

The challenges faced in terms of subjective, environmental and technical problems were addressed from a wide angle consist of the subject, researchers and the laptop or system. Verbal feedback was collected in terms of the face-to-face interview using pen and paper method.

3. Result and Discussion

3.1 Steady-state visual evoked potential (SSVEP) Response

Two elderly subjects with MND were recruited in this study. Their prominent frequencies recorded were: Participant A: 8 and 11 Hz and Participant B: 7, 8, 11, and, 12 Hz. Remaining other frequency hits were 0 respectively. These were weak brain signals recorded as true positive hits for at least once for each subject. A sum of more than 10 trial sessions was conducted for both pre-test and selection practice. These test trial were conducted with the SSVEP selection without the wheelchair. In this test, the subject has experienced the way to select when using a BCI wheelchair. In this trial, both subjects were unable to select the direction they want to move to for the first few attempts. The rest of their attempts land them to the wrong selection. Out of 10 attempts, Participant A has none correctly selected attempt while for Participant B has 2 correctly selected attempts. Many failed attempts made them tired and bored, these were expressed by the subjects after the attempts. As a matter of fact, for smooth control of EEG-based BCI wheelchair, there should be continuous sensorimotor frequency produced in order to select the wheelchair direction. An EEG-based BCI wheelchair does not ease a control to both elderly subjects in this study. Therefore, an alternate device is preferred for a smooth BCI wheelchair control.

3.2 Challenges/ Issues in Wheelchair Implementation

This current research studies on the challenges faced in implementing a BCI wheelchair in a home-care setting and its proposed solution. There are, indeed, challenges faced by many researchers [3,4,8,11] in applying a BCI system in-home setup. The challenges divided into three major factors which are subjective, environmental, and, technical factors.

3.2.1 Subjective Factor. Subject factors result from the research subject themselves. It ranges from 1)unavailability of the subject due to hospital follow-up and in-house physiotherapy session, 2)poor health condition due to disease progress and illness, and, 3)unable to achieve a sufficient control level due to tiredness, discomfort and distraction were to influence the subject's control over the BCI wheelchair navigation. As a matter of fact, the subject felt tired due to their sleepless night due to discomfort caused by pain, getting up earlier in the morning, and, lethargy due to aging as both were geriatric subject's. Also, subjects felt discomfort when the EEG electrode removed together with the surgical tape. Plus, the elderly can be easily distracted with noise, movement by passerby or conversation happens around them. Besides, sweating leads to poor EEG signal as sweat loosens the EEG electrode connection on the subject's scalp. Poor EEG signal results in destination selection difficulty and, thus, the wheelchair control is disturbed which is a major issue for BCI wheelchair implementation in the home-care setting.

3.2.2 Environmental Factor. The environmental factor is best explained by nature. Firstly, bright sunlight was a major issue there as the wheelchair navigation route is under a roofed walkway leaving the surrounding unroofed. In this case, the bright sunlight produced a glare on the laptop screen which acts as an obstacle for the subject to look and select their destination appeared on the screen.

3.2.3 Technical Factor

The technical challenges were contributed by the laptop or system. The current wheelchair only consists of a 2D laser mounted at 170cm and another 2D laser mounted at the height of 90cm. The position of the sensor limits the obstacle sensing capabilities of the wheelchair in a home care setting where there are multiple dynamically moving obstacles such as other patients on a wheelchair, food trolleys, and cats. The path that can be taken by the wheelchair is narrow and dynamic obstacle avoidance in such a small area by the wheelchair is not possible with the current sensor setup.

4. Proposed Solution

In this section, the potential solutions were listed out for the avoidable factors.

4.1 Subjective Problem-Solution

We are proposing solutions as stated: 1)unavailability of subject can be solved by checking the availability of the subject with their caregivers and draft a schedule accordingly, 2)taking a break from being the research participant is the best choice to offer when they are in ill health and tired body, while, and, 3)distraction can be reduced by reminding them to focus to select and once after the wheelchair starts to move they are free to do their own activities such as talking and sightseeing.

While the poor EEG signals made their selection difficult, therefore, switching EEG electrode to an eye tracker will be the best solution. Moreover, an eye tracker eases the subjects to make the selection as the eye gaze on the screen provides visual feedback to them. It also ensures whether the preset destination is focused and selected correctly. In this way, the end user will be certain with their selection.

4.2 Environmental Problem-Solution

On the other hand, environmental factors can be solved by fixing a black sheet on the navigation laser holding bars as it helps to block the direct sunlight from producing glare on the laptop screen.

4.3 Technical Problem-Solution

For the purpose of the experiment, the dynamic pathfinding feature of the wheelchair is disabled. Fixed waypoints are set as the trajectory for the wheelchair. The wheelchair is programmed to stop upon the detection of an obstacle. The wheelchair will resume movement once it detected that the path is cleared.

5. Conclusion

For a successful implementation of a BCI wheelchair in a home-care setting, there are, indeed, many challenges faced from subjective, environmental, and, technical aspects. However, poor EEG signal is a major issue in a BCI wheelchair control. Henceforth, in this article, the best solution proposed to replace the EEG electrode for end users with poor visual frequency is by substitute it with an eye tracker.

Acknowledgements

This research was supported in part by a grant from the Universiti Tunku Abdul Rahman (UTAR) Research Fund 2018 Cycle 2, IPSR/RMC/UTARRF/2018-C2/T01. We would, also, convey our appreciation to Jabatan Kebajikan Masyarakat Malaysia (JKMM), the research subjects and those who helped us both directly and indirectly.

References

[1] Bastos-Filho TF, Cheein FA, Müller SM, Celeste WC, de la Cruz C, Cavalieri DC, Sarcinelli-Filho M, Amaral PF, Perez E, Soria CM, Carelli R. *Towards a new modality-independent interface for a robotic wheelchair.* IEEE Transactions on Neural Systems and Rehabilitation Engineering. 2013 Jun 4;**22**(3):567-84.

[2] Guger C. *Brain Computer Interface, Lecture notes in The 4th International Summer School on Emerging Technologies in Biomedicine: Advanced Methods for the Estimation of Human Brain Activity and Connectivity*, Applications to Rehabilitation Engineering, Patras, Greece, 2008 July 4.

[3] Lin JS, Yang WC. *Wireless brain-computer interface for electric wheelchairs with EEG and eye-blinking signals.* Int. J. Innov. Comput. Inf. Control. 2012 Sep 1;8(9):6011-24.

[4] MND, (2012). *Motor Neuron Diseases Fact Sheet | National Institute of Neurological Disorders and Stroke.* Retrieved from https://www.ninds.nih.gov/Disorders/Patient-Caregiver-Education/Fact-Sheets/Motor-Neuron-Diseases-Fact-Sheet. Accesses 2018 Dec.

[5] Murguialday AR, Hill J, Bensch M, Martens S, Halder S, Nijboer F, Schoelkopf B, Birbaumer N, Gharabaghi A. *Transition from the locked in to the completely locked-in state: a physiological analysis.* Clinical Neurophysiology. 2011 May 1;1**22**(5):925-33.

[6] Nortina S. *More awareness on motor neuron disease needed; four to six M'sians diagnosed each month.* New Straits Times. Retrieved from https://www.nst.com.my/news/nation/2017/11/306794/more-awareness-motor-neuron-disease-needed-four-six-msians-diagnosed-each. 2017 November 23.

[7] Olaronke I, Rhoda I, Gambo I, Oluwaseun O, Janet O. *Prospects and Problems of Brain Computer Interface in Healthcare.* Current Journal of Applied Science and Technology. 2018 Oct 23:1-7.

[8] Wolpaw JR, Birbaumer N, McFarland DJ, Pfurtscheller G, Vaughan TM. *Brain–computer interfaces for communication and control.* Clinical neurophysiology. 2002 Jun 1;**113**(6):767-91.

Study on Pulse Electric Field Exposure Effect on HeLa Cells For Wound Healing Application

Muhammad Mahadi Abdul Jamil[1], Mohamed A. Milad Zaltum[1], Mansour Youseffi[2] and Farideh Javid[3]

[1]Biomedical Engineering Modeling and Simulation (BIOMEMS) Research Group, Faculty of Electrical and Electronics Engineering, Universiti Tun Hussein Onn Malaysia, 86400,
Batu Pahat, Johor, Malaysia
[2]Biomedical and Electronics Engineering Department, Faculty of Engineering and Informatics, University of Bradford, Bradford, BD7 1DP, UK
[3]Pharmacy, School of Applied Sciences, University of Huddersfield, Queensgate, Huddersfield, HD1 3DH, UK

M.Youseffi@bradford.ac.uk

Abstract. This study investigated the effect of pulse electric field (PEF) exposure on cervical cancer cells (HeLa cells) in an *in-vitro* wound repair model. The study mainly focused on the healing time of HeLa cell line wound model. During the experiments HeLa cells were maintained at 37°C in a modified Chamlide EC magnetic chamber where they were exposed to high electric fields. A Nikon inverted microscope (Ti-series) with Metamorph® time lapse software were used to monitor, image and capture photomicrographs and videos of the cells. The tests carried out during this study revealed that pulse electric field enhanced the migration of HeLa cells. Cells exposed to PEF (1kV/cm, 100μs, and single pulse) healed the wound in ~2 hours (from initial wound gap of 54.53μm ± 0.55SD to 0.66μm ± 0.61SD). On the other hand, non-PEF (control) healed the wound in ~10 hours (from initial wound gap of 56.33μm ± 0.57SD to 0.46μm ± 0.45SD). It was therefore found that the healing rate with PEF is ~five times faster than non-PEF group. It is believed that PEF usage on diseased biological cells would enable a novel method for assisting drug free wound repair systems and many other potential biomedical engineering applications such as treatment of neurological disorders including Alzheimers and Parkinsons.

1.Introduction

All living cells have some electrical features and consequently respond to and generate electric fields and currents [1-2]. Electrically focused and organized cell migration plays a vital role in mammalian cell embryonic growth [3], therefore, manipulation of the cell electrical characteristics may give a significant and direct therapeutic option for cancer and wound healing process [1]. Electroporation is a biophysical procedure carried out by the use of high voltage electrical pulse to tissue *in-vivo* or cells *in-vitro* and it has been found to enhance the cell's uptake of molecules by generating transient pores in the cell membrane [4-5]. This study used an *in-vitro* wound repair model to estimate precise cellular level interactions when exposed to pulse electric field (PEF) under controlled environmental conditions. Besides the opening of pores in the cell membrane, electroporation can induce cytoskeletal restructuring and there is evidence that these changes influence cell adhesion and migration [4]. Redistribution of actin cytoskeletal has been found in cell electro-migration studies in response to direct current electric fields [6-8]. Cell migration is an important component in wound healing and growth of an organism [9-10]. In cancer cell development, the ability of the cells to migrate and proliferate is a major disease-causing factor. Though electric field can change the cell physiology by linking with intracellular signaling pathology, little is known about the process. In a previous study [11] it was shown that electric fields have a meaningful influence on the attachment speed and distribution characteristics of the colon cell line (HT29) [11]. Cell adhesion and migration have been closely associated with cellular activities [12], and more investigation is required to study the migration properties of HeLa cells in an electric field. Therefore, in order to examine and quantify HeLa cell migration in response to electric fields, an experimental setup was designed in this work so as HeLa cell migration could be observed and recorded during exposure to PEF [13-14]. Cells were imaged using confocal laser Nikon inverted microscope (Ti-series) with the help of integrated devices for time-lapse live imaging such as the CCD camera, temperature and CO_2 controllers. Commercially available pulse generator was used for these experiments to produce pulse of 1kV/cm for a duration of 100μs [15-20].

2.Materials and Methods

2.1 HeLa cell lines and cell culture procedure

In this study, HeLa cells were used because they are well known for their high replication rates. This high replication rate makes HeLa cells suitable for simple and continuous harvesting of large number of HeLa cells for in-vitro experimental work. Figure 1 shows a micrograph of the cervical cancer cells known as HeLa cells. HeLa cells were cultured in RPMI 1640 media with 10% fetal bovine serum (FBS) and 1% of penicillin and streptomycin solution. Cells were incubated in 5% CO_2 at 37°C until reaching 90% confluency. When the tissue culture flask reached 90% confluency, media was removed from the tissue culture flask, then, 3ml of phosphate buffered saline (PBS) was added to the flask to wash the cells, which was removed after few seconds. 2 ml of TryPle Express from GIBCO was added to detach the cells from bottom of the flask. TryPle Express works best in a warm surrounding, so the flask was incubated for seven minutes at 37°C. After this process, 0.5ml of cell suspension at cell concentrations of 1.6×10^6 cells/ml was transferred to a new 25cm^2 flask along with 5ml of fresh media. Cells were incubated in RPMI containing 10% fetal bovine serum (FBS) and 1% of penicillin and streptomycin solution at 37°C in 5% CO_2 for the next growth phase.

2.2 Plating cell on the coverslip

After 90% confluency, the cells were washed 3 times with PBS and then harvested by incubating them with 2ml of TryPle Express for seven minutes and neutralizing them with an equivalent volume of complete growth medium. 25mm circular cover glasses were placed, one in each well of a six well cell culture plate. 2ml of pre-warmed complete growth medium (RPMI 1640+ 10% FBS + 1% penicillin-streptomycin) was then added to each well along with 200μl of the detachment agent to the cells suspension with a concentration of 1.6×10^6 cells/ml. The cells were then incubated in a humidified incubator at 37°C and 5% CO_2 until reaching 85 to 90% confluency.After HeLa cells reached 90% confluency, the culture medium was aspirated and the cover slips from each chamber were washed once with 2ml PBS. The cover glass was then removed and placed on the bottom plate of the Chamlide EC magnetic chamber. Figure 2 shows the EC magnetic chamber with 10mm gap platinum electrodes. With the help of a 23 gauge sterile needle, a wound model was formed at the center of the coverslip parallel to the two electrodes of the Chamlide EC magnetic chamber. Cells were rinsed twice with 1ml of PBS to remove cell debris. Furthermore, 1ml of complete growth medium was added to cells in the chamber and the chamber was placed on the upper stage of the inverted microscope for live imaging as shown in Figure 3. The CO_2 and temperature of the integrated live imaging system were maintained at 5% and 37°C, respectively. Cells were then finally subjected to PEF (1000V, 100μs and single pulse) during which time cell responses could be visualized simultaneously.

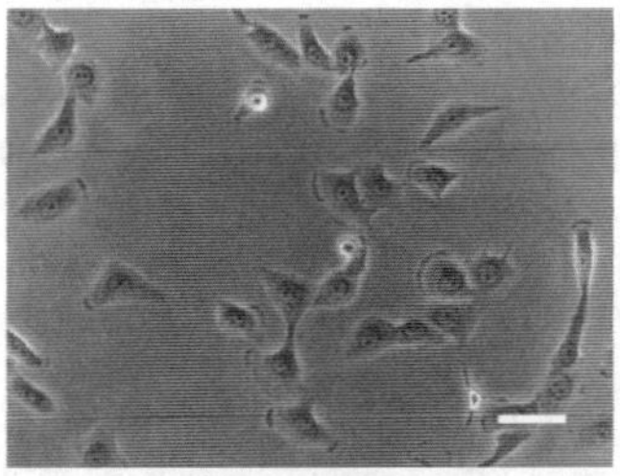

Figure 1. Cervical cancer cells (HeLa Cells) [scale bar=30μm].

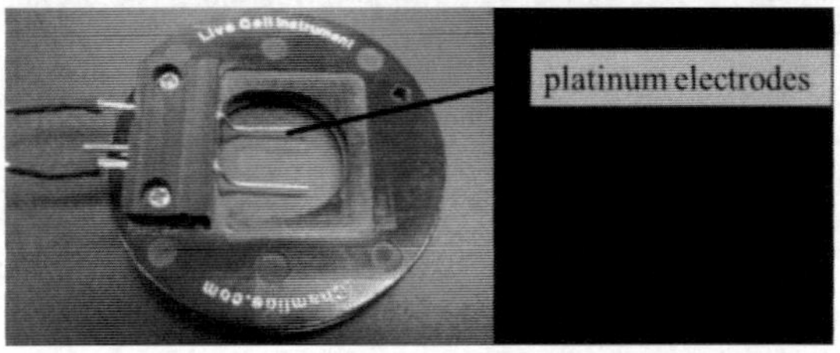

Figure 2. EC magnetic chamber with 10mm gap platinum electrodes.

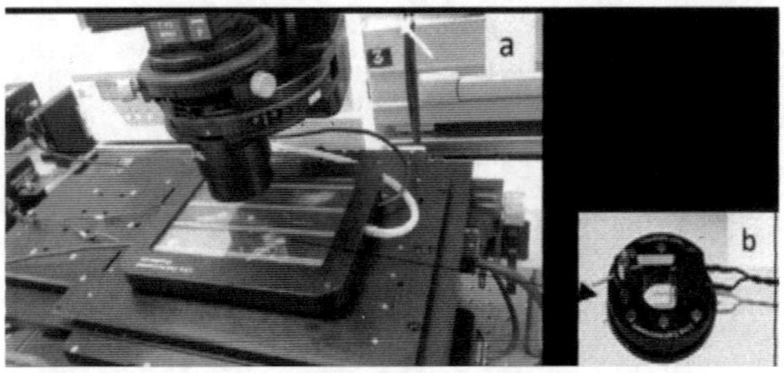

Figure 3. (a) Chamlide stage; (b) Magnetic chamber.

3.Cell imaging

Cell responses to PEF were imaged using a Nikon inverted microscope (Ti series) and monitoring was achieved using Metamorph® time lapse software to observe morphological changes of wound closure between two samples of HeLa cells. Images were taken and recorded at a frequency of 1frame per minute until wound closure. Using the region measurement in the Metamorph® software, five vertical lines to the wound edges were drawn, and their average height was taken as the wound gap at that point in time.

4.Result and Discussion

The HeLa cells cultured in RPMI 1640 at 37°C and 5% CO_2 reached 90% confluency after 72 hours in culture. A manually created wound of width ~55μm was formed on the glass coverslip monolayer which was placed on the stage of the inverted microscope for live imaging. Time-lapse video microscopy was employed for examining the migration speed of HeLa cell line after the wound was created. The experiment conducted was to identify if electroporation can enhance the *in-vitro* healing rate for HeLa cell line. This can further be explored for enabling wound healing process.

The results obtained from this experiment showed the effect of charged PEF (1000V, 100μs and single pulse) on HeLa cell behavior and morphology. Wounded HeLa cell monolayers showed complete healing after approximately 2 hours as can be seen in Figure 4. Figure 5 is the graph showing the average wound width plotted against the healing time, which shows clearly that there is an abrupt reduction in the mean wound width over time for the PEF treated HeLa cell monolayers.

In contrast, the control closure response of the HeLa cell monolayers took approximately 10 hours as shown in Figure 6. The re-epithelialization displayed by the HeLa cells at the start of injury appeared to involve spreading of the cells at the wound border and the migration of cell sheets near the site of wound closure [19]. These findings were further investigated by repeating the experiments five times for each sample with and without PEF. The average vertical distance between the wound gaps recorded over time is given in Table 1.

Interestingly, the rate of wound closure with respect to time can also be seen from Figure 5. ZXC being almost linear in the first two hours and much faster for the PEF treated monolayers as compared to that of the control, and for control sample after two hours is almost three times slower than first two hours that is why the time scale is not linear. The imaging results showed that the time taken by HeLa cells to proliferate towards wound area had significant contribution towards rate of wound closure. From the control sample it was found that complete wound closure was achieved at ~10 hours after the creation of the wound suggesting a much slower healing process (from initial wound gap of 56.33μm ± 0.57SD to 0.46μm ± 0.45SD).

In comparison, PEF exposed cultures achieved complete wound closure after ~2 hours (from initial wound gap of 54.53μm ± 0.55SD to 0.66μm ± 0.61SD). This indicates reduction in the healing time for wound closure. These findings were further investigated by repeating the experiments three times which showed repeatable results. Therefore, this study strongly suggests that the physical activity and morphological properties of HeLa cells are affected significantly ($P<0.01$) by the application of PEF.

The results in this study are in agreement with previous findings of the same authors [11, 13, 16, 20] who demonstrated a similar pattern of HeLa cells wound repair under the influence of electric field using PEF at 2700V/cm, 30μs with single pulse which showed that wounded HeLa cell monolayers were repaired after ~3 hours under electric field treatment. In this study, a PEF of 1000V/cm at 100μs with single pulse was used, and wounded HeLa cell monolayers were repaired in ~2 hours under the influence of pulse electric field treatment.

Table 1. Average wound width over time for both PEF and Non-PEF HeLa cell monolayers.

Experimental	Time (hour)	Mean Wound Gap (μm) ± Std. Deviation (SD)	Std. Error (SE)
PEF Group	0.0	54.53 ± 0.55	0.31
	0.5	41.4 ± 0.45	0.26
	1.0	25.5 ± 0.86	0.5
	1.5	13.66 ± 0.57	0.33
	2.0	0.66 ± 0.61	0.35
Non-PEF Group	0.0	56.33 ± 0.57	0.33
	0.5	51.33 ± 1.52	0.88
	1.0	42.83 ± 0.76	0.44
	1.5	36.23 ± 0.49	0.28
	2.0	29.46 ± 0.55	0.31
	4.0	19.93 ± 0.4	0.23
	6.0	14.03 ± 0.25	0.14
	8.0	9.5 ± 0.5	0.28
	10.0	0.46 ± 0.45	0.26

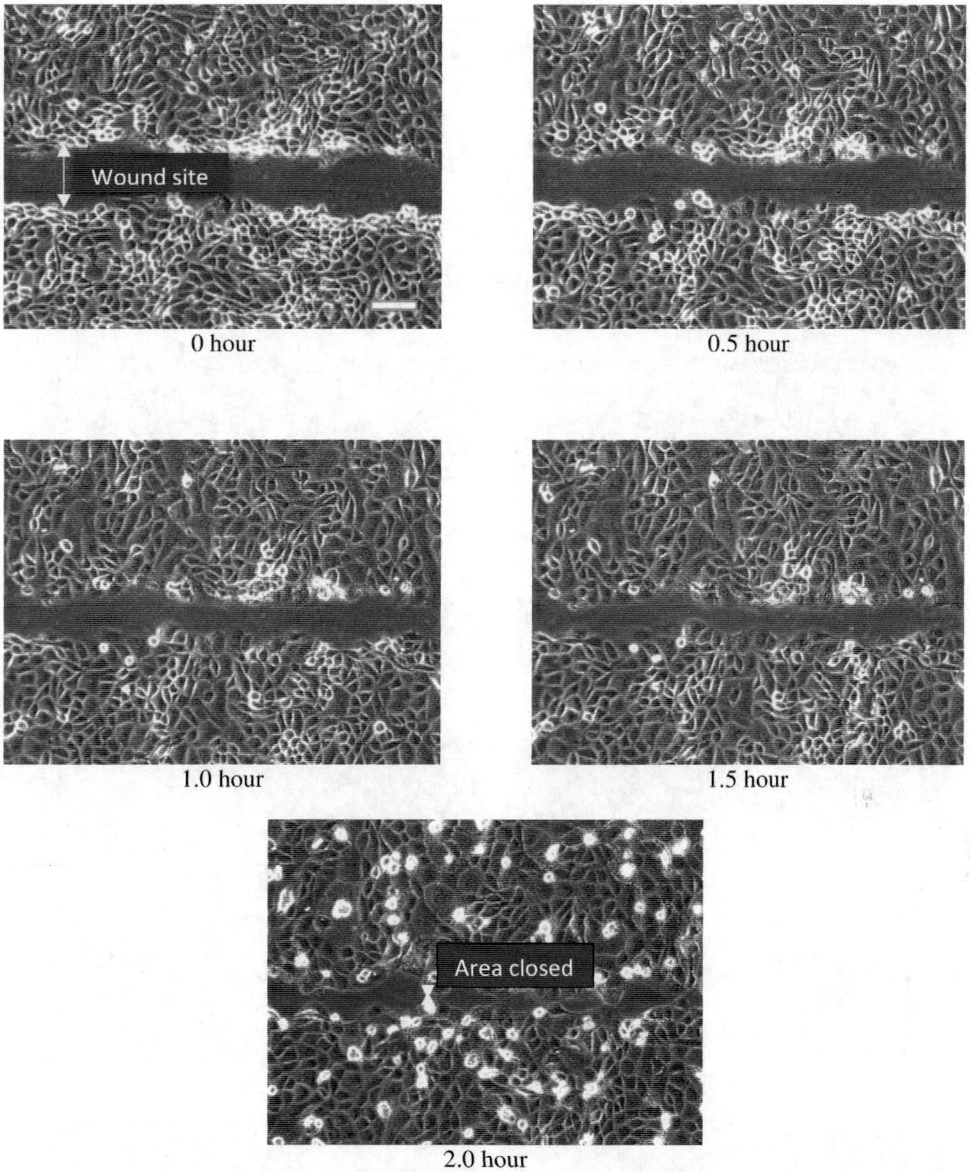

Figure 4. Wound healing closure response with PEF treatment (scale bar = 50μm).

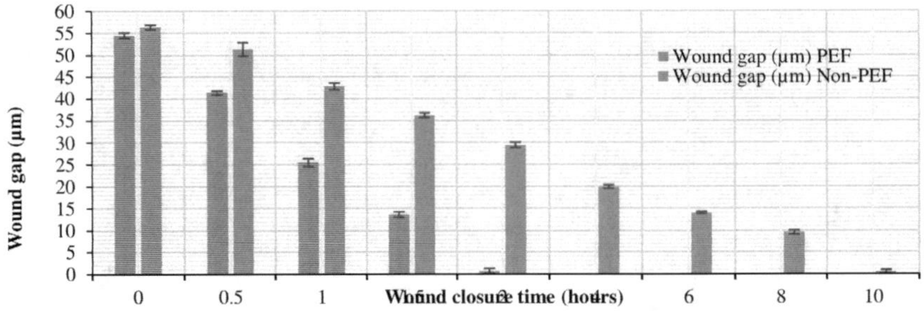

Figure 5. Time evolution of wound gap distance for HeLa cell line for PEF

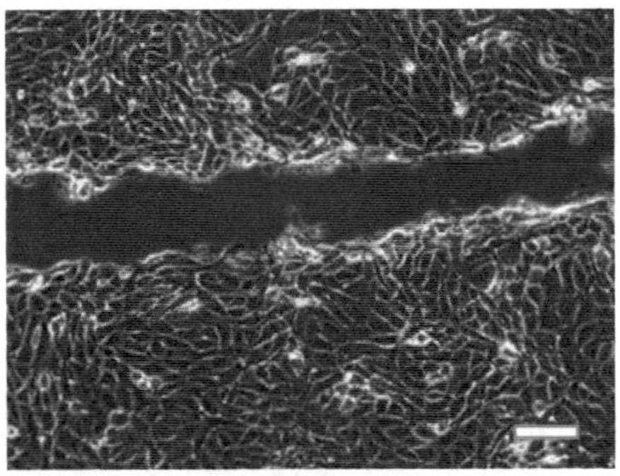
0 hour

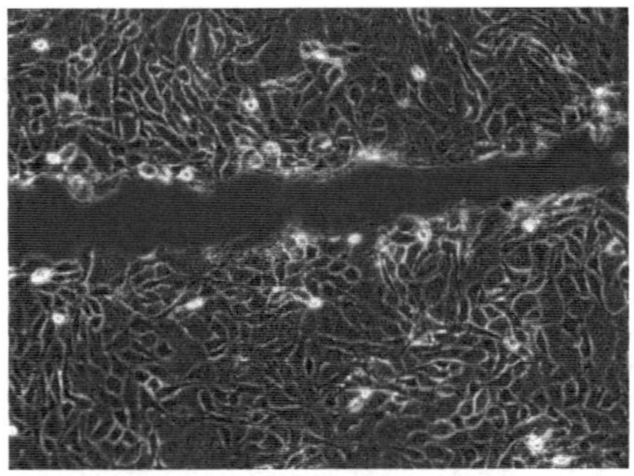
0.5 hour

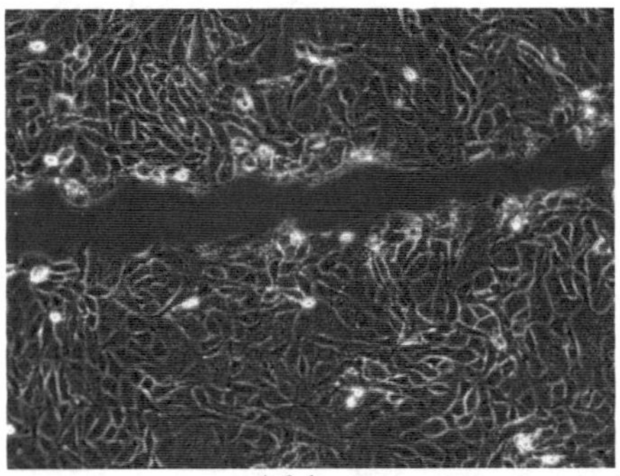
1.0 hour

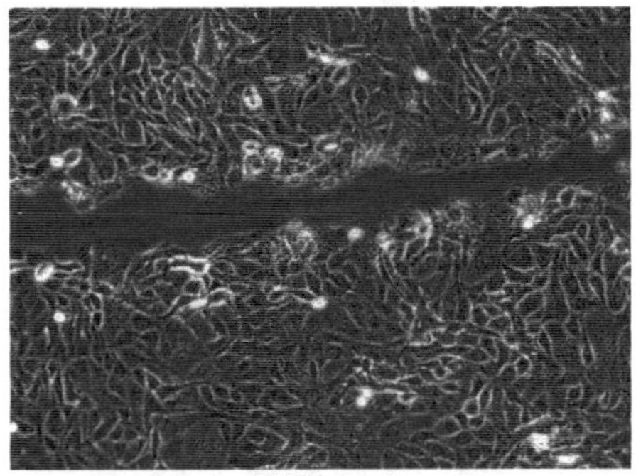
1.5 hour

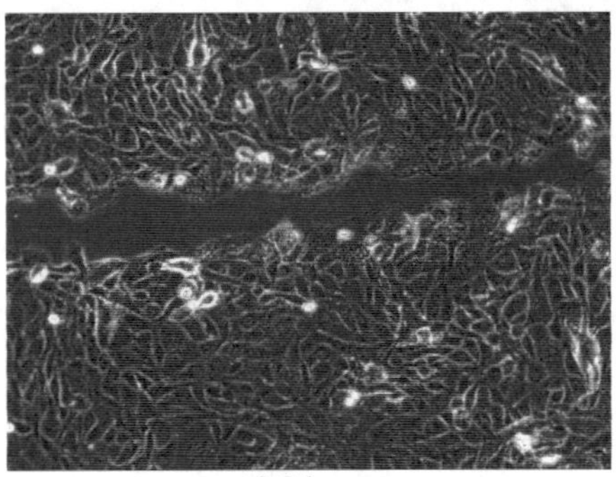
2.0 hour

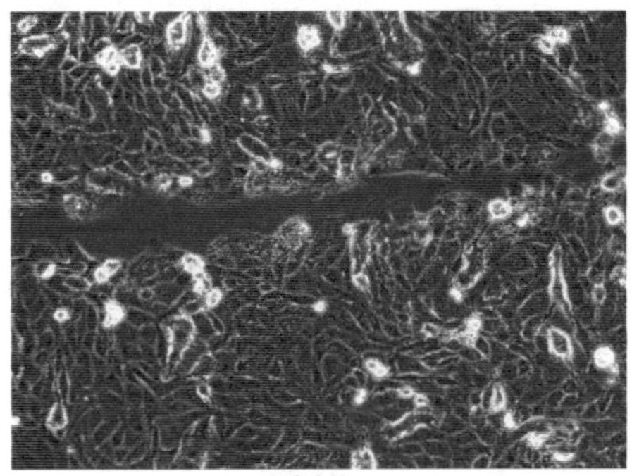
4.0 hour

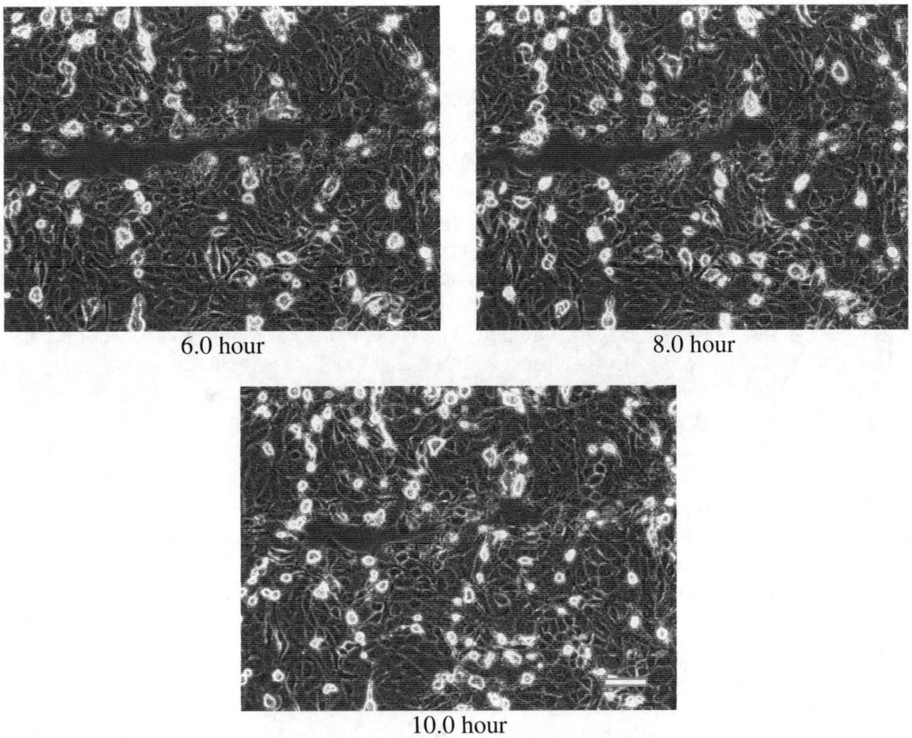

Figure 6. Control wound healing closure response with Non-PEF (scale bar = 50μm).

5. Conclusions

It can be concluded that up to 80% decrease in healing time was achieved with PEF treatment. It is believed that the electric field reduces the adhesiveness [11] of the cells and hence increases the migration ability of the cells to cover the wound gap faster. The PEF stimulates greater cell proliferation and spreading ability of the cells, thus more cells are available to cover the wound gap faster. The reduction in time for wound closure in this study showed a novel contribution by using PEF assisted wound closure response and this research provides a fundamental approach for wound healing applications.

Acknowledgement

The authors would like to thank the Ministry of Higher Education Malaysia and UTHM for the research grant under the Fundamental Research Grant Scheme (FRGS) phase 2/2014, Vot 1488, University Contract Grant Vot U551 for funding and supporting this research work. The continued R&D collaborations with Universities of Bradford and Huddersfield are also very much appreciated.

REFERENCES

[1] Hondroulis, E., Melnick, S. J., Zhang, X., Wu, Z. Z., and Li, C. Z. (2013). Electrical field manipulation of cancer cell behavior monitored by whole cell biosensing device. Biomedical microdevices, 15(4), pp. 657-663.

[2] Saulis, G. (2012). Electrochemical processes occuring during cell electromanipulation procedures. A short review. Electroporation in Laboratory and Clinical Investigations, pp. 99-113.

[3] Pehlivanova, V. N., Tsoneva, I. H., and Tzoneva, R. D. (2012). Multiple effects of electroporation on the adhesive behaviour of breast cancer cells and fibroblasts. Cancer cell international, 12(1), pp.1-9.

[4] Titushkin, I. and Cho, M. (2009). Regulation of Cell Cytoskeleton and Membrane Mechanics by Electric Field: Role of Linker Proteins. Biophysical Journal, 96(2), pp. 717-728.

[5] Miklavcic, D. and Kotnik, T. (2004). Electroporation for electrochemotherapy and gene therapy. Bioelectromagnetic medicine, pp. 637-656.

[6] Weijer, C. J. (2009). Collective cell migration in development. Journal of cell science, 122(18), pp. 3215-3223.

[7] Hamilton, D. W., Wong, K. S., and Brunette, D. M. (2006). Microfabricated discontinuous-edge surface topographies influence osteoblast adhesion, migration, cytoskeletal organization, and proliferation and enhance matrix and mineral deposition in vitro. Calcified tissue international, 78(5), 314-325.

[8] Zaltum, M. A., Adon, M. N., and Mahadi, M. A. (2013). Electroporation Effect on Growth of HeLa cells. In Biomedical Engineering International Conference (BMEiCON), pp.1-4.

[9] Mackay, C. R. (2008). Moving targets: cell migration inhibitors as new anti-inflammatory therapies. Nature immunology, 9(9), pp. 988-998.

[10] Xiao, D., Tang, L., Zeng, C., Wang, J., Luo, X., Yao, C., and Sun, C. (2011). Effect of actin cytoskeleton disruption on electric pulse-induced apoptosis and electroporation in tumour cells. Cell biology international, 35(2), pp. 99-104.

[11] Mamman, H. B., and Jamil, M. M. A. (2015). Investigation of electroporation effect on HT29 cell lines adhesion properties. In 2nd International Conference on Biomedical Engineering (ICoBE), IEEE, pp. 1-4.

[12] Zhao, J., Li, P., Feng, H., Wang, P., Zong, Y., Ma, J., and Lu, A. (2013). Cadherin-12 contributes to tumorigenicity in colorectal cancer by promoting migration, invasion, adhersion and angiogenesis. J. Transl. Med, 11, pp. 288-288.

[13] Nazib, M. Adon, Dalimin, M. N., Jamil, M. M. A., and Kassim, N. M. (2012). Development of high voltage pulse inducement method for biological cell. In International Conference on Biomedical Engineering (ICoBE), pp. 501-503.

[14] Hovis, K. R., Padmanabhan, K., and Urban, N. N. (2010). A simple method of *in-vitro* electroporation allows visualization, recording, and calcium imaging of local neuronal circuits. Journal of neuroscience methods, 191(1), pp. 1-10.

[15] Ušaj, M., Trontelj, K., Miklavčič, D., and Kandušer, M. (2010). Cell–cell electrofusion: optimization of electric field amplitude and hypotonic treatment for mouse melanoma (B16-F1) and Chinese hamster ovary (CHO) cells. The Journal of membrane biology, 236(1), pp. 107-116.

[16] Zaltum, M. A. M., Adon, M. N., Hamdan, S., Dalimin, M. N., and Jamil, M. M. A. (2015). Investigating a critical selection of pulse duration effect on growth rate of HeLa cells. In International Conference on BioSignal Analysis, Processing and Systems (ICBAPS), pp. 33-36.

[17] Li, C., Yao, C., Sun, C., Guo, F., Zhou, W., and Xiong, Z. (2011). Dependence on electric field intensities of cell biological effects induced by microsecond pulsed electric fields. Dielectrics and Electrical Insulation, IEEE Transactions on 18(6), pp. 2083-2088.

[18] Pucihar, G., Krmelj, J., Rebers, M., Napotnik, T., and Miklavčič, D. (2011). Equivalent Pulse Parameters for Electroporation. IEEE Transactions on Biomedical Engineering, 58(11), pp. 3279-3288.

[19] Zahm J. M., Kaplan, H., Herard, A. L., Doriot, F., Pierrot, D., Somelette, P. and Puchelle, E. (1997). Cell migration and proliferation during the *in-vitro* wound repair of the respiratory epithelium. Cell motility and the cytoskeleton, 37(1), pp.33-43.

[20] Adon, M. N. (2015). Pulse Electric Field Exposure Effect on Morphological Properties of HeLa Cells. PhD thesis submitted to the University Tun Hussein Onn Malaysia.

Journal of Physics: Conference Series **1372** (2019) 012022 doi:10.1088/1742-6596/1372/1/012022

Intelligent Fuzzy PD+I Controller with Stabilizer for Nano Satellite Attitude Control System

Hani Hazza Ali Ahmed*[1], Mohd Yusoff Mashor[1], and Mohammed Chessab Mahdi[2]

[1]School of Mechatronic Engineering, Universiti Malaysia Perlis (UniMAP), 02600, Arau, Perlis, Malaysia.

[2]Al-Furat Al-Awsat Technical University, Iraq

*HaniHazza@gmail.com

Abstract. This paper discusses the variation of proportional and derivative gains in a system under the control of PD+I controller that uses fuzzy inference method an effort is made to design the fuzzy proportional-derivative (PD) + I controller for a nonlinear the proposed fuzzy PD + I structure shows superior performance in the Nano-satellite attitude control system. Given that 2-input PD type controllers are commonly used to control Nano satellite attitude systems (Innovative Satellite InnoSAT), This paper analysis the performance of Fuzzy PD+I on satellite system. Fuzzy PD+I controller design can control Nano satellite attitude control system nicely and utilized the Fuzzy PD+I controller which optimize the stability of the Nano Satellite control system. In addition, the scheme is robust to variations in nonlinearities, as well as the steady-state gain of the plant. We illustrate the effectiveness of our scheme In MATLAB / SMULINK simulation environment.

1. Introduction

This paper presents a type of fuzzy PD+I controller that uses fuzzy inference method. Given that 2-input PD type controllers are commonly used to control Nano satellite attitude systems, the structure discussed in this paper suggests that the controllers will achieve the same performance made by the general-purpose fuzzy PID controller, with additional enhances to the control performance while addressing a wide range of input and parameter variations. Complexity of the proposed fuzzy PD+I controller is minimized as possible and only two design variables are used to adjust the rate of variations of the proportional gain and derivative gain. This paper utilized the Fuzzy PD+I controller which optimize the stability of the Nano Satellite control system eradicate the steady state error, while the PI controller might face limitations in optimizing the response to the momentary variations. On the other hand, three-input fuzzy systems to construct a fuzzy PID controller can create a hardship in defining the control rules [1].

In order to address these limitations, the fuzzy PID controllers have adopted the use of multiple two-input fuzzy controllers. For example, using the two-input fuzzy system from Ying's work on fuzzy controllers [2], Misir [3] developed a digital PID controller that used the two "two-inputs" controllers. Further work by Li was made to improve the proportional segment of the PID controller using a 36 inputs regions in a two-input fuzzy systems [4].In addition, Mann [5] examined the inputs and outputs to come up with groupings for the fuzzy PD+I elements to categorize the fuzzy PD+I structures by the types and numbers of its consisting elements. However, advances in this regard and the development of the PID controllers are largely dependent on the background and use of the researchers and developers. Therefore, attempts to execute direct evaluation of the performance made by the traditional PID controllers and the fuzzy PD+I controllers has proven to be challenging. Yet a comparison might still be possible considering that many attempts to tune the traditional controllers have been developed, and consequently the tuned PID factors can lay the foundation for selecting a fuzzy PID controller. The selection will take into consideration the relation between the fuzzy PD+I controllers and equivalent PID controller [6].

For the purpose of this paper, the traditional PID control has been modified with two-input fuzzy controller to include the varying gains of the proportional and derivation signals. The two signals play the greater role in determining the transient response, while the integral signal has a fixed gain for removing the steady state errors. Therefore, the proposed controller will be referred to as the fuzzy PD+I type. For this comparison, the linear inference method will be performed instead of the gravity method or similar general inference methods as proposed by Mizumoto [7], [8]. For this paper, the designed procedure will use membership functions and control rules following implementing a direct output interference method for the PID output surface. When using the traditional nonlinear inference method, the undesired nonlinearities will become apparent on the fuzzy system's output surface. However, by using the product-sum-gravity method the nonlinearities can be created on the output surface and created to enhance the performance of the controller. Using this approach, the added nonlinearities and its effects can be adapted and controlled depending on the rates of nonlinearities introduced to the proportional and derivative parts [9].

To thoroughly examine this approach, this paper will be organized as follows: Section 2 addresses the structure of the proposed fuzzy PD+I and explain its components. Section 3 describes the design of the two-input fuzzy system with modified PD part. Section 4 focuses on the practical structure of the fuzzy PD+I controller for Nano satellite attitude control system. Section 5 demonstrates the results of the simulation for fuzzy PD+I controller for Nano satellite attitude control system.

2. Structure of Fuzzy PD+I controller with stabilizer

The traditional method of PID control is to use the proportional, integral and derivative signals of the errors of the attitude to create a control signal. It should be taken into consideration that the attitude error promptly changes in the same moment when the reference input change, which might cause the derivative signal of the attitude to shock the system. Therefore, to overcome this issue is the PID structures that relies on the derivative signal for the output signal rather than for the attitude error. Hence, for this paper the structure of the fuzzy controller suggested is shown in Figure 1 consisting a fuzzy PD+ linear I structure. In the cases where there is no significant differences between the magnitude of the three gains of the traditional PID controller, it is recommended to use a three-input fuzzy control like fuzzy PI+D or fuzzy PD+I because they will produce more control over the results [10].

2.1. Structure of Fuzzy PD+I controller

Since this paper will only focus on the steady state errors and the transient response, the fuzzy PD+I controller will be used. The fuzzy system used in the proposed structure will only be applied to the proportional and derivative signal of the linear PID controller, in which the integral signal will be relying on the traditional linear method [11]. As explained above the integral signal mainly focuses on eliminating the steady state error. Therefore, the transient outcome is mainly caused by the proportional and derivative signals. To improve the transient response, a two-input fuzzy system will use a verity of gains to control the proportional and derivative parts. The nonlinearities are then added by the fuzzy control rules and membership functions, therefore, accentuating the proportional gain to deal with large attitude errors and promptly decreases the speed of the attitude error. In this case, the nonlinearities of the derivative gain overpowers overshoot and enhances damping until reaching a stability. The fuzzy system in this structure is regulated in response to the maximum range of the reference input signal.

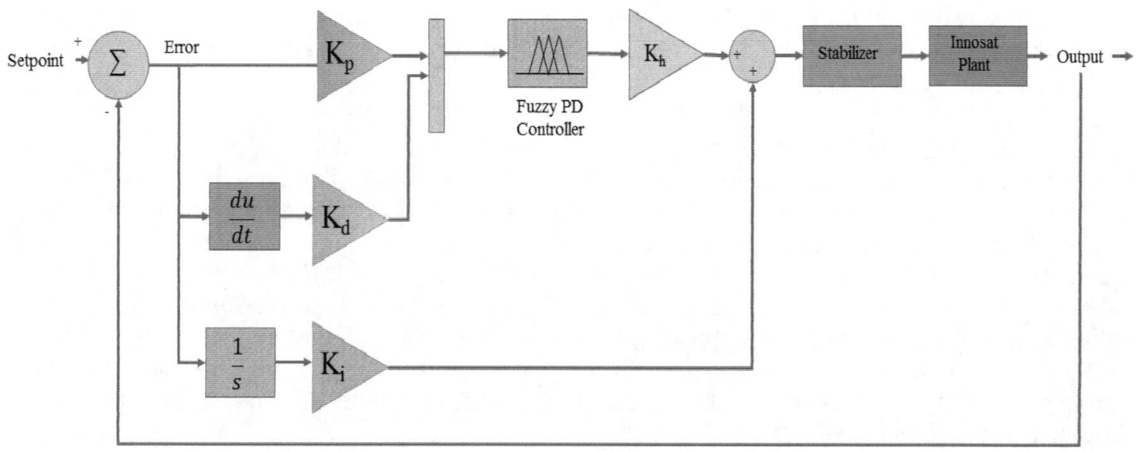

Figure l. Block diagram of fuzzy PD+I controller

The Euler angles and their derivatives result represent the transfer functions of InnoSAT attitude control can be shown in equation (1), [12]

$$\begin{aligned} \phi_{(s)} &= \frac{s^2 + 0.3051s + 0.2040}{s^4 + 1.1050s^2 + 0.1650} \\ \theta_{(s)} &= \frac{1}{s^2 - 7.1138 \times 10^{-3}} \\ \psi_{(s)} &= \frac{s^2 - 0.3023s + 0.8088}{s^4 + 1.1050s^2 + 0.1650} \end{aligned} \tag{1}$$

2.2. Stabilizer

The stabilizer used in the proposed control scheme is based on lead compensator where the parameters for the compensator can be achieved by the method of root locus. If a plant is dynamic, adjusting the feedback gain alone is not enough. Therefore, some modification or compensation must be made in the feedback to achieve the desired specifications. Typically, the transfer function of lead compensator takes the form [13] :

$$H(s) = K\,\frac{s + a}{s + b} \tag{2}$$

The transfer function of stabilizer to stabilize the InnoSAT can be expressed by the following equation:

$$H(s) = \frac{s + 1}{s^2 + 40s + 400} \tag{3}$$

Table 1. Rule base for the controller of roll, pitch and yaw angles

	Change of Error		
Error	**NE**	**ZE**	**PO**
NE	LO	LO	ZE
ZE	LO	ZE	HI
PO	ZE	HI	HI

Three similar fuzzy controllers for the three angles roll φ, Pitch θ and Yaw ψ, are shown in Figure 2. Table 1 show the memberships for inputs Error, Change of Error, and the list of the rules of fuzzy inference structure (FIS) [14] .

3. Fuzzy PD + I Controller Design

The experiment uses computer simulation and is performed using fuzzy PD+I controller for InnoSAT attitude control system. Figure 2 demonstrates the block diagram of the fuzzy PD+I control system. In this system, the transient response of the proportional and derivative actions and their behaviours can be easily recorded and observed. However, the corresponding variations of the integral gain might have limited effect on the transient response, and its associated control rule might not be easily perceived. The proposed controller focuses on improving the transient response. Therefore the considered fuzzy PD+I controller uses two input fuzzy system for the proportional and derivative gains and a normal I controller for the integral gain. In this design, the change of error will be replaced by the derivative of error signal as the derivative signal.

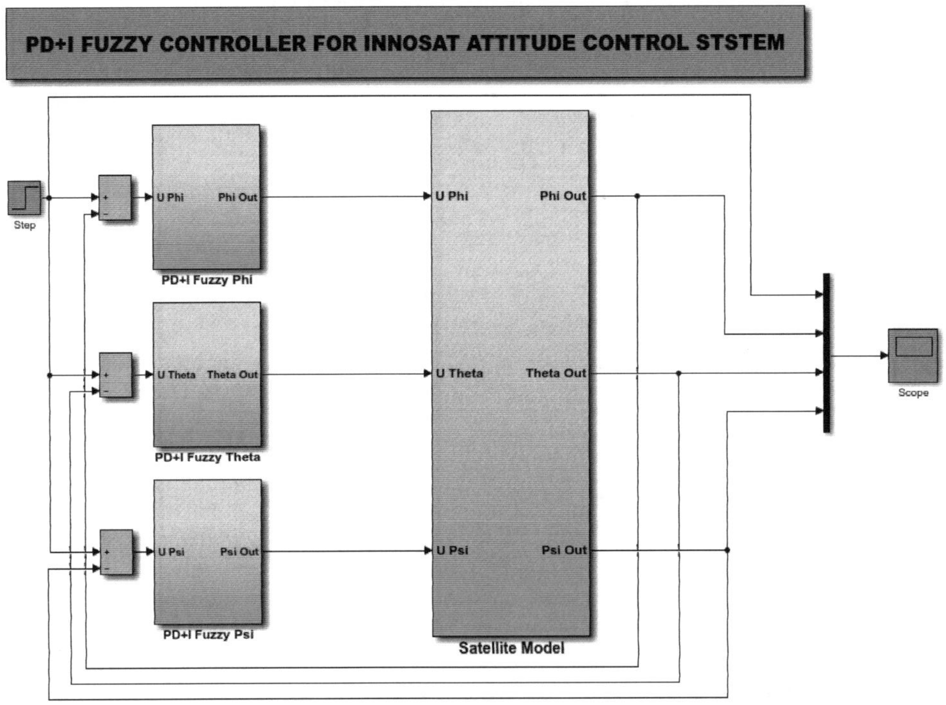

Figure 2. Fuzzy PD+I controller for InnoSAT Attitude Control System.

4. Simulation Results

The simulation was conducted using MATLAB and Simulink. By using the initial Gain Co-efficient parameters for Fuzzy PD+I Controller in Table 2, performance measures of the controller are as shown in Table 3. Under these optimum tuning parameters, the Fuzzy PD+I controller shows a very good performance.

Table 2. Gain Co-efficient Parameter for Fuzzy PD+I Controller

	Controller		
Parameter	K_p	K_d	K_i
Roll $\phi_{(s)}$	50	15	5
Pitch $\theta_{(s)}$	70	14	3
Yaw $\psi_{(s)}$	50	15	5

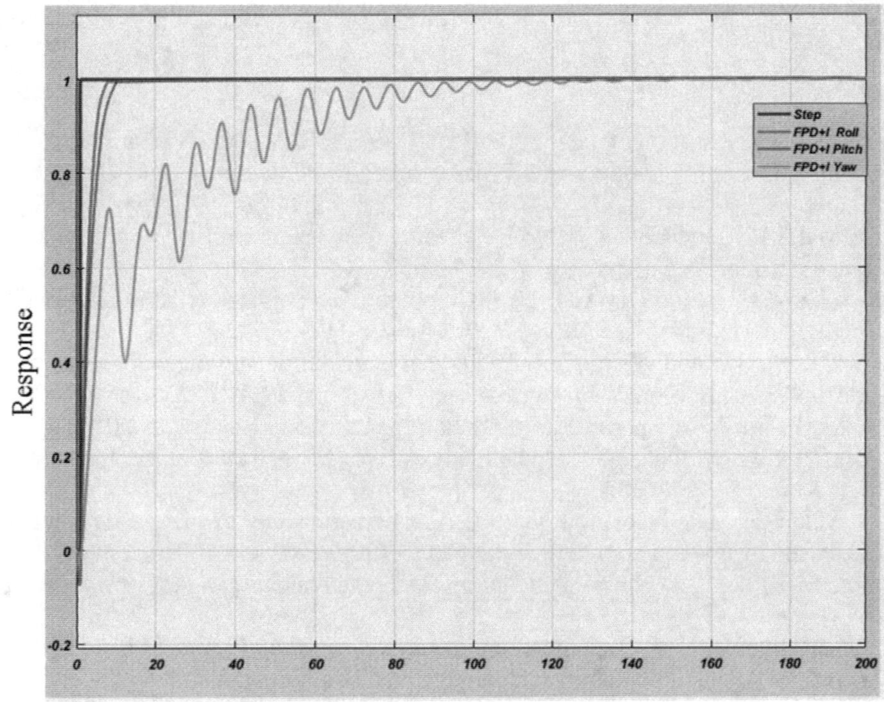

Figure 3. Step response of Fuzzy PD+I with Stabilizer for InnoSAT Attitude Control.

Table 3. Performance measures of the proposed controlled system

Angle	Delay Time (Td) sec	Rise Time (Tr) sec	Peak Time (Tp) sec	Settling Time (Ts) sec	Peak Overshoot PO%	Steady State Error %
Roll $\phi_{(s)}$	3	5.2	10	15	0.2	0.1
Pitch $\theta_{(s)}$	2	3.8	8	10	0.3	0.1
Yaw $\psi_{(s)}$	6	33.9	117	90	0.5	0.2

Response of the Fuzzy PD+I with Stabilizer for InnoSAT Attitude Control System is shown in Figure 3. During the simulation, the fuzzy PD+I control demonstrated very small overshoot and a rapid rise time for Roll and Pitch. Therefore, the suggested fuzzy PD+I control is less sensitive to the differences of the Kp, Ki, Kd gains. However, the response for Yaw is relatively slow due to unstable original response. In this case the controller has to play two roles, which are to stabilise and then control the Yaw dynamic. Thus a longer time is required to compensate the system response.

5. Conclusion

This paper discusses the design of a fuzzy PD+I controller to improve the attitude control performance of InnoSAT. Two input and one-output fuzzy system has been adapted for the proportional and derivative part of the traditional PID controller. The results proved that using this fuzzy PD+I controller gives additional effectiveness compared to the fuzzy inference method. The fuzzy inference method showed that nonlinearities are implemented only by the fuzzy control rules and membership functions. With comparatively large attitude error, the proportional gain is highlighted by the nonlinearities; alternatively with small attitude error, the derivative gain is emphasized. The fuzzy PD+I controller also has a fast rise time and more effective control on overshoot. The results concluded that with those selected gains and parameters, the fuzzy PD+I controller produced good and robust responses.

References

[1] Kumar V, Rana K P S and Gupta V 2008 Real-Time Performance Evaluation of a Fuzzy PI + Fuzzy PD Controller for Liquid-Level Process *Int. J. Intell. Control. Syst.* **13** 2 89–96

[2] Ying H 2000 *Fuzzy control and modeling: Analytical foundations and applications*.

[3] Misir D, Malki H A and Chen G 1996 Design and analysis of a fuzzy proportional-integral-derivative controller *Fuzzy Sets Syst.* vol. **79** 3 297–314

[4] Chopra V, Singla S K and Dewan L 2014 Comparative Analysis of Tuning a PID Controller using Intelligent Methods *Acta Polytech. Hungarica.* **11** 8 235–249

[5] Mann G K I, Hu B G and Gosine R G 1999 Analysis of direct action fuzzy PID controller structures *IEEE Trans. Syst. Man, Cybern. Part B Cybern.* **29** 3 371–388

[6] Bakri F A, Mashor M Y, Sharun S M, Bibi Sarpinah S N and Bakar Z A 2016 Adaptive fuzzy logic controller with direct action type structures for InnoSAT attitude control system *IOP Conf. Ser. Mater. Sci. Eng.* **152** 1

[7] Hu B G, Mann G K I and Gosine R G 2001 A systematic study of fuzzy PID controllers - Function-based evaluation approach *IEEE Trans. Fuzzy Syst.* **9** 5 699–712

[8] Mizumoto M 1988 Fuzzy controls under various fuzzy reasoning methods *Inf. Sci. (Ny).* **45** 2 129–151

[9] Preeti D and Beniwal N 2012 Comparison of Conventional and Fuzzy P/PI/PD/PID Controller for Higher Order Non Linear Plant with High Dead Time *Int. J. Sci. Res.* **2** 8 217

[10] Uakarn C 2016 Attitude Control of Quadcopter Using Fuzzy PD + I Controller *Int. J. Comput. Internet Manag.* **24** 2 69–74

[11] MAHDI M C 2016 FUZZY PID CONTROLLER FOR NANO- *J. Sci. Arts.* **4** 37 407–416

[12] Hazza H A, Mashor M Y and Mahdi M C 2019 Performance of Manual and Auto-Tuning PID Controller for Unstable Plant - Nano Satellite Attitude Control System *2018 6th Int. Conf. Cyber IT Serv. Manag. CITSM 2018*. 7–11

[13] Mahdi M C, Shehab A, and Al Bermani M J F 2014 Direct Fuzzy Logic Controller for Nano-Satellite *Journal of Control Engineering and Technology* **4** 210–219

[14] Alavandar S and Nigam M J 2008 Fuzzy PD+I control of a six DOF robot manipulator *Ind. Rob.* **35** 2 125–132

Preliminary Study on Muscle Force Estimation using Musculoskeletal Model for Upper Limb Rehabilitation with Assistive Device for Home Setting.

Muhamad Fadzli [1], Akihiko Hanafusa[1], Yuji Kubota[1] and Daigo Nishimori[1]

[1] Department of Bio-science and Engineering, Shibaura Institute of Technology, 307 Fukasaku, Minuma-ku, Saitama 337-8570, Japan

nb18110@shibaura-it.ac.jp

Abstract. Post-stroke rehabilitation using assistive device has the potential to cover the need for improvement of the upper limb functionality. Moreover, using a biomechanical model to estimate the muscle activity during the rehabilitation training could improve the training module as well as help understand the target muscle during the motion of body part while using the assistive device. In this study, the author has focused on using a musculoskeletal model of the right arm to estimate the individual muscle force by simulating the movement of the right arm while using the developed assistive device. A developed upper limb assistive device has been investigated for its potential as a rehabilitation device for persons with physical disability of upper limb motion. The muscle force estimation is based on an inverse dynamic method, improved with additional constraints of the joints in order to obtain the muscle's activity from motion capture data. The acquired muscle force data could be used to improve the arm assistive device in rehabilitation training for home setting purpose.

1. Introduction

The number of stroke survivors in this world is quite large, and most of these survivors experience impairment impact on the upper limb function [1]. Patients who suffer from the upper limb impairment usually have difficulty performing daily activities that require using the upper limb, such as feeding, washing, etc. Some patients may recover some functionality of the upper limb function following the rehabilitation. However, most of the high technology assistive device are placed at the rehabilitation centre and must be operated with the help and observation of the therapists. Recently, wearable assistive devices have also started to play an important role as a rehabilitation device [2–4]. The needs of the assistive device and its cost effective and user friendly nature could help patients with rehabilitation training at home. The lack of the need for a therapist number is an added advantage, thus demanding the assistive device to be robust and easy to handle. The development and improvement of the assistive device need to come together with an understanding of the muscle force and muscle activation of the target muscle during the rehabilitation training.

Presently, three main approaches i.e., assessment scales, movement evaluation, and surface electromyography (sEMG) analysis are widely applied to evaluate the upper extremities. As these assessments are mainly viewed and scored by the therapist, the evaluation results are often subjective and general. The movement evaluation method using the motion capture systems can provide data on the physical movement of the upper limb, which can then be used for monitoring the progress of the rehabilitation. However, this method cannot account for muscle characteristics in patients, and the neurological mechanism used to overcome the problems associated with their pathology is also still unknown. Although all these methods are useful in the assessment of upper limb function, they are still inadequate for quantitative evaluation due to the lack of deep muscle's activation information and noise contamination from the movement artefact. Moreover, a direct interaction with the subject is needed in order to gain the information, and this could be a limitation based on the patient's condition, time consumed to setup the system and the cost for the actual test involving many equipment and subjects.

Given this, this study presents a method using a musculoskeletal model focusing on the upper limb to predict muscle force during the elbow flexion with the assistive device. Individual muscle force was investigated during the movement of the elbow flexion with the healthy subject using the assistive device and the results were compared with those of the one not using any assistive device. Through this approach, the specific functional muscle involved during the movement can be known, making it possible to conduct improvement in the assistive device for rehabilitation training purpose. A brief flowchart of the proposed muscle force estimation is shown in figure 1.

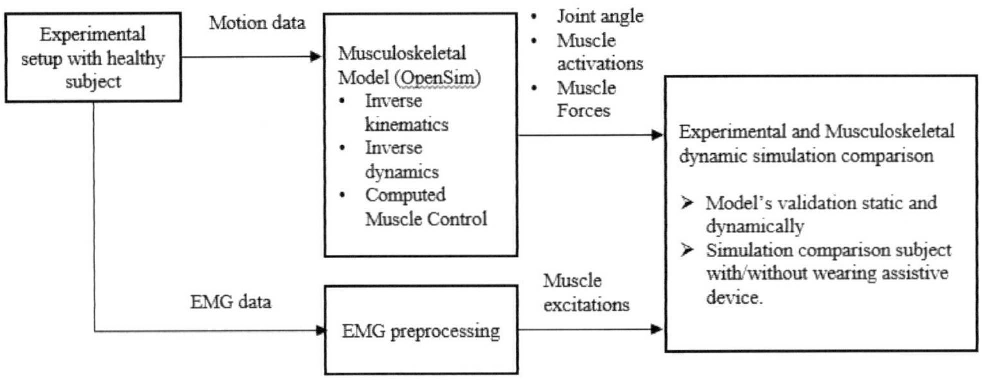

Figure 1: Flowchart of proposed muscle force prediction using musculoskeletal model.

2. Method

In order to study the muscle force based on the given motion data, a musculoskeletal model from the musculoskeletal software, OpenSim [7] was utilized. A suitable pre-existing musculoskeletal model has been used and scaled to match the marker data. Then the model is validated dynamically by comparing the muscle force estimated from simulation and EMG data from the experiment. Finally, an experiment is performed on a healthy subject with assistive device and the result is compared with the simulation data for the same motion.

2.1. Data Acquisition and Experimental Setup for Musculoskeletal Model Validation

2.1.1. Electromyogram (EMG) & Motion Capture Data Recording.

A healthy subject volunteered for this investigation and gave their informed, written consent. The project was approved by the Human Research Ethics committee at the Shibaura Institute of Technology (SIT), Japan. The subject was quietly seated in the chair with their torso kept upright and their right hand keeping at close to a 90-degree elbow flexion. The reach forward motion is designed to obtain motion of the right upper limb. The motion simulated the elbow flexion from 90 degree to forward reaching and return to initial position. The same motion was repeated three times to get three sets of data. The configuration of the EMG recording and marker placement for motion capture system is shown in figure 2. Motion data was acquired using the Mac3D system available in our laboratory. This equipment consists of 10 infrared cameras that are able to capture the 3D position of the different markers over time. During the motion recording, 10 markers were used at specific positions together with marker clusters according to the recommendation on definitions of joint coordinate systems [6].

Six predominant muscles at upper arm activating elbow DoFs were selected to be the muscle of test, as shown in figure 2 (a) and (b). Six channels of bipolar differential amplifier were carefully placed on these muscles based on both the anatomy and hand touch experience. The active EMG electrodes of each channel were positioned at the muscle belly. The skin underneath the electrodes was cleaned with alcohol patch to reduce the resistance between the skin and the electrodes. The motion recording was sampled at 200 Hz and synchronized with the EMG recording through the motion capture system.

Ch.1: Deltoid (anterior part) Ch.4: Triceps (lateral head)
Ch.2: Deltoid (posterior part) Ch.5: Biceps (short head)
Ch.3: Biceps (long head) Ch.6: Triceps (long head)

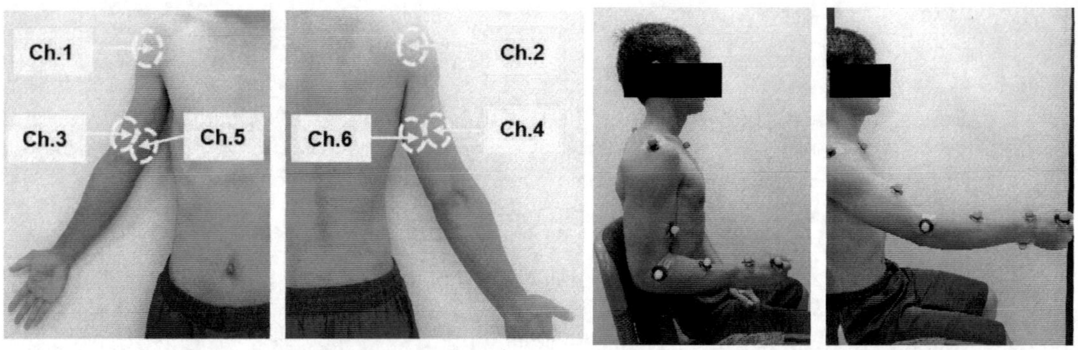

(a) 3 channels (front) (b) 3 channels (back) (c) initial position (d) reach forward motion

Figure 2: The configuration of 6 channels EMG electrodes for upper arm (a) and (b), and a healthy subject performing the reach forward movement is shown in (c) and (d).

2.1.2. Joint angle estimation, Joint torque, and Muscle Force Estimation using OpenSim

Delp et al. [5] have developed an open source platform called OpenSim. This platform allows the dynamic simulation on the musculoskeletal system using provided motion capture data. These simulations use inverse kinematics method to obtain kinematics data such as joint angle of each joint during the movement, which is later used in inverse dynamic simulations to obtain the joint moments. Then, an original algorithm, called Computed Muscle Control (CMC), based on inverse dynamics method is used to compute the muscle forces allowed to obtain the muscle excitation. An upper limb model for the right hand is available in this platform. It has realistic movements and precise muscular topology for the joints. This study utilizes the CMC to validate the OpenSim Model. Then the same approach is used to study the active muscle in the human musculoskeletal model during the motion of the upper arm.

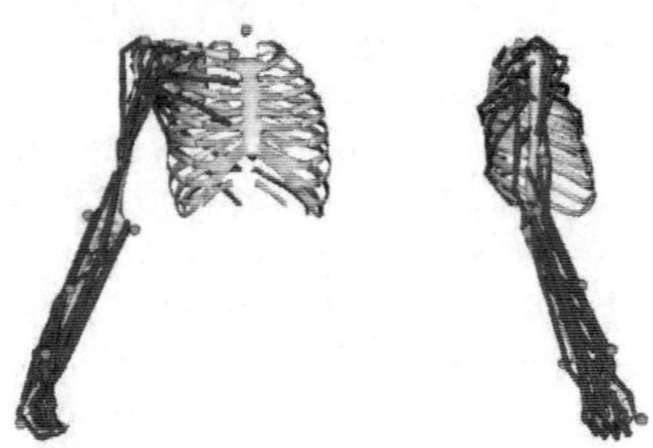

(a) Front view (b) Lateral view

Figure 3: OpenSim upper limb musculoskeletal model. This model was developed by Saul KR [7]. It consists of 7 body segments and 32 muscles across the shoulder, elbow, forearm, and wrist.

2.1.3. Validation Result

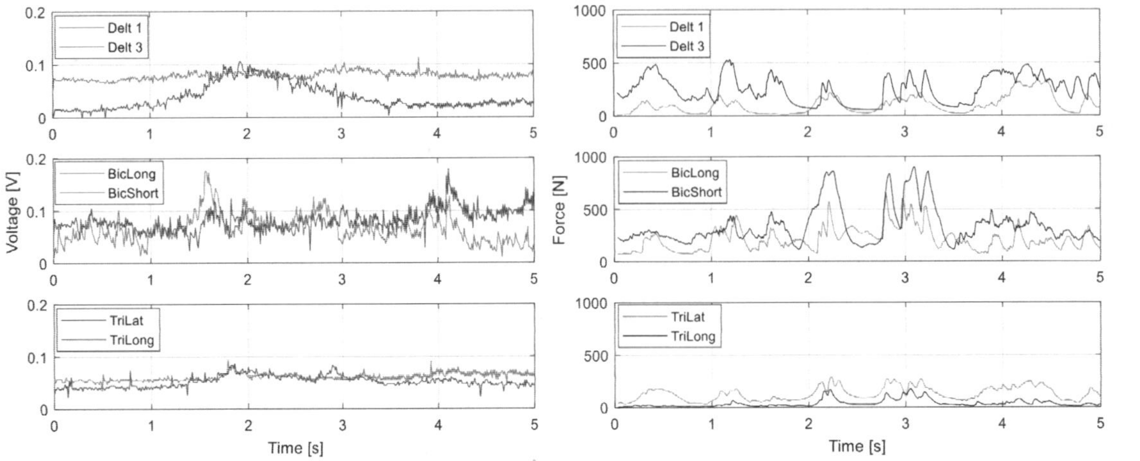

(a) Experimental processed EMG data (b) Estimated muscles force using upper limb model

Figure 4: Experimental muscle excitation and estimated muscle forces for the same reach forward motion.

As mention in section 2.1.1, we selected deltoid anterior part (Delt 1), deltoid posterior part (Delt 3), short head of biceps (BicShort), long head of biceps (BicLong), long head of triceps (TriLong), and lateral head of triceps (TriLat) as 6 muscles of interests. Figure 4 (a) shows the processed EMG data for the muscles taken during the experiment and estimated muscle forces shown in figure 4 (b) from the simulation of the musculoskeletal model. By adopting a neural mapping method [10], we assumed that BicShort has the same activation as BicLong and other two heads of triceps have the same activation. We can see in both experimental EMG data and estimated muscle forces from simulation shows that the Biceps muscles are working during the motion and presents the main activity. On the other hand, the triceps muscle does not show big activity due to its minor role in joint motion. We also can see that the pattern of the most muscle forces (Biceps and Triceps) shows similarity with the recorded EMG pattern. The similarity agreement between the recorded EMG data and the estimated muscle forces for the same motion shows that this model is acceptable to predict muscle activity for upper limb right hand motion.

2.2. Experiment with Assistive Device and Simulation with Musculoskeletal Model

2.2.1. Assistive Device, Experiment & Simulation Protocol

A lightweight assistive device [8], which is wire-driven by 2 servo motors was developed. The device can generate the motions of elbow flexion/extension movement and internal/external rotation movement, performed by pulling the wire hung on a pulley connected to the wrist part. During the experiment, the subjects were standing with their torso kept upright and their right hand kept relaxed. One motion designed had the elbow flexion from the natural position to close to a 100 degree. Two separate motions have been designed where the subject performed the motions with and without the assistive device. Then the same motion was simulated using the musculoskeletal model in OpenSim. The marker data taken is produced using the Motion Capture 3D system available in our laboratory. A torque of 4Nm is applied to the musculoskeletal model, simulating the model with assistive device to achieve the same flexion motion close to 90 degree. A target muscle, which is the bicep (short), is investigated for muscle activity.

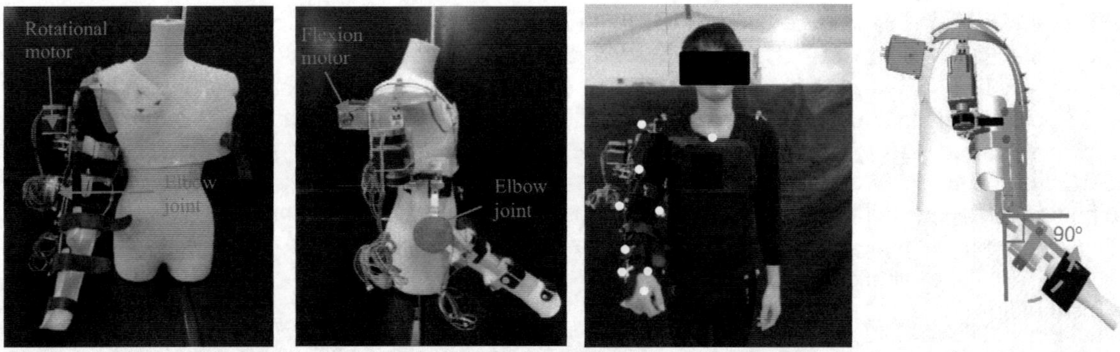

Figure 5: Upper limb right hand assistive device and subject wearing the device performing motion of elbow flexion close to 90 degree.

3. Results and Discussion

The purpose of the current work is to study the muscle activation estimation method via musculoskeletal model when using upper limb assistive device. The muscle estimation was evaluated by simulating the musculoskeletal model using the motion data provided through the experiment of the subject wearing and without wearing the developed assistive device. Figure 6 (a) shows the measured EMG data signal processed with 1Hz cut off low pass filter and figure 6 (b) shows the estimation of the muscle force of the target muscle during the elbow flexion of one subject using and without using the assistive device.

As can be seen from figure 6, the pattern of the muscle force estimation using the musculoskeletal model shows good agreement with the experimental muscle excitation (EMG) pattern. The most significant difference can be observed by magnitude of the muscle activation without device, which is slightly higher when compared to the muscle with assistive movement. Even though it is normal for the assistive part to produce less muscle activities, our proposed estimation method proven can be used and it will be possible for us to get a deeper understanding of the human dynamic movement mechanism while using an assistive device for rehabilitation purpose.

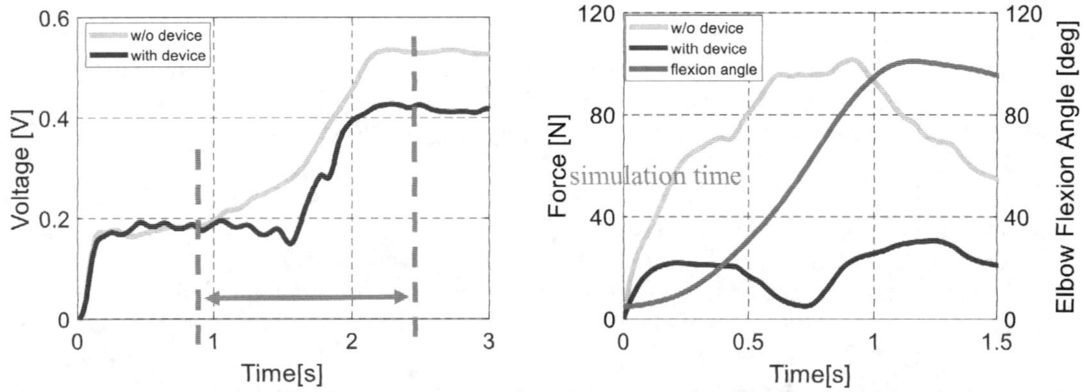

(a) Filtered EMG data (b) Estimated muscle force & elbow flexion angle

Figure 6: EMG data from experimental and estimated muscle force from simulation for the subject wearing/without wearing the assistive device.

Several limitations of our study should be noted. Firstly, the musculoskeletal model used in OpenSim is only scaled based on the marker data from the 3D motion capture system. The best practice in building the musculoskeletal model, the MRI data from the subject should be used to construct the model from scratch. However, the procedures are very difficult and require a lot of time to construct and validate a musculoskeletal model for simulation purpose. Second, the choice of muscles should be investigated carefully. As there may be more muscles wrapping around the elbow, each muscle mechanical function should be first figured out and the main action muscles chosen then as the flexors and extensors as muscles of interest. Lastly, as we only collected one subject's motion and EMG data for one movement, a generalizable criterion cannot be established. Nevertheless, the scaling and validation process make the musculoskeletal model make it suitable to be used for the tasks. This could allow the performance of a variety of movement imitated rehabilitation training while using the assistive device in the future.

4. Conclusion

This work proposed an approach to use a musculoskeletal model of the upper right arm to predict individual muscle activation during the motion of elbow flexion when the subject is using the assistive device. The model is first validated dynamically by comparing the simulation of the movement and experimental EMG data, which shows agreement in the pattern. The validated model is later used to assess the upper limb target muscle when the subject is using the assistive device. This method has great potential to be explored and is reliable to determine the muscle activation that could provide us with deeper understanding of the muscle characteristics during the rehabilitation training when using the device. Future improvement of the mechanical design and movement of assistive device can also be considered aligned with the human dynamic movement data achieved from simulations results.

References

[1] Mozaffarian D, Benjamin E J, Go A S et al. 2015 Heart disease and stroke statistics—2015 update: a report from the American Heart Association *Circulation* **131** 4 pp. e29–e322,.

[2] Jung Y and Bae J 2015 Kinematic analysis of a 5-DOF upper-limb exoskeleton with a tilted and vertically translating shoulder joint *IEEE Trans. on Industrial Mechatronics* **20** 3 pp. 1643–1648.

[3] Perry J C, Rosen J and Burns S 2007 Upper-limb powered exoskeleton design *IEEE Trans. on Industrial Mechatronics* **12** 4 pp. 408–17.

[4] Nishimura M, Hyodo K, Kawanishi M and Narikiyo T 2015 Proof of concept for robot-aided upper limb rehabilitation using disturbance observers *IEEE Trans. on Human-Machine Systems* **45** 1 pp. 110–18.

[5] Delp S L, Anderson F C, Arnold A S, Loan P, Habib A and John C T 2007 Guendelman, E., Thelan, D.G. OpenSim: Open-source software to create and analyze dynamic simulations of movement *IEEE Transactions on Biomedical Engineering* **55** pp 1940–50.

[6] Wu G, F C T van der Helm, H E J, (DirkJan), Veeger M, Makhsous, Van Roy P, Anglin C, Nagels J, Karduna A R, McQuade K and Wang X 2005 ISB recommendation on definitions of joint coordinate systems of various joints for the reporting of human joint motion. Part II. Shoulder, elbow, wrist and hand *J. Biomech* **38** pp. 981–92.

[7] Saul K R, Hu X, Goehler C M, Daly M, Vidt M E, Velisar A and Murray W M 2015 Benchmarking of dynamic simulation predictions in two software platforms using an upper limb musculoskeletal model *Computer Methods in Biomechanics and Biomedical Engineering*. **18** pp.1445–58. 10.1080/10255842.2014.916698.

[8] Hanafusa A, Shiki F, Ishii H, Nagura M, Kubota Y, Ohnishi K and Shibata Y 2018 Development of an active upper limb orthosis controlled by EMG with upper arm rotation. In Intelligent Human Systems Integration – *Proc. 1st Int Conf on Intelligent Human Systems Integration IHSI* 2018.

[9] Shahrol M, Annisa J, Ana S, Zainal A, Mohd S J, Noor A M, Muhamad F A and Helmy H 2015 Development of Upper Limb Rehabilitation Robot Device for Home Setting *Proc 2015 IEEE Int Symp on Robotics and Intelligent Sensors*, pp. 376–80.

[10] Pizzolato, C.; Lloyd, D.G.; Sartori, M.; Ceseracciu, E.; Besier, T.F.; Fregly,B.J. 2015 CEINMS: A toolbox to investigate the influence of different neural control solutions on the prediction of muscle excitation and joint moments during dynamic motor taks. *J. Biomech., **48**, 3929-3936.*

Journal of Physics: Conference Series **1372** (2019) 012024 doi:10.1088/1742-6596/1372/1/012024

A Study of Biomagnetic Stimulation for Transcranial Application

Nor Shahira Ab Ghafar[1], Zulkarnay Zakaria[1*], Anas Mohd Noor[1], Chong Yen Fook[1], Ahmad Nasrul Norali[1], Asyraf Hakimi Abu Bakar[1], Muhammad Khairul Ali Hassan[2], Mohd. Hafiz Fazalul Rahiman[2], Jaysuman Puspanathan[3], Elmy Johana Mohamad[4], Siti Zarina Mohd Muji[4]

[1]*Biomedical Electronic Engineering Department, School of Mechatronic Engineering, Universiti Malaysia Perlis*
[2]*Mechatronic Engineering Department, School of Mechatronic Engineering, Universiti Malaysia Perlis*
[3]*Sport Innovation & Technology Centre (SiTC), Faculty of Engineering, Universiti Teknologi Malaysia, Malaysia*
[4]*Faculty of Electrical and Electronic Engineering, Universiti Tun Hussein Onn Malaysia*

**zulkarnay@unimap.edu.my*

Abstract. Biomagnetic stimulation is a treatment uses the application of magnetic field principle for a therapeutic purpose. This non-invasive therapeutic technique use low frequency electromagnetic field which is less than 10 MHz. In transcranial application, electromagnetic field is use to stimulate the brain tissue for the people affected by weaken neurologic and psychiatric disorders. This paper discusses the modelling and simulation of Transcranial Magnetic Stimulation using Comsol Multiphysics software. In this research four designs have been simulated. The overall analysis indicates that a single coil design has the best performance in stimulating the area of interest at the frequency of 5MHz.

1. Introduction

The human brain is made up of biological tissues with a specific value of conductivity that is different than others biological tissues in human body. Any electrical abnormality that happens to human brain can leads to a neurological disorder such mental illnesses and ischemic [1]. Transcranial Magnetic Stimulation is basically based on Faraday electromagnetic induction principle where a pulse of electric current with a high intensity is applied near a patient head through a coil, non-invasively. The current in the coil in turn will generate a electromagnetic field which penetrates the scalp and skull and the brain tissue. The energy form the field is transferred to the brain tissue and trigger the metabolism of the brain cells and activates the brain activities through depolarising the neurons which eventually producing a different neurophysiological responses and behavioural changes at the targeted region [2].

In the case of glioma which is the interest of this project, it is found that 40% of glioma located in the frontal lobe, 29% in temporal lobe, 14% in parietal lobe, and 3% in occipital lobe and with 14% in the deeper structures. Glioma is a kind of tumor that occurs in brain and spinal cord [3],[4].

2. Modelling and Simulation procedure

COMSOL Multiphysics is used to design the biomagnetic stimulation coil for transcranial which involves brain tissue. The coil that will be used is a circular type coil with 1A [5] and the radius is set up to 25mm with 10 number of turns [6]. The brain model is a 66 years old human head with a radius of 100mm [7]. The distance of stimulating coil from the scalp is fixed to be 5mm [8]. The depth of a cortical surface from the surface of the head is assumed to be 15 mm [9]. Hence, the radius of cortical region of the brain will be 85mm. The conductivity and permittivity value for the brain tissue is set to be 0.292 S/m and 320 respectively [10]. In this project, the frequency is set to be fixed at 5MHz. The material of the coil is made up of copper. The flow chart of methodology is shown Figure 1.

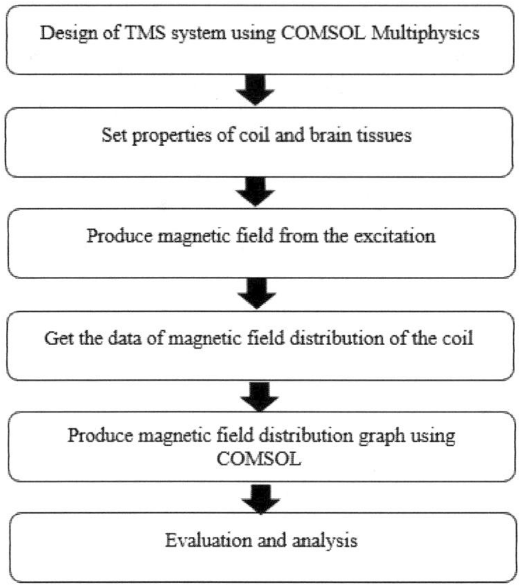

Figure 1. Flowchart of the procedure

3. Result and Discussion

3.1. The modelling of biomagnetic stimulation effect on transcranial application.

Figure 2 shows the model design in Comsol and the magnetic flux density after stimulation. Figure 3 shows a graph for magnetic density versus distance. From this graph, we can conclude that as the distance increase, the value of magnetic density decreases. Hence, this system is valid for a normal size of human brain.

Journal of Physics: Conference Series **1372** (2019) 012024 doi:10.1088/1742-6596/1372/1/012024

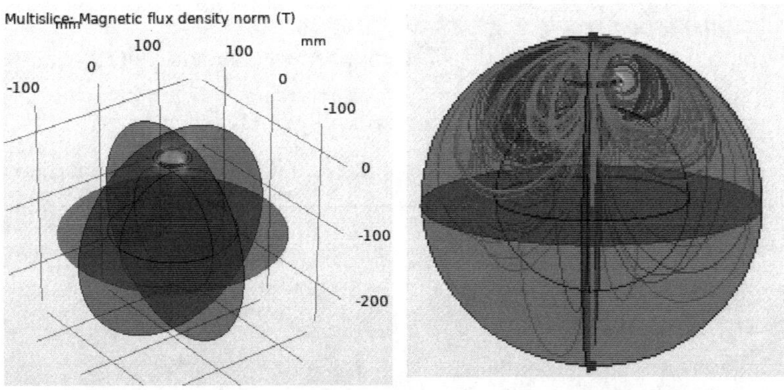

Figure 2. Coil design and magnetic flux density stimulation using COMSOL Multiphysics

3.2. The modelling and stimulation of four different design of coil placement

Table 1. Four different designs of coil placement

DESIGN	 TOP VIEW	 FRONT VIEW
1 COIL	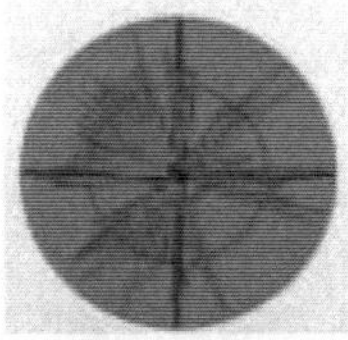	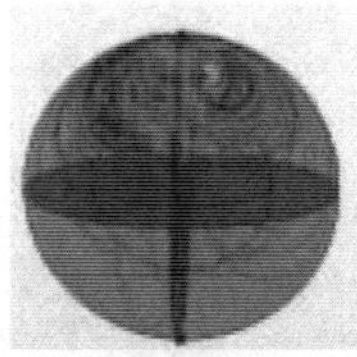
2 COIL	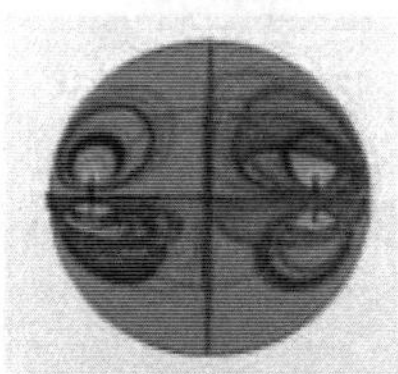	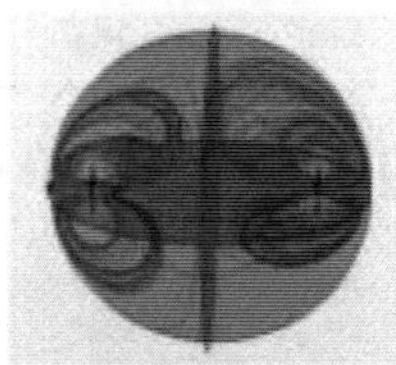
3 COIL	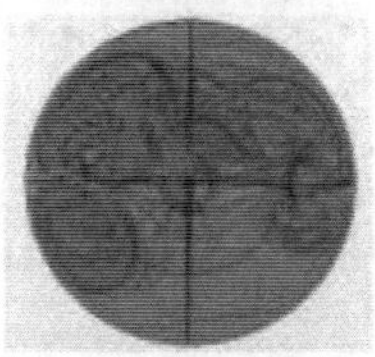	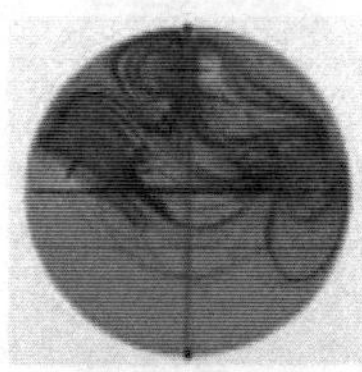
4 COIL	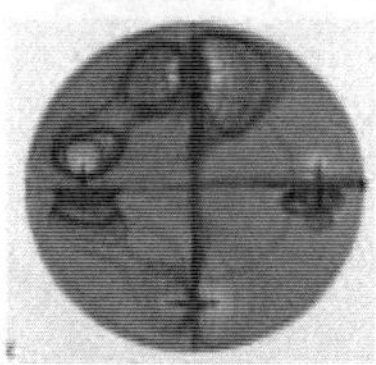	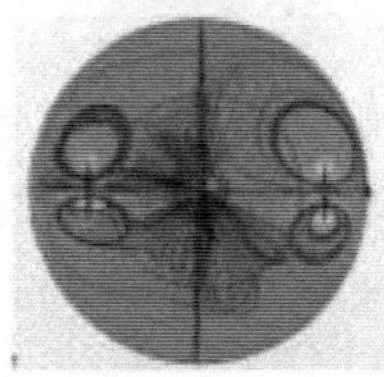

3.3. The evaluation of simulation result

Table 2 shows the graph for magnetic field distribution produced for each design.

Table 2. Magnetic Flux Density (T) vs Distance (mm)

One coil	Two coil
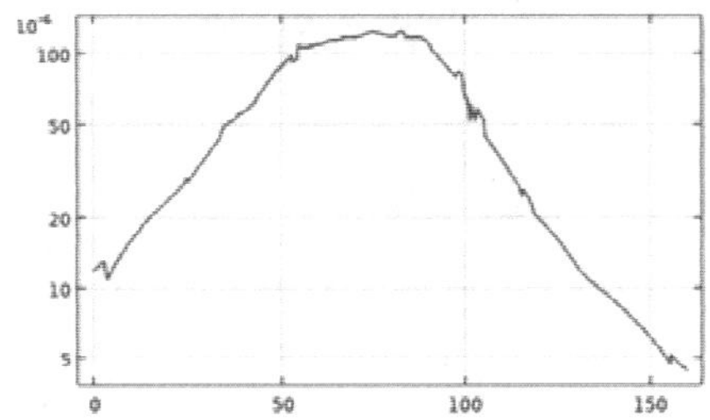	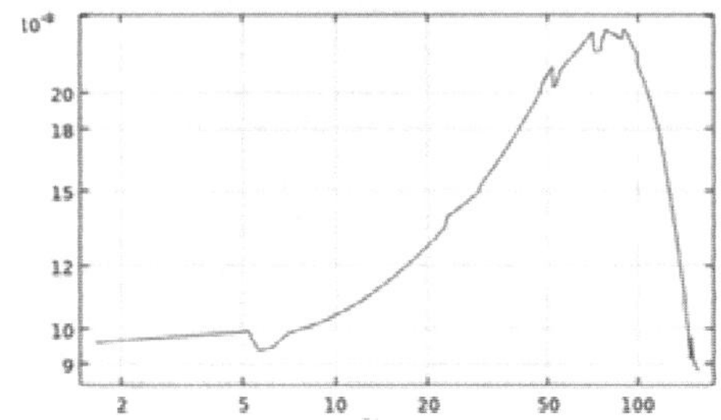
Three coil	**Four coil**
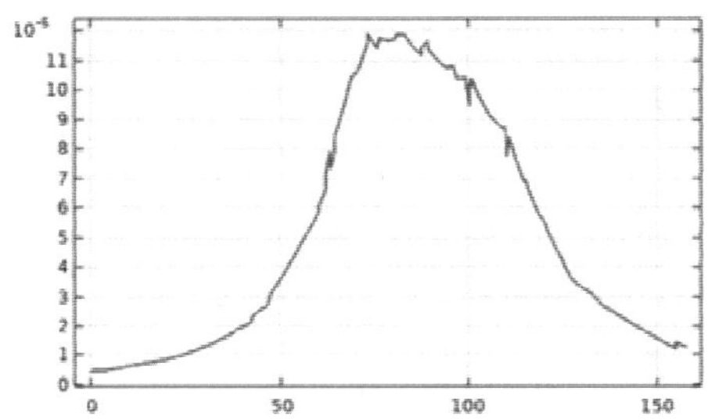	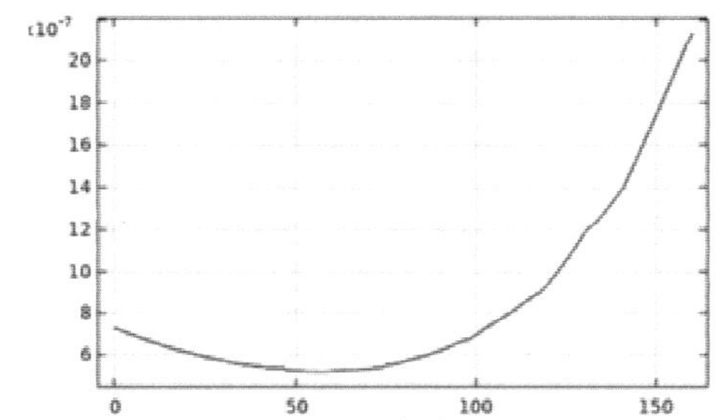

Table 1 show One-Coil and Three-Coil design has distribution of electromagnetic field at the central of brain area whereas for Two-coil and Four-coil design the distribution are towards the end of the line. In term of maximum strength, approximately 120x10^{-4} T, 30x10^{-4} T, 12x10^{-5} and 22x10^{-7} T for One-Coil, Two-Coil, Three-Coil and Four-Coil respectively.

4. Conclusion

The results show that One-coil design has been the best option which produce focus central electromagnetic field distribution and with the highest maximum amplitude.

Acknowledgments

The author would like to thanks University Malaysia Perlis (UNIMAP) for supporting this final year project for undergraduate students.

References

[1] S. U. Ã and M. Sekino, "Biomagnetics and Bioimaging for Medical Applications," *Magn. Magn. Mater.*, vol. 304, pp. 122–127, 2006.

[2] M. Sekino, H. Ohsaki, Y. Takiyama, K. Yamamoto, T. Matsuzaki, Y. Yasumuro, A. Nishikawa, T. Maruo, K. Hosomi, and Y. Saitoh , "Eccentric Figure-Eight Coils for Transcranial Magnetic Stimulation", *Dep. Electr. Eng. Inf. Syst.*, vol. 1, pp. 1–38, 2015.

[3] S. Larjavaara, R. Mäntylä, T. Salminen, H. Haapasalo, and J. Raitanen, "Incidence of Gliomas by Anatomic Location," *Neuro. Oncol.*, vol. 23, pp. 319–325, 2007.

[4] K. Akeret, C. Serra, O. Rafi, V.E. Staartjes, J. Fierstra, D. Bellut, N. Maldaner, Lukas L. Imbach, F. Wolpert, R. Poryazova, L. Regli, and N. Krayenbühl, "Anatomical Features of Primary Brain Tumors Affect Seizure Risk and Semiology," *Neuroimage (Amst).*, vol. 22, pp. 1–9, 2019.

[5] S. Sarkawi, Z. Zakaria, I. Balkhis, J. A. Jalil, M.A.A. Rahim, M.H.F. Rahiman, N. Mustafa, R.A. Rahim, and Z. Shaffie"3D Model Simulation on Magnetic Induction Spectroscopy For Fetal Acidosis Detection Using COMSOL Multiphysics," in *AIP Conference Proceedings*, 2016, vol. 050003, no. 1774, pp. 2–10.

[6] S. M. Goetz and Z. Deng, "The development and modeling of devices and paradigms for transcranial magnetic stimulation," *HHS Public Access*, vol. 29, no. 2, pp. 115–145, 2018.

[7] D. Lazutkin, A. Harkara, and P. P. Husar, "Magnetic Stimulation of the Human Brain with Low-Intensity Field," in *COMSOL conference*, 2010, pp. 1–4.

[8] M.G. Stokes, C.D. Chambers, I.C. Gould, T.R. Henderson, N.E. Janko, N.B. Allen, and J.B. Mattingley,"Simple Metric For Scaling Motor Threshold Based on Scalp-Cortex Distance : Application to Studies Using Transcranial Magnetic Stimulation," *J Neurophysiol*, vol. 94, pp. 4520–4527, 2005.

[9] A. V. P. Zhi De Deng, Sarah H.Lisanby, and A.V. Peterchev,"Electric Field Depth–Focality Tradeoff in Transcranial Magnetic Stimulation: Simulation Comparison of 50 Coil Designs," *NIH Public Access*, vol. 6, no. 1, pp. 1–13, 2013.

[10] R. Pethig, "Electrical Properties of Biological Tissue," *Inst. Mol. Biomol. Electron.*, pp. 94–131, 2014.

[11] Z. Zakaria, S. Sarkawi, J. A. Jalil, I. Balkhis, M. Aliff, A. Rahim, and N. Mustafa, "Simulation of Single Channel Magnetic Induction Spectroscopy for Fetal Hypoxia Detection," *J. Teknol.*, vol. 73, no. 6, pp. 107–110, 2015.

Smart Medical Chatbot with Integrated Contactless Vital Sign Monitor

Wan Muhamad Asyraf Wan Zaki[1], Muhammad Faiz Md Shakhih[1], Muhammad Hanif Ramlee[1,2,] and Asnida Ab Wahab[1,2*]

[1]School of Biomedical Engineering and Health Science, Faculty of Engineering, Universiti Teknologi Malaysia, 81310, Johor, Malaysia
[2]Medical Device and Technology Centre (MEDITEC), Institute of Human Centered Engineering (iHuman), Universiti Teknologi Malaysia, 81310, Johor, Malaysia.

[*]asnida@biomedical.utm.my

Abstract. Artificial intelligence is one of the important fields in modern technologies to help us strive for better life. Healthcare industries nowadays spend a lot of money researching on how artificial intelligence can help improve their services and give the highest satisfaction to their customers. Most healthcare organisations have a passive relationship to their patients when it comes to communication and this situation is often worsened because of a lack of inter-operability between client and provider. Mobile applications on the other hand have become one of the effective strategies in bridging the interaction between provider and end user. In this study, an automated self-learning system is designed to provide conversational healthcare for personalised proactive experience. This system is developed along with the in cooperation of contactless monitoring device using a vision-based real-time monitoring of vital signs which allow patients to monitor their oxygen level, heart rate and respiration rate. This system is also automatically calibrated across patients, allowing precise measurement using highest probability method and natural language processing. Results obtained from the comparative analysis show a promising result with an error of 1.16 for pulse sensor and 2.917 for ECG which are below the threshold error. This allows user to accurately measure vital signs in a non-obtrusive way, and to provide them with the data required to determine to the right timing for any intervention procedure needed. The developed system would also help to bridge the gap of interoperability between client and medical provider

1. Introduction

One of the requirements for general artificial intelligence (AI) is to interact with the machine through natural language. This AI field includes dialog systems, chatbots or spoken dialog systems. The chatbot is also drawing upon ever-growing medical question range, to broaden its already significant wealth of medical expertise. Many seemingly static scenes contain subtle changes that are invisible to the naked human eye. However, it is possible to pull out these small changes from videos through the use of algorithms via motion magnification [1]. Motion magnification gives a way to visualize these small changes by amplifying them and to pull out interesting signals from these videos, such as the human pulse [2]. Motion magnification techniques can be divided into two categories which are Lagrangian and Eulerian approaches [5,4,3]. Lagrangian approach explicitly extracts the motion field (or optical flow) and uses it to move the pixels [10]. Lagrangian perspective is to analyse the angle of motion of the pixels (particles) of interest in the tracking image [6]. However, the Lagrange perspective approach has the following shortcomings where it is necessary to accurately track and estimate the trajectory of particles, which requires more computational resources. Unlike the Lagrangian perspective, the Euler perspective does not explicitly track and estimate the motion of the particle, but instead fixes the perspective in one place, such as the entire image. In 2012, Wu et al. started from this perspective and proposed a method called Eulerian Video Magnification [7]. Eulerian techniques are useful for revealing subtle motions, but they are hand-designed, and do not take into account many issues such as occlusion. Wadhwa et al. improved the technology in 2013 and proposed a phase-based image motion processing technique [8]. In this study, the eulerian technique is used to extract the heart rate of the target person, where occlusion does not affect the end result much. Current systems utilize various sensors which require constant and secure physical contact with the patients during monitoring process and have a passive relationship to their clients when it comes to communication. This often worsens the problem because of a lack of inter-operability between client and provider [9]. The use of personal health assistants on our phone has suddenly become definite responses for certain needs, allowing precise measurement without the need for individual adjustments and contactless vital sign acquisition. However, for patient with specific conditions, a less intrusive monitoring method is critically required to avoid any unnecessary distress to the patient [10]. This study is focus on chatbot operation and process, where the text presented to them by the user (a process known as "parsing"), inferring what user mean and/or want, and determine a series of appropriate responses based on that patient information and vital sign prior to responding to a series of algorithms that interpret and identify the common disease such as common cold (viral rhinopharyngitis), tension-type headache (TTH) and migraine with exception of cancer related disease.

2. Methodology

2.1 Database and Automated Diagnosis

The system encompasses of state of art chatbot interface along with vision based vital sign solutions that offers user to accurately measure vital signs in a non-obtrusive way, and to provide them with the data required to determine the need for any intervention procedure [11]. The method of database recall for disease identification, as shown in Figure 1 which was similar in ideas and formulaa to those of the "maximum likelihood method" and (head, ear, eye, nose and throat) HEETN exam, were applied to the automated diagnosis. The likelihood was weighted with a prior probability of each disease. Clinical data include quantities that are continuous, discrete, and bivariate [12]. Each of the data is therefore considered to be bivariate in terms of quantity either a symptom occurred or not. The format item that is specified here and used directly for automated diagnosis is designed to select from the three: YES, NO, and UNKNOWN. In this way the answer can be chosen.

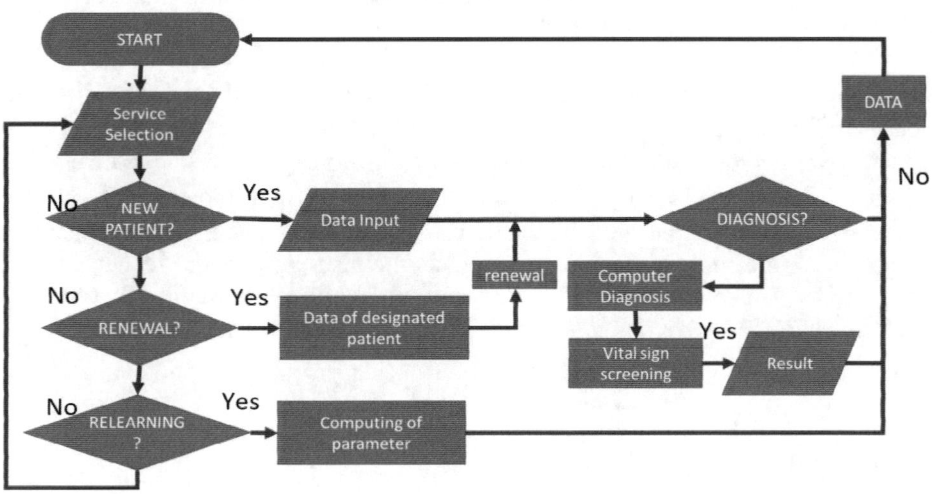

Figure 1. Flowchart of Automated Diagnosis.

2.2 Motion Magnification

Heart rate is measured from subtle change in colour and movement under skin epidermis, due to variation in volume and oxygen saturation of the blood in vessel at every heartbeat. Each cardiac cycle appears in peak within small range e.g. 50 and 90 bpm.

2.3 Proposed Algorithm Pipeline

The automated diagnosis system is integrated with a chatbot and motion magnification for vital sign sensing. The architecture, as shown in Figure 2, consists of user interface and server with different sets of parameters. The left one (machine learning) is called the encoder, while the below one (corresponding logic and rule parser tokens) is called the decoder. The encoder machine learning conceives a sequence of context tokens one at a time and updates its hidden state. After processing the whole context sequence, it produces a final hidden state, which incorporates the sense of context and is used for generating the answer. For this purpose, a SoftMax layer over vocabulary of HEETN exam is maintained in the decoder machine learning [12]. At each time step, this layer takes the decoder hidden state and outputs a probability distribution over all words in its vocabulary.

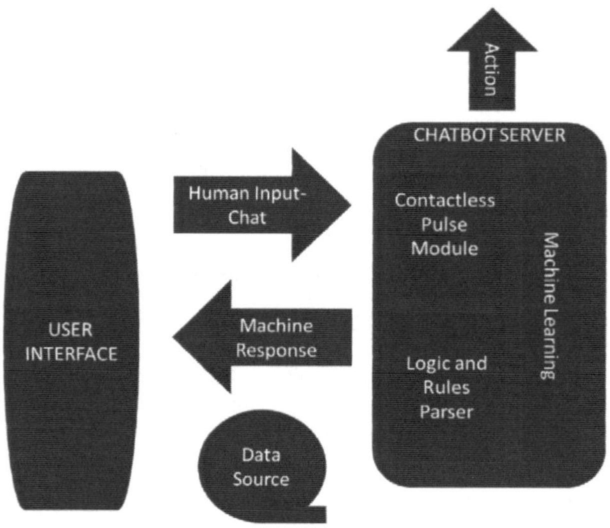

Figure 2. Overview of the system, which compromise of user interface and a chatbot server that contain principle component to realise automated diagnosis.

First, decoder hidden state is initiated with final encoder hidden state, $h_0^{dec} = h_n^{enc}$. Then, logic and rule parser token are passed as first input to the decoder and update first hidden state, $h_0^{dec} = rnn_\theta[h_{n-1}^{dec}, \omega_t]$. Next is to sample (or take one with max probability) first word from first SoftMax layer $\omega_{t+1} \sim p_{t+1} = softmax(g_\varphi(h_t^{dec}))$. Finally, pass this word as input, update hidden state and generate new word. Repeat step 4 until logic and rule parser token is generated or maximum answer length is exceeded. More simply, the network predicts the next word in the sequence by providing it with a correct prefix. Training is performed via maximum likelihood training, which leads to classical cross-entropy loss [12]:

$$L = \sum_{t=1}^{m} \sum_{i=1}^{|V|} I[y_t = i] log\hat{p}_{t,i} = \sum_{t=1}^{m} log\hat{p}_{t,y_t} \quad (1)$$

Where, y_t is a correct word in reply at time step *t*.

2.4 Contactless vital sign monitor

As mentioned above, the first step of image amplification technique is to spatially filter the video sequence to obtain basebands of different spatial frequencies. The topmost image, that is, the image that in contrast with changes in blood volume in a portion of the peripheral microvasculature had the lowest spatial frequency and the highest signal-to-noise ratio, hence can use the maximum magnification, and the magnification of the next layer is sequentially decreased. In addition, it is easy to approximate the heart rate signal where images with higher spatial frequencies, such as the original video image, approximate with Taylor series expansion for reconstruction, as shown in Figure 3. In a nutshell, the contactless vital sign monitor can be visualised in block diagram as follows:

Figure 3. Step for vital sign via Eulerian video magnification.

3. Result and Discussion

Comparative analysis between our developed method and existing methods was performed. Based on Figure 4, vital sign result was based on the Eulerian magnification, specifically the heart rate measurement. All results in this section were processed with temporal filters unless otherwise noted.

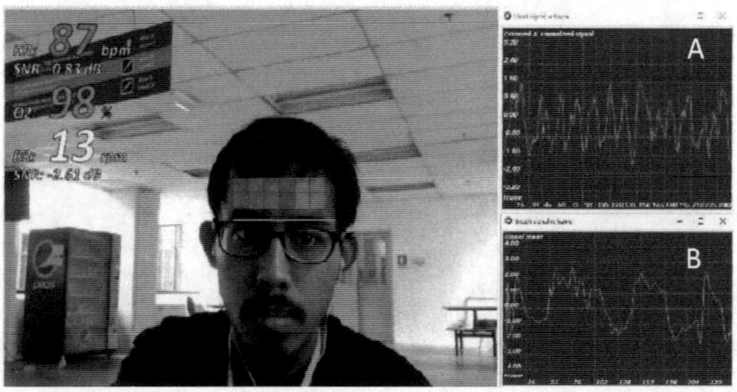

Figure 4. User Interface of the Contactless Vital sign Monitor. Graph A shows the heart rate while graph (B) shows the breath rate.

3.1 Disease identification accuracy

Disease identification metrics such as Dice indexes (DI), Jacard indexes (JI), false positive (FP) and false negative (FN) were used to validate the performance of proposed automated diagnosis of common disease. In this test, performance of the developed system on common cold (viral rhinopharyngitis), TTH and migraine as shown in Table 1 were measured

Table 1. Performance of measured result

Disease	Acc (%)	Precision	Sensitivity	Specificity	F-Measure	DI	JI
C. Cold	96.667	0.9678	0.9670	0.0126	0.9583	0.9753	0.9711
TTH	93.333	0.9357	0.9340	0.0187	0.9096	0.9640	0.9390
Migraine	89.876	0.8850	0.8825	0.0346	0.8550	0.8976	0.8314

Based on Table 1, it can be seen that migraine had lower accuracy due to a lot of subclasses for the machine to pin down exact root symptom. Yet, the system still performs better than threshold accuracy, which is over 96.667% for common flu, and 93.333 for TTH. The overall system had more than 0.85 true positive or sensitive, which provide high probability sample for disease identification.

3.2 Comparison with pulse sensor and electrocardiogram

Based on comparison with pulse sensor and ECG system in Figure 5, it is found out that the developed system had mean difference of 1.25 for pulse sensor and 2.92 for ECG for the heart rate measurement. This is because, our contactless system suffers from outdoor occlusion and changing intensity, but with closed room, the difference might be lower.

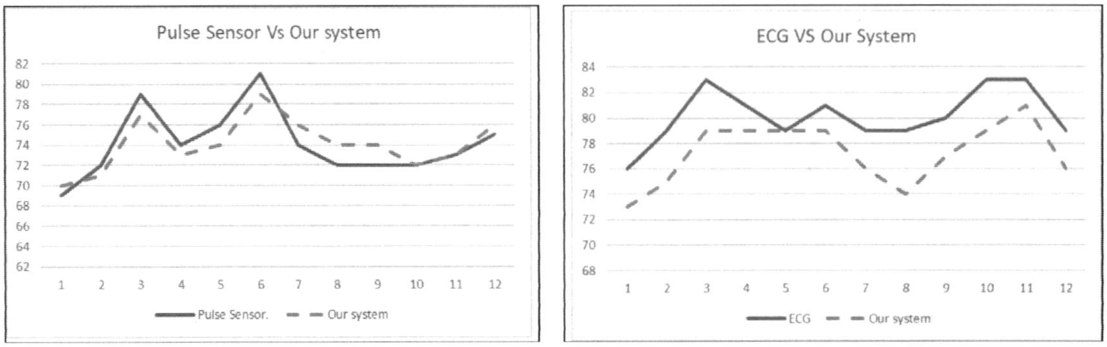

Figure 5. The results of 12 sample readings from pulse sensor, ECG and our system.

Mean average error (MAE) was calculated to determine the difference between two continuous heart rate measurements. This was performed to obtain the error contributes to MAE in proportion to the absolute value of the error. In addition, the feasibility of the system can be observed in ways that disregard the direction of which a measure that place emphasis on this is significant on same scale as the data being measured [13]. The formula to calculate MAE is as follows:

$$MAE = \frac{1}{n}\sum|Actual - Forecast| \quad (2)$$

From the pulse sensor, the MAE attained is 1.167 and for ECG is 2.917. This is obvious because pulse sensor calculates heart rate based on constitutes another means of determining the timing of cardiac cycles via continuous monitoring of changes in blood volume in a portion of the peripheral microvasculature while ECG measure the bio-potential generated by electrical signals that control the expansion and contraction of heart chambers. ECG should have highest accuracy to benchmark for heart rate, but as per monitoring and initial diagnosis, the threshold for error is less than 3 in which the system performs above the threshold [14].

4. Conclusion

In terms of computer diagnosis or medical information system, the diagnostic system with learning capability is very useful. As regards, the automated diagnosis, bivariate in terms of quantity either a symptom occurred or not, using the highest probability method where there are many diseases need to be identified, with 99.667% for common cold, 93.333% for TTH and 89.876% for migraine, method of maximum probability is practical and powerful [15]. With the in cooperation of contactless vital sign monitor, an error of 1.16 for pulse sensor and 2.917 for ECG which are still below threshold error offers user to accurately measure vital signs in a non-obtrusive way, and to provide them with the data required to determine to the right timing for any intervention procedure. This system helps bedridden patients who usually stay at home yet require a constant health monitoring without having to wear or attach to bulky medical equipment [16] which can bridge the gap of inter-operability between client and medical provider.

Acknowledgments

Authors would like to express gratitude to Ministry of Higher Education and Universiti Teknologi Malaysia for supporting this research under the Institutional Research Grants Vote Number (07G22)

References

[1] Zhang, Y., Pintea, S.L., van Gemert, J.C. 2017. Video Acceleration Magnification. *IEEE Conf. on Computer. Vision and Pattern Recognition.* 502-510

[2] Brandtzaeg, P. B., & Følstad, A. 2017. Why People Use Chatbots. *Internet Science Lecture Notes in Computer Science*. **10673**, 377-392.

[3] Wu, H.Y., Rubinstein, M., Shih, E., Guttag, J., Durand, F., Freeman, W. 2012. Eulerian video magnification for revealing subtle changes in the world. *ACM SIGGRAPH 31*. **4,** 65–8

[4] Wadhwa, N., Rubinstein, M., Durand, F., Freeman, W.T. 2014. Riesz pyramids for fast phase-based video magnification. *IEEE Int. Conf. on Computer. Photography.* 1-10

[5] Wadhwa, N., Rubinstein, M., Durand, F., Freeman, W.T. 2013. Phase-based video motion processing. *ACM Trans. Graph*. **4**, 80

[6] Liu, C., Torralba, A., Freeman, WT, Durand, F. and Adelson, E.H. 2005. Motion magnification. *ACM Transactions on Graphics (TOG),* 519–526

[7] Wu, H.-Y., Rubinstein, M., Shih, E., Guttag, J., Durand, F. and Freeman, W. 2012. Eulerian video magnification for revealing subtle changes in the world. *ACM Transactions on Graphics (TOG)*. **4**, 65.

[8] Fujita, H., Selamat, A., & Omatu, S. 2017. New Trends in Intelligent Software Methodologies, Tools and Techniques *Proceedings of the 16th International Conference SoMeT_17.* (Amsterdam: IOS Press, Incorporated). 139 - 152

[9] Wootton, R., & Batch, J. 2005. Telepediatrics: Telemedicine and child health. London: *Royal Society of Medicine Press*. **4**, 286–293

[10] Reisner, A. T.; Shaltis, P. A.; McCombie, D.; Asada, H. H. 2008. Utility of the Photoplethysmogram in Circulatory Monitoring. *Anesthesiology*. **8**, 950-958.

[11] Hibbard JH, Stockard J, Mahoney ER, Tusler M. 2004. Development of the Patient Activation Measure (PAM): conceptualizing and measuring activation in patients and consumers. *Health Serv Res*. **4**, 5 – 26

[12] Alepis, E., and Lambrinidis, C. 2013. M-health: supporting automated diagnosis and electonic health records. *SpringerPlus*, **2**,103.

[13] Hyndman, R. and Koehler A. 2005. Another look at measures of forecast accuracy. *International Journal of Forecasting* **4**, 679-688

[14] Valenza, G., Citi, L., & Barbieri, R. 2017. Linear and Nonlinear Parametric Models in Heart Rate Variability Analysis. *ECG Time Series Variability Analysis*, 87-116.

[15] Yeung, D. 2017. Keynote. *Proceedings of the 2nd Workshop on Deep Learning for Recommender Systems - DLRS 2017.*

[16] Tasli H.E.; Gudi, A.; den Uyl, M. 2014. Integrating Remote PPG in Facial Expression Analysis Framework. *Proceedings of the 16th International Conference on Multimodal Interaction, Istanbul*, 74-75.

Mitotic Cell Detection in Breast Histopathology Image: A Review

Tan Xiao Jian[1], Mustafa Nazahah[1*], Mashor Mohd Yusoff[1], Ab Rahman Khairul Shakir[2]

[1]University Malaysia Perlis, 02600 Arau, Perlis, Malaysia.
[2]Hospital Tuanku Fauziah, Jalan Tun Abdul Razak, 01000 Kangar, Perlis, Malaysia.

*nazahah@unimap.edu.my

Abstract. Mitotic assessment is one of the critical features in the Nottingham Histological Grading (NHG) system for breast cancer grading. In recent years, rapid development in the performance of mounted camera, computer processing power, and data processing speed have fostered the research interest in detection of mitotic cells using image processing techniques. This paper is meant as an introduction of the recent methods and approaches specifically in detection of mitotic cells for the non-experts. It starts with overview of the grading system, online dataset, and the recent proposed methods in each step (i.e., color normalization, nuclei detection, feature extraction, and classification) of the mitotic cell assessment procedure. A brief discussion and the open problems pertinent to the established works were presented at the end of this paper.

1. Introduction

Over the past decade, tremendous growth in image processing algorithms and computational technology have foster the development of computer-aided analytical approaches such as computer-aided diagnosis (CAD) and computer-aided prognosis (CAP) in the medical domain. The recent advent of Whole Slide Imaging (WSI) scanner allows analogue histopathology slides to be converted into digital slides. This scanner demonstrates a significant impact to relieve the pressing need for CAD to reduce the workload among pathologist [1]. Furthermore, the emergence of WSI images allows image analysis and processing approaches to be incorporated to reduce inter- and intra-observer variation in grading [1, 2].

In this paper, we provide an overview for methods that have been published in the field of mitotic cell detection in breast histopathology images. The organization of the paper is as follows: Section 2, Nottingham Histological Grading (NHG) system, Section 3, Assessment of mitotic cell; and Section 4, discussion and conclusion. For a broader overview in breast histopathology images analysis, we refer the recent review in [1, 2, 3].

2. Nottingham Histological Grading (NHG) System

A patient that confined with breast carcinoma is required to perform the grading procedure. Grading of breast carcinoma provides a strong prognosis value and is well recognized as one of the key components of clinical decision-making-tools [4]. The grading assessment was proposed by Patey and Scarff [5] and

Bloom and Richardson [6]. The grading system was modified by Elston and Ellis [7] and becoming more semi-quantitative when compared to the previous methods. The modified grading system by Elston and Ellis is known as the Nottingham Histopathological Grading (NHG) system [4]. The NHG system is nowadays accepted worldwide and is used as a gold standard in performing grading procedure [4, 8]. The overall output grade obtained from the NHG system is dependent on the total score obtained from three important features. The features are glandular formation, nucleus pleomorphism, and mitotic count. Table 1 shows details explanation for each feature.

Table 1. The semi-quantitative histology grading system: NHG system.

Feature	Remark	Score
Glandular formation	Majority of tumor (>75.0%)	1
	Moderate degree (10.0%-75.0%)	2
	Little or none (<10.0%)	3
Nucleus pleomorphism	Small nuclei with little increase in size respect to normal breast epithelial cells, regular outlines, uniform nuclear chromatin and low degree of variation in size	1
	Increase in cell size with open vesicular nuclei, visible nucleoli, and moderate variability in term of shape and size	2
	Vesicular nuclei, often with prominent nucleoli, present significant variation in shape and size, occasionally with very large and eccentric forms	3
Mitotic counts (field diameter: 0.40mm)	≤4	1
	5 - 9	2
	≥10	3
Final grading	Grade 1	Total score: 3-5
	Grade 2	Total score: 6 or 7
	Grade 3	Total score: 8 or 9

Note: Score 1: Low, score 2: moderate, and score 3: high.

The assessment of mitotic count is the most prognostically significant [9]. Assessment of mitotic count is time consuming, tedious, required intensive care and experience [10, 11, 12]. The assessment depends on tissue fixation and slide preparation procedures. The observers must only include clear and definite mitotic cell. Karyorrhectic and apoptotic nuclei that shared similar appearance with the mitotic cells should be neglected. To perform mitotic count, the observers should first delimit the area with the highest mitotic activity. The total number of mitotic cells in the 10 consecutive high power field (HPF) is then recorded. The cut off threshold of the mitotic count is dependent on the field diameter. Thus, it is important to control the calibration of the microscope or the magnification level (40x magnification) during the mitotic count.

3. Assessment of Mitotic Cell

3.1 Online Dataset

There are publicly assessable online datasets which provide a platform for the researchers to compare their proposed methods. The online datasets contain breast carcinoma biopsy slides which were stained with hematoxylin and eosin (H&E) dyes to enhance the contrast between different tissue components. MITOS-ATIPIA2014 [13] dataset consists of breast carcinoma slides with magnification of x20 and x40. These slides were scanned using two scanners: Aperio Scanscope XT and Hamamatsu Nanozoomer 2.0-HT. AMIDA2013 [14] dataset consists of breast carcinoma slides with magnification of 40x. These slides were scanned using Aperio Scanscope XTscanner. Similar to AMIDA2013, the Mitosis-ICPR2012 [15] dataset contains slides at 40x magnification and were scanned using Aperio Scanscope XT and Hamamatsu Nanozoomer 2.0-HT scanners. A summary of the online datasets is shown in Table 2.

Table 2. Summary of the online datasets.

Datasets	Number of frame	Number of mitotic cells	Resolution
MITOS-ATIPIA2014 [13]	1200	749	1539 × 1376
AMIDA2013 [14]	311	550	2000 × 2000
Mitosis-ICPR2012 [15]	50	327	2084 × 2084

In recent year, a new online available dataset, namely TUPAC16 [16] was released to the public. The dataset consists of 500 training and 321 testing data (WSI slides). Only the ground truth of the training dataset was provided to the public to ensure an independent and fair evaluation on the dataset. Overall, there are 383 mitotic cases with score 1, 194 cases with score 2 and 244 cases with score 3. The TUPAC16 [16] provides two tasks to the participants: (1) to predict the mitotic scores and (2) to reproduce the most common method of assessing tumor proliferation (to predict the gene expression based on PAM50 proliferation scores).

3.2 Color Normalization

Color normalization is one of the essential steps in the assessment of mitotic cell, typically in histopathology images. The main purpose of color normalization is to reduce the color inconsistency across different histopathology slides. The heterogeneity properties of the disease, different manufactures, timing of stain absorption, concentration of the reagent, thickness of the biopsy section, and the differences in fixation and staining protocol are some of the main contributions to this problem [2, 17]. Histogram matching [18], color transfer [2], color swapping [19], spectral matching [20], illuminant normalization [21] and color deconvolution [22] are examples of techniques used in color normalization. It is important to emphasis that the selection of color normalization method is closely related to the object of interest. For instance, histogram matching technique is fast and simple method as compared to color deconvolution method. Histogram matching involves in simple mathematic computation but requires careful selection of reference image.

3.3 Nuclei Detection

Nuclei segmentation remains as a challenging task albeit numerous attempts have established to address this problem. Irshad implemented Laplacian of Gaussian (LoG) together with thresholding, morphological and active contour model methods to segment the mitotic cells [23]. Irshad et al. proposed a novel mitotic cell detection based on texture, Scale-Invariant Feature Transform (SIFT) feature and Hierarchical Model and X (HMAX) biologically inspired approach [24]. Wang and Cruz-Roa combined LoG and local dynamic thresholding on a blue ratio space of RGB input images [25]. Maximum Likelihood Estimation (MLE) was used to segment the true mitotic cells [26]. Lu and Mandal implemented Bayesian Modelling and local-region thresholding in the mitotic cells segmentation [27]. Logambal and Saravanan also used local-region thresholding [28]. They performed Otsu method to calculate the local-region thresholding. In recent year, Jian et al. proposed a color thresholding method on color decorrelation stretched image [12] and a K-Mean with guided initialization [11] to segment the nuclei cells. Several studies extended their study by implementing false positive reduction to reduce the number of non-mitotic candidates [29, 30, 31, 32].

3.4 Feature Extraction

The extracted features from the mitotic cells could be characterized into three domains: color, texture, and shape domains. In color domain, color moment (e.g., min, max, mean, median, variance, skewness, kurtosis) and color histogram can be extracted whereas in texture domain, Gray Level Co-occurrence Matrix (GLCM) [33] and Complete Local Binary Pattern (CLBP) [34] can be extracted. In GLCM, 14 textural features can be measured from the probability matrix: energy, entropy, contrast, variance, homogeneity, correlation, sum average, sum entropy, sum variance, difference variance, different entropy, maximum correlation coefficient, information measures of correlation 1, and information measures of correlation 2 [35]. Eccentricity, solidity, area, perimeter, form factor, ratio of minor and major axis, extent, length, roundness, compactness, minor, and major axis are parameters that can be extracted from shape domain. To the best of our knowledge, there is lack of evidence showing which features are best to describe the mitotic cell.

3.5 Classification

Support Vector Machine (SVM) is a widely used classifier. SVM has shown effective classification in mitotic detection [26, 29]. Other classifiers are decision tree [36], nearest neighborhood tree [36], discriminant analysis [36] and ensemble classifier [36]. Some studies used neural network [25, 37] and Model Explanation System (MES) classifier [27] in mitotic cell classification.

4. Discussion and Conclusion

The introduction of WSI scanner has fostered the development of medical image processing specifically in histopathology images. Several leading journals are focused in the publication of works pertinent to automated analysis in medical images. However, there is a need to provide evidence in clinical applicability to further evaluate the established works. It is important for the established works to demonstrate tangible and clear effect in the clinical domain in order to benefit the pathologist in the standard routine procedure.

By far, the publicly available online datasets are limited. Due to the limited clinical annotation and time constraint, the histopathology data are anticipated to undergo a bottleneck in the near future. Comparison across different literature is difficult. Researchers tend to evaluate their results using different metrics and some researchers evaluate their detection system using self-collected data which is confidential. Implementation of CAD and CAP in routine procedure is cost effective. For instance, image-based detection can be used to predict disease occurrence and recurrence [1] as a high mitotic count is correlated with high speed of tumor cells replication and/ or high proliferative activity in tumor cells. Hence, the expensive molecular assays may not be required for the same purpose. It is important to maintain a constant collaboration between researchers from biomedical, computer, and electronic engineering background with the clinical pathologists as the pathologists could provide knowledge in histopathology which may eventually lead to new research ideas.

Acknowledgments

The authors would like to acknowledge the support from the Fundamental Research Grant Scheme (FRGS) under a grant number of FRGS/1/2016/SKK06/UNIMAP/02/3 from the Ministry of Higher Education Malaysia. The protocol of this study has been approved by the Medical Research and Committee of National Medical Research Register (NMRR) Malaysia (NMRR-17-281-34236).

References

[1] Gurcan MN, Boucheron L, Can A, Madabhushi A, Rajpoot N and Yener B 2009 Histopathological Image Analysis: A Review. *IEEE Biomed Eng*;2:147–71. doi:10.1109/RBME.2009.2034865.Histopathological.

[2] Veta M, Pluim JPW, Van Diest PJ and Viergever MA 2014 Breast Cancer Histopathology Image Analysis: A Review. *IEEE Trans Biomed Eng*;61:1400–11. doi:10.1109/TBME.2014.2303852.

[3] Jian TX, Mustafa N, Mashor MY and Shakir K, Rahman 2018 A. Image Processing in Breast Carcinoma Histopathological Image : A Review. *J of Eng Res Educ*;10:57–72.

[4] Lakhani SR, Ellis IO, Schnitt SJ, Tan PH and Vijver MJ Van De, editors 2012 WHO Classification of Tumours of the Breast. 4th ed. *International Agency for Research on Cancer (IARC)*.

[5] Patey D and Scarff R 1928 The Position of Histology in the Prognosis of Carcinoma of the Breast. *Lancet*;1:801–4.

[6] Bloom H and Richardson W 1957 Histological Grading and Prognosis in Breast Cancer; A Study of 1409 Cases of Which 359 Have Been Followed for 15 Years. *Br J. Cancer*;22:36–7. doi:10.1038/bjc.1957.43.

[7] Elston C and Ellis I 1991 Pathological Prognostic Factors in Breast Cancer. I. The Value of Histological Grade in Breast Cancer: Experience From A Large Study with Long-Term Follow-up. *Histopathology*;19:403–10.

[8] Tan XJ, Mustafa N, Rahman KSA, Mashor MY and Ang WC 2018 Simple Landscapes Analysis for Relevant Regions Detection in Breast Carcinoma Histopathological Images. *Int. Conf. Comput. Approach Smart Syst. Des. Appl., IEEE*; p. 1–5. doi:10.1109/ICASSDA.2018.8477610.

[9] Diest PJ Van, Wall E Van Der and Baak JPA 2004 Prognostic value of proliferation in invasive breast cancer: a review. *J. Clin Pathol*;57:675–81. doi:10.1136/jcp.2003.010777.

[10] Li C, Wang X, Liu W and Jan L 2018 DeepMitosis : Mitosis detection via deep detection , verification and segmentation networks. *Med Image Anal*;45:121–33. doi:10.1016/j.media.2017.12.002.

[11] Jian TX, Mustafa N, Mashor MY and Shakir K 2018 An Improved Initialization Based Histogram of K-Mean Clustering Algorithm for Hyperchromatic Nucleus Segmentation in Breast Carcinoma Histopathological Images. *Lect Notes Electr Eng* 2018:1–6.

[12] Jian TX, Mustafa N, Mashor MY and Shakir K 2018 Hyperchromatic Nucleus Segmentation on Breast Histopathological Images for Mitosis Detection. *J Telecommun Electron Comput Eng*;10:27–30.

[13] Available as on 10.06.2014. [Online]. Available, http://mitosatypia-14.grand-challenge.org/dataset/.

[14] Available as on 5.02.2013. [Online]. Available: http://amida13.isi.uu.nl/?q=node/62.

[15] Available as on 8.01.2012. [Online]. Available: http://ludo17.free.fr/mitos_2012/dataset.html.

[16] Available as on 24.08.2016. [Online]. TUPAC16 n.d. http://tupac.tue-image.nl/.

[17] Roy S, Jain A, Lal S and Kini J 2018 A study about color normalization methods for histopathology images. *Micron*;114:42–61. doi:10.1016/j.micron.2018.07.005.

[18] Arai K, Kadoya N, Kato T, Endo H, Komori S, Abe Y and et al 2017 Feasibility of CBCT-based proton dose calculation using a histogram-matching algorithm in proton beam therapy. *Phys Medica*;33:68–76. doi:10.1016/j.ejmp.2016.12.006.

[19] Govindarajan A, Balamuralitharan S, and Sundaresan T 2017 Color Swapping to Enhance Breast Cancer Digital Images Qualities Using Stain Normalization. *IOP Conf. Ser. Mater. Sci. Eng.*, p. 0–7. doi:10.1088/1742-6596/755/1/011001.

[20] Sanjeevi S 2014 Spectral matching approaches in hyperspectral image processing. *Int J Remote Sens*;35:8217–51. doi:10.1080/01431161.2014.980922.

[21] Amal KRG and Arun LCK 2017 A Complete Color Normalization Method On Pathological

Images. *Int J Adv Res Innov Ideas Educ*;2:60–8.

[22] Khan AM, Rajpoot N, Member S, Treanor D and Magee D 2014 A Nonlinear Mapping Approach to Stain Normalization in Digital Histopathology Images Using Image-Specific Color Deconvolution. *IEEE Trans Biomed Eng*;61:1729–38.

[23] Irshad H 2013 Automated Mitosis Detection in Histopathology Using Morphological and Multi-channel Statistics Features. J *Pathol Inform*;4:10. doi:10.4103/2153-3539.112695.

[24] Irshad H, Jalali S, Roux L, Racoceanu D, Naour G, Hwee L and et al 2013 Automated mitosis detection using texture, SIFT features and HMAX biologically inspired approach. *J Pathol Inform*;4:12. doi:10.4103/2153-3539.109870.

[25] Wang H and Cruz-Roa A 2014 Basavanhally A, Gilmore H, Shih N, Feldman M, et al. Mitosis detection in breast cancer pathology images by combining handcrafted and convolutional neural network features. *J Med Imaging (Bellingham, Wash)*;1:034003. doi:10.1117/1.JMI.1.3.034003.

[26] Tashk A, Helfroush MS, Danyali H and Akbarzadeh M 2013 An automatic mitosis detection method for breast cancer histopathology slide images based on objective and pixel-wise textural features classification. *IKT 5th Conf Inf Knowl Technol*:406–10. doi:10.1109/IKT.2013.6620101.

[27] Lu C and Mandal M 2014 Toward automatic mitotic cell detection and segmentation in multispectral histopathological images. *IEEE J Biomed Heal Informatics*;18:594–605. doi:10.1109/JBHI.2013.2277837.

[28] Logambal G and Saravanan V 2015 Cancer Diagnosis using Automatic Mitotic Cell Detection and Segmentation in Histopathological Images. *Proc. Glob. Conf. Commun. Technol.*, p. 128–32.

[29] Nateghi R, Danyali H and Helfroush MS 2017 Maximized Inter-Class Weighted Mean for Fast and Accurate Mitosis Cells Detection in Breast Cancer Histopathology Images. *J Med Syst*;41. doi:10.1007/s10916-017-0773-9.

[30] Jian TX, Mustafa N, Mashor MY and Shakir K 2018 Segmentation Based Classification for Mitotic Cells Detection on Breast Histopathological Images. *J Telecommun Electron Comput Eng*;10:2–5.

[31] Nateghi R, Habibollah D MSH and FPP 2014 Automatic detection of mitosis cell in breast cancer histopathology images using genetic algorithm. *21st Iran. Conf. Biomed. Eng. (ICBME 2014)*, p. 1–6.

[32] Nateghi R, Danyali H, Sadegh Helfroush M and Tashk A 2014 Intelligent CAD System for Automatic Detection of Mitotic Cells from Breast Cancer Histology Slide Images Based on Teaching-Learning-Based Optimization. *Comput Biol J*;2014:1–9. doi:10.1155/2014/970898.

[33] Singh S, Srivastava D and Agarwal S 2017 GLCM and Its Application in Pattern Recognition. 5th Int. *Symp. Comput. Bus. Intell.*, p. 20–5.

[34] Abdul M and Holambe RS 2018 A collaborative representation face classification on separable adaptive directional wavelet transform based completed local binary pattern features. *Eng Sci Technol an Int J*;21:611–24. doi:10.1016/j.jestch.2018.05.008.

[35] Webel J, Gola J, Britz D and Mücklich F 2018 A New Analysis Approach Based on Haralick Texture Features for the Characterization of Microstructure on the Example of Low-alloy Steels. *Mater Charact*;144:584–96. doi:10.1016/j.matchar.2018.08.009.

[36] Gupta V and Bhavsar A 2017 Breast Cancer Histopathological Image Classification : Is Magnification Important ? *IEEE Conf. Comput. Vis. Pattern Recognit. Work.*, p. 1–8. doi:10.1109/CVPRW.2017.107.

[37] Cruz-roa A, Basavanhally A, González F, Gilmore H, Feldman M, Cruz-roa A and et al 2014 Automatic detection of invasive ductal carcinoma in whole slide images with convolutional neural networks. *SPIE Med. Imaging*, vol. 904103, 2014. doi:10.1117/12.2043872.

Segmentation of Irrelevant Regions Using Color Thresholding Method: Application in Breast Histopathology Images

Tan Xiao Jian[1], Mustafa Nazahah[1*], Mashor Mohd Yusoff[1], Ab Rahman Khairul Shakir[2]

[1]University Malaysia Perlis, 02600 Arau, Perlis, Malaysia.
[2]Hospital Tuanku Fauziah, Jalan Tun Abdul Razak, 01000 Kangar, Perlis, Malaysia.

*nazahah@unimap.edu.my

Abstract. Segmentation of irrelevant regions in the breast histopathology image is essentially performed in preliminary or post processing stages. This study presents a color thresholding method to segment the irrelevant regions, specifically blood cells and hemorrhage in the RGB breast histopathology images. The conventional gray level global thresholding method was extended to color images by modifying the thresholding algorithm. The color thresholding method was performed by decomposing the RGB input image into coherent regions based on similarity in color feature. Area Overlap Measure (AOM) and Combined Equal Importance (CEI) are two statistical metrics used to evaluate the performance of the color thresholding method. The obtained AOM and CEI for the overall dataset in blood cells are 0.92 and 0.89, respectively. The color thresholding method is found to be effective and is able to produce plausible thresholding results in segmentation of irrelevant regions in RGB breast histopathology images.

1. Introduction

Image segmentation has a paramount role in meaningful analysis and interpretation of acquired image. It is a fundamental step in many video, image, graphical and computer vision related applications where the subsequence steps are strongly rely on the quality of the segmentation [1]. The main purpose of image segmentation is to decompose the input image into spatially coherent regions sharing similar attribute and homogeneous properties.

Over the last decades, many segmentation techniques have been proposed to address the different segmentation problems. However, a universal segmentation approach that is suitable for all applications (in monochrome and color images) is yet to exist [2]. Segmentation can be performed through bi-level thresholding where the input image is decomposed into two homogeneous regions. The thresholding value can be obtained based on the textural feature, intensity value and relative edges [1]. Gray level global thresholding [3, 4], fuzzy analysis based segmentation [5, 6] and landscape analysis [7, 8] are few typical segmentation approaches. Among these approaches, gray level global thresholding is the most commonly used segmentation technique with many related algorithms have been proposed in recent years [1, 9, 10]. Gray level global thresholding approach can be extended to color images by modifying the thresholding algorithm. The color thresholding defines the regions of interest based on the similarity in color feature (i.e., red (R), green (G) and blue (B) channels).

The main objective of this study is to segment blood cells (Figure 1 (a)) and hemorrhage (Figure 1 (b)) from the background areas on the RGB breast histopathology images. Blood cells and hemorrhage are irrelevant components in the breast histopathology image. Removal of these components is normally performed at the preliminary or post processing stage to ensure the smoothness in the relevant regions (i.e., tumor regions) segmentation. In this study, the gray level global thresholding method was modified for color image segmentation. A set of comprehensive empirical study was performed to select the optimal thresholding values. The organization of this paper is as follows: Section 2, methodology, Section 3, experimental results and Section 4, conclusion.

(a) Blood cells (b) Hemorrhage

Figure 1. Examples of breast histopathology images with blood cells and hemorrhage.

2. Methodology

Figure 2 shows the block diagram of the proposed method. The proposed method consists of three main steps: color normalization, empirical analysis and color thresholding. The details of each sub-step are described in the following subsections.

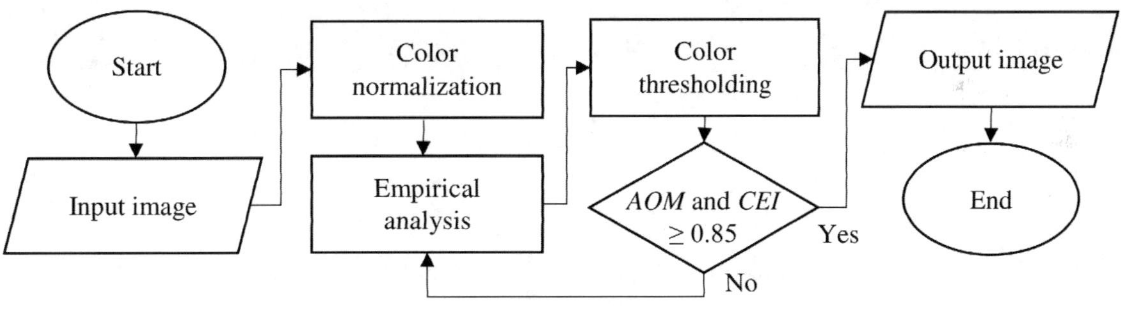

Figure 2. Block diagram of the proposed method.

2.1 Color Normalization

Color normalization step aims to standardize and reduce color inconsistency in the RGB input histopathology images. This step was done using histogram matching algorithm [11]. For this purpose, one breast carcinoma histopathology image was selected as a reference image. The reference image was selected based on the following criterions: (1) the reference image should be chosen from the dominant region and captured under 10x magnification and (2) the reference image should have the common image contents such as tumor regions, non-tumor regions, blood cells, hemorrhage and background to reduce unwanted bias during histogram matching. The R, G and B channels from input image were matched to the R, G and B channels of the reference image.

2.2 Empirical Analysis and Color Thresholding

The conventional gray level global thresholding method [12] in monochrome image is performed by selecting a threshold value from a range of intensity values (i.e., 0 to 255). The equation of the conventional gray level global thresholding method is given in Equation (1), where T is the threshold value, $f(x,y)$ is the original input pixel value and $g(x,y)$ is the output pixel value.

$$g(x,y) = \begin{cases} 0, & f(x,y) < T \\ 1, & f(x,y) \geq T \end{cases} \tag{1}$$

As the RGB input histopathology consists of three color channels, the thresholding process was performed once at a time for each color channel. The Boolean AND operator was used to combine the thresholding process into one rule as given in Equations (2) and (3). To find the suitable values for the thresholding purpose, a set of comprehensive empirical analysis was performed. The study was conducted on five different breast histopathology slides. Three histopathology images were captured from each slide at the dominant area, under 10x magnification. The 15 histopathology images were first pre-processed using the method described in Section 2.1. For each image, the threshold value of R, G, and B were manually adjusted until all the blood cells and hemorrhage were segmented from the background. Table 1 shows the results of the empirical study performed on the 15 histopathology images. The mean value in each color channel was selected as the thresholding value. The blood thresholding values in R (R_{blood}), G (G_{blood}), and B (B_{blood}) are 139, 100, and 130, respectively. The same procedure was applied to the hemorrhage. The hemorrhage has brighter areas (i.e., in terms of intensity values) compared to the blood cells. The hemorrhage threshold values for R ($R_{hemorrhage}$), G ($G_{hemorrhage}$), and B ($B_{hemorrhage}$) are 200, 171, and 171, respectively.

$$g_{blood}(x,y) = \begin{cases} 1, & R(x,y) < R_{blood}\ AND\ G(x,y) > G_{blood}\ AND \\ & B(x,y) > B_{blood} \\ 0, & otherwise \end{cases} \tag{2}$$

$$g_{hemorrhage}(x,y) = \begin{cases} 1, & R(x,y) < R_{hemorrhage}\ AND\ G(x,y) > G_{hemorrhage}\ AND \\ & B(x,y) > B_{hemorrhage} \\ 0, & otherwise \end{cases} \tag{3}$$

Table 1. R, G, and B values obtained from the empirical study performed on the 15 breast histopathological images.

Threshold values	Irrelevant objects					
	Blood cells			Hemorrhage		
	R	G	B	R	G	B
Mean	139	100	130	200	171	171
Maximum	143	112	139	207	179	174
Minimum	128	96	124	192	165	167

2.3 Evaluation Metric

Two statistical metrics were used: Area Overlap Measure (AOM) and Combined Equal Importance (CEI). AOM is referred as a ratio of the intersection to the union of the two areas to be compared. This metric is commonly used to evaluate the performance of the segmentation algorithm. CEI is an equation that combined AOM, over-segmentation, and under-segmentation to provide them an equal importance. The equations of AOM and CEI are given in Equations (4) and (5) [13], where A is to the result obtained from the proposed procedure and B is the ground truth.

$$AOM = \frac{area|A \cap B|}{area|A \cup B|} \tag{4}$$

$$CEI = \frac{AOM + \left(1 - \frac{area|B| - area|A \cap B|}{area|B|}\right) + \left(1 - \frac{area|A| - area|A \cap B|}{area|A|}\right)}{3} \tag{5}$$

3. Experimental Results

3.1 Dataset

The histopathology slides of breast carcinoma were locally collected, in Malaysia. The breast histopathology slides were obtained from the Pathology Department, Hospital Tuanku Fauziah, Kangar, Perlis, Malaysia. A total of five breast histopathology slides were used in this study. These slides were prepared under a standard staining procedure aligned with the guideline of MOH, Malaysia, as in [14], using the Hematoxylin and Eosin (H&E) stain. An Aperio CS2 WSI scanner was used to convert the slides into digital form (i.e., digital histopathology slides). The input histopathology images were captured using the Aperio ImageScope at 10x magnification. The captured histopathology image is an 8-bit RGB color with dimensions of 614x1240 pixels (size of pixel: 0.2521 μm per pixel) and is in tiff file format. For the purpose of evaluation, a total of 50 histopathology images (i.e., 25 images with blood cells and 25 images with hemorrhage) were captured, such that 10 images were captured from each slide which corresponding to the different dominant regions. For evaluation purpose, ground truth images for the 50 histopathology images were annotated by an experienced histopathologist.

3.2 Results and Discussion

Three sample images were selected from each dataset, namely A1, A2 and A3 for blood cells images; and B1, B2 and B3 for hemorrhage images. Figures 3 and 4 show the outputs of color thresholding for the blood cells and hemorrhage images, respectively. In each figure, images (a) to (c) show the original RGB input images and images (d) to (f) show the outputs of color thresholding for the respective images.

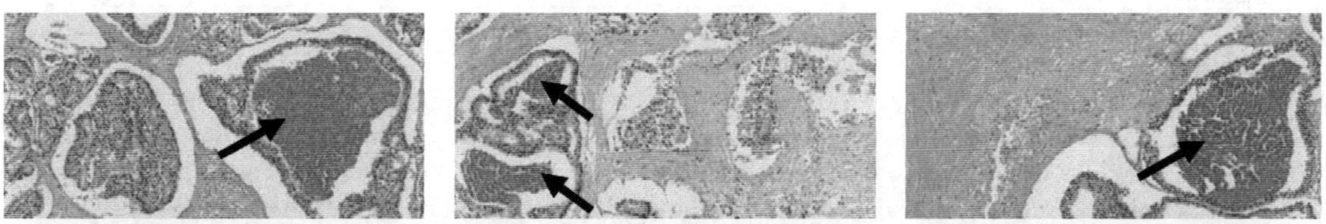

(a) Original image of A1 (b) Original image of A2 (c) Original image of A3

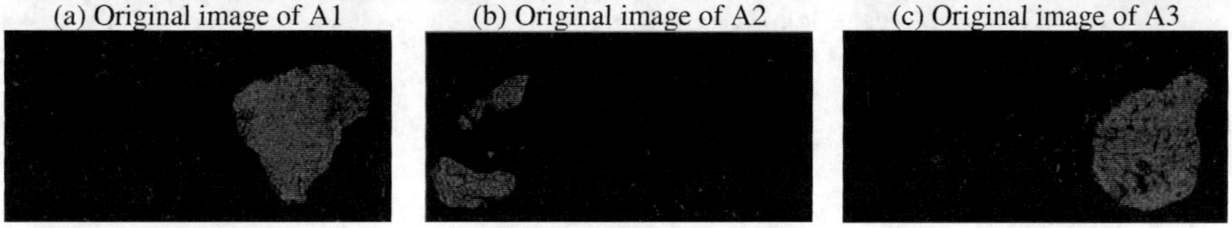

(d) Color thresholding of A1 (e) Color thresholding of A2 (f) Color thresholding of A3

Figure 3. Blood cells images.

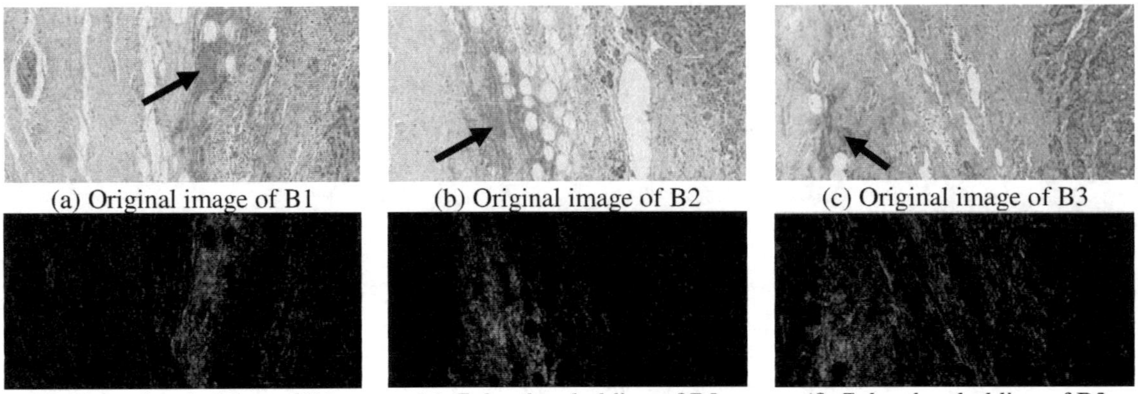

(a) Original image of B1 (b) Original image of B2 (c) Original image of B3

(d) Color thresholding of B1 (e) Color thresholding of B2 (f) Color thresholding of B3

Figure 4. Hemorrhage images.

Table 2 shows the obtained AOM and CEI for the blood cells images. The hemorrhage is complex in structure and is humanly impossible to manually delimit each region in pixel-based accurately. Therefore, for hemorrhage, the performance of color thresholding is solely depends on human intervention. Referring to Figures 3 and 4, based on the human intervention, the color thresholding method is found to be effective and is able to produce plausible segmentation outputs. The color thresholding method is fast in computation as the thresholding algorithm do not involves with complex mathematical equation. The color thresholding method could be used in the preliminary or post processing stages to remove the irrelevant regions (e.g., blood cells and hemorrhage) from the breast histopathology images.

Table 2. The obtained AOM and CEI for blood cells images.

Image	AOM	CEI
A1	0.94	0.90
A2	0.89	0.86
A3	0.91	0.89
Average for 25 images	0.92	0.89

4. Conclusion

In this study, the conventional gray level global thresholding method was extended to color image segmentation. The color thresholding method was used to segment blood cells and hemorrhage from the RGB breast histopathology images. The thresholding values for each color channel were selected based on the comprehensive empirical analysis done on the respective components. Qualitative and quantitative (blood cells images only) evaluations were performed to evaluate the performance of the color thresholding method. The obtained AOM and CEI for the overall blood cells dataset are 0.92 and 0.89, respectively. This study concludes that the color thresholding method is effective in the segmentation of blood cells and hemorrhage by decomposing the input RGB images into coherent regions based on the similarity in color feature.

Acknowledgments

The authors would like to acknowledge the support from the Fundamental Research Grant Scheme (FRGC) under a grant number of FRGS/1/2016/SKK06/UNIMAP/02/3 from the Ministry of Higher Education Malaysia. The protocol of this study has been approved by the Medical Research and Committee of National Medical Research Register (NMRR) Malaysia (NMRR-17-281-34236).

References

[1] Sarkar S, Das S and Chaudhuri SS 2015 A Multilevel Color Image Thresholding Scheme based on Minimum Cross Entropy and Differential Evolution. *Pattern Recognit Lett*;54:27–35. doi:10.1016/j.patrec.2014.11.009.

[2] Harrabi R and Braiek E Ben 2012 Color image segmentation using multi-level thresholding approach and data fusion techniques : application in the breast cancer cells images. *J Image Video Process*;11:1–11.

[3] Jian TX, Mustafa N, Mashor MY and Shakir K 2018 Segmentation Based Classification for Mitotic Cells Detection on Breast Histopathological Images. *J Telecommun Electron Comput Eng*;10:2–5.

[4] Cheremkhin PA and Kurbatova EA 2019 Comparative appraisal of global and local thresholding methods for binarisation of off-axis digital holograms. *Opt Lasers Eng*;115:119–30. doi:10.1016/j.optlaseng.2018.11.019.

[5] Jian TX, Mustafa N, Mashor MY and Shakir K 2018 An Improved Initialization Based Histogram of K-Mean Clustering Algorithm for Hyperchromatic Nucleus Segmentation in Breast Carcinoma Histopathological Images. *Lect Notes Electr Eng* 2018:1–6.

[6] Alzu S, Shehab M, Al-Ayyoub M, Jararweh Y and Gupta B 2018 Parallel implementation for 3D medical volume fuzzy segmentation. *Pattern Recognit Lett*;0:1–7. doi:10.1016/j.patrec.2018.07.026.

[7] Tan XJ, Mustafa N, Rahman KSA, Mashor MY and Ang WC 2018 Simple Landscapes Analysis for Relevant Regions Detection in Breast Carcinoma Histopathological Images. *Int. Conf. Comput. Approach Smart Syst. Des. Appl., IEEE*; 2018, p. 1–5. doi:10.1109/ICASSDA.2018.8477610.

[8] Klonowski W, Stepien R and Stepien P 2010 Simple fractal method of assessment of histological images for application in medical diagnostics. *Nonlinear Biomed Phys*;4:7. doi:10.1186/1753-4631-4-7.

[9] Jian TX, Mustafa N, Mashor MY and Shakir K 2018 Hyperchromatic Nucleus Segmentation on Breast Histopathological Images for Mitosis Detection. *J Telecommun Electron Comput Eng*;10:27–30.

[10] N SriMadhava Raja and V. Rajinikanth 2014 Otsu Based Optimal Multilevel Image Thresholding Using Firefly Algorithm. *Modelling and Simulation in Engineering*;2014:1. doi:http://dx.doi.org/10.1155/2014/794574.

[11] Arai K, Kadoya N, Kato T, Endo H, Komori S, Abe Y and et al 2017 Feasibility of CBCT-based proton dose calculation using a histogram-matching algorithm in proton beam therapy. *Phys Medica*;33:68–76. doi:10.1016/j.ejmp.2016.12.006.

[12] Kulkarni N 2015 Color Thresholding Method for Image Segmentation of Natural Images. *Int J Image, Graph Signal Process*;1:28–34. doi:10.5815/ijigsp.2012.01.04.

[13] Abbas Q, Celebi ME and Garcia IF 2013 Breast mass segmentation using region-based and edge-based methods in a 4-stage multiscale system. *Biomed Signal Process Control*;8:204–14. doi:10.1016/j.bspc.2012.08.003.

[14] Guideline CP, editor 2010 Management of Breast Cancer. 2nd ed. *Ministry of Health Malaysia*.

Segmentation of Relevant Region in Breast Histopathology Images using FCM with Guided Initialization

Tan Xiao Jian[1], Mustafa Nazahah[1*], Mashor Mohd Yusoff[1], Ab Rahman Khairul Shakir[2]

[1]University Malaysia Perlis, 02600 Arau, Perlis, Malaysia.
[2]Hospital Tuanku Fauziah, Jalan Tun Abdul Razak, 01000 Kangar, Perlis, Malaysia.

*nazahah@unimap.edu.my

Abstract. This study proposes a modified initialization approach for the conventional FCM, namely FCM with guided initialization. The FCM with guided initialization was implemented to segment the relevant regions in the breast histopathology images. The initialization method to select initial centers is based on the Cyan (C) channel histogram. Area Overlap Measure (AOM) and Combined Equal Importance (CEI) were used to evaluate the performance of the proposed FCM with guided initialization. The obtained AOM and CEI for the overall dataset achieved promising results: 0.89 in AOM and 0.88 in CEI. When comparing the number of iterations required to complete the proposed FCM clustering algorithm, the FCM with guided initialization is found to be effective in reducing the search space by showing a lower number of iterations.

1. Introduction

Segmentation is often used as a preliminary step in the image processing, specifically in the breast histopathology image. The intrinsic nature of the breast histopathology image such as fuzziness, complexity, and heterogeneous [1, 2] elevate challenges in the segmentation. Segmentation of relevant regions (i.e., tumor regions) is crucial for breast carcinoma grading specifically in the assessment of glandular formation [3, 4]. The relevant regions provide diagnostically important information which hold a paramount role in providing scores to the overall grade.

Clustering is a common technique employed in data mining and data analysis [5]. Clustering technique performs classification or segmentation on objects that share similar characteristics, features and properties into subsets of data known as clusters. Theoretically, clustering algorithms aim to maximize similarity between intra-cluster objects and minimized similarity between inter-cluster objects. Fuzzy C Mean (FCM) is one of the popular approaches in clustering analysis. FCM has been reported by many studies on its superior ability in partition fuzzy dataset into different clusters with similar properties [5, 6]. However, FCM is not flawless. This algorithm is sensitive to initial guess. Dead center, center redundancy and trap in local minima are some typical limitations of FCM [7].

The main objective of this study is to reduce limitations of the conventional FCM algorithm. The conventional approach to obtain the initial centers for each cluster was modified so that the initial centers can be selected systematically from the intensity histogram of the input images. The proposed method for center initialization with FCM algorithm was implemented on the breast histopathology images to obtain relevant regions, non-relevant regions, and background areas (Figure 1). The organization of this study is as follows: Section 2, methodology; Section 3, experimental results; and Section 4, conclusion.

Figure 1. Samples of histopathology images. The relevant regions, non-relevant regions and background are shown by the black, red and blue arrows, respectively.

2. Methodology

Figure 2 shows the block diagram of the proposed method. The proposed method consists of four steps: color normalization, selection of color channel, FCM with guided initialization and post processing. The details of each sub-step are described in the following subsections.

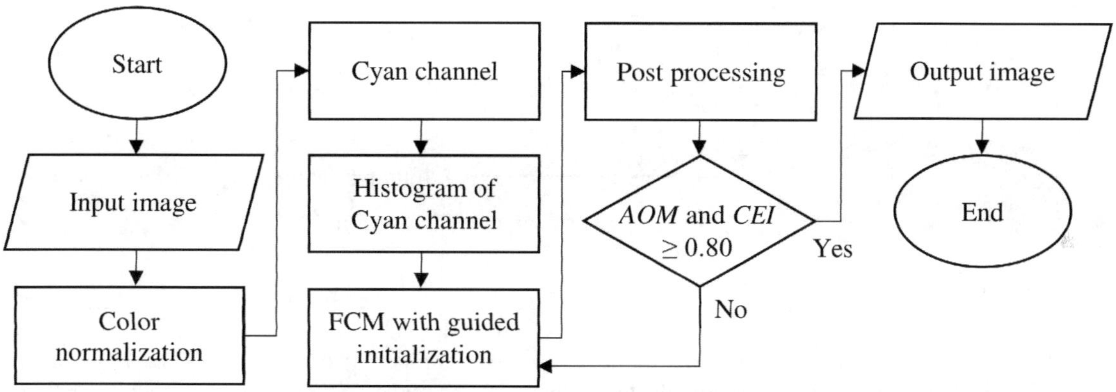

Figure 2. Block diagram of the proposed method.

2.1 Color Normalization

Color normalization step aims to standardize and reduce color inconsistency in the RGB input histopathology images. This step was done using histogram matching algorithm [8]. Histogram matching method was selected as it can provide fast and simple computation on large size of image and it can ensure smoothness of the clustering process in the later sub-step of segmentation.

2.2 Selection of color channel

For the purpose of relevant region segmentation, a comprehensive empirical study was conducted by investigating YCbCr, HSV, and CMY color models. The empirical analysis found that the Cyan (C) channel from the CMY color model shows significant discrimination in color features between the relevant, non-relevant regions, and the background areas. The C channel which complements the R channel (i.e., based on the RGB to CMY color conversion) provides a relatively high C value for the relevant regions, moderate C value for irrelevant regions, and a low C value for the background areas. Hence, the C channel was opted as the input color channel to the consequence sub-steps.

2.3 FCM with Guided Initialization

Based on the domain knowledge, the relevant and non-relevant regions in the breast histopathology image have a distinct degree of stain absorption, therefore, the relevant and non-relevant regions could be discriminated using the color feature [9]. The FCM clustering algorithm was selected to partition the input image into relevant regions, non-relevant regions, and background. The conventional FCM clustering algorithm applied with random initialization method have been reported with several limitations as described in Section 1. To address these limitations, the initialization method was modified to select initial centers based on the C channel histogram. Incorporation of the domain knowledge in color feature, specifically in the selection of initial centers for the relevant regions, non-relevant regions, and background are hypothesized to reduce the search space and avoid the limitations aforementioned. Three initial centers were selected for background (*IC1*), non-relevant regions (*IC2*), and relevant regions (*IC3*). The search for IC1 and IC3 were performed using the local maxima search algorithm (i.e., hill climbing method [10]) to obtain the first and second intensity peaks (Figure 3) whereas *IC2* was obtained using Equation (1).

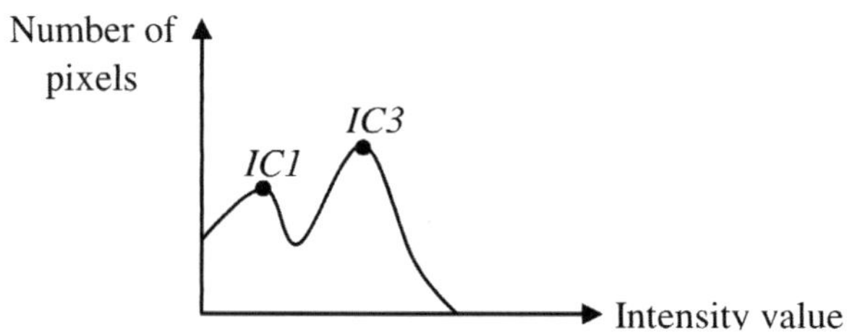

Figure 3. Illustration of C channel histogram.

$$IC2 = IC1 + \frac{IC3 - IC1}{2} \tag{1}$$

FCM clustering algorithm requires two input parameters: number of clusters (n_c) (i.e., n_c=3 in this study) and initial cluster value for each cluster. Consider an image with N number of pixels to be clustered into n_c regions. V_i is i^{th} data and C_j is j^{th} center (i = 1, 2, …, N and j = 1, 2, …, n_c). The steps of the FCM are as follows:

1. Set number of cluster (n_c).
2. Set initial cluster value for each cluster obtained using the proposed initialization method.
3. Calculate the membership (M_{ji}) using Equation (2):

$$M_{ji} = \frac{1}{\sum_{k=1}^{n_c}\left(\frac{d_{ji}}{d_{ki}}\right)^2}; if\ d_{ki} > 0, \forall j, i \tag{2}$$

where $d_{ji} = \left\|V_i - C_j\right\|$

$$\left.\begin{array}{l} M_{ji} = 1 \\ M_{ji} = 0; for\ i \neq k \end{array}\right\} if\ d_{ki} = 0$$

4. Calculate the new center using Equation (3):

$$C_j = \frac{\sum_{i=1}^{N} M_{ji}^2 V_i}{\sum_{i=1}^{N} M_{ji}^2} \tag{3}$$

5. Repeat steps 3 and 4 until the center remain unchanged.

2.4 Post Processing

The small artefacts on the FCM with guided initialization outputs were removed using morphological operation [11]. A "closing" morphological operation with a "disk" structure element (i.e., radius of 1 pixel) was implemented to remove and fill holes in the obtained outputs.

2.5 Evaluation Metrics

Two statistical metrics were used for evaluation: Area Overlap Measure (AOM) and Combined Equal Importance (CEI). AOM is defined as a ratio of the intersection to the union of the two areas to be compared. This metric is commonly used to evaluate the performance of the object region segmentation algorithm. CEI is an equation that combined AOM, over-segmentation, and under-segmentation to provide them an equal importance. The equations of AOM and CEI are given in Equations (4) and (5) [12], where A denotes to the result obtained from the proposed procedure and B denotes the ground truth.

$$AOM = \frac{area|A \cap B|}{area|A \cup B|} \tag{4}$$

$$CEI = \frac{AOM + \left(1 - \frac{area|B| - area|A \cap B|}{area|B|}\right) + \left(1 - \frac{area|A| - area|A \cap B|}{area|A|}\right)}{3} \tag{5}$$

3. Experimental Results

3.1 Dataset

The histopathology slides of breast carcinoma were locally collected, in Malaysia. The breast histopathology slides were obtained from the Pathology Department, Hospital Tuanku Fauziah, Kangar, Perlis, Malaysia. A total of five breast histopathology slides were used in this study. These slides were prepared under a standard staining procedure aligned with the guideline of MOH, Malaysia, as in [13], by using the Hematoxylin and Eosin (H&E) stain. An Aperio CS2 WSI scanner was used to convert the slides into digital form (i.e., digital histopathology slides). The digital histopathology slides were viewed and captured using the Aperio ImageScope software at 10x magnification. The captured histopathology image is an 8-bit RGB colour with dimensions of 614x1240 pixels (size of pixel: 0.2521 μm per pixel) and is in tiff file format. A total of 25 histopathology images were used in this study, such that five images were captured from each slide which corresponding to different dominant regions.

3.2 Results and Discussion

In Figure 4, images (a), (b) and (c) show three RGB breast histopathology images namely, BC1, BC2 and BC3, respectively. Images (d), (e) and (f) show the outputs of the FCM with guided initialization applied on BC1, BC2 and BC3, respectively. The clustering output of *IC3* (i.e., relevant regions) are shown in white areas. The obtained AOM and CEI for BC1, BC2, BC3 and the overall dataset are given in Table 1. From the table, the proposed method is found to be robust and effective in segmenting the relevant regions. The obtained AOM and CEI in the overall dataset show promising results: 0.89 and 0.88, respectively.

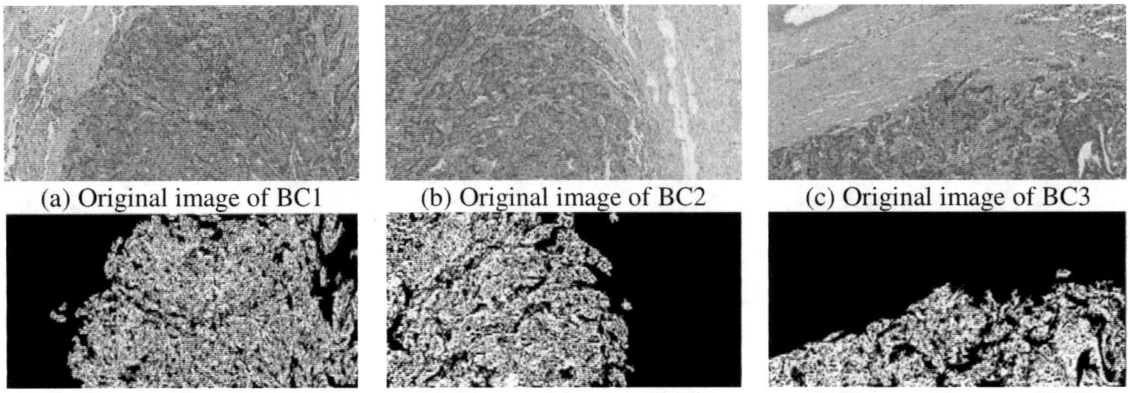

(a) Original image of BC1 (b) Original image of BC2 (c) Original image of BC3

(d) Clustered image of BC1 (e) Clustered image of BC2 (f) Clustered image of BC3

Figure 4. Results of FCM clustering using the proposed initialization method.

Table 1. The obtained AOM and CEI for BC1, BC2, BC3 and the overall dataset.

Images	AOM	CEI
BC1	0.84	0.86
BC2	0.86	0.88
BC3	0.85	0.89
Average for 25 images	0.89	0.88

The number of iterations for the conventional FCM (with random initialization) and FCM with proposed guided initialization were compared and shown in Table 2. The FCM with proposed guided initialization shows its ability to reduce the search space and complete the clustering algorithm with a smaller number of iterations (i.e., shorter computation time). As the initial centers were selected systematically from the histogram of Cyan channel, the limitations such as dead center, center redundancy and local minima could be avoided.

Table 2. The number of iterations for BC1, BC2, BC3 and the overall dataset.

Images	Iteration(s)	
	Conventional FCM	FCM with guided initialization
BC1	12	7
BC2	12	5
BC3	11	6
Average for 25 images	13	6

4. Conclusion

This study presented a modified initialization method for the conventional FCM clustering algorithm, namely FCM with guided initialization. The initial centers of the FCM were not inherent from random generation, but selected systematically from the histogram of the Cyan channel. Domain knowledge in breast histopathology image was incorporated into the algorithm to assist the selection of initial centers from the histogram. The FCM with guided initialization is found to be robust and effective in the segmentation of relevant regions in breast histopathology images (Table 1). The proposed method is able to reduce the search space by completing the clustering algorithm with the lower number of iterations (Table 2). The proposed method is also able to avoid limitations such as dead center, center redundancy and trap in local minima as reported in Section 1. The similar initialization approach could be used in the other cluster analysis such as K-Mean clustering algorithm to address the similar limitations. It is important to emphasize that for different medical images (e.g., magnetic resonance (MR) images), the FCM with guided initialization might require different domain knowledge in selecting the initial center.

Acknowledgments
The authors would like to acknowledge the support from the Fundamental Research Grant Scheme (FRGS) under a grant number of FRGS/1/2016/SKK06/UNIMAP/02/3 from the Ministry of Higher Education Malaysia. The protocol of this study has been approved by the Medical Research and Committee of National Medical Research Register (NMRR) Malaysia (NMRR-17-281-34236).

References

[1] Ramírez-torres A, Rodríguez-ramos R, Sabina FJ, García-reimbert C, Penta R, Merodio J and et al 2017 The Role of Malignant Tissue on the Thermal Distribution of Cancerous Breast. *J Theor Biol*;426:152–61. doi:10.1016/j.jtbi.2017.05.031.

[2] Samuel SM, Varghese E, Varghese S and Büsselberg D 2018 Challenges and Perspectives in the Treatment of Diabetes Associated Breast Cancer. *Cancer Treat Rev*;70:98–111. doi:10.1016/j.ctrv.2018.08.004.

[3] Jian TX, Mustafa N, Mashor MY and Shakir K 2018 Hyperchromatic Nucleus Segmentation on Breast Histopathological Images for Mitosis Detection. *J Telecommun Electron Comput Eng*;10:27–30.

[4] Jian TX, Mustafa N, Mashor MY and Shakir K 2018 Segmentation Based Classification for Mitotic Cells Detection on Breast Histopathological Images. *J Telecommun Electron Comput Eng*;10:2–5.

[5] Cebeci Z and Yildiz F 2015 Comparison of K-Means and Fuzzy C-Means Algorithms on Different Cluster Structures. *J Agric Informatics*;6:13–23. doi:10.17700/jai.2015.6.3.196.

[6] Velmurugan T 2012 Performance Comparison between k-Means and Fuzzy C-Means Algorithms using Arbitrary Data Points. *Wulfenia J*;19:234–41.

[7] Jian TX, Mustafa N, Mashor MY and Shakir K 2018 An Improved Initialization Based Histogram of K-Mean Clustering Algorithm for Hyperchromatic Nucleus Segmentation in Breast Carcinoma Histopathological Images. *Lect Notes Electr Eng* 2018:1–6.

[8] Arai K, Kadoya N, Kato T, Endo H, Komori S, Abe Y and et al 2017 Feasibility of CBCT-based proton dose calculation using a histogram-matching algorithm in proton beam therapy. *Phys Medica*;33:68–76. doi:10.1016/j.ejmp.2016.12.006.

[9] Tan XJ, Mustafa N, Rahman KSA, Mashor MY and Ang WC 2018 Simple Landscapes Analysis for Relevant Regions Detection in Breast Carcinoma Histopathological Images. *Int. Conf. Comput. Approach Smart Syst. Des. Appl., IEEE*; p. 1–5. doi:10.1109/ICASSDA.2018.8477610.

[10] Ganesan P, Kalist V and Sathish BS 2016 Histogram based Hill Climbing Optimization for the Segmentation of Region of Interest in Satellite Images. *IEEE WCTFTR - Proc. World Conf. Futur. Trends Res. Innov. Soc. Welf.*, p. 5–9. doi:10.1109/STARTUP.2016.7583961.

[11] Mohammad S, Hasan A and Ko K 2016 Depth edge detection by image-based smoothing and morphological operations. *J Comput Des Eng*;3:191–7. doi:10.1016/j.jcde.2016.02.002.

[12] Abbas Q, Celebi ME and Garcia IF 2013 Breast mass segmentation using region-based and edge-based methods in a 4-stage multiscale system. *Biomed Signal Process Control*;8:204–14. doi:10.1016/j.bspc.2012.08.003.

[13] Guideline CP, editor 2010 Management of Breast Cancer. 2nd ed. *Ministry of Health Malaysia*.

Voice Pathology Analysis using DT-CWPT and ReliefF Algorithm

Farah Nazlia Che Kassim[1,3*], Vikneswaran Vijean[1], Hariharan Muthusamy[2], Rokiah Abdullah[1,3] and Zulkapli Abdullah[3]

[1] School of Mechatronic Engineering, Universiti Malaysia Perlis, 02600 Arau, Perlis
[2] Department of Electronic Engineering, National Institute of Technology,Srinagar (Garhwal), Uttarakhand India
[3] Pusat Kejuruteraan, Universiti Malaysia Perlis, 02600 Arau, Perlis

*nazlia@unimap.edu.my

Abstract. Voice pathology analysis has been one of the useful tools in the diagnosis of the pathological voice. This method is non-invasive, inexpensive and reduces time required for analysis. This paper investigates the feature extraction based on the Dual-Tree Complex Wavelet Packet Transform (DT-CWPT) with entropies and energy measures tested with two classifiers, k-Nearest Neighbors (k-NN) and Support Vector Machine (SVM). Feature selection using ReliefF algorithm is applied to reduce redundancy features set and obtain the optimum features for classification. Massachusetts Eye and Ear Infirmary (MEEI) voice disorders database and Saarbruecken Voice Database (SVD) are used. This research was done on multiclass and by specific pathology. The experimental results automates the process of voice analysis hence produce promising results of the presence of diseases in vocal folds.

1. Introduction

Voice is the sound produced by humans using the lungs and the vocal folds in the larynx, or voice box. The presence of pathologies or diseases in vocal folds can affect the normal vibratory patterns, which will affect the sound produced. Some of the voice pathologies are like dysphonia, laryngeal paralysis, cancer, nodules, vocal polyps, vocal fold paralysis, leukoplakia (keratosis), ulcers and voice pathologies due to neurodegenerative disorders like Parkinson disease. A doctor can use the available apparatus such as electromyography (EMG), electroglottography (EGG) and video endoscopy to assess and diagnose malfunctioning vocal fold. However, these methods are known as invasive, high cost and require an expert to analyze the human speech signal parameters. Therefore, a significant attention has been paid to find various parameters obtain from voice signal hence produce early detection results of the presence of diseases in vocal folds, low cost and non-invasive diagnostic technique.

Many automatic systems for the voice pathologies detection have been developed based on voice signal processing. Nowadays, feature extraction using wavelet packet analysis seems to have the reliable capability for the identification of vocal fold pathology [1-5]. The wavelet was preferred because it can provide accurate information about the fast fluctuations of the signals in the time domain [1].Wavelet packet transform (WPT) is chosen as speech signal parameterization method because of its ability to analyze a signal at several levels of chosen resolution, considering simultaneously overall information of the spectral band for the speech signal which provides more features and characteristics of the time-frequency characteristics [2].

A typical automatic voice analysis system [4] consist extraction of relevant parameters, decision machines and performance assessment. Feature selection becomes an essential part of a successful data mining pipeline. It is important to discard the irrelevant, redundant and to retain the relevant features to improve the performance of a learning model. At the same time, the time and memory consumed to train a learning model are also reduced. Three major approaches of feature selection methods are filter methods, wrapper methods and embedded methods. The filter methods is renowned to reduced computational burden compared to both embedded and wrapper techniques [4]. Apart from that this methods are independence of the classifier, fast and has good generalization ability. Among the most used filter-based strategies, ReliefF is an iterative, randomized, and supervised approach that estimates the quality of the features according to how well their values differentiate data samples that are near to each other [6]. Other feature selection methods that commonly used in dimensionality reduction technique are fisher method, Laplacian score, genetic algorithms and etc.

In most researches, a two-class classification is done to identify healthy or pathological voice on various databases. Classification is carried out using techniques like vector quantization, artificial neural networks, SVM, hidden Markov model, Gaussian Mixture Model, decision trees, linear classifier, K-means clustering and combined classifiers. Multiclass analysis is less explored due to limited number of available databases with different set of speech tasks and vowel pronounced, labeled in different languages, gender or age and a few type of voice pathologies sample was unevenly distributed across the databases. It is difficult to compare their findings, since their results vary between published papers mostly because of differences in selected voice pathology samples, acoustic features employed, databases and classifiers involved in the research. Selvakumari and Radha [3] concluded in their comparative study between seven researches work mentioned that the acoustic features play the main role in finding the pathology. Therefore, this paper aims are to investigate the use of DT-CWPT for analyzing the voice signals using energy and entropy measures, implementing ReliefF algorithm to select relevant features, performed multiclass classification hence produce results of the presence of diseases in vocal folds.

2. Database

The voice signals from normal person and patients suffering from disorders were acquired from recordings. The voice samples were taken from two databases, MEEI and SVD database for the input signal processing. MEEI database, the most widely used and the only one commercially available become as a benchmark database in the field of pathological speech analysis [7]. The SVD database, a freely downloadable database was recorded by the Institute of Phonetics of Saarland University [8]. Only few studies of voice pathology analysis have been explored on this database [9,10] .Two databases were used as to compare the dependency of voice signal between different language, English and Germany. Table 1 show the number of voice samples uses in the investigations of multiclass analysis for the two different databases. These pathological samples were chosen as they were available and common between the two databases.

The voice signals files contain sustained normal pitch vowel /a/ samples were downsampled to 25 kHz frequency sampling. An adaptive synthetic (ADASYN) sampling approach is applied to these sets of imbalanced class data to balance up the minority sample to achieve better accuracy. The ADASYN approach improves learning by reducing the bias introduced by the class imbalance and adaptively shifting the classification decision boundary toward the difficult examples [11]. After ADASYN, samples for each class have been balance up to give a total of 201 and 592 samples for MEEI and SVD database respectively.

Table 1. Number of voice samples for multiclass pathology analysis database.

Pathology Class	MEEI Database		SVD Database	
1	Vocal nodules	19	Cysts	6
2	Paralysis	67	Paralysis	194
3	Polyp	20	Polyp	44
Total Samples		106		244

3. Methodology

The block diagram of the voice pathology analysis is shown in figure 1 consisting DT-CWPT with energy or entropies as feature extraction, feature selection using ReliefF algorithm and multiclass classification by using k-NN and SVM classifier.

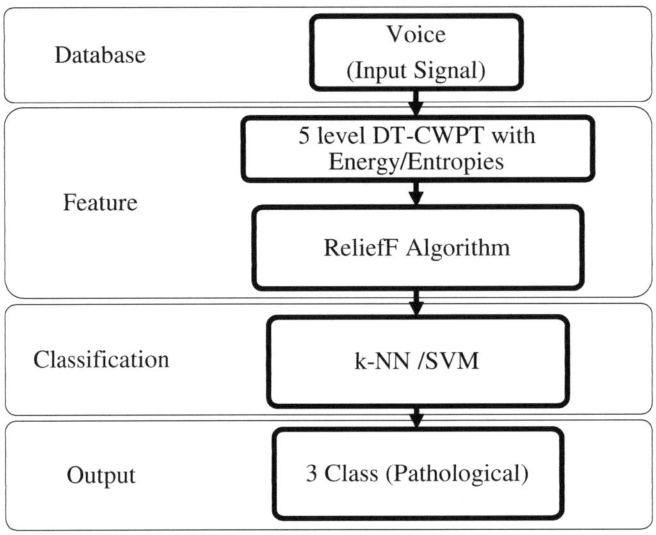

Figure 1. Block diagram of the voice pathology multiclass analysis.

3.1. Dual-tree Complex Wavelet Packet Transform

The DT-CWPT using energy and entropy measures are investigated as the feature extraction. The DT-CWPT provides time-frequency analysis of the voice signals. It is an extended algorithm from Dual Tree Complex Wavelet Transform (DT-CWT), with two band of DWPT operating in parallel [18]. DT-CWPT has the same shift-invariance and good directional selectivity properties of DT-CWT, and it also has fewer energy leakages into its negative frequency bands [12,13]. Each voice signal is decomposed using the fifth level DT-CWPT and further calculated with energy and non-linear entropy measures, namely Shannon and Renyi entropy which produced a total of 64 (2^5 x 2) features.

3.2. ReliefF Algorithm

This algorithm is implemented after feature extraction. The selection of features was based on the ranking and weightage computed by ReliefF algorithm which list the most useful to least useful features. The original relief can deal with nominal and numerical attributes. However, it cannot deal with incomplete data and is limited to two-class problems. ReliefF notably, the 'F' refers to the sixth algorithm variation (from A to F) proposed by Kononenko [14] is extended to respond to deal with multiclass problems, incomplete and noisy data and are more robust. Instead of finding one near miss M from different class, the algorithm finds one near miss M(C) for each different class and averages their contribution for updating estimates W[A]. The average is weighted with the prior probability of each class as in equation (1).

$$W[A]{:=}W[A]-\sum_{j=1}^{k}\frac{\mathrm{diff}(A,R_i,H_j)}{(m.k)}+\sum_{C\neq class(R_i)}\left[\frac{[P(C)]}{1-P(class(R_i))}\sum_{j=1}^{k}\mathrm{diff}\left(A,R_i,M_j(C)\right)\right]/(m.k) \quad (1)$$

For i=1 to m times, ReliefF randomly selects an instance R_i, find k nearest hits H_j and misses $M_j(C)$.The k is tuned to 10 (optimal k for dataset) [14-16]. The values of k is also varies to investigate the stability and reliability of attribute ranks and weights. It updates the quality estimation W[A] for all attributes A depending on their values for R_i, hits H_j and misses $M_j(C)$. The contribution for each class of the misses is weighted with the prior probability of that class P(C).

3.3. Classification

Two common classifiers, k-NN and SVM were used. k-NN is one of the simplest machine learning algorithms, also called as lazy learner [5]. Classification was based on majority of k-Nearest Neighbor's category and use distance measure to decide in the train data. k values were varied between 1 and 10 for a given training and test data. A 10-fold cross validation classification (CVC) scheme was used to increase the reliability of the results [17,18]. The SVM with radial basis function (RBF) as a kernel is used. This classifier have several advantages such as it can provide a good out-of-sample generalization, it can produce a unique solution if the parameters are properly chosen, unlike neural networks, because the optimality problem is convex, and with the help of kernels, it can achieve the flexibility of the form of the threshold separating the classes [11].

Various experiments had been made, aiming to select the best number of features for classification as well as reducing the time taken for classification. First experiment is the classification using all the features from DT-CWPT with energy features, Shannon entropy and Renyi entropy features which produced 64 features each. Next experiment involved reducing the features by selecting top most relevant 10, 20 and 30 features as until almost half of the original features. The accuracy and time taken for each classifier were then observed and compared.

4. Results and Discussion

Table 2 and Table 3 summarises the multiclass results obtained from the experiments for both database. Table 2 shows the accuracy and classification time taken to classify three pathological class MEEI database using DT-CWPT with energy, Shannon and Renyi entropy. Using k-NN classifier, the highest prediction rates were obtained by selecting the top 30 features. The best achieved accuracy for k-NN is 80%, 80.10% and 79.55% for DTCWPT with energy, Shannon and Renyi entropy respectively. These accuracy increases 1-6% compared to using all the 64 features with the advantages of reduced the classification time taken. In contrast using SVM classifier, the accuracy reduced about 2.5% for DT-CWPT with Shannon entropy for chosen 30 features. The prediction rate of DT-CWPT with energy and Renyi entropy increase and are among the highest achieved, 93.63% and 94.03%. ReliefF give the advantage of reducing almost 30% of the classification time.

The classification results for SVD database using DT-CWPT with energy, Shannon and Renyi entropy were shown in table 3. Better accuracy achieved, 85.81- 87.03% about 3% increase using selected 30 features for k-NN classifier. Average calculation time reduce about 3-11% compared to use all features. In contrast to the SVM classifier results, the highest prediction rate is 97.65% by taking all features from DT-CWPT with Renyi entropy but takes a longer time. A slight 1% reduces of accuracy achieved for the selected 30 features but time for calculation become faster.

Comparing both of the classifiers, SVMs give higher accuracy than k-NN. SVM classifier is used most widely due to the fact that SVM classifier performs better for high-dimensional data as well as small sized dataset [5] but SVMs intensive computation takes a longer time to make decision. Both class 2 in k-NN classifier show a low rate maybe due to accuracy of regions declines for higher dimensional data sets. It determines the class by taking the majority vote of class labels among the k-nearest neighbors (uses all features in computing distances).

ADASYN was used to overcome the small samples, this give influence to class 1 SVD database which gives perfect prediction rate. In future studies, the findings from synthetic data could be validated with real samples. Overall, the performances of both classifiers using the dataset after feature selection are better and improved in term of both accuracy and time taken.

Table 2. Comparison of accuracy and classification time for three class MEEI database.

Classifier	Features Experiments	Accuracy (%) ± sd				Time (seconds)
		Average	Class 1	Class 2	Class 3	
KNN	DT-CWPT + Energy					
	30 features	80.00 ± 0.77	97.19 ± 0.66	46.57 ± 1.83	96.29 ± 1.54	1.61
	64 features	79.25 ± 0.47	97.19 ± 0.99	44.78 ± 1.00	95.86 ± 0.45	1.62
	DT-CWPT + Shannon entropy					
	30 features	80.10 ± 0.81	96.88 ± 0.00	46.12 ± 2.38	97.29 ± 1.05	0.63
	64 features	75.82 ± 1.05	96.56 ± 1.23	34.03 ± 2.88	96.86 ± 0.60	0.64
	DT-CWPT + Renyi entropy					
	30 features	79.55 ± 0.95	96.72 ± 1.15	45.22 ± 1.42	96.71 ± 0.96	0.62
	64 features	79.10 ± 0.91	97.03 ± 0.89	44.48 ± 2.97	95.86 ± 0.81	0.63
SVM	DT-CWPT + Energy					
	30 features	93.63 ± 1.04	92.03 ± 2.70	94.03 ± 0.70	94.71 ± 1.18	13.81
	64 features	93.48 ± 0.92	90.63 ± 2.33	97.01 ± 0.00	92.71 ± 1.25	19.37
	DT-CWPT + Shannon entropy					
	30 features	89.90 ± 1.00	91.88 ± 1.77	88.66 ± 1.04	89.29 ± 2.26	13.66
	64 features	92.19 ± 0.78	89.38 ± 1.92	98.36 ± 0.85	88.86 ± 1.31	19.44
	DT-CWPT + Renyi entropy					
	30 features	94.03 ± 0.66	93.44 ± 1.23	94.03 ± 1.00	94.57 ± 1.88	13.60
	64 features	93.53 ± 0.70	90.47 ± 2.01	97.16 ± 0.47	92.86 ± 1.51	19.20

Table 3. Comparison of accuracy and classification time for three class SVD database.

Classifier	**Features Experiments**	**Accuracy (%) ± sd**				**Time (seconds)**
		Average	**Class 1**	**Class 2**	**Class 3**	
KNN	DT-CWPT + Energy					
	30 features	86.42 ± 0.36	100.00 ± 0.00	62.58 ± 0.92	96.18 ± 0.39	1.64
	64 features	83.75 ± 0.40	100.00 ± 0.00	52.58 ± 0.97	97.94 ± 0.45	1.71
	DT-CWPT + Shannon entropy					
	30 features	85.81 ± 0.51	100.00 ± 0.00	59.74 ± 1.25	97.11 ± 0.59	0.63
	64 features	83.85 ± 0.44	100.00 ± 0.00	53.40 ± 1.01	97.45 ± 0.56	0.72
	DT-CWPT + Renyi entropy					
	30 features	87.03 ± 0.54	100.00 ± 0.00	63.35 ± 1.69	97.21 ± 0.61	0.70
	64 features	84.34 ± 0.35	100.00 ± 0.00	54.43 ± 1.24	97.89 ± 0.33	0.78
SVM	DT-CWPT + Energy					
	30 features	96.55 ± 0.35	100.00 ± 0.00	97.53 ± 0.33	92.35 ± 1.23	98.00
	64 features	97.45 ± 0.26	100.00 ± 0.00	98.97 ± 0.00	93.58 ± 0.75	148.31
	DT-CWPT + Shannon entropy					
	30 features	93.77 ± 0.30	100.00 ± 0.00	85.05 ± 0.81	96.13 ± 0.82	98.53
	64 features	94.81 ± 0.28	100.00 ± 0.00	86.60 ± 0.49	97.70 ± 0.40	147.13
	DT-CWPT + Renyi entropy					
	30 features	96.55 ± 0.35	100.00 ± 0.00	97.53 ± 0.33	92.35 ± 1.23	97.48
	64 features	97.69 ± 0.33	100.00 ± 0.00	96.60 ± 0.36	96.52 ± 0.71	149.96

5. Conclusion

In conclusion, ReliefF algorithm can be an effective way to improve the overall performance of the multiclass classification. It can enhance the DT-CWPT performance by selecting relevant features and reduce time taken for accuracy calculation. For future works, the system performance should be tested by increasing the training samples and the used of another database to assess the independence of the developed algorithms. From the promising experimental results, it is hoped that the system can be test rigorously in the medical field and provide the clinicians with a useful early diagnosis for voice pathology.

References

[1] Taib,D., Tarique, M., Islam, R.: Voice feature analysis for early detection of voice disability in children. IEEE International Symposium on Signal Processing and Information Technology (2018) 12-17

[2] Arjmandi, M.K., Pooyan, M.: An optimum algorithm in pathological voice quality assessment using wavelet-packet-based features, linear discriminant analysis and support vector machine. Biomedical Signal Processing and Control 7(1) (2012) 3–19

[3] Selvakumari, N.A.S., Radha, V.: Voice pathology identification: a survey on voice disorder. International Journal of Engineering and Manufacturing 7(2) (2017) 39–49.

[4] Gómez-garcía, J.A., Moro-velázquez, L., Godino-llorente, J.I.: On the design of automatic voice condition analysis systems . Part I : Review of concepts and an insight to the state of the art. Biomedical Signal Processing and Control 51 (2019) 181–199

[5] Hegde, S., Shetty, S., Rai, S., Dodderi, T.: A survey on machine learning approaches for automatic detection of voice disorders. Journal of Voice (2018)

[6] Roffo, G., Melzi, S., Castellani, U., Vinciarelli, A.: Infinite latent feature selection: a probabilistic latent graph-based ranking approach. In: Proceedings of the IEEE International Conference on Computer Vision (2017) 1398-1406

[7] Harar, P., Galaz, Z., Alonso-Hernandez, J.B., Mekyska, J., Burget, R., Smekal, Z.: Towards robust voice pathology detection. Neural Computing and Applications (2018) 1–11

[8] Barry, W.J., Pützer, M.: Saarbruecken voice database. Institute of Phonetics, University of Saarland, http://www. stimmdatenbank. coli. uni-saarland. de. (2007)

[9] Martinez, D., Lleida, E., Ortega, A., Miguel, A., Villalba, J.: Voice pathology detection on the saarbr¨ucken voice database with calibration and fusion of scores using multifocal toolkit. In: Advances in Speech and Language Technologies for Iberian Languages Springer, Berlin, Heidelberg (2012) 99–109

[10] Muhammad, G., Alsulaiman, M., Ali, Z., Mesallam, T.A., Farahat, M., Malki, K.H., Bencherif, M.A.: Voice pathology detection using interlaced derivative pattern on glottal source excitation. Biomedical Signal Processing and Control (2017) 156–164.

[11] He, H., Bai, Y., Garcia, E.A, Li, S.:ADASYN: Adaptive synthetic sampling approach for imbalanced learning. In IEEE International Joint Conference on Neural Networks (IEEE World Congress on Computational Intelligence) (2008) 1322–1328

[12] Bayram, I., Selesnick, I.W.: On the dual-tree complex wavelet packet and m-band transforms. IEEE Transactions on Signal Processing 56(6) (2008) 2298–2310

[13] Selesnick, I.W., Baraniuk, R.G., Kingsbury, N.G.: The dual-tree complex wavelet transform. IEEE signal processing magazine 22(6) (2005) 123-151.

[14] Kononenko, I.: Estimating attributes: analysis and extensions of relief. European Conference on Machine Learning, Springer, Berlin, Heidelberg (1994) 171–182

[15] Roffo,G.:Feature selection library (MATLAB toolbox).arXiv preprint arXiv:1607.01327 (2016)

[16] Robnik-Šikonja, M., Kononenko, I.: Theoretical and empirical analysis of ReliefF and RReliefF. Machine learning 53 (1-2) (2003) 23-69

[17] Abdullah, R., Muthusamy, H., Vijean, V., Abdullah, Z., Kassim, F.N.C.: Real and complex wavelet transform approaches for malaysian speaker and accent recognition. Pertanika Journal of Science & Technology 27(2) (2019) 737–752

[18] Hariharan, M., Paulraj, M.P., Yaacob, S.:Time-domain features and probabilistic neural network for the detection of vocal fold pathology. Malaysian Journal of Computer Science 23(1) (2010) 60–67

Computer Aided Diagnosis of Eye Disease for Diabetic Retinopathy

Wei Sheng Ng[1], Wan Mahani Hafizah Wan Mahmud[1*], Audrey Kah Ching Huong[1], Wan Nur Hafsha Wan Kairuddin[1], Hong Seng Gan[2], Raja Mohd Aizat Raja Izaham[3]

[1]Faculty of Electrical and Electronic Engineering, Universiti Tun Hussein Onn Malaysia, 86400 Batu Pahat, Malaysia.
[2]British Malaysian Institute,Universiti Kuala Lumpur, 53100 Gombak, Malaysia.
[3]Kolej Kemahiran Tinggi MARA Ledang, 84410 Ledang, Malaysia.

*wanmahani@uthm.edu.my

Abstract. Diabetic retinopathy (DR) is one of diabetes complication that could cause the vision loss, where it is caused by the damage of the blood vessels at the back of the eye. Due to this, regular eye check-up and timely treatment is needed. However, the lack of specialized ophthalmologists together with associated higher medical costs makes regular check-up costly. Therefore, any application of the technologies such as Computer Aided Diagnosis (CAD) system that could help in analysing DR efficiently in its early stage may help this current situation. Although CAD systems were developed before, but the graphic user interface for user is not developed for the ease of uses for everyone and not just limited to professional. So, in this study, a system is created in order to help the doctor to reduce their burden on the job daily and the false negative rate for the benefit of the patient. The input of the system is the Retinal fundus images (RFI) from STARE database, and the system was built with the ability to enhance and process the image for confirmation of DR. In addition, the system will help to extract out the important features based on Grey Level Co-Occurrence Matrix (GLCM) and classify it using artificial neural network (ANN) whether the patient is associated with the characteristics of DR. Also, the system will be easy to use to everyone as it will have its own graphic user interface to make it clear to everyone not just professionals so that the image from the RFI can be inserted and the result will come out in a short duration of time. The developed system able to achieve as high as 88% sensitivity, 80% specificity and 84% accuracy.

1. Introduction

Diabetes is now regarded as an epidemic, with the population of patients expected to rise to 380 million by 2025. Tragically, this will lead to approximately 4 million people around the world losing their sight from diabetic retinopathy (DR), the leading cause of blindness in patients aged 20 to 74 years [1]. Theoretically, the blood supply towards all layers of retina is done through micro blood vessels which are susceptible to unrestrained blood sugar level. However, when a large amount of glucose or fructose gathers in blood, the vessels may be crumbled and blocked because of insufficient distribution of oxygen to cells, thus leading to a severe eye injury [2]. As a result, metabolic rate slows down and leads to structural abnormality in vessels which causes DR [3].

DR is a main reason of blindness and its development is at different rates in different persons [2]. Previous study reported 90% of DR cases can be prevented through early detection and treatment [4].

But DR can easily perceive until a severe stage is reached. Therefore, regular eye check-up and timely treatment is needed. However, the lack of specialized ophthalmologists together with associated higher medical costs makes regular checkup costly [5]. That is why there is a huge demand for the latest technologies and methodologies to analyze DR efficiently and correctly in its early stage.

An automated screening system for DR can help in reducing the chances of complete blindness due to DR along with lowering the workload on ophthalmologists. For DR screening, development of low cost and versatile Computer Aided Diagnosis (CAD) systems which can be used in clinical environments, have drawn much more attention in recent years [6-13]. These systems as any machine vision systems include the lighting, the acquisition, the processing and the segmentation/classification steps. The objective of CAD systems mostly is to recognize DR and normal images in utilizing various eye features readable from fundus images like exudates (EXs), microaneurysms (MAs), veins, texture, hemorrhages (HMs), node points, and so forth [6, 7]. DR screening between two classes (DR and normal) was produced by Usher et al. [8], Aptel et al. [9], and Quellec et al. [10]. They utilized clinical components like EXs, veins, MA, Cotton Wool Spots (CWS), and HMs. Usher et al. segmented out dark lesions using moat operator and EXs were extricated using recursive region growing (RRGT) and adaptive intensity thresholding (AIT) [8]. Their methods obtained specificity of 46.3% and sensitivity of 95.1% [8]. Quellec et al. have applied optimal filters to extricate Mas [10]. The artificial neural network (ANN) [8] and Bayesian outline work [9] were used for classification.

Although CAD systems were developed before, but the graphic user interface for user is not developed for the ease of uses for everyone and not just limited to professional. So, in this project, a system is created in order to help the doctor to reduce their burden on the job daily and the false negative rate for the benefit of the patient. The input of the system is the Retinal fundus images (RFI) and the system was built with the ability to enhance and process the image for confirmation of DR. In addition, the system will help to extract out the important feature and classify it whether the patient is associated with the characteristics of DR. Also, the system will be easy to use to everyone as it will have its own graphic user interface to make it clear to everyone not just professionals so that the image from the RFI can be inserted and the result will come out in a short duration of time.

2. Material and Methods

There were several steps that need to be done in order to complete this project, as shown in Figure 1. After acquiring the fundus images, the images will undergo image processing step namely image enhancement and image segmentation. Image enhancement is performed to filter out the noises in the image and intensify the features of the image while image segmentation was performed to remove any irrelevant and unwanted parts. After that, the images were further analyzed by extracting the features, selecting the good features and classify the features selected. Feature extraction was performed using Gray-Level Co-Occurrence Matrix (GLCM) to measure the pixel of the image. Then, extracted features were selected so the dimensional of the image can be reduced to only a subset of image. Next, feature classification by using Neural Network took place where it can visualize, simulate and classify the output images that have been processed. Finally, CAD system was developed.

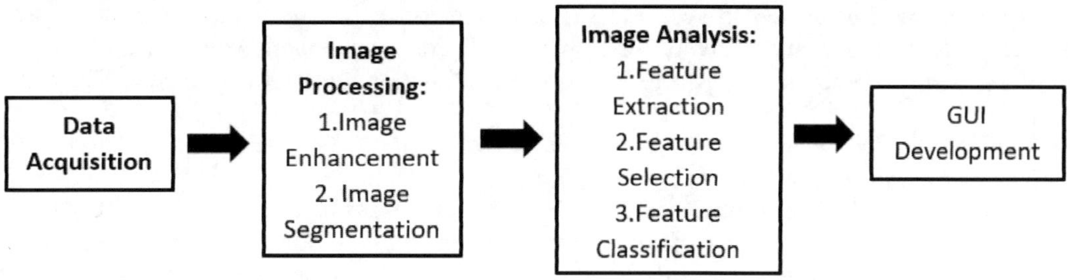

Figure 1. Block diagram of CAD system for DR.

2.1 Data Acquisition
In this study, the data were taken from the public database called Structured Analysis of the Retina (STARE). Images of retina are taken by a device called fundus camera. RFI is the name given to these images. This camera takes images of the internal surface of retina, posterior pole, macula, optic disc, and blood vessels. Figure 2 shows the example of the rgb images taken from STARE database.

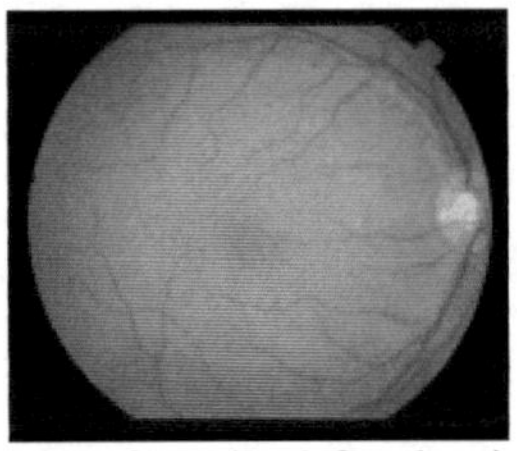

Figure 2. Example of Retinal fundus images (RFI).

In RFI related to DR, most important feature that need to be focused on is the blood vessel from the eyeball as DR will cause the blood vessel to be grown on the back of the eye where the blood vessel is very weak and it might burst and bleed out on the other part of the eye. So as the blood vessel from the image come out as more black (lower intensity pixel value) than the surrounding eyeball, an image processing procedure is needed for better determination of the blood vessel. For this study, 40 images of normal eye and 40 images of DR eye were used.

2.2 Image Enhancement and Segmentation
Image enhancement is usually the first significant step for an image processing in order to create a good quality of imaging. Firstly, the rgb image was converted into grayscale image as the blood vessel in the grayscale image is darker than the other part. After that, the image is filtered with median filter of 3x3 neighborhood in order to clear up all the noises in the images as the median filter uses to make the edge of each pixel 0's to clear up any blurring part in the image. In this study, few different filters were initially tested, but median filter was chosen in the end because it is the most effective than other filters with highest value of Peak Signal to Noise Ratio (PSNR) and lowest value of Mean Squared Error (MSE). Next, the image was darkened to make it more obvious. Subsequently, the image was complemented in order to inverse the image to make the blacker colored blood vessel to become white. Afterwards, in order to have a distinct of an image where the black and white are more intensified, the adjustment have been made to have the input contrast limit to become low and the output contrast limit to become high to make the image have larger difference. Thus, adaptive histogram equalization was used to focus on the part that wanted where it enhanced the contract of every tile to make the blood vessel more obvious.

Next, image segmentation was performed to the output image of adaptive histogram equalization step. In this part, the techniques that were taken into consideration were based on morphological processing where only the important regions were retrieved and selected out for the imaging process. A structuring element was used to sort and select the regions, thus removing those that were not needed. The structuring element always used with morphological open which is the function to remove the structure element that was selected before the function where it can remove the shape that was unwanted.

2.3 Feature Extraction, Selection and Classification
The feature extraction was performed using the Gray-Level Co-Occurrence Matrix (GLCM). GLCM is a texture filter where it can extract the pixel on the image and give out as a statistical data view for the user. In this study, there were four features of GLCM used such as Contrast, Correlation, Homogeneity and Energy. Contrast includes the calculation of the intensity contrast that relate to its nearest pixel it has for the whole image. Correlation is the calculation of the correlation in the image

with the neighboring pixels and relationship between the gray levels and the pixels. Energy on the other hand gives the total number of the square element in the GLCM properties. Lastly, homogeneity is the value that count the distribution of the pixel in one image to measure the elements of the GLCM.

Feature selection was performed using the Statistical Package for the Social Sciences (SPSS). For this study, one-way ANOVA was used. One-way ANOVA is one-way analysis of variance where it is one of the techniques that is used in comparing two or more groups of data or samples. The data that this project looking for is the p-values of the both set of normal eyes and DR eyes where the normal eyes result was stated as "1" and the DR eyes was set to "2". The result in numbering that is closer to the "1" was set to be normal eyes and vice versa. P-value is the probability value that was needed to proof the data is sufficiently inconsistent to make sure that the data would not happen repetitively. When the p-value is less than 0.05%, it can be proven that the group of data were significantly different.

Feature classification is the process where the image is to be identified with various situations and whether the retina is contagious with the diabetic retinopathy disease. The supervised algorithm of Artificial Neural Network (ANN) was used where this ANN works with arrangement of simple and complex cell where these cells are arranged together to each layer to make sure the probability of an object is confirmed by each cell decision. Each layer is decided by the feature that has been selected through one-way ANOVA analysis previously so that the system would better distinguish between normal and diabetic retinopathy eyes. In this step, 80 images were used as training of ANN (40 images for each group of normal and DR), while 50 images were used as testing of ANN (25 images for each group of normal and DR).

2.4 Graphical User Interface (GUI) Development

The Graphic User Interface Development Environment (GUIDE) in the MATLAB system provides the controls and designing option for the user to enhance the coding performed in the MATLAB. The development of Graphic User Interface (GUI) may help users to perform the analysis faster.

3. Result and Discussion

This section includes the result of the image processing steps taken in order to observe the blood vessel of the eye.

3.1 Result of Image Processing

Figure 3 shows the result after the RGB image with DR was converted into grayscale and filtered using median filter. Based on the result, the blood vessel inside the retina can be seen where the pixels were darker rather than the other part. Figure 4 shows the result after the segmentation step using the morphological process. Based on the result, we can say that the output image can highlight the blood vessel better. While in Figure 5, the red circle shows that the patient suffers from diabetic retinopathy where the consecutively black dots shown that the blood vessel burst and spreading around the back of the eyes and this was the cause of vision impairments in DR patient.

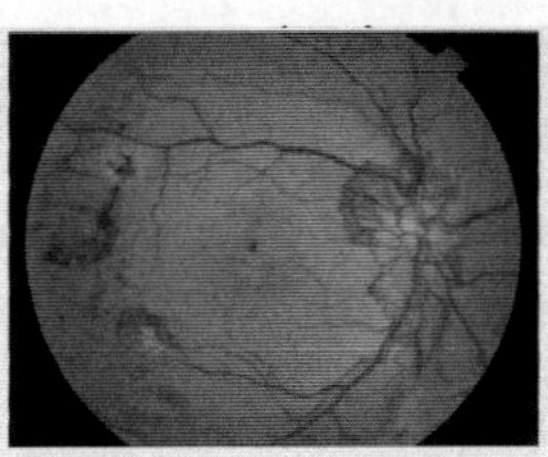

Figure 3. The result of DR image after median filter.

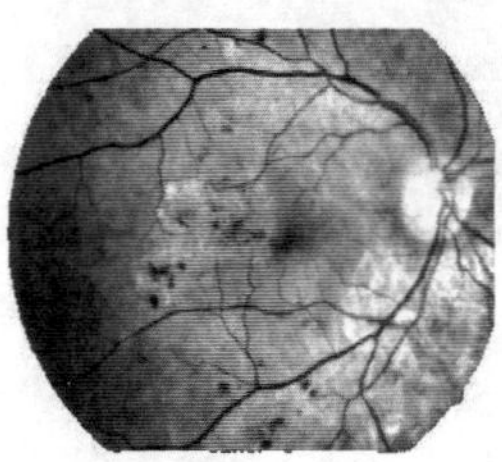

Figure 4. Final output for the image processing.

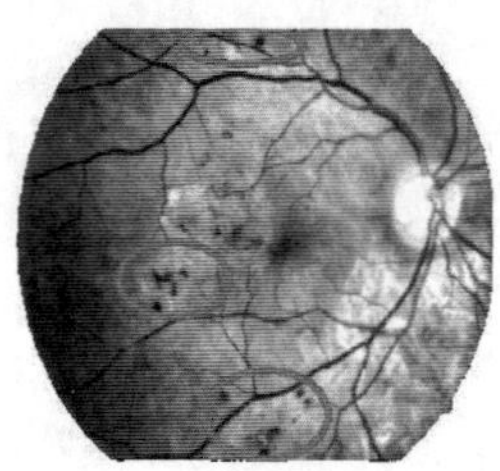

Figure 5. Proof that the patient suffering from diabetic retinopathy.

3.2 Result of Feature Extraction

Figure 6 shows the mean result of the GLCM for each Contrast, Correlation, Energy and Homogeneity for both normal eye and DR eye. As can be seen in the Figure, Contrast has shown a distinguished result between normal and DR eye, followed by Energy, and Homogeneity. Correlation only shows slightest difference between the normal and DR eye.

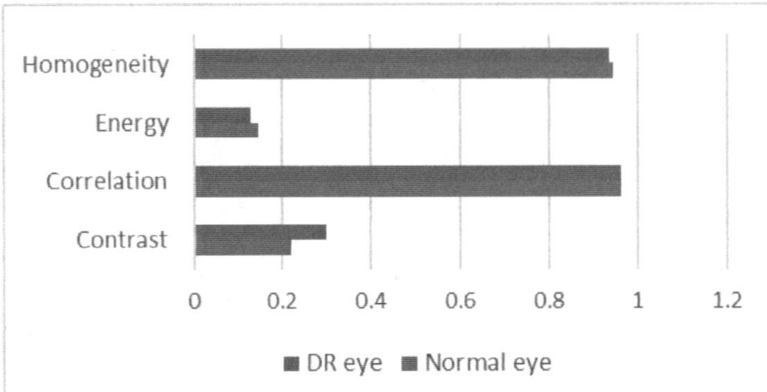

Figure 6. Result of GLCM for both normal eye and DR eye.

3.3 Result of Feature Selection

Based on the statistical analysis using one-way ANOVA, two groups between normal eyes and diabetic retinopathy eyes were being compared and analyzed. According to the result, the p-value of only contrast and energy is 0.000 where it is <0.05 thus both of contrast and energy were used for the classification step using ANN. Homogeneity and Correlation features were excluded to be used for classification purposes.

3.4 Result of GUI Development

Figure 7 shows the complete GUI design for detection of DR that is easy to be used by any medical personnel. The read and process image button can be clicked by user to choose the image that the user wanted to be processed and the system will auto train the system and show the diagnosis result in the panel. If the value was less than 1.5, the image was diagnosed as normal eyes, while if the value was more than 1.5, the image was diagnosed as diabetic retinopathy eyes.

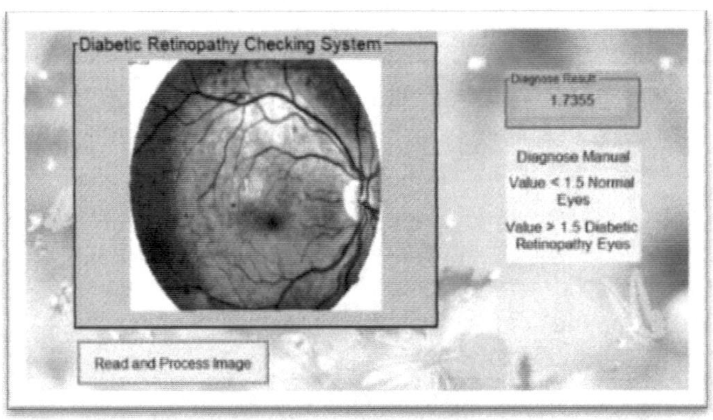

Figure 7. Complete GUI design for detection of diabetic retinopathy.

The accuracy of the system was further tested by using 50 images (25 images of normal eye and 25 images of DR eye). Table 1 shows the result of the accuracy testing. Out of 25 DR eye images, 22 of them were successfully tested as DR eye (true positive (TP)) while another 3 was wrongly tested as normal eye (false negative (FN)). Based on 25 normal eye images, 20 of them were detected as normal (true negative (TN)), while 5 of them were falsely detected as DR eye (false positive (FP)). The sensitivity value of the system was calculated 88%, the specificity value of the system was 80%, and the accuracy value of the system was 84%.

Table 1. The result of accuracy testing.

Tested \ Actual	DR eye	Normal eye
DR eye	22	5
Normal eye	3	20

4. Conclusion

As a brief conclusion, image processing steps including enhancement and segmentation have been successfully implemented, where the output images enable good features to be extracted from the images. The burst out of the blood vessel was clearly pointed out from the multiple image processing stages. With the aid of the GUI system, this system was able to be used at ease for users with quite decent accuracy. As recommendations, this developed system can be improved by collecting the image data personally with collaboration of hospital as the image intensity differs with each machine. With the image consistently produced, the feature can be easily extracted, and the neural network can perform better, thus higher accuracy of the system may be achieved.

Acknowledgments
The authors would like to thank the sponsor for this study. This study was funded by grant TIER 1 H223.

References

[1] Tarr J M, Kaul K, Chopra M, Kohner E M and Chibber R 2013 *ISRN Ophthalmol.* 1-13
[2] Amin J, Sharif M and Yasmin M 2016 *Hindawi Publishing Corporation Scientifica* 1-20
[3] Kayal D and Banerjee S 2014 *Proc. of the 1st Int. Conf. on Sig. Processing & Int. Networks* 141– 44
[4] Kertes P J and Johnson T M 2007 *Lippincott Williams & Wilkins*
[5] Meriaudeau F, Sidibé D, Adal K, Ali S, Giancardo L, Karnowski T P and Chaum E 2013 *11th Int. Conf. on Quality Control by Artificial Vision* 241-47
[6] Trucco El, Ruggeri A, Karnowski T, Giancardo L, Chaum E, Hubschman JP, Al-Diri B, Cheung CY, Wong D, Abràmoff M, Lim G, Kumar D, Burlina P, Bressler NM, Jelinek HF, Meriaudeau F, Quellec G, Macgillivray T, Dhillon B 2013 *Investigative Ophthalmology & Visual Science*
[7] Chua K, Lim C M, Ng E Y K and Laude A 2013 *Comp. in Biol. and Med.* **43** 12 2136–55
[8] Usher D, Dumskyj M, Himaga M, Williamson T H, Nussey S and Boyce J 2004 *Diabetic Med.* **21** 1 84–90
[9] Aptel F, Denis P, Rouberol F and Thivolet C 2008 *Diabetes & Metabolism* **34** 3 290-93
[10] Quellec G, Russell S R and Abramoff M D 2011 *IEEE Trans on Medical Imaging* **30** 2 523–33
[11] Agurto C, Murray V, Yu H, Wigdahl J, Pattichis M, Nemeth S, Barriga E S and Soliz P 2014 *IEEE Journal of Biomedical and Health Informatics* **18** 4 1328–36
[12] Aqeel A F and Ganesan S 2014 *Proc. of IEEE Int. Conf. on Electro/Information Technology* 206–15
[13] Rokade P M and Manza R R 2015 *Int. J of App. or Innov in Eng. & Mgmt* **4** 5 402–10

Smoking status classification by optical spectroscopy and partial least square regression

Audrey Kah Ching Huong[1*], Wan Mahani Hafizah Wan Mahmud[1], Kim Gaik Tay[1], Xavier Toh Ik Ngu[1,2]

[1]Faculty of Electric and Electronic Engineering, Universiti Tun Hussein Onn Malaysia, 86400 Batu Pahat, Johor, Malaysia
[2]Center for Applied Electromagnetics, Universiti Tun Hussein Onn Malaysia, 86400 Batu Pahat, Johor, Malaysia

*audrey@uthm.edu.my

Abstract. Smoking status of individuals is often revealed through self-reported data and interviews. The incidence of false reports severely impairs the proper assessment of the individuals' health conditions and their risk to tobacco associated diseases, delays clinical intervention and treatment services. This paper presents the use of optical technique combined with partial least square (PLS) regression model in the classification of smoking status. The focus of this work is on light absorbance signals (by skin tissue) in the wavelength range of 520 – 600 nm; based on these data dendogram and PLS techniques are used to estimate the number of likely clusters within the considered dataset and to predict smoking status of individuals, respectively. The results from the processing of spectral information of smoking and nonsmoking populations revealed a high misclassification rate of 26.67 % using dendrogram method, but a considerably high accuracy of 90 % evaluated via leave one out cross validation was obtained using PLS component number 4. This study concluded that the spectral oscillation patterns and descending rates corresponded to nonsmoking and smoking individuals could be differentiated and specified using PLS technique in the determination of smoking status.

1. Introduction

Identifying an individual's smoking status has become an increasingly important part of public health's responsibility in this era of civilization, catalysed by the increased environmental awareness and high level of health education among the general public. It was reported by World Health Organization in year 2018 that 14.6 % of population worldwide are smokers [1], these statistics were shown to be affected by a variety of factors such as demographic characteristics, socioeconomic and education level of an individual [2], the revolution in broadcasting industries such as advertisement, film, television and media, and local tobacco control policies [3]. Tobacco use is known as a risk factor for numerous chronic diseases such as cardiovascular and respiratory diseases [4], gene defects, cancers and stroke. Other effects of tobacco use include premature birth and mortality.

For this reason, world leaders, independent organizations and economic major players from different countries are working together towards tobacco-free world by year 2040 [5]. This is evident with the cessation of tobacco production by Dunhill, a widely recognizable name in tobacco industry, and the passing of anti-smoking policy on the discontinuance of new smoking recruits in a Japanese university [6], all in the efforts to mitigate smoking habit and tobacco addiction among smokers. In addition, bans on tobacco advertising, promotion and sponsorship, taxes on tobacco products and graphic pictorial warning to educate and encourage awareness among the smokers are some examples of anti-smoking strategies undertaken in the global level.

The common proxy measures of identifying smokers including through diagnosis of tobacco-use disorder, counselling visits records, anti-smoking prescription history [7], and records of mobile apps that support smoking cessation [8]. However, these data may not be available for those who have not taken initiative to receive intervention for their smoking habit. Meanwhile the validity of records based on self-report approach is subjected to authentication as workers in [9] found a high percentage (of 48 %) of their recruits made false statements about their smoking status.

A rapid and accurate means of determining smoking status of an individual is of the utmost importance to identify the health condition of the person for applications such as in epidemiological studies of smoking, clinical studies of smoking related diseases [4], for effective treatment strategy, and for the evaluation of health insurance coverage. Among the measurable parameters related to tobacco use are traces of nicotine and increase of carbon monoxide (CO) level in blood. The state of the art of noninvasive technologies for the prediction of smoking status include pulmonary function test, urinary cotinine concentration test and exhaled carbon monoxide (eCO) tester, but there are reservations about the sensitivity and specificity of these technologies. Previous reports [9] found contradictory readings obtained using these technologies in a clinical trial involving 144 subjects.

Optical system that uses light waves of different wavelengths and a light sensor, or its advanced forms (e.g. polarization technique) [10], to noninvasively extract the underlying health conditions has seen many of its clinical applications; these are such as in the diagnosis of hyperbilirubinemia [11], hyperoxia [12] and CO poisoning [12, 13]. The imaging capacity of this technology is inherently limited to the skin penetration depth of a few hundred microns (i.e. dermal layers of human skin). Existing commercial available optical sensors to determine CO poisoning include Pulse CO-Oximetry (Masimo, USA), which operation is based on changes of the absorption spectra of hemoglobin derivatives.

Meanwhile with the recent massive growth and progression in the processing and memory capability of a computing unit, artificial intelligence (AI) plays a crucial role in nearly all systems or technologies that required decision making. It is, therefore, the aim of this study to explore the use of optical technology combined with AI system as a means for rapid classification of a person's smoking status based on their blood CO level.

2. Materials and methods

2.1. Spectral data sample

Thirty light absorbance spectra recorded from right index finger of 18 smoking and 12 nonsmoking subjects in [12] were considered in this study. In this preceding research, light reflected from the examined site was spectrally resolved by a spectrometer (model no. USB4000, OceanOptics, Inc) across the wavelength range of 200 – 1100 nm at a resolution of 0.2 nm; the detected light signals was used in the analysis to predict blood oxyhemoglobin and carboxyhemoglobin saturations. The smoking characteristics of 18 smokers are listed in Table 1. The subject index of nonsmoking control group is from 19 to 30 (aged between 20 and 38 years). The difference in the age range of subjects in these two groups was due to the limited availability in non-smoking subjects of other age categories who were willing to participate in the study.

This research considered only data across wavelengths from 520 to 600 nm based on the previous reports of the distinctive optical signature of hemoglobin components, specifically oxyhemoglobin, deoxyhemoglobin and carboxyhemoglobin, across this wavelength range (see reports of [14]). The use of tobacco would increase blood CO level (i.e. carboxyhemoglobin level) rendering changes in light absorbance spectra. This study used unsupervised dendrogram method to provide an estimate of grouping of dataset and a visual summary of similarity among spectra. The latter was inferred based on the height of the link in the dendrogram. The supervised regression analysis was via PLS component number from 2 to 4; the training and testing data were iteratively chosen based on the concept of leave one out cross validation.

Table 1. The information on age and smoking characteristics of smoking individuals considered in this study (Subject index 1 –18).

Subject index	Age	Daily tobacco use (Cigarette sticks)	Number of years smoking
1	22	9	5
2	28	8	10
3	20	5	2
4	21	5	3
5	52	12	37
6	38	10	20
7	20	14	8
8	43	12	20
9	25	2	2
10	20	7	7
11	23	10	5
12	48	20	20
13	43	20	25
14	58	20	15
15	62	36	50
16	51	4	20
17	24	20	10
18	24	5	3

2.2. Partial Least Square Regression Discriminant Analysis (PLS-DA)
This study explored the use of supervised PLS-DA technique in the identification of smoking status of a person using MATLAB (version 2018a). This method predicts the responses, *Y*, in a population via extracting latent variables from sampled factors and responses as followed [15]:

$$X = T * P' + E_0$$
$$Y = U * Q' + F_0 \quad (1)$$

where *X* and *Y* are the predictor and response, respectively, *T* and *U* are score matrices or latent variables, P' and Q' are loading matrices while E_0 and F_0 are residuals. The scores are solved via least square regression as followed:

$$U = TB + F_1 \quad (2)$$

where *B* and F_1 are inner relation coefficient and residuals, respectively, to give the overall partial regression model in formula (3).

$$\tilde{Y} = X * (P * B * Q') + F. \quad (3)$$

$\tilde{Y}$ is a categorical variable represented by 0 (to indicate nonsmoker) and 1 (for smoker). The response loadings with $\tilde{Y} \geq 0.5$ were discriminated as that of a smoker.

3. Results

Light absorbance spectra for smoking and nonsmoking subjects are shown in Figure 1. The dendrogram in Figure 2 shows three distinct groups. While three (nonsmoking) subjects of index 20, 24 and 30 were misclassified, five spectral data, from that of both smoking and nonsmoking groups (index 4, 10, 16, 21, 25), were found in the third cluster. The bottom diagrams of Figure 2 are the spectral data of subject index 3, 23 and 21 from the clustered groups. The subjects' smoking status value predicted using different PLS component number is shown in Figure 3, also shown in this diagram is a dotted horizontal line drawn to indicate the set threshold value of 0.5. The accuracy of 76.67 %, 93.3 % and 90 % was achieved using PLS component number 2, 3 and 4, respectively.

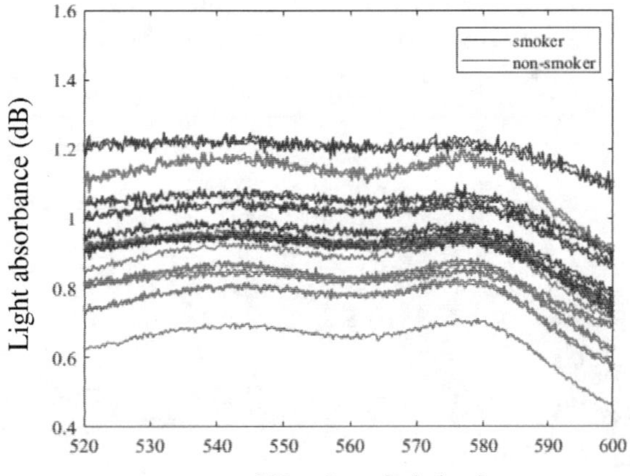

Figure 1. Light absorbance of smoking and non-smoking individuals.

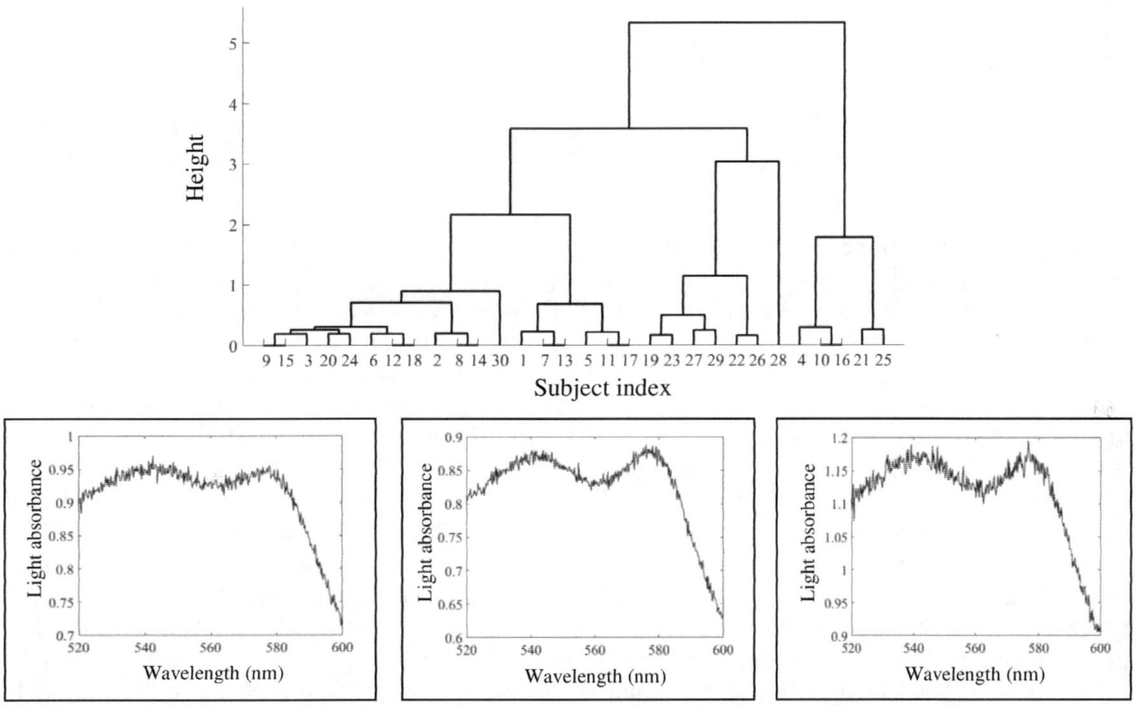

Figure 2. Cluster analysis of spectral data for 30 subjects. (Bottom left to right) Spectral data of subject index 3, 23 and 21 from three distinct groups.

4. Discussion

The dendrogram in Figure 2 showed that the absorbance spectrum for most of the subjects fall into the two distinct clusters, distinctive difference in the visibility of oscillation patterns can be observed in their spectral graphs in the wavelength range of 530-580 nm. Meanwhile several individuals (subject index 4, 10, 16, 21 and 25) from both control and investigation groups were categorized under the third class producing misclassification rate of 26.67 %. An investigation into the absorbance spectrum of these individuals revealed comparative higher magnitudes of light absorbance for individuals in this group, suggesting skin pigmentation as an important feature in spectral classification. It is interesting to note on the striking similarity in the data of several smoking individuals (e.g. subject index 11 and 17). It was found that despite similar absorbance spectrum was measured, their smoking habit and history in Table 1 are relatively different. The main reason for this remained to be clarified, but the authors have not ruled out the possibility of prevalence monotonic effects of CO in blood for smoking individuals.

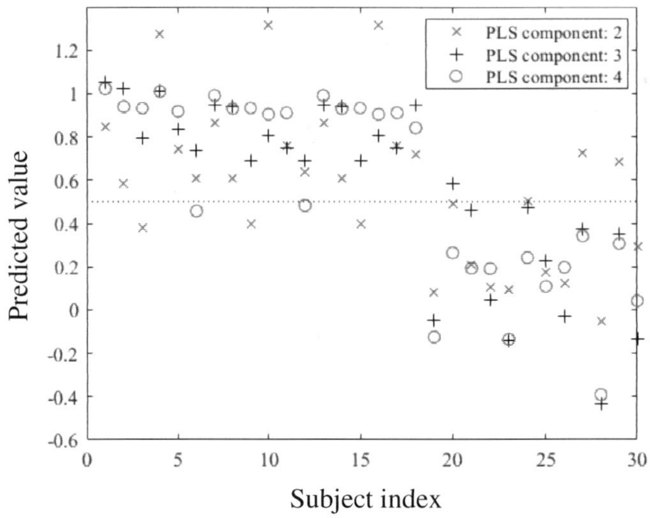

Figure 3. Smoking status value predicted for different subjects using partial least square (PLS) component number from 2 to 4. Value greater than 0.5 was classified as positive.

Even though this study has not considered the effects of variations in nicotine content of cigarette smoke, the lapse of time between the last cigarette smoked and the time of experiment, and the different ability of one's body to expel the CO, the results in Figure 3 showed remarkable consistency in the value predicted for smoking subjects using PLS regression of 4th component. The DA results revealed a slightly lower accuracy of 90 % as compared to that using PLS component number 3, which produced an accuracy of 93 %. The two peaks in the light absorbance spectra, contributed by oxyhemoglobin component, gradually resolved in the investigation group in Figure 1 as compared to that of the control group. This work found that the absorbance spectrum of subject index 6 and 12, which were marginally misclassified in Figure 3, has similar oscillation patterns and magnitudes as that of other smokers, the most apparent disparity is in their spectrum descending rate in between wavelength 580 and 600 nm. The descending pattern observed for these two spectra shows moderate downward movement, it is neither extremely fast nor slow, and is of similar value to that of subject index 27 and 29. Interestingly this misclassification was not observed in subject index 18, whose spectral pattern is considerably similar to that of 12 in Figure 2, however this study has not eliminated the possibility of this being the source of the error.

This study is subjected to limitations including the small number of sample size and the gap in the information of the investigation group, but the results of this study could have implication for future healthcare management and intervention practices. Addressing the above mentioned limitations could be the next steps of this research, which comprises implementation of unsupervised clustering of spectroscopy data before using PLS-DA technique in the prediction of smoking status for new set of data.

5. Conclusion

This work demonstrated the use of PLS-DA in the prediction of a person's smoking status, and a considerably good accuracy of 90 % was obtained using this technique. The findings of this work add to the literature that the prediction using this approach is independent of nicotine content of cigarette use and the last time a person smoke. This study found that, in addition to CO level, skin pigmentation, spectral oscillation patterns and descending rates were among the important determinants in the prediction of smoking status.

Acknowledgments
This work is financially supported in part by the Ministry of Education Malaysia under FRGS (grant no. 1581), PRGS (grant no. K038) and Universiti Tun Hussein Onn Malaysia (B084, U848, H302, H310).

References
[1] World Health Organization WHO 2018 Tobacco Retrieved from https://www.who.int/news-room/fact-sheets/detail/tobacco
[2] Wang Y, Sung H Y, Yao T, Lightwood J and Max W 2017 Factors associated with short term transitions of nondaily smokers: socio-demographic characteristics and other tobacco product use *J. Addict.* **112** 864-872
[3] Pomerleau J, Gilmore A, McKee M, Rose R and Haerpfer C W 2004 Determinants of smoking in eight countries of the former Soviet Union: results from the Living Conditions, Lifestyles and Health Study *J. Addict.* **99** 1577–85
[4] Perez-Warnisher M T, Carballosa de Miguel M P and Seijo L M 2019 Tobacco Use Worldwide: Legislative Efforts to Curb Consumption *Ann. Glob. Health* **85** 9
[5] Beaglehole R, Bonita R, Yach D, Mackay J and Reddy K S 2015 A tobacco-free world: a call to action to phase out the sale of tobacco products by 2040 *Lancet* **385** 1011-1018
[6] Victor D 2019 Japanese university says it wont hired teachers who smoke Retrieved from The New York Times, https://www.nytimes.com/2019/04/25/world/asia/smoking-teachers-nagasaki-japan.html
[7] Desai R J, Solomon D H, Shadick N, Iannaccone C and Kim S C 2016 Identification of smoking using Medicare data- A validation study of claims based algorithm *Pharmacoepidemiol Drug Saf.* **25** 1-8
[8] Heminger C L, Schindler-Ruwisch J M and Abroms L C 2016 Smoking cessation support for pregnant women: role of mobile technology *Subst. Abuse Rehabil.* **7** 15-26.
[9] Stelmach R, Fernandes F L, Carvalho-Pinto R M, Athanozio R A, Rached S Z, Prado G F and Cukier A 2015 Comparison between objective measures of smoking and self-reported smoking status in patients with asthma or COPD: are our patients telling us the truth? *J. Bras. Pneumol.* **41** 124-32.
[10] Huong A, Yip J and Ngu X 2017 Polarized optical system for noncontact imaging of epidermal microcirculation activities *ACM Proc. Int. Conf. on Imaging, Signal Processing and Communication* (Penang, Malaysia) p 59-62.
[11] Cheng N Y, Lin Y L, Fang M C, Lu W H, Yang C C and Tseng S H 2019 Noninvasive transcutaneous bilirubin assessment of neonates with hyperbilirubinemia using a photon diffusion theory-based method *Biomed. Opt. Express* **10** 2969-2984.
[12] Huong A, Philimon, S and Ngu X 2015 Non-invasive estimation of blood oxyhemoglobin and carboxyhemoglobin saturations using cumulant based forward model *APRN J. of Eng. Appl. Sci.* **10** 8421.
[13] Jędrzejewska-Szczerska M, Karpienko K, Wróbel M S, Tuchin V V 2016 *Biosensors for Security and Bioterrorism Applications. Advanced Sciences and Technologies for Security Applications* ed D Nikolelis and G P Nikoleli (Switzerland: Springer)
[14] Zijlstra W G, Buursma A and van Assendelft O W 2000 *Visible and near infrared absorption spectra of human and animal haemoglobin: determination and application* (Netherlands:VSP)
[15] Chiang Y Q, Zhuang Y M and Yang J Y 1992 Optimal Fisher discriminant analysis using the rank decomposition *Pattern Recognit.* **25** 101-111

Journal of Physics: Conference Series **1372** (2019) 012032 doi:10.1088/1742-6596/1372/1/012032

Fundamental design of thermoelectrical nature air conditioning system in small room for living cat (mammalian)

Muhd Nur Rahman Yahya[1*], Muhammad Azam Md Amin[1], Muhammad Izham Ismail[1], Zuradzman Mohamad Razlan[1], Shahriman Abu Bakar[1], Wan Khairunizam[1], I Zunaidi[2]

[1]School of Mechatronic Engineering, Universiti Malaysia Perlis, Arau, Perlis, MALAYSIA.
[2]Faculty of Technology, University of Sunderland, St Peter's Campus, Sunderland, SR60DD, UNITED KINGDOM.

*nurrahman@unimap.edu.my

Abstract. Living mammalian especially European cats will face some problems when traveling from cold ambient in Europe to hot ambient country. Some of the problems are, these cats will become panting when exposed to the high temperature of ambient for a short term. Also, for a long term exposed to high temperature, the cats will experience heat stress and hair loss. The objectives of this study are on cat thermal comfort, cooling load for cat compartment, and analysis of air flow to achieve the desired temperature for the cat compartment using CFD simulation. In this study, three parts of analysis will be considered in CFD simulations which are the thermal distribution for four Peltier plates with aluminium heat sink using ANSYS Transient Thermal, air flow in the ducting to a cold heat sink, and air flow distribution in cat compartment using ANSYS Fluent. The temperature for the Persian cat's thermal comfort is around 20-25°C. Cooling load gain from designed portable cat compartment with the desired space is 177 W. By ANSYS Transient Thermal, time taken for heat sink change from 22°C to 12°C is around 200-250 seconds. Then, by using ANSYS Fluent, the temperature of initial air temperature in the ducting when flow through a cold heat sink was changed from 27.15°C to 20.93°C by simple cooling. Next, the average air temperature in the cat's compartment is 23.83°C that achieved average thermal comfort for mammal or Persian cat which is 20-25°C of dry bulb temperature. However, the study only focuses on steady-state dry air simple cooling analysis only with the laminar flow as a preliminary result by CFD simulation before build the actual portable cat compartment. The dry air simulation can be assumed similar as humid air for simple cooling process only proved by the comparison of the same cooling capacities values towards the different value of absolute humidity.

1. Introduction

Hundred million of cats are kept as pets around the world. Siamese or Oriental Shorthair cats were categorized as a very light breed, the Persian cat as a light breed, the Norwegian Forest and the Siberian Cat as a large breed, and the Maine Coon cat as a giant breed [1]. In general terms, long-haired cats can tolerate much colder temperature than short-haired cats [2]. Pet lovers around the world willing to spend a lot of money on their pet's life care. Nowadays, cat's exhibitions or tournaments are held around the world. Mostly, when they are traveling with their pets from Europe to other hot and humid country, the cat's fur will fall due to expose towards high ambient temperature. Thus, the air conditioning system for a portable cat compartment is needed to overcome these problems. By applying computer analysis software, dry air simulation is an attempt before designing an air conditioning system using thermoelectric to achieve and maintain the desired temperature for the cat's compartment. All the variables and parameters in the simulation can be changed in order to find the best airflow system. Also, to improve the efficiency of the cooling capacity of the thermoelectric to give maximum heat exchange from air inside the confined space to the outside.

Thermal comfort for a mammal is between 18.5 °C to 27 °C of air temperature and 30-70% of relative humidity. According to J. Physiol [3], some experimental study had been done to analyze the reaction of a cat in a cage when the temperature is increased. When the surrounding temperature increase from 25 °C to 38 °C, the cat becomes excited active, pacing about the cage and stretching the wall. Thermoneutral zone is the minimum temperature required to maintain body temperature by using a minimum amount of energy. From estimation, 15-20 °C for long-haired cat breeds and 20-25 °C for short haired cat breeds [4]. Yet most cat housing areas in homes and laboratories are maintained closer to 22 ± 2 °C for long-haired cat [5]. Thermoelectric module (TEM) consists of an arrangement of P-type and N-type of semiconductor which can convert either electric power to thermal power or thermal power to electric power [7]. When there is some temperature difference is placed on the plates, the device can convert heat energy into electrical energy [8]. Cooling or heating energy can be produced through the flowing of electrical current into the module [9]. The cooling capacity is suitable to cover small cooling loads. The coefficient of performance of thermoelectric is less than a conventional air conditioning system. Some popular advantages of thermoelectric are small size, no chemical reaction, low energy consumption, and suitable for small cooling loads [10].

2. Related equations

2.1. Analysis of cooling load

Heat can be transferred in three different modes which are conduction, convection, and radiation. The cooling load can be calculated by adding the rate of heat transfer from conduction from the surrounding through the wall, natural convection from outside of the compartment, and thermal radiation emits from the cat's body.

$$Total\ cooling\ load = \dot{Q}cond + \dot{Q}conv + \dot{Q}rad \quad (1)$$

$$\dot{Q}total = \frac{T1 - T2}{R\ total} \quad (2)$$

$$R_{conduction} = \frac{\Delta x}{kA} \quad (3)$$

$$R_{convection} = \frac{1}{hAs} \quad (4)$$

$$R_{radiation} = \frac{1}{\mathcal{E}\sigma As} \quad (5)$$

Where, T_1= temperature at point 1 (K), T_2= temperature at point 2 (K), k = thermal conductivity (W/m·K), Δx = Thickness (m), A= surface area of the wall (m^2), h = convection heat transfer coefficient (W/m^2·K), A_S= Surface area contact with the fluid (m^2), σ = Stefan Boltzmann constant (W/$m^2 \cdot K^4$) and $\mathcal{E}$ = emissivity (W/$m^2 \cdot K^4$).

2.2. Convection of heat transfer coefficient

The convection heat transfer coefficient h is not a property of the fluid. The value of natural convection heat transfer coefficient of air, the properties of air at the film temperature can be obtained in APPENDIX B. Thus, the natural convection of heat transfer can be calculated in equation 8 based on Rayleigh number and Nusselt number equations.

$$Ra_L = \frac{g\beta(T\infty - Ts){L_C}^3}{v^2}(P_r) \quad (6)$$

Where, g = Gravitational acceleration (m/s^2), β = Volume expansion coefficient (K^{-1}), v = Kinematic viscosity (m^2/s), Pr = Prandtl number, L_C = Characteristic length, Ts = Surface temperature (K) and $T\infty$ = Outside temperature (K).

$$N_u = 0.59(Ra_L)^{\frac{1}{4}} \tag{7}$$

$$h = \frac{k}{L_c} N_u \tag{8}$$

Where, Ra_L = Rayleigh number, h = convection heat transfer coefficient (W/m^2·K) and k = thermal conductivity (W/m·K) and L_c = Characteristic length (m).

2.3. Cooling capacity for thermoelectric

Based on the cooling capacity of thermoelectric, the value of voltage and electric current can be determined by this formulation from equation (9).

$$\dot{Q}in = \propto T_C I - \frac{1}{2} R I^2 - k(T_H - T_C) \tag{9}$$

Where, $\propto$ = Seeback coefficient (*V/K*), k = thermal conductivity (*W/m·K*), R = System resistance (Ω), I = System current (I), T_c = temperature cold (K) and T_H= Analysis of cooling load.

2.4. Lumped system

The formula to determine heat sink temperature after contact with the Peltier plate can be assumed the same due to small thickness by using lump system analysis formulation in equation (10).

$$Bi = \frac{hL_c}{K} \tag{10}$$

Where, h = convection heat transfer coefficient (W/m^2·K) and k = thermal conductivity (W/m·K). To use lump system analysis, the value of Bi must be less than 0.1.

$$b = \frac{hA_S}{\rho V C_P} \tag{11}$$

Where, h = convection heat transfer coefficient (W/m^2·K), A_s= Surface area (m^2), ρ = Density (kg/m^3) and C_P = specific heat (kJ/kg.K).

2.5. Velocity analysis

Value of air velocity needed in this system can be determined by mass flow rate, enthalpy differential from the psychrometric chart, and cooling load calculation. All the formulation can be used as below.

$$\dot{m} \times \Delta h = cooling\ load\ from\ the\ system \tag{12}$$

$$\dot{V} = \dot{m} \times v \tag{13}$$

$$V = \frac{\dot{V}}{A} \tag{14}$$

Where, $\dot{m}$ = mass flowrate (*kg/s*), Δh = enthalpy differential (*kJ/kg.K*), $\dot{V}$ = Volume flowrate (m^3/s), v = specific volume (m^3/kg) and V = air velocity (*m/s*).

3. Methodology

3.1. Cat thermal comfort

From estimation, temperature for cats use minimum energy to maintain their body temperature are around 15-20 °C for long-haired cat breeds and 20-25 °C for short haired cat breeds [4]. Yet most cat housing areas in homes and laboratories are maintained closer to 22 ± 2 °C for long-haired cat [5]. In this case, by taking the average of thermal comfort of mammals which are 20°C to 25 °C of air temperature and 40% to 60% of relative humidity as the thermal comfort to keep prevent hair loss and heat stress for European cat which is Persian cat.

3.2. Total cooling load

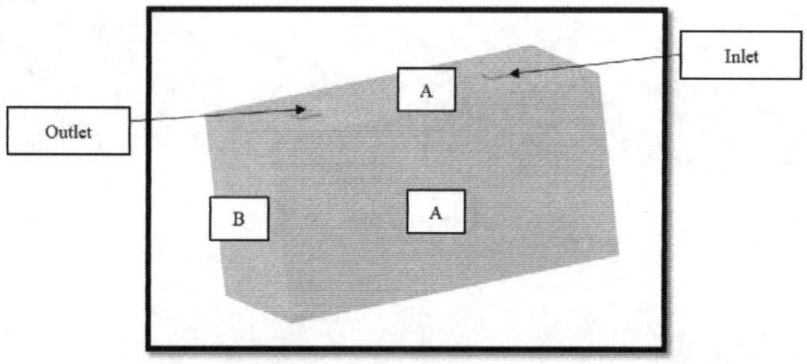

Figure 1. Compartment.

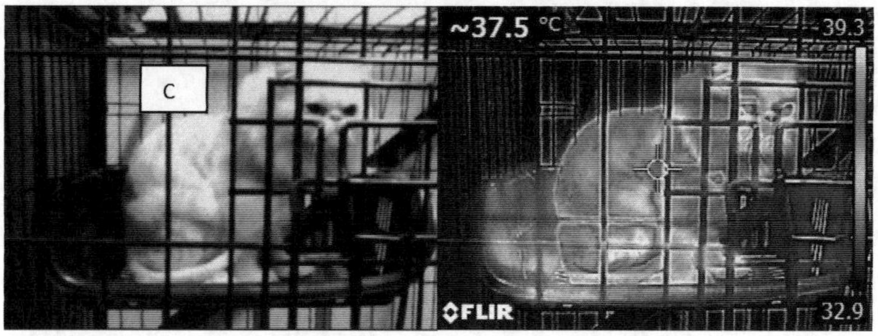

Figure 2. Cat surface temperature.

From Equation 1 in methodology, there are conduction and convection from the surrounding. Then, thermal radiation from the cat's body.

$$Total\ cooling\ load = 111\ W(\dot{Q}cond, \dot{Q}_A) + 34\ W(\dot{Q}conv, \dot{Q}_B) + 32\ W(\dot{Q}rad, \dot{Q}_C) = 177\ W$$

Based on the calculation of subsidiary law in heat transfer. Total cooling load gained is 177 W.

3.3. Velocity at the inlet of the ducting
Mass flow rate: Equation 12, $\dot{m} = 0.0151\ kg/s$.
Volume flow rate: Equation 13, $\dot{V}in = 0.01306\ m^3/s$.
Area of the ducting cross-sectional area, $A = 6.4 \times 10^{-3}\ m^2$

Inlet velocity: Equation 14, $Vin = 2.01\ m/s$
Thus, velocity inlet and outlet of the ducting is 2 m/s each can be assumed.

4. Result and Discussion

4.1. Heat sink
The heat sink thermal distribution simulation using ANSYS Transient thermal analysis.

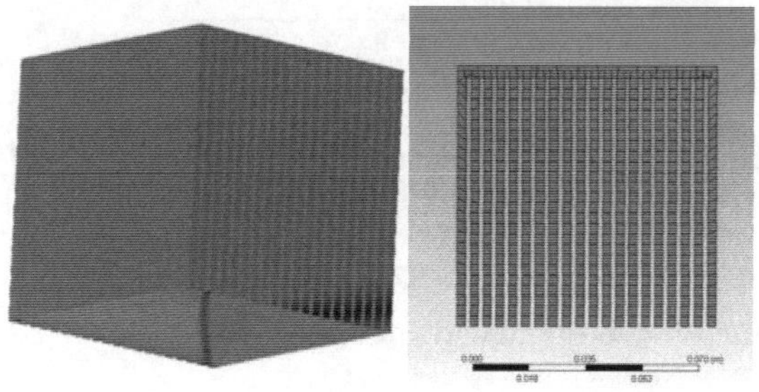

(a) (b)

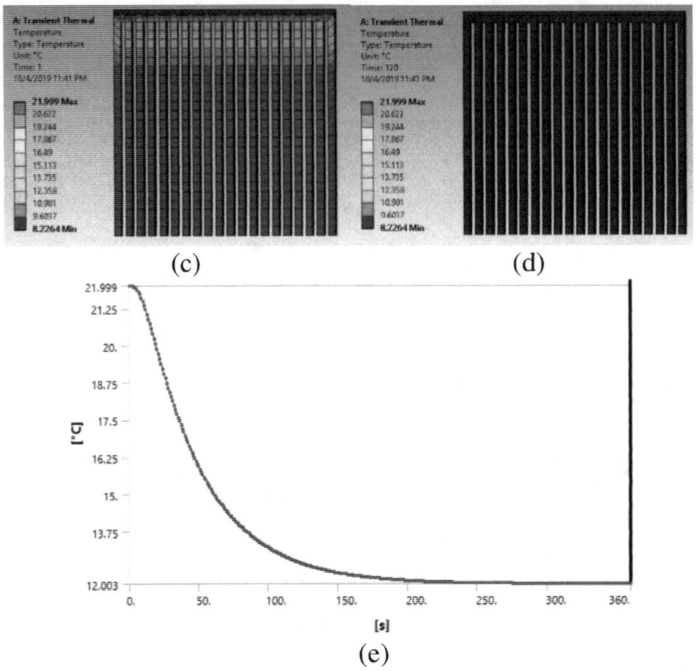

Figure 3. a) Heat sink geometry, b) meshing, c) at 1 second, d) at 120 seconds, and e) graph temperature vs time.

By observing the graph in Figure 3, the behavior of the graph is inversely proportional between temperature and time. The temperature aluminum heat sink from 22 °C at room temperature decrease to 12 °C following the Peltier plate's temperature. Time taken needed for the heat sink become 12 °C is around 200 seconds to 250 seconds only.

4.2. Ducting
Airflow in the ducting simulation using ANSYS Fluent.

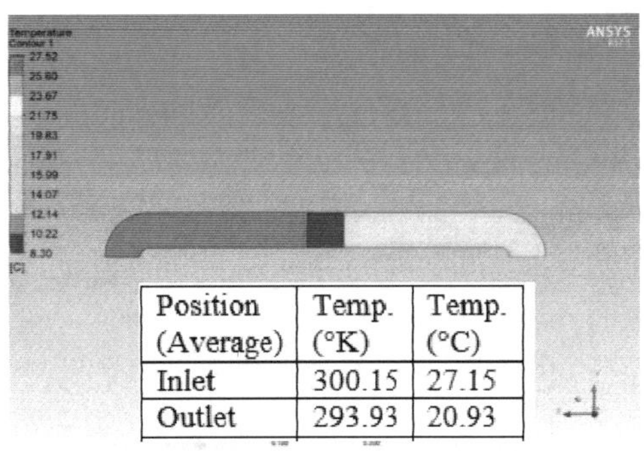

Position (Average)	Temp. (°K)	Temp. (°C)
Inlet	300.15	27.15
Outlet	293.93	20.93

Figure 4. Temperature distribution.

Figure 4 shows the average temperature of dry air in the ducting changed from 27.15 °C to 20.93 °C of average outlet temperature after going through the constant temperature of the heat sink at 12 °C.

4.3. Compartment
Airflow in the ducting simulation using ANSYS Fluent.

Journal of Physics: Conference Series **1372** (2019) 012032 doi:10.1088/1742-6596/1372/1/012032

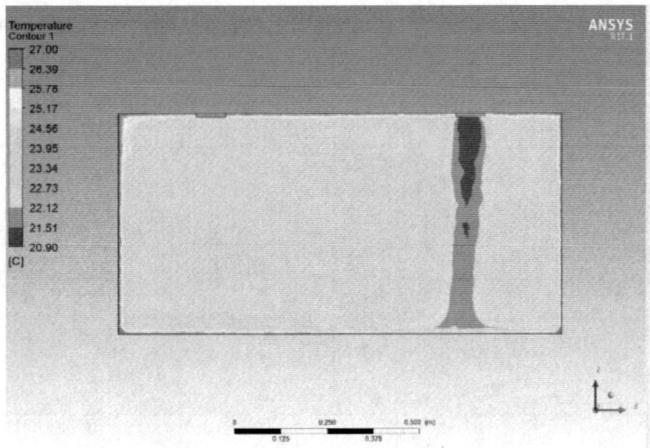

Condition	Temp. (°K)	Temp. (°C)
Average	296.89	23.89
Maximum	299.60	26.60
Minimum	294.08	21.08
Inlet	294.08	21.08
Outlet	297.20	24.20

Figure 5. Temperature distribution.

Then, the average temperature at the compartment is 23.89 °C can be seen in Figure 5. The average inlet temperature of the compartment to the outlet temperature changed from 21.08 °C to 24.04 °C. The average temperature in the compartment achieved the average thermal comfort of mammal or cat which is air temperature between 20 °C to 25 °C.

4.4. Compartment analysis

There are some possible constraints that may affect the values of temperature which are dry air analysis using software with humid air in real life. Referring data of simple cooling at Figure 6, by increasing the absolute humidity from 0 g/kg to 50 g/kg and 100 g/kg, the air temperature at the outlet of the ducting can be determined for same cooling capacity. This data from the calculations will prove the similarity relationship between dry and moist air simulation with the same cooling capacity and differences values of the absolute humidity as long as it is a simple cooling process in Table 1.

From the graph in Figure 7, the similarity of dry air and humid air relationship can be seen when increasing the absolute humidity, the changes of the inlet to outlet temperature are the same. The final temperature did not reach or below than the dew point temperature to prove simple cooling process by this system. By this relationship, this dry air simulation can be used regardless the value of absolute humidity as a preliminary result as long as a simple cooling process before starting to design a cooling system for cat's compartment with actual humid air in the real life.

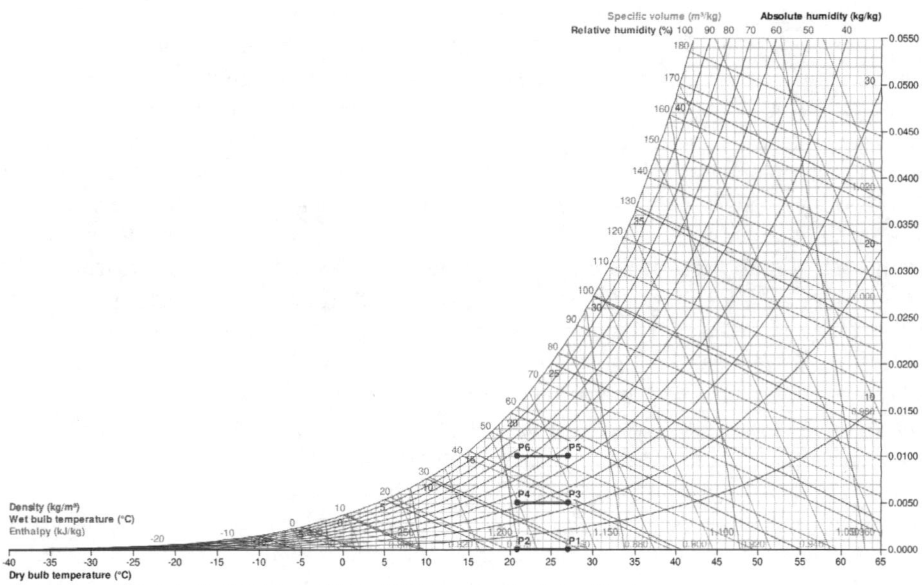

Figure 6. Simple cooling for 0 g/kg, 50 g/kg, and 100 g/kg of absolute humidity.

Table 1. Comparison data gained for different absolute humidity between 0 g/kg, 50 g/kg, and 100 g/kg

Condition from the inlet of ducting	Cooling load	Mass flow rate, $\dot{m}$	Mass flow rate differential, $\Delta\dot{m}$	Enthalphy differential, Δh	Outlet temperature
27.1 °C to 20.9 °C at 0 g/kg of absolute humidity from simulation	94.5 W	0.0150 kg/s	Decrease 0.0001 kg/s	6.3 kJ/kg	20.9 °C
27.1 °C at 50 g/kg of absolute humidity	94.5 W	0.0149 kg/s		6.34 kJ/kg	20.9 °C
27.1 °C 100 g/kg of absolute humidity	94.5 W	0.0148 kg/s		6.38 kJ/kg	20.9 °C

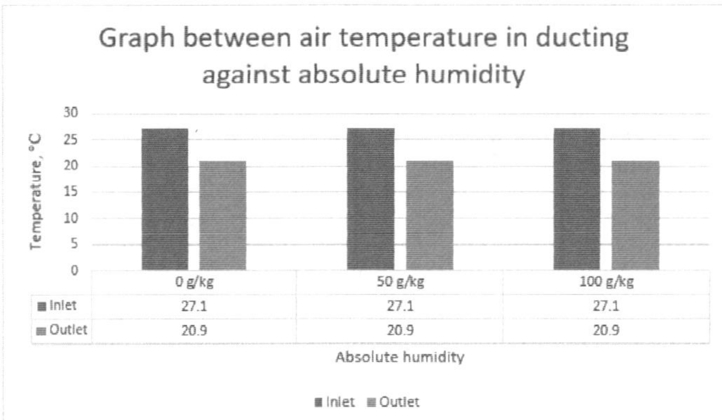

Figure 7. Simple cooling differential temperature with 0 g/kg, 50 g/kg, and 100 g/kg of absolute humidity.

5. Conclusion

The portable cat compartment works to overcome problems such as heat stress and hair loss toward European cats when traveling from cold to a hot environment. Cat thermal comfort is the target point to be achieved and regulated in the portable cat compartment. From the survey at Jabatan Perkhidmatan Veterinar Negeri Perlis, there is no specific temperature and humidity for cat's thermal comfort. A cat can adapt to surrounding temperature when living in a new environment. But, for traveling for a certain period of time in a hot environment, the temperature of the compartment needs to be regulated to maintain the cat's thermal comfort. Thus, by taking the average of thermal comfort of mammals which are 20°C to 25°C of air temperature and 40% to 60% of relative humidity as the thermal comfort to keep prevent hair loss and heat stress for European cat which is Persian cat.

The cooling loads for the portable cat compartment can be calculated by using subsidiary law in heat transfer which are Fourier law of conduction, Newton law of cooling in convection, and Stefan Boltzmann law in thermal radiation. Total cooling load from the cat's compartment obtained by the calculations is 177 W. By these cooling load, the velocity for the inlet ducting with 0.0064 m^2 of the area is 2 m/s. Thus, 4 pieces of Peltier plates will be used with 6 A of current, 6 V of voltage, and around 50-60 W each. The COP of the Peltier module is 1.42 which is likely an efficient device. All fours Peltier plates will be attached to an aluminum heat sink with 8.3 °C.

By ANSYS Transient Thermal, time taken for heat sink change from 22 °C to 12 °C is around 200-250 seconds. Then, by using ANSYS Fluent, the temperature of initial air temperature in the ducting was changed from 27.15 °C to 20.93 °C by simple cooling. Next, the average temperature in the cat's compartment achieved is 23.83 °C which is achieved average thermal comfort for mammal or Persian cat which is 20-25 °C of dry bulb temperature. But this data using steady-state simulation with laminar flow and fully dry air. This dry air simulation same as humid air simulation had been proved in discussion before which is only for simple cooling analysis. All three objectives had been achieved in this study.

References

[1] E. Kienzle and K. Moik 2011 A pilot study of the body weight of pure-bred client-owned adult cats *Br. J. Nutr.*, vol. 106, no. S1, pp. S113–S115.

[2] J. Richards CATS 2013 London: POPULAR UK PTE LTD .

[3] S. Campbell 1948 From the Department of Physiology *J. PHysiol.*, pp. 87–108.

[4] F. Complete and C. P. Food 2017 FEDIAF_Nutritional_Guidelines_2017 no. May.

[5] J. L. Stella and C. C. Croney 2016 Environmental Aspects of Domestic Cat Care and Management: Implications for Cat Welfare *Sci. World J.*, vol. 2016.

[6] Yunus A. Cengel, Afshin J. Ghajar 2015 Heat and Mass Transfer, Singapore: McGraw-Hill Education.

[7] S. Twaha, J. Zhu, Y. Yan, and B. Li 2016 A comprehensive review of thermoelectric technology : Materials , applications , modelling and performance improvement vol. 65, pp. 698–726.

[8] S. Yu, Q. Du, H. Diao, G. Shu, and K. Jiao 2015 Start-up modes of thermoelectric generator based on vehicle exhaust waste heat recovery," *Appl. Energy*, vol. 138, pp. 276–290.

[9] X. F. Zheng, C. X. Liu, Y. Y. Yan, and Q. Wang 2014 A review of thermoelectrics research - Recent developments and potentials for sustainable and renewable energy applications," *Renew. Sustain. Energy Rev.*, vol. 32, pp. 486–503.

[10] Y. Zhou, T. Zhang, F. Wang, and Y. Yu 2018 Performance analysis of a novel thermoelectric assisted indirect evaporative cooling system *Energy* vol. 162 pp. 299–308.

Numerical Analysis on Modelling Thermal Comfort Minor Operation Theatre

Muhd Nur Rahman Yahya[1*], Muhammad Azam Md Amin[1], Muhammad Izham Ismail[1], Zuradzman Mohamad Razlan[1], Shahriman Abu Bakar[1], Wan Khairunizam[1], I Zunaidi[2]

[1]School of Mechatronic Engineering, Universiti Malaysia Perlis, Arau, Perlis, MALAYSIA.
[2]Faculty of Technology, University of Sunderland, St Peter's Campus, Sunderland, SR60DD, UNITED KINGDOM.

*nurrahman@unimap.edu.my

Abstract. An operation theatre (OT) is a special room inside the hospital where medical surgery is carried out by a surgeon with the help of medical personnel. Technical standards or requirements which have been set for heating, ventilation and air conditioning (HVAC) inside an operation theatre is important not only for the comfort of surgeon, patient, and medical personnel, but also to reduce the risk of surgical site infection during surgery. This research focus on Minor Operation Theatre (MOT) which is dental surgery room at Universiti Malaysia Perlis Health Centre with room dimensions of 2.89 m(H)×3.12 m(W)×3.4 m(L) is used for numerical analysis. The air flow supplied to the MOT is from single unit air-conditioning system. Computational Fluid Dynamics (CFD) analysis is a part of an investigation to determine the air flow and temperature distribution inside the MOT. A simulation conducted by using ANSYS Fluent only consider dry air inside MOT. Therefore, the main aim of this research is to compare and analyze the simulation of dry air conducted with previously obtained experimental data of humid air distribution inside the MOT. The comparison of humid and dry air temperature throughout the MOT shows that the difference is 25.3%. The average temperature of humid air inside the MOT is 21.8 °C while for dry air is 16.3 °C. Moreover, the cooling capacity of humid air and dry air are 2.23 kW and 1.64 kW respectively. Thus, the difference between humid air and dry air-cooling capacity is 26.5%. However, the dry air simulation and humid air simulation is the same if only the process is considered as simple cooling.

1. Introduction

Thermal comfort is when a person is satisfied with the thermal environment and does not need to use their thermo-regulator mechanism[1]. In hospital, the needs to predict the thermal comfort perception of patients and medical personnel is essential for building system design and architectural. A healthy and comfortable environment for patient, medical personnel or any other occupant in hospital is important especially in an Operation Theatre (OT). Thus, indoor air quality can be control by installing HVAC system and appropriate filtering system in OT [2]. The risk of hospital infection is occurred in all medical procedures, especially when it comes to surgery, and in certain types of operations it can be particularly serious. The optimal indoor air quality and thermal comfort must be controlled in order to primarily reduce the risk of surgical site infection (SSI) [3]. A serious surgical site infection can often permanently causes crucial pain and affect several aspects of the patient's lives [4]. This research work is based on the requirement to develop a realistic Computational Fluid Dynamic (CFD) model of the Minor Operation Theatre and study the airflow pattern in the MOT at the UniMAP Health Centre. By using a software simulation, an effort is made to model a hypothetical or real situation on a computer to study the working system. Changing variables in the simulation can predict the behavior of the system.

2. Methodology

This section includes all the methods or steps used to solve the problems related to this project. It comprises a few principles, theoretical analysis, and others. Previous researches had already done an experiment to record the temperature, humidity and air velocity based on MOT which is dental surgical room at UniMAP Health Centre [5]. The CFD analysis is a part of this investigation to determine the airflow patterns, temperature distribution and humidity in a MOT.

2.1 Modelling of MOT

Measurement of the MOT is based on actual size of UniMAP Health Care dental room. Dimensions of the room is 3.4 m(Length)×3.12 m(Width)×2.89 m(Height) as shown in Figure 1. The velocity of air outlet conditioner is 2.8 m/s and from analysis of previous data shows the air outlet coil temperature of 9°C which has a size of 650 mm×90 mm. The air return size of air-conditioning system is 990 mm×263 mm. ANSYS Fluent is used for the numerical analysis of MOT [5] [6].

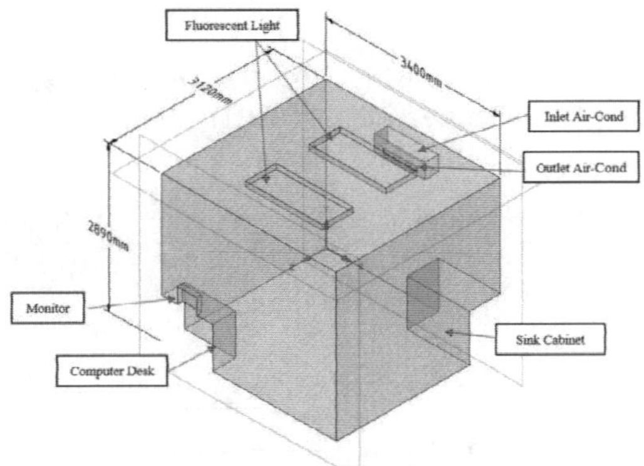

Figure 1. MOT 3D model.

2.2. Meshing

Cut cell method is used to generate the mesh with fine relevance center. Total nodes and elements are 35471 and 32163 respectively. All the specific details are shown in Figure 2.

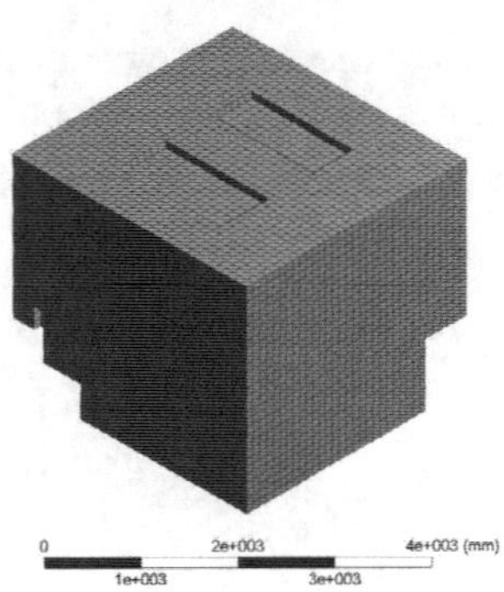

Figure 2. Meshing of MOT.

2.3 Marking Point Inside MOT

Two method is used to collect data in the MOT. First method is by using Colemeter and second method is by using data logger. The second method shows that for each location have four points. The first point starts from 0.5 m above the ground. Second point is 1.0m, third point is 1.5 m and fourth point is 2.0 m from ground. The total marking points inside this MOT is 36 points.

Figure 3 shows nine locations inside the Minor Operation Theatre (MOT). Each location consist of four points. For one location, the first point starts from 0.5 m above the ground. Second point is 1.0 m, third point is 1.5 m and fourth point is 2.0 m from ground. The total marking points inside this MOT is 36 points.

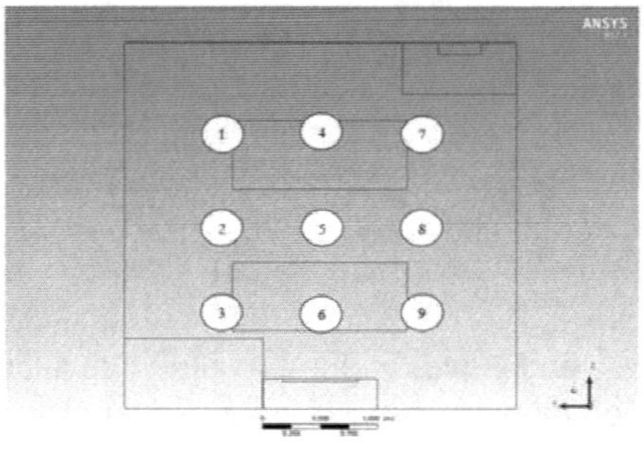

Figure 3. Top view of MOT.

Figure 4 shows 36 marking points inside this Minor Operation Theatre. These points are the same as per experimental data collection points. Therefore, dry air temperature data collected at these points can be compared with experimental humid air temperature data.

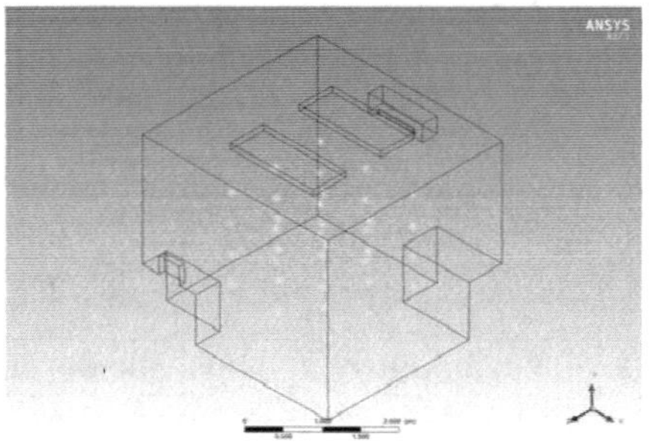

Figure 4. Marking Points in MOT.

2.4 Thermal Image Inside MOT

Inside the MOT, there are a few objects that generates heat. By using FLIR Ex-series infrared cameras, heat generated from objects can be identify. Therefore, these objects must be considered in the simulation process so that the condition of the MOT 3D model is similar as possible as the real MOT.

Table 1. Object and its temperature inside MOT.

Object	Maximum Temperature (°C)
Wall Exposed to Outside	26.7
Fluorescence Lamp	45.4
Computer Monitor	34.2

3. Result and Discussion

3.1. Experimental data (Humid Air)

Tables 2 and 3 show data measured by using Colemeter and Data Logger respectively.

Table 2. Data Measured by using Colemeter.

Point	Temperature °C	Relative humidity %	Absolute Humidity kg/kg
1	22.2	42.9	0.0071
2	22.0	42.7	0.0070
3	22.1	44.0	0.0073
4	21.5	44.8	0.0071
5	21.8	43.3	0.0070
6	20.5	47.2	0.0071
7	21.9	47.2	0.0077
8	22.3	51.7	0.0087
9	22.1	47.6	0.0079

Table 3. Data Measured by using Data Logger.

Point	Temperature °C					
	T1	T2	T3	T4	T5 (Air Return)	T6 (Air Outlet)
1	22.6	22.7	22.2	21.8	26.7	21.6
2	22.8	22.6	22.7	22.5	25.6	21.3
3	22.2	22.2	22.0	23.1	25.4	21.2
4	23.6	21.7	21.5	21.2	25.1	20.8
5	21.4	21.5	21.8	21.4	24.7	21.3
6	21.7	20.3	21.3	21.2	24.5	21.0
7	22.5	21.9	21.5	21.0	24.0	21.3
8	22.5	22.0	21.5	21.0	23.6	21.0
9	22.2	22.0	21.8	21.0	23.2	23.0

3.2. Simulation data (Dry Air)

Data from Table 4 obtained in ANSYS Fluent simulation for dry air only indicates that the minimum and maximum temperature are 15.725 °C and 17.325 °C respectively. Moreover, the average temperature inside the MOT is 16.3 °C. The air outlet coil temperature of air-conditioning system is 9 °C. Meanwhile, the air return temperature is 16.95 °C. Absolute humidity is the measure of moister in the air. Therefore, the absolute humidity is equal to 0 for dry air.

Table 4. Simulation data of temperature distribution.

Point	Temperature (°C)			
	T1	T2	T3	T4
1	15.841	16.369	16.298	16.130
2	15.792	16.382	16.283	16.238
3	16.240	16.043	16.315	16.203
4	15.808	15.804	15.850	15.900
5	15.725	15.949	16.115	16.212
6	15.743	16.082	16.354	16.553
7	16.337	16.382	16.480	16.612
8	16.192	16.924	16.973	16.802
9	16.344	17.056	17.325	17.106

3.3. Comparison Dry Air and Humid Air Temperature Inside Minor Operation Theatre

Figure 5 shows the relationship of humid and dry air temperature distribution inside the Minor Operation Theatre (MOT). The air outlet coil temperature is the same for both dry and humid air which is 9 °C. The graph indicates that the pattern for humid and dry air is the same. But, humid air temperature is higher throughout the MOT compare to dry air. This graph indicates that dry air needs less energy to cool.

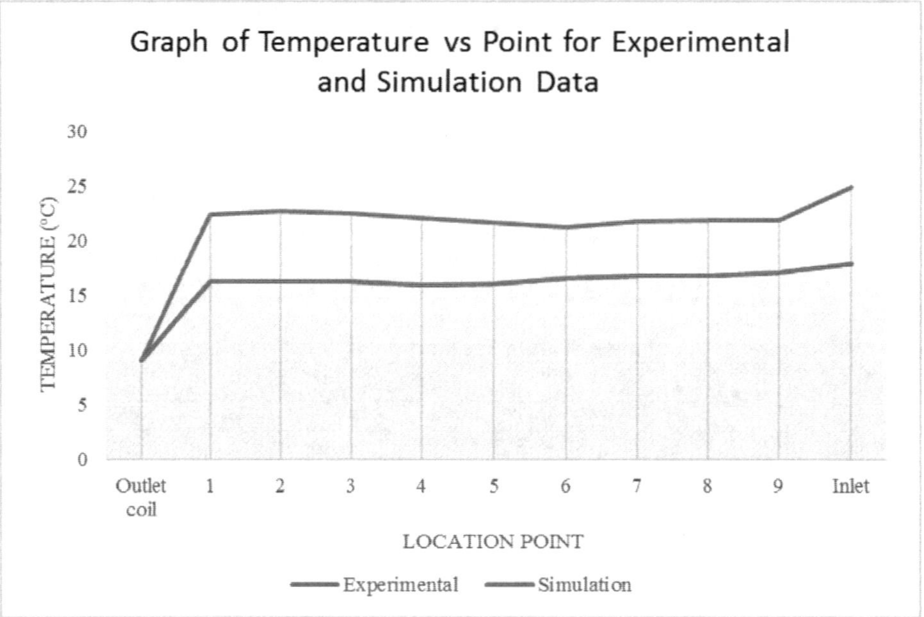

Figure 5. Graph of experimental vs simulation.

The average differences between humid air and dry air temperature is 25.3%. The reason is that the relative humidity of dry air is less than 1% or zero absolute humidity [7]. However, humid or moist air is a mixture of dry air and water vapour [8]. There is higher amount of water vapour when the air is more humid. The specific heat, Cp for air and water is 1.005 kJ/kg.K and 4.18 kJ/kg.K [9]. Therefore, amount of heat needed to raise the temperature of water by one degree celsius is higher than air.

3.4. Relations of Dry Air Simulation Data and Humid Air

The cooling capacity obtained from dry air simulation is 1.64 kW and the air return of air-conditioning is 16.95 °C. The air return size is 990 mm×263 mm with velocity of 0.652 m/s. By increasing the absolute humidity, air outlet of air-conditioning for same cooling capacity can be determine. The volumetric flow rate at air return of air-conditioning system, $\dot{V}$ is $0.1698\ m^3/s$.

Table 5. Increasing absolute humidity in simulation data.

Cooling Capacity	**Air Return Temperature**	**Absolute Humidity**	**Specific Volume, m^3/kg**	**Mass Flow Rate, kg/s**	**Enthalpy Differences, kJ/kg**	**Air Outlet Temperature, °C**
1.64 kW	**16.95°C**	0g/kg	0.822	0.206	7.96	**9.0**
1.64 kW	**16.95°C**	5g/kg	0.828	0.205	8.00	**9.1**
1.64 kW	**16.95°C**	10g/kg	0.835	0.203	8.08	**12.0**

Higher amount of cooling capacity is required for humid air that undergoes process of cooling with dehumidification in order to achieve same amount of temperature differences in simulation of dry air [9]. Besides that, the dry air and humid air for simple cooling process shows the same result. Therefore, the dry air simulation and humid air simulation is the same if only the process is simple cooling.

4. Conclusion

This study has succeeded in achieving all of the objectives of this research. In order to achieve thermal comfort of a MOT, the parameters that must be control are temperature, humidity, pressure, air change per hour (ACH) and air ventilation. According to Ministry of Health Malaysia (MOH), the acceptable temperature is between 18 °C to 21 °C, relative humidity is from 50% to 60%, air change rate is about 15 to 25 ACH and the room must be in positive pressure. A Minor Operation Theatre (MOT) which is Dental Surgery Room at UniMAP Health Centre is used for numerical analysis of air distribution throughout the room. Dimensions of the MOT are 2.89 m (height), 3.12 m (width) and 3.40 m (length). Single unit air-conditioning system is used as cooling system inside the MOT. Previous researcher had conducted an experiment and the data had been analyze shows that air outlet temperature of air-conditioning unit is 9 °C with velocity of 2.8 m/s. These data is used in the ANSYS Fluent to simulate the real MOT conditions. Dry air temperature results obtained from simulation by using ANSYS Fluent is used to compare with humid air temperature distribution inside the MOT. Therefore, the comparison of humid and dry air temperature throughout the MOT shows that the difference is 25.3%. Average temperature of humid air inside the MOT is 21.8 °C while for dry air is 16.3 °C. Moreover, the cooling capacity of humid air and dry air are 2.23 kW and 1.64 kW respectively. Thus, the difference between humid air and dry air cooling capacity is 26.5%. Finally, this research found that higher amount of cooling capacity is required for humid air that undergoes process of cooling with dehumidification in order to achieve same amount of temperature differences in simulation of dry air due to the presents of water vapour. However, the dry air simulation and humid air simulation is the same if only the process is simple cooling.

References

[1] F. Patania, A. Gagliano, F. Nocera, and A. Galesi 2011 Thermal comfort in operating rooms: A case study *WIT Trans. Biomed. Heal.* vol. 15, pp. 105–114.

[2] C. A. Balaras, E. Dascalaki, and A. Gaglia 2007 *HVAC and indoor thermal conditions in hospital operating rooms* vol. 39, pp. 454–470.

[3] S. Sadrizadeh and S. Holmberg 2016 Thermal comfort of the surgical staff in an operating room: a numerical study on laminar and mixing ventilation systems *Proc. 14th Int. Conf. Indoor Air Qual. Clim.*

[4] A. E. Andersson, I. Bergh, J. Karlsson, and K. Nilsson 2010 Patients' experiences of acquiring a deep surgical site infection: An interview study *Am. J. Infect. Control*, vol. 38, no. 9, pp. 711–717.

[5] A. Zamri 2018 Analysis of Air Psychrometry Distribution in Minor Operation Theatre of Hospital *UniMAP*.

[6] M. N. Y. Rahman, Z. M. Razlan, M. Izham, M. I. Omar, and N. A. A. Zambri 2018 A Test of Possibility on Relative Humidity Function in Minor Operation Theatre *IOP Publishing*.

[7] P. Wolkoff 2018 The mystery of dry indoor air – An overview *Environ. Int., vol. 121, no. October, pp. 1058–1065.*

[8] P. Doran 2013 Unit Operations *Second Edi. Elsevier Science & Technology Books.*

[9] Y. A. Cengel 2015 Thermodynamics: An Engineering Approach, *Eight Edit. McGraw-Hill Education.*

Image quality assessment (IQA) using high-frequency and image variance (HFIV) for colour image

Li Chien Tan[1*], Haniza Yazid[2] and Yen Fook Chong[3]

School of Mechatronic Engineering, Universiti Malaysia Perlis, Pauh Putra Campus, 02600, Arau, Perlis, Malaysia.

*LCtan94@gmail.com

Abstract. Image quality is often lost during image acquisition, transmission, and compression. Therefore, image quality assessment (IQA) is crucial in image processing. Currently, image quality can be measured from the frequency domain features, but it only applicable to blurred grayscale images. Nevertheless, noise distortion is also a common problem in digital images, and colour also affects the perception of image quality. Therefore, this paper proposes an enhanced blur and noise specific colour image quality assessment that measures high-frequency components and image variance. The number of high-frequency components is related to the edge and noise. In order to distinguish the distortion of the image, the image variance estimation is included. Experiments on public databases have shown that this method outperforms PSNR and SSIM in terms of noise and blur distortion and has low processing time of 0.0941 s/img.

1. Introduction

In most electronic imaging applications, high-quality images are requested and often required. However, the visual quality of the image may be corrupted during the process of image acquisition, transmission and so on. The process of evaluating the quality of a distorted image with respect to the original image is referred as IQA. IQA is important because if a poor medical image is not identified immediately, misdiagnosis and delayed treatment may occur, which may result in death. On the other hand, when the same situation occurs in the industrial sector, misdetection will result business and reputation losses to the company. Therefore, there is a great need for IQA to measure the image quality.

In general, IQA can be divided into two categories, subjective and objective method. The subjective method evaluates image quality by human observers. Human is the final receivers in most of image processing application [1] and we can easily evaluate images without referring any reference, hence, the subjective method is the most reliable [2]. The popular subjective quality evaluators, such as Mean Opinion Score (MOS) and Difference Mean Opinion Score (DMOS) have been used in various image processing applications. However, these methods always rely on the visual capability of the viewer and the luminance of the environment. Thus, it is inconvenient, time consuming and costly. Also, it is usually too slow to be used in real-world applications.

In consequence, objective IQA methods are mainly used in most applications. They can usually be subdivided into three types: Full-Reference (FR), such as Mean-Squared Error (MSE) and Structural Similarity Index (SSIM), which require a complete reference image and measure the similarity between the test and reference image; Reduced-Reference (RR), only part of the reference image is available; No-Reference (NR), no reference picture is required. Both FR and RR require a reference image. However, in most cases, there is no reference image. Therefore, NR can meet people's requirements at this time.

NR methods are typically designed for specific distortion. In this paper, our work focuses on blur and noise estimation in images, which is the most common type of distortion in digital images. Therefore, let us briefly summarize the relevant existing literature in this regard. The sharpness of the image can be measured by computing the wavelet decomposition of the images [3], measuring the edge spread in the image [4] and also the high-frequency (HF) components in the frequency domain [5]. Both methods assume uniform blur.

Besides, some methods focus on both noise and blur distortion, for example, the metrics proposed by Zhu and Milanfar [6], H metric, can detect blur and noise based upon the singular value decomposition (SVD) of the local image gradient matrix based on the assumption that the noise variance is known or estimable. In 2010, they modified the method by introducing Q metric [7,8]. This method is easy to compute, has a reasonable response to both blur and random noise, but the long processing time is the weak point. In addition, Cohen and Yitzhaky [9] estimated the noise in both spatial and frequency domain while blur is only estimated in the frequency domain. This method worked well with additive white noise, Gaussian blur and de-focus blur. But, the performance of the method may be degraded when the image contains strong periodic patterns and when the blurred image contains high noise power. However, all of the methods mentioned above were only suitable to evaluate grayscale images. Not only that, the accuracy, complexity and processing time of the existing metrics for blur and noise images always directly proportional. Therefore, they are not suitable for integration in real-time systems.

Inspired by [5], we propose an improved frequency domain based colour NR-IQA algorithm for blur and noise images. First, the ratio of the HF components in each colour component (YCbCr) is measured separately and then combined by the weight of the rod and the cone receptor in the human retina. Second, measure the image variance of each set of images. If the minimum image variance of each set of images is less than a predetermined threshold, then that particular set of images is considered blurry, otherwise is noisy. The experiments showed that our proposed method was accurate, effective, well-correlated with the human visual system and has short processing time. The rest of the paper is organized as follows: In Section 2, the methodology of the proposed method is discussed, and in Section 3, we present experimental results and correlations with subjective IQA. Finally, in Section 4, the entire paper will be finalized and possible future work is discussed.

2. Proposed colour NR-IQA

Our proposed method is the hybrid of *HF measurement and image variance estimation (HFIV),* mainly composed of three stages. First, the algorithms described in [5] was implemented in the YCbCr colour space, followed by image variance estimation and lastly the final image quality calculation. The block diagram of the proposed algorithm is shown in Figure 1.

2.1. High-Frequency Measurement in YCbCr colour space

Figure 2 represents the overall flow of the first stage. First, split the colour image into Y, Cb and Cr colour components, and then compute their respective Fourier Transform representation individually. Next, calculate the threshold and measure the ratio of HF components in each colour components. Lastly, combine all together for final image quality score calculation. In this paper, the YCbCr colour space is chosen because it always used as a part of the colour image pipeline in video and digital photography systems, therefore is more suitable for IQA [10]. Y represents the luminance, while Cb and Cr are the blue-difference and red-difference chroma components.

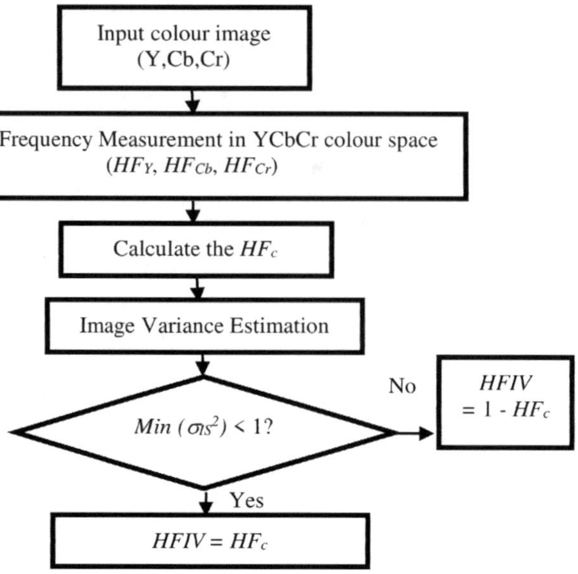

Figure 1. Block diagram of the proposed *HFIV* algorithm.

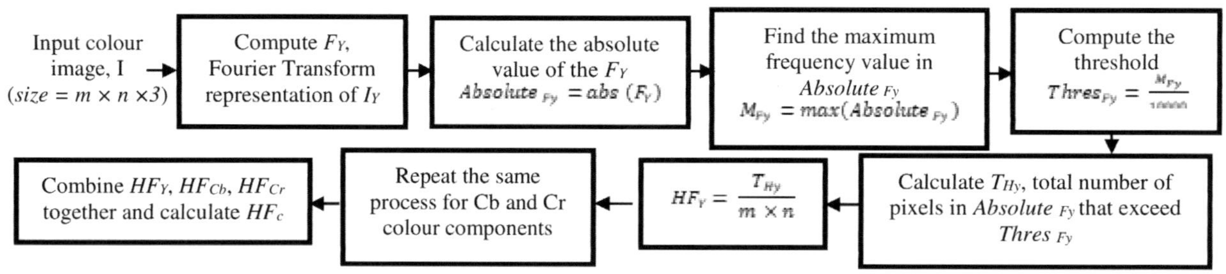

Figure 2. Block diagram of the Stage I (High-Frequency Measurement in YCbCr colour space).

Human retina has two types of photoreceptors, rods and cones. The rod is responsible for scotopic vision, very sensitive to light, and does not mediate colour perception, while on the other hand, cones provide colour perception and the spatial perception of the eye. The fovea part of the human eye has approximately 7 million cones and 120 million rods [11,12]. Therefore, the weight of the luminance component (L_w) and the weight of the colour components (C_w) are calculated as follows, Eq. (1) and Eq. (2). Then, the number of HF component of the colour image (HF_c) can be easily obtained using Eq. (3) [13].

$$L_w = \frac{120{,}000{,}000}{(120{,}000{,}000 + 7{,}000{,}000)} = 0.9449 \quad (1)$$

$$C_w = \frac{7{,}000{,}000}{(120{,}000{,}000 + 7{,}000{,}000)} = 0.0551 \quad (2)$$

$$HF_c = 0.9449\,(HF_Y) + 0.0551\left(\frac{HF_{Cb} + HF_{Cr}}{2}\right) \quad (3)$$

As we know that, as the blur in the image increases, the number of HF components in the image is suppressed. A clearer, the higher quality image will have a greater number of HF components than a blurred image. However, on the contrary, noise is the high frequency in nature. The more the image is contaminated by noise, the greater the number of HF components. Then the problem arises. The image with the highest number of HF components can be the sharpest image or the noisiest image. In order to solve this problem, the image variance estimation phase is added.

2.2. Image Variance Estimation
In this stage, this paper implements the fast noise variance estimation algorithm proposed by John Immerkaer [14]. The reason for choosing this method is that it is very simple. Grayscale images, rather than colour images, are used at this stage to reduce the processing time. Apply a noise estimation operator *N* that is almost insensitive to the image structure to the input grayscale image, I_G.

$$N = \begin{array}{|c|c|c|} \hline 1 & -2 & 1 \\ \hline -2 & 4 & -2 \\ \hline 1 & -2 & 1 \\ \hline \end{array} \quad (4)$$

Next, compute the variance of each output image, $\sigma_{g,n}{}^2$ and find the minimum variance of each group of image, $\sigma_{kmin}{}^2$ using Eq. (5) and (6), respectively.

$$\sigma_{g,n}{}^2 = \left(\frac{\pi}{2}\frac{1}{6\,(W-2)(H-2)}\sum_I |I_G(x,y) * N|\right)^2 \quad (5)$$

$$\sigma_{k,min}{}^2 = \min\left(\sigma_{k,1}{}^2, \sigma_{k,2}{}^2, \ldots\ldots \sigma_{k,n}{}^2\right) \qquad , k = 1, 2, \ldots\ldots g \quad (6)$$

where *g* is the number of image groups, *n* is the number of image per group, *W*, *H* represent the width and height of the input image, $I_G(x,y) * N$ is the output image, and * represents convolution.

When the blurred image's edge is smoothed, the $\sigma_{kmin}{}^2$ is very small, usually below 1. In contrast, when the noise contained in the image increases, the $\sigma_{kmin}{}^2$ will be increased, generally more than 1. However, this method does not apply if there is only one image. The details of the method can be referred in [14].

2.3. Image Quality Measure (IQM)
Noise and blur have a paradoxical effect on image attributes. In the frequency domain, noise affects the entire frequency band and primarily increases the higher frequencies, while blurring primarily suppresses higher frequencies. For this reason, the image quality measurement of noise and blurred images is different.

For a blurry image, the higher the number of HF components, the better the image quality. For noisy images, the HF_c value is inversely proportional to the image quality. Therefore, the final *HFIV* of the blurred image and noisy image can be expressed as Eq. (7), respectively.

$$HFIV = \begin{cases} 1 - HF_c & ; \text{if } \sigma_{kmin}{}^2 \geq 1 \\ HF_c & ; \text{if } \sigma_{kmin}{}^2 < 1 \end{cases} \quad (7)$$

3. Experimental Results
In order to evaluate the performance of the proposed IQM, our experiments used three subjective IQA databases, CSIQ [15], TID2008 [16] and TID2013 [17]. In this paper, only a few types of distortion are included. In the CSIQ database, it includes additive white Gaussian noise (AWGN), Gaussian blur (Blur) and additive pink Gaussian noise (Fnoise). For the TID2008 and TID2013 databases, the types of distortion include additive Gaussian noise, additive noise in colour components, spatially correlated noise, masked noise, HF noise, impulse noise, quantization noise, Gaussian blur, and image denoising. Each of the distorted images in TID2008, TID2013, and CSIQ include a corresponding MOS and a DMOS, respectively, which can be used as a ground truth. Table 1 summarizes the database information used in our experiments.

To illustrate the performance of IQA metrics, two evaluation criteria are used, the Pearson linear *(PLCC)* and the Spearman's rank order correlation coefficients *(SRCC)*. High *PLCC* scores and *SRCC* scores are associated with high accuracy, monotonicity, and consistency of the metrics in the test. The closer the value of *SRCC* and *PLCC* to 1, the better the metric performance.

Table 1. Database with selected distortion for performance evaluation

Database	Reference Image	Selected distortion types	Distortion level	Total distorted image used	Subjective score
CSIQ	30	3	5	450	DMOS
TID2008	25	9	4	900	MOS
TID2013	25	9	5	1125	MOS

The performance of this method is also compared to PSNR and SSIM. The evaluation metrics *SRCC* and *PLCC* are then calculated to present their accuracy and consistency. Tables 2, 3 and 4 show the results of PSNR, SSIM and proposed methods in the CSIQ, TID2008 and TID2013 databases, respectively. A total of 2475 images were used in this evaluation.

As can be seen from Table 2, in CSIQ database, our proposed method can compete with the FR method. Even though the result of our algorithm is slightly lower than PSNR and SSIM, but, the difference between *SRCC* and *PLCC* is less than 0.04. Our result shows a strong negative *SRCC* of 0.9978 and a strong negative *PLCC* of 0.9465 between our calculated score and the DMOS. The relationship between our objective score and DMOS is a strong negative relationship, which means high prediction accuracy and monotonically.

Table 2. Average *SRCC* and *PLCC* with DMOS for different types of distortion in the CSIQ database.

	SRCC			*PLCC*		
	PSNR	SSIM	Proposed method	PSNR	SSIM	Proposed method
Average	-1	-1	-0.9978	-0.9828	-0.9689	-0.9465

From Table 3 and Table 4, it is clear that our proposed method significantly outperforms PSNR and SSIM in the TID2008 and TID2013 databases. Our approach achieves a very strong positive *SRCC* and *PLCC* correlation between subjective ratings MOS, which are very close to 1. This demonstrates that our algorithm can achieve stable and good performance across different noise and blur distortions for the entire three databases.

Table 3. *SRCC* and *PLCC* with MOS for different types of distortion in the TID2008 database.

	SRCC			*PLCC*		
	PSNR	SSIM	Proposed method	PSNR	SSIM	Proposed method
Average	0.8404	0.8404	0.9758	0.8351	0.8352	0.9668

Table 4. *SRCC* and *PLCC* with MOS for different types of distortion in the TID2013 database.

	SRCC			*PLCC*		
	PSNR	SSIM	Proposed method	PSNR	SSIM	Proposed method
Average	0.8680	0.8680	0.9675	0.8656	0.8661	0.9634

In addition, the average execution time of each IQA algorithms was measured to assess the complexity. The operating system used in this experimental environment was Windows 10 64-bit operating system with an i7-6500U CPU @ 2.60GHz CPU and 8.00GB RAM. Table 5 summarizes all simulation results. It shows that PSNR has the fastest execution speed, followed by our proposed method and SSIM. Although PSNR has the fastest speed, it requires a reference image that is not always available. Therefore, our method is still the best way to deal with noise and blurred images. It can process images in 0.1s without reference, so it is suitable for real-time applications.

Table 5. Average execution time (s / img) of PSNR, SSIM and proposed method.

Method	PSNR	SSIM	Proposed method
Average	0.0328	0.3811	0.0941

4. Conclusion

This paper introduces a hybrid of *HF component measurement and image variance estimation* (*HFIV*) for NR-IQA. This method is dedicated to blur and noise distortion, which is the most common type of distortion in digital imaging applications. In addition, we compared our approach to the performance of PSNR and SSIM in the CSIQ, TID2008 and TID2013 databases. Experiments showed that our proposed method has strong *SRCC* and *PLCC* correlation with human visual perception, able to accurately evaluate the image without any reference. In addition, the experiment also demonstrated that our proposed algorithm has a short processing time of 0.0941s/img, so, it can meet the real-time requirement of most current applications. Blocking artifacts caused by compression are also very common in digital imaging. Therefore, for future work, this method can be improved to accommodate JPEG and JPEG2000 distorted images.

References

[1] Yi Y, Yu X, Wang L and Yang Z 2008 Image Quality Assessment Based on Structural Distortion and Image Definition *2008 International Conference on Computer Science and Software Engineering* (IEEE) pp 253–6

[2] Choi M G, Jung J H and Jeon J W 2009 No-Reference Image Quality Assessment using Blur and Noise *Int. J. Electr. Comput. Energ. Electron. Commun. Eng.* **3** 184–8

[3] Abdel-Hamid L, El-Rafei A and Michelson G 2017 No-reference quality index for color retinal images *Comput. Biol. Med.* **90** 68–75

[4] Feichtenhofer C 2003 No-reference Sharpness Metric Based on Local Gradient Analysis *Proceedings International Conference on Image Processing* (IEEE Xplore) pp 1–34

[5] De K and Masilamani V 2013 Image Sharpness Measure for Blurred Images in Frequency Domain *Procedia Eng.* **64** 149–58

[6] Zhu X and Milanfar P 2009 A no-reference sharpness metric sensitive to blur and noise *2009 International Workshop on Quality of Multimedia Experience* (IEEE) pp 64–9

[7] Zhu X and Milanfar P 2010 Automatic Parameter Selection for Denoising Algorithms Using a No-Reference Measure of Image Content *IEEE Trans. Image Process.* **19** 3116–32

[8] Prajapati P, Narmawala Z, Darji N P, Moorthi S M and Ramakrishnan R 2015 Evaluation of Perceptual Contrast and Sharpness Measures for Meteorological Satellite Images *Procedia Comput. Sci.* **57** 17–24

[9] Cohen E and Yitzhaky Y 2010 No-reference assessment of blur and noise impacts on image quality *Signal, Image Video Process.* **4** 289–302

[10] Wang Q, Chu J, Xu L and Chen Q 2016 A new blind image quality framework based on natural color statistic *Neurocomputing* **173** 1798–810

[11] Gonzalez R C and Woods R E 2009 *Digital Image Processing (3rd Edition)* vol 56

[12] Yalman Y and Ertürk I 2013 A new color image quality measure based on YUV transformation and PSNR for human vision system *Turkish J. Electr. Eng. Comput. Sci.* **21** 603–12

[13] Yalman Y 2014 Histogram based perceptual quality assessment method for color images *Comput. Stand. Interfaces* **36** 899–908

[14] Immerkaer J 1996 Fast Noise Variance Estimation *Comput. Vis. Image Underst.* **64** 300–2

[15] E. C. Larson and Chandler D M 2010 Most Apparent Distortion: Full-Reference Image Quality Assessment and the Role of Strategy *J. Electron. Imaging* **19**

[16] Ponomarenko N, Lukin V, Zelensky A, Egiazarian K, Astola J, Carli M and Battisti F 2009 TID2008-A Database for Evaluation of Full-Reference Visual Quality Assessment Metrics *Adv. Mod. Radioelectron.* **10** 30–45

[17] Ponomarenko N, Jin L, Ieremeiev O, Lukin V, Egiazarian K, Astola J, Vozel B, Chehdi K, Carli M, Battisti F and Jay Kuo C-C 2015 Image database TID2013: Peculiarities, results and perspectives *Signal Process. Image Commun.* **30** 57–77

Effects of Electrospinning Voltage and Flow Rate on Morphology of Poly-vinyl Alcohol Nanofibers

Amnah Asiri[1], Rania Hussien Al-Ashwal[2*], Mohd Helmi Sani [1] and Syafiqah Saidin[3]

[1]Department of Bioscience, Faculty of Science, Universiti Teknologi Malaysia, 81310 UTM Johor Bahru, Johor, Malaysia.
[2]Department of Clinical science, School of Biomedical engineering and Health Science, Universiti Teknologi Malaysia, 81310 UTM Johor Bahru, Johor, Malaysia.
[3]IJN-UTM Cardiovascular Engineering Centre, Institute of Human Centered Engineering, Universiti Teknologi Malaysia, 81310 UTM Johor Bahru, Johor, Malaysia.

*rania@utm.my

Abstract. Nanofibers have obtained considerable interest for use in various applications. Polyvinyl alcohol (PVA) has been used to achieve many benefits for diverse pharmaceutical and biomedical applications. We investigated in this study the effects of applied voltage, needle diameter, and flow rate on morphologies of PVA nanofibers. A constant volume of the feeding solutions delivered to the needle at a flow rate of 1and 2 mL/h with high potentials of voltage was applied as they exit the needle. After that, the electrospun fibers collected on the ground connected aluminum foil. The electrical conductivity measurements of feeding solutions performed at room temperature. Characterization of the PVA nanofibers conducted using scanning electron microscopy (SEM) and attenuated total reflectance-Fourier transform infrared spectroscopy (ATR-FTIR). The obtained nanofibers SEM images show beads when using the flow rate at 1 mL/h, whereas increasing the voltage and the flow rate improved the morphology of the nanofibers to uniform without beads. The FTIR results show that O-H and C-O bands are the main attributing to the chemical functionality of PVA nanofibers. As a conclusion, the high voltage and flow rate considered as the most critical parameters that impacted on PVA nanofibers morphology.

1. Introduction

In recent years, focus on nanofibers research has increased and nanofibers have obtained a considerable interest in several applications such as energy storage, healthcare, biotechnology, environmental engineering, defence and security [1-2-3]. Nanofibers could be synthesized from natural polymers (e.g., chitosan, fibronectin, gelatin, collagen, silk, ethyl cellulose) in addition to synthetic polymers (e.g., polylactic acid (PLA), polyglycolic acid (PGA), polylactic-*co*-glycolic acid (PLGA), tyrosine-derived polycarbonates, poly ε-caprolactone (PCL), polyurethane (PU), polyvinyl pyrrolidone (PVP), polyvinyl alcohol (PVA), and their various combinations) [2].

Polyvinyl alcohol is categorized as a polyhydroxy polymer which has been studied intensively due to its excellent film forming and physical properties, with high processability, biocompatibility, hydrophilicity and excellent chemical resistance [4-5-6]. This polymer has been utilized in several areas, such as food chemistry, pharmaceuticals, medicine, and biotechnology [7]. Particularly PVA has high usefulness in diverse pharmaceutical and biomedical applications [8] where it has the Food and Drug Administration (FDA) approval [7-8]. Polyvinyl alcohol has also been used as PVA hydrogels to fabricate contact lenses, wound dressings, coatings for sutures and catheters. It is also prominent to be modified as a water-soluble and biodegradable carrier that beneficial in drug delivery systems [7-9].

Electrospinning is an economical and convenient way to industrialize nanofibers [10]. In electrospinning, a high electrical potential is applied to a polymer solution [10] where fibers are derived by usually charging the fluid versus the ground at a short distance, resulting in charging the injection into the liquid from an electrode [11] The infected charge mark depends on the polarity of electrode; that is, the negative pole produces a negatively charged fluid [11]. The charged liquid is referred to the ground pole from an opposing polarity, forming a so-called Taylor cone at the opening of nozzle where a polymer solution is extracted by an electric field to form a liquid jet [11]. When a polymer jet solvent evaporates, the jet hardens, and polymer fibers are formed [11].

Electrospun membranes are highly potent in tissue regeneration since they can closely mimic the structure of extracellular matrix [12-13]. Therefore, in this study, PVA nanofibers were electrospun at high voltage with two numerals of flow rate and two sizes of needle diameter. The changes in chemical composition and morphology were observed on different parameters of PVA nanofibers through an attenuated total reflectance-Fourier transform infrared spectroscopy (ATR-FTIR) and scanning electron microscopy (SEM) analyses.

2. Materials and Methods

2.1. Sample Preparation

A PVA solution was prepared by dissolving PVA (fully hydrolyzed, MW 30000, Japan) in deionized water at a concentration of 25%. The solution stirred at 80°C for 2 h to obtain a homogeneous solution. Then, the PVA solution electrospun using an electrospinning instrument (NaBond Technologies Co., Limited, Shenzhen, China) which consisted of the syringe needle, ground electrode, aluminum sheet and a high voltage supply.

The needle connected to the high voltage supply, which generates a positive DC voltage and current, up to 35 kV. The electrospinning process then conducted at room temperature with a fixed needle-electrode distance at 12 cm. The positive voltage applied to the polymer solution varied at 15 and 16 kV while the volumetric flow rate varied at 1 and 2 mL/h. The effect of needle diameter was only investigated for the sizes of 18G and 21G.

2.2. Morphology Analysis

The morphology of the PVA nanofibers was investigated by SEM (JEOL, iT300LV, Japan) at an accelerating voltage of 10 kV. Before the SEM observation, the nanostructured membranes of PVA were dried under vacuum (0.1 mbar) and coated with a gold thin film using a gold sputter coater (Q150R, Quorum Technologies Ltd, England). The SEM images were captured at 3000x magnification to visualize the formation of mat nanofibers, possible beads and electrospray fibers.

2.3. Chemical Functionality Analysis

The chemical functionalities of the PVA nanostructured membranes analyzed by ATR-FTIR (L1600301, PerkinElmer, UK) between wave numbers of 500 and 4000 cm^{-1}. The scanning performed at 4 cm^{-1} resolution. The acquired ATR-FTIR spectra were then annotated and analyzed using Originlab software (Origin Pro 8, OriginLab Corporation, USA).

3. Results and Discussion

Electrospinning has been proven to be a potential technique for fabricating polymeric fibers with diameters ranging between micrometers and nanometers due to the ease of use, cost-effectiveness and adaptability [10-14]. In this study, similar parameters of PVA nanofibers were used according to Uyar *et al.* [15] which reported that the PVA (25%, w/w) could be electrospun into finest uniform nanofibers.

Two flow rates (1 mL/h and 2 mL/h) with two sizes of needle diameter (18G and 21G) and two different voltages (15 kV and 16 kV) were applied to determine the best morphology (Figure 1 and Figure 2). Increasing the flow rate to 2 mL/h has displayed smooth surface nanofibers without bead formation compared with the electrospun at 1 mL/h. The results were similar for both sizes of needle diameter and voltage due to the fact that difference both needle diameter and voltage might and might not affected on fibers diameter [11].

The voltage then tailored at 15kV and 16 kV, independently, to remove the tip needle droplets and to acquire homogenous mat nanofibers [11]. Increasing the voltage with the flow rate of 2 mL/h has provided uniform nanofibers without beads formation while some beads appeared on the nanofibers which have been electrospun at 1 mL/h flow rate. The high voltage affects the continuous electric current that was a quasi-stable point where the jet flow rate was equal to feed rate according to Fallahi *et al.* [16].

The nanofibers were homogenously electrospun without beads by using the 21G needle diameter compared to 18G needle diameter, specifically observed on the SEM images at 1mL/h flow rate while the SEM images of 2 mL/h did not show a significant difference when various sizes of needle diameter used at different voltages.

First, the applied voltage was chosen 15 kV and 16 kV with two sizes of needle diameter (18G,21G), and used two flow rates (1mL/h, and 2mL/h) to determine the best parameter of PVA nanofiber. Figures 1, 2 (b) and (d) the results display that the nanofiber of PVA obtained smooth surface compare with Figure 1, 2 (a) and (c) which PVA nanofibers include few beads. Thus, it may be concluded that increasing the flow rate to 2mL/h obtained smooth nanofibers with both size of needle diameter and different voltage due to high voltage could effect on a continuous electric current that was the quasi-stable point where the jet flow rate was equal to the feed rate according to [16].

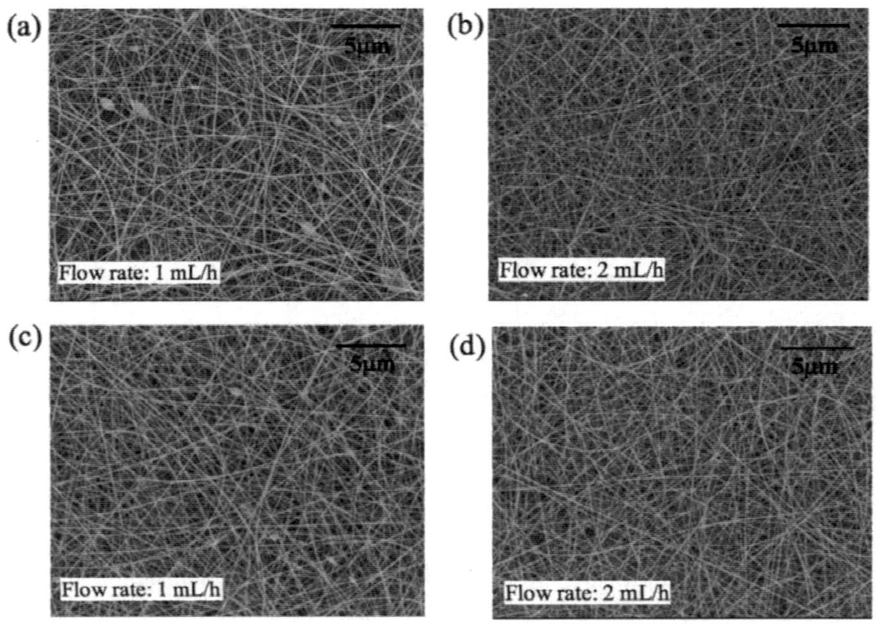

Figure 1. SEM Morphologies of electrospun PVA nanofibers with different flow rates (1 and 2 mL/h), 15 kV and different sizes of needle diameter: (a) 18G, (b) 18G, (c) 21G, (d) 21G.

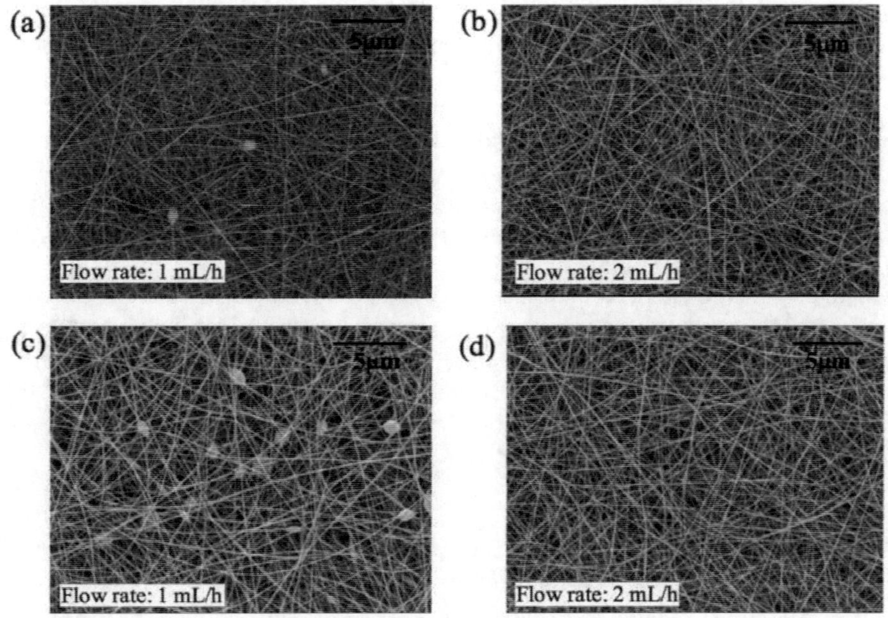

Figure 2. SEM Morphologies of electrospun PVA nanofibers with different flow rates (1 and 2 mL/h), 16 kV and different sizes of needle diameter: (a) 18G, (b) 18G, (c) 21G, (d) 21G.

The ATR-FTIR analysis of nanofibers electrospun from blended solutions was carried out to discover any peak change that referenced to chemical interactions [17]. The ATR-FTIR spectra in Figure 3 shows the functional groups of PVA nanofibers which have been electrospun at 2mL/h flow rate with different voltages and needle diameters. All nanofibers were composed of similar functional groups of PVA. However, the 21G needle has produced greater peak intensity for both 15 and 16 kV compared to the 18G due to better homogeneity of nanofibers distribution without bead disturbance (Figure 2). The 21G needle has projected greater full coverage of mat nanofibers to be detected by the ATR-FTIR analysis. In all nanofibers, the functional groups of PVA were presented at 1000-1320 cm^{-1} caused by the bond of C=O stretching [18]. While the hydroxyl group (O-H bond) of PVA were recorded at 3250-3400 cm^{-1} [19] and 910-950 cm^{-1}. Besides, peaks of C-H-C-H and C=O stretch were observed at 610-700 cm^{-1} and 1665-1710 cm^{-1} [20], respectively. Another variable peaks of C=H bond were also noticed at (2850-3000 cm^{-1}) [21] and 1450-1470 cm^{-1}[20].

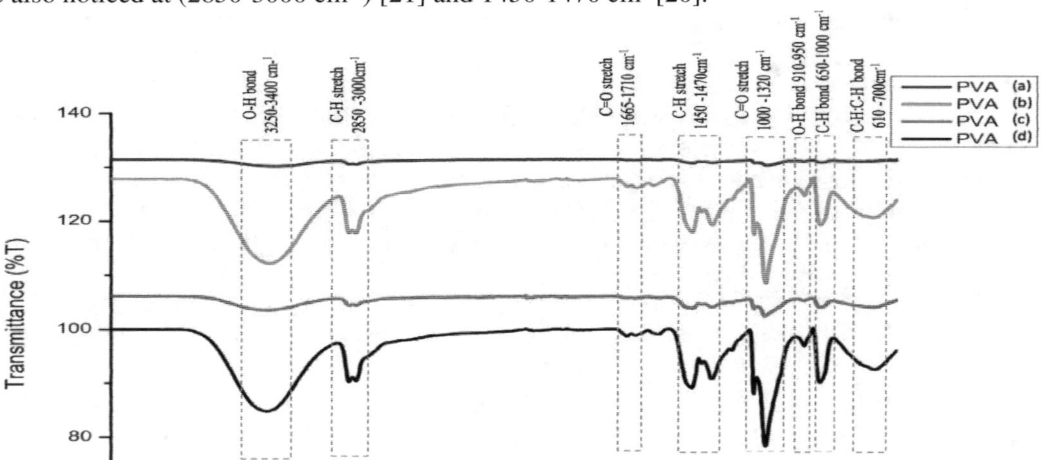

Figure 3. ATR-FTIR spectra of PVA nanofibers at flow rate of 2mL/h, voltage and needle diameter of (a) 15 kV and 18G, (b) 15 kV and 21G, (c) 16 kV and 18G, (d) 16 kV and 21G.

4. Conclusion

Electrospinning process and the factors affecting the properties of electrospun nanofibers have been widely studied for different fields in recent time. An applied voltage, flow rate and needle diameter are some of the affecting parameters during electrospinning. In this study, the formation of nanofibers was improved with the increasing of electrospun voltage and flow rate. However, there was no clear effect of needle diameter on the morphology of PVA nanofibers, in spite of its chemical functionalities. Therefore, tailoring the parameters of electrospinning in necessary to achieve desirable properties of nanofibers for a specific application.

Acknowledgments

The author acknowledges the support from ministry of higher education (MOHE) KPT grant transdisciplinary research grant (TRGS) R. J130000.7845.4L843, UTM university research grant Tier 1 Q. J130000.2545.16H98, the members of Biomedical Materials research group (B-MEMS), the Saudi Arabian Cultural Mission (SACM) and Advanced Diagnostic and Progressive Human Care Research Group.

References

[1] Ramakrishna S, Fujihara K, Teo W E, Yong T, Ma Z and Ramaseshan R 2006 Electrospun nanofibers: solving global issues. *Mater. today.* **9** 40-50

[2] Kamble P, Sadarani, B Majumdar A and Bhullar S 2017 Nanofiber based drug delivery systems for skin: A promising therapeutic approach *J. Drug Deliv. Sci. Technol.* **41** 124-33

[3] Zhu J, Ge Y, Jasper S and Zhang X 2017 Physical characterization of electrospun nanofibers. *In Electrospun Nanofibers.* 207-38

[4] Yang E, Qin X and Wang S 2008 Electrospun crosslinked polyvinyl alcohol membrane. *Mater. Lett.* **62** 3555-7

[5] Rajeswari N, Selvasekarapandian S, Karthikeyan S, Sanjeeviraja C, Iwai Y, and Kawamura J 2013 Structural, vibrational, thermal and electrical properties of PVA/PVP biodegradable polymer blend electrolyte with CH 3 COONH 4. *Ionics.* **19** 1105-13

[6] Gozutok M, Sadhu V and Sasmazel H T 2019 Development of Poly (vinyl alcohol)(PVA)/Reduced Graphene Oxide (rGO) Electrospun Mats *J. Nanosci. Nanotechnol.* **19** 4292-8

[7] Chaouat M, Le Visage C, Baille W E, Escoubet B, Chaubet F Mateescu M A and Letourneur D 2008 A novel cross-linked poly (vinyl alcohol) (PVA) for vascular grafts *Adv. Func. Mater.* **18** 2855-61

[8] Baker M I, Walsh S P, Schwartz Z and Boyan B D 2012 A review of polyvinyl alcohol and its uses in cartilage and orthopedic applications *J. Biomed. Mater. Res. Part B: Appl. Biomater.* **100** 1451-7

[9] Premraj R and Doble M 2005 Biodegradation of polymers. *Indian. J. Biotechnol.* **4** 186-93

[10] Wongkanya R, Chuysinuan P, Pengsuk C, Techasakul S, Lirdprapamongkol K, Svasti J and Nooeaid P 2017 Electrospinning of alginate/soy protein isolated nanofibers and their release characteristics for biomedical applications *J. Sci. Adv. Mater. Devices.* **2** 309-16

[11] Haider A, Haider S and Kang I K 2018 A comprehensive review summarizing the effect of electrospinning parameters and potential applications of nanofibers in biomedical and biotechnology. *Arab. J. Chem.* **11** 1165-88

[12] Yao L, HaasT W, Guiseppi-Elie A, Bowlin G L, Simpson D G and Wnek G E 2003 Electrospinning and stabilization of fully hydrolyzed poly (vinyl alcohol) fibers. *Chem. Mater.* **15** 1860-4

[13] Choi J K, Jang J H, Jang W H, Kim J, Bae I H, Bae J and Park J W 2012 The effect of epidermal growth factor (EGF) conjugated with low-molecular-weight protamine (LMWP) on wound healing of the skin. *Biomaterials* **33** 8579-90

[14] Sill T J and von Recum H A 2008 Electrospinning: Applications in drug delivery and tissue engineering. *Biomaterials* **29** 1989-2006

[15] Uyar T, Çökeliler D, Doğan M, Koçum I C, Karatay O and Denkbaş E B 2016 Electrospun nanofiber reinforcement of dental composites with electromagnetic alignment approach. *Mater. Sci. Eng. C* **62** 762-70

[16] Fallahi D, Rafizadeh M, Mohammadi N and Vahidi B 2008 Effect of applied voltage on jet electric current and flow rate in electrospinning of polyacrylonitrile solutions *Polym. intl.* **57** 1363-8

[17] Safi S, Morshed M, Hosseini Ravandi S A and Ghiaci M 2007 Study of electrospinning of sodium alginate, blended solutions of sodium alginate/poly (vinyl alcohol) and sodium alginate/poly (ethylene oxide) *J. appl. polym. sci.* **104** 3245-55

[18] Kumar A, Negi Y S, Bhardwaj N K and Choudhary V 2013 Synthesis and characterization of cellulose nanocrystals/PVA based bionanocomposite. *Adv. Mater. Lett.* **4** 626-31

[19] Shao L, Li J, Guang Y, Zhang Y, Zhang H, Che X and Wang Y 2016 PVA/polyethyleneimine-functionalized graphene composites with optimized properties. *Mater. Des.* **99** 235-42

[20] Xu Z, Zhou H, Tan S, Jiang X, Wu W, Shi J and Chen P 2018 Ultralight super-hydrophobic carbon aerogels based on cellulose nanofibers/poly (vinyl alcohol)/graphene oxide (CNFs/PVA/GO) for highly effective oil–water separation. *Beilstein J. Nanotechnol.* **9** 508-19

[21] Rajendran S, Sivakumar M, Subadevi R and Nirmala M 2004 Characterization of PVA–PVdF based solid polymer blend electrolytes. *Phys. B Condens. Matter.* **348** 73-8

Characterization of plasma based on the electrode size of atmospheric pressure plasma jet (APPJ)

Jeyashri Veeraiah [1], Siti Khadijah Zaaba[1,2*], Mohammad Taufiq Mustaffa[3]

[1]School of Mechatronic Engineering, Universiti Malaysia Perlis, Arau, Perlis, Malaysia
[2]Centre of Excellence for Advanced Sensor Technology (CEASTech), Universiti Malaysia Perlis
[3]School of Manufacturing Engineering, Universiti Malaysia Perlis, Arau, Perlis, Malaysia

*khadijah@unimap.edu.my

Abstract. Atmospheric Pressure Plasma Jet recognize as non-thermal plasma or cold plasma where known as APPJ which is generated under pressure of the atmospheric environment, currently APPJ using in many kind of application especially in biomedical application to reduce microbial activity. In APPJ, there are various type of configuration can be set to produce plasma discharge. This project began by changing the configuration of electrodes thickness, electrodes distance, and quartz tube's inner and outer diameter to generate ideal plasma discharge. The ideal plasma discharge can be choose from Optical Emission Spectroscopy (OES) analysis by using SpectraWiz software and bacteria inactivation method. This OES analysis used to analyze the charged species and chemical reactive species that present in the plasma plume. The length and reactive species of plasma discharge is all depend on the APPJ configuration used. From the result of OES analysis, the best parameters can choose to perform the bacteria inactivation method. It was found that the bacterial inactivation depend on the oxygen radical was introduced into the plasma jet system. The ideal configuration that was choosen from this research can be applicable in future plasma process to get an effective result.

1. Introduction

Nowadays, development in life science technology becomes more promising and plays a great role in the daily lives of people and making their lifestyle advance. One of the recent science technologies is atmospheric-pressure plasma jets (APPJ). APPJ have attracted major attention due to their widespread applications in plasma medicine, nanotechnology, as well as surface and materials processing [1]. The application of APPJ in plasma medicine is plasma surface modification, plasma bio-decontamination, and therapeutic plasma application [2].

Besides, APPJ technology is driven by electrons, which gain energy from applied voltage and carrier gas. This is due to several orders of differences in mass between electrons makes vigorous collision between energized electron and heavy particle cause excitation, dissociation and ionization. Plasma jet contains neutral particles and electron, ions, molecules, including atom, radicals and UV photons. Carrier gas which commonly for bio-decontamination process is Heliun (He) because it can produce a homogeneous glow discharge much easier at the lower voltage [2]. Helium gas also has long lifetime of meta-stable excited atoms and highest electron energy level among all material, easy dissociation of mixed gas molecules, and generation of ions and radicals [3]. The plasma generated inside the tube and expands into the open air molecules forming reactive species. In this research, plasma are generated by applying a certain voltage to High-Voltage (HV) electrode. When the voltage exceeds a threshold value called the breakdown voltage, the gas ionization becomes self-sustaining, and the kinetic process occurs in the electrodes induced the plasma to generate [4].

A different number of reactive species will obtain in plasma plume with different type of APPJ configuration [5]. In this study, a certain configuration of quartz tube's diameter, electrode thickness of atmospheric APPJ, and electrode distance is examined to determine the optimum plasma parameters for the plasma process. Moreover, to generate and utilize stable plasma process, it is crucial to determine all the parameters that lead to a stable plasma discharge and produce a very reliable disinfection device and surface modification device [6].

This study focuses on determining the ideal plasma parameters by changing the configuration of electrodes thickness, electrodes distance, and quartz tube's inner and outer diameter to generate stable plasma discharge by reducing the usage of carrier gases, applied voltage, and electrodes. To analyze the ideal plasma discharge, we use optical emission spectroscopy (OES) analysis to investigate the spectrum emitted by the plasma jet. The expected result would be if the plume have longer plume with higher intensity and number of reactive species present in it can carry out sterilization and disinfection process effectively. This result can be prove by taking bacteria to perform the plasma treatment with the suggested configuration of APPJ in order to examine the density of bacteria present after the treatment. From this, we can determine the best parameter to perform the plasma treatment in many field by giving best output.

2. Methodology

2.1. Atmospheric pressure plasma jet device configuration

The principle operation of the atmospheric plasma is high voltage supplied to the flowing gas in a glass tube. Firstly need to analyze the requirement of this project by understanding the operation of atmospheric pressure plasma jet (APPJ) in detail and study the effect of different configuration on the plasma discharge. In this work, the development of plasma system consists of plasma generating devices, including electronic instrument, power supply, helium gas, pulse invertor, and plasma jet.

The experimental arrangement is shown in Figure 1. The set up composed of fours mains parts; gas control system, pulse inverter, plasma jet setup, and spectrometer device. The gas control system includes the flow meter system, valve, power system, control system, gases and Labview® (2011) software. Plasma jet setup consists of two electrodes where the ionization occurs. These electrodes are placed 25 mm between each other on the glass tube (diameter 1.50 mm). Helium gas with flow rate 1000 ml/min are used as a main gas source.

The emission line spectrums for plasma jet plume were measured using EPP2000-HR High Resolution spectrometer from Stellar. The plume was collected by using fiber optic probe (UV-VIS-NIR). The fiber optic probe was held 0.5 mm away from plume at 45°. The recorded spectrum line was observed using SpectraWiz OS v5.3(c) 2013 software.

Power supply is an important component in this setup which will affect the production of the plasma. The power supply is a device that supplies the electric power to the electric load. It converts the AC or DC power into the required voltage according to the user setting such 16.5V [7]. In this experiment, pulse inverter was used to convert 8VDC to 16KV AC.

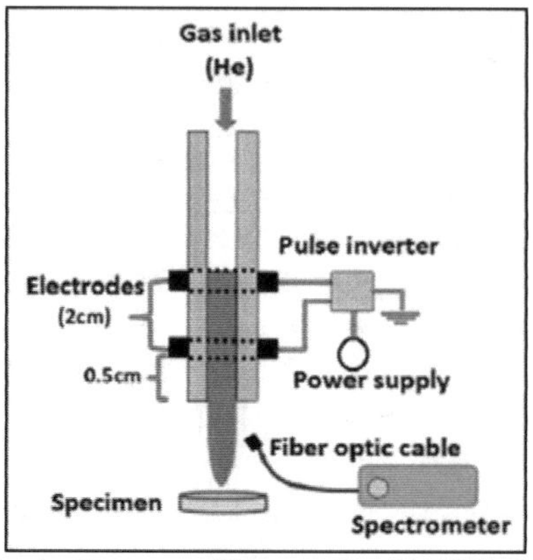

Figure 1. Atmospheric-Pressure Plasma Jet (APPJ) device configuration

2.2. The changing parameters for APPJ configuration

The optimum plasma plume can obtain with proper configuration by following all the parameters as listed in the Table 1 and refer figure 2 which show the parameter need to change as well. Table 1 shows the parameters that need to use for the investigation of distance change between two electrodes. Table 2 shows the parameters that need to consider for the investigation of thickness changes of High-Voltage (HV) electrode by remaining constant of ground electrode's thickness.

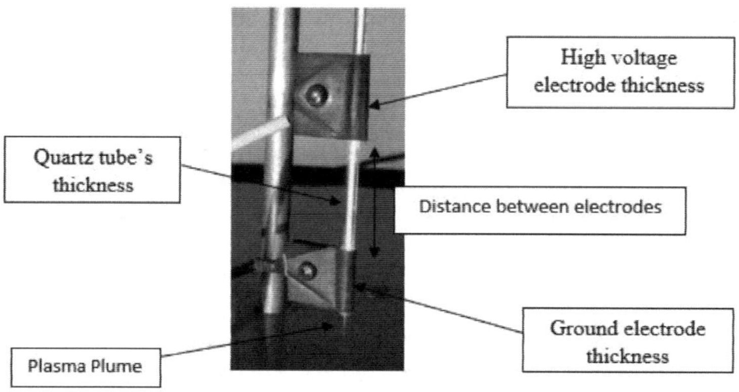

Figure 2. Changing parameters

Table 1. Parameters of distance (mm) between the two electrodes

Constant parameters	Changing parameters
1. HV electrode thickness = 0.5 mm	Distance between the two electrodes; 10 mm, 15 mm, 20 mm, and 25 mm
2. Ground electrode thickness = 0.5 mm	
3. Inlet gas = Helium	
4. Flow rate of gas = 1000 ml/min	
5. Quartz tube's thickness = 1.50 mm	

Table 2. Parameters of thickness (mm) of high voltage (HV) electrodes

Constant parameters	Changing parameters
1. Distance between the two electrodes = 25 mm	HV electrode thickness = 0.5 mm, 0.7 mm, 0.9 mm, and 1.2 mm
2. Ground electrode thickness = 0.5 mm	
3. Inlet gas = Helium	
4. Flow rate of gas = 1000 ml/min	
5. Quartz tube's thickness = 1.5 mm	

Table 3 shows the parameters that need to consider for the study of thickness changes of ground electrode by remaining constant of High-Voltage (HV) electrode's thickness. Other parameters such as constant flow rate, quartz tube's thickness, and type of inlet gas are fixed.

Table 3. Parameters of thickness (mm) of ground electrodes

Constant parameters	Changing parameters
1. Distance between the two electrodes = 25 mm	Ground electrode thickness = 0.5 mm, 0.7 mm, 0.9 mm, and 1.2 mm
2. HV electrode thickness = 0.5 mm	
3. Inlet gas = Helium	
4. Flow rate of gas = 1000 ml/min	
5. Quartz tube's thickness = 1.50 mm	

Table 4 shows the configuration that need to use for the study of thickness changes of quartz tube's thickness by remaining constant of other parameter such as constant flow rate, High-Voltage (HV) thickness, ground electrode, and type of inlet gas are fixed. Quartz tube has two different thickness which is from aspect inner diameter and outer diameter. Both dimension play the important role in producing plasma discharge.

Table 4. Parameters of thickness (mm) of quartz tube

Constant parameters	Changing parameters	
1. Distance between the two electrodes = 15 mm	Outer Diameter (OD)	Inner Diameter (ID)
2. HV electrode thickness = 0.5 mm	3 mm	1.5 mm
3. Ground electrode thickness = 0.5 mm	5 mm	3 mm
4. Inlet gas = Helium	8 mm	5 mm
5. Flow rate of gas = 1000 ml/min	13 mm	10 mm

2.3. Analyzing method of plasma properties

The plasma plume that discharges from the plasma jet can be analyzed using Optical Emission Spectrometry technique. This method can be done by using an EPP2000-HR High-Resolution spectrometer from Stellar [8]. The plume was collected by using a fiber optic probe (UV-VIS-NIR). The fiber optic probe was held 0.5 mm away from plume at 45°. The recorded spectrum line was observed using SpectraWiz OS v5.3(c) 2013 software [9]. The spectrometer will detect the reactive species obtain in the plume and generate a line emission spectrum with the ions intensity and wavelength [10].

3. Results and Discussions

Two different research is carry out in this project which are choose the ideal configuration of Atmospheric Pressure Plasma Jet (APPJ) by varying quartz tube diameter, electrode thickness, and electrodes distance.

The distance between two electrodes at 25 mm give the most stable plasma discharge as shown in Figure 3 (a) due to the stable energy distribution of electrons/ions and many type of reactive species present in the line emission spectrum as in Figure 3 (b). The plasma discharges by using electrodes distance at 25 mm are suitable for bacteria inactivation process and plasma surface modification because of the presence of reactive oxygen radicals and hydroxyl, OH radicals and N_2.

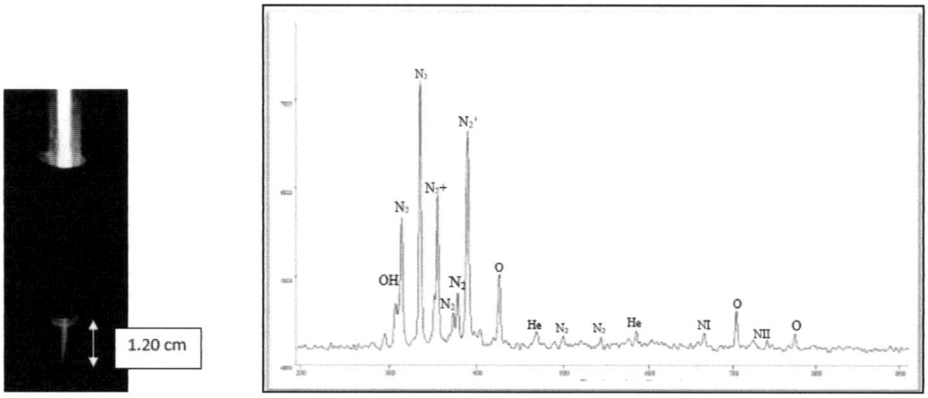

(a) (b)

Figure 3. (a) Plume length and (b) plasma Line Emission Spectrum for the distance changes of electrode at 25 mm

The high-voltage electrode thickness of 0.9 mm with ground electrode thickness of 0.5 mm give the most stable plasma discharge as shown in Figure 4(a), because it drift more electrons/ions to ionize and transmit to ground electrode to discharge through the nozzle of quartz tube[8].

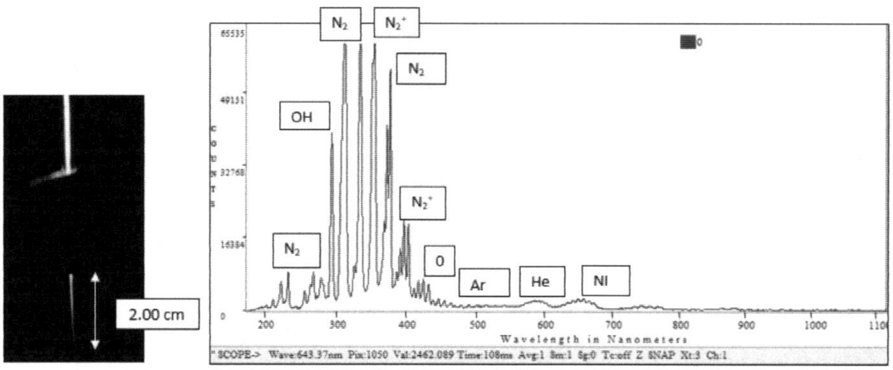

(a) (b)

Figure 4. (a) Plume length and (b) plasma emission when high-voltage electrode thickness is 0.9 mm with ground electrode thickness is 0.5 mm

The ground electrode thickness of 0.9 mm with High Voltage (HV) thickness of 0.5 mm give the most stable plasma discharge as shown in Figure 5 (a) due to the high intensity value and different type of reactive species presented in the line emission spectrum (Figure 5 (b)) among others especially the intensity of OH radical is higher at 6033 W/m2 that can used in bacteria inactivation.

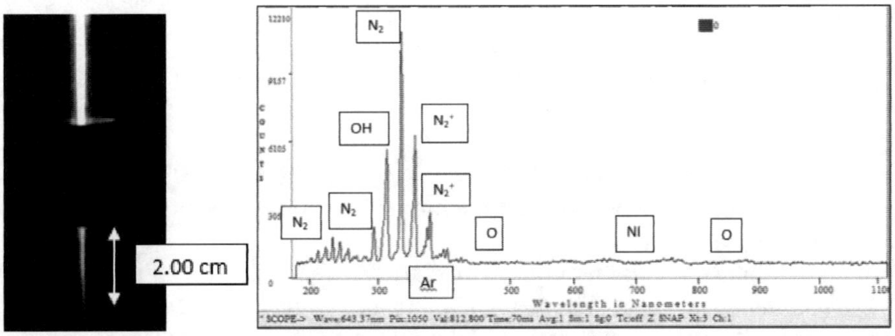

(a) (b)

Figure 5. (a) Plume length and (b) emission Line emission spectrum for the High-Voltage electrode thickness 0.90 mm

The quartz tube thickness of outer diameter 3 mm and inner diameter 1.5 mm give the most stable plasma discharge as shown in Figure 6 (a) due to the small inner space of the quartz tube to discharge the plasma and have different reactive in line emission spectrum (Figure 6 (b)). The other quartz tube size produce a non-stable plume by splitting into two plume and discharge inside the quartz tube because of inadequate flow rate, larger volume, and low collision of electrons and neutral particles.

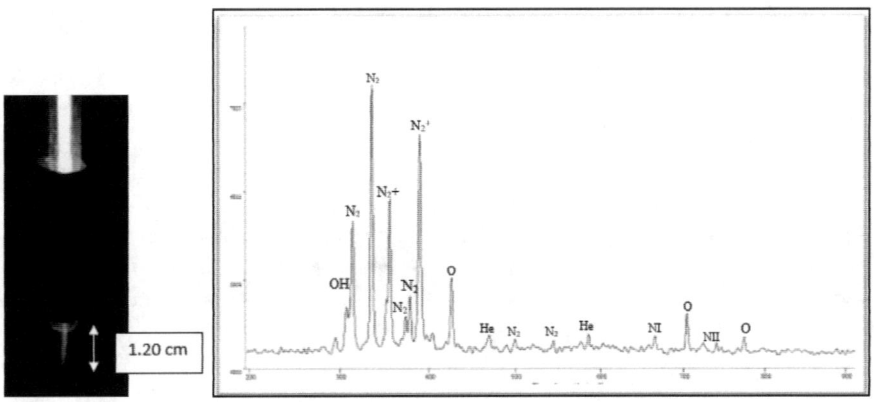

(a) (b)

Figure 6. (a) Plume length and (b) emission Line emission spectrum for the Quartz tube's thickness of (OD= 3 mm & ID = 1.5 mm)

4. Conclusion

The result of present study demonstrated that atmospheric pressure plasma jet with thickness of 0.9 mm High-Voltage electrode and 0.5 mm ground electrode by using flow rate 1000 ml/min, electrode distance 25 mm, and thickness of quartz tube 1.5 mm showing the strongest configuration because it has higher intensity reactive oxygen species and nitrogen species in optical emission lines 300 nm until 400 nm which had an important effect on the bacterial inactivation. These species causes damage by providing abnormal load oxidation process to bacteria [11]. This then, will cause death of bacteria.

The purpose of electrodes folded around quartz tube (dielectric material) in this atmospheric pressure plasma jet (APPJ) is to stop the development of the arc discharge by the reverse charge effect of the real charges in the discharge [11]. Thus, if thicker the copper electrode, the higher the reverse charge effect in producing the plasma discharges. Other elements inside plume such as ions, electrical field and UV plays secondary role but still important in inactivation purpose. The developed atmospheric plasma jet device is comparatively a simple device which possesses good potential as microbial inactivation device since it is able to inactivate the pathogen in short time [12].

Acknowledgement

The author would like to acknowledge the support from the Fundamental Research Grant Scheme (FRGS) under grant number of FRGS/1/2015/SG02/UNIMAP/02/5 from the Ministry of Education Malaysia.

References

[1] Laroussi M and Lu X 2005 Room-Temperature Atmospheric Pressure Plasma For Biomedical Applications, *Appl. Phys. Lett.* **87** 1–3.

[2] Nordin N A, Zaaba S K, Baharuddin B, Zakaria A, Wan K and Shahriman A B 2015 Study on the surface characteristic of PTFE after atmospheric plasma treatment *Proceedings 2015 2nd International Conference on Biomedical Engineering (ICoBE)* 30–31.

[3] Nur M, Kinandana A W, Winarto P and Muhlisin Z 2016 Study of an atmospheric pressure plasma jet of Argon generated by column dielectric barrier discharge, *Journal of Physics: Conference Series* **776**(1).

[4] Saad N A et al. 2015 Optical emission spectroscopy analysis of atmospheric plasma jet plume on bacteria inactivation, *J. Teknol.* **6**(77) 77–81.

[5] Fridman G, Friedman G, Gutsol A, Shekhter A B, Vasilets V N and Fridman A 2008 Applied plasma medicine, *Plasma Process. Polym.* **5**(6) 503–533.

[6] Jha N, Ryu J J, Choi E H and Kaushik N K 2017 Generation and Role of Reactive Oxygen and Nitrogen Species Induced by Plasma, Lasers, Chemical Agents, and Other Systems in Dentistry. *Oxid. Med. Cell. Longev.*

[7] Schütze A, Jeong J Y, Babayan S E, Park J, Selwyn G S and Hicks R F 1998 The Atmospheric-Pressure Plasma Jet: A Review and Comparison to Other Plasma Sources, *IEEE Trans. PLASMA Sci.* **26**(6).

[8] Gorbanev Y, Soriano R, O'Connell D and Chechik V 2016 An Atmospheric Pressure Plasma Setup to Investigate the Reactive Species Formation. *J. Vis. Exp.* **117**.

[9] Donnelly V M 2004 Plasma electron temperatures and electron energy distributions measured by trace rare gases optical emission spectroscopy *J. Phys. D. Appl. Phys.* **37**(19).

[10] Fused Silica/Quartz Glass - Properties and Applications of Fused Silica/Quartz Glass by Goodfellow Ceramic & Glass Division. [Online]. Available: https://www.azom.com/article.aspx?ArticleID=4766. [Accessed: 08-Dec-2017].

[11] Zhao S et al. 2013 Atmospheric Pressure Room Temperature Plasma Jets Facilitate Oxidative and Nitrative Stress and Lead to Endoplasmic Reticulum Stress Dependent Apoptosis in HepG2 Cells *PLoS One* **8**(8) 1–14.

[12] Seo Y S, Mohamed A A H, Woo K C, Lee H W, Lee J K and Kim K T 2010 Comparative studies of atmospheric pressure plasma characteristics between He and Ar working gases for sterilization *IEEE Trans. Plasma Sci.* **38**(10) 2954–2962.

Journal of Physics: Conference Series **1372** (2019) 012037 doi:10.1088/1742-6596/1372/1/012037

Time-frequency EMG features comparison of biceps brachii and erector spinae in evaluating work level categories of RTW patients

Ezreen Farina Shair[1,2*], Siti Anom Ahmad[1,3], Abdul Rahim Abdullah[2], Mohammad Hamiruce Marhaban[1], Shamsul Bahri Mohd Tamrin[4]

[1]Department of Electrical and Electronic Engineering, Faculty of Engineering, Universiti Putra Malaysia, 43400 UPM, Serdang, Selangor, Malaysia
[2]Centre for Robotics and Industrial Automation, Faculty of Electrical Engineering, Universiti Teknikal Malaysia Melaka, Hang Tuah Jaya, 76100 Durian Tunggal, Melaka, Malaysia
[3]Malaysian Research Institute on Ageing, Universiti Putra Malaysia, 43400 UPM Serdang, Selangor, Malaysia
[4]Department of Environmental and Occupational Health, Faculty of Medicine and Health Sciences, Universiti Putra Malaysia, 43400 UPM Serdang, Selangor, Malaysia

[*]ezreen@utem.edu.my

Abstract. Electromyography (EMG) is one of the special tool in assessing human muscle performance. The muscle performance can be evaluated when there exist changes on both the amplitude and frequency simultaneously. However, various time-frequency EMG signal features from literature do not provide easy and direct interpretation of the muscle condition specifically for non-researchers in rehabilitation centers. This study investigates EMG behavior of biceps brachii and erector spinae based on new time-frequency EMG features during functional capacity evaluation (FCE)'s core-lifting task in evaluating work level categories of return to work patients. Surface EMG signals are filtered using band pass filter with cut-off frequency at 5-500 Hz and auto-segmented using spectrogram, while S-transform is used for the features extraction. Result shows that the differences of the proposed EMG features values are significant in indexing muscle performance and are good indicators for the determination of work level categories.

1. Introduction

Functional capacity evaluation (FCE) is a test used to check an individual's physical ability to perform specific task and to decide return-to-work (RTW) patients' readiness to work [1]. The determination of work level categories for RTW patients is vital to decide if the worker is capable of handling the light, medium, or heavy physical requirements of a task. The lopsidedness between recovery and load will prompt maladaptation and on the off chance that maladaptation is prolonged, more severe condition will take over where workers with previous history of musculoskeletal disorders (MSDs) may suffer from injury on the previously injured muscle.

In an FCE, one of the critical task to test an RTW patient's practical capacity is the core-lifting task. The patient needs to perform lifting to a maximal limit and only the therapist's visual perceptions are utilized to decide whether the maximum limit has been reached. Many opinions agreed that a noteworthy constraint inherent to this FCE test is the probability that the therapist's decisions could be affected by information other than visual perceptions [2]. To enhance the reliability and validity of the FCE, electromyography (EMG) signal is considered in this study to determine the work level categories of each RTW patient. EMG signal is a highly complicated non-stationary signal which represents the neuromuscular activation with a contracting muscle. Previously, various amplitude and frequency features of an EMG signal have been introduced separately. Though time-frequency distributions are known to be more accurate as it can provide both time and frequency information simultaneously, time-frequency based features, however, are harder to be interpreted especially for non-researchers.

Therefore, this paper presents EMG behavior of biceps brachii and erector spinae during FCE's core-lifting task. New time-frequency EMG features which are the muscle strength, muscle power, and muscle endurance are plotted to observe the differences of EMG behavior between different work level categories. Other physiological responses such as the maximum load mass lifted as well as heart rate are also recorded.

2. Experimental procedure

2.1. Subject recruitment and preparation

Two different groups of subjects were formed in this study. The first group is the control group which consists of eleven healthy subjects (six male and five female) without previous history of MSD, with a mean age of 25.73±1.74 years, and body mass index (BMI) of 22.47±1.33 kg/m2. They were recruited randomly via notices in public settings. The second group is the test group which consists of subjects with previous history of MSDs either on the upper limb or back and are certified fit to partake the FCE. They were recruited among RTW patients at the Social Security Organization (SOCSO) Tun Abdul Razak Rehabilitation Centre, Melaka. The group also consists of eleven subjects (6 male and five female) with a mean age of 27.36±3.72 years and BMI of 21.76±1.81 kg/m2. Subjects with movement difficulties, suffering from chronic diseases, pregnant, and on medication that could affect the EMG activity were excluded from the study. The subjects are required to answer the self-assessment questionnaires (upper extremity functional index questionnaire and Oswestry low back pain questionnaire) in order to assess the inclusion and exclusion requirements. The study protocol reported in this paper was undertaken in compliance with the ethical principles of the Helsinki Declaration and was approved by the Ethics Committee for Research Involving Human Subjects of Universiti Putra Malaysia. Information consent forms were signed by the subjects before the start of the experiment.

2.2. Core-lifting task protocols

The Joule-Valpar FCE's core-lifting task protocols are employed in this study. It follows the conventions utilized by the SOCSO Tun Abdul Razak Rehabilitation Centre, Melaka to ensure validity and reliability of the results. The core-lifting comprised of six phases in one cycle; P1: waist to waist, P2: waist to waist, P3: waist to floor, P4: floor to waist, P5: waist to above shoulder, and P6: above shoulder to waist. Toward the beginning of the lifting protocol, the subject is required to remain in front of the Valpar-Joule premium package work centre as in Figure 1, where certain anthropometric that are related to lifting was recorded (waist height and above shoulder height).

After one cycle, loads were added to the lifting container. The load progression flip-chart and the T-handled color-coded loads (Figure 1) were utilized to advance the subject through the core-lifting task. The lifting cycle was repeated to the point where the subject experienced symptoms such as lack of energy, blurred vision and dizziness, which speak to the indications of fatigue. Heart rate was observed throughout the experiment using Polar chest strap to ensure that the subjects already reached the maximal limit and needed to stop lifting.

The work level categories' definition utilized at the SOCSO Tun Abdul Razak Rehabilitation Centre are taken form the Dictionary of Occupational Titles. It very well can be divided into five categories dependent on the maximum lifting load. These categories by one means or another differ upon the frequency of lifting work required when the subjects return to the industry. However, this study just focusses on the frequent lifting frequency which includes between 2.5-5.5 hours of lifting in an 8-hour shift. The work level categories for the frequent lifting frequency is characterized as shown in Table 1 [3]. Because of subject constraints, just two work level categories were considered in this study which are medium work (MW) and heavy work (HW) categories.

Table 1. Work level categories for frequent lifting frequency

Light work	Medium work	Heavy work	Very heavy work
Exerting up to 4.5 kg of force to lift.	Exerting up to 4.5 - 11 kg of force to lift.	Exerting up to 11 -23 kg of force to lift.	Exerting in excess of 23 kg of force to lift.

2.3. Signal acquisition

Surface EMG signals were recorded using Shimmer 3 Consensys EMG Development Kit (Shimmer Sensing, Dublin, Ireland). The full system offers signal to noise ratio of 42.13dB, which fulfils the qualities of a good EMG signal. Surface EMG signals of the biceps brachii and erector spinae were recorded and the reference electrode was placed on the wrist of the subject. The signals were sampled at a sampling frequency of 1500 Hz to avoid aliasing [4]. These muscles were selected as they are the most activated muscles during lifting and suffer a higher rate of injury in the industries [5]. Pairs of Ag/AgCl disc electrodes (Kendall Meditrace 200, Covidien, USA) with a 20 mm inter-electrode distance were used to acquire the EMG signals. The procedure for the electrode placements was referred to the Surface EMG for Non-Invasive Assessment of Muscles (SENIAM) guidelines.

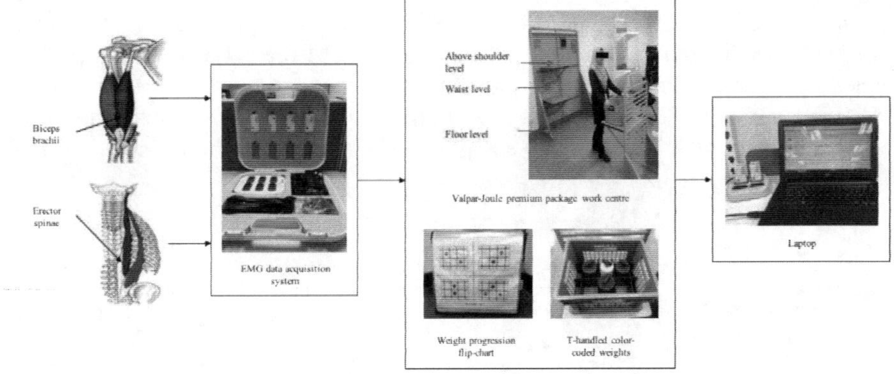

Figure 1. Experimental setup

3. Data analysis

3.1. Pre-processing

Surface EMG signal was filtered using band pass filter with a cut-off frequency at 5-500 Hz. Signals below 5 Hz and above 500 Hz were wiped out as they have insignificant power spectral density of the surface EMG signal. It was likewise filtered so as to reduce the baseline noise and corner frequency caused by unstable and unpredictable fluctuation as recommended by [6]. The EMG signals were then auto-segmented to separate the EMG signal muscle contraction from the baseline/noise. This is to reduce the computation complexity for long recordings of EMG signal as the important information to be analyzed is only available in the muscle contraction. The auto-segmentation algorithm used in this study is based on the thresholding of the instantaneous energy of the EMG signal [7].

3.2. Time-frequency analysis

The time-frequency representation (TFR) of the EMG signal was computed via S-transform. The root mean square (RMS) voltage of the TFR was then measured instantaneously and the average values were calculated for the work level categories evaluation. To provide easier interpretation of the muscle condition, three new time-frequency EMG features were derived from the average RMS voltage plot. The features are:

- Muscle strength – Ability to exert maximal force.

$$Strength = \frac{1}{n}\sum_{i=1}^{n} V_{rms(avg)i} \tag{1}$$

- Muscle power – Ability to exert maximal force within a short period of time.

$$Power = \frac{\sum(x-\bar{x})(y-\bar{y})}{\sum(x-\bar{x})^2} \tag{2}$$

- Muscle endurance – Ability to perform a specific muscular action for a long period of time.

$$Endurance = \max(n) \tag{3}$$

where n is the number of lifting cycles, $V_{rms(avg)_i}$ is the average RMS voltage, x is the load mass, y is the$V_{rms(avg)_i}$, $\bar{x}$ and $\bar{y}$ are the sample means (averages) of load mass and $V_{rms(avg)_i}$, respectively.

4. Result and discussion

4.1. Physiological response analysis

Figure 2 shows the maximum load mass lifted by the subjects. C1 until C11 are the control subjects, while T1 until T11 are the test subjects who participated in the experiment. From the figure, it demonstrates that none of the subjects were able to complete the lifting task which comprised of 39 lifting cycles in total with the maximum load of 22.4 kg. Most of the subjects stopped the lifting when they have reached their maximum strength. A few test subjects needed to stop due to muscle pain that developed on previously injured muscles. Five control subjects and six test subjects satisfy the requirement of the heavy work (HW) level category, while the rest of the subjects fall under the medium work (MW) level category.

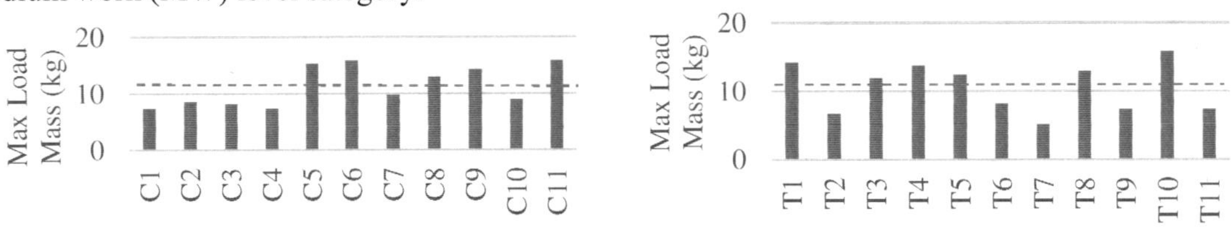

(a) (b)

Figure 2. Maximum load mass lifted by (a) control subjects and (b) test subjects

Maximum heart rate (HRmax) is important to be recorded as it decides the intensity of the training, thus proving that the subjects have reached their maximal lifting limit. Based on Figure 3, all 22 subjects have reached their maximal lifting effort before ending the task and all of them accomplished either hard (80-89%) or very hard (90-100%) intensity zone categories. Subjects experienced heavy breathing and muscular fatigue after the lifting task.

(a) (b)

Figure 3. Maximum heart rate percentage of (a) control subjects and (b) test subjects after completing the core-lifting task

The HRmax of the test subjects are found to be lower compared to the control subjects. Based on this result, it shows that previous history of MSDs may have caused the subjects' strength to degrade, thus forcing them to quit the lifting at a lower intensity.

4.2. Surface electromyography analysis

The EMG signal features which are the muscle power, muscle strength, and muscle endurance were approximately determined based on the plot of the mean average RMS voltage of each subject. Figure 4 and 5 show the plot of the EMG signal features of all 22 control subjects and test subjects for biceps brachii and erector spinae muscles. The biceps brachii of subject C1 is seen to have high muscle power and high muscle strength, yet low muscle endurance, as opposed to subject C11 who has low muscle power and low muscle strength, yet high muscle endurance. Previously, in the study of [8], it was claimed that healthy individuals exerting higher strength during lifting may effortlessly experience fatigue, thus resulting in poor endurance. For good muscle performance, each of the three EMG signal features must be balanced [9].

Behavior of the EMG signal features in evaluating the MW and HW categories on the investigated muscles is compared to the maximum load mass in Figure 3. The result illustrates that the EMG signal features for each subject have comparable behavior and response on the investigated muscles. The EMG signal features under the MW category (subject C1, C2, C3, C4, C7, C10, T2, T6, T7, T9, and T11) tend to have higher muscle power and higher muscle strength compared to the EMG signal features under the HW category (subject C5, C6, C8, C9, C11, T1, T3, T4, T5, T8, and T10).

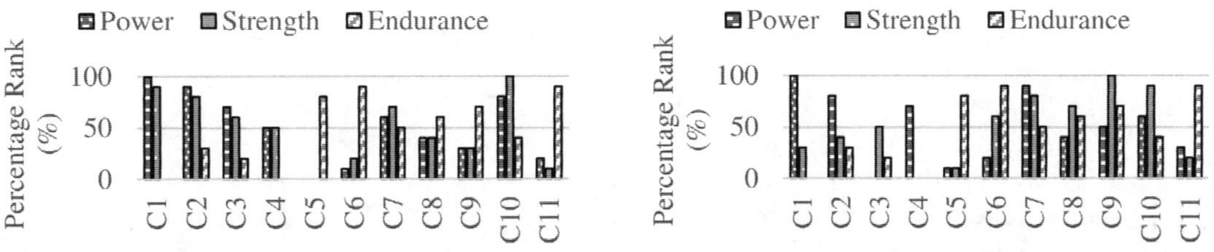

(a) (b)

Figure 4. EMG signal features plot for (a) biceps brachii and (b) erector spinae muscles of control subjects

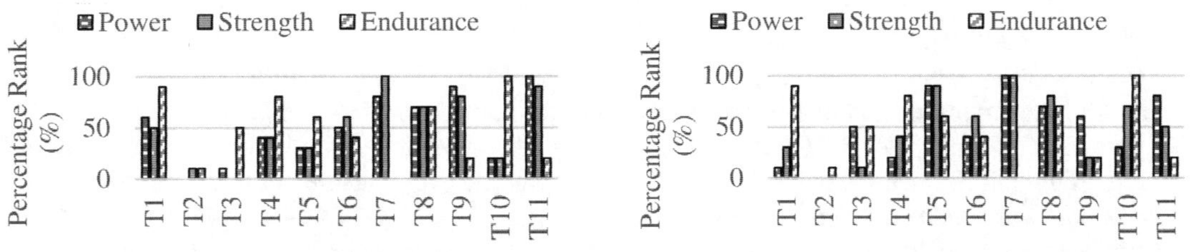

(a) (b)

Figure 5. EMG signal features plot for (a) biceps brachii and (b) erector spinae muscles of test subjects

The higher muscle power and muscle strength under the MW category in Figure 4 and 5 demonstrate that the subjects struggled to lift the load, thereby forcing them to use more effort even at the beginning of the lift [9]. Subjects under the MW category were expected to exert much higher strength as their body required more muscle power to move the object against gravity and also to maintain stability. This result supports the findings by [10] that less fit subjects are normally expected to exert higher levels of muscle power in order to sustain long duration. However, due to the generation of higher muscle power, the supply of oxygen and the metabolic by-product clearance during exertion are reduced which resulted in poor muscle endurance [11].

5. Conclusion

Results presented in this paper show that the proposed time-frequency EMG signal features are significant to evaluate the work level categories of an FCE. The differences found in the work level categories based on the EMG signal features, however, were prominent in muscle with higher activation (biceps brachii) in contrast to less activated muscle (erector spinae). Overall, the result suggests that the differences of EMG signal features of the control and test subjects are significant on both activated muscle. In short, rise and decrease of the muscle power, strength and endurance values indicate valuable information which reflect to the muscle performance of RTW patients treated in rehabilitation centers. In future study, these features are recommended to be fed to a classifier in order to provide a real-time work level categories' classification.

Acknowledgements

The authors would like to acknowledge and thank the Control and Signal Processing group of Universiti Putra Malaysia, Rehabilitation and Assistive Technology research group, and Advanced Digital Signal Processing research laboratory of Universiti Teknikal Malaysia Melaka, and Social Security Organization Tun Abdul Razak Rehabilitation Centre for their continuous support in the research. This research is funded by the Universiti Putra Malaysia under the Geran Putra Berimpak (GPB) grant.

References

[1] Postema S G, Bongers R M, Reneman M F and van der Sluis C K 2018 Functional capacity evaluation in upper limb reduction deficiency and amputation: Development and pilot testing *J. Occup. Rehabil.* **28** 158–69

[2] Reneman M, Jaegers S, Westmaas M and Goeken L 2002 The reliability of determining effort level of lifting and carrying in a functional capacity evaluation *Work* **18** 23–7

[3] United States Department of Labour 1991 *Dictionary of Occupational Titles* 4th ed (Washington DC: United States Government Printing Office) p 1

[4] Shair E F, Tengku Zawawi T N S, Abdullah A R and Shamsudin N H SEMG signals analysis using time-frequency distribution for symmetric and asymmetric lifting 2015 *Proc. Int. Symp. on Technology Management and Emergent Technologies* pp 233–7

[5] Al-Ashaik R A, Ramadan M Z, Al-Saleh K S and Khalaf T M Effect of safety shoes type, lifting frequency, and ambient temperature on subject's MAWL and physiological responses 2015 *Int. J. Ind. Ergon.* **50** 43–51

[6] De Luca C J, Donald Gilmore L, Kuznetsov M and Roy S H 2010 Filtering the surface EMG signal: Movement artifact and baseline noise contamination *J. Biomech.* **43** 1573–79

[7] Shair E F, Abdullah A R, Tengku Zawawi T N S, Ahmad S A and Mohamad Saleh S 2016 Auto-segmentation analysis of EMG signal for lifting muscle contraction activities *J. Telecommun. Electron. Comput. Eng.* **8** 17–22

[8] Shih Y C, Chen W L and Chi C F 2009 The effects of gender, exerting type, and temperature on hand grip/pinch strength under the muscular fatigue 2009 *Proc. Wseas Int. Conf. on Systems Theory and Scientific Computation* pp 50–5

[9] Abdel Kader Hasan N A K, Mohamed Kamal H and Ahmed Hussein Z 2016 Relation between body mass index percentile and muscle strength and endurance *Egypt. J. Med. Hum. Genet.* **17** 367–72

[10] Maffiuletti N A, Ratel S, Sartorio A and Martin V 2013 The impact of obesity on in vivo human skeletal muscle function *Curr. Obes. Rep.* **2** 251–60

[11] Hicks A L, Kent-Braun J and Ditor D S 2001 Sex differences in human skeletal muscle fatigue *Exerc. Sport Sci. Rev.* **29** 109–12

Number of Screws Affecting the Stability and Stress Distributions of Conventional and Locking Compression Plate: A Finite Element Study

Herman Tuminoh[1], Mohamad Hafiz Mohamad Afandi[1], Hong Seng Gan[2], Mohammad Rafiq Abdul Kadir[3,4,] Muhammad Hanif Ramlee[1,4*].

[1]Medical Devices and Technology Centre (MEDITEC), Institute of Human Centered Engineering (i-Humen), Faculty of Engineering, Universiti Teknologi Malaysia, Skudai 81310, Johor, Malaysia
[2]Universiti Kuala Lumpur, British Malaysian Institute, Batu 8 Jalan Sg. Pusu, 53100 Gombak, Selangor, Malaysia
[3]Sports Innovation and Technology Centre (SITC), Institute of Human Centered Engineering (iHumen), Universiti Teknologi Malaysia, Skudai 81310, Johor, Malaysia
[4]Bio-inspired Innovation Group (BIOINSPIRA), School of Biomedical Engineering and Health Sciences, Faculty of Engineering, Universiti Teknologi Malaysia, Skudai 81310, Johor, Malaysia

*Corresponding author: muhammad.hanif.ramlee@biomedical.utm.my

Abstract. The effectiveness of malleolar fracture fixation is still questionable. Internal fixator is the one of the treatment for treating this fracture. However, the analysis of various type of internal fixator is still lacking in the literature in terms of biomechanical characteristics and behaviour. Thus, the aim of the study was to compare the stability of locking compression plate (LCP) and one third tubular plate (OTT) in different configuration of screws. Computed Tomography (CT) images of bone was used to develop 3D model of fibula bone. The plate was constructed in Solidworks software and number of screws used were 3 and 5. Further, finite element study was conducted for both model. For LCP, the highest von Mises stress (VMS) observed at the plate for 3 screws was 484 MPa, whereas for 5 screws plate was 667 MPa. Besides, for OTT, the highest VMS at plate observed for 3 screws was 300.5 MPa, whereas for 5 screws plate was 127.5 MPa. Based on the results, it can be noted that the usage of 3 screws can causes a low VMS at plate compare to 5 screws. However, the relation is valid for LCP. For OTT, 5 screws constructs gave a low VMS than 3 screws constructs.

1. Introduction

From 2007 to 2011, hospital from the United States of America (USA) reported that the ankle has the highest number of fractures at lower limb which was 280 933 cases [1-2]. According to these numbers, it is shown that malleolar fracture has the highest ranking of occurrence specifically at the ankle joint [1-2]. In order to treat malleolar fracture, a surgery should be conducted to restore the fracture into the original formation of the structure. One of the method in restoring the ankle fracture is the open reduction and internal fixation method [3].

Conventional plating methods are based on the use of an adequate number of anchoring screws to press the plate against the bone with high compressive forces, creating a stable bone-implant connection. To date, there is no problem with the conventional plate, which is one third tubular plate (OTTP) if the patient's bone density is in normal condition. Another design that have been used to restore the ankle fracture is the locking compression plate (LCP). To be noted, the use of these two plates is favourable in treating ankle fracture specifically at fibula region [3].

In clinical practises, the complication from the application of plate fixation on the fibula fracture is 6.2% for the LCP fixation and 1.4% for the OTT plate fixation [4]. These cases will require medical surgeons to conduct second surgery for removing the implant. The complications reported in the literature are from various aspects such as implant loosening, infection, mal-union, non-union and misalignment [4-9]. As far as the authors are concerned, there is no study performed to evaluate the biomechanical properties of LCP and OTT in treating fibula fracture. Therefore, this study was conducted to compare the stability and stress distribution of LCP and OTT for better understanding on biomechanics principle of these two designs.

2. Methodology

2.1. Bone reconstruction

Computed Tomography (CT) images data set were used to reconstruct anklebone models which is fibula bone. The bone model images provided by CT scan were used to reconstruct the three-dimensional (3D) model of bone via Mimic software. Next, the 3D images of each bone model was made to undergo meshing process where the mesh was 3.0 mm sized tetrahedral. An opening of 2 mm gap was simulated by removing a cylindrical-shaped of bone from the distal part of the fibula.

The Conventional plate and Locking Compression plate were designed using a Solidworks software. All the parameters (size and distance) were determined following the techniques of MacLeod [8]. Both OTTP and LCP systems have 5 screws and 3 screws orientation while the screw type is kept constant which using cortical screw. Further, the process is continuing with fixing the bone plate to the fracture area using Mimic software.

For finite element analysis, axial loading was chosen to be tested on the model due to standing condition. The force was position axially to the fibula and tibia as the patient body weight was 70 kg, the load is distributed equally between two legs and further the load was distributed to tibia (93%) and fibula (7%) [10]. The load distributed between tibia and fibula was connected with link to imitate ligament. Besides, the model was also fixed at the bottom of talus surface to make sure the model is static. All friction coefficient values were set to 0.3. The analysis was done using commercial finite element software MARC with the equivalent von Mises stress (EVMS) and displacement of the model relative to the fibula. The Figure 1 below showed the finite element study of the model.

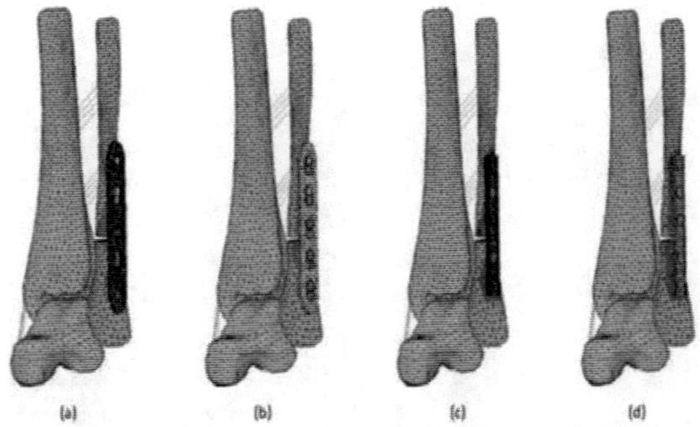

Figure 1: (a) 3 screws (LCP) (b) 5 screws (LCP) (c) 3 screws (OTT) (d) 5 screws (OTT)

3. Results and Discussion

The contour plots of von Mises stress (VMS) of the fibula and different configuration of screw for Locking Compression Plate (LCP) and One Third Tubular Plate (OTTP) are shown in Table 1. In general, the contour plots of all the configurations showed that the bone had a high stress concentration at the pin-bone interface for stance phase.

When the plate was placed towards the fracture side, 5 screws plate gave less stress to the bone in order to maintain the bone union. For LCP, the highest VMS was recorded for 3 screws with 183.7 MPa at the fibula pin-bone interface and then the 5 screws with 70 MPa was the lowest VMS. For OTTP, the highest VMS was observed for 3 screws (141.3 MPa) at the fibula pin-bone interface, whereas for 5 screws was 56 MPa.

The stress distribution at the plates is as shown in Table 1. The highest VMS of LCP plate for 3 screws (484 MPa) was observed lower than 5 screws plate (667 MPa). Besides, the highest VMS of OTTP plate for 3 screws was 300.5 MPa being higher than 5 screws plate (127.5 MPa). Even with the same pattern of stress distribution, the magnitude of each combinations was different.

From the results, a higher stress was seen at bone plate rather than fibula pin-bone interface. The stability of LCP depends on the number of screw plate where the reduction of screw number will reduce the stress distribution of implant. This is due to a low force required to achieve bone contact. The low force will minimize the effect of shielding force because the shielding effect need to be minimized is to reduce the bone resorption process that will weaken the bone structure [4, 7, 8].

The stresses at interface between the screw heads and the plate holes have been ignored due to the difficulty to accurate the modelling the threaded interfaces. The interfaces were not considered as the weakest part of system if screws are well aligned with plate hole and tightened with proper torque and not exceed the ultimate strength [11]. In this study, the use of LCP and OTT for treating malleolar fracture can be accept since the maximum stress at the bone when using LCP and OTT does not exceed the ultimate strength of the bone used (193 MPa) [12].

Figure 2 shows the displacement plot of the fibula bone. Generally, the displacement values were below 4 mm for all plates except for the 3 screws of OTTP (5.649 mm). When simulating the stance phase, the displacement of LCP for 3 screws (3.59 mm) and 5 screws (3.51 mm) were slightly close while the displacement of OTTP for 3 screws and 5 screws were 5.65 mm and 3.72 mm respectively.

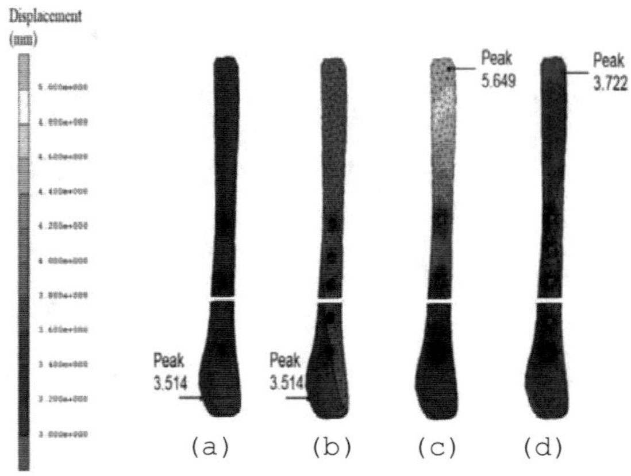

Figure 2: The displacement plot of the fibula bone for different configuration of screw for Locking Compression Plate (a) 3 screws (b) 5 screws and One Third Tubular Plate (c) 3 screws (d) 5 screws.

Table 1: The contour plots of von Mises stress (VMS) of the (1) fibula and (2) plates for different configuration of screw for Locking Compression Plate (a) 3 screws (b) 5 screws and One Third Tubular Plate (c) 3 screws (d) 5 screws.

Contour plot	Maximum von Mises stress (MPa)				Ultimate strength (MPa)	
	(a)	(b)	(c)	(d)	3 screws	5 screws
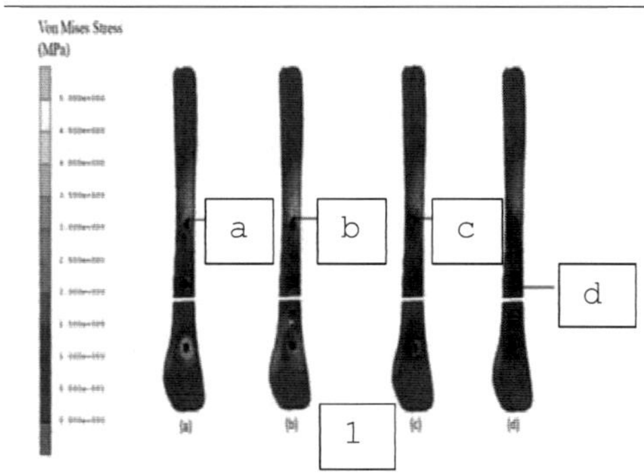	184	70	141	56	193	193
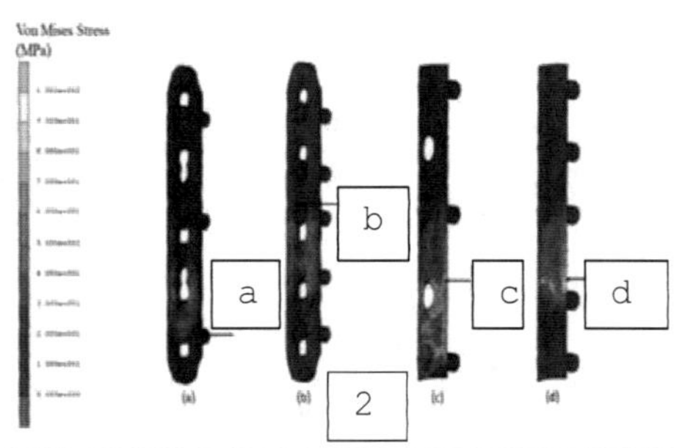	484	667	301	128		

4. Conclusion

Based on the results, it can be noted that the use of 3 screws causing a low VMS at plate compare to 5 screws but the relation is only valid for LCP. However, for OTTP, the 5 screws plate gave a lower VMS than 3 screws construct. Furthermore, the result for VMS at bone is always lowest when using 5 screws construct. The stability in term of displacement, the value is quite close for both 5 screws and 3 screws construct for LCP.

5. Acknowledgement
This work was financially supported by the Universiti Teknologi Malaysia under the Research University Grant (20H20, 4J358 and 20H26) and Ministry of Higher Education Malaysia.

6. References
1. Jacobs, J.J., J.L. Gilbert, and R.M. Urban, *Corrosion of metal orthopaedic implants.* Journal of Bone and Joint Surgery-American Volume, 1998. **80A**(2): p. 268-282.
2. Shibuya, N., M.L. Davis, and D.C. Jupiter, *Epidemiology of foot and ankle fractures in the United States: an analysis of the National Trauma Data Bank (2007 to 2011).* The Journal of Foot and Ankle Surgery, 2014. **53**(5): p. 606-608.
3. Chang, T.-L., et al., *Low profile pelvic fixation: anatomic parameters for sacral alar-iliac fixation versus traditional iliac fixation.* Spine, 2009. **34**(5): p. 436-440.
4. Miller, D.L. and T. Goswami, *A review of locking compression plate biomechanics and their advantages as internal fixators in fracture healing.* Clinical Biomechanics, 2007. **22**(10): p. 1049-1062.
5. Ahmad, M., et al., *Biomechanical testing of the locking compression plate: when does the distance between bone and implant significantly reduce construct stability?* Injury, 2007. **38**(3): p. 358-364.
6. Gervais, B., et al., *Failure analysis of a 316L stainless steel femoral orthopedic implant.* Case Studies in Engineering Failure Analysis, 2016. **5-6**: p. 30-38.
7. Hasenboehler, E., D. Rikli, and R. Babst, *Locking compression plate with minimally invasive plate osteosynthesis in diaphyseal and distal tibial fracture: a retrospective study of 32 patients.* Injury, 2007. **38**(3): p. 365-370.
8. MacLeod, A.R., A.H.R. Simpson, and P. Pankaj, *Reasons why dynamic compression plates are inferior to locking plates in osteoporotic bone: a finite element explanation.* Computer methods in biomechanics and biomedical engineering, 2015. **18**(16): p. 1818-1825.
9. Sommer, C., et al., *Locking compression plate loosening and plate breakage: a report of four cases.* Journal of orthopaedic trauma, 2004. **18**(8): p. 571-577.
10. Varsalona, R. and G. Liu, *Distal tibial metaphyseal fractures: the role of fibular fixation.* Strategies in trauma and limb reconstruction, 2006. **1**(1): p. 42-50.
11. Sommer, C., et al., *First clinical results of the Locking Compression Plate (LCP).* Injury, 2003. **34**: p. B43-54.
12. Ramlee, M.H., et al., *Biomechanical features of six design of the delta external fixator for treating Pilon fracture: a finite element study.* Medical & biological engineering & computing, 2018. **56**(10): p. 1925-1938.

Analysis of genu and splenium of the corpus callosum: comparison between healthy subjects with and without Leukoaraiosis

Siti Nor Fatihah Salbinih[1], Nur Hartini Mohd Taib[1*], Muzaimi Mustapha[2], Ahmad Hadif Zaidin Samsudin[1], Muhammad Nur Salihin Yusoff[3], Ibrahim Lutfi Shuaib[4] and Suzana Mat Isa[4]

[1] Department of Radiology, School of Medical Sciences, Universiti Sains Malaysia, Kubang Kerian, Kelantan, Malaysia.

[2] Department of Neurosciences, School of Medical Sciences, Universiti Sains Malaysia, Kubang Kerian, Kelantan, Malaysia.

[3] Medical Radiation Programme, School of Health Sciences, Universiti Sains Malaysia, Kubang Kerian, Kelantan, Malaysia.

[4] Advanced Medical and Dental Institute, Universiti Sains Malaysia, Kepala Batas, Pulau Pinang, Malaysia.

*nhartini@usm.my

Abstract. The purpose of this study to investigate whether there are any differences in Diffusion Tensor Imaging (DTI) values in subjects that have leukoaraiosis (LA) in the frontal white matter (fWM) and occipital white matter (oWM) compared to subjects that have no LA in those areas in the brain. DTI indices namely axial diffusivity (AD), radial diffusivity (RD), mean diffusivity (MD), fractional anisotropy (FA) were measured via retrospective study. Finding from this study shows that there is no significant difference in AD, RD, MD, and FA values between both groups. It is concluded that there are no significant differences in DTI values of genu and splenium of the corpus callosum between subjects with and without LA. This study shows that subjects with LA does not having degeneration on fiber bundles of the genu and splenium, in which the genu and splenium function in connecting right and left of hemisphere. Also, it might does not affect or not yet affect the genu and splenium fiber bundles. This finding the differences is important to determine whether any changes in AD and RD values.

Keywords- diffusion tensor imaging; white matter; magnetic resonance imaging; leukoaraiosis; fractional anisotropy

1. Introduction

Leukoaraiosis (LA) is defined as diffuse, confluent white matter abnormality which appears low density on computed tomography (CT) scan, and hyperintensity on T2-weighted or FLAIR MRI. LA also have irregular shape and size. The term was derived from the Greek stem "leuko-" which means white and referring to the white matter, while the adjective "araios" means rarefied [1]. LA is very common in elderly and as a risk factor for developing dementia, mobility problems and strokes. It is usually accompanied by the change in corpus callosum morphology [2

Corpus callosum (CC) consists of over 190 million axons, so it is the largest white matter tract. The CC connects the cerebral hemispheres and provides for interhemispheric integration and transfer of information [3]. It plays an important role in brain development, which involves in transferring sensory, motor and cognitive information between hemispheres [4].

Diffusion tensor imaging (DTI) is an imaging technique in MRI that measures three-dimensional (3D) diffusion of water molecules within tissue by applying diffusion gradient in various directions [5]. In this technique, anisotropic diffusion is used to estimate the organization of the axon in the brain. DTI provides quantitative analysis of magnitude and directionality of water molecules. The parameters derived from DTI data includes axial diffusivity (AD), radial diffusivity (RD), mean diffusivity (MD) and fractional anisotropy (FA). Axial is defined as the longest eigen vector, $\lambda_{1,}$ which describes the mean diffusion coefficient of water molecules diffusing parallel to the tract within the voxel of interest. While, radial diffusivity is defined as the average of the two shorter eigen that are (λ_2 and λ_3)/2. It shows the magnitude of water diffusion perpendicular to the tract [6]. Recently, AD and RD have been used to obtain further information on the integrity of white matter (WM). Mean diffusivity describes the average extent of water diffusion, that are $(\lambda_1 + \lambda_2 + \lambda_3)/3$. It is also as a marker of WM integrity. Meanwhile, fractional anisotropy describes the directionality the diffusion tensor [7].

This study is conducted to further analyze the condition of the CC of subjects with and without LA in the frontal white matter (fWM) and occipital white matter (oWM) areas. Analysis was performed specifically to investigate the differences in DTI values between the two groups of subjects.

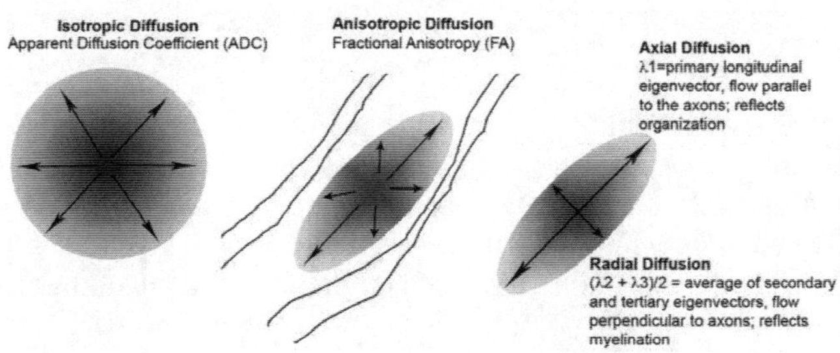

Figure 1. Illustrating Axial Diffusion, Radial Diffusion and Fractional Diffusion [8]

2. Materials and methods

2.1. Data acquisition

This is a retrospective study in which all DTI images were retrieved through USM picture archiving and communication system (PACS) available at Advanced Medical and Dental Institute (AMDI), USM. The data were acquired using 1.5 Tesla MRI System (Signa HDxt, GE Healthcare) available at AMDI. The optimal imaging parameters used were as described elsewhere [9,10]. This study was approved by Human Research Ethics Committee USM (USM/JEPeM/17050265).

2.2. Subjects
Subjects were divided into two groups. The first group consist of those with LA in the fWM and/or oWM areas. This group was labelled as "With LA", whereas the second group consist of those without LA in those areas, but they do have LA in other areas in their brain. This second group was labelled as "Without LA". All subjects did not present neurological and/ or cardiovascular illnesses when the MRI scanning was performed.

2.3. Post-processing
Analysis was carried out using Osirix software version 4.1.2. Diffusion tensor elements and anisotropy at each voxel were computed. MD, FA, and color-coded FA maps were reconstructed.

2.4. Region of interest and quantitative measurements
Region of interest (ROI)s were located on the center, axial slice at the horizontal level of the striatum containing the genu and splenium of the corpus callosum. Single circular ROI with area of 0.30 ± 0.02 cm^2 was drawn manually on the genu and splenium area on the MD maps. All ROIs drawn on the MD maps were copied to the FA maps and pasted on the AD, RD, and FA maps to ensure an accurate and standardized placement of the ROIs on all parametric map. Then, the values displayed on each map were recorded.

2.5. Statistical Analysis
Statistical analysis was performed using Microsoft Excel (version 16.0.11425.20202, Microsoft Corporation). F-Test was performed to compare variances between the two groups. Then, t-test was performed to compare the differences between groups. For all comparisons, $p < 0.05$ was considered statistically significant.

3. Results
3.1 Axial diffusivity
Based on Figure 2., the average AD values recorded for subjects with LA in genu is 1830.95 x 10^{-6} mm^2/s and 1853.35 mm^2/s for those without LA. Whereas the average AD values recorded in splenium for subjects with LA and without LA are 1769.33 x 10^{-6} mm^2/s and 1825.17 x 10^{-6} mm^2/s respectively. The result shows that the AD values in genu for subjects with LA are slightly lower than the subjects without LA. Meanwhile, in splenium, the AD values for subjects with LA is also slightly lower than that of the subjects without LA. Statistical analysis, t-test shows that there is no significant difference in AD for both genu ($p = 0.70$) and splenium ($p = 0.46$) between the groups as descripted in Figure 1.

3.2 Radial diffusivity
The average RD values recorded for subjects with LA in genu are 324.66 x 10^{-6} mm^2/s and 315.39 x 10^{-6} mm^2/s for those without LA. Meanwhile, the RD values for those with LA and without LA in splenium are 272.72 x 10^{-6} mm^2/s and 261.62 x 10^{-6} mm^2/s respectively. As displayed in Figure 3, the graph shows the RD values for those with LA in genu is higher than those who do not have LA in fWM an oWM. Whereas, in splenium, the observed RD values of subjects with LA was found slightly higher than the subjects without LA. Statistical analysis, t-test shows that there is no significant difference in RD for both genu ($p = 0.67$) and splenium ($p = 0.59$) between the groups.

3.3 Mean diffusivity

Average MD values recorded for subjects with LA in genu is 832.89 x 10^{-6} mm^2/s and the subjects without LA is 827.26 x 10^{-6} mm^2/s. Meanwhile, in splenium, the average of MD values for subjects with LA and without LA are 798.08 x 10^{-6} mm^2/s and 779.04 x 10^{-6} mm^2/s respectively. Based on Figure 4, the graph shows the MD values for subjects with LA in genu is slightly higher than subjects without LA. Whereas, in splenium, it shows the MD values for subjects with LA is also higher than subjects without LA. Statistical analysis, t-test shows that there is no significant difference in MD for both genu (p = 0.71) and splenium (p = 0.33) between the groups.

3.4 Fractional anisotropy

The average FA recorded for subjects with LA and without LA in genu are 7984.83 x 10^{-4} and 8057.27 x 10^{-4} respectively. Whereas in splenium, the subjects with LA are 8160.29 x 10^{-4} and 8403.23 x 10^{-4} for those subjects without LA. The graph in Figure 5 shows that the FA values for subjects with LA in genu is a little bit higher than those subjects without LA. In splenium, we observed that the FA values for subjects with LA is lower than those subjects without LA. Statistical analysis, t-test shows that there is no significant difference in MD for both genu (p = 0.67) and splenium (p = 0.34) between the groups.

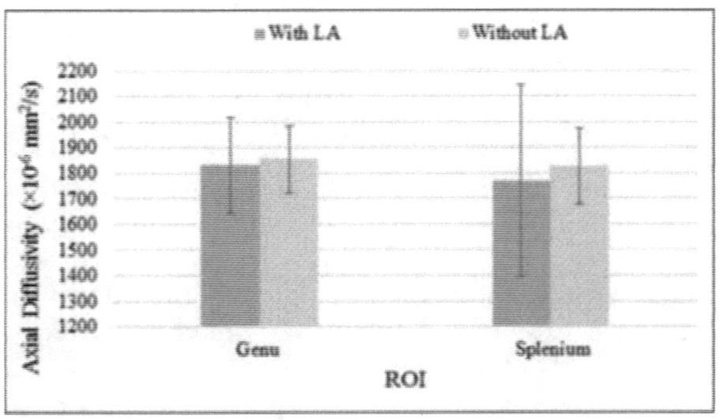

Figure 2. Comparison of AD values for genu and splenium measured in both groups.

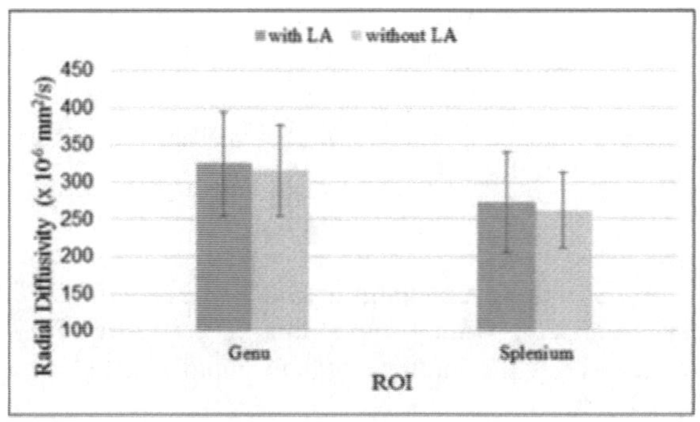

Figure 3. Comparison of RD values for genu and splenium measured in both groups.

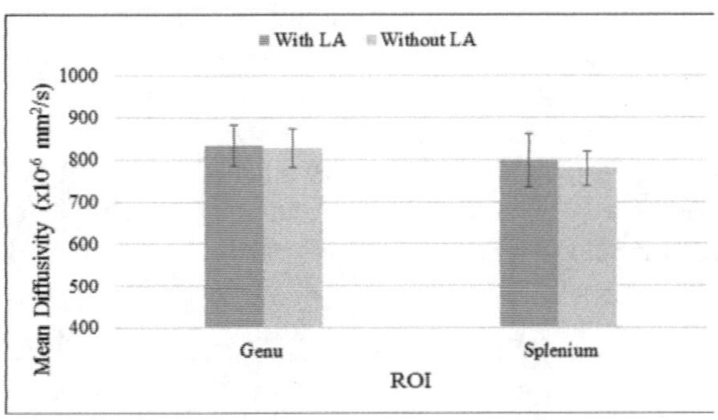

Figure 4. Comparison of MD values for genu and splenium measured in both groups.

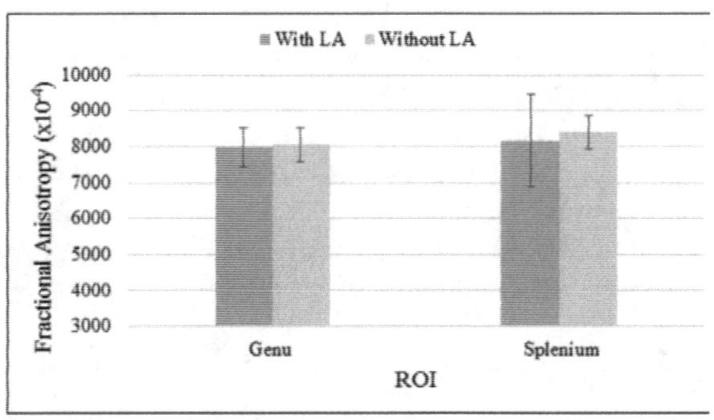

Figure 5. Comparison of FA values for genu and splenium measured in both groups.

4. Discussion

Based on Figure 2 to Figure 5, the greatest difference between subjects with LA and without LA in genu and splenium was seen in fractional anisotropy values, followed by axial diffusivity, mean diffusivity and radial diffusivity value. The major findings in this study is that there are no significant differences in DTI values between those with LA and without LA for both genu and splenium in which these results are not as expected. From this investigation, we expected this study can confirm that eigenvalues of DTI parameters do not indicate the same microstructural characteristics in both groups since the orientation of eigen vector is different. However, this study results opposed our expectation.

4.1 Axial Diffusivity (AD)
AD measures the diffusion of water parallel to axons. In this study, we found decreases AD in both ROIs of the corpus callosum for the group of subjects that are with LA. However, the difference was not significant. The previous study performed by Takahashi et al. showed that AD increases when the later stages of white matter development of rats, the axonal orientation becomes less asymmetry [11]. This means the AD also portrays the axonal degradation. Therefore, axonal orientation is possible decreases since the axons degraded.

4.2 Radial Diffusivity (RD)
Radial diffusivity appears to be closely with myelination. In normal aging, the axons degenerated causes some myelin sheaths breakdown [12]. This would reflect the myelin degeneration is correlated with increased RD. In our analysis, increased RD were observed in group of subjects with LA for genu and splenium, though it is not significant.

4.3 Mean Diffusivity (MD)
We also found an insignificant difference in MD values between both groups although the values of subjects with LA is higher than those without LA. Based on Liu et. al. (2016), the higher MD values reflect the white matter integrity impairment [13].

4.4 Fractional Anisotropy (FA)
This study shows that decreased FA for subjects with LA in both genu and splenium caused by myelin changes. Due to myelin changes which is myelin degradation, the white matter hyperintensities increases. In previous studies, the lower FA because of myelin degradation or axonal degradation [14]. Based on the previous findings also showed decreases FA due to microstructural changes such as myelination loss, microtubules loss and the deep white matter hyperintensities (WMH) seen on T2-weighted images [15]. These microstructural changes are also known as neurodegenerative process. However, we were suggesting that there may be no apparent changes in the microstructural of subjects with LA and without LA in genu. The causes of this difference in FA for those with LA and without LA is not entirely clear. But the possible causes may be the microstructural phenomenon is same as FA also measures intravoxel coherence as stated by Pfefferbaum.

5. Conclusion
In the current study, it was found that there are no significant differences in all DTI values between subjects with LA and those without LA in both genu and splenium. The value of AD and RD indicate the changes in structure of the axons and myelin sheath respectively. In this case, changes in AD value shows that axonal dyemelination, whereas RD shows that myelin sheath degeneration. This study is important to examine whether any changes in AD and RD values.

Acknowledgments

This study was funded by Universiti Sains Malaysia under USM Short Term Grant (304/PPSP/6315095).

References

[1] O'Sullivan M 2008 Leukoaraiosis practical neurology **8** 26–38

[2] Feng L, Jiang H, Li Y, Teng F and He Y 2017 Effects of citicoline therapy on the network connectivity of the corpus callosum in patients with leukoaraiosis *Medicine* **96**(4)

[3] Fabri M and Polonara G 2013 Functional topography of human corpus callosum: An fMRI mapping study *Neural Plasticity* **2013**

[4] Provenzale J M, Isaacson J, Stinnet S and Chen S 2012 Analysis of corpus callosum diffusion tensor imaging parameters in infants *The Neuroradiology Journal* **25** 342–50

[5] Basser P J and Pierpaoli C 1996 Microstructural and physiological features of tissues elucidated by quantitative-diffusion-tensor MRI. *J Magn Reason* **111** 209–19

[6] Winklewski P J, Sabisz A, Naumczyk P, Jodzio K, Szurowska E and Szarmach A 201. Understanding the physiopathology behind axial and radial diffusivity changes—what do we know? *Frontiers in Neurology* **9** 92

[7] Zeestraten E A, Benjamin P, Lambert C, Lawrence A J, Williams O A, Morris R G, Barrick T R and Markus H S 2016 Application of diffusion tensor imaging parameters to detect change in longitudinal studies in cerebral small vessel disease *PLoS One* **119**

[8] Karlsgodt K H, Rosser T, Lutkenhoff E, Cannon T, Silva A and Bearden C 2012 alterations in white matter microstructure in neurofibromatosis *PLoS One* **7**

[9] Mohd Taib N H, Wan Abdullah W A K, Shuaib I L, Magosso E and Mat Isa S 2017. Determination of optimum combination of voxel size and b-value for brain diffusion tensor imaging *Sains Malaysiana* **46** 67–74

[10] Mohd Taib N H, Wan Abdullah W A K, Shuaib I L, Magosso E and Mat Isa S 2015 Diffusion tensor imaging of leukoaraiosis, normal appearing brain tissue, and normal brain tissue. *Malaysian Journal of Medicine and Health Sciences* **11** 1–10

[11] Takahashi M, Ono J, Harada K, Maeda M and Hackney D B 2000 Diffusional anisotropy in cranial nerves with maturation: Quantitative evaluation with diffusion MR imaging in rats *Radiology* **216** 881–85

[12] Inano S, Takao H, Hayashi N, Abe O and Ohtomo K 2011 Effects of age and gender on white matter integrity *American Journal of Neuroradiology* **32** 2103–09

[13] Liu H, Yang Y, Xia Y, Zhu W, LeaR Kk, Wei Z, Wang J and Hu X 2016 Aging of cerebral white matter *Ageing research reviews* **34** 64–76

[14] Alicata D, Chang L, Cloak C, Abe K and Ernst T 2009 Higher diffusion in striatum and lower fractional anisotropy in white matter of methamphetamine users *Psychiatry research* **174**(1) 1–8

[15] Pfefferbaum A, Sullivan E V, Hedehus M, Lim K O, Adalsteinsson E and Moseley M 2000 Age-related decline in brain white matter anisotropy measured with spatially corrected echo-planar diffusion tensor imaging *Magn. Reson. Med.* **44** 259–68

The Effect of Material Characteristic of Shoe Soles during Long Duration Running

Chan Siow Cheng*, J P Tan and Tan Yin Qing

Department of Mechatronics and Biomedical Engineering Lee Kong Chian Faculty of Engineering and Science Universiti Tunku Abdul Rahman Selangor, Malaysia

*chansc@utar.edu.my

Abstract. Long duration running has been associated with musculoskeletal injuries such as patellofemoral pain syndrome and knee osteoarthritis. This study was designed to investigate the effects of two different shoe insoles on kinematic and kinetic performance during long duration running. In total 15 subjects (7 males and 8 females) were recruited to perform 15 minutes continuous running on the treadmill in 3 different shoe insole conditions: (i) no insole; (ii) Ethyl Vinyl Acetate (EVA) insole and (iii) Polyurethane (PU) insole. The result revealed a significant increase in maximum vertical ground reaction force (VGRF) with the PU insole compare with the no insole and EVA insole conditions ($p<0.05$). There is no difference between conditions in kinematic variables: step length, stride length and cadence ($p>0.05$).

1. Introduction

Recreational running is one of the common exercises that can improve our cardiorespiratory function and general well-being [1]. The popularity of running as a form of exercise and recreation in maintaining healthy life-style has grown rapidly due to its low-cost, equipment-free and time independent. The negative impact of running is the increasing incidence of running-related injuries such as patellofemoral pain syndrome, Achilles tendinopathy and medial tibial stress syndrome [2-4] especially those runners just committed into training [5]. One of the strategies to reduce the impact loading on the human locomotor system is insertion of shoe insole in running shoe to provide the shock protection and cushioning comfort [6-7]. They are many commercially shoe soles with different characteristic e.g. softness, flexible, durability, high in tensile strength etc. However, whether the available shoe soles in the market can better protect the runners from injuries is still unknown especially in long duration running.

O'Leary et al [8] has revealed that a cushioned insole (Sorbo Air, Sorbothane, Inc, Kent, Ohio) has significant reduction in mean VGRF peak impact and the loading rate. Another study conducted by Dixon et al [9] showed that running in military boots with new and degraded insoles reduce the peak loading rate as well as the peak ankle dorsiflexion. For the flat insoles, medium EVA was found to have higher peak pressures whereas low and medium density PU were effective in increasing contact area and reducing pressure time integral [10]. Wang et al [11] showed that the foam category and hardness/density affected the biomechanical performance of running shoes. There were some studies determined that the cushioned insole could minimize the risk of encountering injuries [12-14]. On the other hand, there were other research groups showed that it did not provide any protective effect [15-16]. The purpose of this study is to investigate the biomechanical effects during long duration with different shoe insole. It would enhance the understanding of running injury mechanism on lower extremities as well as provide a recommendation on the strategy to select shoe insole for competitive and recreational runners.

2. Methodology

2.1 Subject Recruitment

The target population of this study is 15 young and healthy students (7 males and 8 females) who are free from any injuries in lower extremity and health problem such as varus, valgus, asthma and cardio diseases. The written informed consent form was signed by the participated subject before participation to this study. The physical anthropometry data is shown in Table 1.

Table 1. Anthropometry data.

Variable	Mean ± Standard Deviation
Age	21.67±1.05
Weight (kg)	56.75±8.54
Height (cm)	166.13±8.44

2.2 Shoe Characteristics

Two types of commercially shoe insoles: (i) EVA and (ii) PU were tested in this study (Figure 1). Table 2 summaries the maximum tensile stress for these two types of insole materials.

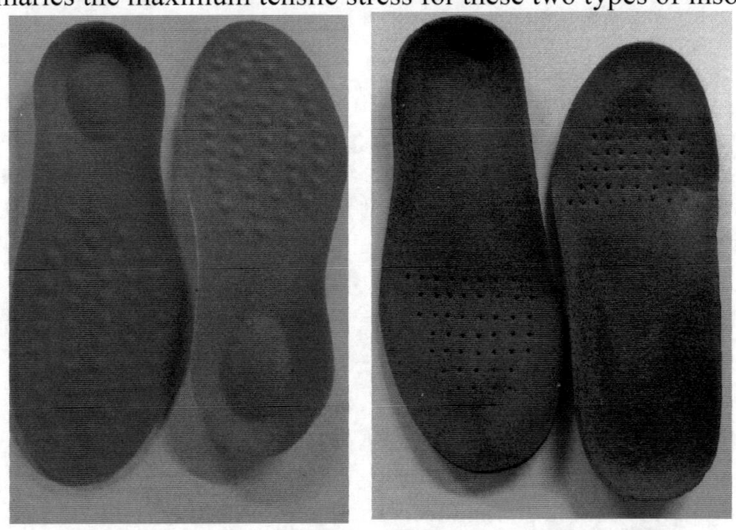

Figure 1. PU insole (left) and EVA insole (right).

Table 2. Tensile test for two types of insoles.

Insole's Material	Maximum Tensile Stress (MPa)	Extension at Maximum Tensile stress (mm)
Polyurethane (PU)	> 0.66002	> 25.69334
Ethyl Vinyl Acetate (EVA)	> 0.01816	> 27.44015

Figure 2 is the plots of tensile stress as a function of applied strain. The slopes of the linear lines denote the relaxation moduli of each insole respectively. PU insole has lower relaxation modulus than EVA insole.

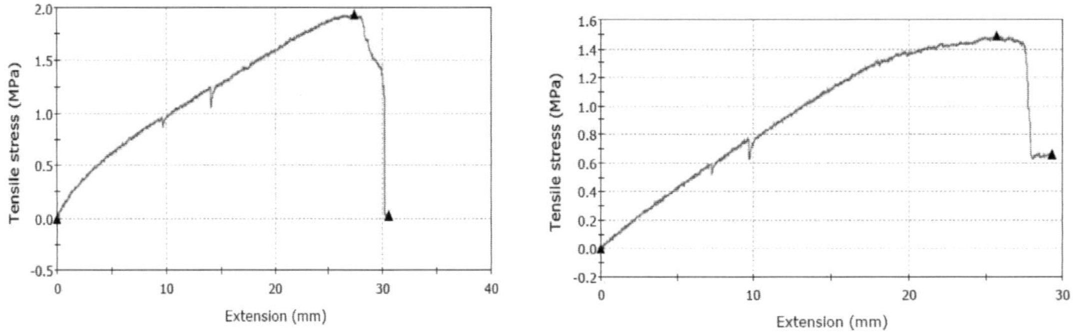

Figure 2. Tensile stress as a function of extension for: EVA insole (left) and PU insole (right).

2.3 Experimental setup and data collection

The physical anthropometric data such as body height, weight and length of lower limb were measured. LED markers were placed at the joints of lower extremity to calculate the angular displacement on sagittal plane. Video was recorded throughout the whole experiment. The subjects were required to run with their preferable speed for 15 minutes on the H/P Cosmos Instrumented Treadmill embedded with force plate (Model: TLA10004681). Subjects were given some practice time to achieve their normal running gait before the data collection. Three trials with three different conditions were performed in three separate days.

2.4 Data analysis

The kinematic and kinetic parameters were normalized based on the anthropometry data of each subject to reduce the bias effect as shown in Table 3 [17].

Table 3. Normalization kinetic and kinematic data.

Parameters	Normalized equation
VGRF, F	$\hat{F} = \frac{F}{mo\ g}$
Step Length/ Stride Length, l	$\hat{l} = \frac{l}{lo}$

Cadence, c	$\hat{c} = \dfrac{c}{60\sqrt{g/lo}}$
Moment, M_o	$M = \dfrac{M_o}{m_o \times g \times l_o}$

Given *mo* is Body Mass (kg), *g* is gravity acceleration (9.81m/s^2), *lo* is leg length and M_o is moment (N.m).

2.5 Mathematical Modeling

Figure 3 shows the running gait cycle for three conditions. There is only single peak VGRF in all three running conditions. The corresponding free body diagram used to calculate knee joint moment is shown in Figure 4.

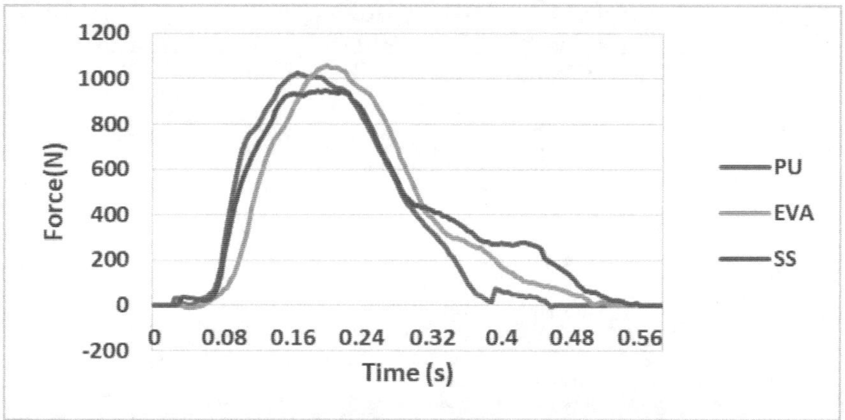

Figure 3. Comparison of VGRF for three different conditions: PU (Polyurethane insole), EVA (Ethyl Vinyl Acetate insole) and SS (Sport Shoe).

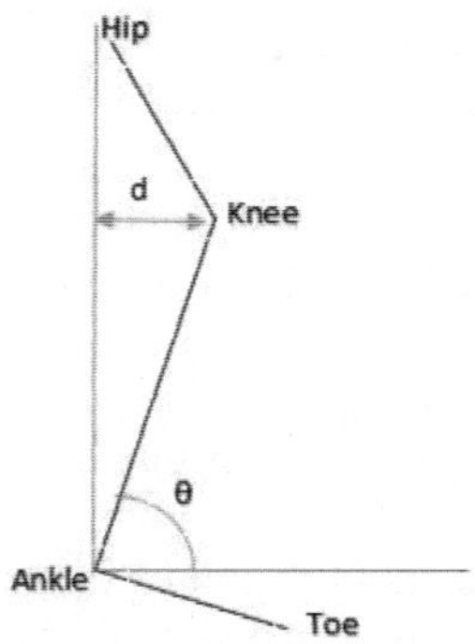

Figure 4. Example of free body diagram.

The angle between shank and horizontal axis was measured using Kinovea whereas the distance between the knee and vertical normal line, d can be calculated via

$$\mathrm{d} = L_{KA}\cos\theta$$

where L_{KA} is the length from knee to joint and θ is the angle between the shank and horizontal normal line. The knee joint moment of the knee joint is denoted as

$$M = F.d$$

where F refer to the maximum ground reaction force.

2.6 Statistical test

All the kinematic and kinetic data were summarized using descriptive statistics of mean and standard deviation. The statistical test, Mann Whitney test was employed with the use of Statistical Package for the Social Sciences (SPSS) for between condition differences in biomechanical parameters comparison.

3. Results and discussion

Both kinematic and kinetic parameters for running gait were summarized in Table 4. The statistical results for running data were shown in Table 5.

Table 4. Kinetic and kinematic data for running gait (mean ± standard deviation).

	Maximum VGRF	Knee Joint Moment	Push Off Peak Force	Step Length	Cadence
SS	1.8598±0.1795	.4006 ± 0.0422	1.5109 ± 0.2824	.6275 ± 0.0779	1.5172 ± 0.1936
PU	1.9176±0.2202	.4043 ±0.0530	1.6157 ± 0.3216	.6511 ± 0.0944	1.5428 ± 0.2889
EVA	1.8930±0.2980	.3988 ± 0.0850	1.6977 ± 0.2984	.6586 ±0.0959	1.5253 ± 0.2677

Table 5. Statistical results for running gait.

	Maximum VGRF	Knee Joint Moment	Push Off Peak Force	Step Length	Cadence
SS vs PU	.00001*	.46812	.19215	.20327	.26763
SS vs EVA	.49202	.25463	.04093*	.18141	.38591
PU vs EVA	.00001*	.38591	.18141	.41683	.37070

From Table 4, it revealed that both insoles did not reduce the maximum VGRF. It showed significant increase in VGRF when comparing PU insole with sport shoe (no extra insole). This finding is being in line with the result from Rauno et al. (2009) which suggest that insole do not attenuate shock or reduce loading rate.

There were no significant differences among insoles for knee joint moment which in line with the previous study [19]. PU insole showed no significant effect on the motion of hip and knee joint on sagittal plane.

Additionally, both insoles have higher push off peak forces. EVA insole was significantly higher than no insole condition. Similar findings have been reported on textured insoles during accelerated running [20].

Some of the earlier studies showed that different insoles would significantly affect the step length due to the increment of knee extension during the heel strike phases [20-21]. However, based on our results, the mean values of step length and cadence did not change significantly for these two types of different insoles. The insoles did not affect kinematic performance of lower extremities. The contrary finding of this study may be due to the insoles used in previous studies were much different from both EVA and PU insoles used in this study.

4. Conclusion

Insoles are often marketed to reduce impact loading force and prevent injury. However, our study did not support this assumption. The results showed that both EVA and PU insoles used in this study did not reduce the peak of VGRF as well as the knee joint moment compared to the conventional insoles furnished in running shoes. The use of these two types of insoles does not seem to protect the lower extremity from injury.

Acknowledgments

This work was supported by a grant from UTAR Research Fund Project No. IPSR/RMC/UTARRF/2018-C2/C01.

References

[1] Bartholomew J B, Morrison D and Ciccolo J T 2005 Effects of acute exercise on mood and well-being in patients with major depressive disorder *Medicine & Science in Sports & Exercise* **37(12)** 2032-2037

[2] Williams III D, McClay I and Hamill J 2001 Arch structure and injury patterns in runners *Clinical Biomechanics* **16** 341-347

[3] Lopes A D, Hespanhol L C, Yeung S S and Costa L O P 2012 What are the main running-related musculoskeletal injuries? *Sports medicine* **42** 891-905

[4] Van Gent R, Siem D, Van Middelkoop M, Van Os, Bierma-Zeinstra S, Koes B and Taunton J 2007 Incidence and determinants of lower extremity running injuries in long distance runners: a systematic review *British Journal of Sports Medicine* **41** 469-480

[5] Macera C A, Pate R R, Powell K E, Jackson K L, Kendrick J S and Craven T E 1989 Predicting lower-extremity injuries among habitual runners *Archives of internal medicine* **149(11)** 2565-2568

[6] Jones B, Thacker S and Gilchrist J 2002 Prevention of lower extremity stress fractures in athletes and soldiers: a systematic review *Epidemiologic Reviews* **24** 228-247

[7] Hinz P, Henningsen A, Matthes G, Jäger B, Ekkernkamp A and Rosenbaum D 2008 Analysis of pressure distribution below the metatarsals with different insoles in combat boots of the German Army for prevention of march fractures *Gait & Posture* **27** 535-538

[8] O'Leary K, Vorpahl K A, and Heiderscheit B 2008 Effect of cushioned insoles on impact forces during running *Journal of the American Podiatric Medical Association* **98** 36-41

[9] Dixon S J, Waterworth C, Smith C V and House C M 2003 Biomechanical analysis of running in military boots with new and degraded insoles *Medicine and science in sports and exercise* **35** 472-479

[10] Healy A, Dunning D N and Chockalingam N 2012 Effect of insole material on lower limb kinematics and plantar pressures during treadmill walking *Prosthetics and orthotics international* **36** 53-62

[11] Wang L, Hong Y and Li J 2012 Durability of running shoes with ethylene vinyl acetate or polyurethane midsoles *Journal of Sports Sciences* **30** 1787-1792.

[12] Schwellnus M, Jordaan G and Noakes T 1990 Prevention of common overuse injuries by the use of shock absorbing insoles: a prospective study *The American Journal of Sports Medicine* **18** 636-641

[13] Mundermann A, Stefanyshyn D and Nigg B 2001 Relationship between footwear comfort of shoe inserts and anthropometric and sensory factors *Medicine and Science in Sports and Exercise* **33** 1939-1945

[14] Milgrom C, Finestone A, Shlamkovitch N, Wosk J, Laor A, Voloshin A and Eldad A 1992 Prevention of overuse injuries of the foot by improved shoe shock attenuation *Clinical Orthopaedics and Related Research* **281** 189-192

[15] Gardner L, Dziados J, Jones B, Brundage J, Harris J, Sullivan R and Gill P 1988 Prevention of lower extremity stress fractures: a controlled trial of a shock absorbent insole *American Journal of Public Health* **78** 1563-1567

[16] Withnall R, Eastaugh J and Freemantle N 2006 Do shock absorbing insoles in recruits

undertaking high levels of physical activity reduce lower limb injury? A randomized controlled trial *Journal of the Royal Society of Medicine* **99** 32-37

[17] Hof A L 1996 Scaling gait data to body size *Gait & posture* **3** 222-223

[18] Ruano C, Powell D, Chalambaga E T and Renshaw D 2009 The effects of tempur insoles on ground reaction forces and loading rates in running *International journal of exercise science* **2** 186

[19] Özmanevra R, Angin S, Günal İ H and Elvan A 2018 Effect of Different Insole Materials on Kinetic and Kinematic Variables of the Walking Gait in Healthy People *Journal of the American Podiatric Medical Association* **108** 390-396

[20] Kelleher K J, Spence W D, Solomonidis S and Apatsidis D 2010 The effect of textured insoles on gait patterns of people with multiple sclerosis *Gait & Posture* **32** 67-71

[21] Hatton A L, Dixon J, Rome K, Brauer S G, Williams K and Kerr G 2016 The effects of prolonged wear of textured shoe insoles on gait, foot sensation and proprioception in people with multiple sclerosis: study protocol for a randomised controlled trial *Trials* **17** 208

Prediction of Parkinson's disease using Ensemble Machine Learning classification from acoustic analysis

Amit Kumar Patra[1], Ratula Ray[2], Azian Azamimi Abdullah[3], Satya Ranjan Dash[4]*

[1]Tata Consultancy Services Limited, USA
[2]School of Biotechnology, KIIT University, Bubaneswar, India
[3]School of Mechatronic Engineering, Universiti Malaysia Perlis, Malaysia
[4]School of Computer Applications, KIIT University, Bubaneswar, India

*sdashfca@kiit.ac.in

Abstract. Parkinson's disease (PD) is a neurodegenerative disorder, which upon progression affects the movements. Tremors associated with Parkinson's disease are the major symptoms to look out for in such cases. This generally results in breakdown of the neurons producing dopamine. Qualitative speech starts to decline as the disease progresses and the variability in the vocal cord vibration (also known as fundamental frequency) starts to occur. People with PD have shown to produce greater variability in frequency as compared to normal people. In our paper, we are focused on comparison of the voice measurement features of patient dataset to understand whether a patient is suffering from PD or not using Machine Learning classifiers. We have implemented Decision Trees, Logistic Regression and K-nearest neighbors as base classifiers and have compared their performance with Ensemble learning classifiers Bagging, Random Forest and Boosting. We have compared the accuracy (%) of the classifiers and discussed which one of them is more accurate at predicting the outcome of the disease. We also found out the most relevant features associated with the classification and ranked them based on feature importance. Our main aim here is the classification of healthy individuals from people suffering from PD by detection of dysphonia (difficulty in speaking due to declining health conditions).

1. Introduction

With increase in the aging population, the people affected with neurological disorders are also increasing rapidly and thus, the need for quick and effective means for monitoring them becomes more and more important. It has been noticed that in almost 90% of the cases, the people suffering from PD experience impairment of the vocal cords, which result in dysphonia. At much more severe stage, articulatory deficit can also manifest along with vocal impairment [1].

Understanding the measurements related to voice detection thus can play an important role in predicting the progression of the disease. According to the traditional sound measurement standards, jitter, shimmer, pitch, noise to harmonics ratio and pressure level of absolute sound have been used effectively. Inspired by nonlinear dynamical systems theory, very recently, few other measures have also been included in the system which can be used for the purpose of sound measurement. These tools used for voice disorders detection include Recurrence Period Density Entropy(RPDE) and Detrended Fluctuation Analysis(DFA)[2].

In our paper, we are aiming to apply the Machine learning classifiers in order to predict the status of health of the patients (that is, if the person is suffering from PD or not) fusing acoustic analysis . We have taken under consideration the various traditional as well as newer non-linear tools for voice measurements as the attributes for classification of the health status of the individuals. So along with classification, we will be also looking into the most important features that are responsible for making the decision and check if the newer attributes are better than the traditional ones.

2. Literature review

Previous studies have shown the implementation of Neural networks in assessment of the risk factors associated with Parkinson's disease. Factors such as genes, age, stroke and diabetes were found out to have higher correlation with the outcome of the prediction [3].

Feature relevance analysis has also been carried out previously using Parkinson's voice dataset. Tensor flow deep learning library has also been used to detect the increase in disease severity in case of Parkinson's disease [4].

Subtype identification of PD has also been done before using both supervised and unsupervised machine learning methods. The progression rate of the disease can be predicted using these learning systems and can be classified as: highly progressive, moderately progressive and mild progressive state of Parkinson's Disease. This has been utilized effectively in order to improve the quality of the counseling of patients, improve clinical trial design and using resources available in a better way [5].

Work has also been carried out on MRI dataset of the brain for identification of medical image related biomarkers. Supervised Machine Learning classifiers such as Support Vector Machine has been utilized in order to understand the progression of the disease and role of the biomarkers in identifying the progression behavior [6].

Gait classification in case of patients suffering from PD has also been studied using Artificial Neural Networking (ANN) and Support Vector Machine classifiers. Kinematic, kinetic and spatiotemporal gait are the 3 basic parameters, which are to be taken under consideration before performing the classification task [7].

3. Methods

3.1. Dataset

We have used the Parkinson's dataset of UCI Repository which comprises of measurements of voices of 31 people, out of which 23 of them suffer from PD and the rest are healthy individuals[8]. The non-linear measurements are also added to the dataset, which are used for classification purpose along with the traditional measurements[9].

3.2. Dataset Attributes

The different acoustic signal measurements chosen as attributes for the classification of the presence and absence of Parkinson's disease are as follows:

- MDVP:Fo(Hz) - Average vocal fundamental frequency
- MDVP:Fhi(Hz) - Maximum vocal fundamental frequency
- MDVP:Flo(Hz) - Minimum vocal fundamental frequency
- MDVP:Jitter(%),MDVP:Jitter(Abs),MDVP:RAP,MDVP:PPQ,Jitter:DDP - Several measures of variation in fundamental frequency

- MDVP:Shimmer,MDVP:Shimmer(dB),Shimmer:APQ3,Shimmer:APQ5,MDVP:APQ,Shimmer:DDA - Several measures of variation in amplitude
- NHR,HNR - Two measures of ratio of noise to tonal components in the voice
- status - Health status of the subject (one) - Parkinson's, (zero) - healthy
- RPDE,D2 - Two nonlinear dynamical complexity measures
- DFA - Signal fractal scaling exponent
- spread1,spread2,PPE - Three nonlinear measures of fundamental frequency variation

3.3. Classifiers

3.3.1. Base Classifiers

- Decision Tree: In case of Decision Tree classifier, the input is split into sub-spaces based upon certain functions. It helps in reaching a conclusion based upon conditional control statements[11].
- Logistic Regression: This classification occurs on the basis of a sigmoid function, known as the logistic function which takes in a real input and gives out a value in between 0 and 1[12].
- k-Nearest Neighbors: It is a "k" biased classification technique where majority voting among the neighborhood data points are taken under consideration for classification, which may or may not rely upon the distance based weighting parameters[13]. We have considered the value of n=5 in our classification problem.

3.3.2. Ensemble Classifiers

- Random Forest Classifier: This is an ensemble learning classifier which randomly generates decision trees and averages the result, thereby reducing over-fitting of the model[14]. It selects only a subset of features randomly and the best feature is used to split at the node.
- Bagging Classifier: This classification technique is almost like Random Forest but differs from it in the sense that it does not consider any subset of features for splitting. Rather, it takes into consideration all the features available for splitting at the node[15].
- AdaBoost Classifier: Also known as Adaptive Boosting classification technique, this ensemble learner works upon fitting the weak classifiers and improve the estimate upon each iteration[16].

3.4. Validation procedure

In the paper, we have used 10- fold cross validation over the entire dataset. In 10-fold cross validation system, the entire dataset is divided into 10 parts with equal number of data in each part. We have to then consider (10-1) to be the training model and the rest is the testing model. This is repeated for 10 times with (10-1) as training set and the rest as validation(testing) set [10].

3.5. The metrics used for this purpose are as follows:

$$\text{Recall}=\frac{\boldsymbol{True\ Positive}}{\boldsymbol{True\ Positive+False\ Negative}}$$

$$\text{Precision}=\frac{\textit{True Positive}}{\textit{True Positive}+\textit{False Positive}}$$

$$\text{F1-score}=\frac{2*(\textit{Recall}*\textit{Precision})}{(\textit{Recall}+\textit{Precision})}$$

$$\text{Accuracy}=\frac{\textit{True Positive}}{\textit{True Positive}+\textit{False Positive}}$$

4. Results and Discussion

We have used three base classifiers whose performance has been compared in Table 1, based on the statistical metrics being used.

Table 1. Performance analysis of the Base classifiers

Base Classifiers	Precision	Recall	F1-score	Accuracy%
Decision Tree	83%	82%	82%	79.95%
Logistic Regression	81%	82%	81%	82.03%
K-nearest neighbors(n=5)	75%	76%	75%	76.31%

The result obtained suggests that the performance has increased substantially for the ensemble classifiers. For the base classification system, Logistic Regression showed good results in terms of accuracy, though Decision Tree shows better stability as a classifier in terms of f1 score.
Ensemble classifiers are the set of classifiers which takes under consideration the performance of individual base classification systems and give the final result. We have used both averaging method (Bagging and Forests of randomized trees) and sequential method (Boosting) here to build upon the result of the base decision tree classifier. Table 2 shows our result for the same.

Table 2. Performance analysis of the Ensemble classifiers

Ensemble Classifiers	Precision	Recall	F1-score	Accuracy%
Random Forest	83%	82%	82%	84.59%
Bagging	80%	81%	80%	81.11%
Boosting(Adaboost classifier)	83%	83%	83%	82.51%

Inappropriate management of PD is very common and can be often misdiagnosed as depression. Acoustic analysis provides a more distinct approach for differentiating between the patients suffering from both of these diseases by considering the temporal measure of the speech. Understanding the most relevant features associated with the status of the patients in case of Parkinson's Disease is also very important and has been depicted in Fig. 1. We have selected the best performing classifier (in this case, Random Forest), and have evaluated the most important attributes associated with the prediction of PD in a set of individuals. For a forest of decision trees, the measure of how much each feature decreases the weighted impurity in a tree is important to note. The mean of the impurity decrease can be used to rank the features accordingly.

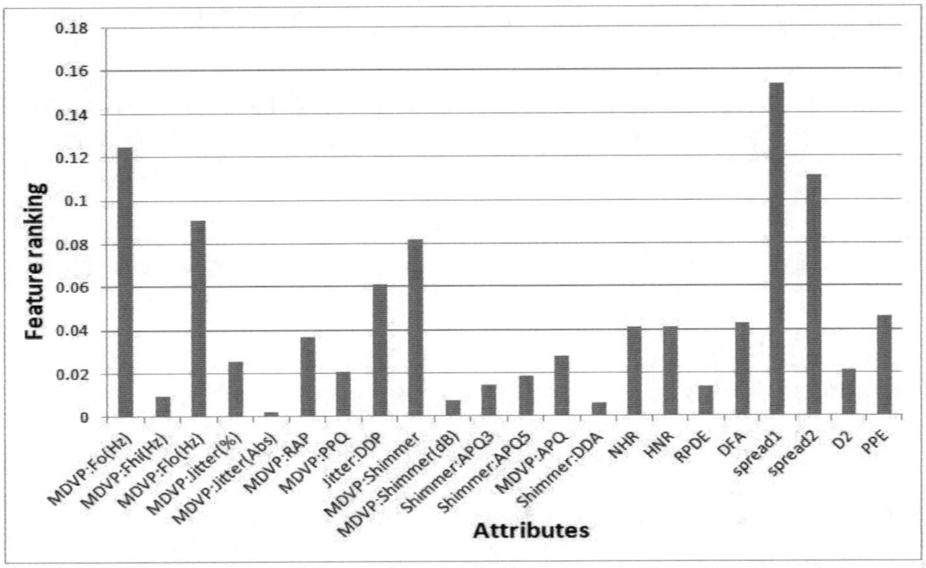

Figure 1. Feature ranking based on the importance of the features associated with the classification

5. Conclusion

Our observations have shown that Ensemble Classifiers show much better accuracy as compared to the base classifiers in most of the cases. Out of them all, Random Forest Classifier gives the best result with an accuracy of 84.59%, though if we consider the f1-scoring system as the standard metrics for classification, Boosting shows a slightly better result (83%). Both of these ensemble methods can thus be used to classify the dataset effectively. Among the base classifiers, Logistic Regression performs quite well with an accuracy of 82%.

Our observations also highlight the most important features that are to be considered for prediction of dysphonia in patients suffering from PD. Non-linear measure of fundamental frequency variations (spread 1 and spread 2) have shown to have much higher importance and is thus thought to be most related in the prediction outcome. Among the traditional methods used, Average vocal fundamental frequency and Minimum vocal fundamental frequency have showing higher ranking, thus showing more correlation to the prediction outcome of the classification.

The main highlight of this paper is to showcase the power of Ensemble classifiers over base classification systems and how combining the power of multiple heterogeneous classifiers can give a more reliable accuracy, free from bias. Using f1 scoring system as the standard metric for assessing the performance of the classifier gives us an opportunity for taking into account both Recall and Precision. Unlike accuracy, it gives a more balanced result by considering the False Negative while computing the final score.

References

[1] Ho, A. K., Iansek, R., Marigliani, C., Bradshaw, J. L., & Gates, S. (1999). Speech impairment in a large sample of patients with Parkinson's disease. *Behavioural neurology*, *11*(3), 131-137.

[2] Little, M. A., McSharry, P. E., Roberts, S. J., Costello, D. A., & Moroz, I. M. (2007). Exploiting nonlinear recurrence and fractal scaling properties for voice disorder detection. *Biomedical engineering online*, *6*(1), 23.

[3] Joshi, S., Shenoy, D., Rrashmi, P. L., Venugopal, K. R., & Patnaik, L. M. (2010, February). Classification of Alzheimer's disease and Parkinson's disease by using machine learning and neural network methods. In *2010 Second International Conference on Machine Learning and Computing* (pp. 218-222). IEEE.

[4] Grover, S., Bhartia, S., Yadav, A., & Seeja, K. R. (2018). Predicting Severity of Parkinson's Disease Using Deep Learning. *Procedia computer science*, *132*, 1788-1794.
[5] Faghri, F., Hashemi, S. H., Leonard, H., Scholz, S. W., Campbell, R. H., Nalls, M. A., & Singleton, A. B. (2018). Predicting onset, progression, and clinical subtypes of Parkinson disease using machine learning. *bioRxiv*, 338913.

[6] Salvatore, C., Cerasa, A., Castiglioni, I., Gallivanone, F., Augimeri, A., Lopez, M., ... & Quattrone, A. (2014). Machine learning on brain MRI data for differential diagnosis of Parkinson's disease and Progressive Supranuclear Palsy. *Journal of Neuroscience Methods*, *222*, 230-237

[7] Tahir, N. M., & Manap, H. H. (2012). Parkinson Disease Gait Classification based on Machine Learning
Approach. *Journal of Applied Sciences*, *12*(2), 180-185.

[8] Little, M. A., McSharry, P. E., Hunter, E. J., Spielman, J., & Ramig, L. O. (2009). Suitability of dysphonia measurements for telemonitoring of Parkinson's disease. *IEEE transactions on bio-medical engineering*, *56*(4), 1015.

[9] Little, M. A., McSharry, P. E., Roberts, S. J., Costello, D. A., & Moroz, I. M. (2007). Exploiting nonlinear recurrence and fractal scaling properties for voice disorder detection. *Biomedical engineering online*, *6*(1), 23.

[10] Sharma, R. C., Hara, K., & Hirayama, H. (2017). A machine learning and cross-validation approach for the discrimination of vegetation physiognomic types using satellite based multispectral and multitemporal data. *Scientifica*, *2017*.

[11] Mashat, A. F., Fouad, M. M., Philip, S. Y., & Gharib, T. F. (2012). A decision tree classification model for university admission system. *Editorial Preface*, *3*(10).

[12] Stylianou, N., Akbarov, A., Kontopantelis, E., Buchan, I., & Dunn, K. W. (2015). Mortality risk prediction in burn injury: Comparison of logistic regression with machine learning approaches. *Burns*, *41*(5), 925-934.

[13] Guo, G., Wang, H., Bell, D., Bi, Y., & Greer, K. (2003, November). KNN model-based approach in classification. In *OTM Confederated International Conferences" On the Move to Meaningful Internet Systems"* (pp. 986-996). Springer, Berlin, Heidelberg.

[14] Kumari, G. P. (2012). A Study Of Bagging And Boosting Approaches To Develop Meta-Classifier.

[15] Freund, Y., Schapire, R., & Abe, N. (1999). A short introduction to boosting. *Journal-Japanese Society For Artificial Intelligence*, *14*(771-780), 1612.

Determination of Fracture Risk on Patient-specific Model of Femur with Osteogenesis Imperfecta

Siti Fatima Zahra' Ahmad[1], Mohd Hanafi Mat Som[1*], Khairul Salleh Basaruddin[1], Ruslizam Daud[1] and Mohd Shukrimi Awang[2]

[1] School of Mechatronic Engineering, Universiti Malaysia Perlis, Pauh Putra Campus, 02600 Arau, Perlis, Malaysia.
[2] Kuliyyah of Medicine, International Islamic University Malaysia, Jalan Sultan Ahmad Shah, Bandar Indera Mahkota, 25200 Kuantan, Pahang.

*mhanafi@unimap.edu.my

Abstract. Osteogenesis Imperfecta (OI) is a group of genetic disorder that affects the bones. It causes bone to break easily. For severe case, it can lead the patient to death. However, doctors still have no quantitative method to predict fractures. Therefore, this study aims to investigate the fracture risk for OI femoral bone under load of activities in daily living and various loading direction. A finite element model reconstructed from radiography CT images was developed. The variations of daily activities were subjected to the femoral model. In standing and walking configuration, the model of OI has the ability to perform without having any fracture. However, during jumping, both configurations cause the OI femoral fractured. We can conclude that when the OI patients more active in their daily life, the higher chances for fracture to occur. The risk of fracture increases with increasing the intensity of the activity. Thus, early prediction of fracture risk could help prevent or treating fracture for OI patients.

1. Introduction

Osteogenesis Imperfecta (OI) or brittle bone disease is a genetic disorder. It is characterized by easily broken bones or 'imperfectly formed bone'. OI's patient has genetic defect that impairs the formation of strong bones [1]. This disease is caused by a mutation in COL1A1 and COL1A2, genes that encode type 1 collagen. Type 1 collagen is the main structural protein of the extracellular matrix of bone, skin, and tendons [1] [2]. A person with skeletal abnormalities tends to have very low bone mass and bone strength, which will lead to an easy and frequent fracture. Generally, OI is divided into 4 main types, Type I, Type II, Type III and Type IV. These types are divided based on radiography structures and clinical [3].

The main issue of Osteogenesis Imperfecta is the prediction of bone fracture. It is important to investigate the ability of the bone to resist the force before bone fracture occurs [4,5]. Currently, there is no established method in determining fracture risk for OI patients [6]. In order to estimate the fracture risk and plan the preventive interventions, it is necessary to determine the strength of bone accurately [7,8]. Finite Element Analysis (FEA) is broadly used in analysing bone structures and mechanical behaviours. The bone structures are divided into finite elements and each of the elements was assigned to proper material properties [6].

Wanna et al. used finite element model of OI to study the effect of bowing angles during standing [9]. A fracture risk assessment model of the femur in children with osteogenesis imperfecta (OI) during gait [6]. The findings showed that the risk of fracture is influenced by various load impact and gait cycle. In finite element, the key parameters of the femoral model comprised the mechanical properties of material, the boundary conditions, and the geometry of the model. These parameters make an important contribution to the analysis of fracture.

In this study, the OI finite element model of a femur was reconstructed based on radiography CT images of a real OI patient. This patient-specific model was used to analyse the distribution of mechanical stress, fracture risk, and fracture load under various loads of daily activities (ADL)—standing, walking, and jumping. The result of this simulation might provide important information in the determination of fracture risk in OI patients.

2. Methodology

2.1. Reconstruction of Femoral Model

A set of 310 CT images was obtained from a 13 years old female patient of IIUM Medical Centre. The patient was diagnosed with type III OI. The bone and soft tissues were segmented using global thresholding method as described by Mansor et al. [10]. A 3D femoral model was created using the segmented bone images on VOXELCON (Quint Corporation, Japan). Each voxel was assigned with 1 mm of width, 1 mm of height, and 1 mm of slice pitch of a pixel, thus resulting in 1 mm^3 of voxel. Figure 1(a) shows the lower limb of the patient where it was apparent that the bones were bowed at a certain angle. Figure 1(b) shows the resulted 3D femoral model created using the VOXELCON software.

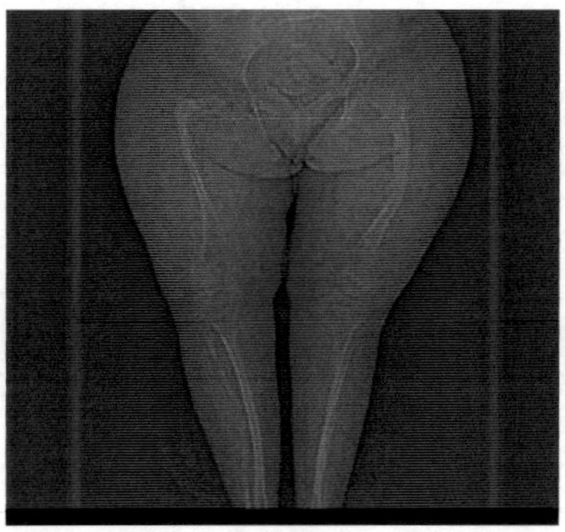

(a)

(b)

Figure 1. (a) CT image showing the lower limb of a type III OI patient. (b) 3D femoral model generated from the segmented images using VOXELCON.

2.2. Material Properties

The material properties of the reconstructed OI bone was assigned as isotropic linear elastic. The femur was assumed to has similar properties of cancellous bone, where Young's Modulus was set to 19 GPa and Poisson ratio of 0.3 [11]. The voxel element size of the femur model was generated based on the pixel size of the original image, thus eliminating the needs for convergence test.

2.3. Boundary and Load Conditions

The boundary condition was set to be in static condition. Boundary constraint was set at the distal of femur as to support the OI femoral model and prevents it from moving. A set of compressive forces were applied on the femoral head simulating the loads of ADL. In standing position, the compressive force was assumed as one-third of body weight. While walking, the femur needs to absorb three to seven times of body weight. Whereas during jumping, the femur takes approximately 15 to 20 times of body weight [12]. Since the body weight was not recorded, it is assumed that patient's weight is 37.5 kg (367.9 N) based on the study by Graff et al. [13]. Figure 2 illustrate the setup of the femur throughout the study.

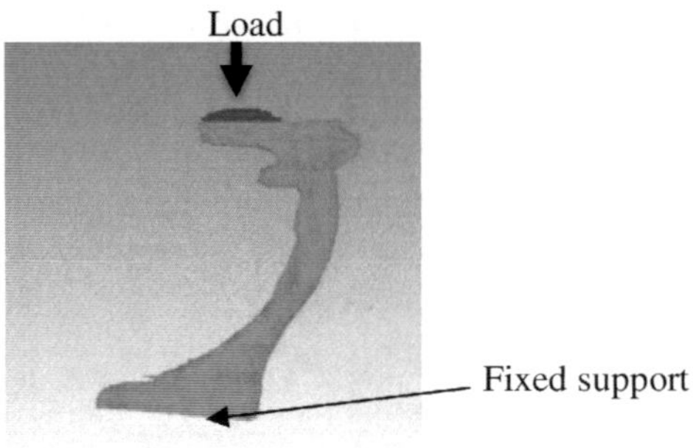

Figure 2. Boundary condition setup on 3D femoral model. The load was applied on the femoral head, whereas the condyles was fixed in its place.

2.4 Fracture Risk Analysis
Five loading conditions were applied to the distal part of femoral model to simulate standing, walking, and jumping. The fracture risk is calculated based on formulation created by Keyak et al. [14]. It is a test to measure the fragility of the bone. Based on nanoindentation test done by Fan et al. [11], the fracture strength was assumed to be 115 MPa. The von Mises stress value obtained through the simulation.

$$\text{Fracture Risk} = \frac{\text{Fracture Strength}}{\text{Von Mises Stress}} \quad (1)$$

3. Result and discussion

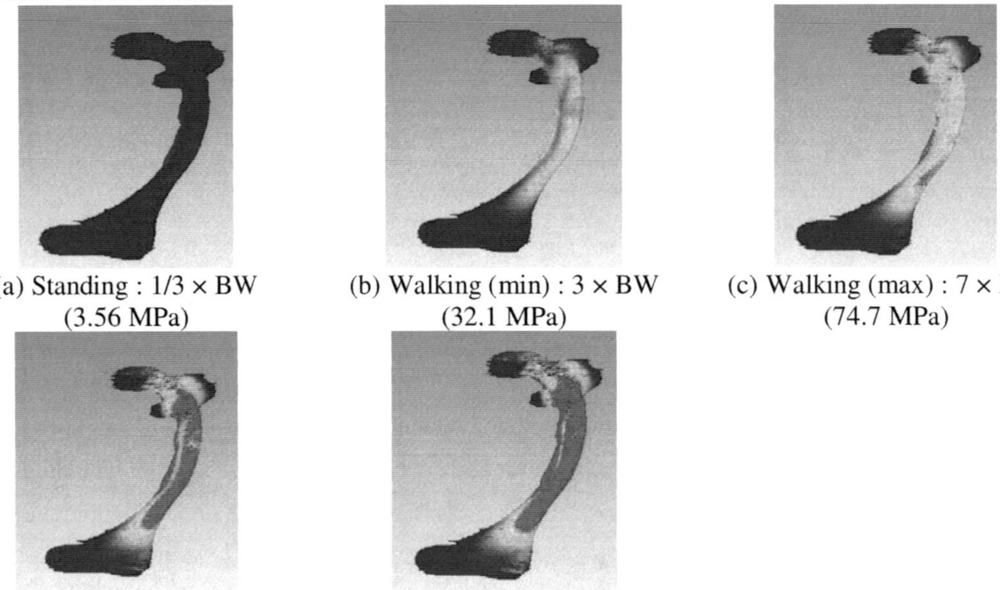

(a) Standing : 1/3 × BW (3.56 MPa)
(b) Walking (min) : 3 × BW (32.1 MPa)
(c) Walking (max) : 7 × BW (74.7 MPa)
(d) Jumping (min) : 15 × BW (159.7 MPa)
(e) Jumping (max) : 20 × BW (213.9 MPa)

Figure 3. Maximum Von Mises stress on the femoral model upon compressive force under different ADL

The results are represented as a contour map of stress distribution on the surface of femur as shown in Figure 3. The minimum von Mises stress was normalized to 0 MPa and maximum 30 MPa and the contour range were applied to all results. The contour plot highlights the location of high fracture risk. The area that appears as red colour indicates high von Mises stress level. Both scenario of jumping resulted in von Mises stresses that exceed the fracture stress of 115 MPa.

Figure 4 shows the plot of fracture risk against the coefficient of body weight based on ADL. The dotted line indicates the point when the fracture occurs. For standing condition, the result shows that OI model able to stand without causing any fracture. Significant decrease of fracture risk values in relative to standing condition were observed for both minimum and maximum walking condition. However, the forces acting on the femur did not cause any fracture as indicated by the fracture risk values of 3.58 and 1.54. This finding is supported by previous study by Wanna et al. where the OI patient can stand and walk on their own without causing any femur fracture [5]. This finding also supported by Fritz et al. where their study proved that all OI patient were able to perform full gait cycle without causing any bone fracture [6]. On the contrary, both minimum and maximum jumping resulted in fracture of femur as implied by the contour plot and fracture risk values calculated. The fracture risk value below 1 indicate that the bone is fractured as suggested by Keyak et al. [14]. Thus, that the more the compressive force is subjected to the OI femur, the higher the chances of femur to fracture. Based on the power regression, it was found that the fracture will be initiated on the femur when the body weight coefficient is 11.

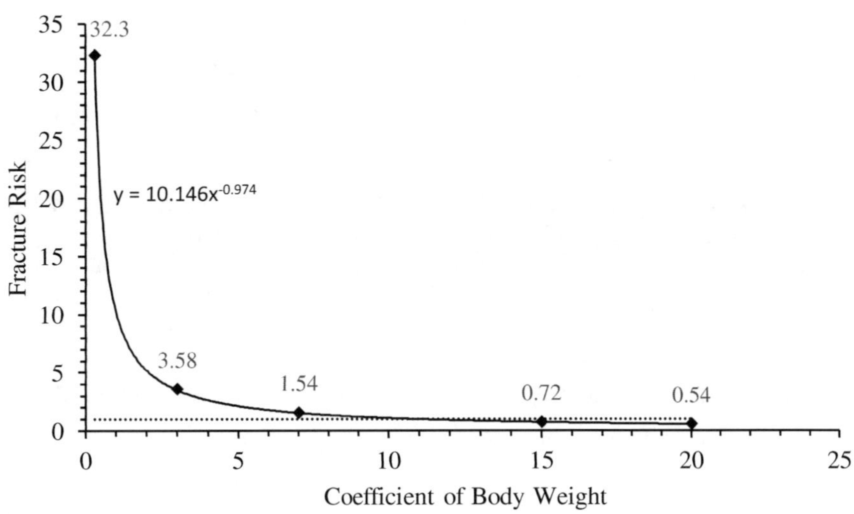

Figure 4. Fracture risk curve against the coefficient of body weight based on ADL

4. Conclusion

The fracture risk analysis of patient-specific OI femur has been performed using various loadings based on ADL. The simulation results showed that the standing and walking activities did not cause pose fracture risk to the femur. However, when the body weight coefficient is higher than 11, the femur started to fracture. Jumping activity have the body weight coefficient between 15-20, hence resulted in femur fracture. As a conclusion, OI patient have to be careful when performing activities in the daily life as excessive forces may result in fracture of the bone.

Acknowledgement

The authors would like to acknowledge the support from the Fundamental Research Grant Scheme (FRGS) under a grant number of FRGS/1/2016/TK03/UNIMAP/02/6 from the Ministry of Education Malaysia.

References

[1] Alharbi S A 2016 A Systematic Overview of Osteogenesis Imperfecta **5** 1–9

[2] Forlino A and Marini J C 2016 Osteogenesis imperfecta *Lancet* **387** 1657–71

[3] Sillence D O, Senn A and Danks D M 1979 Genetic heterogeneity in osteogenesis imperfecta. *J. Med. Genet.* **16** 101–16

[4] Wanna S B C, Basaruddin K S, Mat Som M H, Mohamad Hashim M S, Daud R, Abdul Majid M S and Sulaiman A R 2017 Prediction on fracture risk of femur with Osteogenesis Imperfecta using finite element models: Preliminary study *Journal of Physics: Conference Series* vol 908 p 012022

[5] Wanna S B C, Basaruddin K S, Mat Som M H, Sulaiman A R, Shukrimi A, Khan S F, Abdul Majid M S and Ridzuan M J M 2018 Fracture risk prediction on children with Osteogenesis Imperfecta subjected to loads under activity of daily living *IOP Conference Series: Materials Science and Engineering* vol 429 p 012004

[6] Fritz J M, Guan Y, Wang M, Smith P A and Harris G F 2009 A fracture risk assessment model of the femur in children with osteogenesis imperfecta (OI) during gait *Med. Eng. Phys.* **31** 1043–8

[7] Bessho M, Ohnishi I, Matsuyama J, Matsumoto T, Imai K and Nakamura K 2007 Prediction of strength and strain of the proximal femur by a CT-based finite element method *J. Biomech.*

[8] Sulong I, Basaruddin K S, Mat Som M H and Daud R 2018 Nanoindentation measurement on mechanical properties of Osteogenesis Imperfecta (OI): A review *ARPN J. Eng. Appl. Sci.* **13** 3435–40

[9] Wanna S B C, Basaruddin K S, Som M H M, Salleh A F and Sulaiman A R 2018 Effect of loading direction on fracture of bone with osteogenesis imperfecta (OI) during standing *AIP Conference Proceedings* vol 2030 p 020094

[10] Mansor N, Beh Y M, Mat Som M H, Basaruddin K S, Mustafa N, Yazid H and Salleh A F 2018 Reconstruction of cortical and cancellous bone in tibia with osteogenesis imperfecta *J. Telecommun. Electron. Comput. Eng.* **10** 115–9

[11] Fan Z F, Smith P, Rauch F and Harris G F 2007 Nanoindentation as a means for distinguishing clinical type of osteogenesis imperfecta *Compos. Part B Eng.* **38** 411–5

[12] Pria Bankoff A D 2012 Biomechanical Characteristics of the Bone *Human Musculoskeletal Biomechanics*

[13] Graff K and Syczewska M 2017 Developmental charts for children with osteogenesis imperfecta, type I (body height, body weight and BMI). *Eur. J. Pediatr.* **176** 311–6

[14] Keyak J H, Rossi S A, Jones K A and Skinner H B 1997 Prediction of femoral fracture load using automated finite element modeling *J. Biomech.* **31** 125–33

Variation of Stress Intensity Factor and Strain Energy Release Rate in Human Cortical Bone using Finite Element Analysis

Mohammad Shahril Salim[1*], Anis Najwa Azahari[1], Ahmad Faizal Salleh[1], Ruslizam Daud[1], Hamzah Sakeran[1]

School of Mechatronics Engineering, Universiti Malaysia Perlis, Pauh Putra Campus, 02600 Arau, Perlis, Malaysia.

*mshahrilsalim@gmail.com

Abstract. Finite element modelling and analysis is an alternative way to study human fracture behaviour in biomechanics field compared to experimental test in laboratory. The aim of this study is to analyse the stress intensity factor (*K*) and strain energy release rate (*J*-integral) when three-point bending test and tensile test are applied on a two-dimensional (2D) cortical bone model developed by using finite element software with different applied loads and crack-to-width ratios. Two methods are used to to evaluate the values of *K* and *J*-integral which are Displacement Extrapolation Method (DEM) and CINT method. The values then been compared with other theoretical expression done by other researchers for validation purposes. It has been proved that by using CINT method, the error is very small compared to DEM method. As a result, stress intensity factor values increased as crack-to-width ratio increased and strain energy release rate also increased as crack-to-width ratio and loads increased. It can be concluded that finite element analysis can be used to study the fracture behaviour of human cortical bone. Therefore, researchers can easily study the behaviour of cortical bone using simulation instead of mechanical experiment.

1. Introduction

The bone is a complex hierarchical microstructure that made up of water, collagen and mineral. The bone consists of two layer which are cortical and trabecular bone. Cortical bone is the compact bone and outer layer bone, protecting the inner cavity which is trabecular bone. Human bones are exposed to many possible ways of failure such as bending and tension. It is because, the complex microstructure of human cortical bone has a significant effect on its mechanical and fracture properties [1].

This research was done as finite element modelling of cortical bone is tested with two testing method in a closed fracture. Finite element is a method which is used to solve complex problems using simpler solution [2]. Usually, problems are expressed in partial differential equation solved using analytical method. But, as some of the problems cannot be settled using analytical method, numerical technique is used instead. Finite element modelling, an effective computed method is utilised in solving biomedical engineering problems such as bone mechanics [3], observing how a product behaves under a particular condition. Using this method, 2D of human cortical bone can be formed and tested using finite element software which is Mechanical ANSYS Parametric Design Language (APDL) 14.0. Ansys is one of the computational method software used to solve the finite element analysis (FEA) in working out mechanical problems such as static or dynamic structural analysis [4]. The DEM is a proven numerical method used for modelling particle systems as this numerical technique is able to describe the mechanical behaviour of a material, makes it widely used in biomechanics to study the fracture toughness of bone [5]. Meanwhile, CINT method is the most preferable method to find the SERR for fracture analysis using FE method. This technique is based on the assumption that the energy required to form a crack is equal to the energy needed to close the crack.

Then, Linear Elastic Fracture Mechanics (LEFM) is the basic theory of fracture developed using SIF determined by stress analysis [6]. In fracture mechanics, SIF is one of the most fundamental and a practical parameter developed by George R Irwin (1957) [7]. In a material with a crack, the stress at the crack tip is represented by stress intensity factor (SIF) which is correlated to the crack propagation and the failure of the material. Moreover, strain energy releasr rate (SERR) is introduced by Griffith which also represents fracture toughness, which known as energy dissipated during fracture per unit [8]. When a material is loaded, the crack propagates and produced new fracture surface area by the help of the strain energy. Other than observing the fracture behaviour, two parameters were also have been observed which are stress intensity factor (SIF) and strain energy release rate (SERR).

As the world keeps evolving, the demands for advancement and development in medical field specifically in orthopaedic field is growing. So, many studies are made to feed the needs of the people proving the determination of previous researchers. This scenario somehow seeds the motivation for this research to be done. However, solving this problem using 2D model on simulation software is not yet done by many. Previous studies showed that experiments are done on the fresh bone either animal or human bone as specimen to observe the crack propagation and crack behaviour using various testing method. To obtain the fresh bone specimens are not that easy under certain circumstances including ethics issue that might rise from it. Furthermore, most of them did analyse on stress intensity factor and strain energy release rate in different testing methods. So, this is the suitable platform to observe the behaviours of cortical bone with respect to different testing methods. Thus, this will be a new challenge where closed fracture on human bone with different crack lengths and applied loads were tested by using simulation software. This will be a point where other researchers feel encourage to shift their focus on problem relate to finite element modeling and analysis.

2. Methods

The mechanical properties possessed by cortical bone was defined and been specified where the Young's Modulus and Poisson's ratio is 14 GPa and 0.3 respectively. The single layer model was set in solid plane strain, structural, linear, isotropic quadrilateral elements (8 node PLANE 183) with singularity elements and around the crack tip having 8 nodes throughout the analysis as shown in figure 1.

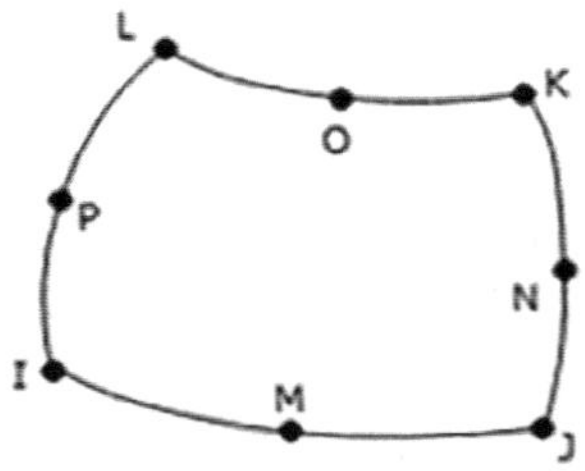

Figure 1. The quad element type 8 node (PLANE 183)

After model was created in Ansys, postprocessing is been done where two mechanical properties will be analysed from the models which are SIF and SERR from both testing methods.

2.1. Stress Intensity Factor, K

SIF values were calculated using two different FEA methods which are DEM and CINT metho. In this section, *K*-values for two different testing methods are analyzed separately. A load of 1000 N was applied in different crack-to-width ratio (*a*/*W*).

For three-point bending test, *K* values obtained is compared between DEM and CINT method with the theoretical expression proposed by Bower (2009) [9] and Tada (2000) [10]. The theoretical expression of Bower is expressed in equation 1 and equation 2 and Tada's in equation 3, where *P* is load, *B* represents specimen thickness and *S* is span length of the specimen.

$$K = \frac{P}{B\sqrt{W}} f(\frac{a}{W}) \tag{1}$$

$$f(\frac{a}{W}) = \frac{3\frac{S}{W}\sqrt{\frac{a}{W}}}{2\left(1+2\frac{a}{W}\right)\left(1-\frac{a}{W}\right)^{\frac{3}{2}}}[1.99 - \frac{a}{W}(1-\frac{a}{W})\{2.15 - 3.93\left(\frac{a}{W}\right) + 2.7\left(\frac{a}{W}\right)^2\}] \tag{2}$$

$$K_I = \frac{4P}{B}\sqrt{\frac{\pi}{W}}[1.6\left(\frac{a}{W}\right)^{\frac{1}{2}} - 2.6\left(\frac{a}{W}\right)^{\frac{3}{2}} + 12.3\left(\frac{a}{W}\right)^{\frac{5}{2}} - 21.2\left(\frac{a}{W}\right)^{\frac{7}{2}} + 21.8\left(\frac{a}{W}\right)^{9/2}] \tag{3}$$

Meanwhile, *K* values obtained from DEM were compared with CINT method for tensile test.

2.2. Strain Energy Release Rate, J-integral

J-integral values were obtained through FE software by using CINT method. Five different loads and eight different crack-to-width ratios are tested onto the cortical bone model by using two different testing methods separately.

3. Results & Discussion

3.1. Analysis of Stress Intensity Factor (SIF), K

The values gained from FE software were compared with theoretical expression and the average percent of error for DEM compared to theoretical expressions is 1.415% while CINT method has average percent of error of 1.173%. Therefore, CINT method is used to compare with DEM in tensile testing as it is proved reliable with percent of error less than 10%. Comparisons of data are tabulated in table 1.

Table 1. *K*-values Obtained Through DEM, CINT, Bower's and Tada's Expression at Different Crack-to-Width Ratio, *a*/*W*

a/*W*	K_{Bower}	K_{Tada}	K_{DEM}	Error % (Bower)	Error % (Tada)	K_{CINT}	Error % (Bower)	Error % (Tada)
0.126	30.981	32.299	31.301	1.033	3.090	31.582	1.941	2.219
0.1926	38.879	39.639	38.879	1.242	1.917	39.163	1.980	1.202
0.2520	45.741	46.317	45.741	1.305	1.244	45.996	1.874	0.693
0.3190	54.467	54.923	54.467	1.447	0.830	54.646	1.780	0.505
0.3780	63.794	64.206	63.794	1.597	0.642	63.846	1.680	0.561
0.5040	92.623	92.748	92.623	1.599	0.135	91.942	0.852	0.870
0.5700	116.860	116.266	116.86	1.761	0.511	115.083	0.214	1.017
0.622	143.800	141.762	143.8	2.847	1.438	140.400	0.415	0.961

For both testing method as indicated in figure 2, the SIF increased as crack length increased. This indicated that the fracture toughness of the cortical bone keeps increasing in order to resist fracture which brings to failure of the bone. Therefore, this property of bone protected the cortical bone when it is exposed to force.

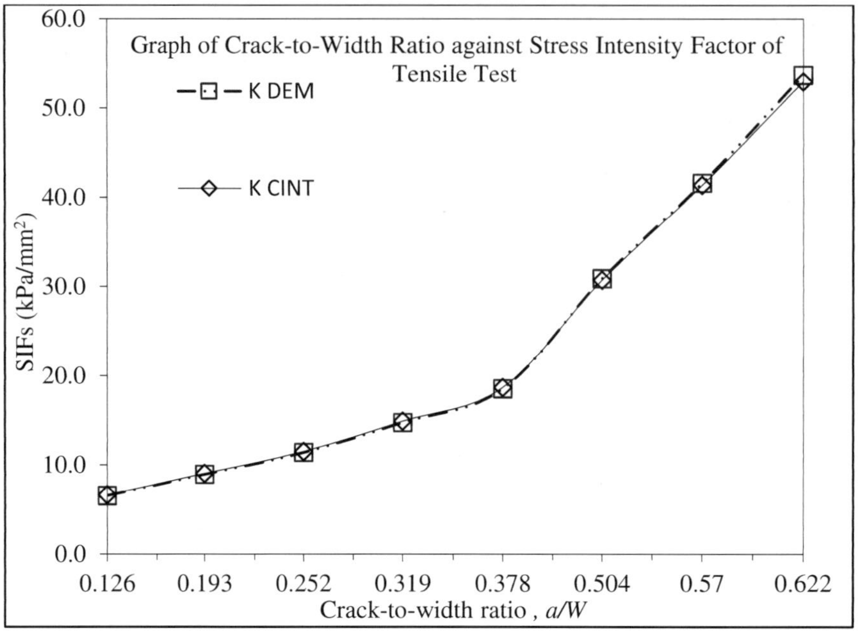

Figure 2. Comparison of *K* values in tensile test with different crack-to-width ratio, *a/W*

3.2. Analysis of Strain Energy Release Rate, J-Integral

For both testing method, it can be seen that as the crack-to-width ratio is raised, the *J*-integral will be escalated too. Besides, as the loads were increased, the *J*-integral values increased too. This drastic change is affected by the fracture toughness of the cortical bone since *J*-integral is related with fracture toughness of a material. The energy release rate is affected by the loads and cracks. That is why, as the crack length increase together with the crack length, SERR increased.

3.2.1. Three-Point Bending Test.

Referring to table 2, a graph in figure 3 is constructed to show the correlation between *J*-integral and crack-to-width ratio. It clearly showed that *J*-integral increase along with crack-to-width ratio and load. No matter what value of load exerted, the lines are displaying nonlinear increment of the relationship.

Table 2. *J*-integral values when different loads applied at different crack-to-width ratio for three-point bending test

Crack-to-Width Ratio, *a/W*	Strain Energy Release Rate, *J*-Integral				
	1000N	2000N	3000N	4000N	5000N
0.126	0.0617	0.2470	0.5557	0.9879	1.5437
0.193	0.0949	0.3798	0.8545	1.5192	2.3737
0.252	0.0976	0.5257	1.1788	2.0956	3.2744
0.319	0.1849	0.7395	1.6638	2.9579	4.6218
0.378	0.2524	1.0095	2.2713	4.0378	6.3091
0.504	0.5235	2.0941	4.7118	83765	13.0883
0.570	0.8224	3.2940	7.4014	13.1583	20.5597
0.622	1.2221	4.9140	10.9980	19.5527	30.5510

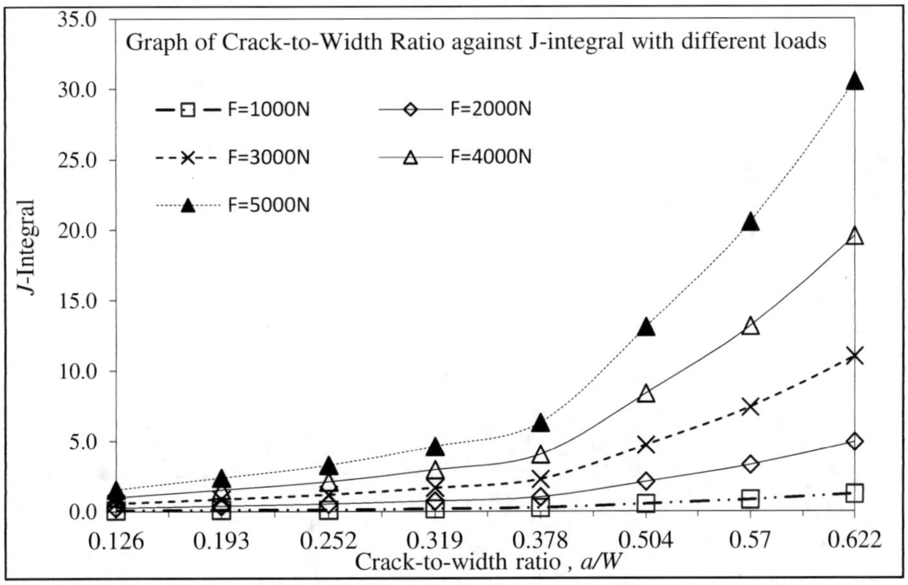

Figure 3. Graph of comparison between *J*-integral of three-point bending test when different loads are applied on different crack-to-width ratio, *a/W*

3.2.2. Tensile Test.

Referring to Table 3, a graph (figure 4) was constructed to show the correlation between *J*-integral and crack-to-width ratio for tensile test. It can be seen that as the crack-to-width ratio is raised, the *J*-integral will be escalated too. Besides, as the loads were increased, the *J*-integral values increased too. This drastic change is affected by the fracture toughness of the cortical bone since *J*-integral is related with fracture toughness of a material. The energy release rate is affected by the loads and cracks. That is why, as the crack length increase together with the crack length, SERR increased. However, *J*-integral values obtained were smaller compared to three-point bending test.

Table 3. *J*-integral values when different loads applied at different crack-to-width ratio for tensile test

Crack-to-Width Ratio, *a/W*	Strain Energy Release Rate, *J*-Integral				
	1000N	2000N	3000N	4000N	5000N
0.126	0.0027	0.0109	0.0246	0.0437	0.0683
0.193	0.0051	0.0203	0.0456	0.0811	0.1268
0.252	0.0082	0.0328	0.0739	0.1313	0.2052
0.319	0.0137	0.0549	0.1236	0.2198	0.3434
0.378	0.0216	0.0865	0.1946	0.3460	0.5407
0.504	0.0592	0.2367	0.5327	0.9470	1.4796
0.570	0.1060	0.3408	0.7667	1.3630	2.0080
0.622	0.1736	0.5020	1.1296	2.0080	3.1377

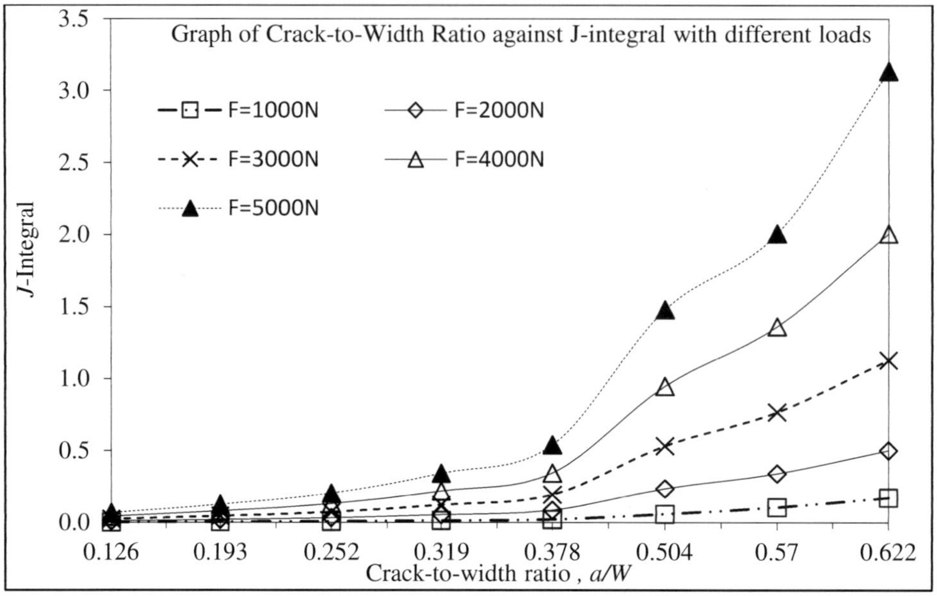

Figure 4. Graph of comparison between *J*-integral of tensile test when different loads are applied on different crack-to-width ratio, *a/W*

4. Conclusion

It is proved that fracture toughness of cortical bone which are SIF and SERR increased as crack-to-width ratio or load applied increased. This indicated that the fracture toughness of the cortical bone keeps increasing in order to resist fracture which brings to failure of the bone since SIF increased with crack-to-width ratio. Besides, FE method were proved as accurate as theory when DEM and CINT method were compared with theoretical expression. Moreover, it can be deduced that both forces exerted and crack-to-width, *a/W* ratio affect the *J*-integral during both tests since as the crack length increase together with the loads, SERR increased. However, SIF and SERR values for tensile test is lower than three-point bending test because of the force distributed by the load, as the force for bending test is directly above the crack while for the tensile test, the force is far from the crack.

5. References

[1] Nalla RK, Kinney JH, Ritchie RO 2003 Mechanistic fracture criteria for the failure of human cortical bone. Nature materials. **2**(3), 164.
[2] Rao SS 2017 The finite element method in engineering. Butterworth-heinemann.
[3] Zimmermann EA, Busse B, Ritchie RO 2015 The fracture mechanics of human bone: influence of disease and treatment. BoneKEy reports 4.
[4] Stolarski T, Nakasone Y, Yoshimoto S 2018 Engineering analysis with ANSYS software. Butterworth-Heinemann.
[5] Qian G, González-Albuixech VF, Niffenegger M, Giner E 2016 Comparison of KI calculation methods. Engineering Fracture Mechanics **156**, 52-67.
[6] Song C, Ooi ET, Natarajan S 2018 A review of the scaled boundary finite element method for two-dimensional linear elastic fracture mechanics. Engineering Fracture Mechanics **187**, 45-73.
[7] Soman S, Murthy K, Robi P 2018 A simple technique for estimation of mixed mode (I/II) stress intensity factors. Journal of Mechanics of Materials and Structures **13**(2), 141.
[8] Karr DG, Akçay FA 2016 A criterion for ductile fracture based on continuum modeling of energy release rates. International Journal of Fracture **197**(2), 201-212.
[9] Bower AF 2009 Applied mechanics of solids. CRC press.
[10] Tada H, Paris P, Irwin G 2000 The analysis of cracks handbook. New York: ASME Press. **1**.

Journal of Physics: Conference Series **1372** (2019) 012044 doi:10.1088/1742-6596/1372/1/012044

Classification of Acute Leukemia Based on Multilayer Perceptron

Nurul Hazwani Abd Halim[1*] Mohd Yusoff Mashor[1] Rosline Hassan[2]

[1]Electronic & Biomedical Intelligent Systems (EBItS) Research Group,
School of Mechatronics Engineering, University Malaysia Perlis,
02600 Kampus Pauh Putra, Perlis, Malaysia.
[2]Department of Hematology, School of Medical Science, University Science Malaysia,
Kubang Kerian, Kelantan, Malaysia.

[*]nurul.hazwani43@yahoo.com.

Abstract. In general, various artificial neural network have been applied in many areas such as modelling, pattern recognition, signal processing, diagnostic and prognostic. In this paper, artificial neural network are used to detect and classify the white blood cell (WBC) inside the acute leukemia blood samples. There are 25 features have been extracted from segmented WBC, which consist of shape, color and texture based features. Then, it have been fed up as the neural network inputs for the classification process in order to classify the segmented regions into two classes either B or T. The training algorithm for MLP network is Levenberg-Marquardt (LM). The MLP network achieves the highest testing accuracy of 96.99% for 4 hidden nodes at state of 5 by using the overall 25 input features. Thus, MLP network trained by using LM algorithm is suitable for acute leukemia cells detection in blood sample.

1. Introduction

Leukemia is a blood cancer, which can affect both children and adults. Acute leukemia can be divided into two types: Acute Lymphoblastic Leukemia (ALL) and Acute Myelogenous Leukemia (AML) [1]. It is wise to identify the types of leukemia early and rapidly assist the patients by giving appropriate treatments. In addition, the process to classify the type of leukemia is depending on the skills and experience of the hematologist. However, the manual microscopic examination has several drawbacks. Even for an expert, this traditional method can be very tedious and furthermore time-consuming [2].

Several development of semi and automated detection system related to leukemia have been proposed. Shafique *et*.al [3] proposed blood detection that utilizes thresholding based on Zack algorithm and shape and color based features were extracted from the segmented white blood cell (WBC).It able to achieve accuracy of 93.70% by using support vector machine (SVM) in order to classify either normal or blast cells. Meanwhile, Asadi *et*.al used Backpropagation Neural Network to classify ALL and AML, which produced 86.66% of accuracy [4]. Rajpurohit *et*.al proposed various classifiers such as convolutional neural network (CNN), feed-forward neural network (FNN), support vector machine (SVM) and K-Nearest Neighbor (KNN) with their following accuracies respectively, 98.33%, 95.40%, 91.40% and 93.30%. CNN provided a highest accuracy compare to other classifiers [5]. In this study, a multilayer perceptron by training using Levenberg-Marquardt (LM) has been proposed in order to classify the acute lymphoblastic leukemia either it is B or T. The motivation to do this classify because the hematologist face some difficulty to distinguish these cells unless do some specific test such as flow cytometry. The following section will be discussed about several image processing and how to implement this classifier effectively.

Journal of Physics: Conference Series **1372** (2019) 012044 doi:10.1088/1742-6596/1372/1/012044

2. Methodology

The block diagram below shown the proposed classification method for acute leukemia images which consist of several image processing techniques such as image enhancement, color thresholding, feature extraction and classification.

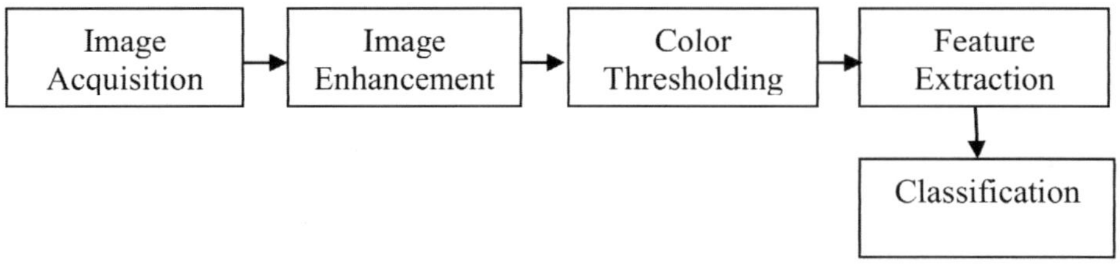

Figure 1: The proposed method for classification of acute lymphoblastic leukemia.

2.1. *Image Acquisition*

Figure 2 (a) and (b) were some examples of the captured images of acute lymphoblastic leukemia for type B and T, respectively, which, were captured under 100X magnification and at a resolution of 1280×960 pixels in Hospital Universiti Sains Malaysia (HUSM).

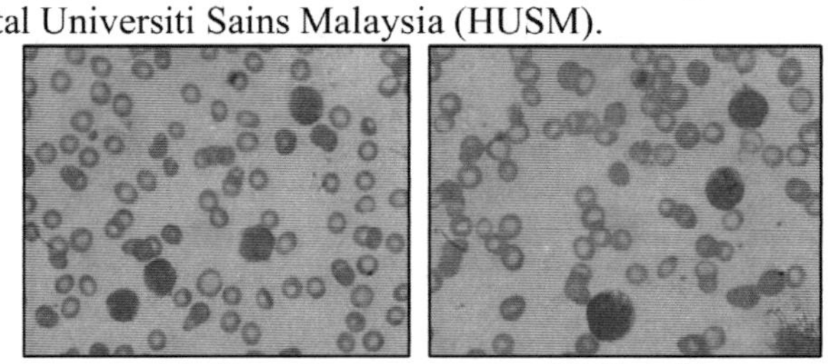

(a) Type B (b) Type T

Figure 2. Sample images of acute lymphoblastic leukemia (ALL).

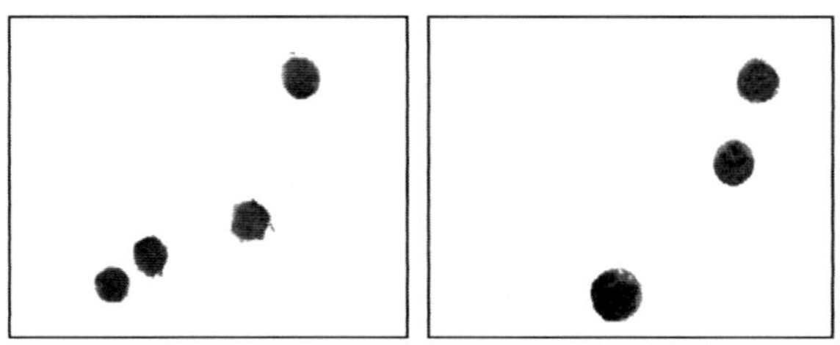

(a) Type B (b) Type T

Figure 3. Segmented images of acute lymphoblastic leukemia (ALL).

2.2. *Image Processing*

The effectiveness of the image enhancement process will make it easier for image segmentation, features extraction and classification of the blood sample slide images to identify leukemia. Thus, local contrast stretching has been implemented on B and T types of ALL images. By implementing this algorithm, each red, green and blue color space will be distributed linearly over the whole histogram so that the dynamic range of the histogram is fulfilled (0 – 255)[6][7]. After that, automatic color thresholding based on HSI (Hue, Saturation, Intensity) color space has been applied. In order to segment white blood cell (WBC) from the background and red blood cell (RBC), Hue component has been extracted from HSI color space. Based on the previous study, Hue component can provided a fully information about WBC while saturation component contains information about nucleus only [8]. The threshold value is set to 0.5. Any region which has greater than 0.5 will be considered as the WBC and the rest will be eliminated from the images as shown in Figure 3.

(a) Type B (b) Type T

Figure 4. Zooms of Segmented images of acute lymphoblastic leukemia (ALL).

2.3. *Feature Extraction*

Figure 4 shown some individual cells from the segmented images for each category. In order to distinguish WBC either it is type B or type T, it must have fulfilled the characteristic as below [9]:

- Type B has rather small, uniform blast cells with scanty cytoplasm and rounded with usually a single nucleolus and have a smudgy homogenous chromatin without prominent nucleoli.
- Type T has varied lymphoblasts, including numerous larger blasts with more open chromatin, prominent nucleoli and abundant cytoplasm.

After that, feature extraction is used to measure the properties of WBC. A number of approaches have been developed for feature extraction in acute leukemia identification system, such as the geometrical features [10][11], texture features [12][13], and the combination of geometrical ,texture and color features[14][15][16]. Before that, all of the WBC should be crop manually in order to extract features efficiently. There are some shape and geometrical based features that have been extracted such as area, perimeter, convex area, eccentricity, solidity, circularity and Affine Moment Invariant. For texture-based features such as contrast, correlation, energy and entropy have been extracted from segmented WBC images. Finally, color-based features are also extracted such as standard deviation and mean of RGB (Red, Green,Blue) color space. In total, there are 25 features have been extracted from the segmented WBC of both types of ALL which were then fed up as the neural network inputs for the classification. Beforehand, the features must be normalized between 0 to 1 in order to achieve high performance of classification. Noted that these features have been implemented on whole images in this study and the choice of the features has been driven by suggestions of the experts in HUSM and validate by them.

2.4. *Classification using Multilayer Perceptron*

Multilayer Perceptrons are gaining popularity in classification task due to its flexibility, robustness and high computational rates [17]. The classification performance of the MLP network depends on the structure of the network and training algorithm. The MLP network consists of an input layer that accepts the input data used in the classification, hidden layers and an output layer as shown in Figure 5. In this study, the Levenberg-Marquardt (LM) was used as a training algorithm due to it has a much better learning rate and can keep the relative stability.Detailed about LM algorithm can be found here [18].

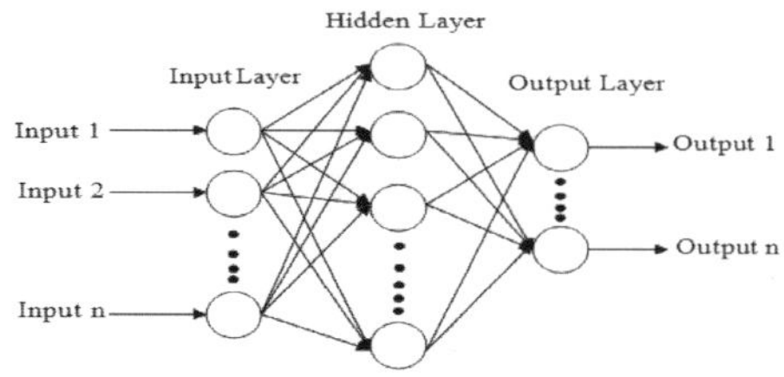

Figure 5. Schematic diagram of MLP model with 1 hidden layer

For the classification of overall WBC inside acute leukemia blood samples, there are a total of 1062 WBCs have been segmented from B-type and T-type images. The data division is set into 70% for training phase while 30% for the testing phase as shown in Table 1 below. In this study, the comparison has been done in five states in order to obtain the optimum number of hidden nodes for training and testing data. For the hidden node, the numbers used during the training phase are varied from 1 to 30 nodes, with the interval of 1 until the best classification result is achieved.The states are referring to the initial value assigned to the function when a random number generator (RNG) is called. The structure of the MLP network is set to 25:10:1 (input node: hidden node: output node). The other parameters are set such as training algorithm = LM, goal = 0.001, the number of epochs is set to 1000 and activation function is tan-sigmoid. Noted that,the evaluation is based on accuracy only due to the analysis is based on positive samples,which are B and T, hence, sensitivity and specificity are neglected in this context.

Table 1. Input data division for classification method

Overall Dataset	Training Dataset (70%)	Testing Dataset (30%)
B = 601 T = 461	B = 421 T = 323	B = 180 T = 138
Total = 1062	Total = 744	Total = 318

3. Results&Discussions

In this section, the classification performance based on overall features will be elaborated here. For this purpose, a total of 25 input features have been fed into the MLP network. There are two different analyses that have been conducted which are analysis of finding the best number of states and the best number of hidden nodes. All of this can be obtained when the MLP network achieved the highest testing result. The states refer to the initial values assigned to the function when a random number of generator is called. Different initial random values will produce different results. Hence, the testing has been done in five states to choose the best structure for MLP network.

. In order to avoid the problem of over-fitting, it is necessary to determine the best number of hidden nodes. Table 2 and Figure 6 show the analysis of number of hidden nodes for classification between B and T using MLP(LM) network.Based on Table 2, the best classification performance is obtained at state 5 and number of hidden nodes of 4 with testing accuracy of 96.99%.The results also show that the testing accuracy better than 90% has been archieved for other state.

Table 2. Analysis of number of hidden node for classification between T-type and B-type WBC using MLP_LM network

State	Number of Hidden Node	Training Accuracy (%)	Testing Accuracy (%)
1	29	99.61	95.32
2	29	99.48	95.65
3	30	99.48	95.97
4	21	99.74	94.31
5	**4**	**97.51**	**96.99**

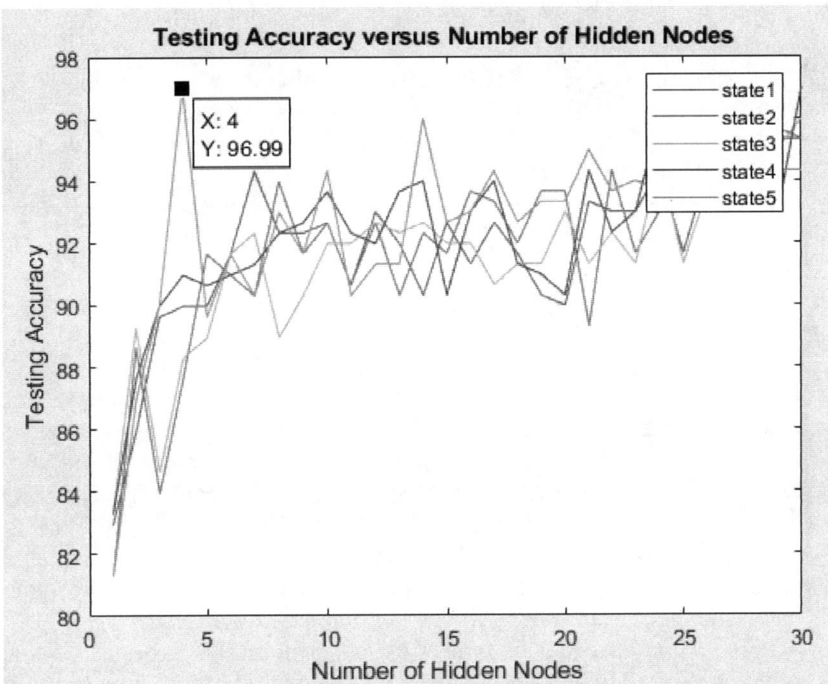

Figure 6. Analysis of Number of Hidden Nodes for classification between B and T using MLP(LM) network

4. Conclusion

It is shown that MLP_LM network produces the best classification performance of 96.99% at state of 5 using 4 hidden nodes. Hence, MLP network is suitable for detection of acute leukemia cells in order to classify either it is B or T type of leukemia.

Acknowledgments
The authors gratefully acknowledge to the National Science Fund (NSF) team members and University Science Malaysia (USM).

References

[1] Rajendran S 2007 Image Acquisition and Retrieval Systems for Leukaemia Cells *Master's thesis, Department of Computer Science, UM.*

[2] Thung F and Suwardi I S 2011 Blood parasite identitication using feature based recognition *International Conference on Electrical Engineering and Informatics (ICEEJ)* 1-4.

[3] Shafique S, Samabia T, Anas T 2019 Computer-assisted Acute Lymphoblastic Leukemia detection and diagnosis *2nd International Conference on Communication, Computing and Digital Systems (C-CODE)*, 184–189.

[4] Asadi F, Putra F M, Sakinatunnisa M I and Marzuki I 2017 Implementation of Backpropagation Neural Network and Blood Cells Imagery Extraction for Acute Leukemia Classification. *5th International Conference on Instrumentation, Communications, Information Technology, and Biomedical Engineering (ICICI-BME)* 106–110.

[5] S.Rajpurohit, S.Patil, & N.Choudhary(2018). Microscopic Blood Image Using Image Processing, (Cll), 2359–2363.

[6] Aimi Salihah A N, Mashor M Y, Harun N H, Abdullah A A and Rosline H 2010 Improving colour image segmentation on acute myelogenous leukaemia images using contrast enhancement techniques *Proceedings of 2010 IEEE EMBS Conference on Biomedical Engineering and Sciences, IECBES* 246–251.

[7] Mat Isa N A, Mashor M Y and Othman N H 2003 Contrast Enhancement Image Processing on Segmented Pap Smear Cytology Images *International Conferences on Robotic, Vision, Information and Signal Processing* 118 – 125.

[8] Huang C Y and Wu M J 2006 Image Segmentation *Final Report, University of Winconsin-Madison*.

[9] Miwa S 1998 Atlas of Blood Cells. Tokyo, Japan: *Bunkodo Co., Ltd.*.

[10] Chatap N and Shibu S 2014 Analysis of blood samples for counting leukemia cells using Support vector machine and nearest neighbour *IOSR Journal of Computer Engineering (IOSR-JCE)* **16** 79-87.

[11] Faivdullah L, Azahar F, Htike Z Z and Naing W N 2015 Leukemia Detection from Blood Smears *Journal of Medical and Bioengineering* **4** 488-491.

[12] Jyoti Rawat, Annapurna Singh, Bhadauria HS, Jitendra Virmani and Jagtar Singh Devgun 2017 Computer assisted classification framework for prediction of acute lymphoblastic and acute myeloblastic leukemia *Biocybernetics and Biomedical Engineering* **37** 637–654.

[13] Morteza Moradi Amin, Saeed Kermani, Ardeshir Talebi and Mostafa Ghelich Oghli 2015 Automatic recognition of acute lymphoblastic leukemia cells in microscopic images using k-means clustering and multiclass support vector machine classifier *Journal of Medical Signals & Sensors* **5**.

[14] Srisukkham W, Zhang L, Neoh S C, Todryk S and Lim C P 2017 Intelligent leukaemia diagnosis with bare-bones PSO based feature optimization *Applied Soft Computing* **56** 405–419.

[15] Lim H N, Francis E U, Mashor M Y and Hassan R 2017 Classification of bone marrow acute leukemia cells using multilayer perceptron network *3rd International Conference on Electronic Design, ICED 2016* 486–490.

[16] Kumar P and Udwadia S M 2017 Automatic detection of Acute Myeloid Leukemia from microscopic blood smear image *2017 International Conference on Advances in Computing, Communications and Informatics, ICACCI 2017* 1803–1808.

[17] Lim T Y, Ratnam M M and Khalid M A 2007 Automatic Classification of Weld Defects using Simulated Data and An MLP Neural Network *Journal INSIGHT, Learned & Professional Society Publisher* **49**(4) 154-159.

[18] Nasir A S A, Mashor M Y and Rosline H 2011 Detection of Acute Leukaemia Cells Using Variety of Features and Neural Networks *5th Kuala Lumpur International Conference on Biomedical Engineering 2011. IFMBE Proceedings* **35** 40–46.

Influence of simulated leg length discrepancy on the spinal kinematics during stance phases

Siti Mariam Mad Jusoh, Ahmad Faizal Salleh*, Khairul Salleh Basaruddin, Mohammad Shahril Salim, Muhammad Juhairi Aziz Safar, Hamzah Sakeran

School of Mechatronic Engineering, Universiti Malaysia Perlis, Perlis, Malaysia
*ahmadfaizal@unimap.edu.my

Abstract. Exploring the walking of LLD are clinically helpful in understanding which LLD level contributed to the changes at the upper limb. The aim of this work was to determine which LLD level during single leg stance and double leg stance contributed to the spinal changes. Eighteen healthy male subjects were induced with 0.5 cm to 4.0 cm height of insoles with interval 0.5 cm accordingly and their spine motion during the stance phase was captured by 5 Oqus cameras and evaluated by using Qualisys Track Manager (QTM) software. The effect of LLD height on spinal kinematics were assessed using spinal curvature parameters which are the lateral bending angle at lumbar (LBAL), the shoulder- girdle and pelvic line bending angle (SPBA) and the shoulder-girdle line and pelvic line rotation angle (SPRA). 4 cm LLD height at double stance phase created a relatively larger LBAL when compared with normal walking. The influence of LLD on SPBA and SPRA was very minimal.

1. Introduction

Leg length discrepancy (LLD) or anisomelia affected 40 to 70 percent of the human population [1-5]. According to Burke Gurney, Jan W. Raczkowski, Barbara Daniszewska and Krystian Zolynski in their review article stated that, leg length discrepancy (LLD) is defined as the pair of lower leg did not have same inequalities that lead to the difficulties in maintaining the stability of body posture during walking, running, standing and sitting [4,6,7,8]. That difficulties were causes by the changes in the segment of spine such as lumbar and pelvic during movement of the lower limb [4, 9].

Liu et al and Walsh et al stated that LLD cause the changes in the kinematic in the lower limbs and pelvis when it reaches the severity of 2 cm and above [4,10,11]. This is because higher severity of the LLD can lead to the disease related to spinal shape such as lumbar scoliosis and low back pain [12]. Not only that, the continuous pelvic tilt in LLD patients may lead to the structural changes in spine [1-6]. As we already know, spine is the important system that served to support the human body to bear the weight, to facilitate daily movement and to protect the spinal cord to carry their vital functions [12].

When the magnitude of the lateral bending exceeds the maximum, the lumbar scoliosis may become permanent disease of patient with LLD [12]. The difficulties are causes by the changes in the segment of spine such as lumbar and pelvic [1-6]. Other difficulties related to LLD are scoliosis, change in human gait, low back pain because of the inequalities of lower limb will make the weight of body been supported more toward the short legs [2,4,10,12].

During walking there are several movements of our body such as the rotation of the hip and shoulder, the lateral bending of spine and the movement of muscle at ankle joint, knee joint and hip joint [6]. Walking which compromise of eight phases become awkward in LLD patient [10]. This is because the distribution of their body weight being support more toward short leg [1-6]. Figure 1 shows the position of normal human without LLD and with LLD during walking. This shown that there are compensating of the LLD patient compared to normal person [10].

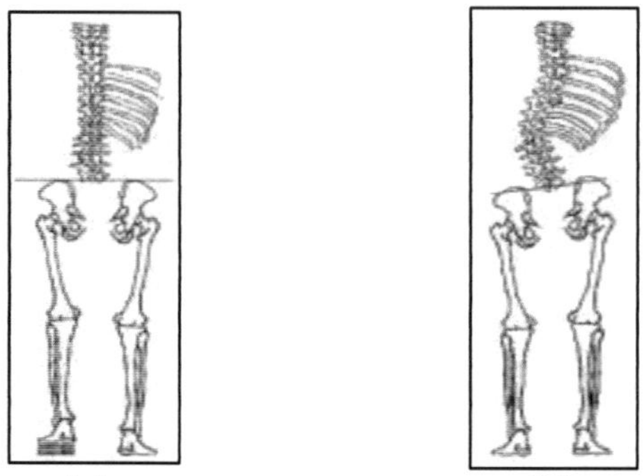

(a) (b)

Figure 1: (a) Normal person (b) LLD person [4]

LLD can be classified into four groups which are functional LLD, structural LLD, environmental LLD and different severity level of LLD [2]. Functional LLD (apparent LLD) is the condition of actual leg length changes which depends on the movement of the lower limb. Meanwhile, structural LLD (structural LLD) due to different level of femur head and calcaneus [6]. Environmental LLD usually found in athlete who stepped on non- smooth track for a long period of time, while the level of LLD could be classified whether it was mild, moderate and severe [2,7]. LLD with less than 3 cm is mild, 3 cm to 6 cm is moderate and more than 6 cm is severe [2]. Plus, LLD can further divide whether the persons experienced LLD since kids and generate LLD later at certain time of their age [13].

The main objective of this study was to assess the effect of LLD on the lateral bending angle of the lumbar (LBAL), shoulder-girdle and pelvic line bending angle (SPBA) and shoulder- girdle and pelvic line rotation angle (SPRA). All of these parameters were analysed during single leg stance and double leg stance. These was to know which LLD level affected the most on the LBAL, SPBA and SPRA more specifically during single leg stance and double leg stance.

2. Method

2.1 Subjects

Eighteen male subjects (age 22±2 years, mass 62 ± 8 kg, height 1.66 ± 0.05 m, BMI 22.5 ± 1.7 kgm^{-2}) with no history of back pain and no gaits disorders volunteered to participate in the study. Informed consent was obtained from all participants prior to participation in the study. The study was conformed to the local ethics committee guidelines.

2.2 Experimental Protocol

The subjects were instructed to wear tight sport shirt. 30 round reflective skin markers, 12.5 mm in diameters were placed on the subject's body as shown in Figure 2. They had to wear the sandal with induced insoles inserted under the right foot started from 0.5 cm, 1.0 cm, 1.5 cm, 2.0 cm, 2.5 cm, 3.0 cm, 3.5 cm and 4.0 cm accordingly.

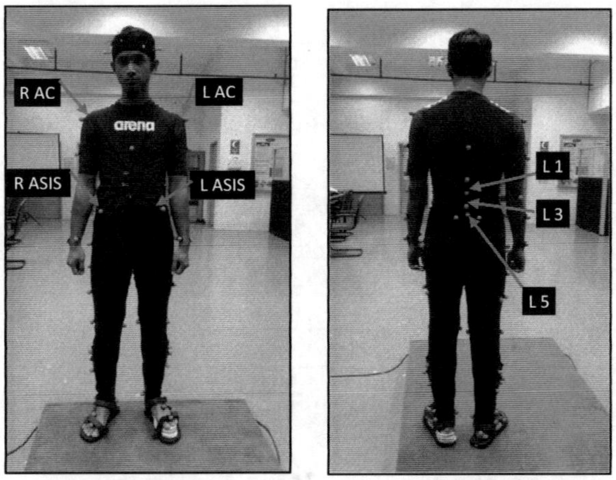

(a) Front (b) Back

Figure 2: Marker placements

Three-dimensional kinematics and kinetics data were obtained using 5 cameras motion capture system (100 Hz; Qualisys AB, Gothenburg, Sweden) and two rectangular Bertec force platforms (FP4060-10-2000, Bertec Corporation, Columbus, USA) operated at 1000 Hz. The two force platforms were installed in series in the middle of 7m walkway to record the gait cycles. Participants walked three to five time on 7 m platform to familiarize with the artificially created LLD. Subject completed 5 trials and one-minute rest between trial was provided to prevent fatigue.

2.3 Data and Statistical Analysis

The data was analysed using Qualisys Track Manager (QTM) software (Gothenburg, Sweden). The LBAL, SPBA and SPRA were defined based on the study done by *kakushima et al* [10]. The LBAL was defined as the angle formed by markers placed at Lumbar 1, 3 and 5 and viewed in coronal plane. The SPBA was defined as an angle formed by shoulder girdle line and pelvic line and viewed from coronal plane. The SPRA was defined as an angle formed by shoulder girdle line and pelvic line and viewed from transverse plane. The LBAL, SPBA and SPRA were evaluated during single stance of right leg (SR), single stance of left leg (SL) and double stance (DS). A *two-way* repeated measure ANOVA were applied to test the effect of artificially created LLD heights and Stance Phases on spinal curvature parameters. Statistical analyses were completed using SPSS 16 (SPSS, Chicago, IL)

3. Result

3.1 Lateral Bending Angle of Lumbar (LBAL)

In each stance phases, comparison of LBAL created at different LLD height with normal gait (0 cm LLD) were performed. No significant differences were found for LBAL created at SL and SR when compared with normal gait. However, significantly larger LBAL value was shown when comparing 4.0 cm LLD height with normal gait at DS, $F\,(1,17) = 6.544$, $r = 0.52$.

In each LLD height, comparison was made between SL, SR and DS stance phases. No significant differences of LBAL values were found during normal gait in all stance phases. There was a significant main effect of LBAL values when comparing between stance phases at 4 cm LLD height, $F\,(2,34) = 3.887$, $p < 0.05$; however, it did represent a medium-sized effect, $r = 0.32$. Contrast revealed that value of LBAL of DS stance phase was higher than SL and SR stance phases, $F\,(1,17) = 6.917$, $r = 0.53$ and $F\,(1,17) = 5.252$, $r = 0.49$, respectively.

Table 1 Means and SD of SL, SR and DS for Lateral Bending Angle of The Lumbar with and without insoles

Level of LLD (cm)	Lateral Bending Angle of Lumbar (◦)		
	SL	SR	DS
0.0	5.2±2.0	4.4±1.3	4.7±2.3
0.5	4.3±1.5	4.9±1.0	4.7±3.2
1.0	4.7±1.7	5.5±2.0	4.6±2.3
1.5	4.2±2.3	4.1±1.8	3.8±2.6
2.0	5.5±2.0	4.4±2.6	4.5±2.6
2.5	3.9±2.4	4.9±2.8	6.0±3.3
3.0	4.1±2.7	4.7±2.2	4.6±2.3
3.5	3.7±2.3	3.6±2.5	6.0±3.2
4.0	5.1±2.1	4.4±1.7	*6.4±1.2

*$p<0.05$

3.2 Shoulder- girdle and Pelvic line Bending Angle (SPBA)

The SPBA at each LLD height was compared with the SPBA at normal gait in each stance phases. As shown in table 2, there were no significant LBAL created at SL, SR and DS when compared with normal gait. Comparison of SPBA between stance phases for each LLD height were performed. No significant differences of SPBAs between stance phases were found during normal gait and walking with 4 cm LLD height.

Table 2 Means and SD of SL, SR and DS for SPBA with and without insoles

Level of	Shoulder-girdle and Pelvic Lines Bending Angle (◦)		
LLD (cm)	SL	SR	DS
0.0	3.8±0.9	3.8±1.0	3.4±1.3
0.5	3.3±1.6	3.4±1.8	3.2±1.4
1.0	3.3±1.8	4.6±2.2	3.1±1.9
1.5	2.8±1.0	3.6±1.4	2.8±1.6
2.0	3.5±2.6	4.4±1.3	2.7±1.0
2.5	3.9±2.8	4.2±1.1	2.5±2.2
3.0	2.3±0.8	4.4±1.6	2.1±1.9
3.5	4.0±2.1	3.5±1.9	2.6±1.9
4.0	4.7±2.5	4.4±0.6	3.8±2.3

3.3 Shoulder- girdle and Pelvic line Rotation Angle (SPRA)

Comparison of LBAL created at different LLD height and at different stance phases with normal gait were performed. As shown in table 3, there were no significant LBAL created at SL and DS when compared with normal gait. However, significantly larger SPRA value was shown when comparing 4.0 cm LLD with normal gait at SR, $F\,(1,16) = 5.801$, $r = 0.52$.

In each LLD height, comparison was made between SL, SR and DS stance phases. No significant differences of LBAL values were found between stance phases during normal gait and walking with 4 cm LLD height.

Table 3 Means and SD of SL, SR and DS for Shoulder- girdle and Pelvic Lines Bending Angle with and without insoles

Level of LLD (cm)	Shoulder-girdle and Pelvic Lines Rotation Angle (◦)		
	SL	SR	DS
0.0	5.4±1.9	4.4±1.5	7.3±1.6
0.5	5.9±1.7	3.9±1.6	7.5±1.5
1.0	6.4±1.9	5.2±2.0	7.6±1.4
1.5	6.0±1.1	4.8±1.6	6.7±1.8
2.0	5.4±1.8	3.7±1.3	6.9±1.9
2.5	5.3±2.3	4.0±1.6	6.9±2.0
3.0	4.8±1.3	2.9±2.0	5.9±1.9
3.5	3.9±1.7	5.7±2.5	5.1±1.2
4.0	3.8±1.9	5.4±2.5	5.3±2.3

4. Discussion

In this study, the effects of LLD during gait on the spinal kinematics were evaluated in healthy male adults. Analysis were performed to identify any significant angle values created at different LLD height when compared with normal walking in each stance phases. During normal walking, there was no significant difference observed in LBAL, SPBA and SPRA value in single stance of left and right leg. However, a relatively larger LBAL value was observed at 4 cm LLD height during double stance phase.

Thus, LLD at double stance phase is likely to expose the spine to a larger amount of lateral bending stresses. *Kakushima et al* performed similar studies using 3 cm insoles on the right foot and found that the Maximum Lateral Bending Angle of the Lumbar (MLBAL) of right side was significantly larger in LLD than in normal gait. However, it was not clear at which gait event the MLBAL was assessed. *Bargenter et al* investigated the effect of LLD on spinal gait kinematics in patients with structural LLD of 20-60 mm. They did not report any significant differences for Lumbar Lateral Bending Angle viewed in sagittal plane. However, patients with LLD showed larger frontal plane lumbar bending angle to the longer side (convex on the short side) during initial contact and early mid-stance. In the current study, the frontal plane lumbar bending angle was not considered for analysis. In the case of SPBA and SPRA, none of the LLD height give significant impact to the spine kinematics. Some errors in determining SPBA and SPRA could be the reasons for these insignificant results. Further investigation is necessary to find the changes of LLD on SPBA and SPRA.

5. Conclusion

4 cm LLD during double leg stance contributed to the change of the spinal kinematic. LLD patients may find difficulties during physiological and physical activities of their body. Thus, the treatment for LLD may be helpful in preventing the physical and physiological of the body from being aggravated.

References

[1] Betsch M, Wild M, Große B, Rapp W and Horstmann T 2012 The effect of simulating leg length inequality on spinal posture and pelvic position: a dynamic rasterstereographic analysis *Eur. Spine J.* **21** pp. 691–697

[2] Khalifa A A, 2017 Leg Length Discrepancy: Assessment and Secondary Effects *Ortho & Rheum Open Access* **6**. Pp 1-5

[3] Murray K J, Molyneux T, Le Grande M R, Castro Mendez A, Fuss F K, and Azari M F 2017 Association of mild leg length discrepancy and degenerative changes in the hip joint and lumbar spine *J. Manipulative Physiol. Ther.* **40** pp. 320–329

[4] Raczkowski J W, Daniszewska B and Zolynski K 2010 Functional scoliosis caused by leg length discrepancy *Arch. Med. Sci.* **6** pp. 393–398

[5] Gurney B, 2002 Leg length discrepancy *Gait Posture* **15** pp. 195–206

[6] Eliks M, Ostiak-Tomaszewska W, Lisiński P and Koczewski P 2017 Does structural leg-length discrepancy affect postural control? Preliminary study *BMC Musculoskelet. Disord.* **18** p. 346

[7] Oxland T R 2015 A history of spine biomechanics *Unfallchirurg* **118** pp. 80–92

[8] Shelburne K B, Pandy M G, Anderson F C and Torry M R 2004 Pattern of anterior cruciate ligament force in normal walking *J. Biomech.* **37** pp. 797–805

[9] Resende R A, Kirkwood R N, Deluzio K J, Morton A M and Fonseca S T, 2016 Mild leg length discrepancy affects lower limbs, pelvis and trunk biomechanics of individuals with knee osteoarthritis during gait *Clin. Biomech.* **38** pp. 1–7

[10] Kakushima M, Miyamoto K, and Shimizu K 2003 The effect of leg length discrepancy on spinal motion during gait: three-dimensional analysis in healthy volunteers *Spine* **28** pp. 2472–2476

[11] Izzo R, Guarnieri G, Guglielmi G, and Muto M 2013 Biomechanics of the spine. Part II: Spinal instability *Eur. J. Radiol.* **82** pp. 127–138

[12] Mishima K, Kitoh H, Kadono I, Matsushita M, Sugiura H, Hasegawa S, Kitamura A, Nishida Y and Ishiguro N 2015 Prediction of clinically significant leg-length discrepancy in congenital disorders *Orthopedics* **38**, pp. e919–e924

[13] Yamato Y, Hasegawa T, Kobayashi S, Yasuda T, Togawa D, Arima H, Oe S, Iida T, Matsumura A, Hosogane N and Matsumoto M 2016 Calculation of the target lumbar lordosis angle for restoring an optimal pelvic tilt in elderly patients with adult spinal deformity *Spine* **41**, pp. E211–E217

[14] Bargenter C., Romkes J., Lorenzetti S., Krieg A.H., Hasler C.C., Brunner R., Schmid S. 2019 What are the biomechanical consequences of a structural leg length discrepancy on the adolescent spine during walking? *Gait & Posture* **68**, pp. 506-513

Encapsulation of *C. odorata* extracts for antimicrobial activity

Siti Nur Amylina Azmi[1], Che Wan Sharifah Robiah Mohamad[1*], Khairul Farihan Kasim[2]

[1]Biomedical Electronic Engineering Program, School of Mechatronic Engineering, Universiti Malaysia Perlis, 02600 Arau, Perlis
[2]School of Bioprocess Engineering, Universiti Malaysia Perlis, 01000 Jejawi, Perlis

*robiah@unimap.edu.my

Abstract. The purpose of this study is to investigate the relationship between the methycellulose (MC) and ethylcellulose (EC) of encapsulated *C. odorata* extracts (methanol and ethanol) and their antimicrobial properties against *E. coli* and *Staphylococcus aureus*. The encapsulated of *C. odorata* extract was prepared by solvent displacement method through dialysis tube cellulose membrane and it was freeze-dried to obtain the suspension of encapsulated *C. odorata* extract. Then, antimicrobial properties of encapsulated *C. odorata* extracts was performed by broth dilution method on target bacteria. A few drops of hydrochloric acid (HCl) were added into growth media until pH 4 to mimic the acidic conditions in human stomach. The results showed the encapsulated methylcellulose of methanol extract (MC-M) inhibit both bacteria types.

1. Introduction

Nowadays, natural antimicrobial has become very important in global health because of their potential as antibiotics for many diseases [1]. Many types of antimicrobials have been created in order to destroy and inhibit the growth of bacteria since last half of the previous century. But now it is become more complicated because of resistance of bacteria for certain vaccination or antibiotic are developing increasingly. A study shown the nano-capsule represent a target of new compositions for inhibiting the growth of bacteria in medical field [2].

Chromolaena odorata (C. odorata) belongs to the family Asteraceae and is known locally in Nigeria through many common names as siam weed, Elizabeth weed, obirakara, olorohuru and independence weed [3]. The anti-microbial properties have made it a popular choice in disinfecting and treating open wounds [4]. The wider use as an effective therapy against diarrhoea, malaria fever, tooth ache, diabetes, skin diseases, dysentery and colitis has been severally documented [4-5]. The leaves could be ground and the extracted juice taken to alleviate fever or the treatment of diabetes [6]. This study is aimed at providing preliminary studies on *C. odorata* leaves and its potential as antimicrobial activity to combat infectious disease.

Cellulose have been famous in the food and pharmaceutical industries as a coating material due to a few characteristics that allows the increase of active ingredients ability and biodegradability. Other than that, cellulose does not react with the material to be encapsulated. Then, cellulose also help to seal and hold the active material within its structure during processing or storage. It also provided maximal protection of the active against environmental conditions. The encapsulation membrane/shell (wall material) around active compound/phytochemical were functioning as a barrier or main protective role [7]. Figure1 shows the schematic representative of matrix encapsulation system.

However, there are lack of information about the potential encapsulated of *C. odorata* as antimicrobial activity since its behaviour has not yet been scientifically investigated. This study was to examined the capability of encapsulated of *C. odorata* extract by used cellulose for microbial activity using broth dilution method.

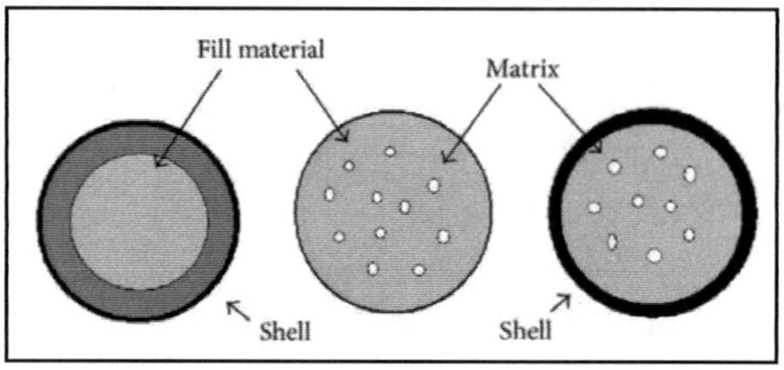

Figure 1: Schematic representative of matrix encapsulation [7].

2. Methodology

2.1. Materials

C. odorata leave were collected from Changlun, Kedah. Fresh *C. odorata* leave was washed and dried at 40ºC for 24 hours. After dried, the leaves are grinded into powder using mechanical blender.

2.2. C. odorata extraction

The Soxhlet extractor apparatus have been used to extract *C. odorata* leave. 30 g of fine *C. odorata* sample added into 250 ml of ethanol was used for minimum 6 hours to ensure optimum extraction of bioactive compound in the leave. The solvent then evaporated using EYELA rotary evaporator for 1 hour and the ethanol crude extracts obtained were kept in 4°C until further use. The extraction process was repeated using methanol as solvent

2.3. Encapsulation of C. odorata extract

Encapsulation of *C. odorata* extract using EC and MC were prepared by solvent displacement method through dialysis tube cellulose membrane. 0.1 g of *C. odorata* leave extract and 0.1 g cellulose (EC or MC) at ratio of 1:1 (w/w) was co-dissolved in 25 ml of solvent (ethanol or methanol) and 2.5 ml tween 20 (a polysorbate-type non-ionic surfactant). The solution was transferred into dialysis tube cellulose membrane and dialyzed against 250 ml distilled water. After 2 hours dialyzes process, 50 mL encapsulated sample was transferred into centrifuge tube follow with 4000 rpm for 30 minutes centrifugation. The obtained aqueous suspension was freeze dry and stored in -20°C.

2.4. Broth dilution methods

Typical Soy Broth (TSB) is used as a medium in broth dilution methods. Nine grams of TSB is dissolved in 300 ml distilled water until completely dissolve. The TSB medium was sterilized by autoclaving for 15 minutes at 121°C. The inoculums bacteria were prepared by used 0.5 McFarland bacterial. Overnight culture bacteria were transferred into 5 ml sterile distilled water and vortex for 30 seconds to dissolve the bacteria. Then, the turbidity of bacterial suspension was adjusted according to 0.5 McFarland or (1-2) x 10^8 CFU/ml. After that, 10 mL of 2×10^8 CFU/ml microbial suspension was transferred into 30 ml of TSB. Then, 9 mL of bacterial broth was transferred into 1 mL of *C. odorata* leave extract. The set of test tubes containing a mixture of bacteria and the extracts were incubated at 37 ºC for 24 hours. After 24 hours, both bacteria and extract solution were measured at 600 nm respectively.

2.5. Antioxidant activity

The antioxidant activity was determined using DPPH method. 2.0 ml of extract was mixed with 2.5 ml of 0.1 mM of DPPH solution. The mixture was incubated in the dark for 40 min. after that, the solution was measured using spectrophotometer at 517 nm. The percentage inhibition was calculated using equation below:

$$DPPH\ activity(\%) = \left[1 - \frac{absorbance\ of\ sample}{absorbance\ of\ DPPH}\right] \times 100$$

3. Result and Discussion

3.1. Antimicrobial activity of C. odorata leave extract

4 mg/ml of encapsulation sample with is methanol-methylcellulose (MC-M), methanol-ethylcellulose (EC-M), ethanol-methylcellulose (MC-E) and ethanol-ethylcellulose (EC-E) respectively.

Table 1: Antimicrobial activity of the *C. odorata* extract

Sample (4.0 mg/ml)	Concentration of bacteria (CFU/ml)	
	E. coli (G-)	*Staphy (G+)*
Crude *C. odorata*_ethanol (E)	2.56 X 10^8	2.49 X 10^8
Crude *C. odorata*_methanol (M)	2.52 X 10^8	2.48 X 10^8
MC-E	2.17 X 10^8	1.73 X 10^8
EC-E	2.38 X 10^8	1.94 X 10^8
MC-M	2.01 X 10^8	1.68 X 10^8
EC-M	2.32 X 10^8	1.74 X 10^8

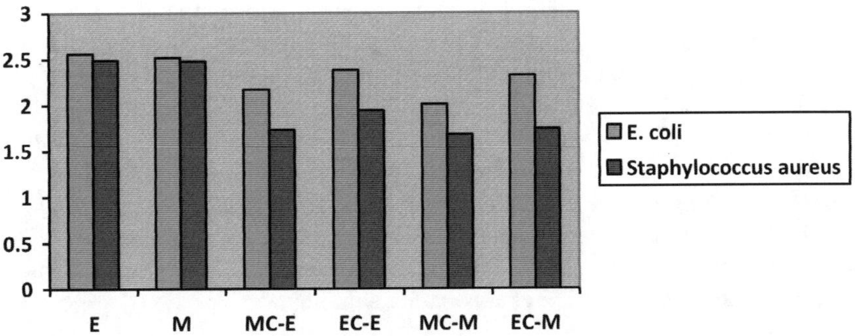

Figure 2: Antimicrobial value for *E. coli* and *Staphylococcus aureus*.

Table 1 and Figure 2 was shown that MC-M more effective in antimicrobial activity towards to *E. coli* (gram-negative) and *staphylococcus aureus* (gram-positive) with the inhibition of bacteria growth $2.01\ x\ 10^8$ and $1.68\ x\ 10^8$. In addition, for EC-E showed least antimicrobial activity against *E. coli* $(2.38\ x\ 10^8)$ and *staphylococcus aureus* $(1.94\ x\ 10^8)$.

3.2. Antioxidant (AOA)

The free radical scavenging activity of the extracts was evaluated based on the ability to scavenge the DPPH. This assay is highly important to provide information about the reactivity of organic compounds with stable free radicals, because of the odd number of electrons.

The result of the study reveals that the extract of the leaves of *C. odorata* scavenged the free radicals by DPPH method. EC-methanol it shows a strong absorption band at 517 nm in visible spectrum (deep violet color). Figure 3 shows the bleaching of DPPH absorption indicates the capacity of the test drugs to scavenge the free radicals. Where, the ethanol (8.8%), EC-ethanol (27.04%), MC-ethanol (16.16%), methanol (21.26%) and MC-methanol (54.9%) have lowest antioxidant activity. Further studies were carried out to find out the correlation between *C. odorata* extract and encapsulated *C. odorata* extract with EC and MC in antioxidant and antimicrobial activity. It shows, methanol has interaction with antioxidant which it contributed to antimicrobial properties [8,10-12]. In another studies, *C. odorata* extract had a positive correlation between the antioxidant activity and total phenolic content. It could be capable in antimicrobial and anti-inflammatory activity of infectious disease [9, 13-14].

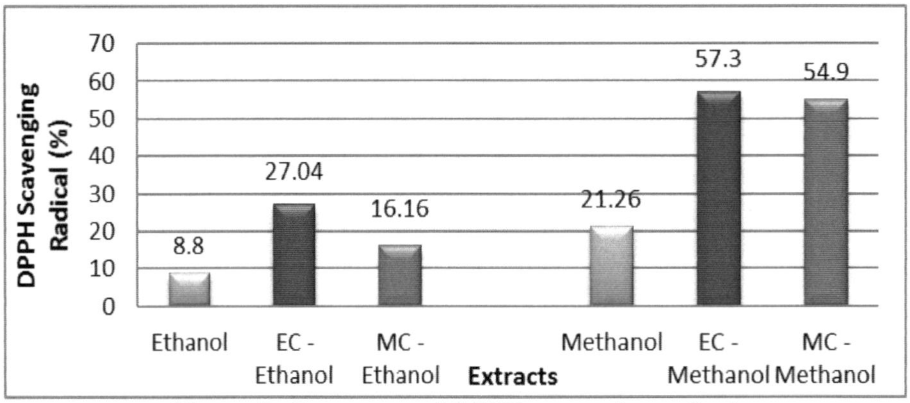

Figure 3: Percentage of DPPH scavenging radical

4. Conclusion

As a conclusion, *C. odorata* leave methanol extracts was shown high activity against antibacterial activity compared *C. odorata* leave ethanol extracts. Ethanol and methanol solvent were used to *C. odorata* leave extract then followed with encapsulated of MC and EC by solvent displacement method through dialysis tube cellulose membrane. The antimicrobial activity of both encapsulated extracts was tested using broth dilution method at concentration 2×10^7 CFU/mL with slightly modification by adding a few drops of HCl into growth media. This modification was representing the acidic environment in human stomach at pH 4. After 24 hours of incubation, MC-M showed the positive result against *E. coli* and *Staphylococcus aureus* with the highest inhibition towards bacteria *S. aureus* as the concentration of bacteria observed is the lowest compared *E. coli* bacteria. Meanwhile, both crude *C. odorata* extract showed the lowest inhibition efficiency by showing the highest reading of target bacteria concentration compared to encapsulated sample.

Acknowledgments

This work was supported by School of Mechatronic and School of Bioprocess, Universiti Malaysia Perlis, Malaysia.

References

[1] Manaia C M 2013 A brief multi-disciplinary review on antimicrobial resistance in medicine and its linkage to the global environmental microbiota *Front. Microbiol.* **4** 1–14

[2] Bates B L, Hiles M C, Johnson C E 2013 Biofilm-Inhibiting Medical Product US Patent 8, 343, 536

[3] Stanley M C, Ifeanyi O E, Nwakaego C C and Esther I O 2014 Antimicrobial effects of Chromolaena odorata on some human pathogens *International Journal of Current Microbiology and Applied Sciences*. vol.3, no.3 1006-1012

[4] Omokhua A G, McGaw L J, Finnie J F and Van Staden J 2016 *Chromolaena odorata* (L.) RM King & H. Rob. (Asteraceae) in sub-Saharan Africa: A synthesis and review of its medicinal potential *Journal of ethnopharmacology* vol.183112-122

[5] Nath L R, Gorantla J N, Joseph S M, Antony J, Thankachan S, Menon D B and Anto R J 2015 Kaempferide, the most active among the four flavonoids isolated and characterized from *Chromolaena odorata* induces apoptosis in cervical cancer cells while being pharmacologically safe *RSC Advances*. vol.5, no.122 100912-100922

[6] Atindehou M, Lagnika L, Guérold B, Strub J M, ZhaoM., Van Dorsselaer and Sanni A 2013 Isolation and identification of two antibacterial agents from *Chromolaena odorata L.* active against four diarrheal strains *Advances in Microbiology*. vol.3, no.01, 115

[7] Lauryna P, Mindaugas M, Valdas J, Liudas I, Dalia M K and Jurga B 2019 Microencapsulation of Elsholtzia ciliata Herb Ethanolic Extract by Spray-Drying: Impact of Resistant-Maltodextrin complemented with Sodium Caseinate, Skim Milk, and Beta-Cyclodextrin on the Quality of Spray-Dried Powders *Molecules*. vol 24, 1461-1483

[8] Thophon S H S, Waranusantigul P, Kangwanrangsan N and Krajangsang S 2016 Antimicrobial activity of Chromolaena odorata extracts against bacterial human skin infections *Modern Applied Science*, vol.*10 no*. 2

[9] Vinceković M, Viskić M, Jurić S, Giacometti J, Kovačević D B, Putnik and Jambrak A R 2017 Innovative technologies for encapsulation of Mediterranean plants extracts *Trends in Food Science & Technology*. vol.69, 1-12

[10] Usunomena U and Efosa E G 2015 Phytochemical Analysis, Mineral Composition and In Vitro Antioxidant Activities of Chromolaena Odorata Leaves *Journal of Pharmaceutical Sciences*. vol.2, no.2, 16-20

[11] Goy R C, Morais S T and Assis O B 2016 Evaluation of the antimicrobial activity of chitosan and its quaternized derivative on E. coli and S. aureus growth *Revista Brasileira de Farmacognosia*. vol.26, no.1, 122-127

[12] Khan I, Miskeen S Khalil A T, Phull A R Kim J and Oh D H 2016 Foodborne Pathogens: Staphylococcus aureus and Listeria monocytogenes An Unsolved Problem of the Food Industry *Pakistan J. Nutr*. vol. 15, no. 6, 505–514

[13] Sadgrove N, Greatrex B and Jones G L 2015 α - Cyclodextrin Encapsulation Enhances Antimicrobial Activity of Cineole Rich Essential Oils From Australian *Species of Prostanthera* (*Lamiaceae*). vol.2, no.2, 30–38

[14] Trivedi M K, Branton A, Trivedi D, Nayak G, Mishra R and Jana S 2015 Characterization of Physicochemical and Thermal Properties of Biofield Treated Ethyl Cellulose and Methyl Cellulose *International Journal of Biomedical Materials Research*. vol.3, no.6, 83-91

Internet of Things (IoT) Fall Detection using Wearable Sensor

Loh Mei Yee[1], Lim Chee Chin[1*], Chong Yen Fook[1], Maslia Binti Dali[2], Shafriza Nisha Basah[1], Lim Sin Chee[3]

[1] School of Mechatronics Engineering, University Malaysia Perlis, Perlis, Malaysia
[2] Hospital Tunku Fauziah, Kangar, Perlis, Malaysia
[3] First City University College, Bandar Utama, Petaling Jaya, Malaysia.
*cclim@unimap.edu.my

Abstract. The IoT fall detection system detects the fall through the data classification of falling and daily living activity. It includes microcontroller board (Arduino Mega 2560), Inertial Measurement Unit sensor (Gy-521 mpu6050) and WI-FI module (ESP8266-01). There total ten (10) subjects in this project. The data of falling and non-falling (daily living activity) can be identified. The falling is the frontward fall, while the daily living activity includes standing, sitting, walking and crouching. K-nearest neighbour (k-NN) classifiers were used in the data classification. The accuracy of k-NN classifiers were 100% between falling and non-falling class. The feature was selected based on the percentage of accuracy of the k-NN classifier. The features of the Aareal.z (97.14%) and Angle.x (97.24%) were selected due to the good performance during the classification of the falling and non-falling class. The performance of the Aareal.z (58.41%) and Angle.x (57.78%) were satisfactory during the sub-classification of the non-falling class. Hence, the feature of Aareal.z and Angle.x were selected as the features which were implemented in the IoT fall detection device.

1. Introduction

A patient falling will cause bone fracture, haemorrhage and even death. The severity of the injury was divided into four levels, such as minor, moderate, major and death. It can be reduced to the minor level by detecting the patient fall. The fastest the fall detected, the lowest the risk of patient falling [1]. The Internet of Things (IoT) fall detection system is developed to overcome some problems. For instance, limited number of the healthcare staffs in the rural or city based healthcare sector, increasing number of patient fall and limitation of the current systems [2]. The problem of limited number of the healthcare staffs in the rural or city based healthcare sector occurs when the healthcare staff is not enough to provide professional care to the increasing number of patients [3], [4], [5]. The problem of the increasing number of patient fall occurs when the patient fall is unable or hardly to be detected. Lastly, there is some limitation of the current systems. The current systems include a camera-based fall detection system [6], 3D fall detection system [7], and multiple sensor fall detection system [8]. This type of system is heavy in size, non-portable and unable detects fall in real-time [9]. Hence, IoT fall detection system is needed to be developed to overcome these limitations.

2. Methodology

2.1 Structural design of the IoT fall detection

An overview of an IoT fall detection device consists of three main parts including data acquisition, data concentration processing and a back-end system is shown in Figure 1(a). A sensor node of an

IoT-based fall detection system is responsible for acquiring motion data and transmitting the data via a wireless communication protocol to a smart gateway. Depending on particular fall detection systems, the collected data can be pre-processed or kept intact before being transmitted. In this system, the collected data (raw data) is processed by complex algorithms or methods like fall detection based on k-nearest neighbor algorithm. Then, the processed data will be classified into falling and non-falling condition. The data and result will be stored in cloud processing analytics. The visualization of data is in the mobile application.

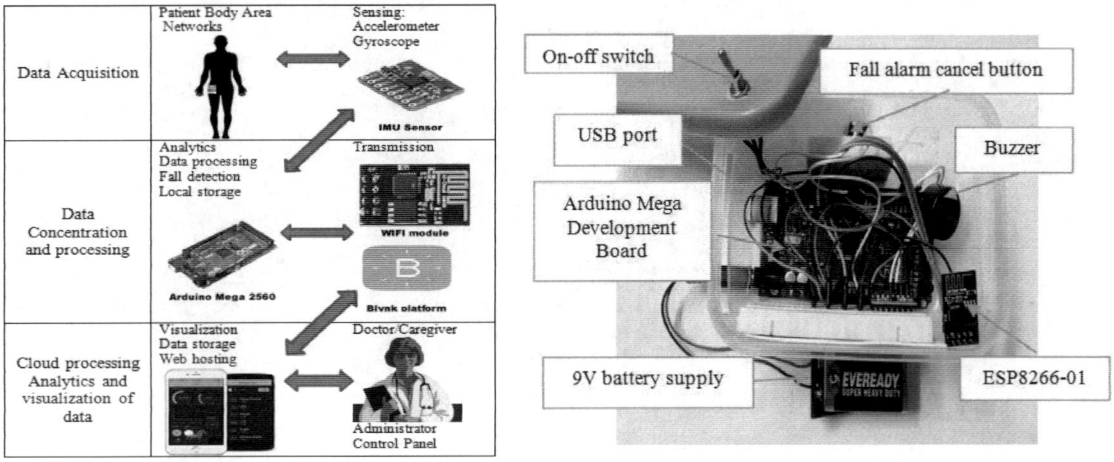

(a) (b)

Figure 1. An IoT fall detection device (a) Overview system (b) Prototype

This system includes of six components, such as microcontroller board (Arduino Mega Development Board (AMDB)), IMU sensor (Gy-521 MPU6050), WI-FI module (ESP8266-01), buzzer alarm, on-off switch and push button as shown in Figure 1. Arduino Mega Development Board was equipped with the ATMEGA2560 processor, which features a maximum clock rate of 16MHz. It was embedded with 54 general-purpose digital I/O pins and 6 communications TX/RX pins that used in the connection with WI-FI module (ESP8266-01). It also supports serial communication of I2C and SPI bus which was used in the connection with the IMU sensor (Gy-521 MPU6050) [10].

The construction of WI-FI module based on IEEE 802.11, is able to support transparent transmission mode and multiple network protocol [11]. It is a self-contained SOC with integrated TCP/IP protocol stack that can allow any microcontroller access to the WI-FI network. It features 802.11 b/g/n, WI-FI Direct or WI-FI peer-to-peer (P2P) is capable of hosting an application of offloading all WI-FI networking functions from another application processor. The module has an integrated 32-bit CPU, which can be used as application processor, and 1MB flash memory [12]. ESP8266-01 is low energy consumption and suitable for the application in mobile devices, wearable electronics and networking application design [13]. ESP8266-01 is selected as the IoT device based on the following strength and advantages, such as low power consumption, easy to compute, light weight and cost-effective.

The piezoelectric buzzer contains the positive and negative pin. The positive pin of the piezoelectric buzzer is connected to the digital pin 13 on the AMDB while the negative pin of the piezoelectric buzzer is connected to the ground of the AMDB. The 3 pins toggle switch acts as on-off switch. It contains pin 1, 2, 3, 4 and 5. When the switch turns to pin 1, the circuit will become "ON" state, while the circuit will become "OFF" state when the switch turns to pin 2. The pin 3, 4, 5 are connected to the ground pin, digital pin 3 and 5V on the AMDB, respectively. The pushbutton act as fall alarm cancel button. When the button is pressed, the fall alarm will be cancelled. The pin 1 is connected to the digital pin 4 on the AMDB and the pin 2 is connected to the ground. The prototype of the IoT fall detection device is shown in Figure 1(b).

The IMU sensor in small size and can be placed in different position of the human body. IMU is unaffected by the external signals or reference [14-15]. Gy-521 MPU6050 is a device that combines the 3-axis gyroscope and a 3-axis accelerometer on the same silicon die together with an on-board

Digital Motion Processor (DMP) as in Figure 2. R represents resulted distance of the Rx, Ry and Rz and calculated by Equation 1. The angle of Axr, Ayr and Azr are calculated by using Equation 2, 3 and 4.

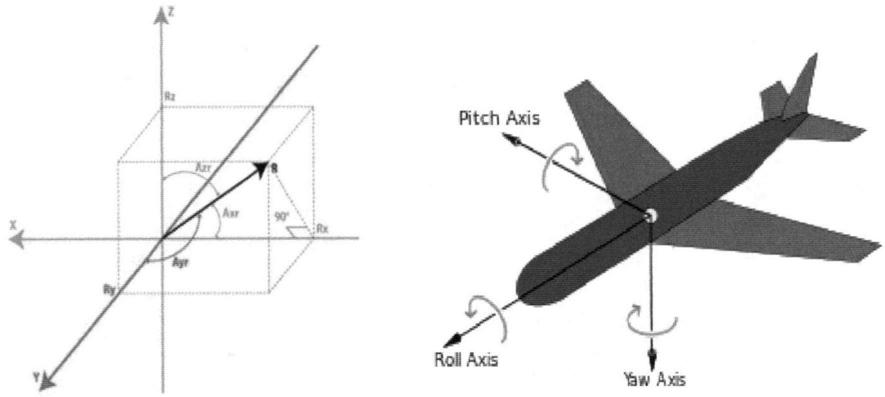

(a) (b)

Figure 2. IMU sensor characteristic: (a) Tri-axial accelerometer [16]; (b) Yaw, pitch and roll orientation [17]

$$\mathrm{R} = \sqrt{(Rx)^2 + (Ry)^2 + (Rz)^2} \quad (1)$$

Where, R = Resulted distance
Rx = Distance along the x-axis
Ry = Distance along the y-axis
Rz = Distance along the z-axis

$$\mathrm{Axr} = \cos^{-1}\frac{Rx}{R} \quad (2)$$

$$\mathrm{Ayr} = \cos^{-1}\frac{Ry}{R} \quad (3)$$

$$\mathrm{Azr} = \cos^{-1}\frac{Rz}{R} \quad (4)$$

Where, Axr = angle between R and the x-axis
Ayr = angle between R and the y-axis
Azr = angle between R and the z-axis

The gyroscope was used to determine the yaw, pitch and roll orientation of the subject. The yaw, pitch and roll orientation are shown in the Figure 2 (b), which can be calculated by using Equation 5, 6 and 7. A Quaternion is represented in the form of $q_0 + q_1 i + q_2 j + q_3 k$. Quaternion number is used in the calculations that involving three-dimensional rotations, such as yaw, pitch and roll [18].

$$Roll = \phi = \arctan\left(\frac{2(q_0q_1 + q_2q_3)}{1 - 2(q_1^2 + q_2^2)}\right) \quad (5)$$

$$Pitch = \theta = \arcsin\left(2(q_0q_2 - q_3q_1)\right) \quad (6)$$

$$Yaw = \psi = \mathrm{atan}\left(\frac{2(q_0q_3 + q_1 + q_2)}{1 - 2(q_2^2 + q_3^2)}\right) \quad (7)$$

2.2 Data Extraction

The I2C clock of the IMU sensor is set up as 400 kHz and the serial communication is initialized as 115200 bits per second. The Arduino interrupt pins are set up or configure to fetch the IMU sensor data. The Digital Motion Processing (DMP) is initialized. Then, the Arduino interrupt detection enabled. The data will be stored in the first in, first out (FIFO) buffer [19]. FIFO buffer is temporary data storage and store the data that arrive to a microcontroller peripheral asynchronously. The interrupt status is set as 0x02 memory location to read the data. Then, the tri-axial accelerometer and gyroscope data were read from the FIFO and displayed on the serial monitor. The parameters of the tri-axial distance and angle were extracted.

2.3 Data Testing and Validation
The tri-axial distances were tested in the range from the 0 m to 0.5 m, while the angles were tested from 0 degrees to 180 degrees. The efficiency of the prototype was validated by calculating the average time taken for the IMU sensor calibration process. Ruler, tape, and projector were used during testing and validation process. Each axis was tested and validated by using the same procedures. Firstly, the IMU sensor was placed on the 0 cm of the ruler as the origin of the x-axis. Then, the IMU sensor was moved along the fixed axis of ruler and stopped at the 10 cm of the ruler. The distance travelled was identified by the IMU sensor data and displayed on the serial monitor of the Arduino IDE software. Next, the testing process for the 20 cm, 30 cm, 40 cm, and 50 cm was carried out by using the same procedures.

For the testing and validation of the angle, the IMU sensor was placed horizontally beside the origin of the projector, and then the IMU sensor was elevated to 30 degrees along the semicircle of the projector. The measured angle was identified through the observation of the serial monitor of the Arduino IDE software. The testing for the degree of 60, 90, 120, 150 and 180 was using the same procedure. The IMU sensor data were validated by calculating the average of the total percentage error in the five trials [20] by using Equation 8.

$$\%\,E = \left|\frac{Actual\ value - Measured\ value}{Actual\ value}\right| * 100\% \tag{8}$$

2.4 Data Collection
Data was collected from 10 (5 males and 5 females) University Malaysia Perlis (UniMAP) students. The average age of subjects was 24 ± 1 year. The average weight was 60.1 kg and the average height was 166.1 cm. Each subject performed 4 different Activity Daily Living (ADL) tasks and frontward falling as shown in Figure 3. The 4 different ADL tasks were the sitting, standing, walking and crouching. For the falling protocol, the subject will perform frontward fell on a mattress with a height of 9.5 cm. For sitting protocol, the subject will sit on a chair with height of 45 cm. For standing protocol, the subject will stand steadily about 2 seconds. For crouching protocol, subject will perform a vertical crouched. For walking protocol, the subject will walk with distance of 120 cm. Each task will be performed 5 times. The features of the acceleration, distance and angle in the x, y and z-axis will be recorded.

(a) (b) (c) (d) (e)

Figure 3. Protocol: (a) Falling; (b) Sitting; (c) Standing; (d) Crouching; (e) Walking;

2.5 Features Extraction
Eight features had been extracted, which are Aareal.x, Aareal.y Aareal.z, Dist.x, Dist.y, Dist.z, Angle.x and Angle.z. The Aareal.x, Aareal.y and Aareal.z were the acceleration in the x, y and z-axis respectively. Then, the Dist.x, Dist.y and Dist.z were the distance of x, y and z-axis respectively. The

Angle.x was angle between subject's motion axis and x-axis; while angle.z was the angle between subject's motion axis and z-axis. The acceleration in the x, y and z-axis were calculated by using the DMP of IMU sensor. The function of mpu.getGetQuaternion (&q, fifoBuffer) and mpu.dmpGetGravity were used to extract the quaternion angle and gravity vectors. The quaternion angle and gravity vectors were used for the calculation of acceleration in the x, y and z-axis.

The distance in x, y and z-axis were calculated through the subtraction between current acceleration and previous acceleration. The result was not multiplied by the time taken because the time taken was set as constant. The angle in the x and z-axis were calculated by the equation 3 and 5. Angle.y was not selected because it was too small and be neglected. Since the motion of the subject was on the reference axis (Aareal.y) during the ADL activities and falling action, hence, the angle between the subject's motion and the reference axis was too small and be neglected.

2.6 Data Classification
The data had been classified by using the k-NN. k-NN classifier is a type of instance-based learning that is simple to implement in the data classification. The working principle is based on the measuring the distance or similarity between the tested data and the training data. The k-value is ranged from 1 to 10 and needs to be varied in order to find the match class between the testing and training data. The original sample will randomly partition into k equal sized of subsamples. One of the subsamples will be retained as the validation data for testing of the model, while the remaining k-1 subsamples used as training data. All data will be used for both training and validation by using the k-fold cross-validation method [10, 21]. The performance of the k-NN classifier depends significantly on the distance used [22]. The distances of the Euclidean, city-block, cosine and the correlation are used.

3. Result and Discussion
The result and discussion had been divided into four parts, such as testing and validation, classification between the falling and non-falling class, sub-classification of the non-falling class, and feature selection.

3.1 Testing and Validation
There are only five features out of eight features were selected for testing and validation, which were Dist.x, Dist.y, Dist.z, Angle.x and Angle.z. However, Aareal.x, Aareal.y and Aareal.z were not selected because acceleration is calculated from the differentiation of distance with respect to time.

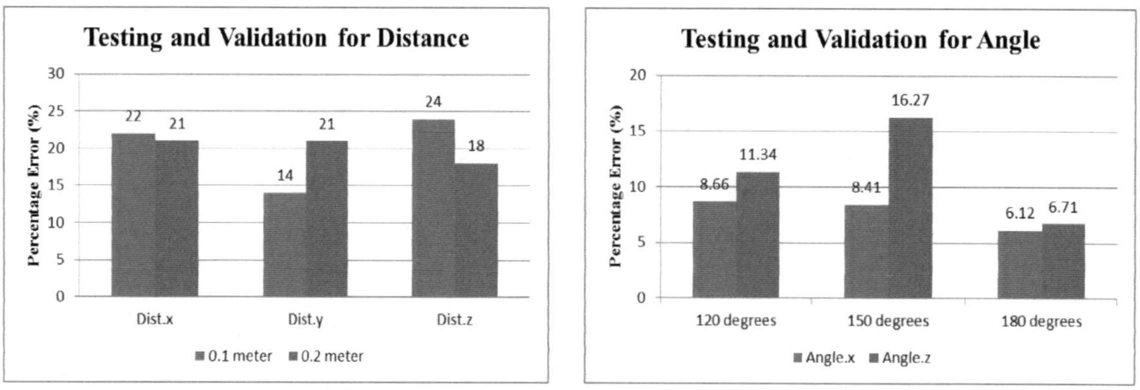

(a) (b)

Figure 4. Testing and Validation of (a) Distance (b) Angle

The result of testing and validation of distances for the 0.1 meter and 0.2 meter is shown in Figure 4(a). There are two distances (0.1 and 0.2 meter) were used because average height of the subject is within 1.66 cm. The result of testing and validation of angles for the 120 degrees, 150 degrees and 180 degrees is shown in Figure 4(b). The 120 degrees, 150 degrees and 180 degrees are used because the values within this range represent the angle between subjects' motions with relevant axis. The results are the average of percentage error for the parameters in five trials. The percentage error of the

Angle.x is lower than the Angle.z is due to the subject was fell toward the x-axis during the frontward fall. Hence, percentage of accuracy for the Angle.x will be higher than the Angle.z.

3.2 Feature Selection

Figure 5(a) show the performance of k-NN classification between falling and non-falling of the eight features using four distances. The features of the Aareal.z (97.14%) and Angle.x (97.24%) show the highest percentage of accuracy among the total four distances. This is because the Aareal.z is affected by the height of the subject while the Angle.x is affected by the angle between the subject's motion axis and the x-axis. The value of the Aareal.z and the Angle.x in the falling class is different in polarity with the value of the Aareal.z and the Angle.x in the non-falling class. Hence, the difference of distance measured between these two classes is large and the percentage of accuracy is high. As the result, Aareal.z and the Angle.x are the best features among the eight features.

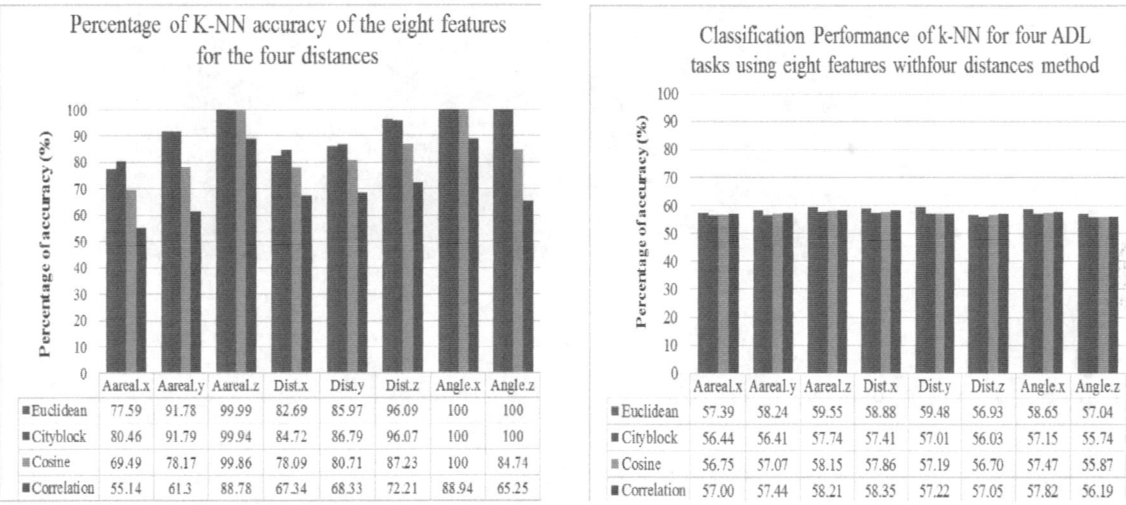

(a) (b)

Figure 5. k-NN classification performance four distances for each eight features. (a) Falling vs. Non-falling classification (b) Four activities daily living (ADL) classification

Based on Figure 5(b), the average accuracy for four distances of features Aareal.z (58.41%) and the Angle.x (57.78%) is the highest among the others eight features. This is because the polarity of the Aareal.z and Angle.x is same in four ADL subclasses. Hence, the difference in distance measured between the input data and the dataset is relatively small. The accuracy for four ADL subclasses is lower as compared with the classification between falling and non-falling class. This is because the feature value of the dataset in the falling and non-falling class is big in different. Thus, the dataset is easily in classification and high percentage of accuracy is resulted. In conclusion, the features of the Aareal.z and Angle.x were selected as the best features that implemented in the Arduino hardware.

3.3 Classification between falling and non-falling

The falling and non-falling class has been classified by using the k-NN classification algorithm in Arduino hardware. There are on using Aareal.z and Angle.x. The k-fold cross-validation method is used for k-NN classifier. A total of 250 data had been folded into 10 sets of the data. The highest the k-fold, the highest the number of the training set obtained. There had obtained 250 testing data and 2250 training data, when 250 data are folded in 10-fold cross-validation. Hence, it aims to optimize the percentage of accuracy by reducing the percentage of error [23],[24]. Data was classified based on the k-value (1 to 10). If the k-value was set at 10, the effect of the noise was reduced during the data classification. The four different types of the distance were applied which produces 100% accuracy.

3.4 Sub-classification of non-falling class

Based on the Figure 6, the percentage of mean accuracy of the sub-classification of non-falling class for the four distances with k-NN classifier is shown. The data for the sub-class is the average of the 10 k-value. The percentage of accuracy of the standing sub-class for the correlation distance is the highest (66.72%). Then, the percentage of accuracy of the crouching (85.91%) and walking (36.61%) sub-class for the cosine distance is the highest. The percentage of accuracy of the sitting sub-class for the Euclidean distance is the highest (48.42%). Different type of distance is suitable for different sub-class. Hence, the suitability of the sub-class with the distance is indicated by the percentage of accuracy. The Euclidean distance is the best computation method because the overall percentage of accuracy is the highest (58.73%) among the distances.

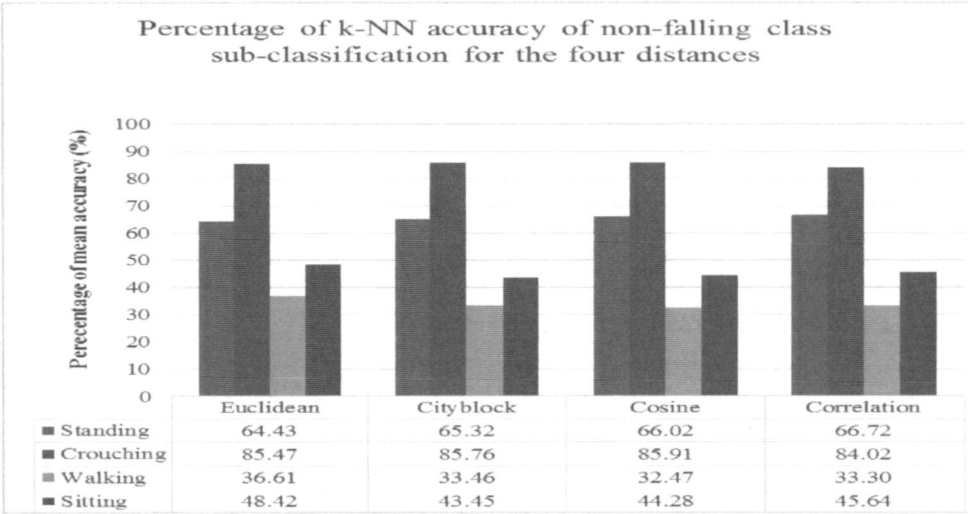

Figure 6. Percentage of accuracy of the sub-classification of non-falling class for the four distances with k-NN classifier.

4. Conclusion

In conclusion, this project aims to develop an IoT fall detection system by using wearable sensor. The IoT fall detection system had be developed with the microcontroller board (Arduino Mega 2560), IMU sensor (Gy-521 mpu6050) and WI-FI module (ESP8266-01). The parameters of IoT fall detection system include tri-axial distance and tri-axial angle had been tested and validated by calculated the percentage of error. The percentage error for the distances in x, y and z-axis are less than 25%, while the percentage error for the angle in x and z-axis is less than 20%. Thus, the IoT fall detection device is a good device that is capable to detect the fall with high accuracy. The dataset had been constructed through the data collection of the 10 UniMAP students. Eight features had been extracted from the dataset and classified through the k-NN. The features of the Aareal.z and Angle.x were selected based on the good performance during the classification. The data of the falling and daily living activity had been differentiated through the classification of the falling and non-falling. Then, the percentage of accuracy for the k-NN classifier by using k-values (1-10) and distances (Euclidean, city-block, cosine and correlation) are 100%. The non-falling class had been further classified into the sub-classes of standing, crouching, walking and sitting.

References

[1] Bouldin E L, Andresen E M, Dunton N E, Simon M, Waters T M, Liu M, Daniels M J, Mion L C and Shorr R I 2013 Falls among adult patients hospitalized in the United States: prevalence and trends *J Patient Saf* **9(1)** 13-17.

[2] Dykes P C, Carroll D L, Hurley A, et al. 2010 Fall Prevention in Acute Care Hospitals: A Randomized Trial *JAMA* **304(17)** 1912–1918.

[3] Campbell D 2014 NHS staff shortages pose risk to patients, warns watchdog | Society | The Guardian https://www.theguardian.com/society/2014/oct/17/nhs-staff-shortages-risk-patients-cqc-report.

[4] Jacobson R 2015 Widespread Understaffing of Nurses Increases Risk to Patients - Scientific American https://www.scientificamerican.com/article/widespread-understaffing-of-nurses-increases-risk-to-patients/.

[5] Chin L C, Basah S N, Affandi M, Shah M N, Yaacob S, Juan Y E and Din M Y 2017 Home-Based Ankle Rehabilitation System: Literature Review and Evaluation *Jurnal Teknologi* **79(6)** 1-21.

[6] Chin L C, Basah S N, Yaacob S, Juan Y E and Kadir A K A 2015 Camera Systems in Human Motion Analysis for Biomedical Applications *AIP Conference Proceedings* **1660(1)** 090006.

[7] Shanyu C, Chin L C, Basah S N and Azizan A F 2019 Development of Assessment System for Spine Curvature Angle Measurement *Proceedings of the 2019 8th International Conference on Software and Computer Applications (ICSCA '19), ACM* 397-402.

[8] Delahoz Y S and Labrador M A 2014 Survey on fall detection and fall prevention using wearable and external sensors *Sensors (Switzerland)* **14(10)** 19806–19842.

[9] Todd C and Skelton D 2004 What are the main risk factors for falls among older people and what are the most effective interventions to prevent these falls? *Copenhagen, WHO Regional Office for Europe (Health Evidence Network report)* http://www.euro.who.int/document/E82552.pdf , accessed 29 June 2019.

[10] Fook C Y, Hariharan M, Yaacob S, and Adom A H 2012 Malay speech recognition in normal and noise condition *IEEE 8th International Colloquium on Signal Processing and its Applications (CSPA 2012)* 409–412.

[11] Juang H S and Lurrr K Y 2013 Design and control of a two-wheel self-balancing robot using the arduino microcontroller board *IEEE Int. Conf. Control Autom. ICCA* 634–639.

[12] Sharma M L, Kumar S, Mehta N 2017 Smart Home System Using IOT *International Research Journal of Engineering and Technology (IRJET)* **4(11)** 1108–1112.

[13] Ilapakurti A, Vuppalapati J S, Kedari S, Kedari S, Chauhan C, and Vuppalapati C 2017 IDispenser-big data enabled intelligent dispenser *3rd IEEE International Conference Proceeding of Big Data Comput. Serv. Appl. BigDataService* 124–130.

[14] Datasheet E 2004 ESP8266 Serial Esp-01 WIFI Wireless *ESP8266 Ser. Esp-01 WIFI Wirel.* https://components101.com/sites/default/files/component_datasheet/ESP8266%20Datasheet.pdf, accessed 30 June 2019.

[15] Chin L C, Basah S N, Ali M A, Fook C Y 2018 Wearable Posture Identification System for Good Sitting Position *Journal of Telecommunication, Electronic and Computer Engineering (JTEC)* **10** 135-140.

[16] Gadget Gangster 2017 Accelerometer & Gyro Tutorial: 3 Steps http://www.instructables.com/id/Accelerometer-Gyro-Tutorial, accessed 30 June 2019.

[17] Nancy Hall 2017 Aircraft Rotations https://www.grc.nasa.gov/WWW/K-12/airplane/rotations.html, accessed 30 June 2019.

[18] Riedesel P 2016 Tilt sensing with low-cost inertial measurement units (IMUs) : Sensor calibration, accuracy specifications and application range http://urn.kb.se/resolve?urn=urn:nbn:se:hig:diva-22497, accessed 12 July 2019.

[19] Choi J, Boyle D K, Dunton J, and Staggs N 2013 Safety & Outcomes *J. Nurs. Adm. Med. Care J. Nurs. Care Qual.* **43(271)** 586–591.

[20] Stratify Labs 2013 A FIFO Buffer Implementation https://stratifylabs.co/embedded design tips/2013/10/02/Tips-A-FIFO-Buffer-Implementation, accessed 13 July 2019.

[21] Aijuan S, Guangyuan S I, and Qiongchan G U 2016 Study on Remote Medical Monitoring System Based on MSP430 AND CC2530 *2016 Chinese Control and Decision Conference (CCDC), Yinchuan* 2415–2418.

[22] Dangeti P 2017 Statistics for Machine Learning *Packt Publishing Ltd.* https://books.google.com/books?id=C-dDDwAAQBAJ, accessed 13 July 2019.

[23] Prasath V B S, Alfeilat H A A, Lasassmeh O and Hassanat A B A 2017 Distance and similarity measures effect on the performance of k-nearest neighbor classifier—a review. arXiv preprint arXiv:1708.04321

[24] Ghosh A K 2006 On optimum choice of k in nearest neighbor classification *Comput. Stat. Data Anal* **50(11)** 3113–3123.

Influence of Diclofenac Sodium Loading on Physicochemical and Mechanical Properties of Dual Layer Polyvinyl Alcohol Transdermal Patch

Shafizah Saadon[1], Saiful Izwan Abdul Razak[2*, 3], Al Emran Ismail [4], Nadirul Hasraf Mat Nayan[5, 6] and Khalida Fakhruddin[1]

[1]School of Biomedical Engineering and Health Science, Faculty of Engineering, Universiti Teknologi Malaysia (UTM), 81310 Skudai, Johor, Malaysia

[2]Centre for Advanced Composite Materials, Universiti Teknologi Malaysia (UTM), 81310 Skudai, Johor, Malaysia

[3]Medical Devices and Technology Group, School of Biomedical Engineering & Health Science, Faculty of Engineering, Universiti Teknologi Malaysia (UTM), 81310 Skudai, Johor, Malaysia

[4]Faculty of Mechanical and Manufacturing Engineering, Universiti Tun Hussein Onn Malaysia (UTHM), 86400 Batu Pahat, Johor, Malaysia

[5]Microelectronic and Nanotechnology-Shamsuddin Research Centre (MiNT-SRC), Universiti Tun Hussein Onn Malaysia, 86400, Parit Raja, Johor, Malaysia

[6]Faculty of Engineering Technology, Universiti Tun Hussein Onn Malaysia, KM1 Jalan Panchor, 84600, Pagoh, Johor, Malaysia

*saifulizwan@utm.my

Abstract. Transdermal drugs delivery offers many advantages over conventional routes of drug delivery that led to this burgeoning interest in this route in recent years. The aim of this work is to investigate the influence of diclofenac sodium (DS) loaded dual layer PVA patch on physicochemical and mechanical properties. The influence of DS-loaded dual layer PVA patch was employed and characterized using Scanning Electron Microscopy (SEM) to observe the polymeric interaction between PVA cryogel and electrospun nanofiber, the FTIR were used to study the interaction of DS in dual layer PVA patch and tensile strength were tested to study the mechanical properties of the DS-loaded dual layer PVA patch. The results show the polymer interaction between PVA cryogel and electrospun nanofiber were observed. The DS spectra were found separately from PVA matrix. As for mechanical strength the curves of DS-loaded were decreased as the percentage of DS loading increases. These results shows that the dual layer PVA patch can be used as the carrier for transdermal DS delivery.

1. Introduction

Transdermal drug delivery system (TDDS) is adhesive drug-containing devices of defined surface area that delivers the specific amount of active drug across the skin, into the systemic circulation [1]. TDDS have an advantages over conventional pharmaceutical dosage formulations, such as circumventing gut and hepatic first-pass metabolism, improving drug bioavailability, reducing administration frequency, and stabilizing drug delivery profiles [2]. For the past few years, transdermal drug delivery systems has gain great interest with the development of various types of transdermal therapeutic systems for several drugs such as, Estradiol for steroidal delivery [3], combination patches containing Gestodene and Levonorgestrel for contraception and hormone replacement [4], Mafenide Acetate for antimicrobial delivery [5], Carvedilol for anti-hypertensive[6], Titanocene Dichloride for antitumor [7]. Insulin for anti-diabetic delivery [8] and Meloxicam delivery systems for anti-inflammatory [9].

Polyvinyl Alcohol (PVA) is one of the polymeric drug carriers that are a semi-crystalline polymer with excellent chemical and thermal stability and demonstrated great interest because of its biocompatibility, non-toxic; non-carcinogenic and water permeable polymer with great electrospinnability [10]. According to Baker *et al.,* [11], PVA shows higher tensile strength and elongation before breaking than hydrogels such as polyhydroxy ethyl methacrylate, making PVA a suitable hydrogel for soft contact lenses, extending wearing time without inducing hypoxia to the cornea. Furthermore, High mechanical strength, rubber-like elasticity and no adhesion to surrounding tissue make PVA gels potential materials for soft tissue replacements, artificial cartilage and other artificial organs [11-15]. PVA hydrogels have been investigated for replacement of damaged cartilage due to their high water content, as well as their elastic and compressive mechanical properties.

Recently, electrospinning techniques gain interest in transdermal drug delivery. This supported by Rives *et al.*, [16] in their studies emphasized that one of the apparent advantages of the electrospinning process compared to conventional film-casting techniques is the highly porous structure of electrospun nanofiber membrane, which exhibits a much higher surface area that could potentially allow drug molecules to diffuse from the matrix more effectively.

Diclofenac sodium (DS) is nonsteroidal anti-inflammatory drugs used to relieve the inflammation, swelling, stiffness and joint pain associated with rheumatoid arthritis, osteoarthritis (the most common form of arthritis) and ankylosing spondylitis (arthritis and stiffness of the spine), toothache, trauma, wound, burn and dysmenorrhea treatment because of their analgesic, antipyretic, and anti-inflammatory roles [17]. According to Satyabrata *et al.,* [18] when DS administered orally, only about 50% of the dose is biologically available because of the first-pass hepatic metabolism, and it possesses short biological half-life.

To the author's knowledge, the combination cryogel and electrospun nanofiber with improved properties as a carrier for DS for TDDS have not been done yet. The aim of this work is to investigate the influence of DS loading on physical and mechanical properties of dual layer PVA for potential prolonged transdermal DS delivery.

2. Methodology

2.1 Fabrication of DS-loaded dual layer PVA patch

2.1.1 Preparation of PVA electrospun nanofibers membrane

A weighed amount of PVA powder was dissolved in distilled water at 80° C for 3 h to prepare a PVA solution at a fixed concentration of 10% w/v. After that, the solution was cooled down to room temperature (25° C). Electrospinning of the as-prepared solutions was carried out by connecting the emitting electrode of positive polarity from a Gamma High-Voltage Research ES30PN/M692 high voltage DC power supply to the solutions contained in a standard 5-ml syringe. The open end of which was attached to a blunt gauge-23 stainless steel needle (outer diameter = 0.91 mm), used as the nozzle, and the collection plate laminated with aluminium foil (dimension = 15 cm x 15 cm), used as the fiber-collection device. A fixed electrical potential of 20 kV was applied across a fixed distance of 15 cm between the tip of the nozzle and the outer surface of the collector plate (i.e., the electrostatic field strength of (20 kV /15 cm). The feed rate of the solutions was controlled to about 1 ml h^{-1} utilizing a syringe pump.

2.1.2 Preparation of PVA solution loaded with DS

The fabrication of freezing-thawing PVA solution was prepared based on the procedure referred in section 2.1.1. The polymer was entirely dissolved, and the obtained transparent solutions were slowly cooled to room temperature. Same methods were applied for DS loaded in PVA solution. Different mass% of DS (i.e., 1.0, 1.5and 2.0 wt% of PVA) was loaded by dissolving in 5 ml ethanol in a separate beaker and then, adding it slowly to the above polymer solution by heating gently to avoid re-precipitation.

2.1.3 Combined techniques of dual layer PVA patch

The aqueous PVA solutions (from section 2.1.2) were then poured on the surface of the electrospun nanofibers membrane (from section 2.1.1) that has been placed inside the specifically designed mould with dimensions: length x width x thickness: 10cm x 10 cm x 1.5cm. Freezing-thawing of dual layer PVA patch was obtained by subjecting the unloaded and DS loaded PVA aqueous solutions with corresponding DS concentrations to repeated 5 cycles of freezing-thawing, consisting a 24 h freezing step at -20 °C followed by a 2 h thawing step at room temperature.

2.2 Scanning Electron Microscopy (SEM) of dual layer PVA patch

The freeze-dried dual layer PVA samples were cut into small dimension (5 mm x 5 mm), and the samples were directly sent to Auto Fine Coater Machine for a sputtered thin layer of gold on its surface at 25 mA plasma current and 2 Pa of chamber pressure to make them conducting samples. The function of the coating is to make sure the insulating freeze-dried dual layer PVA samples are electrically conductive during high-resolution electron imaging applications. The dual layer PVA samples were then examined by using SEM of JEOL-JSM6380LA (Japan) operates at 15 kV at 50 and 100 μm magnifier under high vacuum.

2.3 Fourier-Transform Infrared Red (FTIR) spectroscopy of dual layer PVA patch

DS-loaded dual layer PVA samples were cut into a small cube (10 x 10x 10 mm) and place it at FTIR sample holder. FTIR spectra were recorded in the range of 400 to 4000 cm^{-1} collecting 35 scans with 4 cm^{-1} resolution, in the transmittance mode. It is used to study the interaction between DS and dual layer PVA patch.

2.4 Mechanical testing of dual layer PVA patch

Tensile strength, elongation at break and strain at break of both (unloaded and DS loaded) dual layer PVA patch were measured by Universal Testing Machine (LLOYD Instruments LR30K), with load range; 10N. Each specimen was cut according to ASTM D 412. The crosshead speed and the gauge length were 50 mm/min and 33 mm respectively. A minimum of four samples was analyzed to obtain an average result, and all specimens were tested at room temperature.

3. Results and discussions

3.1 Morphological structure of dual layer PVA patch

For fabrication of DS-loaded dual layer PVA patch process mentioned on section 2.1.3. The combined methods (freezing-thawing cryogel and electrospun nanofiber) has been successfully conducted. As for Figure 1 (a) and (b), represent the surface of unloaded PVA patch and DS-loaded dual layer PVA patch with 5 cycles of freezing-thawing respectively.

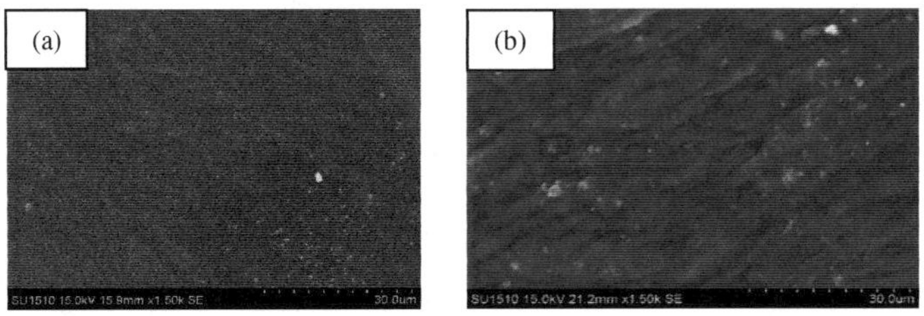

Figure 1. Surface morphology of 5 cycles of (a) Unloaded dual layer PVA patch and (b) DS-loaded dual layer PVA patch.

The micrograph shows rough surface of cryogel after finishing freezing-thawing cycles. This might be due to the presence of the drug's molecule incorporate with cryogel that slightly altered the linkage during freezing-thawing process. DS is partially insoluble in nature, therefore small particles of DS can be observed scattered on the surface of cryogel and nanofibers. Though the presences of DS slightly effect the polymeric structure, number of cycles somehow does not show significant different. On physically observation, the dual layer PVA patch have phase distinction between gel and nanofiber. This claim can be supported by cross-section of the dual layer PVA patch showed in Figure 2.

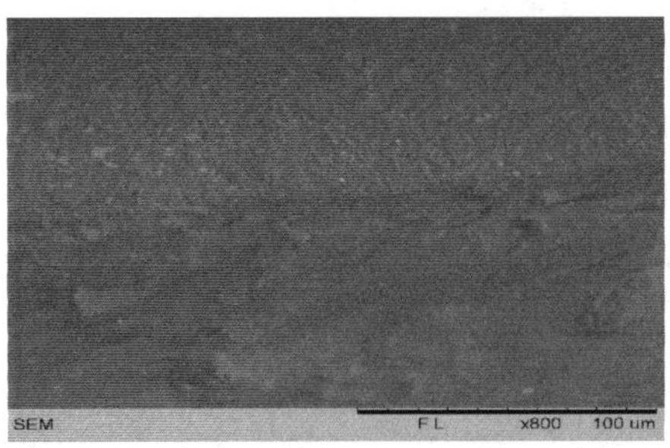

Figure 2. SEM micrograph for cross-section of dual layer PVA patch.

3.2 FTIR spectroscopy of dual layer PVA patch

It is well known that the infrared spectrometer is a powerful instrument for the investigation of the bonding behavior. The FT-IR absorption spectra of unloaded and DS-loaded dual layer PVA patch are shown in Figure 3. 2L-3ml-5C is represent the 5 cycles freezing-thawing of dual layer with 3 ml (volume of PVA spinning) thickness of electrospun nanofiber. While 1.0, 1.5 and 2.0 DS represent the percentages of the drug loading in the dual layer PVA patch.

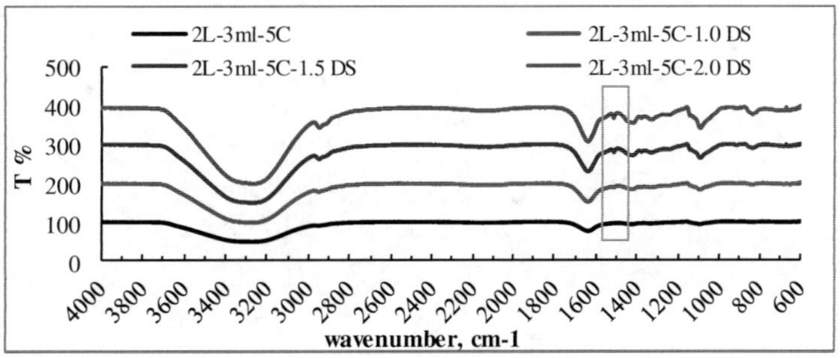

Figure 3. FTIR spectra for unloaded and DS-loaded dual layer PVA patch with different percentage drug loading.

The most characteristic bands of PVA and their respective assignments are observed between range of 3500 and 3100 cm^{-1} and are ascribed to the O-H stretching from the intermolecular and intra molecular hydrogen bonds. The absorption band at 2944 cm^{-1} and a shoulder at 2900 cm^{-1} correspond to the stretching of -CH2- and -CH-, respectively. Further, the band at 1750 cm^{-1} and the shoulder at 1641 cm^{-1} correspond to the unhydrolyzed acetate groups. The band at 1420 cm^{-1} is attributed to -CH2- bending. The band at 1091 cm^{-1} corresponds to C-O unbounded and it could be associated with PVA crystallinity.

Furthermore, the present of DS in dual layer PVA patch (Figure 3), the spectra shows the distinctive peaks at 1566 cm^{-1} (carboxylate COO- asymmetric stretching) and 1515 cm^{-1} (dichlorophenyl ring) [19]. It also can be observed that the intensity of DS peak was increased when the percentage of DS loading increases. Therefore, presence of drug was confirmed in samples as their respective spectra revealed characteristic peaks of DS. These peaks appeared either in separate form, or fused with the characteristic peaks of dual layer PVA patch, confirmed the presence of the drug in the cryogel that did not interact with any of the components in polymer matrix [20].

3.3 Mechanical properties of dual layer PVA patch

The freezing-thawing cycles and DS loading percentage are an important factor determining physical properties, so the mechanical properties of dual layer PVA patch with different percentage were compared Figure 4, the curve represent the stress strain curve for 2L-3ml-5C with 3 different percentage of DS loading.

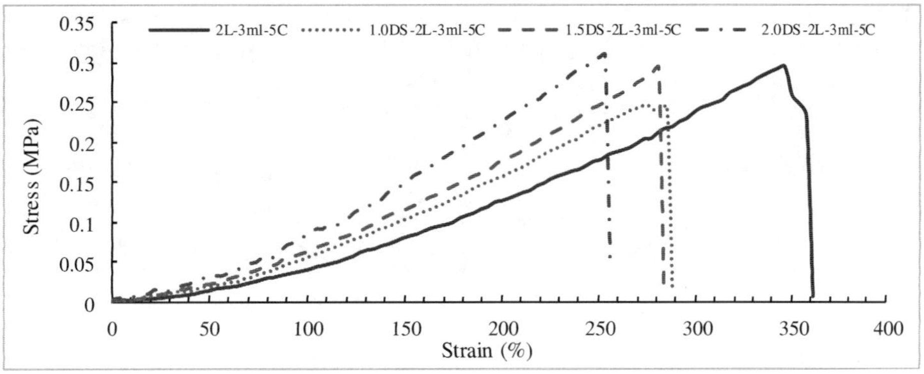

Figure 4. Typical stress–strain curves for unloaded and DS-loaded dual layer PVA patch for 5 cycles.

As shown in Figure 4, the tensile strength were improved as the percentage of DS-loaded in dual layer PVA patch increases (Table 1). However, when comparing with unloaded dual layer PVA patch, the strain percentage of unloaded dual layer PVA shows greater than DS-loaded dual layer PVA patch. This might be due to hydrogen bonding or intermolecular interaction between the drug model and PVA molecules, resulting in its relatively reduced crosslinking density, thus lowering the elongation at break [21].

Table 1. Mechanical Properties of 5 cycles of unloaded and DS-loaded PVA patch with different DS percentages loading.

Cryogel samples	Tensile strength, σ (MPa)	Elongation at break (%)	Young Modulus, *E*
2L-3ml-5C	0.30 ± 0.06	360.92 ± 30.99	0.23 ± 0.06
1.0DS-2L-3ml-5C	0.26 ± 0.02	288.01 ± 41.66	0.29 ± 0.04
1.5DS-2L-3ml-5C	0.30 ± 0.04	283.89 ± 21.44	0.23 ± 0.06
2.0DS-2L-3ml-5C	0.32 ± 0.04	255.70 ± 28.31	0.31 ± 0.06

4. Conclusions

The influence of DS loading in the dual layer PVA structures as well as its physicochemical properties and mechanical strength of the dual layer PVA patch were conducted. It can be observed that after the DS-loaded dual layer PVA patch freezing-thawing cycles is ended, the surface of dual layer cryogel was found uneven compared to control dual layer PVA patch. While the FTIR spectra of DS-loaded dual layer PVA patch, DS peak were observed separately in the spectra indicates that the DS does not interact in PVA matrix during freezing-thawing cycles. The intensity of DS peak also found increases with increasing percentage loading. While for the tensile strength and the young modulus, the result shows that the expansion and elongation at break was found to decrease when the percentage DS loading increases. Overall, the fabrication of dual layer PVA patch using combined methods has improved the physicochemical as well as mechanical properties for further drug release assessment.

Acknowledgments

The authors would like to acknowledge Ministry of Higher Education Malaysia for funding through Fundamental Research Grant Scheme.

References

[1] Rastogi V, Pragya P and Upadhyay 2012 *Interational Journal of Pharmaceutical Sciences and Research* 3(**7**) 1955-1970.
[2] Bennet D and Kim S A 2013 *J. Biomater. Sci., Polym. Ed* **24** 185-209.
[3] Wyrwa R, Möller S, Döhler K, Heppe A and Schnabelrauch M 2015 *BioNanoMat* **16** 184-193.
[4] Galzote R M, Rafie S, Teal R and Mody S K 2017 *International Journal of Women's Health* **9** 315-321.
[5] Abbaspour M, Makhmalzadeh B S, Rezaee B, Shoja S and Ahangari Z 2015 *Jundishapur J Microbiol* **10** e24239.
[6] Namrata V, Lin S and Madan P L 2013 *Asian journal of pharmaceuticalsciences* **8** 28 -38.
[7] Chen P, Wua Q S, Ding Y P, Chu M, Huang Z M and Hue W 2010 *European Journal of Pharmaceutics and Biopharmaceutics* **76** 413-420.
[8] Sadhasivam L, NDey N, Francis A P and Thiyagarajan, D T 2015 *Int J Pharm Pharm Sci*, **5** 84-88.
[9] Yu Q, Wu X, Zhu Q, Wu W, Chen Z and Li Y 2017 *Asian Journal of Pharmaceutical Sciences* 1-34.
[10] Shen X, Xu Q, Xu S, Li J, Zhang N and Zhang L 2014 *J Nanosci Nanotechno.*, **14** 5258-5265.
[11] Baker M I, Walsh S P, Schwartz Z and Boyan B D 2012 *Journal of Biomedical Materials Research Part B: Applied Biomaterials 100B* **5** 1451-1457.
[12] Kadri N A, Raha M G and Pingguan-murphy B 2011 **8** 1489-1493.
[13] Karimi A and Navidbakhsh M 2013 *Materials Technology* **2** 90-100.
[14] Holloway J L, Spiller K L, Lowman A M and Palmese G R 2011 *Acta Biomater* **7** 2477–2482.
[15] Maruyama H, Yokota Y, Hosono K and Arai F 2019 *Sensors* **5** 1102.

[16] Rives M F, Osuna Á A B , Tejedor J A G and Ribelles J L G 2017 *Material* **10** 1448.
[17] Hanan E, Fakhry G and Hanaa E S S 2011 *Asian J Pharm* **5** 33-40.
[18] Satyabrata B P, Kishore K, Muvvala S and Arun K D 2013 *J. Adv. Pharm. Edu. & Res* **3** 248-259.
[19] Wong R S H and Dodou K 2017 *Polymers* **9** 286.
[20] Pal K, Banthia A K and Majumdar D K 2007 *AAPS PharmSciTech* **1** e142-146.
[21] Kim J O, Choi J Y, Park J K, Kim J H, Jin S G, Chang S W 2008 *Biol Pharm Bull.* **31** 2277-2282.

Energy saving mechanism in fog assisted IoT enabled health monitoring framework

Luina Pani[1], Chinmaya Misra[2], Azian Azamimi Abdullah[3], Satya Ranjan Dash[4]*
[1]School of Computer Engineering, KIIT Deemed to be University
[3]School of Mechatronic Engineering, Universiti Malaysia Perlis, Malaysia
[2,4]School of Computer Applications, KIIT Deemed to be University

[1]luinapani@gmail.com,[2]cmisra@yahoo.com, [3]azamimi@unimap.edu.my,
[4]*sdashfca@kiit.ac.in

Abstract. This proposed research work discusses fog assisted IoT enabled health monitoring system performance. This system support local processing nearer the client machine for faster service and removing redundant data. This architecture is suitable for heavy traffic health data monitoring system like old or child monitoring or fitness care monitoring system. After processing huge data, only filtering data will be forwarded to cloud for processing. To transfer the huge amount of health monitoring data form IoT layer to Data Management and Processing Layer (DMPL) consume lots of energy. To minimize the energy requirement in the fog assisted DMPL layer, we adopt sleep mode operation and batch transferring data packet transfer technology. This paper studies the effect of sleeping mode on mean delay, number of data packets in the buffer and probability of blocking. The results show that proposed approach saves energy and effective framework for fog assisted health monitoring framework.

1. Introduction

Internet of Things plays a significant role in development of medical information systems and health care systems. It is a network of devices that are capable of collection and exchange of data among themselves [1]. In medical care the different diagnostic and imaging devices, sensors and other smart devices constitute the principal chunk of the IOT. Application of IOT in health care empowers real-time alerting, tracking, and screening, hence ensure and increase the safety of patients [2]. The structured and unstructured electronic health records, multi-dimensional image data of ultra sound, FMRI etc, and data from wearable sensors are produced continuously. This enormous amount of health data, from different sources and in various formats added in real time referred as Big Data [3]. Big data analytics monitors these data for knowledge extraction and pattern understanding. The IOT devices comprise limited speed non powerful CPU. The medical big data consist of very large and complex electronic data sets, and are difficult to analyse with the existing hardware and software tools [4]. Organisations deploy cloud services to store, process and analyse these data. Cloud has the capability for storage of large volume of data, analysis of that data, power of decision making and generation of reports as per user's need

Usage of cloud in healthcare improves the quality of services as well as it reduces over all medical cost. Health organisations share various medical records like prescriptions, X ray and MRI test results, references and availability of doctors using cloud that can be accessed from any geographical location by authorized individuals at any time. Even though cloud computing provides substantial benefits like elastic resources, great accessibility and better storage at a lower cost, still there are issues concerning security, latency, real time computation of highly sensitive heath care data. In medical emergency, time plays a crucial role, sending data to centralized cloud and from distant cloud to hospital or doctors causes delay. There is chances of failure of the network or a cloud data centre. In order to handle these issues, a distributed architecture called fog computing was incorporated. The fog computing comprises of low-power embedded computers capable of providing local computation close to clients or cloud [5]. Fog nodes are the gateways that integrates all of the clients and simultaneously accomplishes the task of linking with the cloud [14]. These devices are clever enough to filter and find the appropriate data that requires further analysis in cloud This process consumes much less band width [6]. Apart from this, using fog computing the security and privacy issues regarding the processing of highly sensitive medical data can be handled. It inhibits the unauthorized use of data by allowing limited access of data based on user's requirement.

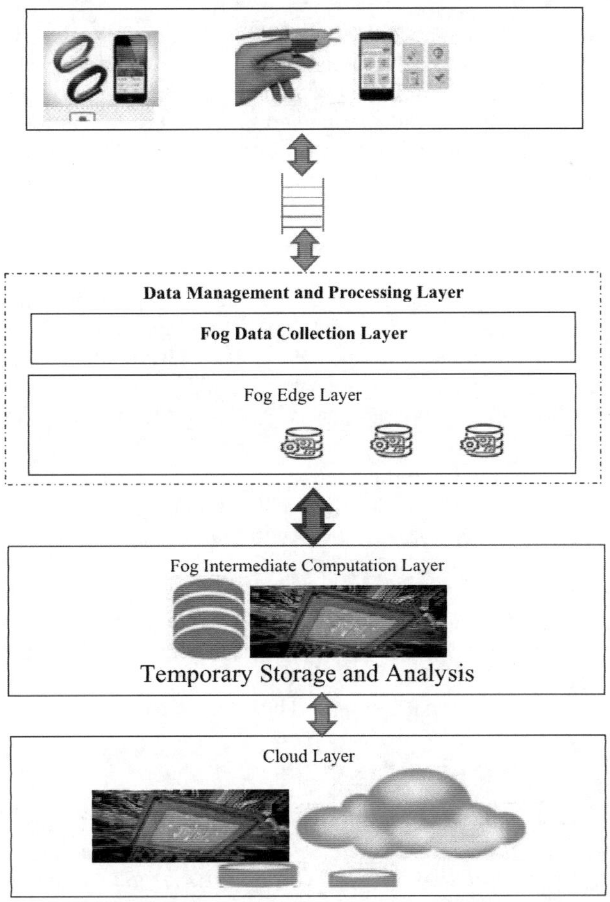

Figure 1. Proposed Fog Assisted Architecture for Medical Big data

As the data storage and computation brought closer where the data are generated, fog computing considerably increases the throughput as well as decreases the latency and round trip delay of the system [7][15]. This IOT paradigm that uses both cloud and fog computing referred as Fog of Things.

The volume and verity of medical big data increasing day by day. These data require efficient processing and systematic analysis for prevention of disease, improvement in care, diagnosis and much more. Involvement of cloud computing in healthcare provide solutions for effective data management and storage in a cost-effective manner.

Automation of data collection of patients with the help of sensors attached medical instruments was discussed by Carlos Oberdan Rolim et al. The data thus collected was uploaded in the cloud from where it can be easily accessed by medical professionals [8]. Yang CT et al[9] implemented the access of Electronic Medical Record on cloud. This system is very helpful for hospitals, patients and care givers to store, exchange and share medical image file for better decision making [10]. Even if cloud computing offers several benefits it has certain draw backs which includes a requirement of high speed steady Internet connectivity, low latency, security and privacy risks etc [11],[12],[13]. Subhadeep Sarkar et al. [14] carried out a comparative performance analysis of fog computing over cloud computing. It was found that the fog computing beat cloud computing in terms of low latency, low CO2 emission and less power consumption.

Tuan Nguyen Gia et al. [15] enhanced the monitoring of health system by employing Fog computing concepts at gateway devices. Electrocardiogram (ECG) features like P and T wave, heart rate were extracted and were analysed at the smart gateways. The resulted system succeeded in achieving better band width consumption and low latency. Amir M. Rahmani et.al [16] utilized the concept of fog computing in healthcare and built a prototype UT-GATE, a smart gateway having enhanced energy efficiency, reliability and interoperability. IOT enabled Fog assisted remote monitoring of patients, they used the techniques like notification services, distributed storage concepts in the smart gateways, the decisions were based on health care data that was collected in real time.

The proposed architecture comprised of a cloud layer, a fog layer which is further subdivided into intermediate computation layer, edge layer, data collection layer and an IOT layer. The cloud layer is mainly concerned with the permanent storage and global analysis of medical data, whereas the processing task is carried out in Fog layers. The healthcare big data generated from various IOT devices are sent to the data collection fog layer and the edge layer which process and manage the data.

In this research work, we emphasize that the large amount of health data is collected from millions of customers. If we continuously transfer this huge amount of data between IoT layer and Data Management and Processing Layer (DMPL) then enormous energy will be needed. We find that all these devices are primarily dependent on battery or external power storage. To minimize the power consumption, we have implemented the sleep mode mechanism with this health monitoring framework. This research work will predict the effect of sleep mode on the Quality of Service (QOS) like delay, packet drop probability. This paper is organized as follows. In section 2 we have defined the proposed model description. In section 3 we have discussed the model analysis result part. In section 4 and 5 we have shown the numerical results and conclusion part.

2. Model Description

The model switch over between sleep mode and busy mode [16]. When the system is not in busy mode, fog assisted processor or DMPL module starts to sleep for a fixed time interval. This is known as sleep interval. The system will be in wake up mode after a fixed time and check if the client side (IoT devices) has any data packet are in buffer for transmission to DMPL layer. If in the buffer there is no data packet to transfer, the system will be in sleep mode again by adjusting the sleep duration. Every time it will check the status of the buffer which is residing between IoT sensor and DMPL, when it will be in wake up state. The sleep interval is varying by some standard protocol but usually it will be 2 times of the earlier interval but certainly there is some threshold value.

In this proposed model we have adopted batch service which will be in sleep mode or busy mode. There is changeover time for switching from sleep to busy mode. Let $\boldsymbol{a}$ and $\boldsymbol{b}$ values indicates the minimum values of packet in buffer for starting the fog service processing and capacity of the batch, respectively. Whenever the fog module completed the transfer of packets, it switch over to sleep. The sleep timer ($\boldsymbol{\beta}$) is exponentially distributed. The fog assisted DMPL awakes after sleep interval whether buffer has some pending transmission left. In the intermediate buffer less than j packets are there and values of $\boldsymbol{j}$ $(\boldsymbol{0 \leq j \leq a-2})$ packets, it switch over to sleep mode. If number of packet is a-1 then it will wait some time and switchover to busy mode that is known as changeover time ($\boldsymbol{\alpha}$). In busy mode it will use more power for transferring the packets. The changeover time is exponentially distributed with parameter $\boldsymbol{\alpha}$.

3. Analysis of the Fog Assisted Health Monitoring Model

In this framework we consider a queuing system $\{(i,j)|0 \leq i \leq N, j = 0,1\} \cup \{(a-1,2)\}$where i specify the size of the intermediate buffer between IoT layer and processing layer and j indicate the state of the fog assisted processor. The state $(i,0)$ represent i number of packet in buffer and the DMPL module in sleep state or inactive state. The state $(i,1)$ represent i packet in buffer but in active or busy state. State $(a-1,2)$ shows $a-1$ data packets in the buffer and the fog server is waiting in the system. The equation can be derived at steady state

$$\lambda P_{i,0} = \lambda\psi(1 \leq i \leq a-2)P_{i-1,0} + \mu P_{i,1},\ 0 \leq i \leq a-2, \tag{1}$$

$$(\lambda+\beta)P_{i,0} = \lambda P_{i-1,0} + \alpha\psi(i = a-1)P_{a-1,2},\ a-1 \leq i \leq N-1, \tag{2}$$

$$\beta P_{N,0} = \lambda P_{N-1,0}, \tag{3}$$

$$(\lambda+\mu)P_{0,1} = \lambda P_{a-1,2} + \mu\sum_{s=a}^{b} P_{s,1} + \beta\sum_{s=a}^{b} P_{s,0}, \tag{4}$$

$$(\lambda+\mu)P_{i,1} = \lambda P_{i-1,1} + \beta\psi(1 \leq i \leq N-b)P_{i+b,0} + \mu\psi(1 \leq i \leq N-b)P_{i+b,1},\ 1 \leq i \leq N-1, \tag{5}$$

$$\mu P_{N,1} = \lambda P_{N-1,1}, \tag{6}$$

$$(\lambda+\alpha)P_{a-1,2} = \mu P_{a-1,1} + \beta P_{a-1,0}, \tag{7}$$

Deriving the equations condition $\sum_{i=0}^{N} P_{i,0} + \sum_{i=0}^{N} P_{i,1} + P_{a-1,2} = 1,$ (8)

$$P_{N,1} = \left[\frac{\lambda T_5 + \frac{\beta\mu}{\lambda}\sum_{j=2}^{a-1} T_2 j + \mu\beta T_2(0)}{\alpha T_3 + \frac{\mu^2}{\lambda}\sum_{j=0}^{a-2} T_1(j)}\right] P_{N,0} \tag{9}$$

4. Performance measures

To measure the system performance we measure the average number of data packets in the buffer (L_q) , average waiting time of the data packets in the buffer (W_q) residing between IoT layer and DMPL and the probability of blocking (PBL).

They are given by

$$L_q = \sum_{i=1}^{N} iP_{i,0} + \sum_{i=1}^{N} iP_{i,1} + (a-1)P_{a-1,2} \tag{10}$$

The PBL is given by $PBL = P_{N,O} + P_{N,1}$. The Lq in the buffer using Little's rule is $W_q = L_q / \lambda'$, where $\lambda' = \lambda(1 - PBL)$.

In Figure 2 (a) and 2 (b) we show the impact of Lq for varying the value of a and b. The Lq decreases first for arrival rate increase but after a certain arrival speed it increases exponentially. The Lq will increases if a value increases. We have seen for constant λ the buffer length decreases as b increases.

Figure 3 (a) shows the impact of Lq on λ and μ. Here we can see that for constant μ the Lq decreases as the λ increases. We have implemented the proposed Fog based model using Raspberry Pi, Pocket Beagle and Arduino Mega. We have shown the CPU load for three different single board system and shown the load balance for different scenarios in Figure 3 (b). Here the arrival rate of data packet is 10 packets per second. The batch size for start the processing is 3 and maximum batch size is 7.

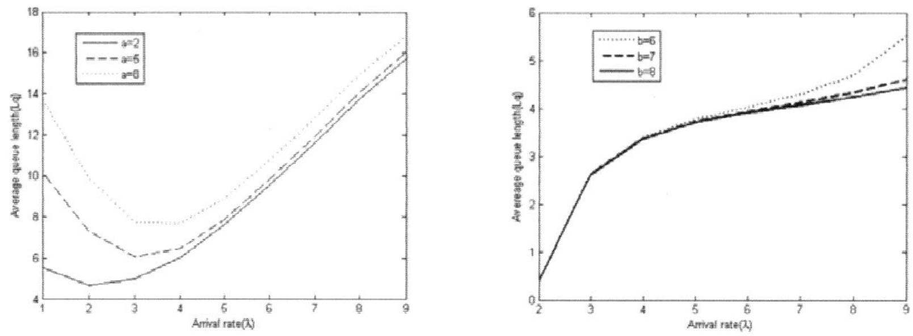

Figure 2. (a) Effect of λ on Waiting Time with varying a (b) Effect of λ on Lq with varying b.

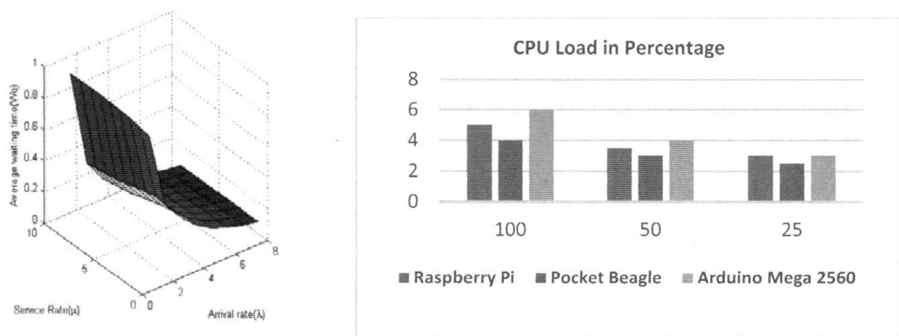

Figure 3. (a) Effect of λ and μ on Wq (b) CPU load in Percentage

5. Conclusion

In this paper, we have designed the energy saving mechanism of fog assisted IoT enabled framework. In this model different impact of QoS parameters like Lq, Wq and PBL are shown for adopting sleep windows. Our proposed model explain the fog assisted IoT enabled health monitoring system which is delay tolerant in nature. In future we extend our work for cost optimization of energy requirement. In this model we show the impact of Lq and Wq on traffic arrival speed and processing speed in case of batch processing. Here we consider internal sleep timer and changeover time for more precise analysis of Lq and Wq.

References

1. Atzori, L., Iera, A., & Morabito, G. (2010). The internet of things: A survey. Computer networks, 54(15), 2787-2805.
2. Gope, P., & Hwang, T. (2015). BSN-Care: A secure IoT-based modern healthcare system using body sensor network. IEEE sensors journal, 16(5), 1368-1376.
3. Baker, S. B., Xiang, W., & Atkinson, I. (2017). Internet of things for smart healthcare: Technologies, challenges, and opportunities. IEEE Access, 5, 26521-26544.
4. Manogaran, G., Thota, C., & Kumar, M. V. (2016). MetaCloudDataStorage architecture for big data security in cloud computing. Procedia Computer Science, 87, 128-133.
5. Dastjerdi, A. V., Gupta, H., Calheiros, R. N., Ghosh, S. K., & Buyya, R. (2016). Fog computing: Principles, architectures, and applications. In Internet of things (pp. 61-75). Morgan Kaufmann.
6. Varshney, P., & Simmhan, Y. (2017, May). Demystifying fog computing: Characterizing architectures, applications and abstractions. In 2017 IEEE 1st International Conference on Fog and Edge Computing (ICFEC) (pp. 115-124). IEEE.
7. Li, J., Zhang, T., Jin, J., Yang, Y., Yuan, D., & Gao, L. (2017, November). Latency estimation for fog-based internet of things. In 2017 27th International Telecommunication Networks and Applications Conference (ITNAC) (pp. 1-6). IEEE.
8. Rolim, C. O., Koch, F. L., Westphall, C. B., Werner, J., Fracalossi, A., & Salvador, G. S. (2010, February). A cloud computing solution for patient's data collection in health care institutions. In 2010 Second International Conference on eHealth, Telemedicine, and Social Medicine (pp. 95-99). IEEE.
9. Yang, C. T., Chen, L. T., Chou, W. L., & Wang, K. C. (2010, December). Implementation of a medical image file accessing system on cloud computing. In 2010 13th IEEE International Conference on Computational Science and Engineering (pp. 321-326). IEEE.
10. Daman, R., Tripathi, M. M., & Mishra, S. K. (2016). Cloud Computing for Medical Applications & Healthcare Delivery: Technology, Application, Security and Swot Analysis.
11. Yeboah-Boateng, E. O., & Essandoh, K. A. (2014). Factors influencing the adoption of cloud computing by small and medium enterprises in developing economies. International Journal of Emerging Science and Engineering, 2(4), 13-20.
12. Padhy, R. P., & Patra, M. R. (2013). Managing IT operations in a cloud-driven enterprise: Case studies. American Journal of Cloud Computing, 1(1), 1-18.
13. Kaur, K., & Rai, A. K. (2014). A comparative analysis: Grid, cluster and cloud computing. International Journal of Advanced Research in Computer and Communication Engineering, 3(3), 5730-5734.
14. Barik, R. K., Dubey, H., Mankodiya, K., Sasane, S. A., & Misra, C. (2019). GeoFog4Health: a fog-based SDI framework for geospatial health big data analysis. Journal of Ambient Intelligence and Humanized Computing, 10(2), 551-567.
15. Barik, R. K., Misra, C., Lenka, R. K., Dubey, H., & Mankodiya, K. (2019). Hybrid mist-cloud systems for large scale geospatial big data analytics and processing: opportunities and challenges. Arabian Journal of Geosciences, 12(2), 32.
16. Barik, R., Dubey, H., Sasane, S., Misra, C., Constant, N., & Mankodiya, K. (2017, July). Fog2fog: augmenting scalability in fog computing for health GIS systems. In 2017 IEEE/ACM International Conference on Connected Health: Applications, Systems and Engineering Technologies (CHASE) (pp. 241-242). IEEE.

Surface inclination effects on muscle and joint contact force during walking: A systematic review

Noor Arifah Azwani Abdul Yamin[1*], Khairul Salleh Basaruddin[1], Ahmad Faizal Salleh[1], Mohd Hanafi Mat Som[1] and Muhammad Juhairi Aziz Safar[1]

[1]School of Mechatronic Engineering, Universiti Malaysia Perlis, Pauh Putra Campus, 02600 Arau, Perlis, Malaysia

*ifahazwani@gmail.com

Abstract. Outdoor surfaces are generally uneven as the surface topography diverges in terms of its inclination. Due to variation of inclination, there are risks of injury caused by muscle response and force reaction during gait. The objective of this review is to characterize the effects of incline surface on muscle and forces during walking that contribute to injury. The search strategy was carried out from Worldcat.org database. Majority of the previous studies agree on the fact that muscle response and force significantly change according to the changes of surface inclination. Future studies involving ligament force should be performed in order to obtain a valid result on muscle response and force reaction during inclined gait.

1. Introduction

Outdoor surfaces are generally uneven as the surface topography diverge in terms of its inclination. The varies of unevenness is might be due to unavoidable human error in manmade work or consequence of the environmental in the case for ramps to access the building or pavement on hills respectively [1].

Muscle are the important component in the incline surface investigation and plays the key role in joint contact forces [2]. It has been reported that, a greater post-exercise muscle soreness and damage during downhill movement which resulting from the increase of muscle eccentric [3] may cause mechanical changes on lower limb posture that lead to future injury [4]. Furthermore, reaction force is reported to be correlated with running surface [5] and increase in lower extremity joint loading in increment of slope compared to level surface gait have been founded [6]. Accumulated high impact forces and joint loading lead to the early onset of fatigue [7], development of stress fractures [8] and patellofemoral pain in runners [9].

Whatever the origin, in order to prevent fall and injury, incline surface require a particular different demand compared to level surface. Therefore, the objective of this paper is to review and characterize the effects of incline surface on muscle and forces during walking that contribute to injury.

2. Method

2.1. Search engine

This review had been completed by performing electronics search in March 2019 involving a database WoldCat.org. The selected articles were limited for past ten years studies only that is from 2009 until 2019. The keyword for the search included ‘inline surface’, ‘slope’, ‘gait’, ‘muscle’ and ‘force’. A manual search on reference list of reviewed articles is completed to ensure no overlooked articles. The articles in English only were individually selected by authors. Articles that met the following criteria were considered: (1) investigation on inclined surface gait, (2) muscle as dependent variable, and (3) force as dependent variable.

2.2. Review process
Based on eligibility criteria, the articles were first screened on the title and abstract. A full-text evaluation is performed if the title and abstract could not provide adequate information of the chosen articles. Re-screened process is also completed to avoid any overlooked information.

2.3. Data extraction
The information collected from the reviewed articles are related to the main research theme that is objective, method implement, independent and dependent variables, as well as findings of the research.

3. Result

3.1. Effect of incline surface on muscle
A considerable amount of literatures has been published on the effect of incline surface to muscle response. In overall, muscle responses to surface inclination during gait can be divided into six measurement; electromyography (EMG) signals, EMG amplitude, muscle timing, muscle geometry, muscle force and muscle efficiency as shown in Table 1.

Based on review done, EMG signals which displayed muscle activities was assessed by majority of the previous studies. All of the articles, monitored EMG signals using surface electrode application. Bavdek et al.[10] and Mohammadi and Phadke [11] applied the surface electrode according to SENIAM method, instead, other previous study does not mentioned method used. However, each of these articles has specific software or method applied in order to analyze and process the EMG signals. The software used are including Labchart 7 [10], LED-synchronized video [12], Biometric Ltd [11], and Delsys EMG works [13]. While, the method implemented in investigating EMG signals are such as telemetry system [14] and trapezoidal method [15]. In addition, musculoskeletal modelling and inverse dynamic was used to investigate muscle geometry and muscle force [2]. Other than that, Saito et al.[16] and Rozumalski et al.[17] extracted muscle synergies using non-negative matrix factorization (NMF).

In overall, by merging all findings of previous studies in this review, muscle activity is reported to be affected and influenced by surface inclination except for Slaughter et al. [13] who found no significant difference between these two variables. On the question of muscle response to incline surface, Kwee-Meier et al. [12] highlighted that peak activities of lower limb are age-related especially for m. gastrocnemius medialis (GAS) and m. soleus (SOL) in incline walking. Furthermore, foot strike pattern and leg dominancy of an individual should be considered to obtain an optimum prescription of walking as both influenced the muscle activity as reported by Ohira et al. [14] and Mohammadi & Phadke [11] respectively. Another important finding was that, muscle synergies extracted from lower limb were consistent in response to different slope and speed of walking [16], [17].

Table 1. Effects of Incline Surface on Muscle

No	Author	Objective	Independent variable	Dependent variable	Method used	Findings
1	Kwee-Meier et al.[9]	To investigate the muscle activation when walking on sloped surface among young and older adult.	Surface inclination 0°, 7°, and 14°	EMG – muscle activity signals	LED-synchronised video stream - analyses EMG signal	Peak activities of GAS and SOL are more restricted in older adult compared to young adult. They compensate these limitations by 'pull' themselves instead of pushing during uphill walking

No	Author	Objective	Independent variable	Dependent variable	Method used	Findings
2	Bavdek et al.[8]	To study the characteristic of peroneal muscles during walking with different style.	Walking types Flat surface (NORM) Medial incline ramp with a 30° inversion (FULL) – cross-slope walking Medial incline ramp with a 30° & maintain the lateral part of the foot elevated	EMG – muscle activity signals Kinematic parameters *gait speed *stride length *contact time *foot position	Surface electrode; according to SENIAM method Labchart 7 – process EMG signal Stalker's type hyper-frequency radar – measure gait speed	Peroneal muscle was affected during incline or ramp walking. Incline walking possibly implement as a strengthening peroneal muscle exercise.
3	Saito, Tomita, Ando, Watanabe, & Akima [15]	To assess the muscle synergies of lower limb during level and uphill treadmill walking.	Surface inclination 0% and 10% grade of treadmill Walking speed 60, 80, and 100 m/min	EMG – muscle activity signals	Surface electrode Non-negative matrix factorization (NNMF) – muscle synergies extraction	The muscle synergies extracted from lower limb showed a consistent pattern during walking on different inclination and speed.
4	Mohammadi & Phadke [11]	To study the effects of incline surface and walking speed to ankle muscle activity among middle-age individual.	Surface inclination 0°,3°, and 6° Walking speed self-selected self-selected +20% self-selected+40%	EMG amplitude Muscle co-contraction Muscle timing	Surface electrode; according to SENIAM method Biometrics Ltd software – EMG processing and analysis	Muscle in dominant leg showed a bias activity compared to the non-dominant leg might be due to age-related adaptation to overcome challenging walking.
5	Rozumalski, Steele, & Schwartz, [14]	To investigate the relation of synergies and gait changes during treadmill walking with different inclinations and speeds in developing children.	Surface inclination and walking speed combination speed: free speed, 110% of free speed, 120% free speed Slope:0°,6°, 12°	EMG – muscle activity signals Kinematics data Kinetics data	Custom software (Matlab) – Ground reaction force collection Surface electrode NNMF – muscle synergies & synergies activation	The muscle synergies remain unchanged in response to changing of biomechanical task constraint show that neuromuscular control strategies do not change. The significant different shown in kinematics and kinetics data indicate different in movement pattern

No	Author	Objective	Independent variable	Dependent variable	Method used	Findings
6	Alexander & Schwameder,[2]	To study muscle force of lower limb during different inclination walking using musculoskeletal model.	Surface inclination (0°, ±6°, ±12° & 18°)	Mass Moment of inertia Muscle sites/geometry Muscle forces	AnyBody Modelling System (musculoskeletal model) Inverse dynamics	Lower limb muscle forces during incline compared to level walk were significantly changed except for Gluteus Medius muscle force.
7	Ortega & Farley[16]	To study the relation of metabolic cost, muscular efficiency and leg muscle co-activation of young and older person during level and uphill walking.	Surface inclination Level and uphill	Metabolic cost Muscular efficiency Leg muscle activation	Rates of oxygen consumption & carbon dioxide production = indirect calorimetry surface EMG signals	Metabolic cost during uphill walking was increased due to increment of leg muscle co-activation and decrement of muscular efficiency among older adult but are unrelated to one another.
8	Ohira et al.,[12]	To study the response of leg muscle during walking on treadmill with different speed and inclination.	Surface inclination 0%, 6%, 10% and 15% Walking speed Foot strike pattern	EMG – muscle activity Kinematics data – knee and ankle angle Time spent for each step	Bipolar surface electrode – monitor muscle activity Telemetry system	To obtain an optimum prescription walking that associated with speed and inclination, the foot strike pattern of an individual should be considered.
9	Silder, Besier, & Delp[17]	To investigate muscle activation patterns, physical criteria and gait characteristic that best estimate metabolic cost during incline walking.	Surface inclination (0°, 5°, 10°)	Kinematic data (marker data) Ground reaction force Moments EMG signal (muscle activity) Joint work	Whole body dual-energy x-ray absorptiometry Surface electrode – monitor muscle activity	Metabolic cost during incline walking can be estimated based on muscle activities and biomechanical gait patterns for healthy adults walking with flexed knees, carrying loads at various speed.
10	Slaughter et al.[11]	To study the EMG activity of lower extremity skeletal muscle at knee and rotational differences between the leg and thigh during walking on different surfaces.	Surface condition Side-slope (5°, 10°) Incline treadmill (10°) Flat (0°) Ascended stairs	EMG signal Maximum voluntary contraction (MVC) Acceleration, rotation & magnetic field data – IMU device Angular velocity	Surface electrode – monitor muscle activity Delsys EMGworks program - analysis	There is little difference in knee joint loading when rotation about vertical axis and EMG activity was considered.

No	Author	Objective	Independent variable	Dependent variable	Method used	Findings
11	Wall-Scheffler CM et al.13]	To analyse the effect of surface inclination and speed to muscular activity of thigh and hip during walking and running.	Surface inclination 0% (0°), 10% (5.7°), 15% (8.6°) and 20% (11.5°) Walking speed 1.2, 1.5, 1.8, 2.7, and 3.6 ms−1	EMG signal (muscle activity)	Surface electrode – monitor muscle activity Trapezoidal method	Walking and running on level treadmills showed differences and the differences decrease as the locomotor intensity (incline, speed and their interaction) increases.

3.2. *Effect of incline surface on joint contact force (JCF)*

Based on literature reviewed for past ten years, only two articles analyze JCF in investigating the surface inclination [3], [6]. Park et al.[3] compare the JCF during moderate and steep downhill running with level running. While, Alexander & Schwameder (2016) investigate JCF during walking at different slope.

As shown in Table 2, both articles have different approach in assessing the JCF. Park et al.[3] generally mentioned that inverse dynamics were performed to calculate joint force and moment. Instead of that, Alexander & Schwameder [6] also characterize JCF by performing inverse dynamics, but with addition of the third order polynomial muscle recruitment criterion applied in musculoskeletal model using AnyBody Modelling system.

Both articles in agreement on the findings state that, ankle joint contact force was decreased either during downhill running or walking which indicate decreased positive ankle work. Instead of that, hip, tibiofemoral and pattelo-femoral joint contact forces are increased with the increased of declination while uphill walking increased all lower joint forces.

Table 2. Effects of Incline Surface on Joint Contact Force

Num	Author	Objective	Independent variable	Dependent variable	Method used	Findings
1	Park et al.[3]	To compare the effect of incline surface on kinematic and kinetic parameters during moderate and steep downhill running with level running.	Surface inclination (Level 0°, Moderate -6°, Steep -9°)	Joint forces 3D joint angles Moments Powers Work	Motion capture system – collect kinematic and kinetic data Visual 3D software – calculate three-dimensional joint angles, moments, powers and work Inverse dynamic – calculate joint force and moment	Runner alter their running mechanics and led to greater kinetic demand on knee to absorb load and energy during downhill running.
2	Alexander & Schwameder[4]	To investigate the lower limb joint compression forces and tibiofemoral joint shear force during incline walking at different slope.	Surface inclination (0°, ±6°, ±12° & 18°)	Joint compression for lower limb (hip, tibiofemoral, patellofemoral & ankle)	Vicon Nexus – process kinematic data AnyBody Modelling System (musculoskeletal model) – calculate lower limb joint force	Downhill walking resulting to increased hip, tibiofemoral & patello-femoral joint forces and decreased ankle joint compression forces, while uphill walking increased all lower limb joint forces.

4. Discussion

Muscle reaction and joint force are related to each other and susceptible to injury [20]. Overcorrecting on certain injury might detrimental on bone, cartilage, musculoskeletal limb and posture. In order to further understand injury risk, this review is completed to compile and outline the published researches that related to biomechanical effects of incline gait in terms of muscle and joint force from the year 2009 until 2019.

Based on the review, there is variability of methodological application selected. However, all findings of the previous studies in this review are consistent which suggested that muscle response and force significantly changes according to the changes of surface inclination except for findings reported by Slaughter et al. [13]. It may be the case therefore that these variations of methodological application showed the flexibility of muscle and joint force measurement.

There are some limitations related to this review. This review only including English-based published articles. Thus, some articles may have been neglected. Other than that, the involvement criteria of the selected articles limited to muscle and force findings only. Thus, other biomechanics effects related findings articles were excluded.

5. Conclusion

As conclusion, there are different methodological applications have been introduced in investigating muscle response and joint force respectively so far. One of the issues that emerges from this review is limited studies assess both muscle response and joint force in an experiment, even though both are the main contributor to injury. Future studies which take these variables into account are therefore recommended with in addition of ligamentous force determination. This is because force validation that including muscle force which implement generic musculoskeletal model (not including ligament element) are not fully validated [21]. The study should be done in order to ensure a valid joint force determination in investigating injury risk due to surface inclination.

Acknowledgement

The authors would like to acknowledge the support from the Fundamental Research Grant Scheme (FRGS) under a grant number of FRGS/1/2016/TK03/UNIMAP/02/6 from the Ministry of Education Malaysia.

References

[1] Vieira M F, Rodrigues F B, Sá e Souza G S, Magnani R M, Lehnen G C, Campos N G, and Andrade A O 2017 Gait stability, variability and complexity on inclined surfaces *J. Biomech* **54** 73-9.

[2] Alexander N and Schwameder H 2016 Effect of sloped walking on lower limb muscle forces *Gait Posture* **47** 62-7.

[3] Park S K, Jeon H M, Lam W K, Stefanyshyn and Ryu J 2019 The effects of downhill slope on kinematics and kinetics of the lower extremity joints during running *Gait Posture* **68** 181-6.

[4] Azizan N A, Basaruddin K S, and Salleh A F 2018 The Effects of Leg Length Discrepancy on Stability and Kinematics-Kinetics Deviations: A Systematic Review *Appl. Bionics Biomech* **2018** 1–22.

[5] Yamin N A A A, Amran M N A, Basaruddin K S, Salleh A F, and Rusli W M R 2017 Ground Reaction Force Response during Running on Different Surface Hardness *ARPN J. Eng. Appl. Sci.* **12(7)** 2313–8.

[6] Alexander N and Schwameder H 2016 Lower limb joint forces during walking on the level and slopes at different inclinations *Gait Posture* **45** 137-42.

[7] Mizrahi J, Verbitsky O, and Isakov E 2001 Fatigue-induced changes in decline running Clin. *Biomech* **16** 207-2.

[8] Milgrom C 1989 The Israeli elite infantry recruit: a model for understanding the biomechanics of stress fractures *Journal Royal College of Surgeons of Edinburgh* **34** S18-22.

[9] Stefanyshyn D J, Stergiou P, Lun V M Y, Meeuwisse W H and Worobets J T 2006 Knee angular impulse as a predictor of patellofemoral pain in runners *Am. J. Sports Med.* **34** 1844-51.

[10] Bavdek R, Zdolšek A, Strojnik V, and Dolenec A 2018 Peroneal muscle activity during different types of walking *J. Foot Ankle Res.* **11** 1-9.

[11] Mohammadi R and Phadke C P 2017 The impact of incline and speed of treadmill on ankle muscle activity in middle-aged adults *J. Body. Mov. Ther* **21** 306-13.

[12] Kwee-Meier S T, Mertens A, and Jeschke S 2018 Age-induced changes in the lower limb muscle activities during uphill walking at steep grades *Gait Posture* **62** 490-6.

[13] Slaughter S, Butler P, Capozzella H, Nguyen A, and Hutcheson L 2012 The comparative gait effects of select walking surfaces using kinetic and EMG analyses *Hum. Mov* **13** 198-203.

[14] Ohira T, Okabe H, Kawano F, Fujita R, Ohira T, Ohira K, Oke Y, Nakai N, Roy R R, Edgerton V R and Ohira Y 2015 Speed and/or inclination-dependent mobilization of human leg muscles during walking with a unique exception *Acta Astronaut.* **116** 237-46.

[15] Wall-Scheffler CM, Chumanov E and Steudel-Numbers K and Heiderscheit B 2011 EMG activity across gait and incline : The impact of muscular activity on human morphology *Am J Phys Anthr.* **143** 601-11.

[16] Saito A, Tomita A, Ando R, Watanabe K, and Akima H 2018 Similarity of muscle synergies extracted from the lower limb including the deep muscles between level and uphill treadmill walking *Gait Posture* **59** 134-9

[17] Rozumalski A, Steele K M, and Schwartz M H 2017 Muscle synergies are similar when typically developing children walk on a treadmill at different speeds and slopes *J. Biomech.* **64** 112-9.

[18] Ortega J D and Farley C T 2015 Effects of aging on mechanical efficiency and muscle activation during level and uphill walking *J. Electromyogr.* Kinesiol **25** 193-8

[19] Silder A, Besier T, and Delp S L 2012 Predicting the metabolic cost of incline walking from muscle activity and walking mechanics *J. Biomech* **10** 1842-9.

[20] Othman N F, Basaruddin K S, Som M H M, Abdul Majid M S and Sulaiman A R 2019 The effect of leg length inequality on joint contact forces of lower limbs during walking *Acta Bioeng. Biomech* **21(1)** 55-62.

[21] Ding Z, Nolte D, Tsang C K, Cleather D J, Kedgley A E, and Bull A M 2016 In Vivo Knee Contact Force Prediction Using Patient- Specific Musculoskeletal Geometry in a Segment-Based Computational Model *J. Biomed. Eng* **138** 1-9

Evaluation on Machine Learning Algorithms for Classification of Autism Spectrum Disorder (ASD)

Azian Azamimi Abdullah[1*], Saroja Rijal[1] and Satya Ranjan Dash[2]
[1]School of Mechatronic Engineering, Universiti Malaysia Perlis, Pauh Putra Campus, 02600 Arau, Perlis, Malaysia
[2]School of Computer Applications, KIIT University, India

[*]azamimi@unimap.edu.my

Abstract. Autism Spectrum Disorder (ASD) was characterized by delay in social interactions development, repetitive behaviors and narrow interest, which usually diagnosed with standard diagnostic tools such as Autism Diagnostic Observation Schedule (ADOS) and Autism Diagnostic Interview-Revised (ADIR-R). Previous work has implemented machine-learning methods for the classification of ASD, however they used different types of dataset such as brain images for MRI and EEG, risk genes in genetic profiles and behavior evaluation based on ADOS and ADI-R. Here a trial on using Autism Spectrum Questions (AQ) to build models that have higher potential to classify ASD was developed. In this research, Chi-square and Least Absolute Shrinkage and Selection Operator (LASSO) have been selected as feature selection methods to select the most important features for 3 supervised machine learning algorithms, which are Random Forest, Logistic Regression and K-Nearest Neighbors with K-fold cross validation. The performance was evaluated in which results Logistic Regression scored the highest accuracy with 97.541% using model with 13 selected features based on Chi-square selection method.

1. Introduction

Autism Spectrum Disorder (ASD) is a neurodevelopmental disability that refers to slow development in verbal skills, limited passion and monotonous behaviour [1]. With the increase in commonness of this problem recently, new pre-screening technology was encouraged [2]. The only reliable and non-invasive diagnosis can be done through screening tool nowadays was Autism Diagnostic Interview-Revise (ADI-R) and Autism Diagnostic Observation Schedule (ADOS). Nowadays there are many problems related to ASD detection where it was hard to find scalable biomarkers for early detection due to the wide variability of the disability and need of simpler measurements that could be applied routinely [3]. Behaviour test based on a clinical tool which was Autism Diagnostic Observation System (ADOS) was employed in this research where they used a machine learning approach to evaluate the data obtained [4]. MRI and EEG approach on ASD classification was in purpose to allow for early detection of ASD and cut down the administration time as follow in ADOS and ADI-R screening tools. The latest research done by Murat Gok by specifying on ASD risk genes of a long non-coding RNA (IncRNA) in order to observe whether the gene characteristic can accurately classify ASD or not [5].

2. Materials and Methods

2.1 *Exploratory Data Analysis (EDA)*

Data consist of 704 instances and 20 features with 1 output. Features consist of ten shortened version Autism Spectrum Quotient (AQ) self-administered questionnaire which assessed on behaviour traits which were represented in binary data (0 and 1) [6]. Python software applied to the raw dataset to visualize the dataset. The dataset observed to have 13.5% missing data. The contribution of individual diagnosed with ASD was 26.85% (189 patients) and not diagnosed with ASD was 73.15% (515 patients). The datasets undergo conversion from characters format into binary and numeric form using "Label encoder" function from Python library.

2.2 *Data Pre-Processing*

Data pre-processing was a crucial step in data mining techniques and it is the initial step after data exploration [7]. As for this project datasets, it contain missing values which it needs proper technique to handle it. Missing values create difficulties for researchers due it can cause a decrement in efficiency during selection or extraction process if not properly handled [8]. The amount of 13.5% missing data was removed from the datasets by dropping the empty column using Python commands. The removal of null datasets giving a total of 609 datasets left with 20 features consisting 19 features and 1 output.

2.3 *Feature Selection*

Feature selection was a technique used for data pre-processing and it was preferable when to perform it before fitting model in machine learning. Selection by mean here was by choosing variable and attributes or variable subset in a dataset to be fit into model and tested for performance. Chi-Square, LASSO and Random Forest method were applied in this study.

2.3.1 *Chi-Square*

The parameter used was α value, which was 0.01 and 0.05, which indicates 99% and 95% confidence interval (CI). The α values as a bench level for choosing the significant features. Any feature falls into the insignificant region will be expelled.

2.3.2 *LASSO*

The same parameter used in LASSO, which was α value but the usage was different from the Chi-Square method. The α value used in LASSO indicates the regularization value, which was the strength of penalization to choose features. The α parameter was selected by using Python "GridSearchCV" function to obtain the best value.

2.3.3 *Random Forest*

The used of "SelectFromModel" and "get.support" function was implemented to obtain the importance of each features.

2.4 *Supervised Machine Learning Method*

In supervised learning, it is defined as the trial to assess the relationship between input features and target features where it was provided with input and output datasets. Few machine learning algorithms used in this study, which was Random Forest, Logistic Regression and K-nearest neighbors.

2.4.1 *Logistic Regression*

Logistic regression has the nature of estimating probability whether it was success or failure even. This algorithm relates the categorical data dependent variables X and branched categorical output of Y [9].

Parameter C equals to 10, which regularization strength was applied in this study, which was selected using "GridSearchCV" function.

2.4.2 Random Forest

Consists of tree predictors, which each tree depends on the random values, which sampled independently. This algorithm was done by aggregating trees and the majority will be chosen [10]. Few parameters were used along this project which were the number of estimators, maximum number of features, maximum depth, minimum sample leaf, minimum sample split and boostrap which was selected by using hyper-parameter tuning "GridSearchCV".

2.4.3 K-Nearest Neighbors

The computation of the closest distance between neighbors which represented as K value. In order to use this algorithm, we have to take count of a few elements, which are the initial subjects set, the K number (nearest neighbors) and the standard approximation between subjects. In this project the parameter K and distance of neighbors was considered.

2.5 Evaluation of Performance Metric

In this research, all performance result was obtained through 5-fold cross-validation instead of using a common split method. This method is applied to make sure the tests are less biased and will better predict the error estimation where it will split the dataset into 5 pieces and run the algorithm 5 times. Different piece of data as a test set and the other 80% of the data is the training set [11].The classifier performance will be classified into few metric measurements which are accuracy, sensitivity, specificity and area under receiving operating characteristics (AUC) which obtained from confusion matrix. In accuracy, it can be observe and identify the accurate prediction made based on the overall number of tests [12]. The sensitivity metric define as the ratio of test diagnosed with ASD which is in true positive (TP) rate indicates how many correct predictions have been made based on the test cases. While specificity is defined as the ratio of test that is not diagnosed with ASD which indicate as true negative (TN) rate [12]. This indicates how many false predictions are made based on the test cases.

3. Results and Discussion

3.1 Feature Selection

After the pre-processing stage, we have 20 features and 1 output datasets. In the LASSO method, a trial on a few penalization values was tested in order to observe the number of features selected for each penalization coefficient. Trials have been performed for different set of alpha, α which was has been chosen earlier. Table 1, Table 2, Table 3, Table 4 and Table 5 show the selected features for each selection methods.

Table 1. LASSO

Feature with non-zero coefficient for α=0.001 (17 features)	
A9_Score	A6_Score
A3_Score	A5_Score
A7_Score	A8_Score
A2_Score	A1_Score
A4_Score	A10_Score
Age	Ethnicity
Relation	Jundice
Used_app_before	Contry_of_res
Gender	-

Table 2. LASSO

Feature with non-zero coefficient for α=0.01 (16 features)	
A9_Score	A6_Score
A3_Score	A5_Score
A7_Score	A8_Score
A2_Score	A1_Score
A4_Score	A10_Score
Age	Ethnicity
Relation	Jundice
Used_app_before	Contry_of_res

Table 3. Chi-square

Features with level of significance / threshold α=0.05 (17 features)	
A9_Score	A6_Score
gender	A5_Score
A7_Score	A8_Score
A2_Score	A3_Score
A4_Score	A1_Score
age	ethnicity
relation	jundice
contry_of_res	A10_Score

Table 4. Chi-square

Features with level of significance / threshold α=0.01 (13 features)	
A9_Score	A6_Score
A10_Score	A5_Score
A7_Score	A8_Score
A2_Score	A3_Score
A4_Score	A1_Score
contry_of_res	ethnicity
relation	-

Table 5. Feature Importance

Feature Selected
A9_Score
A6_Score
A5_Score
A4_Score
A3_Score

3.2 Supervised Machine Learning Method

3.2.1 Logistic Regression

Based on Table 6, it can be observed that the highest performance goes to model LR4-CV with accuracy 97.541%, 96.591% specificity and 100% sensitivity. This model was able to discriminate the ASD from those were healthy by 97.541%. While in term of specificity, it was able to classify 96.591% individual, which was not possessed to ASD case. The sensitivity of the model was 100% where it was able to classify the individual with ASD class correctly. The performance of the models improved as the number of features reduced from 19 to 13 features by using Chi-Square selected features.

Table 6. Logistic regression performance

Models	C value	Accuracy (%)	Specificity (%)	Sensitivity (%)	Features
LR-CV	10	94.687	93.894	94.205	19 (default)
LR1-CV (LASSO = 0.001)	10	95.082	95.455	94.118	17
LR2-CV (LASSO = 0.01)	10	97.541	95.455	100	16
LR3-CV (Chi-square = 0.05)	10	95.082	96.591	94.118	17
LR4-CV (Chi-square = 0.01)	**10**	**97.541**	**96.591**	**100**	**13**

3.2.2 Random Forest

In Random Forest classifier, it was found that the model with Chi-square selection method score the highest among the other models with 96.721% accuracy, 98.864% specificity and 85.294% sensitivity % respectively as shown in Table 7. This performance obtained using 32 numbers of trees in the forest, 2 minimum samples leaf, and 12 maximum depths and with boostrap. The performance trend for this classifier was increased as the features decreased for the two-selection method but not for Random Forest selection method. The performance for Random Forest selected features decreased in term of overall performance metric into 94.262% accuracy, 97.727% specificity and 79.412% sensitivity. The decrement might due to elimination of useful features of datasets.

Table 7. Random Forest performance

Models	Accuracy (%)	Specificity (%)	Sensitivity (%)	Features
RF-CV	93.442	97.727	85.353	19 (default)
RF1-CV (LASSO = 0.001)	90.164	97.727	79.412	16
RF2-CV (LASSO = 0.01)	93.443	98.864	79.412	16
RF3-CV (Chi-square = 0.05)	95.082	98.864	85.294	17
RF4-CV (Chi-square = 0.01)	**96.721**	**98.864**	**85.294**	**13**
RFI (Feature Importance)	94.262	97.727	79.412	5

3.2.3 K-Nearest Neighbors

In K-Nearest Neighbors classifier it was observed that the highest performance goes to model KNN4-CV with 95.902% accuracy, 97.727% specificity and 88.235% sensitivity with K=2 and hamming distance parameters. The result shows that this model has the ability to distinguish correctly patients, which does not possess ASD disorder since it has higher specificity percentage compared to sensitivity. From Table 8 it can be observed that a smaller K value and least number of features giving the best performance out of the 5 models.

Table 8. K-NN performance

Models	K-value	Distance	Accuracy (%)	Specificity (%)	Sensitivity (%)	Features
KNN-CV	5	Hamming	95.082	97.727	88.235	19 (default)
KNN1-CV (LASSO=0.001)	14	Hamming	95.902	95.455	97.059	17
KNN2-CV (LASSO = 0.01)	5	Hamming	81.967	89.773	61.765	16
KNN3-CV (Chi-square=0.05)	16	Hamming	95.902	96.591	94.118	17
KNN4-CV (Chi-square= 0.01)	**2**	**Hamming**	**95.902**	**97.727**	**88.235**	**13**

3.2.4 Comparison on classifier

Comparison on the chosen models for each classifier will be discussed further among the three classifiers in term of accuracy, specificity, sensitivity and area under curve (AUC). Based on previous results, it can be observed that Logistic Regression score the highest with 97.541% accuracy, which it used 13 selected features based on Chi-Square selection method. RF4-CV features scored 96.721% accuracy followed by KNN2-CV selected features scored 95.902% accuracy also with Chi-Square selection method. This means the models are able to differentiate correctly patients, which does not possess ASD disorder. In term of AUC, the AUC score for the all models was summarized in Table 9. The highest AUC curve for Logistic Regression was 0.96 comes from model LR4-CV. While for Random Forest, the highest AUC was from model RF4-CV with 0.95 AUC. For K-Nearest Neighbors, the highest AUC score was 0.946 from model KNN-CV. The closer the AUC score to 1 the least misclassified case occurs, which means the higher dependability of classification.

Table 9. AUC Performance of All Classifiers

Classifiers	Models					
Random Forest	RF-CV	RF1-CV	RF2-CV	RF3-CV	RF4-CV	RFI
	0.93	0.91	0.92	0.94	0.95	0.94
Logistic Regression	LR-CV	LR1-CV	LR2-CV	LR3-CV	**LR4-CV**	-
	0.93	0.93	0.94	0.93	**0.96**	-
K-Nearest Neighbors	KNN-CV	KNN1-CV	KNN2-CV	KNN3-CV	KNN4-CV	-
	0.947	0.94	0.779	0.656	0.946	-

4. Conclusion

Resulting from the selection methods, there are 6 models developed for Random Forest, 5 models for Logistic Regression and 5 models for K-Nearest Neighbors. Logistic Regression LR4-CV model gave highest performance, which 97.541% accuracy, 96.591% specificity and 100% sensitivity. While for Random Forest, the model RF4-CV scored 96.721% for accuracy, 98.864% specificity and 84.294% sensitivity. Moreover, K-Nearest Neighbors with model KNN4-CV gave highest accuracy of 95.902%, 97.727% specificity and sensitivity 88.235%. Based on the 16 models created, it was observed that the model with 13 features selected using Chi-Square method was suitable to be implemented in the classifiers especially with Logistic Regression classifier. The Logistic Regression machine learning performance promotes a high accuracy for behavior observation purpose, which can be implemented for future detection of ASD.

References

[1] C. Lord, M. Elsabbagh, G. Baird, and J. Veenstra-Vanderweele, 2018, "Autism spectrum disorder," *www.thelancet.com*, vol. 392.

[2] M. Duda, R. Ma, N. Haber, and D. P. Wall, 2016, "Use of machine learning for behavioral distinction of autism and ADHD," *Transl. Psychiatry*, vol. 6, p. e732.

[3] R. K. Mariette Awad, "Machine Learning and Knowledge Discovery," .

[4] T. C. McFayden, J. Albright, A. E. Muskett, and A. Scarpa, 2018, "Brief Report: Sex Differences in ASD Diagnosis—A Brief Report on Restricted Interests and Repetitive Behaviors," *J. Autism Dev. Disord.*, vol. 0.

[5] M. Gök, 2018, "A novel machine learning model to predict autism spectrum disorders risk gene," *Neural Computing and Applications*, pp. 1–7.

[6] S. S. Abu-Naser, I. M. Nasser, I. M. Nasser, and M. O. Al-Shawwa, 2019, "Artificial Neural Network for Diagnose Autism Spectrum Disorder electronic Human Resources Management in Palestinian Universities View project Requirements of using Decision Support Systems as an Entry Point for Operations of Re-engineering in the Universities (Applied study on the Palestinian universities in Gaza Strip) View project Artificial Neural Network for Diagnose Autism Spectrum Disorder".

[7] S. Vijayarani and M. P. Research Scholar, "Preprocessing Techniques for Text Mining-An Overview."

[8] S. García, S. Ramírez-Gallego, J. Luengo, J. Manuel Benítez, and F. Herrera, 2016, "Big Data Analytics Big data preprocessing: methods and prospects".

[9] K. Basu, 2018, "Machine learning approaches to the classification problem for autism spectrum disorder," pp. 1–24.

[10] P. Kassraian-Fard, C. Matthis, J. H. Balsters, M. H. Maathuis, and N. Wenderoth, 2016, "Promises, pitfalls, and basic guidelines for applying machine learning classifiers to psychiatric imaging data, with autism as an example," *Front. Psychiatry*.

[11] B. Van Den Bekerom, 2017, "Using Machine Learning for Detection of Autism Spectrum Disorder," *Twente Student Conf. IT*.

[12] F. Thabtah, 2018, "Machine learning in autistic spectrum disorder behavioral research: A review and ways forward," *Informatics Heal. Soc. Care*, pp. 1–20, Feb. 2018.

Design and Fabrication of Biodegradable Microneedle Using 3D Rapid Prototyping Printer

Nur Hazwani Azyan Mansor[1], Marni Azira Markom[1*], Erdy Sulino Mohd Muslim Tan[2] and Abdul Hamid Adom[1]

[1]School of Mechatronic Engineering, Universiti Malaysia Perlis, Kampus Pauh Putra, 02600, Arau, Perlis, MALAYSIA.

[2]Center of Diploma Studies, Universiti Malaysia Perlis, Kampus Unicity, 02000 Padang Besar, Perlis, MALAYSIA

*marni@unimap.edu.my

Abstract. Microneedle is known as transdermal drug delivery (TDD) devices that uses to deliver biological fluid into veins and needle that use to tear skin to collect blood sample. It involves with various parameters and designs. This device gain attention as its benefit can eliminate pain and more convenient compared to intravenous injection due to its micron size. Typically, microneedle is fabricated using MEMS technology. However, this technology requires few processes such as deposition, etching and moulding, as well as consume much times. This paper presents a work of design and fabrication of solid microneedle using 3D rapid prototype printing. A few types of microneedles are designed and they are analysed in terms of stress and force characteristics. Also, it will study the ability of a few materials to withstand stress while force exerted on it. The selected materials are polyvinyl alcohol (PVA), polylactic acid (PLA), polyester resin and acrylonitrile butadiene styrene (ABS). The results show that PVA has the highest ability to withstand force compared other materials. As conclusion, the design and fabrication of microneedle using 3D rapid prototyping printer is succeed using PVA material and real post-analysis can be conducted to test the capability for medical practice.

1. Introduction

Intravenous injection is a traditional method to deliver drug, but limitations of injection can cause pain and tissue trauma [1-4]. Hence, microneedle is invented as alternative method which may give the same function. Microneedle receive an attention among medical expertise and patients as the penetration of microneedle is painless and convenient since it is smaller in size compared to intravenous injection [5]. Previously, this device is fabricated using micro-electro-mechanical system (MEMS) [6-8]. Although, microneedle receive an attention among medical expertise and patients, this device is not widely used in medical field due to the long fabrication period. Recently, there are studies on microneedle fabrication that implement 3D printer, since it has capable to print in micron size [9-10]. Hence, this study presents a work on solid microneedle design using biodegradable material and will fabricate using 3D rapid prototyping printer.

2. Microneedle Design and Fabrication

Basically, microneedle can be categorised into two types; i. hollow microneedle, and ii. solid microneedle. Hollow microneedle is designed to be used to deliver liquids such as medicines to blood vessel. While, solid microneedle is designed to be similar as lancet to drawing blood. In design studies, there are a few parameters that can be considered, such as the height of needle, shape of needle, materials and the tip. All the parameters need to be analysed before undergo fabrication process.

Previously, microneedle is made up from SU-8, silicon, metals, polymers and glass and silicon dioxide through MEMS technology [7]. Despite that, fabrication period using MEMS technology is too long due to many processes that should be conducted, such as deposition, etching and etc [8]. Nowadays, the 3D printer technology has upgraded in terms of its resolution and accuracy that allow the printer to print using many materials and size in micron. The printing size is suitable with size of microneedle. Hence, many researchers introduce the use of 3D printer to fabricate microneedle.

3. Biodegradable Materials

Biodegradable material can be described as a material that can be decomposed rapidly by action of microorganisms. Material with high ability to withstand force is most likely to be chosen in fabricating microneedle. Four material tested in this study are Acrylonitrile Butadiene Styrene (ABS), Polyester Resin, Polylactic Acid (PLA) and Polyvinyl Alcohol (PVA). Table 1 shows the properties of each materials [11-12].

Table 1. Materials and its properties.

Material / Properties	ABS	Polyester Resin	PLA	PVA
Elastic Modulus (GPa)	2.00	4.41	3.50	41.30
Poisson's Ratio	0.394	0.403	0.500	0.140
Shear Modulus (GPa)	0.32	1.40	2.40	1.90
Mass Density (g/cm^3)	1.02	1.2	1.3	1.31
Tensile Strength (MPa)	30	103	50	71.1
Thermal Conductivity $(W/K.m)$	0.2256	0.1700	0.1300	0.3100
Specific Heat $(J/K/kg)$	1386	565	1800	5000

4. Methodology

This work has a few stages in order to complete the fabrication process. It consists of design the microneedle (MN) using a few parameters, analyse the design using Solidworks simulation, determination of suitable parameters and material and fabrication process using 3D printer. Fig. 1 shows the flow of the works.

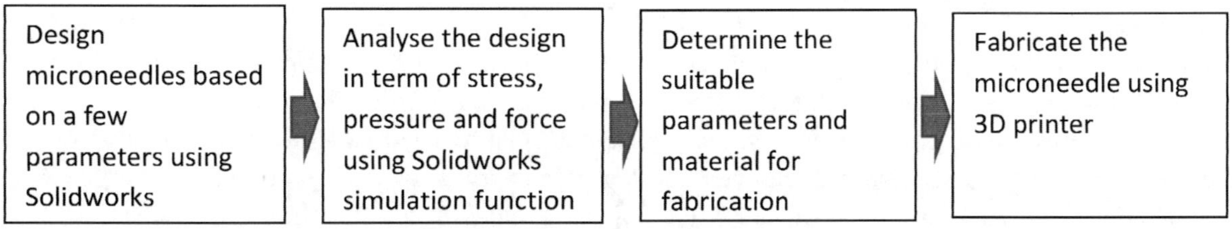

Figure 1. Methodology of research.

4.1 Design

A SolidWorks 2017 is used to design the microneedle. This study focuses on conical and pyramid design as shown in Fig. 2. Table 2 shows the proposed parameters that will be analysed in order to determine the suitable parameters to be printed for microneedle. The parameters consist of shape, number of needles, base size and height of microneedle as well as materials. The value of pressure and force are set to be constant during the simulation analysis.

(a) (b)

Figure 2. The 3D design of the microneedle; (a) conical design (b) pyramid design.

Table 2. Overall design variables of microneedle

Microneedle Design Variable	**Value**		**Type of Variable**
Shape of microneedle	Conical	Pyramid	Manipulated
Material	ABS, Polyester Resin, PLA, PVA		Manipulated
Base are of microneedle (μm)	36π 49π	36×10^4 49×10^4	Manipulated
Height of microneedle (μm)	350 450		Manipulated
Number of needles on patch	Single and triple		Manipulated
Pressure applied on tip of microneedle	3.18 MPA		Constant
Force applied on base of microneedle patch	10kPa		Constant

4.2 Simulation

SolidWorks 2017 provide a simulation features to test the microneedle that implemented with different materials. Human skin exerts 3.18 MPa resistance, which means each needle undergoes a test to resist the pressure exerted by the skin. A force varies from 10 kPa to 100 kPa is placed on the base of the microneedle patch to allowed penetration. However, this paper is focused on 10 kPa force applied. A simulation performed for each design by fixed these two values. The penetration displacement of each design also obtained from the simulation.

4.3 Determination of the suitable parameters and material

From the simulation, the results are coming out according to the selected materials. This stage determines the suitable parameters for conical and pyramid microneedle design that can be applied during microneedle fabrication.

The design that satisfy the ability to withstand stress and penetration of human skin is compared among the fourth materials. The result from study show design with 450 µm height, 49π µm base area and triple needle on a patch for conical design and 450 µm height, 49×10^4 µm base area and triple needle for pyramid design has meet the requirement to penetrate human skin. Table 3 and 4 shows the final comparison of conical and pyramid design for each material.

Table 3. Comparison among the best design of each material for conical design

Material	ABS	Polyester Resin	PLA	PVA
Maximum stress able to withstand (N/m^2)	2.024e+006	1.995e+006	1.500e+006	2.906e+006
Displacement (µm)	56.66	2.514	28.13	353.9

Table 4. Comparison among the best design of each material for pyramid design

Material	ABS	Polyester Resin	PLA	PVA
Maximum stress able to withstand (N/m^2)	2.202e+006	2.191e+006	2.070e+006	2.890e+006
Displacement (µm)	49.35	2.213	24.47	394.2

4.4 Fabrication

SLA 3D printer is chosen to fabricate a microneedle. The mechanism of this printer is it shines UV light into a vat full of selected material which is PVA to cure each layer. This printer capable of printing layer thinner than 10 microns, making it far more detailed compared to FDM printers. Besides, this printer is a low-cost printer and require minimum thirty minutes for every session for printing the microneedle. Fig. 3 shows the printed microneedle using SLA 3D printer.

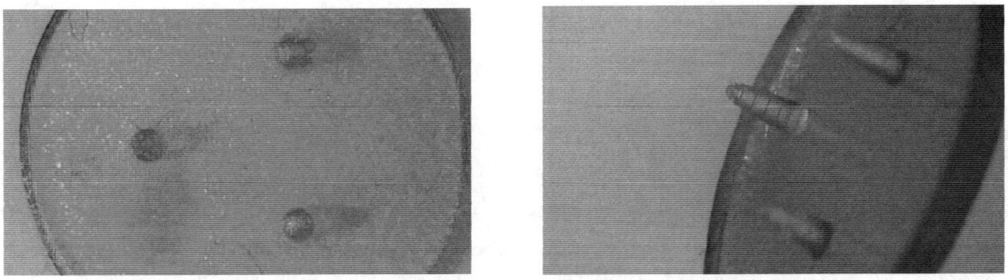

Figure 3. Fabricated microneedle using SLA 3D printer

5. Result and Discussion

In this study, an ability of material to withstand stress and displacement changes after force applied on the design are focused. 10 kPa force is placed on the base of the microneedle patch and 3.18 MPa pressure is positioned on the top of microneedle. The design undergoes a simulation features using SolidWorks 2017 software. PVA is selected as the practicable material to fabricate a microneedle. Data gain from the simulation shows that PVA has the highest ability to withstand stress (force per unit area). From the simulation features, the displacement shown indicating the potential of each design to penetrate human skin. PVA has shown that it can penetrate the deepest and suitable among the rest materials. PVA able to withstand high stress due to its properties which is high tensile strength and high elastic modulus compared to the rest materials. Final stage of this study is to fabricate the microneedle design using SLA 3D printer. Current limitation of microneedle fabrication is lengthy manufacturing period. SLA 3D printer is one of the latest technologies that allowed fabrication using biodegradable material that low-cost and completed in short period.

6. Conclusion

This study constructed based on the comparison of the design, ability to withstand stress and material selected in fabricating a microneedle that can fulfil the requirement to penetrate human skin and at the same time not cause damage towards skin and the needle itself. The simulation result show that PVA is the most trustworthy material in manufacturing microneedle.

References

[1] Ma, G. and Wu, C., 2017. Microneedle, bio-microneedle and bio-inspired microneedle: A review. Journal of Controlled Release, 251, pp.11-23.

[2] McAllister, D.V., Wang, P.M., Davis, S.P., Park, J.H., Canatella, P.J., Allen, M.G. and Prausnitz, M.R., 2003. Microfabricated needles for transdermal delivery of macromolecules and nanoparticles: fabrication methods and transport studies. Proceedings of the National Academy of Sciences, 100(24), pp.13755-13760.

[3] Meites, S., 1988. Skin-puncture and blood-collecting technique for infants: update and problems. Clinical Chemistry, 34(9), pp.1890-1894.

[4] Fruhstorfer, H., Schmelzeisen-Redeker, G. and Weiss, T., 1999. Capillary blood sampling: relation between lancet diameter, lancing pain and blood volume. European Journal of Pain, 3(3), pp.283-286.

[5] Blicharz, T. M., Gong, P., Bunner, B. M., Chu, L. L., Leonard, K. M., Wakefield, J. A., ... & Marlin, S. L. 2018. Microneedle-based device for the one-step painless collection of capillary blood samples. *Nature Biomedical Engineering*, *2*(3), 151.

[6] Kalaiarasi, A. R., Deepika, M., Pooja, D. R., & Preethi, G. S. 2018. Design and Development Mems Based Microneedle for Transdermal Drug Delivery System. *Engineering Reports*, *1*(2), 103-110.

[7] Lin, L. and Pisano, A.P., 1999. Silicon-processed microneedles. Journal of Microelectromechanical Systems, 8(1), pp.78-84.

[8] Song, L., Xia, Y., Li, X., Chen, J., & Ma, S. 2018, January. Fabrication of polymer microneedle array based on tungsten mold and optimized replication method. In *2018 IEEE Micro Electro Mechanical Systems (MEMS)* (pp. 491-493). IEEE.

[9] Economidou, S.N., Lamprou, D.A. and Douroumis, D., 2018. 3D printing applications for transdermal drug delivery. International journal of pharmaceutics, 544(2), pp.415-424.

[10] Farias, C., Lyman, R., Hemingway, C., Chau, H., Mahacek, A., Bouzos, E. and Mobed-Miremadi, M., 2018. Three-Dimensional (3D) Printed Microneedles for Microencapsulated Cell Extrusion. Bioengineering, 5(3), p.59.

[11] Mogusala, N.R., Devadasu, V.R. and Venisetty, R.K., 2015. Fabrication of Microneedle Molds and Polymer Based Biodegradable Microneedle Patches: A Novel Method. American Journal of Drug Delivery and Therapeutics, ISSN, pp.2349-7211.

[12] Material properties. Solidworks Help. Retieved at http://help.solidworks.com 2012/English /SolidWorks/ cworks/c_Material_Properties.htm on 18 December 2018.

Silane treatment of coated carbonate apatite scaffold affects bioactivity and cell viability

Normahira Mamat[1]*, Mariatti Jaafar[2], Zuratul Ain Abdul Hamid[2] and Badrul Hisham Yahaya[3]

[1]School of Mechatronic Engineering, Pauh Putra Campus, Universiti Malaysia Perlis, 02600 Arau, Perlis, Malaysia
[2]Biomaterials Niche Area Group, School of Materials and Mineral Resources Engineering, Engineering Campus, Universiti Sains Malaysia, 14300 Nibong Tebal, Pulau Pinang, Malaysia
[3]Regenerative Medicine Cluster, Advanced Medical and Dental Institute (AMDI), Universiti Sains Malaysia, Bertam, 13200 Kepala Batas, Pulau Pinang, Malaysia

*normahira@unimap.edu.my

Abstract. The aim of this study is to determine effects of surface treatment on coated-carbonate apatite (CO_3Ap) scaffold using silane coupling agent of 3-(trimethoxysilyl) propyl methacrylate. The effect of coating properties with and without surface treatment were investigated through *in vitro* bioactivity and biocompatibility. Firstly, CO_3Ap scaffold was fabricated by transforming ß-TCP scaffold via hydrothermal treatment. CO_3Ap scaffolds were treated in 2% silane solution before coating in chitosan solution (0.5%, 1% and 2%) by dipping method. The coated scaffold without silane treatment denoted as UTR while with silane treatment denoted as TR. It was found that, the presence of Si-OH group enhanced apatite growth on the coated scaffold. TR0.5 has capability to induce apatite growth for 7 days immersion in HBSS compared to UTR0.5. Rougher surface and different surface charge properties have influenced cell viability on the coated CO_3Ap scaffold. To be concluded, it is important to note that bioactivity and cell activity were forced by surface energy or topography, composition and electrostatic interaction.

1. Introduction

In developing bone scaffold, various forms of hydroxyapatite (HA) have been used in orthopaedic, dental or maxillofacial surgery. HA shows excellent biocompatibility and good osteoconductivity, however, the key drawback of HA is its stability in the bone as foreign substance. In other words, HA is difficult to be resorbed in the bone defect. The apatite found in human bone is not stoichiometric of $Ca_{10}(PO_4)_6(OH)_2$ but consisted by other ions mainly carbonate (CO_3^{2-}) content and traces of Na^+, $Mg^{2+}Fe^{2+}$, Cl^- and F^- [1]. Although the presence of these ions is low, they play an important role in the biochemical reactions of bone metabolism. As a result, CO_3^{2-} containing HA has gained much attention due to its chemical composition being closer to bone mineral and favourable for bone growth.

Brittleness of ceramic scaffold can be strengthen using polymer as coating material. Even though the strength will improve, the necessity of biodegradable polymers is also important to be absorbed in host tissue. However, the concern in polymer-coated scaffold fabrication is the interfacial adhesion between bioceramics (inorganic phase) and polymer phase (organic phase). The interfacial bond strength within these dissimilar materials totally depends on the mechanical interlock. Due to chemical bonding does not exist in most bioceramics scaffold, surface treatment using silane coupling agents are often used for providing a strong chemical link in coating system.

Silane is the most commonly used due to its reactivity and ability to yield good and durable adhesion between the relevant substrates [2]. In this regard, the most popular silane coupling agent for bone reconstructions is the 3-(Trimethoxysilyl) propyl methacrylate (also called 3-methacryloyloxy propyl trimethoxysilane) (MPS). Its organofunctional group matches the chemical structure and reactivity of the inorganic/organic systems [3]. Silane coupling agents are often used for ceramic scaffolds to provide a strong chemical link between the oxide groups on the scaffold surface and the polymer molecules on the coating layer [4]. However, limited studies have been conducted to investigate the performance of coated carbonate-substituted apatite scaffold. Most previous researches focused on the silane treatment on composite-based bioceramic [4]. Therefore, this study aimed to investigate the effects of silane treatment on chitosan coated CO_3Ap scaffolds. Bioactivity and cell behaviours on the coated CO_3Ap scaffold with and without silanisation were investigated.

2. Materials and method

2.1. Materials

ß-TCP, $Ca_3(PO_4)_2$, powder with grade purum p.a, >96.0% was used as a main component to prepare ß-TCP scaffold. Polyethylene-imine (PEI, 50 w/v%) was used as a dispersant in ß-TCP slurry. Sodium dodecyl sulphate (SDS) powder was used as a pore forming agent. PVA acts as a binder in the ceramic slurry. These materials were purchased from Sigma Aldrich. Polyethylene glycol (PEG) in liquid form was purchased from Merck (Germany). PEG was used as a plasticizer. Denacol was supplied by Nagase Chemtex, Japan. Denacol was used as a gelling agent and crosslinker in ß-TCP slurry. For transforming ß-TCP scaffold to CO_3Ap scaffold, disodium carbonate (Na_2CO_3) solution was used as a carbonate (CO_3^{2-}) source. Medium molecular weight of chitosan (CS, 75-85% deacetylation) was used as a natural polymer solution for coating purpose. CS was purchased from Sigma Aldrich. 3-(Trimethoxysilyl) propyl methacrylate (MPS)-98% or silane A174 is a silane coupling agent for the surface treatment. Silane was purchased from Sigma Aldrich. Methanol, 99.9% was purchased from JT Baker. Methanol was used as a silane solvent. Acetic acid (100% anhydrous) was used as a CS solvent while glutaraldehyde, GA, 50% solution in water was used as a crosslinker to bind amine group in CS.

2.2. Methods

2.2.1 Fabrication of CO_3Ap scaffold. ß-TCP scaffold was fabricated using gelate-freeze casting method. ß-TCP slurry was prepared by mixing 60 wt.% of ß-TCP powder and 40 wt.% distilled water based solution. Firstly, 4 wt.% of PVA was dissolved in the distilled water. After that, PEG was mixed in the PVA solution. ß-TCP powder was added in the mixture as ceramic slurries. The slurries were homogeneously mixed by using mechanical stirrer at 600 rpm. Then the slurry was added with PEI and continuously stirred. After that, SDS was added under 900 rpm stirring until the slurry 1.5 times expanded. Denacol was lastly added to maintain the slurry expansion. The casting process of the slurry was done by pouring ceramic slurries into mold. The slurry was dried at room temperature for 1 h to complete the gelling phase. Then, the molds were transferred into -30 °C freezer for 3 h. After that, the molds were placed in freeze drier using Freeze Dry System (Labconco) for 24 h. The green bodies were sintered in an air furnace at a temperature of 1115 °C. Fabrication of CO_3Ap scaffold was done using hydrothermal treatment method. ß-TCP scaffold was used as precursors. The precursors were soaked in 5 mol/L of Na_2CO_3 solution in hydrothermal vessel. The vessels were kept at 200 °C in oven for 5 days.

After that, the fabricated CO3Ap scaffolds were soaked in boiling distilled water to wash the excessive sodium in the scaffold. Then, the scaffolds were dried at 80 °C for 1 h.

2.2.2 Chitosan-coated CO_3Ap scaffold without and with surface treatment. The silanisation process was performed by soaking the CO_3Ap scaffolds in the silane solution for 120 min. The silane solution was prepared by depositions from an aqueous methanol solution. This process included a 95% methanol/5% water solution adjusted to a pH 4.5–5.5 with acetic acid. The silane was added while stirring to yield a final concentration of 3%. After soaking for 120 min, the scaffolds were heat-treated in the oven at 160ºC for 2 h. In this study, this is denoted as silanised-CO_3Ap scaffolds. The silanised-CO_3Ap scaffolds were cross-linked with 2% (v/v) GA in ultrapure water overnight with slow agitation. Each sample was rinsed three times with deionized water, then heated and stored for the coating process. CS solutions were prepared by dissolving the powder in distilled water (250 mL) to produce three different concentrations (0.5%, 1%, and 2%) of chitosan, with 2% (v/v) acetic acid. The solution was stirred to dissolve the chitosan completely. Silanised-CO_3Ap scaffolds were then soaked into the chitosan solution. Vacuum was then used for 4 h to force the CS solution to infiltrate the porous structure of the CO_3Ap scaffolds. Scaffolds were dried in a vacuum oven at 80ºC for 24 h. This coated-CO_3Ap scaffolds has been denoted as treated scaffold. The same coating procedures were carried out for CO_3Ap scaffolds without the silane treatment, which is denoted as untreated. The composition and designation of the coatings with the respective scaffolds are summarized in Table 1.

Table 1. Composition and designation of CO_3Ap scaffolds used in this study.

Sample designation	Silane treatment	Chitosan (CS) concentration (w/v%)
Neat CO_3Ap	-	-
Silanised- CO_3Ap	√	-
UTR0.5	-	0.5
TR0.5	√	0.5
UTR1	-	1
TR1	√	1
UTR2	-	2
TR2	√	2

2.3. Characterisations

Hanks' balanced salt solution (HBSS, Gibco), was used as a supporting solution for the CO_3Ap scaffolds in an *in vitro* bioactivity test. The cylinder-shaped of scaffolds were immersed in 25 mL of HBSS to elicit a mineralisation reaction. After soaking, the scaffolds were taken out and rinsed with distilled water to remove residual HBSS and then immediately dried at 50 °C for 24 h. The morphology and composition of the mineral layers formed on the scaffolds were characterised by energy dispersive X-ray spectroscopy (EDX, EDAX Genesis). Cell viability using human osteoblast cell (hFOB). The viability assay was carried out using PrestoBlue™ Cell Viability Reagent using an automated microplate reader according to standard protocol provided by the manufacturer after 1, 2, 3, 5 and 7 d of culture.

3. Results and discussions

3.1 Bioactivity

The capability of UTR and TR scaffolds to induce and form bonelike apatite growth were determined by the changes of Ca/P ratio as presented in Table 2. For 7 days immersion, the Ca/P of the TR scaffolds have higher ratio than those of UTR scaffolds. It revealed a Ca-rich amorphous calcium phosphate

(ACP) with Ca/P ratio of 1.98, 1.67 and 1.58 on TR0.5, TR1 and TR2 respectively. The formation of the Ca-rich ACP was assisted by functional group of Si-OH. The Ca-rich ACP was assumed to take place in consecutive accumulation of the calcium ions, which makes positive charge increased. The Ca-rich ACP on the TR scaffold therefore interacts specifically with the negative phosphate ions in the fluid to form a Ca-deficient (poor) ACP. This type of Ca-deficient (poor) ACP was observed as a precursor, which eventually crystallizes into bonelike apatite that lowered the Ca/P ratio of the TR scaffolds after 28 days immersion, i.e; 1.65, 1.64 and 1.55 for TR0.5, TR1 and TR2, respectively.

Table 2. The molar Ca/P ratio of UTR and TR scaffolds after 7 and 28 days immersion in HBSS solution

Scaffold	Ca/P ratio		
	Immersion time (day)		
	0	7	28
Neat CO_3Ap	1.66	1.78	1.58
UTR0.5	-	1.85	1.86
TR0.5	-	1.98	1.65
UTR1	-	1.65	1.68
TR1	-	1.67	1.64
UTR2	-	1.36	1.68
TR2	-	1.58	1.55

The mechanism of apatite formation starts when the TR scaffold is immersed in the HBSS, Ca^{2+} in the scaffold exchange with H+ in the SBF solution. This leads to the formation of silanol (-Si–OH) in the surface layer at the scaffold–HBSS, and eventually the production of a negatively charged surface with the functional group (-Si–O$^-$). The Ca^{2+} in the HBSS solution are first attracted to the interface between coating and solution to form Ca-rich ACP. Consequently, the ionic activity product of the apatite in the interface is high enough to precipitate apatite on the TR surface [5]. Thus, the results suggest that the changes of apatite-forming behaviour were affected by the surface compositional of CO_3Ap scaffold. The bioactive materials with a superior *in vitro* apatite-forming ability determine the directly bonding to living bone *in vivo*[6].

3.2. *Relation between surface charge and cell proliferation*

Both UTR0.5 and TR0.5 scaffolds were used to be further characterized using zeta potential and cell study. Figure 1 shows the zeta potentials for all CO_3Ap scaffolds at pH 7 in HBSS. All scaffolds are negatively charged, and the absolute zeta-potential comparison sequence was indicated as, silanised-CO_3Ap < neat CO_3Ap < UTR0.5 < TR0.5. Considering CO_3Ap scaffold, the surface was found to be negatively charged represented from the presence of OH- group. After silanisation, CO_3Ap became less negatively charged, as could be expected from the presence of silanol groups. This was related to the nonpolar group of CH^{3+}. Zeta potential of coated CO_3Ap were portrayed more negatively charged *i.e*: -14 mV (TR0.5) and -12.95 mV (UTR0.5) than those of neat and silanised-CO_3Ap scaffolds. It is known that, CS possesses a higher positive surface charges due to more extensive protonation of its amino groups at pH below its pKa,~6.5. However, this explanation is disputable whereby HBSS at pH 7 used in the present study caused the decrement of zeta potential upon neutralization of the respective solution used. This finding is in agreement with results reported by Loh *et al.* [7]. They studied that CS exhibited negative charges at pH 7.4 and positive charges at pH 6.0 as the zeta potentials measured in HBSS. In addition, higher negatively charge of TR0.5 might be resulted from well-dispersed of CS coating on the TR0.5 surface due to chemical link bonding compared to UTR0.5

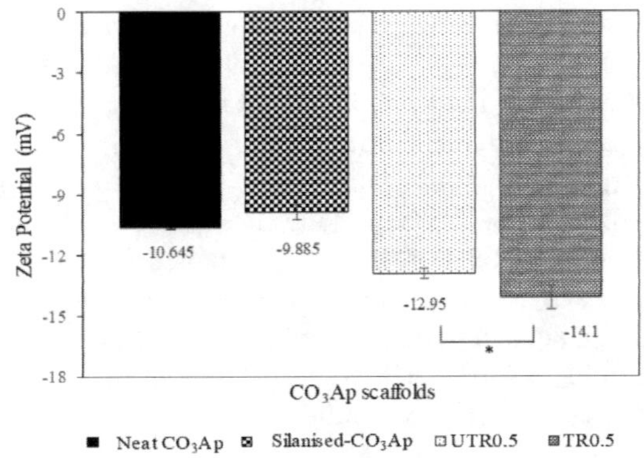

Figure 1. Zeta potential of different surface CO_3Ap scaffolds. * indicates $p < 0.05$ is significantly difference

Figure 2 shows cell proliferation of hFOB cells on the neat CO_3Ap, silanised-CO_3Ap, UTR0.5 and TR0.5 scaffolds Cells seeded on a tissue culture plate without scaffold were taken as a control. The silanised-, UTR and TR CO_3Ap scaffolds showed an increase of cell proliferation than control. The silanised-CO3Ap scaffold showed an increase in cell proliferation compared with others all the time. Rougher surface by silanised-CO_3Ap scaffold (as indicated in Figure 3) is believed has provided higher adhesion of hFOB cells. The rougher surface increased the surface energy which resulted in better chemical bonding with cells.

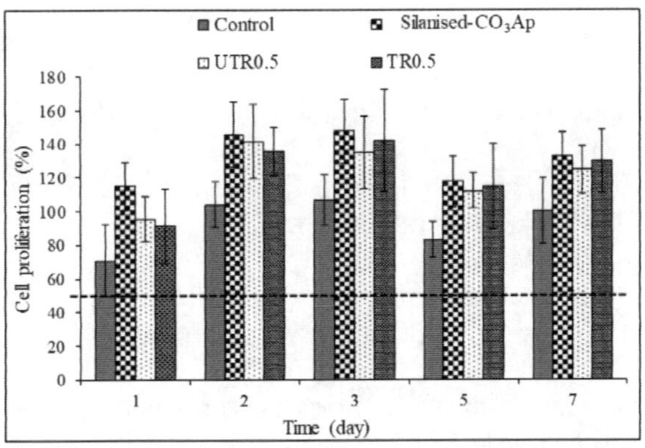

Figure 2. The results of PrestoBlue assay for silanised-, UTR0.5 and TR0.5 CO_3Ap scaffolds in the presence of hFOB cells for 7 days incubation

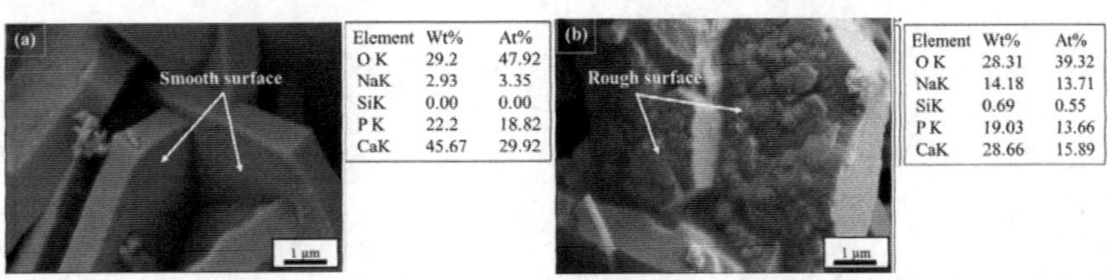

Figure 3. SEM images and EDX spectra shows elemental composition of (a) neat CO_3Ap and (b) silanised-CO_3Ap scaffolds

As indicated, the silanised-CO_3Ap scaffold has less negative charge but the proliferation was superior among others. According to Qin *et al.* [8], osteoblast cells tended to adhere and proliferated more when the magnitude of negative zeta potential increased. Therefore, in this study cells were able to proliferate more on silanised-CO_3Ap surface with a less negative zeta potential (-9.885 mV) could be related with rough surface of silanised-CO_3Ap, which provided adhesion promotion with more micro-mechanical retention. In addition, the silane treatment does not have intrinsic toxicity to be used for cell growth [9]. As already known, cell surface is negatively charged. Due to same negatively charged between the cell and the material no direct electrostatic interaction occurred. Thus, why cells can proliferate on negative surface charge could be explained as followed. Whenever materials encounter a biological environment, the protein adsorption process happened within a few seconds. The protein is amphoteric polyelectrolyte with both acidic and basic peptide. Therefore, it behaves as a bridge between positive or negative groups on cell surfaces or modified CO_3Ap surface. To be concluded, these results provided the outcomes of *in vitro* studies which can be manipulated to select the most optimal performance of surface functionality for *in vivo* studies.

4. Conclusion

The functional group of Si-OH by treated CO_3Ap scaffolds induced the precipitation bonelike apatite in HBSS. Rougher the scaffold surface exhibited by silane treatment has increased cell viability. Utilization of silane treatment on the coated scaffold provided mechanism of enhancement in bioactivity and cell growth for successful bone regeneration.

Acknowledgement

The authors acknowledge the support from the Ministry of Education Malaysia through the Trans Disciplinary Research Grant Scheme (TRGS. 6761004 and 6761002). The authors also gratefully acknowledge support from the Universiti Malaysia Perlis (UniMAP) and Universiti Sains Malaysia (USM).

References

[1] Ishikawa K 2010 Bone substitute fabrication based on dissolution-precipitation reactions *Materials (Basel)* **3** 1138–1155

[2] Zakir M, Ashraf U, Tian T, Han A, Qiao W, Zhang M et al. 2016 The Role of Silane Coupling Agents and Universal Primers in Durable Adhesion to Dental Restorative Materials - a Review *Curr. Oral Heal. Reports* **3** 244–253

[3] Haider A, Haider S, Han S S and Kang I K 2017 Recent advances in the synthesis, functionalization and biomedical applications of hydroxyapatite: a review *RSC Adv.* **7** 7442–7458

[4] Cisneros-Pineda O G., Herrera Kao W, Loría-Bastarrachea M I, Veranes-Pantoja Y, Cauich-Rodríguez J V and Cervantes-Uc J M 2014 Towards optimization of the silanization process of hydroxyapatite for its use in bone cement formulations *Mater. Sci. Eng. C.* **40** 157–163

[5] Oyane A, Kawashita M, Nakanishi K, Kokubo T, Minoda M *et al.*, 2003 Bonelike apatite formation on ethylene-vinyl alcohol copolymer modified with silane coupling agent and calcium silicate solutions *Biomaterials* **24** 1729–1735

[6] Kokubo T, Matsushita T, Takadama H and Kizuki T 2009 Development of bioactive materials based on surface chemistry *J. Eur. Ceram. Soc.* **29** 1267–1274

[7] Loh J W, Saunders M and Lim L Y 2012 Cytotoxicity of monodispersed chitosan nanoparticles against the Caco-2 cells *Toxicol. Appl. Pharmacol.* **262** 273–282

[8] Qin L, Dong H, Mu Z, Zhang Y and Dong G 2015 Preparation and bioactive properties of chitosan and casein phosphopeptides composite coatings for orthopedic implants *Carbohydr. Polym.* **133** 236–244

[9] Ying C, Lung K and Matinlinna J P 2012 Aspects of silane coupling agents and surface conditioning in dentistry : An overview *Dent. Mater.* **28** 467–477

Journal of Physics: Conference Series **1372** (2019) 012055 doi:10.1088/1742-6596/1372/1/012055

Differentiate Characteristic EEG Tobacco Smoking and Non-smoking

Lim Chee Chin[1*], Asmiedah Muhamad Zazid[1], Chong Yen Fook[1], Vikneswaran Vijean[1], Saidatul Ardeenawatie Awang[1], Marwan Affandi[2], Lim Sin Chee[3]

[1] School of Mechatronic Engineering, University Malaysia Perlis (UniMAP), Campus Ulu Pauh, Arau, Perlis, Malaysia

[2] Faculty of Engineering, Rajamangala University of Technology Srivijaya (RUTS), Songkhla, Thailand

[3] First City University College, Bandar Utama, Petaling Jaya, Malaysia.

*cheechin10@gmail.com

Abstract- Electroencephalogram (EEG) signal is non-stationary signal that have low frequency component and amplitude compared to stationary signal. Therefore, present of unwanted substance (nicotine) in Tobacco smoking will alter the brain electrical activity. This paper is proposed to investigate the changes of EEG signal with the present of nicotine and identify the difference brain signal between smoker and non-smoker. There are 20 males (10 smokers, 10 non-smokers) are selected. The subjects are chosen based on inclusion criteria (abstained from smoking within 6 hours before experiment, and do not take any medication and caffeine). The recorded EEG signal contain a lot of noise such as head moving, muscle movement, power line, eyes blinks and interference with other device. Butterworth filter are implemented to remove the unwanted noise present in the original signal. Bandpass filter is used to decompose the EEG signal into alpha, theta, delta and beta frequency. Then, eight features (mean, median, maximum, minimum, variance, standard deviation, energy and power) have been extracted by using Fast Fourier Transform (FFT) and Power Spectral Density (PSD) method. Then, four different type of kernel function ('Linear', 'BoxConstraint', 'Polynomial' and 'RBF') of SVM classifier are used to identify the best accuracy. As a result, PSD (97.50%) have higher performance accuracy than FFT (97.33%) by using Radial Basis Function (RBF) of Support Vector Machine (SVM). Smoking activity caused slightly increase theta and delta frequency. Smoking is activated of five electrode channels (Fp1, Fp2, F8, F3 and C3) and caused additional emotion such as deep rest, stress releasing and losing attention. The attention of smokers can be measure by using stroop test. After smoking activity, smokers become more energetic and increase the time response (1.77 s) of stroop test compared to non-smokers (2.96 s). The result is calculated by using statistical analysis (t-test). The p-value is 0.037 which is less than 0.05. Thus, the null hypothesis is rejected and conclude there is significant different between smokers and non-smoker performance before and after smoking task.

1. Introduction

Brain is an important organ of human body that function as control centre that monitor and coordinate all activity. Cerebrum is the largest region in human brain, it is controls most of function such as speech, emotion, learning and control movement. Cerebrum divided into four lobe which is frontal lobe (F), occipital lobe (O), parietal lobe (P), and temporal lobe (T) [7]. Presenting the harmful substance such as nicotine, drug and alcohol will alter the brain signal activity. Smoking is bad habit that attributes to increase number of chronic diseases and death smokers due to the presence of nicotine. The statistical has been approved 1017 per 100,000 death smokers in Malaysia due to smoking- related illness [1]. Generally, most of smokers aware about the effect of smoking but they cannot stop it. There are strong reasons why smoker addicted to smoke due to present of nicotine [2]. Nicotine is an addictive substance that will affect the EEG in transferring and processing the EEG brain signal [10]. The present of nicotine in human brain caused nicotine bind more effectively with receptor rather than neurotransmitter. Therefore, the cell become overstimulated compare to neurotransmitter. The cell no longer excited when there is no nicotine in body this caused smoker addicted to smoke to get the same effects [11]. In addition, smoking produce psychostimulant profile that effect the power reduction in theta and delta. Smoking one or two cigarettes causes a decrease in the delta in central-posterior head regions [12]. The objectives in this paper is to design the protocol to record EEG signal for smokers and non-smokers based on the previous research, to apply suitable EEG signal processing algorithm for extracting significant features and to compare the topographic EEG pattern before and after smoking task.

2. Literature Review:

In previous study, Edward F. Domino et. al study about tobacco smoking between smokers and non-smokers caused changes of dominant alpha electroencephalographic (EEG) frequency [12]. Subjects were asked to abstain 1 hour before the experiment. Two smoking task had been conducted which is sham smoke (non-smoker) and real smoke (smoker). The result of the experiment is sham smoking slightly increased dominant alpha frequency in Fp2, Fz, F4, F8 electrodes. The right hemisphere of non-smoker undergoes slightly change (± 0.2 Hz) in alpha frequency after sham smoke due to brain response to smoking activity. Catherine P. Canamar et. al. studied about the effect of cigarette smoking toward attention bias by conducting smoking stroop test. There are 51 smoker's subject (age between 18 to 51 years old). The subjects need to perform two tests which are conduct a stroop test after 13-16 hours abstained and the second task conduct a stroop task after 1 hour abstained. Result of the experiment is increase time response speed of stroop test after smoking task [13]. The result is calculated by ANOVA to compare the performance of smokers before and after smoking task.

Bandpass filter is the most preferred filter used in previous study to remove unwanted noise present in raw signal [14][15][11]. Based on Edward F. Domino et. al, bandpass filter (5.0-6.0 Hz) is used to minimize the extraneous artefacts induced by body movement and eyes blinking [14]. Rass et. al study about EEG Signal and personality in daily and non-daily smokers. The author sampled the EEG signal at 1000 Hz sampling rate by using 0.1 until 200 Hz bandpass filter [15]. Murugesan and R.Sukanesh used PSD to extract the features in EEG signal by assuming short time stationary and energy content on EEG signal in conventional EEG frequency band such as alpha (8- 13 Hz), beta (13 - 40 Hz), delta (0-3.5 Hz) and theta (4 -7.5 Hz). The objective of using PSD to identify the power distributed. Spectral analysis in statistics is the process that decomposes a time series into a spectrum of cycles of varying lengths [16]. Goos et. al. applied FFT method to extract the EEG signal and classified the signal by using SVM with 100% accuracy. SVM is the best classifier to classified of two classes of EEG signals [16][17][18][19]. Murugesan and R. Sukanesh found that RBF is the best kernel function in classifying the two classes of EEG signal [16].

3. Methodology

Ten non-smokers and ten smokers were recruited from the local community surrounding University and paid for participation. Participants must satisfy of inclusion criteria of healthy and age range is 20 to 40 years old. Participants were excluded for a history of electroconvulsive therapy, neurological illness and chronic disease. Additional exclusion criteria included of abstain from take any caffeine and stop smoking within 6 hours before the experiment. Hardware EEG truscan and software EEG TruScan Explore was used to recorded brain signal. Thus, non-invasive electrode scalp 19 electrodes are used to obtain EEG signal. Total data of EEG signal is 18430 in one seconds. This data points were saved in ASCII file. After that, it was converted to text file to load into MATLAB software for filtering process. The sampling frequency of data is 1024 Hz. Then, smokers need to fill in the form of Fergerstrom Test Nicotine

Dependence (FTND) to identify the level of nicotine dependency of smoker. Based on FTND, there are 10% light, 20% heavy, 30% is low moderate and 40% is moderate smokers in this experiment.

3.1 Experiment Protocol

The experiment protocol is started with setup the electrode on subject scalp. Then subjects are asked to practice stroop test before start the EEG recording. The light is switch off within 3 minutes to ensure subject really relax and ready form EEG recording. Subjects are asked to close eyes within 3 minutes while EEG recording is started. All of the rest tasks are conducted with eyes closed. After complete rest task, subject required to perform the stroop test in the same time EEG is recorded. Next task is rest task (3minutes). After that is smoking task which is sham smoke (non-smoker task) and real smoke (smoker task). Then, subject asked to rest while EEG is recorded (3 minutes). Finally, second stroop test is conducted after smoking task.

3.2 Signal Processing

The pre-processing technique is used to remove the unwanted signal present during data collection. Head moving, muscle movement, power line, eyes blinks and interference with other device [2] [3] will be removed by using Butterworth Filter. 4th order Butterworth filter is used due to it has linear response. The cutoff frequency is 0.5 to 30 Hz. The delta frequency (below 0.5 Hz) is eliminated because it consider as noise. Second filter have been used in the research is bandpass filter. Bandpass filter is used to decompose the signal in form alpha (8-12Hz), beta (12-30Hz), delta (1-4Hz) and theta (4-8Hz). The signal is segmented from 60 to 120 seconds because during this state is optimum state [8] which subjects are really concentrate on task given thus this is an accurate time to measure EEG signal on given task.

Feature extraction method is the next step after pre-processing method. Feature extraction method is function to extract the signal value and to reduce the loss of important information of signal [20]. In this research, FFT and PSD are used to extract the waveform of EEG brain signal. There are total of eight features are extracted in FFT method which is the mean, median, variance, standard deviation, power, energy, minimum and maximum value. FFT is important to convert time domain to frequency domain to reduce the operation time and increasing the speed. The FFT is can compute the result 0 (N log N) operation [21]. N is mean by the length of the vector. The Equation (3.1) show the FFT functions and Equation (3.2) is function of Inverse Fast Fourier Transform (IFFT)

$$X(k) = \sum_{j=1}^{N} x(j)\omega_N^{(j-1)(k-1)} \tag{3.1}$$

$$x(j) = (1/N)\sum_{k=1}^{N} X(k)\omega_N^{-(j-1)(k-1)} \tag{3.2}$$

Another feature extraction method is by using PSD method is generally known as P-Welch method that implemented to transform the EEG signal form time domain to the frequency domain. In this research, hamming window technique is applied to extract eight significant features. Function of Hamming window is it acts as smoothing value known as an apodization of EEG signal (“removing the foot”, i.e. smoothing discontinuities at the beginning and end of the EEG signal) [9].

$$P_{xx}(\infty) = \frac{1}{2\pi}\sum_{m=-\infty}^{\infty} R_{xx}(m)e^{-jcom} \tag{3.3}$$

$$w(n) = 0.54-0.46\cos(2\pi n/M-1) \qquad 0 \leqslant n \leqslant M-1 \tag{3.4}$$

Support Vector Machine (SVM) is used to classify the EEG signal. Generally SVM can be implemented in two-classifiers due to large margin and mapping data into a higher dimensional space. SVM separate between smoking and non-smoking by using linear line (hyperplane) [20].

4. Result and Discussion

P-Welch is the best feature selection method due to the accuracy is 97.50% compare to FFT accuracy of 97.33% as in Figure 1. P-Welch method provides better performance than FFT method by using RBF kernel function which is 98.33% accuracy compare to FFT 61.50%. This is because P-welch method have high spectrum estimation signal to noise ratio and reduce frequency resolution [22]. P-Welch's method provides good resolution which is the ability to discriminate the feature. There are eight features had been extracted which is mean, medium, maximum, minimum, standard deviation, power and energy.

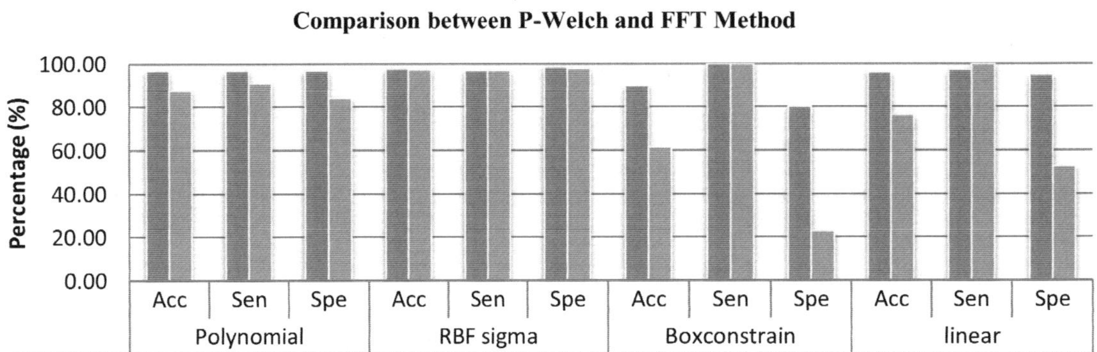

	Polynomial Acc	Polynomial Sen	Polynomial Spe	RBF sigma Acc	RBF sigma Sen	RBF sigma Spe	Boxconstrain Acc	Boxconstrain Sen	Boxconstrain Spe	linear Acc	linear Sen	linear Spe
P-Welch	96.67	96.67	96.67	97.50	96.67	98.33	90.17	100.00	80.33	96.17	97.33	95.00
FFT	87.50	91.00	84.00	97.33	96.67	98.00	61.50	100.00	23.00	76.33	100.00	52.67

Figure 1. Comparison the performance of P-welch method and FFT method
(Note: Acc – Accuracy, Sen – Sensitivity, Spec – Specificity)

4.1 Feature Selection

There are 19 electrodes are used to record EEG signal. Each electrode provides 4 different types of decomposed signals which is alpha, beta, theta and delta frequency. Then, each four type of decomposed signal is extracted into eight features which is the value of mean, medium, maximum, minimum, standard deviation, power and energy. Thus, there are total 608 individual features is extracted from the EEG signal. Since there are too many features, the features selection needs to be done in order to select the best features for classification. All these 608 features will undergo individual classification using SVM and compare each other. The twenty-nine features have been selected as the best feature due to the accuracy is the highest among the others which is more than 57.45%. These twenty-nine features are contributed from five electrodes channels which are Fp1, Fp2, F8, F3 and C3 as show in Figure 2. All of these 29 features are activated during smoking activity and very important in identifying smoking behaviour. Smokers become more energetic after smoking task compare to non-smokers. Smoking behaviour resulted in increasing the theta and delta frequencies due to over stimulate of hormone such as neurotransmitter, melatonin and others [23]. This is have been proved by previous research smoking caused slightly increase the theta and delta frequency [4]. Moreover, smoking caused additional emotion benefits such as reduce stress and anxiety, less feeling of depression and deep rest [5] [6].

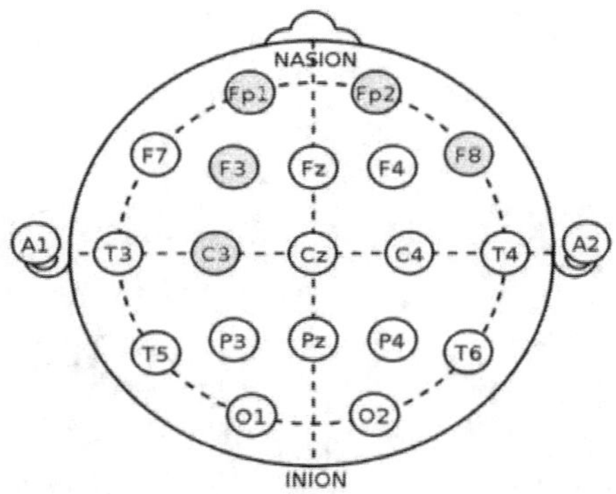

Figure 2. Selected EEG Channels of Topographic

4.2 Classification

Classification between the smoker and non-smoker is done by using SVM classifier as shown in Figure 3. The input of the classifiers either is 19 combined channels or 5 selected channels. The combination channels are combined of 19 electrodes channels while selected channels are combined of five electrodes channel (Fp1, Fp2, F8, F3 and C3). The combination channels 98.33% have higher accuracy compare to selected channels 97.17%.

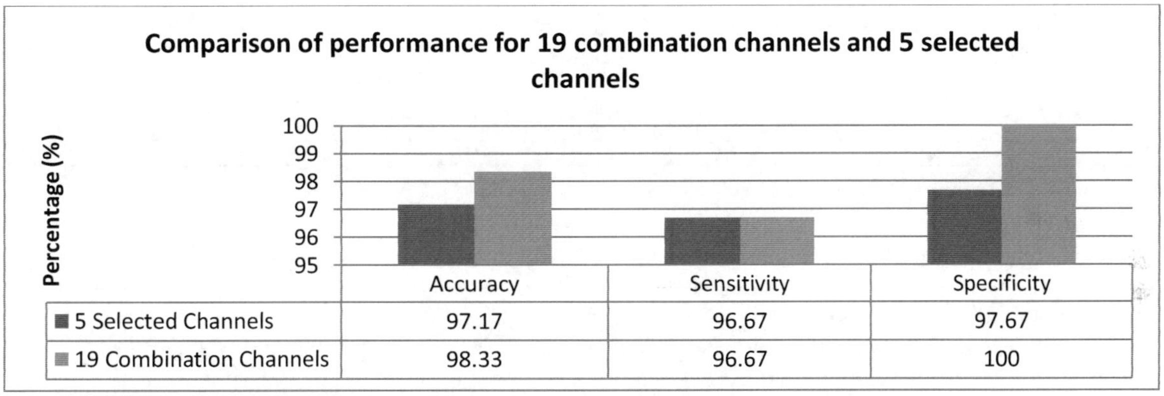

Figure 3. Comparison of performance for 19 combination channels and 5 selected channels

During abstained stage (before smoking task) smokers have low time response of stroop test (12.8 s) compare to non-smokers (12.51 s). After smoking task smokers (11.63 s) have fast time response compare to non-smokers (12.37 s). Smokers perform more number of trials (trials = 37) after smoking and increase speed of response compared to non-smokers (trials = 31) as shown in Figure 4(b). The increasing of number of trial for smokers (trials = 37) lead to increase the time spending to complete the stroop test (430.26 s). The present of nicotine caused smokers become more energetic and reduce the attention level. Therefore, the tendency of smoker to do mistake is higher than non-smoker. The present of nicotine in human will increase the speed of response. The present of nicotine in human body caused over stimulated neurotransmitter in brain. In additional, the present of nicotine can reduce the attention biases of smokers. The over stimulated the Neurotransmitter caused smokers lost their attention to answer stroop test after smoking.

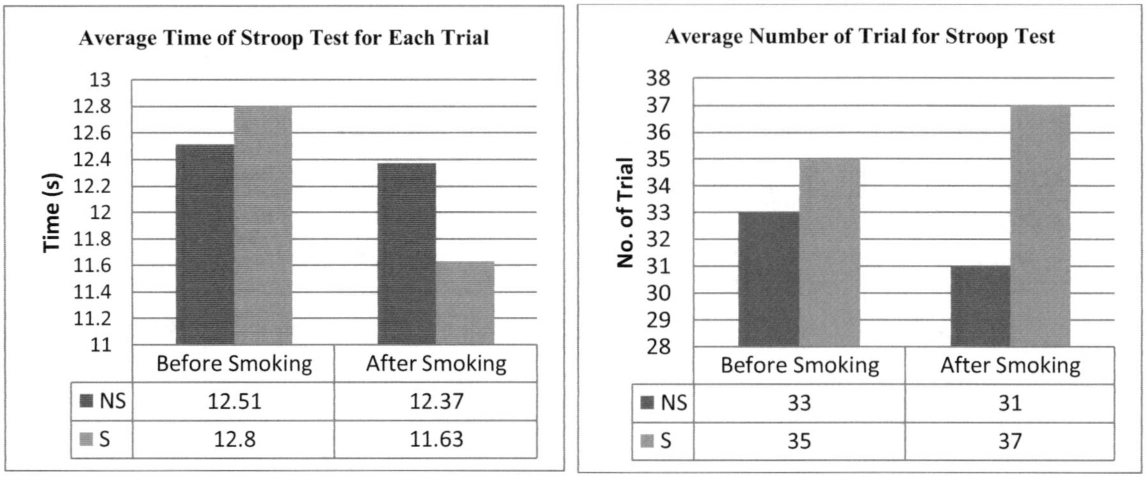

(a) *(b)*

Figure 4. (a) Average response time of stroop test for each trial and (b) Average number of trials for stroop test

Statistical Analysis is calculated to identify the significant difference between smokers and non-smokers. From Table 1, the p-value is less than 0.05 which is 0.037. Thus can conclude there is significant different between smoker and non-smoker performance before and after smoking task.

Table 1. Statistical analysis of significant difference between smoker and non-smokers

	Paired Differences					t	df	Sig. (2-tailed)
	Mean	Std. Deviation	Std. Error Mean	95% Confidence Interval of the Difference Lower	Upper			
Pair 1 non-smokers - smokers	-2.998350	8.778229	1.387960	-5.805764	-.190936	-2.160	39	.037

5. Conclusion

The EEG characteristic between tobacco smoking and non-smoking was successfully differentiated with the P-values is 0.037. The EEG signal is successfully recorded by using 10-20 electrodes placement system with 1024 Hz sampling frequency. The raw EEG signal is filtered by using bandpass and Butterworth filter. Moreover, the features of filtered signal are extracted by using FFT and PSD. Then, signal is classified by using four different type of kernel function of SVM classifier. For the feature extraction method, PSD has higher performances (97.50%) by using RBF kernel function method compare to other types of kernel function ('BoxConstraint', 'Linear' and 'Polynomial'). The present of nicotine can alter the human brain signal, which caused the response time is increased by 1.17 s and reduced in attention the trial number is averagely increased by twice.

To improve the result performance, the numbers of participants need to be increased to contribute large sample size of data. Then, the experiment is suggested to conduct in sound proof room in order to reduce the unwanted noise that will interrupt the recording of EEG signal. Some of the noise cannot be remove completely such as environmental noise, lighting of the room and noise due to muscle movement. Next, increase the abstaining duration which is more than 6 hours to examine the effect of abstain activity toward stroop test. There were will be large significant different between before and after abstain task. Additional of classifier such as Linear Discriminant Analysis (LDA), K-Nearest Neighbor (KNN) or Artificial Neural Network (ANN) to identify which classifiers is able to produce the higher accuracy.

References

[1] Jaswal R 2016 Brain Wave Classification and Feature Extraction of EEG Signal by Using FFT on Lab View *International Research Journal of Engineering and Technology (IRJET)* **3(7)** 1208–1212.

[2] Bang J W, Choi J S, Park K R 2013 Noise reduction in brainwaves by using both EEG signals and frontal viewing camera images *Sensors (Switzerland)* **13(5)** 6272–6294.

[3] Palendeng M E 2011 Removing noise from electroencephalography signals for BIS based depth of anaesthesia monitors.

[4] Li X, Ma R, Pang L, Lv W, Xie Y, Chen Y, Zhang P, Chen J, Wu Q, Cui G, Zhang P, Zhou Y, Zhang X. 2017 Delta coherence in resting-state EEG predicts the reduction in cigarette craving after hypnotic aversion suggestions **7(1)** 1–8.

[5] Kumar J S and Bhuvaneswari P 2012 Analysis of electroencephalography (EEG) signals and its categorization - A study *Procedia Engineering* **38** 2525–2536.

[6] Donovan M 2017 Smoking and Emotional Health | Healthy UNH", *University of New Hampshire* https://www.unh.edu/healthyunh/blog/tobacco/2017/11/smoking-and-emotional-health, accessed 2 July 2019.

[7] Mayfield 2018 "Anatomy of The Brain", *Mayf. Brain Spine*, 1–7.

[8] Xu X and Domino E F 2000 Effects of tobacco smoking on topographic EEG and Stroop test in smoking deprived smokers *Prog Neuropsychopharmacol Biol Psychiatry* **24(94)** 535–546.

[9] Sphinx "scipy.signal.hamming — SciPy v0.19.1 Reference Guide", *The Scipy community*, 2008. https://docs.scipy.org/doc/scipy-0.19.1/reference/generated/scipy.signal.hamming.html, accessed 2 July 2019.

[10] Pritchard W S 1991 Electroencephalographic effects of cigarette smoking *Psychopharmacology (Berl)* **104(4)** 485–490.

[11] Rass O, Ahn W, and Donnell B F O 2016 Clinical Neurophysiology Resting-state EEG, impulsiveness , and personality in daily and nondaily smokers *Clin. Neurophysiol.* **127(1)** 409–418.

[12] Domino E F, Ni L, Thompson M, Zhang H, Shikata H, Fukai H, Sakaki T and Ohya, I 2009 Tobacco smoking produces widespread dominant brain wave alpha frequency increases *International Journal of Psychophysiology* **74(3)** 192–198.

[13] Canamar C P and London E 2012 Acute cigarette smoking reduces latencies on a Smoking Stroop test *Addict. Behav.* **37(5)** 627–631.

[14] Öberg M, Jaakkola M S, Woodward A, Peruga A and Prüss-Ustün A 2011 Worldwide burden of disease from exposure to second-hand smoke: A retrospective analysis of data from 192 countries *Lancet* **377(9760)** 139–146.

[15] Streppel M T, Boshuizen H C, Ocké M C, Kok F J and Kromhout D 2007 Mortality and life expectancy in relation to long-term cigarette, cigar and pipe smoking: The Zutphen Study *Tob. Control* **16(2)** 107–113.

[16] Murugesan M and Sukanesh R 2013 Towards Detection of Brain Tumor in Electroencephalogram Signals Using Support Vector Machines *Int. J. Comput. Theory Eng.* **1(5)** 622–631.

[17] KavitaMahajan M and Rajput M 2012 A Comparative study of ANN and SVM for EEG Classification *Int. J. Eng.* **1(6)** 1–6.

[18] Kumar N, Alam K and Hasan Siddiqi A 2017 Wavelet Transform for Classification of EEG Signal using SVM and ANN *Biomed. Pharmacol. J.* **10(4)** 2061-2069.

[19] Jer L W, Hariharan M, Vijean V, Yazid H, Chin L C 2018 Infant Cry Classification Using Dual Tree Complex Wavelet Transform Features *Advanced Science Letters* **24(3)** 1741-1744.

[20] Al-Fahoum A S and Al-Fraihat A A 2014 Methods of EEG signal features extraction using linear analysis in frequency and time-frequency domains *ISRN Neuroscience Hindawi Publishing Corporation* **2014** 730218.

[21] Zhilu W, Xuexia W, Zhenzhen G and Guanghui R 2005 Automatic Digital Modulation

Recognition Based on Support Vector Machines *International conference on neural networks and brain* **2** 1025–1028.

[22] Gupta H R and Mehra R 2013 Power Spectrum Estimation using Welch Method for various Window Techniques *International Journal of Scientific Research Engineering & Technology (IJSRET)* **2(6)** 389–392.

[23] Kumar J S and Bhuvaneswari P 2012 Analysis of electroencephalography (EEG) signals and its categorization - A study *Procedia Engineering* **38** 2525–2536.

Electrical Stimulation in Combating Pressure Ulcer for Immobilize Subjects: A Conceptual Framework

Amelia Wong Azman[1*], Fawwaz Eniola Fajingbesi[1,5], Zuraida Ahmad[2], Yasir Mohd Mustafah[3], Rashidah Funke Olanrewaju[1] and Nur Azah Hamzaid[4]

[1]Department of Electrical and Computer Engineering
[2]Department of Manufacturing and Materials Engineering
[3]Department of Mechatronics Engineering, International Islamic University Malaysia
[4]Department of Biomedical Engineering, Faculty of Engineering, University Malaya
[5]Department of Physics Electronics and Earth Science, Fountain University Osogbo, Nigeria

*amy@iium.edu.my

Abstract. A paradigm shift in health system did intensify the need for research on wearable electronics. There are numerous reasons that could lead to skin ulcers. The most common are lesion due to constant pressure, incontinence control, excessive humidity and postural changes. Over the years, there have been numerous interventions ranging from specific skin treatment, introducing prevention programs as well as to mechanical intervention. On the rise in such clinical rehabilitation program is the use of electrical stimulation (ES) at a therapeutic level, wound care and muscle function restoration. This paper presents a concept development and implementation process of a new pressure ulcer intervention system which can be used for both chairbound and bedridden patients. The concept here is to integrate the use of electrical stimulation, interface pressure mapping and redistribution with antimicrobial protection not only to act as therapeutic but at the same time as a means of prevention of pressure ulcers by improving muscle tone and decreasing atrophy among immobilized subjects.

1. Introduction

It is but a common medical knowledge today how devastating Pressure Ulcer (PU) could be mostly on immobilized subject. Usually, PU development occurs in subjects when the interface pressure rises above 32 mmHg for an extended period roughly within 2 - 6 hours without timely relief [1]. Although PU pathophysiology has been extensively mapped out and readily available, over the years there have been little to no reports on the reduction in PU prevalence [2]. Similarly, PU diagnosis, treatment and management are classified as a national burden in developed nations, thus one can only imagine the case for developing or underdeveloped ones? [1-3]. As shown in Figure 1, a summary of literature and practice consensus for prevention and management of PU cases can be coined under the acronym S.K.I.N. Keeping proper Diet and Nutrition as constant, any or a combination of this key procedure along with some tailored PU risk assessment tools are mostly employed in today's PU intervention system.

A deeper look into the acronym S.K.I.N. where:

S : represent the stimulation of human skin,
K : represent the keen observation,
I : represent the interface pressure,
N : represent the neutralization of microorganism.

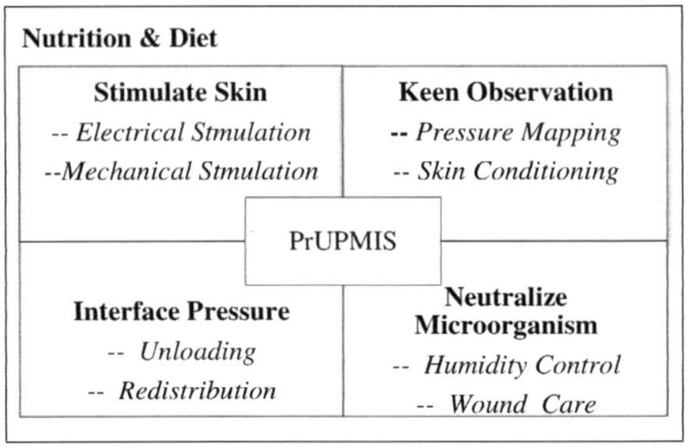

Figure 1. PU Prevention and Management

Therefore, a holistic Pressure Ulcer Prevention and Management System (PrUPMIS), options under skin stimulation would include but not limited to the use of electrical or mechanical stimulation. A keen observation would be maintained either via constant pressure mapping or visually monitoring skin condition. Other factors would include interface pressure unloading or redistribution process, the neutralization of microorganism achieved with proper humidity control and efficient wound care. When these parameters are merged, a holistic PU system can thus be guaranteed.

Table 1 thus provides highlights of selected innovations which can be integrated into a PU management scheme while Table 2 is more on PU dedicated devices some readily available while others are still undergoing clinical testing. It is thus evident from the innovation or devices in Table 1 and Table 2 that none of them could provide a combination of all the S.K.I.N features on a single device system thus the deficiency in combating PU prevalence.

Table 1. Selective highlight on Pressure Ulcer relative innovations [4-7].

		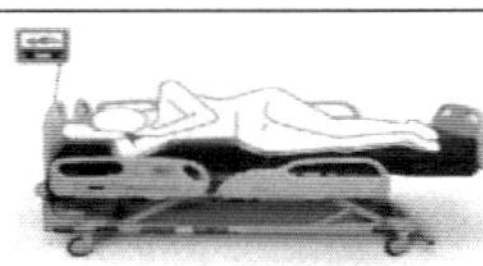	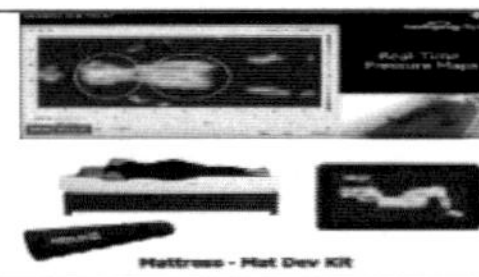	
Product	CONFORMat™ & BPMS™	ForeSite PT ™	Mattress.Mat & Seating.Mat	SmartMat
Company	Tekscan, Inc	XSENSOR$^{(R)}$ Technology Corporation	Sensing.Tex$^{(R)}$	Performance Research, Inc
inventors/ Conceived by	Tekscan, Inc	XSENSOR Technology Corporation	Miguel Ridao Granado; Luis Miguel Gomez Anta; Elkin Diaz Garzon; Francisco Paredes Rivas	Mazir Sadri; Amy Lombardo; Max Bodhisuwan; Paul Hodgson; Leanne Beesley; Chris Gormley
Technology	Thin-film sensor (resistive), 2D & 3D data, USB, WLAN	Capacitive Pressure Sensor, Disinfectble mattress cover, 2D & 3D data, Bluetooth & WLAN	Textile Sensor, Electronic Ink and Paste Materials, a resistive sensor, Bluetooth & USB	Piezoresistive sensor, Bluetooth Interface (iOS& Android), PVC
Strength	1 Sensel/cm^2, 195.07 x 42.67 [cm], 8,064 sensors,	30" x 74", 52 x 118 Sensor Array @ 0.625" by 1mmHg Res; 6000 flex sensors,	85 x 190 [cm], 1152 sensor points, stretchable textile, customizable with SDk kit	71"x26", 21000 sensor, water resistance, withstand 110F heat, non-invasive sensors
Weakness/ Limitation	PU intervention only on pressure redistribution	PU intervention only on pressure redistribution	PU intervention only on pressure redistribution	Not designed for PU intervention

Table 2. Selective highlight on Pressure Ulcer innovations. [8-9, 3, 10-11]

	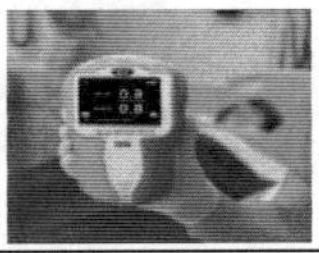	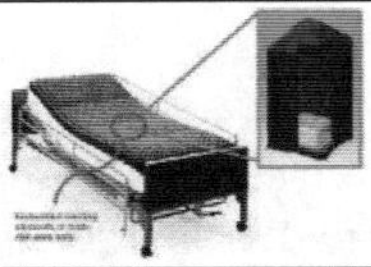			
Product	SEM Scanner™	Pro-TECT™	Project ActiveRest	Smart-e-Pants	PrUPMIS (Proposed)
Company	Bruin Biometrics, LLC	Coulter Translational Research Partners Program, Uni. Pittsburgh	FEDER funds through the COMPETE 2020 Programme	Alberta Innovates Health Solutions & Prev Biotech, Inc	IIUM RMC under MESTECC Research Grant
inventors/ Conceived by	Barbara Bates-Jensen (Ph.D., RN, CWOCN, FAAN) and team.	David Brienza, Alan Murdock	Pereira S., Fonseca J., Almeida J., Car valho R., Pereira P. and Simoes R	Vivian Mushahwar and team.	Corresponding Author Research Team
Technology	Bioimpedance (tissue capacitance) Measurement of Sub epidermal moisture	ROHOTM dry Flotation overlay with embedded Targeted Colling element (gel cushion for SCSCP).	Commercial piezoresistive sensors (FSR) embedded in a textile substrate	Electro-Stimulator, fabric electrodes and anti-bacterial garment	Piezoresistive sensor, Capacitive sensor, web and mobile app-based interactive GUI interface, fabric stimulator electrodes, organic antimicrobial elements
Strength	3-10 day early detection prior to physical changes; high accuracy, Detect DTI and PU, non-invasive	Embedded Colling elements with adjustable dry air redistributive pressure overlay design.	Pressure sensing and monitoring system and can be integrated with an actuation system.	Easy to use and reuse, mild stimulation mimicking fidgeting to prevent PU and increase blood flow	Holistic PU intervention, non-invasive, easy adoption, rapid prototyping cost reduction,
Weakness/ Limitation	PU intervention only on early detection of Risk Marker (SEM)	PU Intervention only at fixed assume high-risk target area.	Developmental phase	Clinical trial phase	Development phase

Looking back at PU pathophysiology and human anatomy, people with severe motor limitations (immobility) mostly have a decreased sensitivity in their body areas in contact with the respective support surfaces, hence are more at risk of developing PU while doing the same actions (siting for 2-6hrs) compared to their counterparts who had either knowingly or subconsciously fidget to relief interface pressure from time to time [8,10].

The National Pressure Ulcer Advisory Panel (NPUAP), European Pressure Ulcer Advisory Panel (EPUAP) and Pan Pacific Pressure Injury Alliance (PPPIA) in her most recent report, *Prevention and Treatment of Pressure Ulcers: Quick Reference Guide*; compiling extensive research with evidence of common preventive strategies and new innovation in the field of combating PU menace to formulate more effective guidelines and point out evidence-based direction where further research is required such as modalities for electrical stimulation in PU cases [1].

This integration and Implementation of Electrical stimulation are expected to be used for mimicking the natural fidgeting movements that people with healthy nervous systems perform subconsciously to alleviate discomfort, improve blood flow, which helps to keep soft tissues healthy while sitting or sleeping for extended periods [1- 3, 8].

Since an Effective PU prevention starts with risk identification and nutritional assessment, followed by the implementation of support surfaces, prophylactic dressings, and repositioning regimens for appropriate patients. A multidisciplinary patient-centered approach is always encouraged for PU management [11]. Hence, we present a concept architecture for PU intervention system (see section 2) that integrates active pressure mapping, electrical stimulation with antimicrobial inhibition on a fabric overlay to achieve a holistic yet cost-effective solution to PU cases.

2. System overview

In order to design a holistic system for PU management, major factors to be considered in such system are its ability to conduct skin-risk assessment and reassessments, manage moisture, minimize mechanical stress (pressure, share and friction) and create a patient, family or caregiver easy access of engagement without physically discomforting either. The novel solution proposed here is thus seen in Figure 2 where real-time active pressure mapping is systemically combined with electrical stimulation.

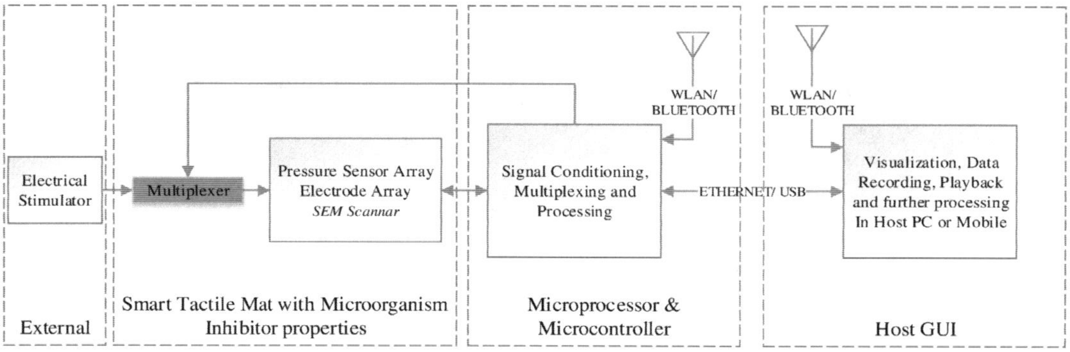

Figure 2. Proposed System architecture

As shown in Figure 2, The proposed system is subdivided into 4 major blocks consisting of an external electrical stimulator device, a fabric mat which an array of electrode is printed on using proprietary conductive inks with antimicrobial properties, a microcontroller block for signal generation, processing and control and finally a communication channel where processed signals are displayed for instant and post-processing. The electrodes embedded in the fabric mat are done systematically to provide both tactile properties for constant pressure mapping and stimulation terminals for the external electrical stimulator signal. Following a specified algorithm uploaded into the microprocessor, the device constantly measures interface pressure for a predefined threshold. Once this value is meet, the electrical stimulator is activated thereby sending stimuli pulse to the localized area above the preset pressure threshold. This ensures a continuous cycle of pressure relief and skin stimulation, improve blood flow, increase muscle tone and decrease atrophy. Similarly, due to the fabric material of the mat and the antimicrobial properties of the collective electrode array, moisture and microbes are easily controlled hence preventing the development or further regression of PU cases.

3. Conclusion

It can be imagined as it still puzzles the public that despite the advances in biotech, medicine, and related research, the prevalence of Pressure Ulcer both in and out of the hospital patients are alarming both in cost and labour hours put into its management and the quality of life of such patients. Hence, we concluded from our related survey of PU reports comprising of clinical research, nursing home reports, Pressure Ulcer Advisory Panels guides and the major problem that stood out in Combating PU cases is the lack of holistic solution rather the focus on specificity in an aspect as shown in Figure 1. Hence, this research proposes yet a novel approach that encompasses existing sparse solution into a single system with a device that would not only reduce costs but also save caregivers work time and as well improve the quality of life in PU patients most especially the at-risk Immobilize categories. Future reports would include the developmental process, procedure, test and benchmarking.

Acknowledgments

This research is fully sponsored under the Malaysian Ministry of Energy, Science, Technology, Environment and Climate Change (eScienceFund Project No. 06-01-08-SF0324).

References

[1] National Pressure Ulcer Advisory Panel, European Pressure Ulcer Advisory Panel and Pan Pacific Pressure Injury Alliance 2014 *Prevention and Treatment of Pressure Ulcers: Quick Reference Guide* ed Emily Haesler (Osborne Park, Western Australia: Cambridge Media)

[2] Tran J P, McLaughlin J M, Li R T and Phillips L G 2016 *J. American Society of Plastic Surgeons: Plastic and Reconstructive Surgery*, **138**(3S), 232S–40S.

[3] Pereira S, Fonseca J, Almeida J, Carvalho R, Pereira P and Simoes R 2019 *Proc. 12th Int. Joint Conf. on Biomedical Engineering Systems and Technologies* 22-24 Feb. 2019 Prague, Czech Republic Vol 1 p 291-6.

[4] Tekscan^{TM} *Body Pressure Distribution | Body Pressure Measurement System (BPMS) - Research* https://www.tekscan.com/products-solutions/systems/body-pressure-measurement-system-bpms-research (accessed April 14, 2019)

[5] Wong H, Kaufman J, Baylis B, Conly J M, Hogan D B, Stelfox H T, Southern D A, Ghali W A and Ho C H 2015 Efficacy of a pressure-sensing mattress cover system for reducing interface pressure: Study protocol for a randomized controlled trial *Trials* **16** 434.

[6] Sensing.Tex *Bedding Mat Platform - Sensing Tex Pressure Mapping* (n.d.). http://sensingtex.com/sensing-mats/mattress-mat/ (accessed April 14, 2019)

[7] INDIEGOGO *SmartMat: The World's First Intelligent Yoga Mat* https://www.indiegogo.com/projects/smartmat-the-world-s-first-intelligent-yoga-mat#/ (accessed April 15, 2019)

[8] Bates-Jensen B M, McCreath H E and Patlan A 2017 Subepidermal moisture detection of pressure-induced tissue damage on the trunk: The pressure ulcer detection study outcomes *Wound Repair and Regeneration*, **25**(3) 502–11

[9] Brienza D, Murdock A 2014 *Pro-TECT: Mattress overlay to prevent pressure ulcers*, Coulter Transl. Res. Partners II Program, Univ. Pittsburgh.

[10] Topfer L A 2015 Smart-e-Pants: Using Intermittent Electrical Stimulation to Prevent Pressure Ulcers *Issues in emerging health technologies,* Issue 132 (Ottawa: Canadian Agency for Drugs and Technologies in Health)

[11] Fajingbesi F E, Azman A W, Olanrewaju R F, Ahmad Z, Mustafah Y M 2019 A Glimpse at Functional Electrical Stimulation for Combating Pressure Ulcers in Immobilize Subjects *Late Breaking IFESS Poster at RehabWeek2019* 25-28 June 2019 Toronto Canada.

Assessment of an Arterial Stiffness to Demonstrate AVF Performance: Modelling Approach

T'ng Chien Mien[1], Shuhaida Yahud[1*]

[1]School of Mechatronic Engineering, Universiti Malaysia Perlis, Malaysia

*shuhaidayahud@unimap.edu.my

Abstract. Arteriovenous fistula (AVF) defined as shunting in the arterial system and the venous conduit is a preferred vascular access used in haemodialysis. The AVF provide a sufficient increase in blood flow rates for effective dialysis. However maintaining AVF patency become the greatest challenge for dialysis patient. Complications arise from this includes stenosis, aneurism and thrombosis. In previous studies, several methods are used to evaluate the performance of AVF. This study will focus on the assessment of arterial stiffness to demonstrate the performance of a side-to-side AVF. The condition of arterial stiffness which alter haemodynamic of AVF, in fact, will affect its performance and functionality. To gain insight and understanding for the haemodynamic condition, a healthy AVF (without arterial stiffening) has been modelled in the computational fluid dynamic (CFD) framework. The aim of this project is therefore to develop a modelling approach, suitable for the analyses of blood flow field to predict the performance and patency of AVF. A 3D CFD method is proposed to analysis flow field or haemodynamic in terms of blood flow velocity, blood pressure and wall shear stress (WSS) along the AVF. A laminar flow was introduced for the simulation of healthy AVF. Correlations in these haemodynamic parameters to assess the vascular failure supports hypothesis suggesting flow field as predicator for AVF performance. Haemodynamic effects lead to AVF failure were determined. Significant flow disturbances with abnormally high or low WSS obtained from the simulation is used as a primary indicator.

1. Introduction

End stage renal disease (ESRD) is also known as one of the common disease related to public health. It is the last stage of chronic kidney disease (CKD) where kidneys only functioning lower than 10 percent of normal function. Patients suffer from ESRD are unable to utilise the function of kidney to eliminate waste product. Haemodialysis is the primary treatment modality for ESRD. A vascular access is created to ensure efficient and successful haemodialysis. Three types of vascular access commonly use for haemodialysis; arteriovenous fistula (AVF), graft, and catheter. In Malaysia, it is reported that AVF is the most preferred access among haemodialysis patients at 85%, followed by catheter at 13.6% and only 1.4% choose graft [1]. However, in the past 10 years (2007 – 2016) the number of patients undergoing haemodialysis using fistula has consistently and gradually reduced, likely due to the absence of functioning fistula upon initiation of haemodialysis, and the increasing number of elderly and diabetic patients with no vascular access.

AVF is regard as the most preferred type of vascular access due to its superior patency rates and it is surgically created by shunting of the artery and vein. The performance of AVF is essential to determine the effectiveness of haemodialysis. Blood flow rate in range of 350 – 400 ml/min is essential to ensure efficient haemodialysis and minimise access recirculation [2]. The performance of AVF is highly affected by the haemodynamic conditions.

The arterial stiffness is assessed to evaluate the haemodynamic conditions and performance in AVF. Arterial stiffness which define the compliance of artery to expand can be related to the internal diameter of artery. Change in the internal diameter of artery, which regulated by the stiffness of artery will alter the hemodynamic conditions such as blood flow velocity, pressure of blood flow, and wall shear stress (WSS) exerted on the blood vessel [3].

Wall shear stress (WSS) which is a measure of force of friction from blood acting along the inner wall of blood vessel is proposed as the primary predictor to the AVF performance. The WSS which is extremely high or low will alter the haemodynamic and prone to complications which affect the performance of AVF. The blood flow can be disrupted and become turbulent in the extremely high level of WSS (> 35 Pa). However, abnormal low level of WSS is predicted as a high risk region to induce complications [2].

Computational fluid dynamic (CFD) framework is established to identify and analyse the hemodynamic conditions in the complex AVF geometry [4]. The aim of this study is to develop a 3D CFD simulation for the analysis of blood flow condition which reflect the performance of AVF The flow field conditions in term of velocity, pressure and WSS are determined in order to identify the correlation between the haemodynamic and prevalence of patency in AVF.

2. Methods

2.1. Physic Defined

A laminar physics was used to simulate the blood flow along the healthy AVF. Laminar flow was defined when Reynolds number was approximated to lower than 2300. The normal blood flow condition is laminar flow, which characterises the movement of blood fluid in parallel along the distance of blood vessel. The simulations of AVF performance in this project will set the fluid as blood properties with dynamic viscosity of 0.0035 Pa.s and density of $1050\ kg/m^3$ [5].

2.2. Geometry Design

The CFD simulation was based on a computer aided design (CAD) model created by AutoCAD 2007. The model of AVF mimic the real fistula was designed. The geometry of healthy AVF was constructed with diameters in 6 mm for vein and 4 mm for artery. Past research shows that the minimum diameter of vein and artery for AVF to achieve maturation was 2 mm [6]. The length of anastomosis was approximated as 12 mm with a total length of 90 mm vein and artery .The anastomosis was created as side-to side geometry. The 3D model of healthy AVF was designed as shown in Fig.1.

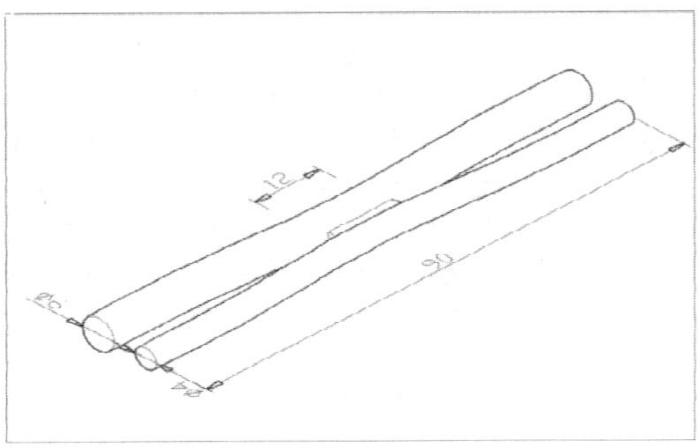

Figure 1. 3D model of healthy AVF

2.3. Fluid Dynamics

The blood flow within the AVF model is governed by an incompressible Navier-Stokes equations as:

$$\rho\frac{\delta u}{\delta t}+\rho(u\cdot\nabla)u=\nabla\cdot[-pI+\mu(\nabla u+(\nabla u)^T)]+F \tag{1}$$

and the continuity equation :

$$\frac{\delta\rho}{\delta t}+\nabla\cdot(\rho u)=0 \tag{2}$$

where u represent the velocity, ρ is the density, p as the pressure, μ is the dynamic viscosity, I as identity matrix, and F is volume force field. Equation (1) formulated as conservation of momentum and the continuity equation formulated as conservation of mass. The both corresponding equations governed the simulation to solve the velocity of blood flow and pressure along the AV.

2.4. Boundary Conditions

The proximal artery (PA) and distal artery (DA) were set as inlets of flow while proximal vein (PV) and distal vein (DV) as outlets of flow. A time related flow parameters for the inlets were specified by equation (3) and (4).

$$V_{PA}(t)=0.1+(\sin(\pi t)\,0.8) \tag{3}$$

$$V_{DA}(t)=0.1+(\sin(\pi t)\,0.5) \tag{4}$$

The PV and DV were both defined as outlets with velocity of 0.0314 m/s [7] set at the DV to indicate flow of deoxygenated blood back to the heart, while venous pressure of 2.6 kPa was set at the PV [8].

3. Results and Discussion

The haemodynamic within a healthy AVF model was examined, in order to demonstrate the performance of AVF and predict it patency rate. The results of haemodynamic in terms of velocity of blood flow, pressure of blood flow and WSS of blood exerted on the wall of AVF were analysed and validated with theoretical studies.

3.1. Velocity Magnitude of Blood Flow

The results of velocity profile of blood flow from the time dependent simulation are depicted in Fig. 2. The velocity of blood flow along the AVF at time, t equals to 0 s, 0.5 s and 1 s were discussed.

The results show that velocity magnitude of blood flow along the artery is higher than velocity magnitude at the venous side along the time of simulation. This phenomenon of flow velocity can be explained theoretically by the principle of conservation of mass:

$$Q=A_1V_1=A_2V_2 \tag{5}$$

where Q is flow rate, A indicates cross section area of blood vessel, V is the velocity of blood flow, V_1 represent the artery and V_2 represent the vein. The equation show that for a constant flow rate, the area of blood vessel is inversely proportional to the velocity.

In Fig. 2(a), it can be observed that the maximum velocity magnitude of blood flow appeared at the bifurcation of artery, which located along the two proximal and distal sites of the anastomosis. This can be explained as maximum velocity at the centreline of parabolic profile had moved towards the bending sites of proximal and distal anastomosis when the both input velocity is low. Therefore an extremely high velocity was found at both sides of bifurcation of artery to the anastomosis which leads to disturbances of flow.

High velocity of blood flow was located at the anastomosis where large pressure gradient between artery and vein induced a blood flow with greater velocity magnitude. This indicate the anastomosis as site which caused disturbances of smooth blood flow to chaotic Turbulent flow.

Fig. 2(b) and (c) show that velocity of blood near the DV side was lower compared to the velocity of blood flow through the PV side. This is due to the converged blood flow at the anastomosis will diverged and flow faster towards the side of PV. The initial flow of deoxygenated blood from DV to PV leads to a greater resistance for blood to flow from the artery to the side of DV [9].

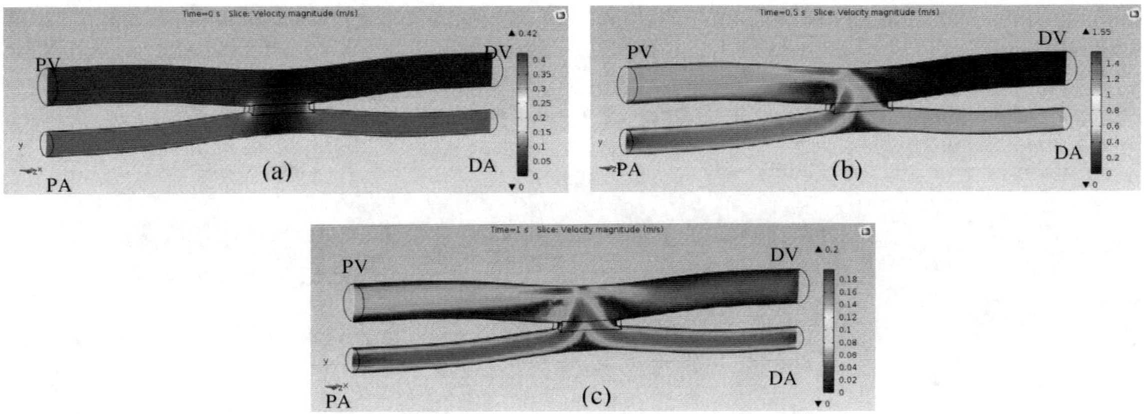

Figure 2. Velocity magnitude (m/s) at (a) t = 0 s, (b) t = 0.5 s and (c) t = 1 s

3.2. Pressure of Blood

The pressure level along the side-to-side AVF was is an important haemodynamic parameter to assess the performance of the AVF. Fig. 3 shows the contour pressure of blood flow in AVF when time, t=0 s, 0.5 s and 1 s. The results of simulation demonstrate that the pressure of blood flow exerted along the artery show greater value compare to pressure along the vein. This result can be explained by the inverse relationship between the cross sectional of blood vessel and the blood pressure as:

$$\boldsymbol{p} = \frac{\boldsymbol{F}}{\boldsymbol{A}} \quad (6)$$

where p is the pressure, F is the force, A is the cross sectional area of vessel.

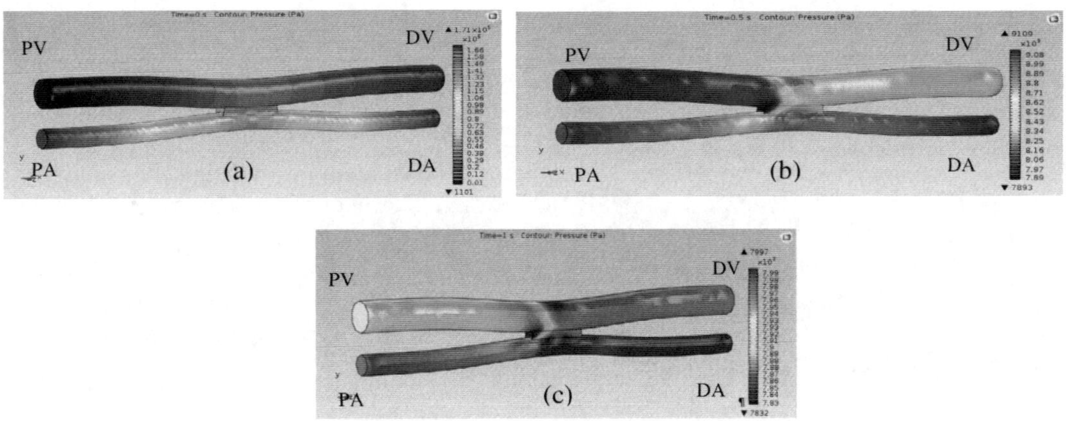

Figure 3. Pressure contour of blood flow at (a) t= 0s, (b) t = 0.5s and (c) t = 1s

The simulation showed that, after the construction of AVF, the blood pressure at vein elevated to value above 20 mmHg (2.6 kPa) which is the physiological level of human systemic venous pressure [10]. The increase of venous pressure by shunting of artery and vein is to ensure adequate blood flow during dialysis [11]. The pressure of blood flow is higher at the DV than the PV due to the explanation that there are a portion of blood return to the heart directed from the DV to PV.

3.3. Wall Shear Stress (WSS) exerted on AVF

The wall shear stress (WSS) play an important role in haemodynamic studies of AVF. It can be used as predictor of AVF performance in haemodialysis. The changes in WSS affect the remodelling of AVF, may influence it patency rate. The results of WSS obtained in the post-processing of simulation was discussed at three different stages; at time, t = 0 s, t = 0.5 s and t = 1 s.

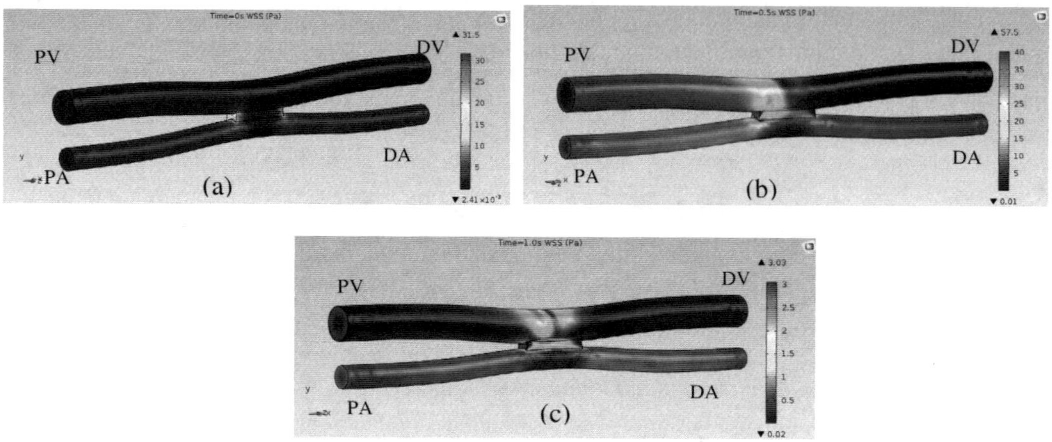

Figure 4. Distribution of WSS at (a) t = 0 s, (b) t = 0.5 s and (c) t = 1 s

The distribution of WSS along the AVF was depicted in Fig. 4. The AVF which functions to induce a greater amount of blood flow to the venous system has brought the WSS distribution to a higher level [12]. This is due to the relationship between the WSS and blood flow rate which defined in equation below:

$$\tau = \frac{32 \cdot \mu \cdot Q}{\pi \cdot d^3} \tag{7}$$

where τ as wall shear stress (WSS), μ is dynamic viscosity, u is the velocity of fluid, Q is flow rate and d is vessel diameter.

The results of simulation show that WSS is high at the bifurcation of artery into the anastomosis. Indicate that the site where blood flow is highly induced into Turbulent flow and causing degradation and damage to elastic mechanism of vessel wall.

The contour plot shows that location of low WSS between 0 to 2 Pa is the segment of DV and the region of artery just below the anastomosis. This is the location of blood vessel which highly promote the platelet lesion and induce common vascular disease such as stenosis, thrombosis and atherosclerosis to affect the performance of AVF for haemodialysis.

4. Conclusion

In recent years, maintaining AVF patency continues to be the greatest challenge for dialysis patients. Arterial stiffness which defined the compliance of blood vessel will influence the vascular function which reflect the patency of AVF. The project is aim to develop a modelling approach to predict the patency rate and performance of AVF. A finite element method with a computational fluid dynamic (CFD) framework was conducted to visualise, study and analyse the flow field and haemodynamic condition of AVF. The haemodynamic condition in terms of velocity of blood flow, blood pressure, and

WSS exerted on the wall along the AVF were focused in order to evaluate the performance of AVF for haemodialysis.

The simulation of healthy AVF was proposed as the initial stage to demonstrate and predict the condition in AVF. This method featuring an insightful analysis of relationship between flows filed conditions and vascular complication in AVF. The assessment of arterial stiffness by employing a CFD analysis to evaluate the haemodynamic conditions had provided a basic demonstration of AVF performance.

References

[1] Wong H S and B L Goh 2018 Twenty Forth Report of the Malaysian Dialysis and Transplant Registry 2016 ISSN 1675-8862.

[2] Lwin N, Wahab M, Carroll J, and Barber T 2014 Experimental and Computational Analysis of a Typical Arterio-Venous Fistula.

[3] Niemann A, Udesen J, Thrysoe S, Nygaard J, Fründ E, Petersen S and Hasenkam J 2010 Can sites prone to flow induced vascular complications in a-v fistulas be assessed using computational fluid dynamics? *Journal of Biomechanics*, vol. 43, no. 10, pp. 2002-2009.

[4] Lorenzini G and Casalena E 2008 CFD analysis of pulsatile blood flow in an atherosclerotic human artery with eccentric plaques *Journal of Biomechanics*, vol. 41, no. 9, pp. 1862-1870.

[5] Siebert M W and Fodor P S 2009 Newtonian and non-newtonian blood flow over a backward-facing step–a case study *COMSOL Conference*, pp. 1-5.

[6] Glass C, Johansson M, DiGragio W, and Illig K A 2009 A meta-analysis of preoperative duplex ultrasound vessel diameters for successful radiocephalic fistula placement *Journal for Vascular Ultrasound*, vol. 33, pp. 65-68

[7] Carroll G, McGloughlin T, Burke P, Egan M, Wallis F and Walsh M 2011 Wall Shear Stresses Remain Elevated in Mature Arteriovenous Fistulas: A Case Study *Journal of Biomechanical Engineering*, vol. 133, no. 2, p. 021003.

[8] Attaran R and Ochoa Chaar C 2016 Compression therapy for venous disease", *Phlebology: The Journal of Venous Disease*.

[9] Meissner M *et al.* 2007 The hemodynamics and diagnosis of venous disease", *Journal of Vascular Surgery*, vol. 46, no. 6, pp. S4-S24.

[10] "The Cardiovascular System Blood Vessels", Class midlanstech.edu, 2016. [Online]. Available://classes.midlandstech.edu/carterp/Courses/bio211/chap19/chap19.html. [Accessed: 10- April- 2016].

[11] Stolic R 2013 Most Important Chronic Complications of Arteriovenous Fistulas for Hemodialysis *Med Princ Pract*, vol. 22, no. 3, pp. 220-228.

[12] Hammes M 2015 Hemodynamic and Biologic Determinates of Arteriovenous Fistula Outcomes in Renal Failure Patients *BioMed Research International*, vol. 2015, pp. 1-7.

Atrial Fibrillation Identification through ECG Signals

Ng Joe Yee[1], Vikneswaran Vijean[1*], Saidatul Ardeenawatie Awang[1], Chong Yen Fook[1], Lim Chee Chin[1]

[1] Biosignal Processing Research Group (BioSIM), School of Mechatronic Engineering, Universiti Malaysia Perlis, Arau, Perlis, Malaysia

*vikneswaran@unimap.edu.my

Abstract. This paper presents an algorithm formulated to identify the atrial fibrillation complications through electrocardiogram (ECG) signals. The ECG data for the study was retrieved from Physio Net which consists of normal, atrial fibrillation and other rhythms. The Discrete Wavelet Transform (DWT) was used to remove baseline wanders. Pan Tompkins algorithm was utilized to detect the P, Q, R, S and T peak and thus the ECG signals were segmented based on each cycle. The morphological features were extracted directly from the time-series while statistical features were extracted after Stockwell transform (S- transform) was applied to the data. Genetic Algorithm (GA) and reliefF algorithm have been applied separately to select the optimum features for classification purpose. Bagged Tree ensemble algorithm, Decision Tree and k-Nearest Neighbour (KNN) algorithm were used as classifiers to identify atrial fibrillation through ECG signals. The classification results with and without feature selection techniques are presented. Prior to the feature selection, Bagged Tree is the classifier best performing classifier with 86.50% of accuracy, 84.38% of sensitivity and 91.94% of specificity. After feature selection, all the three classifiers have almost the same performance which is nearly 100% of accuracy, sensitivity and specificity. This shows that the proposed combinations of algorithms are reliable and able to improve the identification rate of the normal, atrial fibrillation and other rhythms using lesser number of features.

1. Introduction

The statistical report from the Ministry of Health Malaysia illustrated that the number of Malaysians suffering from cardiovascular disease has incremental trends. The number of patients with arrhythmia recorded as 2.8% in government hospital [1]. The incidences of atrial fibrillation have been reportedly increasing with age. The percentage of diagnosis related to atrial fibrillation disease within age range of 50 to 59 is 0.35%, whereas 2.32% is reported for those in the age range of 60 to 69. The age group of 70 to 79 has the highest percentage for atrial fibrillation disease which is 2.59% [1]. Therefore, the diagnosis of cardiovascular diseases with the use of computer aided tools for complications such as atrial fibrillations is a worthy challenge to be addressed. ECG signals play an important role in this identification process because it can effectively depict the condition of the heart. In this paper, an attempt have been made to classify the ECG signals into normal, atrial fibrillation and other rhythms using information retrieved from morphological features and statistical features of the data. MATLAB platform was used to execute the pre-processing, segmentation, feature extraction, statistical analysis, feature selection and classification steps. Each of the process are detailed in section 2.

Journal of Physics: Conference Series **1372** (2019) 012058 doi:10.1088/1742-6596/1372/1/012058

2. Method

This section describes and discusses the flow of the research work as follows:

2.1. Data Source

The ECG signals used in this study were retrieved from the 2017 PhysioNet/Computing in Cardiology Challenge. The single short ECG lead recordings have a length in between 30 seconds and 60 seconds. Only the training set containing 8482 single lead ECG recordings labeled as normal, atrial fibrillation and other rhythms were used in the study as the testing data is inaccessible to the public. The ECG signal were sampled at 300Hz and bandpass filtered with a bandwidth of 0.5Hz to 40Hz [3].

2.2. Pre-processing

Pre-processing is an important step to remove the unwanted artefacts and noise in the signals.In this study, DWT was used to correct the baseline wandering which approximately 0.5Hz from the ECG signal [2]. Db5 wavelet was used to decompose the ECG signal into eight levels and the baseline artefact frequency was removed from the signals. Db wavelet family was chosen as it has a similar structure with the QRS complex and the energy spectrum is concentrated at low frequency. Since Figure 1 is the original ECG signal (A00001). Figure 2 is the ECG signal (A00001) after baseline correction.

2.3. Segmentation

Before segmentation is carried out, Pan Tompkins algorithm was utilised to detect P, Q, R, S and T peak. This algorithm will detect the peak based on the analysis of the slope, amplitude and width of ECG signals [4]. After P, Q, R, S and T peak are detected, the ECG signal were segmented based on each cycle. Figure 3 is the first segment of ECG Signal (A00001) with peak labelling.

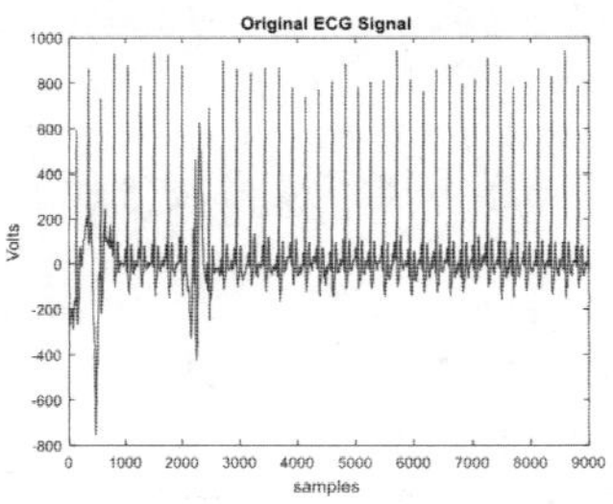

Figure 1. Original ECG Signal (A00001).

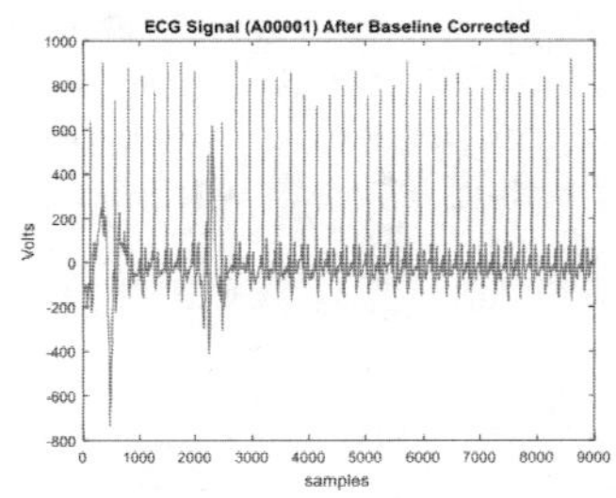

Figure 2. ECG Signal (A00001) after Baseline Correction.

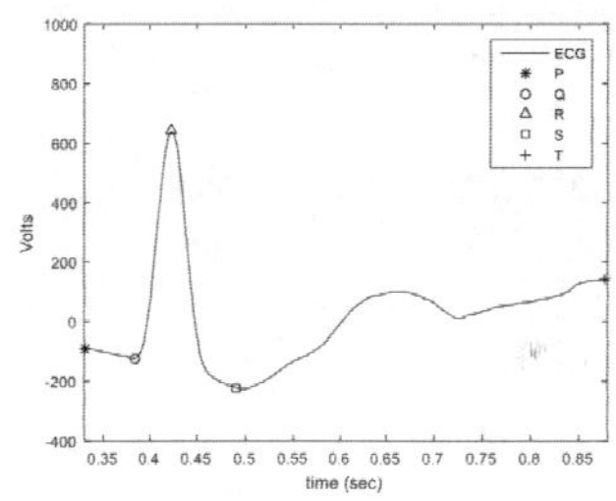

Figure 3. First Segment of ECG Signal (A00001) with Peak Labelling.

2.4. Feature Extraction

Feature extraction is a crucial step where the significant information from the ECG traces are extracted and used as a parameter to classify the ECG signals into normal, atrial fibrillation and other rhythms. The morphological features and statistical features from S-transformed data were used in the study.

The morphological features were extracted directly from the signals after the peak detection process. Features such as heart rate, PR interval, QRS complex interval, RT interval, average RR interval and root mean square of successive RR interval differences (RMSSD) were extracted in the process. Mean of RR interval, RMSSD and heart rate were calculated using Equation (1), (2) and (3).

$$\text{Mean of RR interval, } \bar{I} = \frac{1}{N-1}\sum_{n=2}^{N} I\,(n) \quad (1)$$

$$\text{RMSSD} = \sqrt{\frac{1}{N-1}(\sum_{i-1}^{N-1}((R-R)_{i+1}-(R-R)_i)^2} \quad (2)$$

$$\text{Heart Rate (beats per minute)} = \frac{Number\ of\ RR\ interval}{Totaltime\ (minute)} \quad (3)$$

Stockwell-transform was used to transform the time-domain ECG signals into time-frequency representation [5]. The features that were extracted from Time maximum Amplitude plot (TmA-plot) are the maximum and minimum value of TmA-plot, and energy for the maximum and minimum value of TmA-plot. From Time-Frequency contour (TF-contour), the average maximum frequency was extracted. Energy and power for maximum frequency were therefore calculated. The average skewness and kurtosis values of TF-contour were also measured. Skewness is a measure of the symmetry of the distribution while kurtosis corresponds to the degree of steepness of a probability distribution function. Skewness and kurtosis are defined in Equations 4 and 5.

$$\text{Skewness} = \frac{1}{n-1}\sum_{i=1}^{n}\frac{(x_i-\bar{x})^3}{SD^3} \quad (4)$$

$$\text{Kurtosis} = \frac{1}{n-1}\sum_{i=1}^{n}\frac{(x_i-\bar{x})^4}{SD^4}\text{-}3 \quad (5)$$

2.5. Feature Selection

Feature selection is an important process in order to obtain a smaller feature set so that a simpler and effective model can perform well in classification. In this study, the Genetic Algorithm (GA) and reliefF algorithm were implemented to select the features that can optimally classify the ECG signals into normal, atrial fibrillation and other rhythms. There were 16 morphological and statistical features extracted for the analysis.

2.5.1 Genetic algorithm

The GA was applied using a tournament-based selection strategy. The size of the chromosome was set as equal to the number of features. The purpose of applying GA was to look for a chromosome that represents a set of features which has the highest accuracy in classification. Initially, a random population is created, representing different points in the search space [6]. Each of the initial population was evaluated according to the fitness function [6]. In GA, the reproduction, crossover and mutation are the important steps to produce the next generation from the present generation. Through the repetition process, GA eliminates the low fitness chromosomes and keeps the high fitness chromosomes to the next generation until a good chromosome is obtained. The goal of genetic feature selection stage was to reduce the dimensionality of the feature before the classification process.

2.5.2 ReliefF algorithm

ReliefF algorithm is a filter-based feature selection method. It is a simple and effective algorithm which is able to evaluate the feature's importance by considering dependence from other features. The primary concept of this algorithm is to provide weight to every feature based on the performance of distinguishing the samples that pertain to the same group. A feature with high rank indicates that it can discriminate two near samples which are in the different groups. On the other hand, the feature with low rank expresses as it only can differentiate two near samples in the same group [7].

2.6. Classification

During this stage, Bagged Tree ensemble classifier, Decision Tree and k-Nearest Neighbour (KNN) were employed to classify the ECG signals as normal, atrial fibrillation and other rhythms.

2.6.1. Bagged Tree ensemble classifier

The principal of ensemble classifier is that a group of weak learners combined together to form a strong learner. In this study, Bagging (bootstrap aggregation) is the ensemble learning technique adopted in the classification process. For Bagged Tree ensemble classifier, it has combined several decision trees to produce a better predictive performance as compared to a single decision tree [8].

2.6.2. Decision Tree

Decision Tree is a non-parametric supervised learning method used for classification and regression. It is one of the selections for fast classification and which is also easy to analyze. In a Decision Tree, there are present with three types of nodes: root node, internal node and terminal nodes. A root node represents a single input variable and a split on that variable. Each terminal node contains an output variable which is used to make a prediction.

2.6.3. k-Nearest Neighbour (KNN)

KNN classifier is one of the most basic and non-parametric lazy learning algorithm. The classification is performed by finding the shortest distance from a data set which contains the input or training data and dataset which contains the reference values [9-11].

In this study, tenfold cross-validation was applied to the training and testing data. This indicates that the training dataset was divided into ten sets and the nine datasets will be used as training while the remaining one dataset will be used as testing. The process was repeated ten times such that each dataset has the chances to become a validation set [12]. Consequently, the mean accuracy for the three groups (normal, atrial fibrillation and other rhythms) was calculated and the standard deviation was determined. The mean sensitivity and specificity for the three groups were also investigated by exploring the confusion matrix of the classifier results.

3. Results and Discussion

The Bagged Tree ensemble classifier, Decision Tree and KNN algorithms were used to evaluate the classification accuracy of the proposed work using the selected features. The classification results acquired are presented and discussed in Tables 1, 2 and 3. There are two groups of classification results: (a) before feature selection and (b) after feature selection.

Table 1. Results of Bagged Tree before and after Feature Selection.

	No. of learners	Accuracy (%)	Sensitivity (%)	Specificity (%)
Before feature selection	60	86.50±0.44	84.38±8.45	91.94±7.05
After feature selection	30	99.95±0.03	99.95±0.09	99.96±0.06

From Table 1, it can be inferred that before feature selection, 60 learners were required to provide optimal classification accuracy of 86.50% ± 0.44%, sensitivity as 84.38% ± 8.45% and specificity as

91.94% ± 7.05%. After feature selection, 30 learners were able to provide 99.95% ± 0.03% accuracy, 99.95% ± 0.09% sensitivity and 99.96% ± 0.06% specificity. This demonstrates that the use of feature reduction technique was able to improve the classification results of Bagged tree while requiring lesser number of learners. This is because the dimension of the feature matrix has greatly reduced from 37080 x 16 to 37080 x 3 after applying GA and reliefF algorithm. Therefore, the smaller dimension of the feature matrix needed less number of learners to train the classifier. Through feature selection methods, the accuracy, sensitivity and specificity of Bagged Tree ensemble classifier in classifying normal, atrial fibrillation and other rhythms have improved because it has reduced the redundant feature set.

Table 2. Results of Decision Tree before and after Feature Selection.

	No. of splits	Accuracy (%)	Sensitivity (%)	Specificity (%)
Before feature selection	1400	79.75±1.05	77.93±9.13	88.31±8.10
After feature selection	1300	99.93±0.04	99.94±0.10	99.96±0.07

From Table 2, it can be observed that before feature selection, 1400 splits were required to produce 79.75% ± 1.05% accuracy, 77.93% ± 9.13% sensitivity and 88.31% ± 8.10% specificity. After feature selection, 1300 splits provided the accuracy of 99.93% ± 0.04%, sensitivity of 99.94% ± 0.10% and specificity of 99.96% ± 0.07%. The results of the Decision Tree have improved as compared between before and after feature selection. Although the dimension of the feature matrix had been reduced, the number of splits required in Decision Tree does not have much different. This indicates that the classifier still needed almost similar amount of splits for training purpose although improvement can be seen in overall classification results.

Table 3. Results of k-Nearest Neighbour before and after Feature Selection.

	k-value	Accuracy (%)	Sensitivity (%)	Specificity (%)
Before feature selection	13	46.82±0.61	35.75±27.97	68.12±26.92
After feature selection	4	99.95±0.04	99.95±0.10	99.96±0.07

From Table 3, it can be deduced that before feature selection, k-value of 13 provide the accuracy of 46.82% ± 0.61%, sensitivity of 35.75% ± 27.97% and specificity of 68.12% ± 26.92%. After feature selection, the k-value of 4 provided the accuracy of 99.95% ± 0.04%, sensitivity of 99.95% ± 0.10% and specificity of 99.96% ± 0.07%.The results of KNN have greatly improved as compared to before the application of feature selection process. This is because KNN is a non-parametric lazy learning algorithm. When too many inputs applied to it, the solution of KNN becomes too complex to handle and results in false identification. After the application of feature selection, the dimension of feature matrix was reduced. Therefore, the problems become simpler to be handled by KNN's nearest neighbourhood based majority voting method. For the desired k-value after feature selection, it is smaller than the desired k-value before feature selection. This is because when the dimension of the features matrix decreased, the k-value needed to set as small value so that the probability of getting error decreased.

4. Conclusion

Before feature selection, Bagged Tree with the number of learners equal to 60 has better performance as compare to Decision Tree and KNN. The accuracy, sensitivity and specificity of Bagged Tree are 86.50% ± 0.44%, 84.38% ± 8.45% and 91.94% ± 7.05% respectively prior to feature selection. After feature selection, Bagged Tree, Decision Tree and KNN have similar results which are almost 100% of accuracy, sensitivity and specificity. The results denote that feature selection is an important step in order to improve the performance of classifiers. In this study, GA and reliefF algorithm have been implemented to select the optimum features that able to differentiate between the three classes. This process has greatly reduced the redundant features and improved the performances of classifiers.

References

[1] W. Jin Shyan, R. Bau, F. Ahmad, R. Ismail, and A. F. Yean Yip, 2013, "Prevalence of asymptomatic atrial fibrillation in Malaysian patients with hypertension," *Med. J. Malaysia*, vol. 68, no. 2, pp. 141–143.

[2] K. S. Kumar, B. Yazdanpanah, and P. R. Kumar, 2015, "*Removal of noise from electrocardiogram using digital FIR and IIR filters with various methods*," 2015 Int. Conf. Commun. Signal Process. ICCSP 2015, pp. 157–162.

[3] G. D. Clifford et al. 2017, "AF Classification from a Short Single Lead ECG Recording: the PhysioNet/Computing in Cardiology Challenge 2017," Comput. Cardiol. (2010)., vol. 44, pp. 1–4.

[4] L. Sathyapriya, L. Murali, and T. Manigandan, 2014, "*Analysis and detection R-peak detection using Modified Pan-Tompkins algorithm,*" in 2014 IEEE International Conference on Advanced Communications, Control and Computing Technologies, 2014, pp. 483–487.

[5] V. Vijean, M. Hariharan, and M. N. Mansor, 2016, "Analyzing selected visual anomaly through ST-based multi-resolution VEP decomposition," Proceeding - 2016 IEEE 12th Int. Colloq. Signal Process.its Appl. CSPA 2016, no. March, pp. 64–67.

[6] H. Li, D. Yuan, X. Ma, D. Cui, and L. Cao, 2017, "Genetic algorithm for the optimization of features and neural networks in ECG signals classification," *Sci. Rep.*, vol. **7**, no. 1, p. 41011, Dec. 2017.

[7] F. G. Silva Teodoro, S. M. Peres, and C. A. M. Lima, 2017, "*Feature selection for biometric recognition based on electrocardiogram signals*," in 2017 International Joint Conference on Neural Networks (IJCNN), 2017, pp. 2911–2920.

[8] C. Paper, A. Mert, and B. Teknik, "*Ecg Signal Classification Using Ensemble Decicion Tree*," in Trends in the Development of Machinery and Associated Technology, 2016, no. September 2012, pp. 459–462.

[9] S. Jayalalith, D. Susan, S. Kumari, and B. Archana, 2014, "K-nearest Neighbour Method of Analysing the ECG Signal (To Find out the Different Disorders Related to Heart)," *J. Appl. Sci.*, vol. 14, no. 14, pp. 1628–1632, Dec. 2014.

[10] V. Vijean, M. Hariharan, and S. Yaacob, 2011, "Discrimination of vision impairments using single trial VEPs," *Proc. - 2011 IEEE Int. Conf. Control Syst. Comput. Eng. ICCSCE 2011*, pp. 182–186, 2011.

[11] Q. W. Oung, H. Muthusamy, S. N. Basah, H. Lee, and V. Vijean, 2018, "Empirical Wavelet Transform Based Features for Classification of Parkinson's Disease Severity," J. Med. Syst., vol. 42, no. 2.

[12] M. Hariharan, V. Vijean, C. Y. Fook, and S. Yaacob, 2012, "Speech stuttering assessment using sample entropy and Least Square Support Vector Machine," Proc. - 2012 IEEE 8th Int. Colloq. Signal Process. Its Appl. CSPA 2012, pp. 240–245.

International Conference on Biomedical Engineering (ICoBE) IOP Publishing
Journal of Physics: Conference Series **1372** (2019) 012059 doi:10.1088/1742-6596/1372/1/012059

Software solution for Testing Image Processing Algorithm on Conveyor-Based Vision Systems

Azman Muhamad Yusof[1]*, Ali Yeon Md Shakaff[2] and Saufiah Abdul Rahim[2]

[1] Dept. of Electronic Engineering Technology, Faculty of Engineering Technology, UniMAP.
[2] School of Mechatronic Engineering, UniMAP.

E-mail: *azman@unimap.edu.my

Abstract. Vision systems have been used in many applications that intends to reduce the need for human operators. This is especially true for tasks that are simple but repetitive in nature, which is largely applicable to most manufacturing and agriculture's post-harvest processes. Many such processes utilize conveyor-based systems where the objects being processed are placed on a conveyor belt that runs through multiple processing stations. Implementing a vision system to capture images of an object that is moving usually requires setting up an imaging device to a working conveyor system. Getting a working conveyor system to be ready can take some time and consequently delay development work on the vision system itself, especially those involving image processing algorithms. This paper proposes a software solution that can be used to expedite initial work on such systems. The solution is written in C and is therefore easily ported to any development machine. A basic image processing library has also been developed so that it does not depend on any development library or suite, which is usually huge in size. Thus, the solution can easily be compiled and run on embedded development boards like Raspberry Pi - for a more portable solution.

1. Introduction

Vision systems are getting more attention among researchers and developers. This is understandable from a system point of view, since an image can produce a huge amount of information compared to normal sensing elements. In theory, vision capabilities enable a system to identify objects and their relative positions [1]. Consequently, this also means that vision systems require a more complex processing system with powerful processing elements [2].

Many processing tasks, especially the ones related to manufacturing and agriculture, are now turning to vision systems for simple but repetetive tasks that were traditionally done by humans. As imaging devices and processing elements get faster, real-time image processing applications of many such tasks becomes feasible and realiable for system implementations. Such applications usually involve conveyor systems, where the imaging system is installed at a processing station. Figure 1 shows a block diagram for a generic structure of a conveyor-based vision system.

In manufacturing, vision systems can be used to identify products (i.e. target object) that are being processed either for classification or quality assurance. In [3], such system is being used to detect bottles and to increase the precision of filling the bottles with intended liquid. In another case [4], a vision system is even used to detect defect in on the conveyor belt itself.

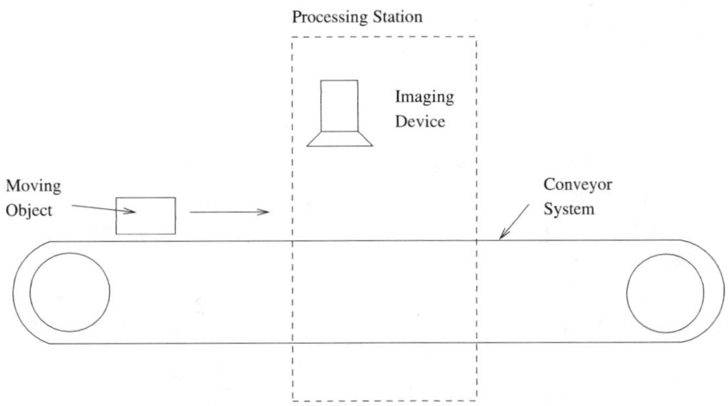

Figure 1. Generic Visualization of a Conveyor-Based Vision System

Agriculture post-harvest processing is another application that is utilizing the usefulness of having vision systems as part of their overall system. Tasks like fruit-sorting [5] and classification (or grading) [6] are actually taking advantage of the availability of such system.

In the next section, a couple of existing implementations of conveyor-based vision systems will be discussed. The following section will cover the proposed software solution for testing image processing algorithms on a conveyor-based system. Consequently, some analysis on practical implementations of the proposed solution will be presented, before more advanced features are discussed in the following section.

2. Existing Work

Most papers do not disclose any information on the methods used at the initial stage of the project when the conveyor is still not ready. Therefore, this section will mostly review a few recent papers that have a conveyor-based vision system setup and analyze the setup of such systems.

2.1. Simple Laboratory Setup

Some systems do not need such tight requirements, either in terms of size or equipments, that it can virtually be built anywhere and still easily moved to another location. This makes it easier to have a mock setup in a laboratory environment, where the researchers can simply start their work.

In [5], a mango sorting system consisting of conveyor system, web camera and a personal computer (PC) has been built. With a dimension of about 1 meter wide, 2 meters in length and height, it can easily be built inside a laboratory environment. Hooking up a web camera and having a desktop computer nearby completes the whole setup. In this work, it seems that all the image processing work was done after the setup was completed, which might have taken a few days if not weeks.

2.2. Industrial-level Setup

Some systems are actually too complex or too tedious to build, that it might as well be built in its intended industrial environment. In [7], such system was built as a novel industrial robot sorting technology. Using industrial manipulator robot arms and intelligent camera, the setup could not possibly be built in a laboratory. The size and setup of the manipulator robot arms themselves would make it impractical to be setup anywhere else. The system uses Cognex In-sight 7000 series industrial intelligent camera. Work to set up such a system would have taken quite a considerable amount of time and would require the software development team to work on site if no simulators are available.

3. Proposed Solution

In order to develop a software solution involving images, a basic image library that defines the image format and provides basic image manipulation functions is required. For that purpose, an open-sourced software *my1imgpro* has been developed and is used as the base for this solution. It is written in C programming language, which makes it highly portable. It has, in fact, been ported to a Xilinx Vertex-II Pro development board, which has a dual PowerPC core running on scarcely-available internal RAM-blocks. This simply shows that it does not require a huge amount of resources to run. The proposed solution involves developing codes that are capable of creating conveyor-like video data, on top of *my1imgpro* library.

3.1. Features of my1imgpro

To represent an image, the *my1imgpro* software uses a data structure named *my1image_t* with 5 members: *width*, *height*, *data*, *length* and *mask*. The first two members, *width* and *height*, are obviously integever values used to store image dimensions. Meanwhile, *data* is a pointer to a dynamically allocated memory of an integer array that holds image pixel values. The size of this pixel array location is saved in *length*. Although *length* seems redundant at first, it can save many multiplication operations required when trying to process the *data* as a one dimensional array.

The *my1imgpro* software was initially developed to process grayscale images. However, as time goes by, it is found that having colour representation is a very useful feature to have. That is the reason for *mask*, the last member in the *my1image_t* image data structure. It can be used to indicate whether the pixel values are actually 8-bit grayscale values or 24-bit RGB (Red-Green-Blue) values. The library also provides functions to manipulate pixels and image regions, functions to load and save image files, along with functions to view the image in Graphical User Interface (GUI). The GUI-related functions are provided by the *GTK* library.

In addition to those mentioned above, a video data structure *my1video_t* has also been defined. It is also a generic data structure, and can theoretically be used with any video capture library and provide a standard interface for applications to analyze the image stream within the video. In order to prove that this structure is adequate, the *ffmpeg* audio-video library has been used as the backend interface to the capture medium, which can be either a file or a live camera feed.

The *my1imgpro* software is available at *https://github.com/azman/my1imgpro* as an open-source software.

3.2. Conveyor-like video library

With a library like *my1impro* already available, work on creating a conveyor-like video generator becomes a lot easier. Instead of have two possible sources for an image stream, a third option has been added whereby the software reads all image files from a designated path and present them as scrolling images that will make any objects in the image look like they are moving on a conveyor system instead. The general idea is as shown in Figure 2.

From implementation point of view, the code basically finds all image files in a path through simple verification process and saves the file names of valid images in a queue list. Then, through iteration, the first two images from the queue are loaded into memory and their file names are reinserted at the back of the queue. This obviously gives an infinite supply of images for the conveyor-like video.

Focusing on code simplicity rather than faster loading time, all operations mentioned earlier are currently implemented in a single thread. As soon as the first image goes out of the intended viewport, the next image file in queue will be loaded into memory and attached to remaining

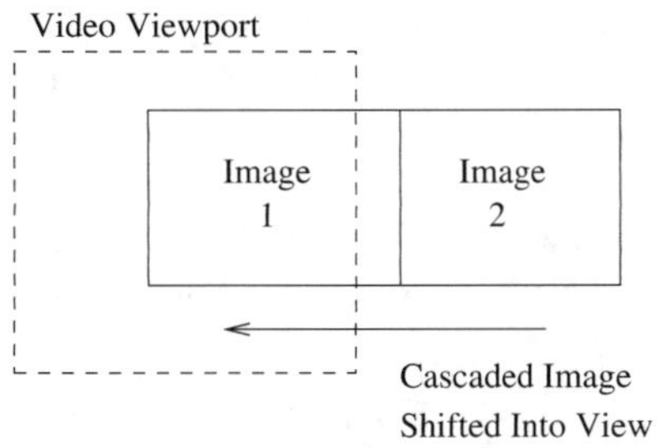

Figure 2. Shifting Image into View for Conveyor-like Effect

image in view. This might create a tolerable slight pause, depending the loading time of the an image.

The main challenge of implementing the conveyor-like image stream is sequencing the image data, i.e. shifting the image, into view. The two-dimensional image data is actually stored in a single-dimensional array, starting with the first row going into the next. Clearly, accessing the rows is not a problem because pixels in a row are already in sequence. However, accessing a column, which is what we need to do in our case, requires an alternative strategy.

While shifting into view, a variable is used to keep track of the next column available. Using this value as an base index, the column pixel data are obtained by offsetting the index by image width, which would get us to the next pixel in the column. Obviously, the process increases processing time but that can be tolerated at the moment since the code implementation is simpler this way and the output are still acceptable. Despite all this, the application using this code so far has shown sufficiently satisfying results.

The code for conveyor-like video generator will also be released as open-source software in the near future.

4. Implementation Analysis

The proposed solution has been used in a project meant for fruit sorting application. At the initial stage, building a conveyor system can take some time and getting fruit samples for testing algorithms can be a challenge, especially for seasonal fruits. With the proposed solution, having just 2 sample pictures of the target fruit would be enough to get things started. As can be seen in Figure 3, the target fruit for that project are mangoes.

Figure 3. Two Sample Pictures of Mangoes on Conveyor System

In the mentioned project, two pictures of mango which incidently are taken while on a conveyor system, were used as source images. This may not seem much, but with some image manipulation like rotation using 90 degrees steps and mirroring, we can easily get 16 samples instead. However, even in its original form, the two images can provide endless video image stream for any algorithm testing.

Figure 4 shows how the two images can be used to generate a conveyor-like video. The upper row shows the generated video image stream, while the lower row shows filtered video image stream. As can be seen from there, the filter actually detects mangoes only when it can be seen as a whole mango.

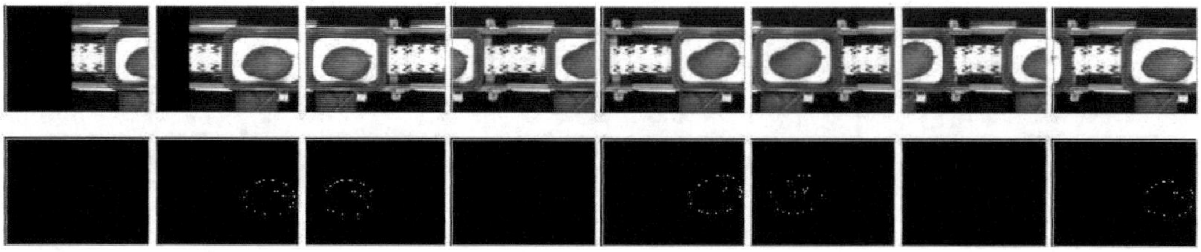

Figure 4. Top row shows the original stream and the bottom row shows filtered stream

Practically, the proposed solution can be used in two different setups. First, a target application with the algorithms being tested can be built together with the conveyor-like video generator code into a single binary. Basically, that means that it can be run together on the same platform (most probably a desktop computer). This is exactly what have been done in the sample demo presented above.

In another setup, the conveyor-like video generator code can be run on a desktop computer and displayed on standard monitor. Meanwhile, the target application with the algorithms being tested can be built separately on the actual intended target platform that has access to an imaging device that points to the monitor. This setup should be more similar to the intended application and therefore provide a more accurate test results compared to the first setup. Figure 5 shows the video generator code running on a desktop PC, while the application code was running on a Raspberry Pi 3 board with 7" display.

5. Conclusion

A software solution to test image processing algorithms on conveyor-based vision system has been presented. It would be very useful in purely research environment where building a conveyor-based system is time-consuming or maybe even not practical. Even without a conveyor system, development work related to vision systems in general or specifically the image processing algorithms can be initiated.

Being written in C programming language makes the proposed solution portable to any platform that has a C compiler - which means, almost all platforms. Since the *my1imgpro* library does not require any huge dependencies, other than *GTK* (for GUI display) and *ffmpeg* (for interface to imaging device and video files), the proposed solution is easy to build. It is also especially useful in embedded systems implementation. The core video generator code can easily be built on Raspberry Pi for a portable vision system solution.

The effect of loading delay of an image when running the conveyor-like video can be reduced by introducing multi-thread implementation. A separate thread can be created to handle loading of the third image while running the first two images. This should create a smoother flow for the conveyor-like video.

Other than the objects on a conveyor system, the conveyor system itself sometimes requires similar setup where it needs to inspect the conveyor belts for any damage periodically. This

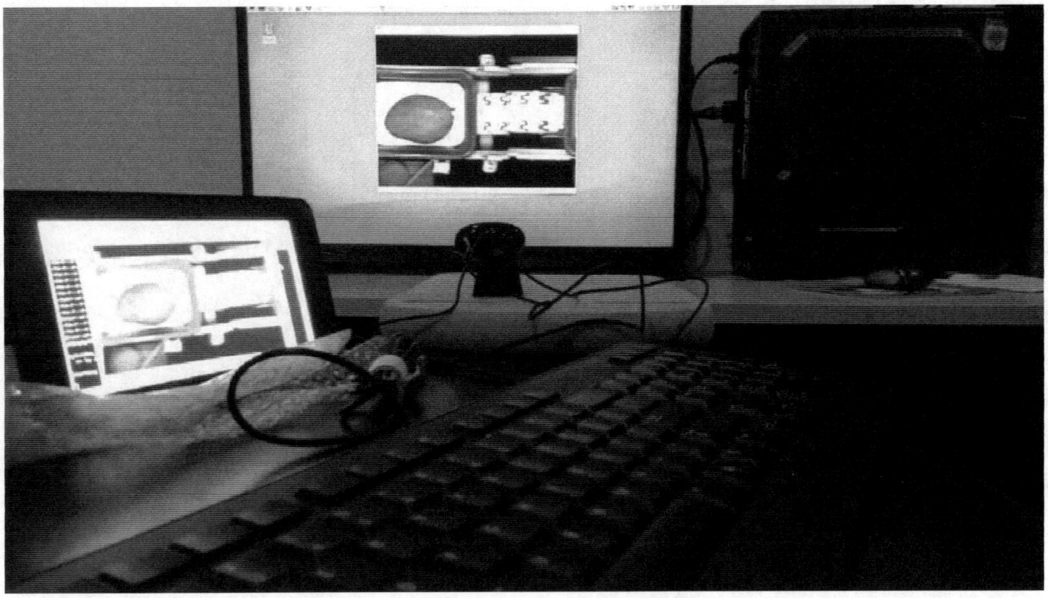

Figure 5. Running image capture and video generator on separate platforms

solution can help in the development of such algorithms without having to have actual damage on the conveyor belts.

Looking further, any applications that involves a vision system looking at objects moving within a certain path could be benefitted by the proposed solution. For example, it can be applied to biomedical analysis tool that is used for blood stream imaging, which is similar to conveyor-like flow.

References

[1] Marr D 1982 *Vision : A Computational Investigation into the Human Representation and Processing of Visual Information* (W.H. Freeman, San Francisco) ISBN 0716712849 0716715678

[2] Yusof A M, Shakaff A Y M and Rahim S A 2018 *Journal of Telecommunication, Electronic and Computer Engineering (JTEC)* **10** 123–129

[3] Thiyagarajan K, Meenakshi R and Suganya P 2016 *2016 International Conference on Advances in Human Machine Interaction (HMI)* pp 1–3

[4] Yang Y, Zhao Y, Miao C and Wang L 2016 *2016 IEEE International Conference on Signal and Image Processing (ICSIP)* pp 315–318

[5] Rojas-Cid J D, Prez-Bailn W, Rosas-Arias L, Romn-Ocampo D B and Lpez-Tello J A 2018 *2018 IEEE International Autumn Meeting on Power, Electronics and Computing (ROPEC)* pp 1–6 ISSN 2573-0770

[6] Su Q, Kondo N, Li M, Sun H, Riza D F A and Habaragamuwa H 2018 *Computers and Electronics in Agriculture* **152** 261 – 268 ISSN 0168-1699

[7] Li C, Ma Y, Wang S and Qiao F 2017 *2017 9th International Conference on Modelling, Identification and Control (ICMIC)* pp 902–907

An Automated Segmentation and Counting of Ki67 Cells in Meningioma Using K-Means Clustering Technique

Fahmi Akmal Dzulkifli*[1], Mohd Yusoff Mashor[1], and Hasnan Jaafar[2]

[1]School of Mechatronic Engineering, Universiti Malaysia Perlis (UniMAP), 02600, Arau, Perlis, Malaysia.

[2]Department of Pathology, School of Medical Sciences, Health Campus, Universiti Sains Malaysia, 16150, Kubang Kerian, Kelantan, Malaysia

*fahmiakmaldzulkifli@gmail.com

Abstract. Meningioma is a type of primary brain tumours. The meningiomas account for about one-third of all primary brain tumours. Image segmentation plays an important role in image analysis, especially detecting the tumours or cancerous areas in medical images. The output images from the segmentation prominently affect the system in detecting the tumour cells. Currently, the pathologists use the 'eye-balling' estimation technique to count the Ki67 cells. This technique was known as a time-saving measure. However, it has poor reliability and accuracy in counting the Ki67 cells. This paper proposed an automatic Ki67 cell counting in meningioma by using k-means clustering approach. The k-means clustering was used to segment the Ki67 cells and then the cells were classified into positive and negative Ki67 cells. The proposed system has been tested on 12 histopathological meningioma images. The proposed system is compared to the manually segmented images that have been validated in prior by the pathologists. The results show that the proposed system was able to segment the Ki67 cells with an average accuracy of 95.29%. The sensitivity and specificity of the proposed system were also high with an average of 93.56% and 97.39%, respectively.

1. Introduction

A brain tumour is referring to a collection of abnormal cells, which often forms a mass of tissue inside the brain. Generally, the brain tumour is divided into two categories, which are primary and metastatic brain tumour [1]. Meningioma is a type of primary brain tumour that arises in the three thin layers of tissue called meninges. According to CBTRUS statistical report, meningioma was the most frequently reported histology among the primary brain tumours, which accounts for 37.1% [2]. Based on the World Health Organization (WHO) grading scale, the meningioma is usually benign and grows. Due to this fact, these tumours cells are difficult to identify in some locations unless the size of the tumour is very large to be discovered. Ki67 is a nuclear antigen which response to a monoclonal antibody, MIB-1. Ki67 is known as an independent prognostic factor for survival rates, which consists of all stages and grade categories [3]. The Ki67 labelling index (LI) is defined as the percentage of immunoreactive tumour cell nuclei. The Ki67 LI is one of the measurement technique used to measure the normal and abnormal proliferation. Jonat and Arnold [4] classified the Ki67 LI into three stages, namely low (LI < 15%), intermediate (16%-30%), and highly proliferating (LI > 30%).

There are various image segmentation techniques were reported in the previous studies such as thresholding, wavelet transform, fuzzy c-means, quick shift technique, and watershed model [5]. K-means clustering is an unsupervised clustering algorithm used for dividing the images into few regions based on image colour property [6]. The clustering can be classified into two types, which are Hard Clustering (Exclusive Clustering) and Soft Clustering (Overlapping Clustering) [7]. For a hard clustering method, each data point belongs to a single cluster and the properties of this element cannot distribute to another cluster [7]. In soft clustering, each data point can belong to more than one cluster. In this study, K-means clustering was applied to partition the histopathological image into two groups namely the positive and negative Ki67 cells. This paper is organized as follows: Section 2 reviews previous studies used k-means clustering technique in medical imaging. Section 3 discusses the procedures and methodology for segmenting the Ki67 cells. Section 4 presents the experimental results obtained from the proposed technique. Section 5 concludes the paper.

2. K-Means Clustering

Nowadays, the use of image segmentation especially in medical fields is more focusing on segmenting tissues and organs. The application also includes tumour detection, surgical planning, brain development study, heart segmentation, and analysis of the cardiac images. A proper segmentation will simplify analysis to yield a more accurate result.

Sharma *et al.* [8] listed the problems existing in computed tomography (CT) and MR images. These include the partial volume effect, different type of artefacts, and noise due to sensors and related electronic system. Thus, the application of image segmentation will be useful in segmenting of an image. Over the years, many approaches have been proposed to segment medical images image by using k-means clustering technique.

Karmilasari *et al.* [9] developed a system that can determine the stage of breast cancer by using k-means clustering in the mammogram images. Three categories of mammogram images were used in the study which is dense-glandular, fatty, and fatty-glandular. At the beginning stage, the system will determine the region of interest (ROI). The proposed system cropped the input images to reduce the size of the input images. Otsu's thresholding was applied to separate the background and foreground of the images. Next, the system segmented the images by using a region growing technique. Then, the calculation of the suspected area size was performed based on the segmented images. K-means clustering was used to partition the suspected area into three stages (I, II, and III). Based on the results, the system can determine the stage of breast cancer based on the calculation of the area of the suspected object.

Bombale and Patil [10] proposed an algorithm to detect lung nodule from CT images by using k-means clustering. The segmentation was done to extract the part of the lungs from the images for detecting the probable cancerous area. Four types of nodules were selected in the study which are Well-circumscribed, Juxta-vascular, Pleural Tail, and Juxta-pleural. The classification process was to determine whether the nodule is benign or malignant. The classification was based on shape and size, appearance and growth rate. Based on the result, the proposed system was able to detect 494 samples from 525 samples of chest CT images. The proposed system obtained a good detection with an overall accuracy of 94.09%. This paper applied the k-means clustering technique to partition the positive and negative Ki67 cells to calculate the Ki67 LI.

3. Automated Ki67 Counting

In this study, the proposed automated Ki67 counting was operated on histopathological images of meningioma.

3.1. Image Acquisition

A total of 12 histopathological images were captured at the Department of Pathology, Hospital Universiti Sains Malaysia (HUSM). In this study, the positive Ki67 cells were stained using diaminobenzidine (DAB), while the haematoxylin was used to stain the negative nuclei. These images were captured under 40x magnification using an Olympus BX51 microscope and Cell^F software that works as an interface to the digital camera that attached with the microscope. The resolution of the captured images were 1360×1024 pixels and the images were saved in (*.jpg) format.

3.2. Image Enhancement

In this study, the image enhancement is essential for increasing the contrast and brightness of the Ki67 cells. In contrast enhancement process, the Ki67 images were enhanced by using Contrast-Limited Adaptive Histogram Equalization (CLAHE). To increase the contrast of an object, the original image must be converted into a colour space that has the image luminosity components. In the beginning, the Ki67 images were converted from RGB to L*a*b* colour space. The RGB colour space needs to be converted into CIEXYZ colour space first and followed by the L*a*b* colour space. The conversion from RGB colour space into CIEXYZ colour space can be calculated as follows [11]:

$$\begin{bmatrix} X \\ Y \\ Z \end{bmatrix} = \begin{bmatrix} 0.4125 & 0.3576 & 0.1804 \\ 0.2127 & 0.7152 & 0.0722 \\ 0.0193 & 0.1192 & 0.9502 \end{bmatrix} \begin{bmatrix} R \\ G \\ B \end{bmatrix} \quad (1)$$

Next is the conversion from CIE XYZ to L*a*b* colour space by using the equations below [12]:

$$L = 116 \left(\frac{Y}{Y_n}\right)^{\frac{1}{3}} - 16 \quad (2)$$

$$a^* = 500 \left[\left(\frac{X}{X_n}\right)^{\frac{1}{3}} - \left(\frac{Y}{Y_n}\right)^{\frac{1}{3}} \right] \quad (3)$$

$$b^* = 200 \left[\left(\frac{Y}{Y_n}\right)^{\frac{1}{3}} - \left(\frac{Z}{Z_n}\right)^{\frac{1}{3}} \right] \quad (4)$$

where X_n, Y_n, and Z_n are the tristimulus value of the reference white. X, Y, and Z describe as any colour that can be perceived by an average human observer. Next step was the luminosity value selection. The range of the values needs to be scaled into [0 1] since the luminosity component image was type double. For the best contrast enhancement result, the luminosity value was selected at 0.87. This value was the minimum value that has been selected manually based on analysis of 12 images. Then, the CLAHE technique is applied to the L component and the results of the L image was multiplied with the luminosity value. The output image from the contrast enhancement process was then retrieved back into RGB colour space. For increasing the brightness of the image, a constant value was selected and added to each of the pixels in the image. After performing analyses using all the captured images, it has been found that the best value was 20 since this value produced the best segmentation results. If the constant value is higher or lower than 20, it will degrade the segmentation results.

3.3. Image Segmentation

In this study, the k-means clustering technique was used to classify between the positive and negative Ki67 cells. At first, the proposed system will remove the background of the output Ki67 image from the enhancement technique. The background removal was done using Otsu's thresholding method. The purpose of this step is to remove the background of that image so that only the objects are remained. Then the resultant image was converted back to the original image.

At this moment, there were two visible objects left in the image which are the positive and negative Ki67 cells. It is known that the positive Ki67 cells appeared to be in granular brown colour while the negative Ki67 cells appeared in blue colour. These two colours were useful in k-means clustering, where the number of clusters needs to be specified at first before the segmentation process. Next step was to fill in the current image with 'dummy' pixels. The reason for adding the 'dummy' pixels is to control any input images that execute in the proposed system so that the system able to separate the objects correctly based on the specified clusters. The 'dummy' pixel contained the colour information of positive Ki67 cells. In this study, the 'dummy' pixel was set to R = 170, G = 130, and B = 60 that represents the brown colour. Before proceeding further process, the proposed system will add a 'dummy' pixel for every 100 pixels to the image. Due to the presence of the ‘dummy’ pixels inside the image, it certainly does not disturb the k-means clustering process since these pixels can be removed later using image filtering method. Afterward was the implementation of k-means clustering technique to segment the Ki67 cells.

At first, the captured image was converted into L*a*b* colour space using Equation (1), (2), (3), and (4). Next was to choose k points into the space represented by the objects that are being clustered. For this study the number of clusters was $k = 2$. This is because the current image (output image after the background removal) only has two colours which are brown and blue. Then was the calculation of the Euclidean distance, d by using the equation [6]

$$d = \|x_i - c_i\| \quad (5)$$

where x_i was the pixel in an image and the c_i was initial group centroids. Then was to assign each pixel to the group that has the closest centroid. When all objects have been assigned, recalculate the positions of the k centroids. The process will be repeated until the centroids are no longer move. This produces a separation of the objects into groups from which the metric to be minimized can be calculated. Next step was to label every pixel in the image using the results from k-means clustering. 'Dummy' pixels were added by using the RGB value that mention earlier in the previous paragraph. After adding the 'dummy' pixels, the process was continued to classify the cluster 1 for positive Ki67 cells and cluster 2 for negative Ki67 cells. Lastly, the noises and the 'dummy' pixels were removed using area opening method.

3.4. Feature Extraction
In this study, there were three main types of features that were selected based on the features of Ki67, namely circularities, area, and solidity based features. The area was expressed by the area of fully segmented Ki67 cells which includes the positive and negative Ki67 cells. The circularities features used to differentiate between the Ki67 cells with unwanted objects like artefacts. Solidity was used to measure the density of the cell. These three features will be the inputs for the proposed algorithm to determine whether the segmented objects were categorized in Ki67 cells or unwanted objects. Those objects will be discarded from the resultant images when it does not meet the requirement of the range value. The extracted image of positive and negative Ki67 cells was later saved into a bitmap (*.bmp) file extension.

3.5. Counting Ki67 Cells
In this step, the binary image from the feature extraction process will be used to count the Ki67 cells. Currently, the binary image only has two possible values (0 or 1) for each pixel. The '0' represented as the background while the '1' referred to Ki67 cells. The Sobel filter was applied to find the edges of the objects. After that, each detected object was labelled with different numbers from 1 to n. This can be performed by calculating the centre of gravity (centroid) to determine the position of the segmented cells and label them.

3.6. Calculation of Ki67 LI
Equation (6) below presents the formula to calculate the Ki67 LI. The equation was used after obtaining the number of positive and negative Ki67 cells. By following the WHO Classification of Tumours of the Central Nervous System, the tumours was labelled as grade I (LI < 7.99%), grade II (8% to 19.99%), and grade III (LI > 20.00%) [13].

$$\text{Ki67 Index} = \frac{\text{No. of positive Ki67 Cells}}{\text{No. positive Ki67 Cells} + \text{No. of negative Ki67 Cells}} \times 100\% \quad (6)$$

4. Results And Discussion
The proposed system was developed using MATLAB software version 2018. A personal computer which runs on an Intel Core i7-5500U, 2.4 GHz, processor with 16.0 GB RAM that operates on Microsoft Windows 10 Pro was used to develop the system. In order to test the performance of the segmentation method, the proposed segmentation algorithm was compared with manual segmentation. The manual segmentation was carried out using an open source image analysis software known as ImageJ. This software was used to segment the Ki67 cells manually and the resultant images were validated by HUSM pathologists. Quantitative analysis was performed to measure the performance of

the proposed system in segmenting Ki67 cells. The evaluation of the performance depended on the calculation of segmentation accuracy, sensitivity, and specificity. These analyses are calculated by:

$$Accuracy = \left(\frac{TP+TN}{(TP+TN+FP+FN)}\right) \times 100\% \quad (7)$$

$$Sensitivity = \left(\frac{TP}{(TP+FN)}\right) \times 100\% \quad (8)$$

$$Specificity = \left(\frac{TN}{(TN+FP)}\right) \times 100\% \quad (9)$$

where TP was defined as the number of pixels that were correctly segmented as positive KI67 cells. FP indicated as the number of pixels that incorrectly segmented as positive Ki67 cells. TN was referred to the number of pixels that were correctly segmented as a negative Ki67 cell, while FN signified the number of pixels that incorrectly segmented as negative Ki67 cells. Table 1 shows the segmentation results of the proposed system.

Table 1. Segmentation results of the proposed system

Input Image	Accuracy (%)	Sensitivity (%)	Specificity (%)
Cell 01.bmp	97.08	95.41	98.88
Cell 02.bmp	96.88	95.75	98.07
Cell 03.bmp	97.32	96.55	98.11
Cell 04.bmp	95.94	92.75	99.64
Cell 05.bmp	94.73	90.77	99.53
Cell 06.bmp	94.26	90.08	99.42
Cell 07.bmp	95.49	96.10	94.90
Cell 08.bmp	95.20	95.99	94.44
Cell 09.bmp	94.35	96.56	92.34
Cell 10.bmp	94.47	91.21	98.29
Cell 11.bmp	93.39	89.61	97.95
Cell 12.bmp	94.36	91.91	97.11
Average (%)	**95.29**	**93.56**	**97.39**

Based on Table 1, the results showed that the proposed system was able to segment the Ki67 images with an average accuracy of 95.29%. All of the images shown to have good accuracy with more than 90%. The results also showed that the proposed system able to detect the Ki67 cells with an average sensitivity value of 93.56%. Next, Table 2 shows the comparison between the proposed automated counting system with the manual counting system for positive and negative Ki67 cells. The table also showed the comparison between the Ki67 index between the proposed counting system with manual counting system. The manual counting was calculated by the pathologists.

Table 2. Comparison between the proposed automated counting with manual counting

Input Images	Manual Counting			Proposed Automated Counting System		
	Immunopositive Ki67	Immunonegative Ki67	Ki67 LI (%)	Immunopositive Ki67	Immunonegative Ki67	Ki67 LI (%)
Cell 01.bmp	51	407	11.14	61	365	14.32
Cell 02.bmp	62	340	15.42	67	328	16.96
Cell 03.bmp	70	351	16.63	78	292	21.08
Cell 04.bmp	21	305	6.44	18	392	4.39
Cell 05.bmp	25	312	7.42	17	411	3.97
Cell 06.bmp	35	373	8.58	28	509	5.21
Cell 07.bmp	41	316	11.48	82	241	25.39
Cell 08.bmp	37	295	11.14	50	265	15.87
Cell 09.bmp	48	287	14.33	62	266	18.90
Cell 10.bmp	79	345	18.63	78	461	14.47
Cell 11.bmp	97	361	13.32	76	415	15.48
Cell 12.bmp	168	308	35.29	133	265	33.42

Based on Table 2, it showed that the proposed system by using the k-means clustering technique was able to calculate Ki67 cells. From this table also, the proposed system showed a good result in counting positive Ki67 cells. For negative Ki67 cells, there was a large gap of counting between the proposed automated counting system and the manual counting. The existing problem happened may be due to the automated number of clusters (k) that was specified earlier in the segmentation process. The average execution time to count Ki67 cells for one image was 17 seconds.

5. Conclusion

This paper proposed an automatic Ki67 cell counting for histopathological meningioma images. The proposed system consists of six steps which are image acquisition, image enhancement, image segmentation, feature extraction, counting Ki67 cells, and calculation of Ki67 LI. The Ki67 images were enhanced using the CLAHE technique and followed by the segmentation process using k-means clustering. Three features were selected to extract the Ki67 cells, which includes of circularity, area, and solidity. A quantitative analysis was used to measure the performance of the proposed system. Based on the result, the proposed system was capable to segment and count the Ki67 cells. The proposed system was able to obtain an excellent segmentation result with an average accuracy of 95.29%. The proposed system also showed a high sensitivity and specificity values with an average of 93.56% and 97.39 & respectively. For future developments, the proposed system may be remodelled to improve the accuracy of the algorithm especially in detecting the Ki67 cells.

Acknowledgments

The author would like to thank the pathologists from Department of Pathology, University Science of Malaysia (HUSM) for helping and contributing to this study.

References

[1] Agravat R R and Raval M S 2016 Brain Tumor Segmentation Towards a Better Life *CSI Commun.* **40** 31–35

[2] Ostrom Q T, Gittleman H, Truitt G, Boscia A, Kruchko C and Barnholtz-Sloan J S 2018 CBTRUS statistical report: Primary brain and other central nervous system tumors diagnosed in the United States in 2011-2015 *Neuro. Oncol.* **20** 1–86

[3] Li L T, Jiang G, Chen Q and Zheng J N 2015 Ki67 is a promising molecular target in the diagnosis of cancer (Review) *Mol. Med. Rep.* **11** 3 1566–1572

[4] Jonat W and Arnold N 2011 Is the Ki-67 labelling index ready for clinical use? *Ann. Oncol.* **22** 3 500–502

[5] Suganya and Menaka 2014 Various Segmentation Techniques in Image Processing: A Survey *Int. J. Innov. Res. Comput. Commun. Eng.* **2** 1 1048–1052

[6] MacQueen J 1967 Some Methods for Classification and Analysis of Multivariate Observations *5th Berkeley Symposium on Mathematical Statistics and Probability 1967*.

[7] Bora D J, Gupta A K and Khan F A 2015 Comparing the Performance of L*A*B* and HSV Color Spaces with Respect to Color Image Segmentation *Int. J. Emerg. Technol. Adv. Eng.* **5**, 2 192–203

[8] Sharma N and Aggarwal L M 2010 Automated Medical Image Segmentation Techniques *J. Med. Phys.* **35** 1 3–14

[9] Karmilasari, Hanum Y , Widodo S, Hermita M, Putri N and ETP L 2014 Sample K-Means Clustering Method for Determining the Stage of Breast Cancer Malignancy Based on Cancer Size on Mammogram Image Basis *Int. J. Adv. Comput. Sci. Appl.* **5** 3 86–90

[10] Bombale A and Patil C G 2017 Segmentation of Lung Nodule in Ct Data Using K-Mean *Int. J. Electr. Electron. Data Commun.* **5** 2 36–39

[11] Kaur A and Kranthi B 2012 Comparison between YCbCr Color Space and CIELab Color Space for Skin Color Segmentation *Int. J. Appl. Inf. Syst.* **3** 4 30–33

[12] Tkaleie M and Tasie J F 2003 Colour spaces - Perceptual, Historical and Applicational Background *IEEE Reg. 8 EUROCON 2003. Comput. as a Tool.* **2** 304–308

[13] Louis D N, Perry A, Reifenberger G, von Deimling A, Figarella-Branger D, Cavenee, W K, Ohgaki H, Wiestler O D, Kleihues P, and Ellison D W 2016 The 2016 World Health Organization Classification of Tumors of the Central Nervous System: a summary *Acta Neuropathol.* **131** 6 803–820

Journal of Physics: Conference Series **1372** (2019) 012061 doi:10.1088/1742-6596/1372/1/012061

Classification of Microorganism Species Based on Volatile Metabolite Contents Similarity

Azian Azamimi Abdullah[1*], Amira Najihah Abdul Aziz[1], Shigehiko Kanaya[2] and Satya Ranjan Dash[3]

[1]School of Mechatronic Engineering, Universiti Malaysia Perlis, Pauh Putra Campus, 02600 Arau, Perlis, Malaysia
[2]Graduate School of Information Science, Nara Institute of Science and Technology, Ikoma, 630-0192, Nara, Japan
[3]School of Computer Applications, KIIT University, India

[*]azamimi@unimap.edu.my

Abstract. Microorganism species can become pathogenic and cause bacterial infection, result from the imbalance in microbial ecosystem between host and microbe. Microorganisms emit secondary metabolites, known as volatile metabolites or organic compounds (VOCs) for various functions such as intra- or inter-species interactions, defense and attraction. Currently, VOCs are widely used as a biomarker for human diseases. This research is aimed to identify the relationship between microorganism species and volatile metabolite compound from the collected species and VOCs emitted organism data by using unsupervised machine learning approaches such as hierarchical clustering and graph-clustering method. Supervised machine learning methods also been used to classify the microorganism pathogenicity such as support vector machine (SVM) and random forest (RF). These data are collected from KNApSAcK and mVOC database where it provides most of the microorganism species and metabolites contents. From the collected data, there are in total of 1088 VOCs emitted by 517 microorganism species. As a result, the application of machine learning methods enable us to identify the relationship of species with their emitted VOCs and classify the microorganism species into their own pathogenicity.

1. Introduction

Microorganism species can become pathogenic and cause bacterial infection, result from the imbalance in microbial ecosystem between host and microbe [1]. Presence of specific genes that contain pathogenic characteristic may cause even the non-pathogenic species became virulence through gene transfer [2]. A variety of microorganism species can cause disease or infection where there are five main types of pathogenic organisms, which are viruses, bacteria, fungi, protozoa and worm. For a particular organism, metabolites are essential to the metabolism process where it promotes in growth, reproduction and defense against pathogen [3]. Basically, metabolites can be classified into primary metabolite, which involved in growth and development and secondary metabolite, which is important for ecological function such as antimicrobial activities and defense against pathogens.

Volatile organic compounds (VOCs) comprise a chemically diverse class of low molecular weight organic compounds having an appreciable vapor pressure under ambient conditions [4]. Because of this characteristic, it allows them to serve as signaling molecules act as identification in plant, food, predators or habitat [5]. The VOCs produced by plant will attract pollinators to disperse the seed to pollinate and provide defense mechanism against disease [6]. For humans and animals, volatiles act as scents and contribute to flavorants of food and as disease biomarkers [7], [8].

The recent classification study are based on metabolite contents is regarding the plant classification. The study focused on the global metabolite content of plants which resulting in classification with natural phylogenetic and chemosystematics pattern [9]. Therefore, a new study should be done focusing on the classification of microorganism species based on the volatile metabolite contents to distinguish the pathogenic and non-pathogenic microorganism instead being done by the previous research. Machine learning has become one of the popular approach in solving complex problem and applications such as self-driving cars, fraud detection and medical prediction. Machine learning can be divided into supervised and unsupervised machine learning. Unsupervised machine learning is when the input of similar types is grouped together without the use of training data to specify which group it belongs to. Previously, this technique has been applied to analyze the relationships between 3D structure and biological activities by using network-based approach [10]. Supervised machine learning is task of training using function that requires input data and target representing the desired output.

In this study, the microorganism species will be classified based on the volatile metabolite contents similarity to determine the pathogenicity of the species. Clustering analysis method such as hierarchical clustering and graph clustering is used to determine the relationship between the species and their VOCs. For the classification process, supervised machine learning is used to distinguish the pathogenicity between species based on their VOCs by using support vector machine (SVM) and random forest (RF) as the classifier.

2. Proposed Methods

2.1. Data extraction

The microbial volatile metabolite data were acquired from KNApSAcK Core DB, which is a part of the KNApSAcK Family DB Pre-processing and mVOC Database, which is based on extensive literature search for microbial volatile organic compounds (mVOCs) [11]. The KNApSAcK Core DB provides a comprehensive species-metabolite relationship, which is connected via the scientific name of each species and this database can be accessed online [12]. The mVOC Database is an online search platform database, which contains information about mVOC and their emitting organism [13], [14].

2.2. Clustering of species based on VOCs similarity

2.2.1 Hierarchical Clustering

Hierarchical agglomerative clustering method is implemented, which it start by putting each observation into its own separate cluster where all the distances between all the observation are being examined and pairs together with the closest ones to form a new cluster and this process continues until only single cluster remain [15]. The result of the clustering is represented in the form of dendogram. The following distance matrix is used to determine the Euclidean distances between species. Let this matrix be called M and $M_{ik}=1$ if the species *i* is related to the *k*th VOC or otherwise $M_{ik}=0$ [7]. Based on the Euclidean distance, the R, an open source programming language is used to analyse the clustering. In this method, four different ways is performed to determine the distance between two clusters. Complete linkage clustering is used to find the possible maximum distance between two points, which belong to two different clusters. The average and all possible pairwise distances for points belonging to two different clusters, mean linkage clustering is applied. The centroid linkage clustering is used to find the centroid of each cluster and the distances between centroids of two clusters. Ward linkage clustering is used to minimize the total variance within the cluster where the cluster with minimum distance between pair of cluster is merged.

$$d\,(i,j) = \sqrt{\textstyle\sum_{k=1}^{n}\left(M_{ik} - M_{jk}\right)^2} \qquad (1)$$

2.2.2 Graph Clustering Based on DPClus

The DPClus is a graph clustering software, which has been developed based on a clustering algorithm that mainly focused on detection of protein complex in interaction networks [16]. This graph clustering software can extract high-density connected nodes as a cluster [7]. The input to the algorithm is the associated matrix of an interaction network and the outputs are protein complexes. The complexes are determined by way of finding clusters [16]. The DPClus presents an algorithm for detection of large interaction network of protein complex. In this network, a node represents a protein while the edge represents an interaction [17]. There are three important inputs in the DPClus algorithm, which are the network, a value of minimum density (d_k) and a minimum value for cluster property (cp_{nk}). The value for density, d_k and cluster property, cp_{nk} should be within the range of:
$0 < d_k \leq 1$ and $0 < cp_{nk} \leq 1$.

2.3. Classification of species based on VOCs Similarity

Supervised machine learning approach is the task of training using function that requires input data and target representing the desired output. For this classification process, microorganism species and their VOCs will represent the input data while the pathogenicity status will represent desired output. During classification process, supervised machine learning is used to separate the pathogen based on binary VOC input data. To ensure the accuracy of the classifier, two machine-learning methods were applied, which are random forest and support vector machine. The RF and SVM algorithm was used to generate classifiers and estimate the classification performance. For a given pathogen class, the classifiers' sensitivity, precision and mean accuracy towards that class are given by (2), (3) and (4), respectively:

$$Sensitivity = \frac{number\ of\ correct\ predictions\ in\ that\ class}{number\ of\ experiments\ in\ that\ class} \times 100\% \quad (2)$$

$$Precision = \frac{number\ of\ correct\ predictions\ in\ that\ class}{number\ of\ predictions\ as\ that\ class} \times 100\% \quad (3)$$

$$Mean\ accuracy = 1 - \frac{number\ of\ incorrectly\ predicted\ pathogens}{total\ number\ of\ experiments} \quad (4)$$

2.3.1 Support vector machine (SVM)

SVM is one of the supervised machine learning methods where the purpose is for classification and regression, which are based on the concept of decision planes or hyperplane that define decision boundaries [18]. Separating line defines boundary in decision plane where it is the one that separates between a set of object having a different class of membership.

2.3.2 Random forest (RF)

Random forest is supervised machine learning method where every observation will be fed into a large number of decision tree. This method operate by constructing a multiple decision tree at training time and put into class or mean prediction of the individual trees. Random subspace method is the algorithm used for this machine learning. In RF classifier, the higher the number of trees in the forest gives the high accuracy results [19]. Variable importance can be used to rank the random forest ranking the importance variables in a regression or classification problem.

3. Results and Discussion

3.1. Clustering of species based on VOCs similarity

From the data extracted, 1088 VOCs emitted by 517 microorganisms are being analyzed to confirm the relationship between them. Two unsupervised learning methods were implemented to determine the relation, which are hierarchical clustering and graph clustering.

Hierarchical clustering is used to analyze the relationship between the microorganism species with their VOCs. There are four different linkage methods in hierarchical clustering were used, which are Ward linkage, Complete linkage, Centroid linkage and Mean linkage clustering to analyze and observe which one out of the four methods will group the pathogenicity of microorganism based on their VOCs correctly. From the result, Ward Linkage Clustering are more preferable to be used in this study to obtain the relation between species and their VOCs. This is because Ward clustering grouping the microorganism by minimizing the variance between cluster and merged the cluster together if they have minimum distance, where in this case the similarity between VOCs causing the distance to be minimized and the microorganism being grouped according to their VOCs similarity. The Euclidean distance is used to calculate the distance between species based on VOCs where the information of microorganism species and its emitted VOCs has been converted into 517×1088 binary matrix. Here, the dendrogram is cut into 50 clusters to represent each microorganism into their own cluster.

To obtain more information regarding the relation of the microorganism with the mVOCs, high-density cluster is applied where the graph-clustering algorithm DPClus is used. Figure 1 shows the graph clustering result where the green node represents the cluster of microorganism while the red edges represent the interaction between clusters. DPClus clustering is set to the following setting: density value, din was set to 0.6, cluster property value, cpin was set to 0.5, the minimum cluster size was set to 2, and the mode is set to non-overlapping. From Figure 1, the size of a green node is proportional to the number of microorganism species in the cluster while the thickness of red edges are proportional to the number of similarity-based relation.

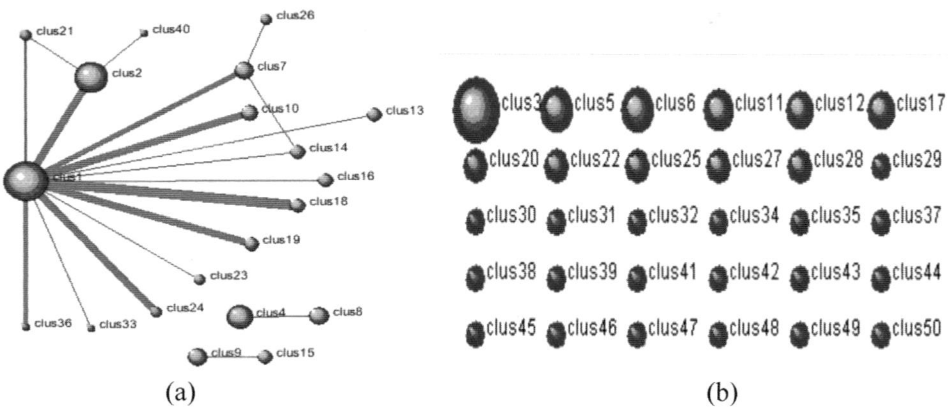

Figure 1. Graph clustering based on DPClus (a) Connected node (b) Independent node.

From the clustering of DPClus, there are 50 clusters are generated where 20 cluster are connected nodes to each other (Figure 2(a)) while the other 30 clusters are independent clusters (Figure 2(b)). From this result we can conclude that the pathogenicity of microorganism can be connected by the similarity of VOCs emitted. Pathogenic microorganism such as Escherichia coli, Klebsiella pneumonia and Staphylococcus aureus that comes from cluster 1 are connected to the other pathogenic cluster for example cluster 2. The results of this graph-clustering and hierarchical clustering are similar because both results show that the VOCs based classification of microorganism is consistent with their classification based on pathogenicity. However, through DPClus we can see the relation between each microorganism to VOCs emitted by their species since DPClus can show the interaction being generated between clusters.

3.2. Classification of species using supervised machine learning methods

The classification was done by using supervised machine learning approach by separating the pathogens based on binary VOC data extracted. The binary VOC data set represents the input while pathogenicity status represents the desired output. The extracted data contains the 1088 VOCs data and their predicted pathogenicity for 517 microorganisms acts as target. The data are being splitted into testing and training with 30% and 70% respectively.

From the result obtained, SVM provide highest accuracy with 99% and 100% and 95% for sensitivity and precision respectively. From Table 1, it is predicted that 131 species is tested non-pathogenic while 22 species are correctly predicted as pathogenic. However, one species is falsely predicted as pathogenic. This incorrect classification may result from the presence of volatile compound. In training confusion matrix, there are 3 species was incorrectly classified as pathogenic while another 308 species are non-pathogenic. However, 53 species are correctly classified as pathogenic species.

Table 1. Confusion Matrix for SVM

	Reference			
	Testing		Training	
Prediction	Non-pathogenic	Pathogenic	Non-pathogenic	Pathogenic
Non-pathogenic	131	1	308	2
Pathogenic	0	22	0	53

For the comparison between classifier, RF was used beside SVM to choose the most reliable classifier among those two. The number of trees (ntree) selected for the random forest algorithm were set to 100. The accuracy obtained by using RF is high with 95%. It also has 100% sensitivity but only 70% specificity. From both testing and training, we can confirm that 65 microorganisms are pathogenic while 13 species which being incorrectly predicted as non-pathogenic but turn up being pathogenic.

Table 2. Confusion Matrix for RF

	Reference			
	Testing		Training	
Prediction	Non-pathogenic	Pathogenic	Non-pathogenic	Pathogenic
Non-pathogenic	131	7	308	6
Pathogenic	0	16	0	49

Figure 2 shows the ROC curve for RF classifier. The rate of true positive is high similar with the rate of false positive. From both classifier of supervised machine learning being tested, SVM shows the highest accuracy with 99% compared to RF (95%). This result confirmed that SVM are more reliable compared to RF even though the accuracy difference is small.

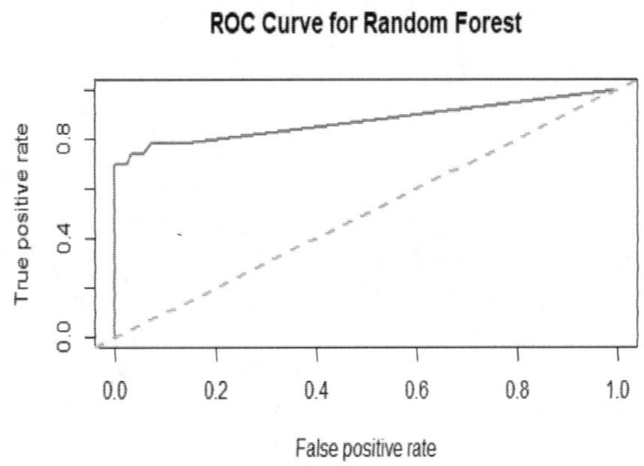

Figure 2. ROC curve for the rate accuracy between species and VOCs.

4. Conclusion

In worldwide healthcare, the development of technologies for medical purposes is growing in fast rate to prevent infectious diseases that endanger human population. A non-invasive monitoring tool is preferable for the detection of volatile metabolite. Through this study, VOCs can act as a potential biomarker to detect the presence of volatile microorganisms. From the study involving 1088 VOCs emitted by 517 microorganisms proved that the classification of microorganism species can be done based on volatile metabolite contents similarity. Based on DPClus, we can see the relation between each microorganism to VOCs emitted by their species since DPClus can show the interaction being generated between clusters. For the supervised learning method, SVM shows the highest accuracy with 99% compared to RF for the classification of microorganisms species based on VOC similarity.

References

[1] G. D. Ehrlich, N. L. Hiller, and F. Hu, "What makes pathogens pathogenic," Genome Biol., vol. 9, no. 6, p. 225, 2008.

[2] K. A. Bliven and A. T. Maurelli, "Antivirulence genes: Insights into pathogen evolution through gene loss," Infect. Immun., vol. 80, no. 12, pp. 4061–4070, 2012.

[3] Q. Wu, M. Chen, H. Zhou, X. Zhou, and Y. Wang, "Metabolite profiles of Populus in response to pathogen stress," Biochem. Biophys. Res. Commun., vol. 465, no. 3, pp. 421–426, 2015.

[4] D. D. Rowan, "Volatile Metabolites," Metabolites, vol. 1, no. 1, pp. 41–63, 2011.

[5] E. Pichersky, J. P. Noel, N. Dudareva, and A. Arbor, "Biosynthesis of Plant Volatiles: Nature ' s Diversity and Ingenuity PV Biosynthetic Pathways Branch Off from Primary Metabolism," vol. 311, no. 5762, pp. 808–811, 2006.

[6] M. Heil and J. Ton, "Long-distance signalling in plant defence," Trends Plant Sci., vol. 13, no. 6, pp. 264–272, 2008.

[7] N. Yusuf, A. Zakaria, M. I. Omar, A. Y. M. Shakaff, M. J. Masnan, L. M. Kamarudin, N. Abdul Rahim, N. Z. I. Zakaria, A. A. Abdullah, A. Othman, and M. S. Yasin, "In-vitro diagnosis of single and poly microbial species targeted for diabetic foot infection using e-nose technology," BMC Bioinformatics, vol. 16, p. 158, 2015.

[8] Yusuf, N., M. I. Omar, A. Zakaria, A. I. Jeffree, R. Thriumani, A. A. Abdullah, A. Y. M. Shakaff et al. "Evaluation of E-nose technology for detection of the causative bacteria in different culture media on diabetic foot infection." In 2014 IEEE Conference on Biomedical Engineering and Sciences (IECBES), pp. 67-70. IEEE, 2014.

[9] K. Liu, A. A. Abdullah, M. Huang, and T. Nishioka, "Novel Approach to Classify Plants Based on Metabolite Content Similarity," Biomed Res. Int., vol. 2017.

[10] Ohtana, Y., Abdullah, A.A., Altaf-Ul-Amin, M., Huang, M., Ono, N., Sato, T., Sugiura, T., Horai, H., Nakamura, Y., Morita, A. and Lange, K.W., 2014. Clustering of 3D-Structure Similarity Based Network of Secondary Metabolites Reveals Their Relationships with Biological Activities. Molecular informatics, 33(11-12), pp.790-801.

[11] A. A. Abdullah et al., "Development and Mining of a Volatile Organic Compound Database," Biomed Res. Int., vol. 2015, 2015.

[12] F. M. Afendi et al., "KNApSAcK family databases: Integrated metabolite-plant species databases for multifaceted plant research," Plant Cell Physiol., vol. 53, no. 2, pp. 1–12, 2012.

[13] M. C. Lemfack, J. Nickel, M. Dunkel, R. Preissner, and B. Piechulla, "MVOC: A database of microbial volatiles," Nucleic Acids Res., vol. 42, no. D1, pp. 1–90, 2014.

[14] M. C. Lemfack, B. Gohlke, S. M. T. Toguem, S. Preissner, B. Piechulla, and R. Preissner, "mVOC 2.0: a database of microbial volatiles," no. November, pp. 1-5, 2017.

[15] A. A. Abdullah, T. Nishioka, T. Katsuragi, and N. Ono, "Analysis of Volatile Metabolites Emitted by Various Species to Reveal Their Roles in Chemical Ecology and Healthcare," no. March, pp. 30–31, 2015.

[16] M. Altaf-Ul-Amin, H. Tsuji, K. Kurokawa, H. Asahi, Y. Shinbo, and S. Kanaya, "DPClus : A density-periphery based graph clustering software mainly focused on detection of protein complexes in interaction networks," J. Comput. Aided Chem., vol. 7, pp. 150–156, 2006.

Journal of Physics: Conference Series **1372** (2019) 012061 doi:10.1088/1742-6596/1372/1/012061

[17] M. Altaf-Ul-Amin, Y. Shinbo, K. Mihara, K. Kurokawa, and S. Kanaya, "Development and implementation of an algorithm for detection of protein complexes in large interaction networks," BMC Bioinformatics, vol. 7, p. 207, 2006.
[18] D. K. Srivastava and L. Bhambhu, "Data classification using support vector machine," J. Theor. Appl. Inf. Technol., vol. 12, pp. 1–7, 2009.
[19] B. López, F. Torrent-Fontbona, R. Viñas, and J. M. Fernández-Real, "Single Nucleotide Polymorphism relevance learning with Random Forests for Type 2 diabetes risk prediction," Artif. Intell. Med., pp. 3–9, 2017.

Classification of Benign and Malignant Breast Cancer using Supervised Machine Learning Algorithms Based on Image and Numeric Datasets

Ratula Ray[1], Azian Azamimi Abdullah[2], Debasish Kumar Mallick[3], Satya Ranjan Dash[4]*

[1]School of Biotechnology, KIIT Deemed to be University, Bubaneswar, India
[2]School of Mechatronic Engineering, Universiti Malaysia Perlis, Malaysia
[3,4]School of Computer Applications, KIIT Deemed to be University, India

[4*]sdashfca@kiit.ac.in

Abstract: Breast cancer has been identified as the second leading cause of death among women worldwide after lung cancer and hence, it becomes extremely crucial to identify it at an early stage, which can considerably increase the chances of survival. The most important part in cancer detection is to be able to differentiate between benign and malignant tumors and this is where the work of Machine Learning comes in. Taking all the dependent features upon consideration, Supervised Machine Learning methods allow for classification with higher degree of accuracy and improve upon the misdiagnosis of the physicians, which might occur almost 20% of the time. In our paper, we are focusing towards understanding the shortcomings of digital mammograms in detection of breast cancer and utilize Machine Learning classifiers for the classification of benign and malignant tumors using image analysis. Apart from this, we are also looking into implementing Supervised Machine Learning classifiers such as Decision Tree, K Nearest Neighbour (KNN), Random Forest and Gaussian Naive Bayes classifiers for assessing the risks involved with breast cancer by analyzing the biomarkers that are involved with it. Our aim is to provide a comprehensive view on prediction of breast cancer through Machine Learning through both image and data analyses, which can play a pivotal role in prevention of misdiagnosis in future. Fig. 1. gives a layout for the breast cancer prediction using Supervised Machine learning classifiers.

1. Introduction

There are majorly two types of breast cancer: invasive and non-invasive breast cancer. Invasive form of breast cancer is more dangerous and it metastasizes from one location and spread to the other organs. Ducts and Lobes are the two most common sites where the breast cancer invades and it can rapidly progress from that point on. Based on these two sites, there are about four different types of breast cancer:

- **Ductal carcinoma in-situ (DCIS):** It is a non-invasive form of breast cancer, where the cancer is restricted to the milk ducts. There is always a chance of recurrence of the cancer even if it has subsided.

- **Lobular carcinoma in-situ (LCIS):** Like DCIS, this form of cancer is also non-invasive but is restricted to the milk producing glands of the breast or lobules. LCIS is associated with abnormal growth of cells on the lobules of the breast.
- **Invasive ductal carcinoma (IDC):** This is a metastatic form of breast cancer and also the most common type. This affects not only the milk ducts but also the surrounding layers of fatty tissues and travels further into the lymphatic system. In our paper, we will deal with this form of breast cancer classification in greater detail.
- **Invasive lobular carcinoma (ILC):** Similar to IDC, this form of cancer is also not restricted to the site of occurrence. Though it originates at the lobules of the breasts, it also rapidly metastasizes to the nearby tissues, affecting them considerably.

Histology can be defined as the study of the anatomy of the tissue and the preparation steps of the histology slide involves the steps such as: fixing, processing, embedding, sectioning and staining [1]. Histopathological analysis of breast cancer refers to the analysis of the biopsies of the breast tissues to identify the region where the cancer is concentrated at and this work is mostly performed by a pathologist. Classification tasks using Machine Learning tools gives the computer the ability to differentiate the entire dataset into multiple classes based upon the features they are being associated with. The dataset that we use can either be in the form of numeric data or images (which need to be transformed into vector forms before it can be used for the classification task). In our paper, we are concerned about classification based on two separate aspects:

a. Classification of the presence of breast cancer based on important biomarkers (numeric data classification)
b. Classification of the histopathological images of breast tissue biopsies to identify particular regions of IDC affecting the tissues to confirm presence of breast cancer (image analysis)

2. Literature review

In our literature survey, the first paper we studied deals with the detection of micro calcifications on breast tissues for aiding in the diagnosis of breast cancer by studying the mammogram database collected from the University of Wisconsin. The image dataset used for this purpose adjusts the contrasts by analyzing the histograms of the respective images, which is followed by application of morphology filtering. Finally through K-means clustering, in association with fuzzy rule based classifiers, the classification process is carried out. The suggested algorithms led to the goal of detection of the cancer type [2].

Literature also suggests medical image analysis performed for the breast cancer dataset, which is followed by the meta-training model. For this they have set up a SOTA for weakly supervised breast screening problem. Several baseline and multiple tasks trained by DenseNet and some fine-tuning by using MIL framework have been applied. But MIL's performance does not provide improved results. In another proposed meta-training of same dataset model, CL sampling yields more accurate result than random sampling, whose performance is closer to that of DenseNet [3].

Another paper also reveals case–control studies of 1171 breast cancer cases with 1659 controls. The digitized film-screen analysis is done by checking for Mammographic breast density and 46 breast texture features are taken into account which includes first and second order feature, fractal analysis and Fourier transformation. The logistic regression model is used to evaluate the features involved in detection of breast cancer after adjustment of the details. Fractal dimensions for the thresholds around 10% and 15% indicate increasing of the risk, whereas fractal dimensions of 60% to 95% indicates decreasing of risk of cancer. Overall, the result of the risk involved in this case depends upon the density of mammographic image [4].

One of the most significant drawbacks of mammographic images is the high rate of false positive for breast cancer diagnosis. Ultrasound data yields more accurate results compared to mammograms in this case. Paper suggests the use of computer-aided system (CAD), which can be used to find out the area affected by cancer more accurately. Utilizing FSVM (Fuzzy Support Vector Machine) with ultrasound dataset in association with feature extraction and selection techniques have also been developed. For this purpose, results have shown that he accuracy of ANN classifier is 88.51%. Moreover, it has also been seen that when SVM classifier yields an accuracy of 87.36%, in that case, FSVM yields 94.25% accuracy [5].

Computer aided detection system is also used to identify the position of the area of the breast affected by cancer using the sum-of-square error function, which improves the efficiency and robustness in identification of the micro calcification clusters. By using artificial neural network, convolution neural network and linear discriminant analysis, selection of the features has been done in cluster method. The result was found by analyzing 96 medical cases with 192 images dataset collected from University of Michigan, and it yielded an accuracy of 96% [6]. Method for development for reduction of false positive results have been developed using computerized detection based on analysis of bilateral information. The bilateral computer aided detection can be used for reducing false positives by training [7]. XR mammography detected 31 patients who were classified for breast cancer according to BIRADS (Breast Imaging-Reporting and Data System), along with their biopsy size and fractal dimension. The parenchyma fractal dimension lesions area showed complex breast calcifications that were consistent with their morphological pattern. The Fractal dimension analysis was used for getting result [8].

3. Implementation

3.1. Dataset

We have utilized the Breast Cancer Coimbra dataset for our prediction analysis of the numeric dataset, which takes into consideration the different biomarkers for breast cancer prediction [9]. For image analysis, we have used the Kaggle dataset for Breast Histopathology images [10].

3.2. *Methodology*

Figure 1 gives a layout for the methodology used for classifying benign and malignant breast cancer using Supervised Machine Learning classifiers.

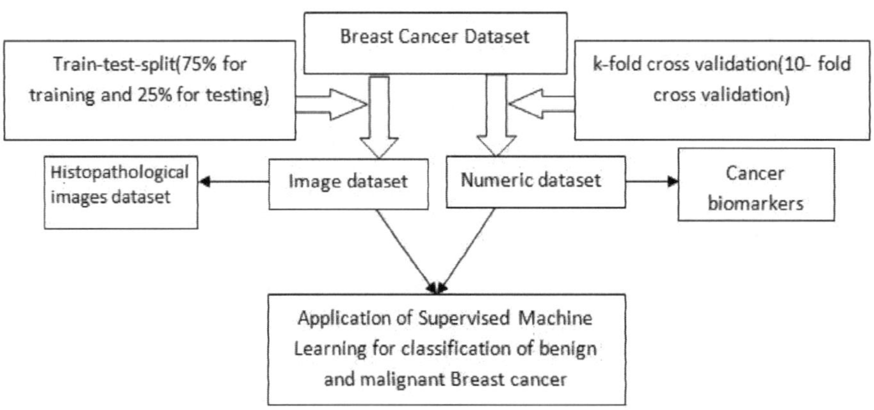

Figure.1. Flowchart of the methodology

3.3. Results

We have used the Precision, Recall and F1-score as the metrics for statistical analysis of the datasets. Table 1. gives the result for the numeric data used for analysis of breast cancer based on different biomarkers such as Age, BMI, Glucose, Insulin, HOMA, Lectin, Adiponectin, Resistin, MCP.1.

The metrics used for this purpose are as follows:

Recall = $\frac{True\ Positive}{True\ Positive + False\ Negative}$

Precision = $\frac{True\ Positive}{True\ Positive + False\ Positive}$

F1-score: $\frac{2*(Recall*Precision)}{(Recall+Precision)}$

Table 1. Statistical results determining the performance of the classifiers for the numeric dataset

Classifiers	Precision	Recall	F1-score
Gaussian Naive Bayes Classifier	66%	60%	59%
Decision Tree Classifier	70%	70%	70%
K-nearest neighbors Classifier(n=5)	48%	48%	48%
Random Forest Classifier	71%	71%	70%

Out of the different classifiers used, we find Decision Tree and Random Forest classifiers showing promising results with an accuracy score of 70%. We hence used Random Forest classifier to check the most important features linked with the prediction of the end result. Figure 2 shows the graphical representation of the ranking of the biomarkers with their relevance to the presence of breast cancer. Out of them, BMI(Body Mass Index) and Resistin show the most dependency on the end result.

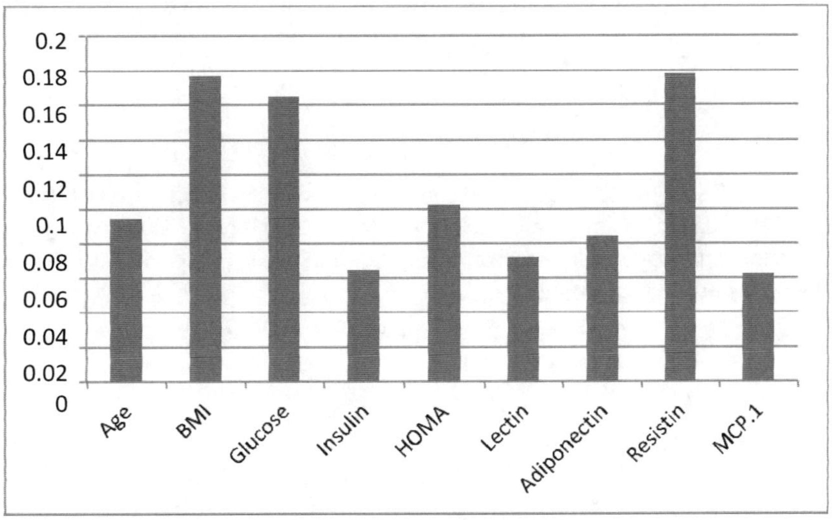

Figure 2. Graphical representation of the feature ranking according to their importance with the data label

The performance of the same sets of classifiers taken for histopathological image analysis for breast cancer tissues are depicted in Table 2. The analysis was done by taking into account the variability of color change upon staining.

Table 2: Statistical results determining the performance of the classifiers for the histopathological image of breast tissues dataset

Classifiers	Precision	Recall	F1-score
Gaussian Naive Bayes Classifier	86%	83%	84%
Decision Tree Classifier	87%	87%	87%
K-nearest neighbors Classifier (n=5)	88%	88%	88%
Random Forest Classifier	91%	92%	92%

4. Conclusion

Histopathological image analysis is the microscopic analysis of breast tissue derived from biopsy specimens. In our case, we have used segmentation of the cellular components on the tissue samples based on color of the staining. Identification of the cancer nuclei is the basic aim of this computational based analysis and this can be done by sectioning of the region of interest from the background information. In case of H and E staining of breast tissue samples, blue color denotes the nuclei and the cytoplasm, whereas the stroma is stained with pink and the RBCs are stained in red. Our result depicts that Random Forest Classifier shows the best result with F1-score of 92%, followed by K-nearest neighbors classifier. Future prospects in this field is huge and there are many aspects of segmentation of the images that can be carried out, like region-based segmentation, shape-based segmentation and texture-based segmentation, which can give much improved and reliable results. Computational constraint for using large dataset for image classification is the reason we have used train-test split as the validation method instead of k-fold cross validation in this case. For the numeric dataset, we have seen that Random Forest and Decision Trees give comparable results. Our findings also suggest that BMI and Resistin are the most important biomarkers that affect the breast cancer results. BMI acts as a major biomarker for breast cancer and it is linked differently in pre and postmenopausal women. It is basically due to the difference in production of estrogen and progesterone and thus in case of pre-menopausal women, higher the BMI; lower is the risk of developing breast cancer. However, in case of post-menopausal women, the risk of breast cancer is positively linked with increasing BMI. Resistin is an adipocytokine and is known to be positively related with the increasing risk of breast cancer.

In this paper, we have highlighted the different classifiers powered by Machine learning techniques in order to detect the breast cancer using both numeric and image datasets. We are hopeful that these techniques will assist the pathologists and doctors to make much more precise and informed decisions and improve the detection process significantly.

References

[1] Aswathy, M. A., & Jagannath, M. (2017). Detection of breast cancer on digital histopathology images: Present status and future possibilities. Informatics in Medicine Unlocked, 8, 74-79.

[2] Guardado-Medina, R. O., Lepe, R. R., & Magaña, B. O. (2012). Intelligent detection of microcalcifications in mammograms based on fuzzy techniques.

[3] Maicas, G., Bradley, A. P., Nascimento, J. C., Reid, I., & Carneiro, G. (2018, September). Training medical image analysis systems like radiologists. In *International Conference on*

Medical Image Computing and Computer-Assisted Intervention (pp. 546-554). Springer, Cham.

[4] Malkov, S., Shepherd, J. A., Scott, C. G., Tamimi, R. M., Ma, L., Bertrand, K. A., ... & Norman, A. (2016). Mammographic texture and risk of breast cancer by tumor type and estrogen receptor status. *Breast Cancer Research*, *18*(1), 122.

[5] Shi, X., Cheng, H., & Hu, L. (2006, October). Mass detection and classification in breast ultrasound images using fuzzy SVM. In *9th Joint International Conference on Information Sciences (JCIS-06)*. Atlantis Press.

[6] Ge, J., Sahiner, B., Hadjiiski, L. M., Chan, H. P., Wei, J., Helvie, M. A., & Zhou, C. (2006). Computer aided detection of clusters of microcalcifications on full field digital mammograms. *Medical Physics*, *33*(8), 2975-2988.

[7] Wu, Y. T., Wei, J., Hadjiiski, L. M., Sahiner, B., Zhou, C., Ge, J., ... & Chan, H. P. (2007). Bilateral analysis based false positive reduction for computer aided mass detection. *Medical physics*, *34*(8), 3334-3344.

[8] Verma, G., Luciani, M. L., Palombo, A., Metaxa, L., Panzironi, G., Pediconi, F., ... & Todde, V. (2018). Microcalcification morphological descriptors and parenchyma fractal dimension hierarchically interact in breast cancer: A diagnostic perspective. *Computers in biology and medicine*, *93*, 1-6.

[9] Patrício, M., Pereira, J., Crisóstomo, J., Matafome, P., Gomes, M., Seiça, R., & Caramelo, F. (2018). Using Resistin, glucose, age and BMI to predict the presence of breast cancer. *BMC cancer*, *18*(1), 29.

[10] Janowczyk, A., & Madabhushi, A. (2016). Deep learning for digital pathology image analysis: A comprehensive tutorial with selected use cases. *Journal of pathology informatics*, *7*.

Pulse electric field effect on cancer cell properties: Reversible electroporation study

Nur Adilah Abd Rahman[1], Muhammad Mahadi Abdul Jamil[1], Mohamad Nazib Adon[1] and Tengku Nadzlin Tengku Ibrahim[2]

[1]Biomedical Modelling and Simulation Research Group, Faculty of Electrical and Electronics Engineering, Universiti Tun Hussein Onn Malaysia, Batu Pahat, Johor.

[2]Center for diploma studies, Universiti Tun Hussein Onn Malaysia, Hab Pendidikan Tinggi Pagoh, KM1, Jalan Panchor, Pagoh, Muar, Johor.

*drmuhammadmahadi@gmail.com

Abstract. Controlling cell function by using pulses of electrical fields have been used widely in chemotherapy application. The pulse electric field induced will create pores through the cell membrane and causes other substance around it to be absorbed into the cell. This method has led to a variety of medical applications, particularly in cell studies. In this preliminary, a range voltage of the pulse electric field used will be demonstrated. Theoretically, pulse electric field does give variety of effect on cancer cell. Which can either be a reversible electroporation or irreversible electroporation. The effect can either heal or kill the cells induced. Thus, will lead towards either cell cancer treatment factors plus the presence of electric field as the stimulator for aggressive adsorption of the anti-cancer agent into the cells or as enhancer towards wound healing applications. The outcome of this techniques will contribute towards understanding the cancer cell treatment factors which may lead towards a new method for cancer cell treatment or wound healing applications. This study will focus on applying pulse electric field on Breast cancer cell to investigate the effect of electroporation process. In addition, this study will discuss the effectiveness of low and high pulse electric field parameter, to look at anti-proliferation effect on the cancer cell towards cancer treatment applications.

1. Introduction

Electroporation or electro-permeabilization is a phenomenon where the lipid bilayer and proteins on cell membrane were momentarily disrupted and permeable to ions and macromolecules when it induced to short but high voltage field pulses [1–3]. There are two types of electroporation, which is reversible (temporal pores) and irreversible electroporation (permanent pores). Temporal open pores means that the cell survive for few microseconds after removing the electric field [8, 9]. While permanently open pores, can cause cell die due to lysis. Square wave electric field has shown to be effective in electroporation method to the cells [16]. It was introduced due to better control of electric field parameter due to the concept of capacitor discharge where the power switch is capable of fast switching and the voltage supply constantly charges the capacitor. The square wave pulse is produced by a partial discharge of a large capacitor, which requires the interruption of high currents against high voltages. Typically, voltage, pulse length, and number of pulses are all directly set on the instrument. Square wave pulses have well defined electric field amplitude, are both effective and also relatively mild to cells to yield higher viability. The membrane potential is the voltage between the external and internal surrounding of living cells. The typical value of this membrane potential is in the range of -30 to -90mV with respect to the exterior of the membrane [17].

In the cell membrane, there are protein channels, pores, and pumps present. The proteins are dependent on the transmembrane potential cause the closing and the opening of these channels. When an electric field induces, many protein channels that sensitive to voltage will open due to the lipid molecules change in orient resulted in creating hydrophilic pores [5]. The pores size could be varied dependent on time range from hundreds of microseconds to milliseconds and will stand through the duration of pulses, shorter pulse durations are thought to create much smaller pores that will allow ions but not large molecules to pass through [12]. But if those parameters exceed the optimal values it can lead to permanent pores (irreversible electroporation) occur on the cell membrane which can cause the cell to die because of lysis [10]. Electroporation is a simple technique that considered to be a non-invasive and nonchemical method [14]. Where this technique does not change the function of the target cell and the biological arrangement itself. The factors that effects the efficiency of this technique are electric field parameters, the chemical composition of media and cell characteristic. Among these factors, the most influential is electric field parameters which consist of amplitude, duration and number of the pulse. The optimal electric pulse parameters were selected based on the selected application [15]. There is an application that requires reversible such as foreign molecule insertion and some other application require irreversible electroporation such as tissue ablation. As for foreign molecule insertion, the parameter selection depends on the type of the molecule. As for small molecule, a repetitive of short pulses with a time duration in the range of micro to millisecond is sufficient. While for large molecule, longer pulses in the range of few millisecond or combination of higher voltage, short pulse duration and low voltage, the longer pulse duration is used. An opening pores, allowed transporting or introducing of therapeutic materials from small to large molecules [6]. Which can be applied in many medical applications. Current applications of this technique include transfer of genes and drug delivery into cells [11], food processing or bacterial decontamination in food technology, micro biotechnology and cancer treatment. It's depends on the parameter applied used. As elaborated, Electroporation method are still in research and investigations on various field and study on electroporation method were performed, focusing on the parameter used for various applications as shown on table 1.

Table 1. Previous study on electroporation technique with different parameter used

References	Cell Type	Electrical Parameter	Pulse Width	Pulse Number	Purpose
[19]	Endothelial cells	1000 V/cm	12.8 ms	1	Endothelial cell seeding of prosthetic surfaces
[20]	Liver (live rat)	50V	50 ms	8	Determine the optimum conditions for local gene therapy
[11]	Jurkat cells	~100kV/cm (100V/cell diameter)	60ns	1	Monitor change in membrane potential in response to nsPEF
[12]	HeLa	500 – 3kV/cm	30μs	1	Wound Healing application
[21]	HeLa	2kV/cm	30μs - 600μs	1	Wound healing applications and treatment applications
[22]	MCF7	1008 V and 1280 V	20 μs	1	Combination therapy for clinical applications
[18]	HT29	600V/cm	500μs	1	Wound healing application
[23]	hippocampal neurons	3 kV/cm (range: 1.9–4 kV/cm)	200ns	1	VGSC activation AP induction
[24]	HeLa	2kV/cm	30μs	1	Wound Healing Applications

2. Literature Review

2.1. Breast cancer cell

Breast cancer is one of the most common cancers develop among female in Malaysia with the highest percentage of patient who died because of it with 52% death among Malaysian people [25]. Breast cancer cell developed from breast tissue with the sign symptom of change in breast shape, an existence of lump in the breast, fluid coming from the nipple, dimpling of skin or red scaly patch of skin. The risk factor is female, obesity, lack of physical activities, drinking alcohol, hormone replacement therapy during monopose, ionizing radiation, and early age at first menstruation, having child late or not at all, older age and family history. It commonly develop in cells from lining of milk ducts and the lobules that supply the ducts with milk. The provided treatment is surgery, radiation therapy, chemotherapy, hormonal therapy and targeted therapy. Breast cancer cell is a metastatic disease, it can spread beyond the original organ such as bone, liver, lung and brain.

2.2. Electroporation (EP)

Electroporation is a method that usually used to apply electric field onto the cells. This phenomenon causes the cell to permeable to ions and macromolecules when inducing with short high voltage pulses [26]. This is due to the voltage breakdown happen on the cell membrane which causes the lipid bilayer to be folded, this phenomenon causes an open pathway on the cell membrane which called an opening pore on the cell membrane. The opening pore on the cell membrane can be used for many applications such as introduction protein, large and small molecule and also cell fusion. A temporary open pore is called reversible electroporation. But if the voltage inducing is too high it can cause cell destruction, which called irreversible electroporation. The study presented in this paper focus on the reversible electroporation

2.2.1. Reversible Electroporation (RE)

Reversible electroporation is a temporal open pore in membranes and cell survival after the removal of the electric field. Reversible electroporation is primarily used for delivery of molecules into the cell. RE is often used to introduce substances into cells, such as dyes, drugs, protein and nucleic acids [27, 28]. Creating small pores in biological and artificial membranes can be done by applying an electric pulse of sufficient amplitude and duration. In the case of reversible, low-amplitude and short-duration pulses is sufficient to produce the pore on the cell membrane and the pores will close within milliseconds to minutes. Which safer in the inducement of the electrical field. This type of electroporation was chosen to see the cell response of breast cancer cell with the reversible range of electroporation method.

3. Material and Methods

3.1. Breast cancer cell (MCF7) culture

Breast cancer cell line is an immortal cell line that mostly used for scientific research. It is the oldest and most frequently used human cell line. The main advantage of this cell is as long as the basic cell survival conditions are met, it can be divided into an unlimited number of times. Therefore, breast cancer cell was used as the primary cell type. These breast cancer cell samples were acquired from the animal laboratory cell cultures (Kulliyyah of Allied Health Sciences, IIUM).

3.2. Experimental setup for electroporation method

The equipment used for electroporation system are consist of High Voltage Pulse Generator (ECM 830) and Nikon inverted microscope TS100 that connected with Dino Camera and Dino capture 2.0 Software. The cell were suspended inside the 4 mm gap size of cuvette and attach to the safety stand that connected to ECM 830. While the parameter that will be set into the pulse generator was 100 – 1000 V/cm (with 100V/cm interval) of electric field intensity, 30 μs of pulse duration and a single number of pulse.

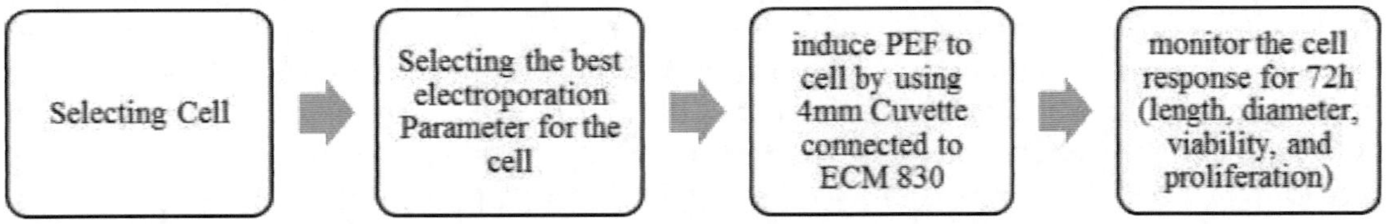

Figure 1. Flow of work of these study

4. Preliminary Result

4.1. Cell Culture Method

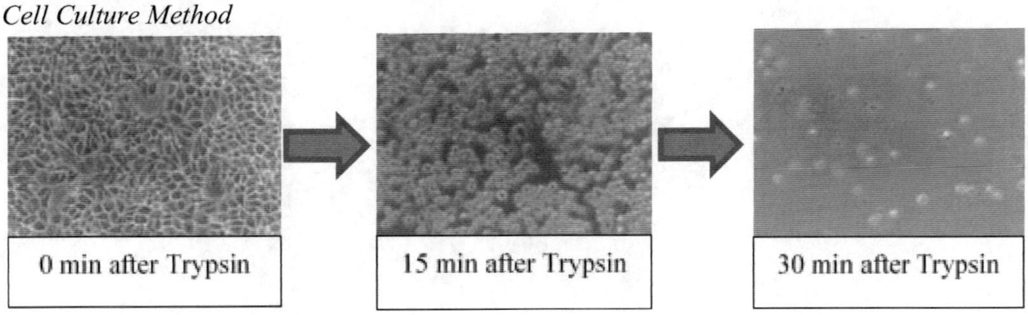

Figure 2. Process of Trypsinization of MCF7

These are the preliminary result for this study. Where the Procedure of splitting cell were done. In order to do the splitting procedure, Cancer cells must be 80 – 90% of confluence in 25 cm^2 flask as shown in figure 2. In order to split the cells, first cancer cells were washed once with 5 ml of PBS and aspirated the PBS. The 2 ml of trypsin was added enough to cover the surface of the 25 cm2 flask, in order to detach the cancer cells from the surface of the 25 cm2 flask. The flask was sealed and incubated for 30 minutes. After the cells fully detached, the flask examined under a high-resolution microscope, a fully trypsinized cell appeared rounded and no longer attached to the surface of flask as shown in Figure 2. After the cells were trypsinized, 2 ml of RPMI media was added to neutralize the trypsin inside the flask. After that, 0.5 – 1 ml of harvested cells were taken using pipette and seeded into a new 25 cm^2 flask that contained 5 ml of new media RPMI + 10% FBS + 1% Pen/Strep. The flask were sealed and incubated until they reached 80 – 90% confluency or 5 – 6 days for the next splitting process.

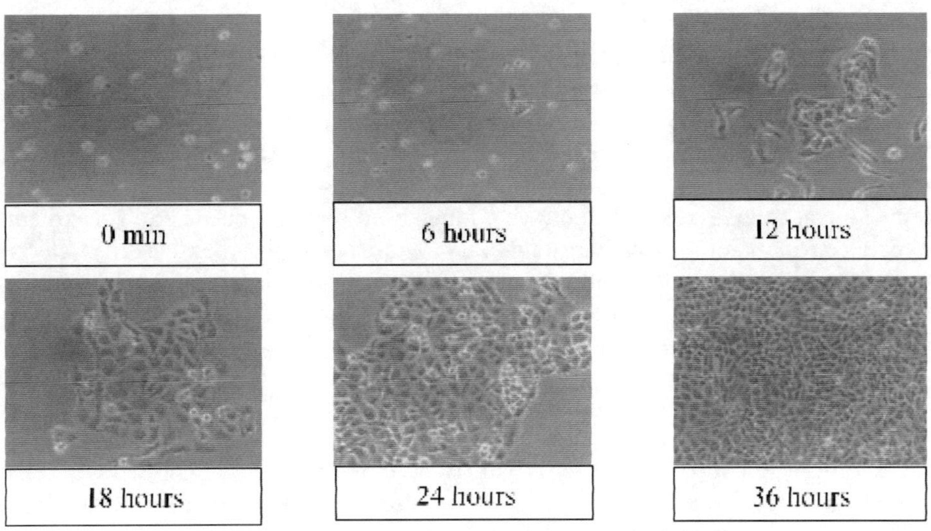

Figure 3. Cancer cell proliferation rate in 36 hours

Figure 3 shows how cancer cell grew in 36 hours. In order to maintain a controlled cells culture method, reagents protocol and sterile work area are important factor that influenced the growth rate of cancer cells. The same splitting technique was applied in every experimental in order to get the consistency of morphology and proliferation rates of cancer cells when exposed to high pulse electric field. These cells will be tested with the series of low electric field range which is 100 – 1000 V/cm and pulse duration of 10, 30 and 100 μs with single number of pulse. In order to look at the pulse electric field effect on the cell properties which is anti-proliferation, cell viability and cell length towards cancer treatment application.

4.2 Pulse Electric Effect on cell length

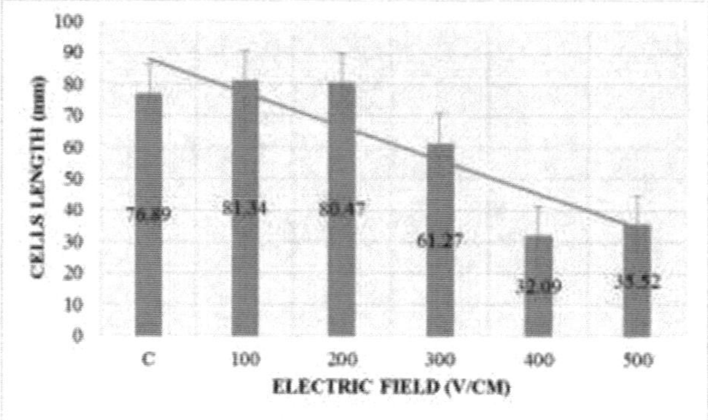

Figure 4. Low range pulse electric field effect on cell length

The preliminary result for this study as shown on the graph in figure 4. It show the cell length after induce with several low ranges of pulse electric field compared with the untreated cells. As shown above, it shows that cell well expands when induced with 100 and 200 V/cm with 81.34±1.36 mm and 80.47±1.15 mm and cell are shorter on cell length when inducing with 400 V/cm with a mean of 32.09±0.60 mm of cell length. This result showed an effect of pulsed electric field on the cancer cell. Which the result is shown and expansion of cell length after 48 hours incubated. The result showed a reversible electroporation effect where the cell survives after the pulse electric field inducement and continue to grow and shown a variance of cell length. The importance of investigating the cell length is due to the greater of cell length could contribute on the enhancement of cell to covering or fulfilling's the wound area for the faster wound healing process [8].

5. Conclusion

In conclusion, Electroporation method is a convenient and non-invasive technique that does not require any chemical method in applying to the cells. It is a phenomenon where the cell permeable to any surrounding molecule due to an opening pore that causes by the voltage breakdown on the cell membrane. These happened due to the electric field effect causes the lipid molecule to change orientation and resulted in creating hydrophilic pores. The pulse electric field inducement influence the cell morphological changes dependent on what range used for the inducement. As explained before there is 2 type electroporation, reversible (temporary) and irreversible (permanently) electroporation which can be used either killing or healing the cell. This study will be more focused on the reversible electroporation which is temporal pores occur on cell membrane and cell survive for few microseconds after removing the electric field. The reasons for choosing reversible electroporation is due to the temporary open pore and the cells survive after the inducement. Thus, it may not harm the normal cell surround it and can be used to enhance the adsorption of the anti-cancer agent into the cancer cells without harming the other normal cell. The aim is to test the low range of pulse electric field which is starting from 100 – 1000 V/cm (with 100 V/cm interval) on breast cancer cells, monitor the cell responds and find the best parameter of pulse electric field inducement, which may enhance the anti-proliferative of the cancer cell. The effects of this range of pulse electric field will be monitored by examining proliferative properties and cell viability of the cells. Thus, the optimum parameter of pulse electric field identified may have great implication for future biomedical applications such as cancer treatment or wound heling applications.

Acknowledgments

Authors wishing to acknowledge the Ministry of Education Malaysia for financial support of this work through Geran Penyelidikan Pascasiswazah (GPPS), VOT U949, Universiti Tun Hussein Onn Malaysia.

References

[1] Andreas Blicher 2011 Electrical aspects of lipid membranes (University of Copenhagen)
[2] Abd Rahman N A, Buhari M H and Jamil M A 2015 an Overview: Investigation Of Electroporation Technique On Cell Properties Cultured On Micropatterned Surface J. Teknol. **6** 61–5
[3] Abd Rahman N A 2017 Fundamental study of pulse electric field effects on HeLa cell cultured over extracellular matrix protein micropatterned surface (Universiti Tun Hussein Onn Malaysia)
[4] Cooper G M 2000 Cell Membranes The Cell: A Moleculer Approach (Sunderland: Sinauer Associates)
[5] Tarek M 2005 Membrane electroporation: a molecular dynamics simulation. Biophys. J. **88** 4045–53
[6] Casciola M and Tarek M 2016 A molecular insight into the electro-transfer of small molecules through electropores driven by electric fields Biochim. Biophys. Acta - Biomembr. **1858** 2278–89
[7] Buchmann L, Frey W, Gusbeth C, Ravaynia P S and Mathys A 2019 Effect of nanosecond pulsed electric field treatment on cell proliferation of microalgae Bioresour. Technol. **271** 402–8
[8] Subhra Santra T, Wang P-C and Gang Tseng F 2013 Electroporation Based Drug Delivery and Its Applications
[9] Jiang C, Davalos R V. and Bischof J C 2015 A review of basic to clinical studies of irreversible electroporation therapy IEEE Trans. Biomed. Eng. **62** 4–20.
[10] Tarek M 2005 Membrane Electroporation: A Molecular Dynamics Simulation Biophys. J. **88** 4045–53.
[11] Frey W, White J A, Price R O, Blackmore P F, Joshi R P, Nuccitelli R, Beebe S J, Schoenbach K H and Kolb J F 2006 Plasma membrane voltage changes during nanosecond pulsed electric field exposure Biophys. J. **90** 3608–15
[12] Adon M N 2015 Pulse Electric Field Exposure Effect on Morphological (Universiti Tun Hussein Onn Malaysia)
[13] Warindi, Hadi S P, Berahim H, Suharyanto, Raz-prag D, et. al. 2017 Effect of cell electroporation on the conductivity of a cell suspension Biophys. J. 1828 14012
[14] Meacham J M, Durvasula K, Levent Degertekin F and Fedorov A G 2014 Physical Methods for Intracellular Delivery: Practical Aspects from Laboratory Use to Industrial-Scale Processing HHS Public Access J Lab Autom **19** 1–18
[15] Sadiq A A, Zaltum M A M, Mamman H B, Adon M N, Othman N B, Dalimin M N and Jamil M M A 2015 An overview: Investigation of electroporation and sonoporation techniques Proc. - 2015 2nd Int. Conf. Biomed. Eng. ICoBE 2015
[16] Jordan D W, Gilgenbach R M, Uhler M D, Gates L H and Lau Y Y 2004 Effect of pulsed, high-power radiofrequency radiation on electroporation of mammalian cells IEEE Trans. Plasma Sci. **32** 1573–8
[17] Hanna H, Denzi A, Liberti M, André F M and Mir L M 2017 Electropermeabilization of Inner and Outer Cell Membranes with Microsecond Pulsed Electric Fields: Quantitative Study with Calcium Ions Sci. Rep. **7**
[18] Buhari M H 2017 Optimization of low aplitude pulse electric field exposure on colon cancer cell for wound healing application (Universiti Tun Hussein Onn Malaysia)
[19] Kotnis R A, Thompson M M, Eady S L, Budd J S, James R F L and Bell P R F 1995 Attachment, replication and thrombogenicity of genetically modified endothelial cells Eur. J. Vasc. Endovasc. Surg. **9** 335–40
[20] Suzuki T, Shin B C, Fujikura K, Matsuzaki T and Takata K 1998 Direct gene transfer into rat liver cells by in vivo electroporation FEBS Lett. **425** 436–40
[21] Milad Zaltum M A, Adon M N, Hamdan S, Dalimin M N and Abdul Jamil M M 2015 Investigation a critical selection of pulse duration effect on growth rate of HeLa cells 2015 Int. Conf. BioSignal Anal. Process. Syst. ICBAPS 2015 33–6
[22] Sree V G, Muthuraman C and Sundararajan R 2015 Anti-proliferation control of breast cancer cells using electric pulses Proc. IEEE Int. Conf. Prop. Appl. Dielectr. Mater. 2015-Octob 160–3

[23] Pakhomov A G, Semenov I, Casciola M and Xiao S 2017 Neuronal excitation and permeabilization by 200-ns pulsed electric field: An optical membrane potential study with FluoVolt dye Biochim. Biophys. Acta - Biomembr. **1859** 1273–81

[24] Rahman N A A and Jamil M M A 2017 Investigation of pulsed electric field on cancer cell cultured on patterned surface Proc. - 6th IEEE Int. Conf. Control Syst. Comput. Eng. ICCSCE 2016 25–7

[25] Yong Chee Meng 2018 The latest treatments for advanced ovarian cancer Star2 Health

[26] Neumann E, Sowers A E and Jordan C A 1989 Electroporation and Electrofusion in Cell Biology (Boston, MA: Springer US)

[27] Garcia P A, Rossmeisl J H and Davalos R V. 2011 Electrical conductivity changes during irreversible electroporation treatment of brain cancer 2011 Annual International Conference of the IEEE Engineering in Medicine and Biology Society (IEEE) pp 739–42

[28] Sundararajan R, Salameh T, Camarillo I and Campana L 2011 Effective use of electrical pulses on cancer cells to control proliferation 2011 - 14th International Symposium on Electrets (IEEE) pp 231–2.

PCA-Polynomial-ELM Model Optimal for Detection of NS1 Adulterated Salivary SERS Spectra

Nur Hainani Othman[1], Khuan Yoot Lee[1, 2, *], Afaf Rozan Mohd Radzol[2,3], Wahidah Mansor[1, 2], Nur Amanina Yusoff [1]

[1]Faculty of Electrical Engineering, Universiti Teknologi MARA, Shah Alam, Selangor DE, Malaysia
[2]Computational Intelligence Detection RIG, Pharmaceutical & Lifesciences Communities of Research, Universiti Teknologi MARA, Shah Alam, Selangor DE, Malaysia
[3]Faculty of Electrical Engineering, Universiti Teknologi MARA, Cawangan Pulau Pinang, Permatang Pauh, Pulau Pinang, Malaysia

*leeyootkhuan@uitm.edu.my

Abstract. Of recent, there has been much interest in the application of Surface Enhance Raman Spectroscopy (SERS) analysis in the detection of diseases such as dengue. Early diagnosis of dengue affords early intervention, greater chance of cure and prevention of mild dengue progressing into life threatening stage. SERS produces, on the interaction of photons from laser beam with saliva samples, a spectral image of its composition here. In the case of dengue fever, Non–Structural Protein 1 (NS1), being its biomarker, is the biochemical fingerprint to be revealed by SERS. NS1 presents in body fluid such as blood and saliva of patients since day one of infection, that makes NS1 a favourite alternative to antibody types of biomarker. However, the concentration of NS1 in saliva is low, yielding a low intensity SERS spectrum. In addition, the spectrum is usually interfered with undesirable noisy features. Extreme Learning Machine (ELM) is a fast algorithm with its strength in data pattern generalization. It has been applied in pattern recognition and machine learning for classification and regression, with encouraging performance. Our work here intends to determine an optimal polynomial-ELM model in classifying SERS spectra of saliva samples adulterated with NS1, amongst the different models subject to three different termination criteria of Principal Component Analysis (PCA). Performance of '100%' is attained for accuracy, sensitivity, specificity and precision, while '1' for kappa, by combining the cumulative percent of total variance (CPV) termination criterion and polynomial-ELM model of power 2 and constant 0.5.

1. Introduction

Dengue fever (DF) is a viral disease spread by female Aedes mosquitoes. When a mosquito bites a person carrying dengue virus in their blood, the mosquito becomes infected and transmits the virus to another person. Dengue virus is one of the flavivirus genome, a member of the Flaviviridae family [1]. It has three structural proteins and seven non-structural proteins, in which NS1 is one of the seven. NS1 plays an important role in viral replication and is essential to the existence of the virus [2]. The advantage of using antigen such as NS1 is that, detection at early stage of DF, without having to wait for the antibodies to form, is possible. DF is the most common mosquito-borne disease and a major health concern in tropic and subtropical regions. According to World Health Organization, more than 40% of the world's population, representing 2.5 billion people, are at risk of dengue infection [3]. There are several methods to detect NS1 such as Eenzyme-Linked Immunosorbent Assay (ELISA), Reverse Transcription Polymerase Chain Reaction (RT-PCR), with blood as the diagnostic medium. As such, they are prone to blood-related infections [4]. Detection of NS1 is one of the many techniques to diagnose dengue virus, such as virus isolation, detection of Ribonucleic acid (RNA), immunohistochemistry, Immunoglobulin M (IgM), Immunoglobulin G (IgG) [5].

Raman spectroscopy (RS) yields information on atomic vibration, which allows test recognizable proof and quantization based on inelastic scattering of monochromatic light [6]. Signals produced by RS are so weak that limit its application till the discovery of SERS. SERS provides greatly enhanced RS signals from Raman active analyte that have been adsorbed onto nano-metal surfaces [7]. The intensity of RS signals has been observed to increase by 10^4 at the least, to as high as 10^{14} [6][8]. The electromagnetic and chemical mechanism contribute to two-third of the enhancement in intensity [7]. The electromagnetic enhancement owes to the roughness of nano-metal surface, while the chemical enhancement is a cause effect of adsorbate electronic states as a result of chemisorption of the analyte [7]. The discovery of SERS encourages RS to be broadly acknowledged as potential diagnostic method for a wide-ranging applications such as, detection of viruses in sputum [9], diabetes in serum [10], melamine in milk [11] and gastric cancer in blood plasma [12]. Applications of RS for detection of DF from NS1 have been reported, yet using blood serum as the diagnostic medium [13][14].

In this paper, the application of SERS for biochemical analysis of saliva samples for detection and classification of NS1-DF is explored. PCA and polynomial-ELM are used to extract features and identify salivary SERS spectra adulterated with NS1. To the best of our knowledge, besides our research group, this is the first report on SERS analysis of saliva samples to reveal NS1-DF for purpose of detection and classification. Section 2 presents theory on PCA and polynomial-ELM classifier. Section 3 elaborates on the NS1 adulterated dataset and methodology employed in this study. Section 4 discusses results from classification of NS1 salivary SERS spectra with different PCA termination criteria and polynomial-ELM vector parameters.

2. Theoretical Background

2.1. Principal Component Analysis, PCA

PCA is probably the oldest multivariate analysis technique introduced by Pearson (1901) [15]. This technique is popularly in use after the introduction of electronic computer. Currently, its algorithm resides in almost every statistical computer package [15].

Being a data reduction technique, PCA reduces the dimensionality of the variables in a large dataset, yet maintains the important features as much as possible [15][16]. In other words, with the use of mathematical procedure, PCA reduces high dimensional data to low dimensional data by transforming the correlated variables into a smaller number of uncorrelated variables, known as principal components (PCs) [17]. Eigenvalue of dataset is a measure of variability in the sample distribution that reflects the variance. PCs are ranked according to variance of datasets. The first place is occupied by PC with the largest variance, while the last place by PC with the least variance. The formula to measure the variability for average squared deviation of each sample from its mean is as follow [16],

$$\text{Variances, } s^2 = \frac{\sum(sample - mean)^2}{Total\ sample - 1} \tag{1}$$

To facilitate data analysis and interpretation of the original data, only significant PCs are selected after ranking using PCA termination criteria. The three termination criteria considered in this study are Eigenvalue-One Criterion (EOC), Cattell's Scree test and Cumulative Percent of Variance (CPV) [18].

2.2. Extreme Learning Machine, ELM

ELM is first developed by Huang in 2004 and published in 2006. It consists of a three layer neural network with sigmoid activation function [19]. It is suitable for implementation on a single-hidden layer feedforward neural networks (SLFNs) [20]. ELM is a learning algorithm well-known for high learning speed, good pattern generalization, least of human intervention and ease of implementation. It not only delivers smaller training error but also better performance.[19][20]. The advantages of ELM to conventional gradient based learning methods are, (i) Stopping criteria, learning rates, learning epochs and local minima are auto-optimized; (ii) Learning algorithm is fast, adopting one pass without reiteration. The learning phase typically completes in seconds or less. It is also suitable for all activation functions [19] [20].

The polynomial kernel is a non-stationary kernel. It is well suited for problems where all the training data are normalized. Polynomial is commonly defined in mathematical form as follow [20]:

$$k(u,v) = (u.v + n)^{power} \quad (2)$$

where n is constant and *power* is the power for the function, while u and v are vectors in the input space computed from training or test samples. The polynomial power starts from two, according to the polynomial function principle. The polynomial *constant* and polynomial *power* are varied to obtain optimized vector parameters.

3. Methodology

Figure 1 illustrates steps in optimizing the selection of PCA termination criteria [Scree, CPV, EOC] and parameters of polynomial-ELM classifier [constant, power], for use with SERS spectra of NS1 adulterated saliva.

The algorithm consists of two parts: feature extraction by PCA and feature classification by polynomial-ELM. SERS spectra of 64 samples each from the control group and the NS1 adulterated group, are acquired from the UiTM-NMRR-12868-NS1-DENV database. The control group data are contributed by healthy volunteers between 23-24 years old, while the NS1 adulterated data are acquired by mixing the control group saliva with pure NS1 protein at different concentrations [2]. Prior to feature extraction and feature classification, the spectra are pre-processed to eliminate spurious noise, background noise, fluorescent effect, signal drifts and spikes from cosmic ray [6][21].

PCA with different termination criteria, i.e. Scree test [128 x 5], CPV [128 x 70] and EOC [128 x 115] are served as inputs to the polynomial-ELM classifiers. Then, polynomial constant n and polynomial *power,* as defined in Equation (2), are varied and their effects recorded. In the first case, the polynomial constant n is varied, as the polynomial *power* is kept constant; vice versa in the second case. The range of variation for polynomial constant n is from 0.000001 to 10000, while that for polynomial *power* is from 2 to 50. For both cases, performance of the classifier models is analysed using indicators such as accuracy, precision, sensitivity, specificity and kappa value. Based on this, the best classifier model is determined.

4. Result and Discussion

This section presents results from classification of NS1 adulterated salivary SERS spectra by using PCA termination criteria integrated with polynomial kernel of ELM classifier. Altogether 3381 polynomial-ELM classifier models, of 69 different polynomial n-values and 49 polynomial *power*-values, are assessed.

With reference to Figure 2, Figure 3 and Figure 4, the best performance in [Accuracy, Precision, Kappa] is [97.37%, 94.74%, 0.95], [100%, 100%, 1.0] and [100%, 100%, 1.0] for Scree-, CPV- and EOC-Polynomial-ELM classifier model with [polynomial constant, polynomial power] of [0.000001, 9], [0.5, 2] and [0.5, 2] respectively. Figure 5 shows the ROC graphs for these best performed classifiers. The nearest point with the straight line indicates the best performance of the ROC graph. CPV and EOC-polynomial-ELM classifier models are observed with the best sensitivity and specificity, which lies at the upper left corner of the ROC graph, with coordinate [0,1].

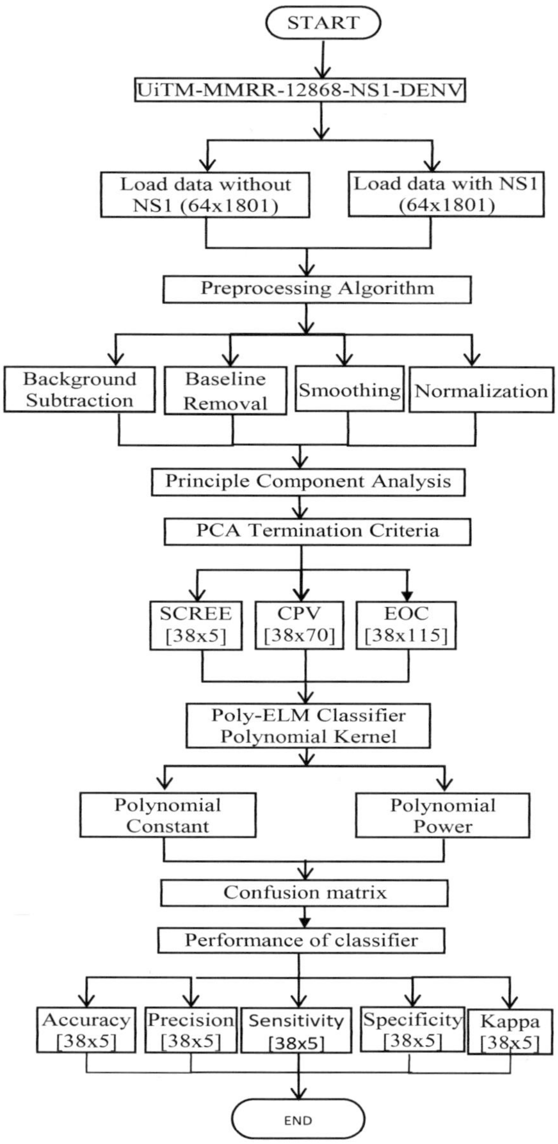

Figure 1. Procedural flowchart to implement the PCA-Polynomial-ELM classifier

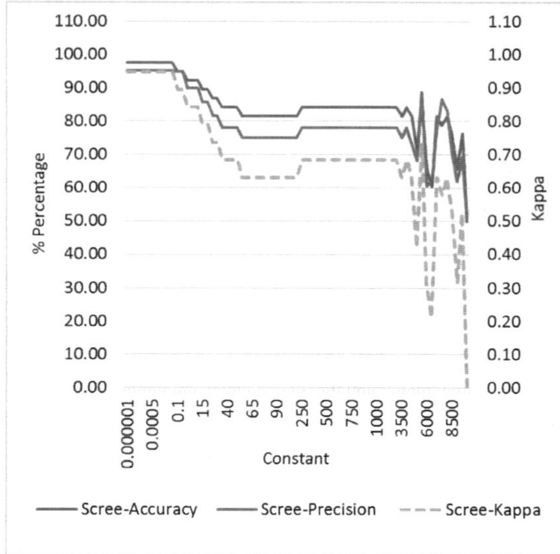

Figure 2. Optimal Polynomial-ELM classifier for Scree termination criterion

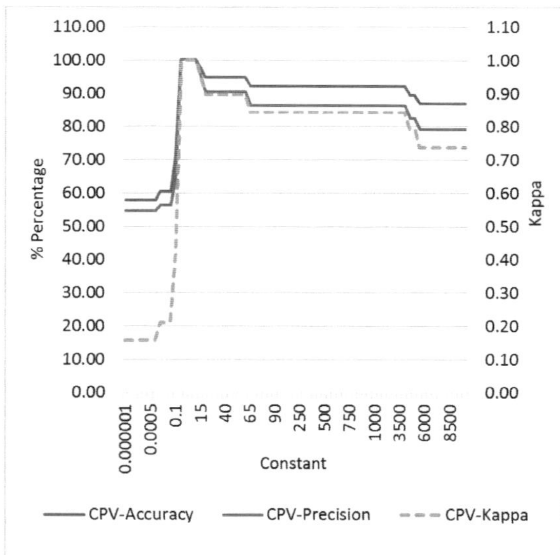

Figure 3. Optimal Polynomial-ELM classifier for CPV termination criterion

As observed from Figure 2, Figure 3 and Figure 4, trending in the accuracy performance is tracked by kappa and precision. When accuracy trends up, the kappa value and precision increases, and conversely. This is because the kappa algorithm is dependent on accuracy. The kappa algorithm compares observational probability of agreement, which is the classifier accuracy, to the hypothetical expected probability of agreement under an appropriate set of baseline constraints such as, total independence of observer classification, which is also known as chance agreement [22]. Theoretically, the precision algorithm does not depend on accuracy. The classifier can be precise but not accurate, and vice versa. However, in this study, the classifier is found precise when it is accurate; imprecise when it is inaccurate, which just shows that the classifier is reliable.

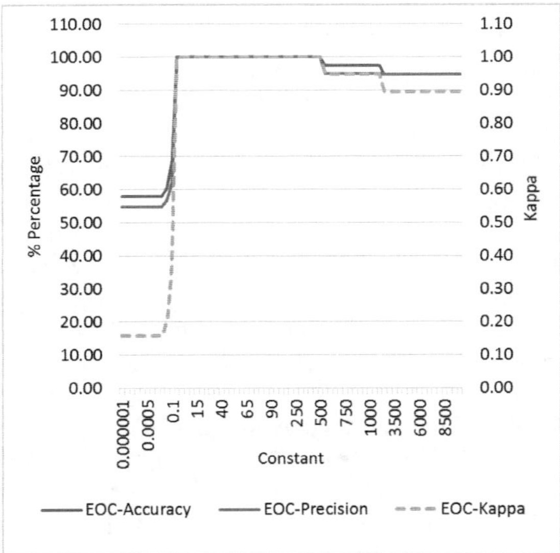

Figure 4. Optimal Polynomial-ELM classifier performance for EOC termination criteria

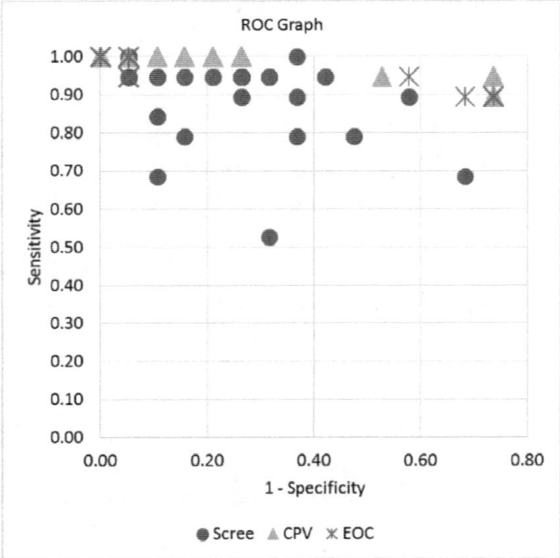

Figure 5. ROC of Polynomial-ELM classifier for different PCA termination criteria

5. Conclusion

Here, the effect of PCA termination criteria and polynomial-ELM parameters in identifying NS1 from SERS spectra of saliva are investigated. As far as our knowledge, research from our group has been the first to investigate the presence of salivary NS1 with SERS, which offers favorable circumstances such as, early, non-invasive and simple sample collection and sample preparation. This is intended as an alternative technique in distinguishing NS1 from saliva, which will profit diagnosis of NS1-DF related infections in the future. For our study here, of the PCA termination criteria, the CPV and EOC criteria which transform the original signal into 70 PCs and 115 PCs respectively, reported the best performance with '1' for kappa; '100%' for accuracy, precision, sensitivity and specificity and sensitivity-specificity pair at coordinate [0,1] of the ROC graph. With further consideration to the number of PCs, CPV uses less PCs in comparison to EOC, which could reduce computational time and load. Even though both CPV and EOC criteria yield similar performance, CPV integrated with polynomial-ELM classifier of polynomial constant 0.5 and power 2 is reckoned with the most optimized classification performance.

Acknowledgments

The author would like to thank the Research Management Institute, for providing the research funding 600-IRMI/REI 5/3 (02 1/2018); Faculty of Electrical Engineering, Universiti Teknologi MARA, Selangor, Malaysia, for the support and assistance given to the authors in carrying out this research.

References

[1] Chen H, Lai Y and Yeh T 2018 Dengue virus non-structural protein 1 : a pathogenic factor, therapeutic target and vaccine candidate *J. Biomed. Sci.* **25** 1–11

[2] Radzol A R M, Lee K Y and Mansor W 2014 Classification of salivary based NS1 from Raman Spectroscopy with support vector machine *Proc. Annual International Conference of the IEEE Engineering in Medicine and Biology Society* (Chicago, IL, USA) p 1835–1838

[3] Bhatt S et al 2013 The global distribution and burden of dengue *Nature* **496**, 504–507

[4] Radzol A R M, Lee K Y and Mansor W 2013 Nonstructural protein 1 characteristic peak from NS1-saliva mixture with Surface-Enhanced Raman spectroscopy *Annu. Int. Conf. IEEE Eng. Med.*

Biol. Soc. (Osaka, Japan) p 2396–2399

[5] Hussain A, Rehman S U, Aslam S, Javed N and Abbas Z 2014 Nano-gold particles mediated detection of NS1; an early diagnostic marker of dengue virus infection *J. Anim. Plant Sci.* **24** 1110–1115

[6] Radzol A R M, Lee K Y, Mansor W and Azman A 2015 Optimization of Savitzky-Golay smoothing filter for salivary surface enhanced Raman spectra of non structural protein 1 *Proc. IEEE Reg. 10 Annu. Int. Conf.* (Bangkok, Thailand) p 9–14

[7] Radzol A R M, Lee K Y, Mansor W and Yahaya S R 2012 Surface-enhanced raman spectral analysis of substrates for salivary based disease detection *Proc. - 2012 IEEE 8th Int. Colloq. Signal Process. Its Appl. CSPA 2012* (Melaka, Malaysia) p 505–509

[8] Fan M, Andrade G F S and Brolo A 2011 A review on the fabrication of substrates for surface enhanced Raman spectroscopy and their applications in analytical chemistry *Anal. Chim. Acta* **693** 7–25

[9] Wu X, Chen J, Li X, Zhao Y and Zughaier S M 2014 Culture-free diagnostics of Pseudomonas aeruginosa infection by silver nanorod array based SERS from clinical sputum samples *Nanomedicine Nanotechnology Biol. Med.*, **10** 1863–1870

[10] Han H W, Yan X L, Dong R X, Ban G and Li K 2009 Analysis of serum from type II diabetes mellitus and diabetic complication using surface-enhanced Raman spectra (SERS) *Appl. Phys. B Lasers Opt.* **94** 667–672

[11] Siak F, Lee K Y, Mansor W, and Radzol A R M 2012 Raman spectra of drop coating deposition Raman for melamine solution *IEEE-EMBS Conf. Biomed. Eng. Sci. IECBES* (Langkawi, Malaysia) p 815–819

[12] Feng S et al. 2010 Gastric cancer detection based on blood plasma surface-enhanced Raman spectroscopy excited by polarized laser light *Biosens. Bioelectron.* **26** 3167–3174

[13] Mahmood T et al. 2018 Raman spectral analysis for rapid screening of dengue infection *Spectrochim. Acta - Part A Mol. Biomol. Spectrosc.* **200** 136–142

[14] Bilal M et al 2016 Raman spectroscopy based discrimination of NS1 positive and negative dengue virus infected serum *Laser Phys. Lett.* **13** 1–7

[15] Jolliffe I T 2002 *Principal Component Analysis, Second Edition* (Spinger-Verlag New York)

[16] Radzol A R M, Lee K Y, Mansor W and Azmin N H R 2015 SVM-RBF Model PCA Criterion Selection for Detection of NS1 Molecule From Raman Spectra of Salivary Mixture *IET International Conference on Biomedical Image and Signal Processing* (Beijing, China) p 1–6

[17] Sahak R, Lee K Y, Mansor W, Yassin A I M and Zabidi A 2010 Detection of asphyxiated infant cry using support vector machine integrated with principal component analysis *Proceedings of 2010 IEEE EMBS Conference on Biomedical Engineering and Sciences* (Kuala Lumpur, Malaysia) p 485–488

[18] Radzol A R M, Lee K Y, Mansor W and Tawi F M T 2016 Signal Processing for Raman Spectra for Disease Detection *Int. J. Pharm. Pharm. Sci.* **8** 4–10

[19] Ismaeel S, Miri A and Chourishi D 2015 Using the Extreme Learning Machine (ELM) Technique for Heart Disease Diagnosis *IEEE Canada International Humanitarian Technology Conference* (Ottawa, ON, Canada) p 1–3

[20] Zhou H, Lan Y, Soh Y C, Huang G B and Zhang R 2012 Credit risk evaluation with extreme learning machine *IEEE Int. Conf. Syst. Man, Cybern.*, (Seoul, South Korea) p 1064–1069

[21] Radzol A R M, Lee K Y, Mansor W and Saadun N 2014 Baseline Correction Customized for NS1 Salivary Raman Spectra with Piecewise Cubic Hermite Interpolation *IEEE Symp. Ind. Electron. Appl.* (Sabah, Malaysia) p 1–5

[22] Landis J R and Koch G G 1977 The Measurement of Observer Agreement for Categorical Data *Biometrics* **33** 159–174

Prediction of Alzheimer's disease (AD) Using Machine Learning Techniques with Boruta Algorithm as Feature Selection Method

Lee Kuok Leong[1] and Azian Azamimi Abdullah[1*]

[1] Biomedical Electronic Engineering Programme, School of Mechatronic Engineering, Universiti Malaysia Perlis (UniMAP), Pauh Putra Campus, 02600 Arau, Perlis Malaysia.

*azamimi@unimap.edu.my

Abstract. Alzheimer's disease (AD) is the most frequent incurable neurodegenerative disease, a general term for memory loss and other cognitive abilities. Early detection of AD can help with proper treatment and prevent brain tissue damage. Traditional medical tests are time consuming, fail to recognize early signs and lack of diagnosis sensitivity and specificity. To achieve promising prediction accuracy, the best predictive machine learning model is selected based on initial pre-processing step followed by vital attributes selection and performance evaluation for five proposed supervised machine learning algorithms. In the pre-processing, all the missing values have been removed since the overall percentage only covered 5.63%. Boruta algorithm as feature selection method resulted Atlas Scaling Factor, Estimated Total Intracranial Volume, Normalized Whole-brain Volume, Mini-Mental State Examination and Clinical Dementia Rating must be included as primary features. With Boruta algorithm, it has been shown that Random Forest Grid Search Cross Validation (RF GSCV) outperformed with 94.39% of accuracy, 88.24% sensitivity, 100.00% specificity and 94.44% AUC among other 12 models that includes conventional and fine-tuned models even for the small OASIS-2 longitudinal MRI dataset. Finally, our developed Graphical User Interface (GUI) prediction tool was evaluated through prediction over OASIS-1 cross-sectional MRI dataset containing 216 samples of imaging sessions that have been pre-processed. Prediction results were closed with the dementia status provided in OASIS cross-sectional data fact sheet.

1. Introduction

Alzheimer's disease (AD), is a type of dementia which is incurable and affecting millions of people worldwide [1], characterized as progressive neuropsychological disorder that have a dramatic consequences on thinking ability and behaviour functions which greatly impact the quality of patient's life with AD and undergo dramatic personality changes [2]. The current issue mainly highlighted that the detection of the AD is still not accurate until the patient reaches the sensible stage of AD. Thus, the accuracy of the early stage diagnosis will directly relate the validation of neuropathological, which should detect the fundamental characteristics neuropathology changes before to the onset of clinical symptoms where with the sensitivity and specificity of no less than 80% [3].

The relationship between combinations of features with the accuracy of prediction model have been concerned in this recent years specifically by Schouten et al. [4], presented a combination of different neuropathological features to prove be a method for improving the model performance. Machine learning techniques have been extensively developed with the aim of automatically classifying AD, producing encouraging results (e.g., Sarraf et al. [5]; Lebedev et al. [6]; Deepika Bansal et al. [7]; Amoroso et al. [8]). This paper mainly focused on detailed study of binary classification (demented or non-demented) and diagnosis of AD which includes initial pre-processing step and conduct of the imperative attributes selection by Boruta algorithm, evaluation of different classifiers and decide the best classifier for prediction on Open Access Series of Imaging Studies (OASIS) imaging datasets. Finally, a high degree of accuracy prediction tool that can help clinicians to evaluate and classify AD with normal individuals in early stage using MRI features based data.

2. Methodology

An overview of the research workflow for this project is shown in Figure 1 below.

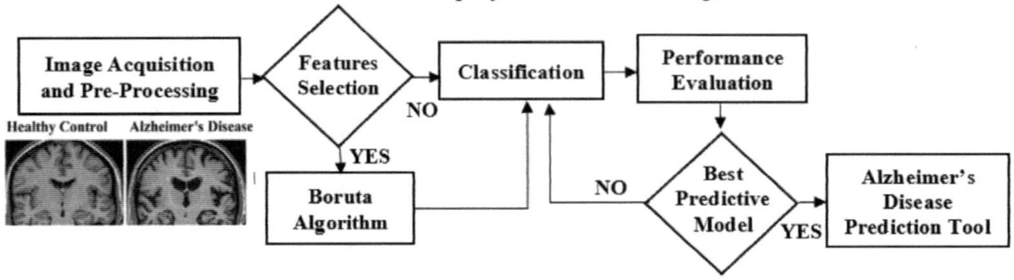

Figure 1. Research workflow

2.1. Data Acquisition

The MRI related data was generated by the Open Access Series of Imaging Studies (OASIS) project. The released data for OASIS-1 cross sectional and OASIS-2 longitudinal MRI datasets were utilized for training and testing with various machine learning models. The subjects from longitudinal dataset include both men (n=62) and women (n=88) with all right-handed dominance and for each subject, at least three or four acquired T1- weighted MRI scan brain imaging images took in single MRI scan session. For a total of 373 imaging sessions in the longitudinal dataset, each subject was scanned on two or more separate instances with separated by at least one year, with an average delay of 719 days (standard delay period within 183 to 1707 days) between visits. Table 1 shows all the features that included in the longitudinal data set.

Table 1. Feature measures counted in the dataset

Age	Age at time of image acquisition (years).
MR Delay	Delayed T1 post contrast
Gender	Gender (M or F)
Education	Years of education
SES	Socioeconomic status classified into categories from 1 (highest status) to 5 (lowest status)
MMSE	Mini-Mental State Examination score (range is from 0 [worst] to 30 [best])
CDR	Clinical Dementia Rating. (0 = no dementia, 0.5 = very mild AD, 1 = mild AD, 2. = moderate AD)
ASF	Atlas scaling factor (unit less). Computed scaling factor that transforms native-space brain and skull to the atlas target.
eTIV	Estimated total intracranial volume (cm^3)
nWBV	Normalized whole brain volume, using automated tissue segmentation process to expressed as a percent of all voxels in the atlas-masked image that are labelled as grey or white matter

2.2. Data Pre-processing

It can be summarized that there are few preferable alternatives that can easily handle the missing values such as (i) remove the data row or column, (ii) use a global constant to fill in for missing values and (iii) use feature mean (or median/ mode/ mean) for all samples fitting to the same class depending on the percentages of missing values in the datasets.

2.3. Boruta Algorithm as Feature Selection Method

2.4. In this paper, Boruta algorithm [9] is used as feature selection method, which is able to select optimal features carrying usable information for model's prediction. Boruta algorithm basically works with Scikit- learn dependency which act as interface medium on Random Forest (RF) classifier which utilizes the importance measure generated by the original algorithm. It is based on the implementation of Python's random forest which runs on a single code to identify all the importance features in the dataset with respect to an outcome variable (categorization of demented or non-demented). In this paper, number of estimators is set to be 'auto' since Boruta algorithm in Python offers automatic number of estimator's selection instead of finding a possibly compact subset of features in dataset since it's applicable to all type of features. To achieve the minimal optimal set of features, Boruta algorithm is used with the following non-default parameters: depth of each tree in the forest is set be 5 since recommended depth in Boruta algorithm is using pruned trees with a depth between 3 to 7 and verbose is set be 2 to controls the verbosity of output. Other than that, parameters such as percentile (perc), alpha, two step correction for multiple testing (two_step) and maximum iterations (max_iter) are set be default where perc=100, alpha=0.05, two_step=True and max_iter=100. However, for the attribute's decision, support and ranking parameters are executing to mask the confirmed selected features as True and assign the features ranking correspond to the ranking position of ith of features (e.g. the best features assign as rank 1 and tentative features as rank 2 etc) respectively. The coded Boruta programme with pre- processed OASIS-2 longitudinal dataset is running over Intel Core i5-7200U 2.5GHz with Turbo Boost up to 3.1GHz and NVIDIA GeForce 940MX with 2GB dedicated VRAM installed with 120GB 530MB/s Solid State Drive.

In conclusion, Boruta algorithm approach is highly recommended for the exploration of high-dimensional data. The formula for model training before the feature's selection is: Group ~ Gender + Age + Visit + Hand + Educ + MR Delay + SES + CDR + MMSE + eTIV + nWBV.

2.5. Supervised Machine Learning Method

2.5.1. Deep Neural Network (DNN)

DNN model is created using Tensorflow. The model will have 11 and 5 fully connected hidden layer with the same number of neurons as input variables for two models, without feature selection (11 features) and with Boruta feature selection (5 features). Thus, an input data is being trained by DNN with difference hidden layer and activation function. The proposed of two DNN models are trained over 100 training epochs with batch size of 10. All the parameters to sett up DNN models are summarized as shown in Table 2. Finally, binary cross-entropy (logarithmic loss function) is used during training, the ideal loss function for binary AD classification.

Table 2. Proposed default estimator for each Deep Neural Network

	DNN	
Feature Selection	With Boruta algorithm	Without Boruta algorithm
Activation Function	Input layer: ReLU Hidden layer: ReLU Output layer: Sigmoid	Input layer: ReLU Hidden layer: ReLU Output layer: Sigmoid
Hidden Layer	11	5
Hidden Neuron	None,11	None,5
Epochs	100	100
Input Dropout Ratio	0.1	0.1
Hidden Dropout Ratio	0.1	0.1

2.5.2. Random Forest (RF)

RF can work well with the Boruta algorithm, as selection is based on the feature's importance. Two types of cross validation techniques are used such as grid search cross validation (GSCV) and randomized search cross validation (RSCV) to search and obtain the best combination of hyperparameters to train RF models. The pre- defined arrays of ideal values for the grid parameters are set to search, specifically declared as Table 3.

Table 3. Set of hyperparameters for RF, RF GSCV and RF RSCV

Grid Parameters	RF (Default)	RF GSCV	RF RSCV
max_depth	None	[2, 4, 6, 8, 10, 12, 14]	[2, 4, 6, 9, 11, 14, None]
max_features	"Auto"	Without Boruta algorithm: [11] With Boruta algorithm: [5]	Without Boruta algorithm: [11] With Boruta algorithm: [5]
min_samples_split	2	[6, 8, 10,12]	[6, 8, 10, 12]
min_samples_leaf	1	[2, 4, 6, 8, 10]	[2, 4, 6, 8, 10]
bootstrap	True	[True, False]	[True, False]
criterion	"gini"	["gini", "entropy"]	["gini", "entropy"]
n_estimators	10	[1, 2, 4, 8, 16, 32, 64]	[1, 2, 4, 8, 16, 32, 64, 100, 200]

2.5.3. Gradient Boosting Machines (GBM)

The most crucial parameters used to define a tree such as max_depth and min_samples_split is tune and maintained the learning rate as default (learning rate= 0.1). Max_depth boundaries the maximum depth of the tree; deeper the tree, more data information is captured and min_samples_split sets the minimum number of samples to split. Table 4 below shows the initial estimates for boosting parameters and range of tree-based tuned parameters which will do a grid search.

Table 4. Initial estimates and range for tuning tree-based parameters

Tune Parameters	GBM (Default)	GBM PT
max_depth	3	[4, 6, 8, 10, 12, 14, 16]
min_samples_split	2	[200, 400, 600, 800, 1000]
Default estimators	• learning_rate= 0.1, • n_estimators= 100, • max_features= None, • subsample=1.0	• learning_rate= 0.1, • n_estimators= 100, • max_features= “auto”, • subsample=1.0

2.5.4. Support Vector Machine (SVM)

As shown in Table 5, three types of SVM models have been proposed with different parameters for linear kernel (LK), radial basis function (rbf) kernel (RBFK), GSCV method and set the C parameter at different values which suite for the function optimization. The GSCV model is proposed to explore combinations of parameters of kernel, C, gamma, decision_function_shape, degree, coefo and shrinking to minimize the misclassifications.

Table 5. Default parameters and tune parameter values of proposed SVM models

Grid Parameters	SVM LK	SVM RBFK	SVM GSCV
kernel	"linear"	"rbf"	['linear', 'poly', 'sigmoid']
C	1	1	[0.0001, 0.001, 0.01, 0.1, 1, 10]
gamma	'auto'	0.01	[0.0001,0.001, 0.01, 0.1, 1]
decision_function_shape	'ovr'	'ovr'	['ovo', 'ovr', None]
degree	3	3	[2, 4]
coefo	0	0	[0, 1, 2, 3]
shrinking	True	True	[True, False]

2.5.5. Logistics Regression (LR)

Two LR models were proposed where the first model is designed with default parameters as shown in Table 6. The main purpose to find the optimal C in the hyperparameter space is to control underfitting and overfitting of the model since large value of C can lead to overfitting And, the second model is mainly target on tuning of parameter C as the regularization parameter or regularization strength.

Table 6. Default parameters and tune parameter values of proposed LR models

Grid Parameters	LR (Default)	LR PT
C	1.0	[0.00001, 0.0001, 0.001, 0.1, 1, 10, 100]
Default estimators	• penalty='l2', • max_iter=100, • multi_class='ovr', • max_iter=100, • random_state=None	• n_splits=10, • shuffle=True, • random_state=42

2.6. Performance Evaluation and Comparison

The classification results were calculated by means of three metric measurements, which are used for quantitative valuation and evaluation, including accuracy, sensitivity (recall) and specificity. Additionally, numerous optimal approaches such as receiver operating curve (ROC) and Area Under the Curve (AUC) are calculated as well.

Fundamentally, the binary classification will produce four outcomes such as True positive (TP), true negative (TN), false positive (FP) and false negative (FN). Terms associated with confusion matrix:

1. True positive (TP): Correct positive prediction

The case where the patient is actually having dementia (1) and the prediction model classified the patient as demented (1) comes under True Positives.

2. True negative (TN): Correct negative prediction

The case where the patient is actually NOT having dementia (0) and the prediction model classified the patient as non-demented (0) comes under True Negatives.

3. False positive (FP): Incorrect positive prediction

The case where the patient is actually NOT having dementia (0) and the prediction model classified the patient as demented (1) comes under False Positives. False (model has predicted incorrectly) and Positive (class predicted was a positive (1))

4. False negative (FN): Incorrect negative prediction

The case where the patient is actually having dementia (1) and the prediction model classified the patient as non-demented (0) comes under False Negatives.

3. Results and Discussion

In this study, the missing values were removed by dropping the 19 rows and 2 rows with missing values in the column of SES and MMSE respectively since in overall, only 5.63% of the values in the dataset were missing. Since the prediction of dementia status are totally relies on complex neuroimaging data, thus replace missing values with sensible values (constant values or column' mean or median value) might affect the accuracy of the dementia classification.

With Boruta algorithm, first five with the highest importance features were selected such as MMSE, CDR, eTIV, nWBV and ASF for the training and testing model later. The formula for model training with the feature's selection is: Group ~ SES + CDR + MMSE + nWBV. With the implementation of Boruta algorithm in the proposed work, 12 proposed machine learning algorithms performed better which shown with higher accuracies, sensitivities, specificities and AUCs as compare that without the feature selection. This is a positive sign for Boruta algorithm, which is suitable for OASIS-2 longitudinal MRI dataset to provide a combination set of features as optimal features for classification. RF GSCV with features selection is showing the best classification performance with accuracy of 94.39%, 88.24% sensitivity, 100.00% specificity and 94.44% AUC but without applying attribute selection, RF GSCV algorithm only perform with accuracy of 91.59%, 85.42% sensitivity, 96.61% specificity and 92.21% AUC as shown in Table 7. High sensitivity and specificity can promise high true positive and true negative rates respectively. And, in case of neurodegenerative diseases it is important to have a high true positive rate (sensitivity) so that all patients with Alzheimer's can identified as early as possible. When compared the experimental results with other recent classification experiments (binary or multi-class classifications) done with OASIS datasets, the current experimental results have shown a good sign in terms of classification accuracy, especially proposed RF GSCV algorithm with 94.39% and 94.44% of accuracy and AUC respectively for binary AD/HC (demented/ non- demented) classification, based on MRI data [10]–[12].

Table 7. Best performance of proposed machine learning algorithms

Model	Feature Selection	ACC (%)	SEN (%)	SPE (%)	AUC (%)
RF GSCV	No	91.59	85.42	96.61	92.21
	Proposed method	**94.39**	**88.24**	**100.00**	**94.44**

RF GSCV model works best with OASIS-2 longitudinal dataset and the fine-tuned parameters are preserved for the predictive model in the GUI prediction tool. Knowing that the performance of best prediction algorithm is based on 94.39% of accuracy rather than percentage lower than 90%, thus it is persuasive that can give more evidence and confidence for making the AD classification diagnosis. Whereas, with the developed prediction tool, the classification results were presented with 83 (38.43%) predicted as demented samples, while 133 (61.57%) as non-demented samples. The predicted subjects with demented were closely to the statement in OASIS fact sheet (rev. 2007-8-20) provided by Marcus et al. [13]. Lastly, three of the subjects with different classes were randomly selected from the OASIS-2 longitudinal MRI dataset for the prediction validation purpose. Five input values (MMSE, nWBV, CDR, eTIV and ASF) were manually key-in to the entries of proposed GUI prediction tool and the predicted classes of the patient were correctly matched with the expected status of patient as shown in Table 8.

Table 8. Randomly selected patients' information from OASIS-2 longitudinal MRI dataset

Subject ID	MRI ID	Group	MMSE	nWBV	CDR	eTIV	ASF	Prediction
OAS2_0001	OAS2_0001_MR1	**Non-demented**	27	0.696	0	1987	0.883	**Non-demented**
OAS2_0002	OAS2_0002_MR2	**Demented**	23	0.736	0.5	1678	1.046	**Demented**
OAS2_0018	OAS2_0018_MR1	**Converted**	30	0.715	0	1406	1.248	**Demented**

4. Conclusion

This study is based on the comparison and evaluation of different machine learning algorithms with pre-processing and Boruta algorithm features selection in the prognosis and prediction of AD. In the proposed method, the missing values were cleansed by columns deletion in the column of SES and MMSE respectively. The experimental results validate that a combination of pre-processing and experimental results with Boruta features selection is a trustworthy technique for early prediction of AD. RF GSCV with Boruta algorithm is the best classifier outperformed among all proposed algorithms, with the accuracy of 94.39%. RF model has better feature propagation and provides better classification result even for the small dataset. The prediction results from the developed GUI prediction tool provided the prediction on the status of dementia based on the available patient information, including demographics, clinical and MRI derived anatomic volume features able to cover the limitations pointed out from the previous researches.

References

[1] M. Prince, R. Bryce, E. Albanese, A. Wimo, W. Ribeiro, and C. P. Ferri, 'The global prevalence of dementia: A systematic review and metaanalysis', *Alzheimer's and Dementia*. 2013.

[2] C. Dillon, C. M. Serrano, D. Castro, P. P. Leguizamón, S. L. Heisecke, and F. E. Taragano, 'Behavioral symptoms related to cognitive impairment.', *Neuropsychiatr. Dis. Treat.*, vol. 9, pp. 1443–55, 2013.

[3] S. Teipel *et al.*, 'Multimodal imaging in Alzheimer's disease: Validity and usefulness for early detection', *The Lancet Neurology*. 2015.

[4] T. M. Schouten *et al.*, 'Combining anatomical, diffusion, and resting state functional magnetic

resonance imaging for individual classification of mild and moderate Alzheimer's disease', *NeuroImage Clin.*, 2016.
[5] S. Sarraf, G. Tofighi, and for the A. D. N. Initiativ, 'DeepAD: Alzheimer′s Disease Classification via Deep Convolutional Neural Networks using MRI and fMRI', *bioRxiv*, 2016.
[6] A. V. Lebedev *et al.*, 'Random Forest ensembles for detection and prediction of Alzheimer's disease with a good between-cohort robustness', *NeuroImage Clin.*, vol. 6, pp. 115–125, 2014.
[7] A. Khan and M. Usman, 'Early Diagnosis of Alzheimer's Disease using Machine Learning Techniques', *2015 7th Int. Jt. Conf. Knowl. Discov. Knowl. Eng. Knowl. Manag. (IC3K).*, 2015.
[8] N. Amoroso *et al.*, 'Deep learning reveals Alzheimer's disease onset in MCI subjects: Results from an international challenge', *J. Neurosci. Methods*, 2018.
[9] M. B. Kursa and W. R. Rudnicki, 'Feature Selection with the Boruta Package', *J. Stat. Softw.*, 2015.
[10] G. Awate, S. Bangare, G. Pradeepini, and S. Patil, 'Detection of Alzheimers Disease from MRI using Convolutional Neural Network with Tensorflow', Jun. 2018.
[11] J. Islam and Y. Zhang, 'A Novel Deep Learning Based Multi-class Classification Method for Alzheimer's Disease Detection Using Brain MRI Data', 2017, pp. 213–222.
[12] J. Islam and Y. Zhang, 'Brain MRI analysis for Alzheimer's disease diagnosis using an ensemble system of deep convolutional neural networks.', *Brain informatics*, vol. 5, no. 2, p. 2, May 2018.
[13] D. S. Marcus, T. H. Wang, J. Parker, J. G. Csernansky, J. C. Morris, and R. L. Buckner, 'Open Access Series of Imaging Studies (OASIS): Cross-sectional MRI data in young, middle aged, nondemented, and demented older adults', *J. Cogn. Neurosci.*, 2007.

Detection of Pulmonary Nodule using Shape-Based Feature Descriptor and Neural Network

Nurfarhana Hazwani Jusoh[1], Haniza Yazid[2*], Shafriza Nisha Basah[3], Saufiah Abdul Rahim[4]

[1,2,3,4]School of Mechatronic Engineering, Universiti Malaysia Perlis, 02600 Arau, Perlis.

*hanizayazid@unimap.edu.my

Abstract. This research aims to detect the pulmonary nodule presented in lung Computed Tomography (CT) scan images. Generally, a Computer-Aided Diagnostic (CAD) system was designed and developed to aid the radiologists in medical imaging department to reduce the time and to obtain faster and better results for lung nodules diagnosis of a patient. Four major stages involve in this paper which are pre-processing, segmentation, features extraction and classification. The images that were utilized were acquired from LIDC-IDRI database that available publicly for CT scan lung images. Initially, the median filter was employed in pre-processing to filter and remove the noises, unwanted distortions and artifacts presented in the images during scanning process. For the second stage, the implementation of Otsu thresholding (thresholding-based method) and watershed algorithm (region-based method) were used to segment the nodules (Region of Interest (ROI)). Manual cropping method was implemented to segment the nodule for further process. The main contribution of this paper is the extraction of the features based on shape descriptor. 10 features were extracted from the segmented nodules. Finally, all extracted features from the segmented nodules were classified into nodule candidates and non-nodule candidates using Back Propagation Neural Network (BPNN). Based on the experiment, it can be observed that the proposed approach works well with CT scan images and segmented the interested nodules with the accuracy of 94%. This semi-automated approach is fast compared with the conventional approach used by the radiologists in current time being.

1. Introduction

Lung cancer is a disease of abnormal cells that develop in the area of lung. The lung cancer become the dominant cause of death among the other types of cancer. The estimation of the new cases of the lung cancer are higher and the number of death due to the lung cancer also increase for both sex; male and female in year 2017. The most common cancer that affected in men are prostate, lung, bronchus and colorectal cancer because it shows 42% out of all cases. For women, breast, lung, bronchus and colorectal cancer show about one-half of all cases present in 2017 [1].

Lung nodules do not always indicate as lung cancer. Additionally, the pulmonary nodules can either considered as cancerous or non-cancerous. A cancerous nodule is referred as malignant and a non-cancerous nodule is referred as benign. Benign tumor will not spread to the other parts of the body unless it agitates the function of the others organs or tissues and cause damage to it. On the other hand, the malignant is built up by cancer cells and it can spread to the nearby tissues by moving into the bloodstream or lymph nodes and it is known as metastasis [2]. Late detection of the pulmonary nodule can cause the delay in treatment and the cancer cells can only be detected at the advanced stage of the disease. Hence, for lung cancer diagnosis, the most effective way is to use the CT scan to detect the pulmonary nodule due to its low cost, high quality, high sensitivity, high image resolution and details and robustness [3]. Therefore, the Computer-Aided Detection (CAD) system is introduced to assist the radiologist to evaluate the nodule in the CT images. It is proven that the CAD system improve the nodule detection and increase efficacy of the scanning [4]. The CAD system plays an important role in detection of cancer cells in the body as the system ease the work of the medical experts and radiologists.

2. Methodology

In this research, there are several stages including image acquisition, pre-processing of the lung CT images, segmentation, feature extraction and classification using neural network. Figure 1 shows the block diagram of the proposed approach.

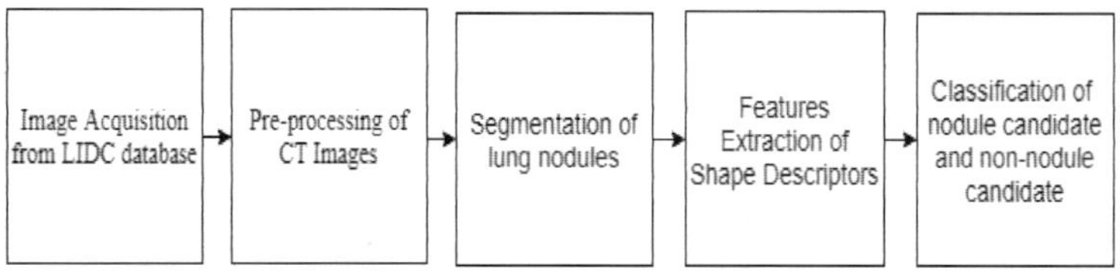

Figure 1. Block Diagram of the proposed system

2.1. Image Acquisition

LIDC-IDRI is the database in which the lung images are acquired for the input of this research. The images are in DICOM format. A total of 200 patients' slices are downloaded from this database for this work [5].

2.2. Pre-Processing Technique on Lung CT Slices

In this stage, the smoothing filtering was used to suppress the noises that present in the images. This smoothing process also blurred all sharp edges that was important information about the images. The noises that may present in the CT images include impulse noise, Gaussian noise and speckle noise. In this research, the median filter, average filter, min filter, max filter and Gaussian Filter were applied to the image to compare and find the best filtering method that can be implemented. The best filtering method was finalized and implemented on the images. The best method was selected based on the quality of the image produced after filtering process using three parameters: Mean Square Error (MSE), Peak Signal-to-Noise Ratio (PSNR) and Structural Similarity Index (SSIM) [6].

2.3. Segmentation of Lung Nodule

The filtered images were then undergo the segmentation process. The segmentation process is essential for segmenting the nodules from the lung CT scan images. The segmentation methods that implemented in this research were Otsu's thresholding and watershed transform. Otsu segmentation method is one of the global thresholding method [7]. It is a nonlinear operation and consecutively converts a grey-scale image into a binary image. Other than that, the watershed transform segmentation was used after using Otsu's thresholding for the images. Watershed segmentation is an edge based continuous domain segmentation method. Watershed is a ridge that splits areas drained by different river structures. A structural region that draining into a river or reservoir is known as catchment basins [8].

2.4. Shape-based Feature Extraction

The features extraction stage is used to extract the description of the nodule from the images after segmentation process. The nodules, non-nodules and background were found in the segmentation process. In this work, the features extraction was implemented using shape-based descriptor. The shape can be represented by moment, polygonal approximation, spatial interrelation feature and etc. These representations can be used for making measurements of the shape, matching shapes or for object recognition. The shape representation and specification techniques can be classified as in Figure 2:

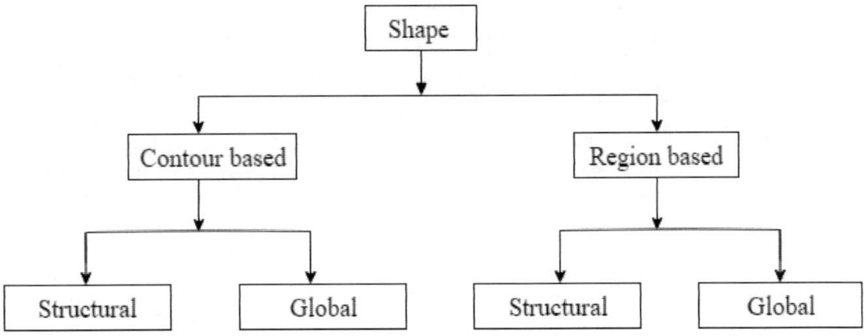

Figure 2. Classification of shape representation and specification techniques [9]

The classification of shape representation is divided into two categories which are contour based methods and region-based methods. Each method consists of two approaches, structural approach and global approach. Both of the approaches are related to a whole shape or segmented shape of the region of interest [9]. The region-based methods consist of geometric features and intensity features. The methods based on region computed all the pixels within the shape region compared to the method based on contour which it need extraction of boundary information [10]. The area, perimeter, major and minor axis length, eccentricity, solidity, centroid, equivalent diameter, circularity and convex area are categorized in region based shape representation and these parameters were extracted from the images to determine the shape of the nodule in the lung [10].

2.5. Neural Network Classification

A suitable neural network is necessary to efficiently determine the efficiency of the proposed system. Artificial neural network system (ANN) is a form of artificial intelligence system and classification method that is commonly used in image processing techniques. It is a collections of mathematical models that emulate the real neural structure of the human brain. This classification is essential for the radiologists to produce correct diagnosis result from the proposed system and necessary treatment can be imposed on the patients. Back Propagation Neural Network (BPNN) is a supervised algorithm of ANN [11].

3. Result & Discussion

3.1. Median Filtering

Pre-processing stage is crucial in any image processing processes as the images were incorporated with noises and any other forms of artifacts during the image acquisition process. Hence, there is a need to implement the filtering. The filtering techniques that had been applied on the images including Gaussian Filter, Median Filter, Average Filter, Min Filter, Max filter and Average filter. In this work several images were used to test the suitable filtering method. Based on the result, it showed that median filter is the best filter that can be implemented on the lung CT scan. This is observed based on the PSNR result produced when compared to the other filtering methods. Table 1 shows the result in term of MSE, PSNR and SSIM for 3 x 3 windowing size. The MSE needs to be low meanwhile PSNR and SSIM need to be high in order to obtain the best filtered image. Based on the results, it can be seen that Median filter produces the best result when compared to min filter, max filter, average filter and Gaussian filter.

Table 1. Performance comparison on different filtering methods

Types of filter	MSE	PSNR	SSIM
Median filter	7.0723	**39.6352**	**0.9539**
Min filter	270.6243	23.8071	0.8423
Max filter	278.5906	23.6811	0.8436
Average filter	24.3175	34.2716	0.9408
Gaussian filter	**3.6336**	34.3955	0.9421

3.2. Lung CT Images Segmentation

After the pre-processing stage, the lung CT scan images undergo segmentation stage. Segmentation stage is important to segment the nodules from background. The extracted nodules will be the determinant factor for diagnosing the lung cancer among the potential patients. Through separation and difference in the threshold values between nodules, non-nodules, background and blood vessels, the objects which were having the same threshold value with the nodules were segmented. Initially, the image is segmented using Otsu thresholding as shown in Figure 3. It can be observed that, the nodules were segmented along with the boundary. Then, the resulting image after thresholding undergoes segmentation process once again using watershed segmentation. Watershed can segment the nodules as the watershed ridge lines produced by this method perfectly encapsulated the outer boundary of the nodules, non-nodules and background as shown in Figure 4. This result supported by the theory that watershed segmentation uses the ridge lines and catchment basis principles in segmenting the objects from background. Figure 5 shows the sample segmentation results comparing with the ground truth image.

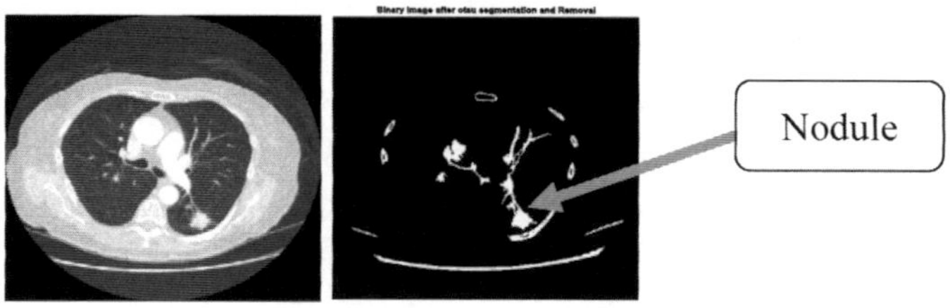

Figure 3. (a) Original image (b) Binary image after Otsu thresholding

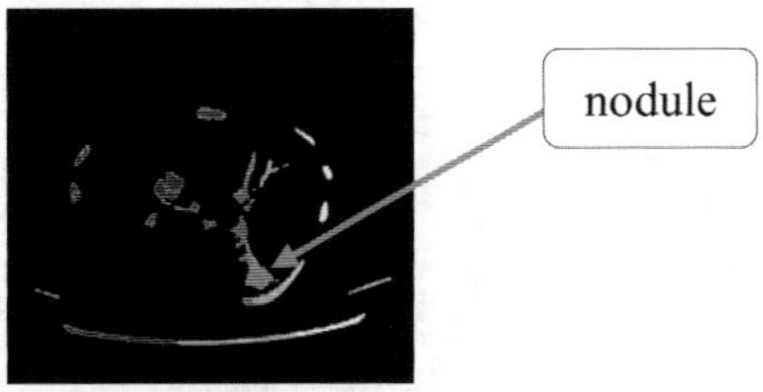

Figure 4. Watershed segmentation

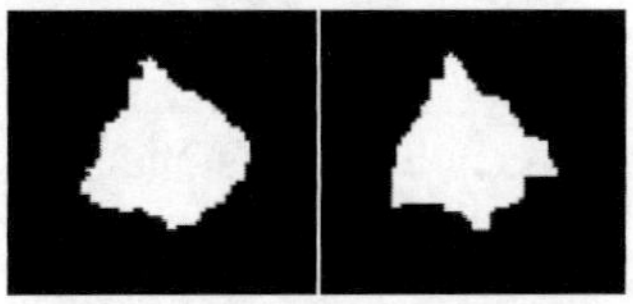

Figure 5. (a) ground truth image from LIDC-IDRI database (b) segmented image using watershed

The performance of the segmented nodules can be measured by calculating the sensitivity, specificity and accuracy. This can be done using the ground truth image and the segmented result from watershed segmentation.

3.3. Extraction of Shape-based Features on Lung Nodule

The third major stage in this project is features extraction stage. After manual cropping process, the important features of the cropped nodule were extracted. The features that were extracted from the cropped nodule included area, centroid, major axis length, minor axis length, eccentricity, convex area, equivalent diameter, solidity, perimeter and circularity. These 10 features determined the development of that particular nodules in subsequent scanning. During the first scanning, all the values of these 10 features were examined and recorded. In subsequent scanning, if any of these important features indicated increment, there indicated the higher probability of developing lung cancer on that patients. Besides, the features that were extracted were important as being the input to the classifier.

3.4. Classification of Lung Nodule Candidate and Non-Nodule Candidate

Next, the extracted features were given to the classifier. The classifier being used in this project was Back Propagation Neural Network (BPNN). The algorithm that was used for classifier was Levenberg-Marquart algorithm, with 10 input features for 200 samples, 12 hidden layers and two output data. The output was set such that the segmented objects were being nodules and non-nodules. The proportion of training and testing were set to 80/20. Three default values were set after the division of data into testing and training stage. The three default parameters were show, epochs and goal, which were set to 10, 200 and 0.5e-2 respectively. The goal parameter is usually the default setting parameter. The epochs value indicates the number of iterations that the system should be reached before the training and testing of data ends. Next, the neural network was trained and the system was simulated. The network was tested after the previous process. After the network has been trained, one could use it to compute the network outputs. Finally, the calculation of sensitivity, specificity and accuracy could be done by using the confusion matrix obtained from the network. The errors and overall performance of the system was calculated.

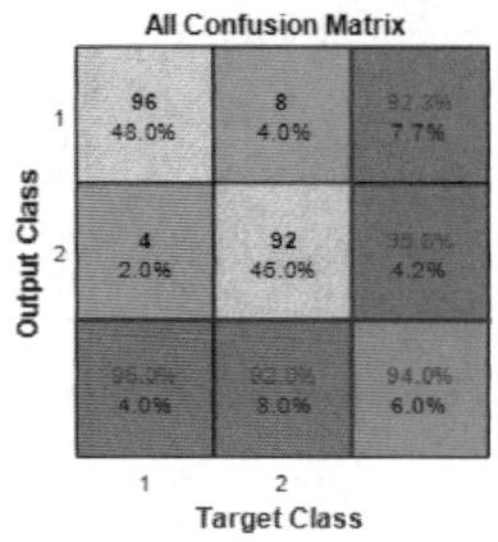

Figure 6. Confusion matrix

From the above confusion matrix as shown in Figure 6, it can be learnt that the overall accuracy of this project is 94% with error at 6%. This accuracy level obtained shows that that the proposed system could detect the segmented objects as nodules at an accuracy level of 94% and minimum error of 6%. Conclusion

This work can assist the radiologist or imaging department of current hospital systems to diagnose the lung cancer. The increment in the major features of the cropped nodules such as circularity, area, perimeter and minor axis length during subsequent scanning can indicate the development of the nodules. The development or increment in the features of the nodule indicate the high possibility of lung cancer on the patients. Therefore, early detection of this cancerous development could be conducted and this will improve the life quality of the patients.

References

[1] R. Siegel, K. D. Miller, and J. Ahmedin, 2017 , "Cancer Statistics," *Ca Cancer J.*, vol. 67, no. 1, pp. 7–10.

[2] R. J. Deberardinis, J. J. Lum, G. Hatzivassiliou, and C. B. Thompson, "Cell Metabolism The Biology of Cancer: Metabolic Reprogramming Fuels Cell Growth and Proliferation."

[3] S. Magalh, Aes, B. Netto, A. Fanes Corrêa Silva, R. A. Nunes, and M. Gattass, 2012, "Automatic segmentation of lung nodules with growing neural gas and support vector machine," *Comput. Biol. Med.*, vol. 42, pp. 1110–1121.

[4] W. J. Choi and T. S. Choi, 2014, "Automated pulmonary nodule detection based on three-dimensional shape-based feature descriptor," *Comput. Methods Programs Biomed.*, vol. 113, no. 1, pp. 37–54.

[5] M. F. Mcnitt-Gray *et al.*, "The Lung Image Database Consortium (LIDC) Data Collection Process for Nodule Detection and Annotation."

[6] J. V L and R. Gopikakumari Professor, "IEM: A New Image Enhancement Metric for Contrast and Sharpness Measurements", *International Journal of Computer Applications*, *79*(9).

[7] H. J. Vala and A. Baxi, 2013, "A Review on Otsu Image Segmentation Algorithm", *International Journal of Advanced Research in Computer Engineering & Technology (IJARCET)*, *2*(2), 387-389.

[8] I. Levner and H. Zhang, 2007, "Classification-Driven Watershed Segmentation," *IEEE Trans. Image Process.*, vol. 16, no. 5.

[9] M. N. Patel, 2016, "A Survey on Feature Extraction Techniques for Shape based Object Recognition," *Int. J. Comput. Appl. (0975 – 8887)*, vol. 137, no. 6, pp. 16–20.

[10] A. Amanatiadis, V. G. Kaburlasos, A. Gasteratos, and S. E. Papadakis, "Evaluation of Shape Descriptors for Shape-Based Image Retrieval," *IET Image Process.*, pp. 1–11.

[11] E. E. Nithila and S. S. Kumar, 2017, "Automatic detection of solitary pulmonary nodules using swarm intelligence optimized neural networks on CT images," *Eng. Sci. Technol. an Int. J.*, vol. 20, no. 3, pp. 1192–1202.

Design and Optimize Solid Microneedle using Genetic Algorithm

Nurul Fatihah Zaharin[1], Marni Azira Markom[1*], Erdy Sulino Mohd Muslim Tan[2] and Norhayati Sabani[3]
[1]School of Mechatronic Engineering, Universiti Malaysia Perlis, Kampus Pauh Putra, Arau, Perlis, Malaysia.
[2]Centre of Diploma Studies, Universiti Malaysia Perlis, Kampus Unicity, Padang Besar, Perlis, Malaysia
[3]School of Microelectronic Engineering, Universiti Malaysia Perlis, Kampus Pauh Putra, Arau, Perlis
*marni@unimap.edu.my

Abstract. Microneedle is an example of Transdermal drug delivery (TTD) device. It is known for eliminating pain and inconvenient intravenous injections. This study presents a work on design microneedle and optimize the designs using Genetic Algorithm (GA). At first, a few solid microneedles are designed with difference parameters, and the considered parameters involve are shape of the needle, materials, size of array, base of microneedle, height of microneedle and number of needles on the microneedle's array base. The microneedle is designed to meet the output requirement which are the total deformation and equivalent stress. Pressure on the tip and size of microneedle base are set to be constant which are 3.18 MPa and 2500 μm x 2500 μm x 50 μm, respectively. Then, optimisation process is conducted on the designs using Genetic Algorithm. Based on the optimisation results, it showed that the optimum structure for the studied microneedles is canonical type, diameter and needle height are 500 μm and 500 μm, respectively, PVA material and involve with 3 number of needles. The results showed that GA is capable to achieve the output requirement and able to optimize the parameter of the microneedle structure.

Keyword: Microneedle, Genetic algorithm, Optimisation and Medical device

1. Introduction

Microneedle is a micron-scale device which has been introduced to increase the ability to deliver a drug while improving the patient compliance. In medical field, microneedle patch is designed to provide an alternative method to conventional needle and syringe [1]. The advantage of this micron-scale device is painless, able to reduce the risk of diseases transmissions through the needlestick injury and wrong way to dispose the needle also to increase the accessibility of the drug into the skin. There are a few types of microneedle such as solid microneedle used for skin penetration, coated microneedle with drug that will dissolves into the skin, polymer microneedle that encapsulated with a drug and fully dissolve in the skin and lastly hollow microneedle which for the drug to infuse into the skin as shown in the Fig. 1 [2].

The studies on Microneedle mostly are of are hollow types, which were designed to be used to deliver liquids such as medicines to blood vessel [3-4]. However, there are solid type of microneedle that also can be applied in medical practice and the studies of it are still lacking. The solid microneedle is more related to the current implementation which is very similar to lancet to drawing blood. Both adults and children use the same size lancets as there is no lancets specific for baby and adult. A typical stainless-steel lancet has a diameter of 0.3–0.8 mm and penetrates 0.7–2.3 mm, still, the pain is real [5-6]. According to [7], 35 from 37 patients reported that they feel pain when using their own lancet.

The study presents an optimisation work of a few microneedle designs. The designs consist of a few parameters such as shape of microneedle, materials, diameter and height of microneedle. The optimisation is conducted by implementing Genetic Algorithm (GA). GA is a computational model which is inspired by the natural evolution. This algorithm can generate a better solution to the optimisation problem [8-9].

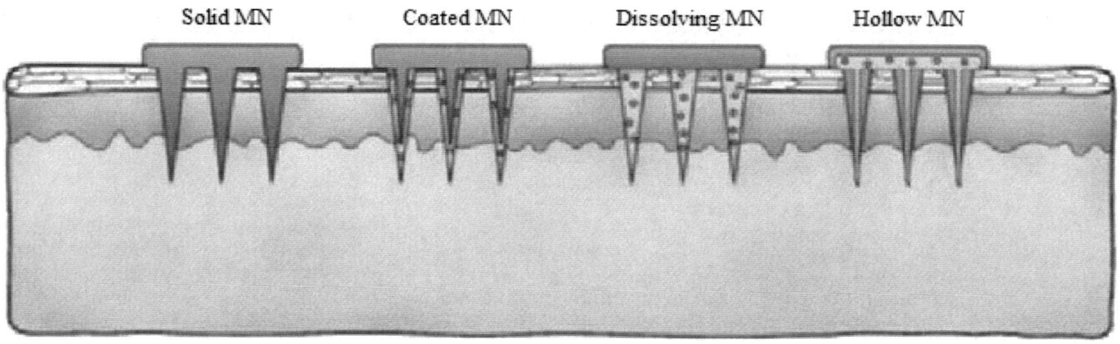

Figure 1. Types of microneedle [2]

2. Description of genetic algorithm

Genetic algorithm (GA) is a computing search technique to find approximate solution to optimisation and search problem. It is inspired by the process of natural selection that belongs to larger class of evolutionary. Two main requirements in GA algorithm; i. genetic representation and, ii. fitness function. Usually, the evolution starts from a population of randomly generated individuals through an iterative process. The population in each iteration is called a generation. The population consists of a set of chromosomes that carry the behaviour and characteristic of the system and design. In each generation, the fitness of every individual in the population is evaluated. The fitness is usually the value of the objective function in the optimization problem being solved. The more fit individuals are selected from the current population, and each individual's genome is modified to form a new generation. The new generation of candidate solutions is then used in the next iteration of the algorithm. Commonly, the algorithm terminates when either a maximum number of generations has been produced, or a satisfactory fitness level has been reached for the population. Below is the typical step of GA algorithm [9]:

i. input GA algorithm
ii. generate initial chromosomes (parents)
iii. measure fitness of parents
iv. GA operation process (mutation, crossover and cut-and-paste)
v. measure fitness of offspring and compare with parents
vi. retain the best population of the chromosome
vii. output of best solution

3. Methodology

The work on optimisation of microneedle has three stages; i. Design the microneedle using a few parameters and variables, ii. Conduct analysis on output requirement which is included total deformation and equivalent stress, and iii. Perform genetic algorithm at the proposed parameters of microneedle. The description methods are illustrated as follow.

Journal of Physics: Conference Series **1372** (2019) 012067 doi:10.1088/1742-6596/1372/1/012067

3.1 Microneedle parameters design

Six manipulated parameters such as shape of microneedles, materials, array of the needles, microneedle base, microneedle height and number of needles on the microneedle's array base are considered to be analysed as shown in Table 1. These parameters are typical parameters that used when design microneedle [3,9]. While, pressure applied at the tip and microneedle array base parameters are set up to be constant. The constant pressure is used because it is included in the range of the value of penetration pressure of needle at skin [10]. These parameters are considered in order to achieve the output requirement as stated in Table 2.

Table 1. Microneedle parameters structure

Microneedle Design Variables	**Range**		**Types of Variables**
	Minimum	**Maximum**	
Shape of microneedle	Canonical, Pyramidal		Manipulated
Materials used	ABS, PVA, Polyester resin		Manipulated
Size of array	1x1, 2x2		Manipulated
Diameter of the microneedle's base(μm)	500	1000	Manipulated
Height of the needle(μm)	500	1000	Manipulated
Number of the needle	1	3	Manipulated
Pressure applied at the tip of microneedle	3.18 MPa		Constant
Microneedle's array base	2500 μm x 2500 μm x 50 μm		Constant

3.2 Output requirement

In this study, microneedle is required to have a low value of the total deformation and equivalent stress. If these parameters values are higher than strength of microneedle material, hence, it will contribute to the microneedle failure. Table 2 shows the output requirement for the microneedle. Scale weight indicates the level of the necessary to the parameters, level 1 indicates the lowest necessary, and 10 indicates the highest necessary. In Table 2, it shows that the total deformation of necessary is about 6 from 10, and the equivalents stress is set to be 8 from 10.

Table 2. Output requirement.

Parameter	**Range**		**Optimization**	**Scale Weight**
	Minimum	**Maximum**		
Total Deformation (μm)	-	-	Minimize	6/10
Equivalent Stress (MPa)	-	-	Minimize	8/10

3.3 Genetic algorithm optimization set up

The genetic algorithm optimization is started with the initial population. 30 chromosomes are initialized randomly and a data set is created. In genetic algorithm, each chromosome in the population represent a possible solution for the setting of microneedle design variables. The generated candidate solution is being fed to the Solidwork simulator in order to get the output specification which are the total deformation (μm) and the equivalent stress (MPa). Both requirements are represented as *F1* and *F2* as shown in Equation (1) and (2).

$$F1 = \text{Total Deformation} \quad (1)$$

$$F2 = \text{Equivalent Stress} \quad (2)$$

Each of the measurement have different unit, to overcome this problem, a normalized approach is used to divide the fitness for each specification with the average specification fitness. Equation (3) shows the normalized approach for overall fitness function.

$$F_{tot} = \sum_{m=1}^{z} Wm \frac{Fm}{Fmaverage} \quad (3)$$

where,

m = specification (*F1, F2*);
z = total specification (which = 2);
Wm = weight for output requirement m;
Fm = fitness value for the output requirement m;
Fmaverage = average value of output requirement m at first iteration

The GA control parameter have been proposed based on the former research as shown in Table 3.

Table 3. The proposed GA control parameter.

Control Parameter	Value
Population Size	30
Maximum Iteration	10
Generation Gap	0.9
Crossover Probability	0.7
Mutation Probability	0.3
No. of Crossover	3
No. of Mutation	3

4. Result and discussion

According to [9], the best result is chosen based on the lowest fitness values which is obtained from Equation (3). If the value of F_{tot} is low, it shows better microneedle structure. Table 4 shows the results of optimization. In the table, it consists the best of six individuals that have been processed for optimization. Among these six sets, the second set of individuals is the best optimised result with the total deformation, stress equivalent and F_{tot} values are 0.011 μm, 0.740 MPa and 0.406, respectively.

Fig. 2 shows that distribution of the chromosomes in a GA model. The figure shows the chromosome's distribution is collected from the generation 2, 5, 7 and 10. The maximum generations of this study is 10. Each of the generations consists of 30 chromosomes and in the figure, it is indicated as red dot. It can be clearly seen that at the second generations, the chromosomes are distributed randomly. As the number of the generations increase, the chromosomes will be started to move at the same point. At the second chromosome, it shows that only one of the chromosomes achieve the maximum fitness value meanwhile at the generation 7, there are three chromosome that achieve the maximum fitness value. Lastly, it is observed that almost all the chromosomes are being joined towards the centre of the solution space, which should be converged optimum solution.

5. Conclusion

MEMS device are involved multiple variables which is difficult to be optimized. The improvement of the structure of microneedle have been shown in terms of total deformation and equivalent stress that have been minimized. GA shows that it is capable in finding a satisfactory solution during a suitable amount of time. Furthermore, it is designed to solve the optimization problems. In order to deal with a problems and constraints, future work will propose on the other optimiser and hybridization to solve more complex microneedle structure such as particle swarm optimisation (PSO).

Table 4. Microneedle Structure which have been optimized using GA (Six best individuals).

Design Variable	GA Optimized Value					
	1st	2nd	3rd	4th	5th	6th
Shape of microneedle	Canonical	Canonical	Canonical	Canonical	Pyramidal	Canonical
Materials used	PVA	PVA	Polyester Resin	ABS	ABS	ABS
Size of array	1x1	2x2	1x1	1x1	2x2	1x1
Diameter of microneedle (µm)	500	500	1000	500	500	1000
Height of microneedle (µm)	500	500	1000	500	500	1000
Number of needle on the microneedle's array base	1	3	3	1	3	1
Total deformation (μm)	0.016	0.011	0.049	0.011	0.097	0.028
Equivalent stress (MPa)	0.344	0.740	1.370	0.500	0.217	0.329
F_{tot}	0.418	0.406	0.414	0.409	0.524	0.442

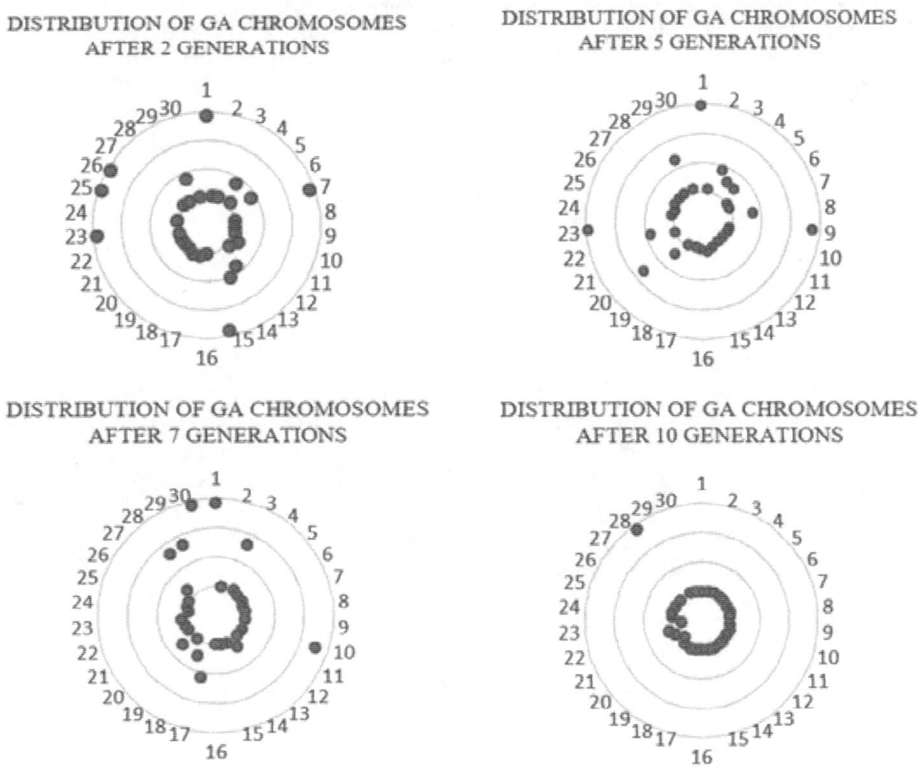

Figure 2. Distribution of the GA model after the 2, 5, 7 and 10 generations.

References

[1] Ma, G., & Wu, C. (2017). Microneedle, bio-microneedle and bio-inspired microneedle: A review. *Journal of Controlled Release*, *251*, 11-23.

[2] Ita, K. (2015). Transdermal delivery of drugs with microneedles—potential and challenges. *Pharmaceutics*, *7*(3), 90-105.

[3] Aoyagi, S., Izumi, H., & Fukuda, M. (2008). Biodegradable polymer needle with various tip angles and consideration on insertion mechanism of mosquito's proboscis. *Sensors and Actuators A: Physical*, *143*(1), 20-28.

[4] Pere, C. P. P., Economidou, S. N., Lall, G., Ziraud, C., Boateng, J. S., Alexander, B. D., & Douroumis, D. (2018). 3D printed microneedles for insulin skin delivery. *International journal of pharmaceutics*, *544*(2), 425-432.

[5] Fruhstorfer, H., Schmelzeisen-Redeker, G., & Weiss, T. (1999). Capillary blood sampling: relation between lancet diameter, lancing pain and blood volume. *European Journal of Pain*, *3*(3), 283-286.

[6] Bloomgarden, Z. T. (2002). Treatment issues in type 1 diabetes. *Diabetes care*, *25*(1), 230-238.

[7] Kim, S. (2010). A pain-free Lancet with a small needle for Glucose Measurement. *Clinical Medicine Insights: Endocrinology and Diabetes*, *3*, CMED-S3775.

[8] Kumari, V. S. R., & Kumar, P. R. (2016). Classification of cardiac arrhythmia using hybrid genetic algorithm optimisation for multi-layer perceptron neural network. *International Journal of Biomedical Engineering and Technology*, *20*(2), 132-149.

[9] Ismail, N. A., Neoh, S. C., Sabani, N., & Taib, B. N. (2015). Microneedle Structure Design And Optimization Using Genetic Algorithm. *Journal of Engineering Science and Technology*, *10*(7), 849-864.

[10] Roxhed, N., Gasser, T.C., Griss, P., Holzapfel, G.A. and Stemme, G., 2007. Penetration-enhanced ultrasharp microneedles and prediction on skin interaction for efficient transdermal drug delivery. *Journal of Microelectromechanical systems*, *16*(6), pp.1429-1440.

Three Dimensional Finite Element Modelling and Analysis of Human Knee Joint- Model Verification

Nur Afikah Zainal Abidin[1], Mohammed Rafiq Abdul Kadir[2] and Muhammad Hanif Ramlee[1,2*]

[1] Medical Device and Technology Centre (MEDITEC), Institute of Human Centered Engineering (i-Humen), Universiti Teknologi Malaysia, 81310, Johor, Malaysia.
[2] Bioinspired Innovation Group (BIOINSPIRA), School of Biomedical Engineering and Health Sciences, Faculty of Engineering, Universiti Teknologi Malaysia, 81310, Johor, Malaysia.

*muhammad.hanif.ramlee@biomedical.utm.my

Abstract. Modelling a three dimensional (3D) model of a human knee joint by extracting the region of interest accurately is one of the main constraints. Oversimplified bone models from previous studies that could affect the accuracy of analyses have become current concerns. An approach to minimize the issue consisting of several steps were done. This study aim to analyse a more precise human knee joint model using finite element technique. Reconstruction of 3D knee models were done by segmenting Computed Tomography (CT) data of a healthy male. Cancellous and cortical bones were segmented based on the Hounsfield unit (HU). The model of knee consists of femur and tibia bones, cartilages and ligaments. Construction of cartilages were done by extracting and offsetting bone layers. Linear spring elements were used to model four ligaments at the knee joint. In order to verify the models, finite element analyses were carried out. Forces ranging between 100 until 1000 N were axially applied on the proximal femur. The results in this study were in an agreement with previous literature reports with maximum peak VMS of 2.928 MPa and 3.25 MPa respectively at articular cartilages. It can be concluded that the knee models were verified.

1. Introduction

One of the largest and complex joint in the lower extremities is knee which is an important load-bearing joint [1]. It is located in between distal femur and proximal tibia bones [2]. There are six different motions of the knee joint that include flexion-extension, internal-external, varus-valgus, medial-lateral, proximal-distal and anterior-posterior [3]. Forces measured at the knee joint during walking were approximately 180 % to 280 % of body weight (BW) and the forces peak up to 350 % BW for stair ascending and descending [4]. The knee is vulnerable to injuries that include fracture and tear of ligaments when excessive forces acted on the joint [5, 6].

Finite element models of human bones and joints to be used in biomechanical analysis have been recognized significantly and widely used since they are capable in providing the understanding of biomechanical behaviours under different types of configurations and loadings [7]. This method could predict and measure local parameters such as internal stress, strain and displacement that would be difficult to be obtained from experimental works [8, 9]. Its efficiency in terms of cost, time and availability is supported by its ability to produce reliable insight and prediction of analysis.

Several finite element studies on knee joints have been reported previously and some of them modelled the bones to be one bone layer (cortical bone only) instead of two bone layers (cortical and cancellous) [10, 11]. Due to that, the material properties for both cortical and cancellous bones were assigned to one single value rather than two different values. These oversimplified models as stated from previous literature could affect the accuracy of analysis [12, 13]. Therefore, a step by step method was carried out to design a more reliable and accurate model of knee joint. The method of developing knee models consists of several steps that includes segmentation of cortical and cancellous bone layers and modelling soft tissues such as cartilages and ligaments. Material properties were assigned to the knee models accordingly and simulated with compression loads ranging from 100 to 1000 N. The results were then compared with other previous study.

2. Materials and Methods

2.1. Three-Dimensional Reconstruction of Bones
Computed Tomography data of a healthy male volunteer with no radiological sign of pathology, 27 years old with height of 169 cm and 75 kg in weight were used to reconstruct 3D models of human knee joint. Mimics (software version 10.01, Materialise, Leuven, Belgium) was used in the development of bones at which different HU values were applied to segment the region of interest autonomously. The HU for segmentation is in a range of 200 to 750 HU for cancellous bone and 750 to 3071 HU for cortical bone [14]. In order to obtain an accurate shape of bone, the layers of CT data were then segmented manually using 'erase and draw' functions. By using the layers that have been segmented, a 3D knee joint that consists of femur and tibia with cortical and cancellous bones were reconstructed as illustrated in Figure 1.

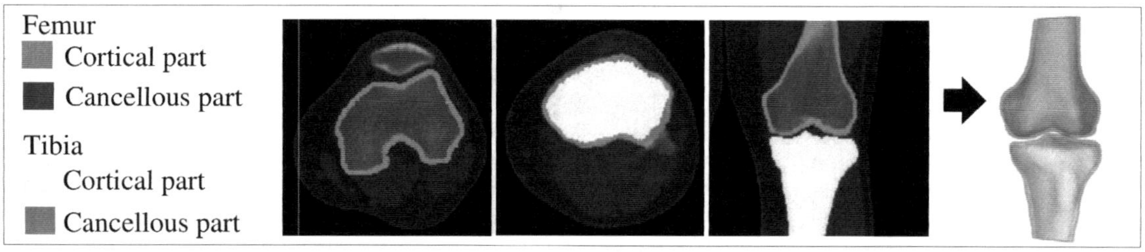

Figure 1. Segmentation of cortical and cancellous bone layers of femur and tibia.

2.2. Development of Cartilages
Cartilages at the knee joint cover the end of distal femur and proximal tibia that function as cushioning effects for shock absorption and minimize the frictions between the bones during knee movement [15]. The cartilages are considered to be linear elastic and isotropic since the loading time for time constant of viscoelastic properties of cartilages approaching 1500 seconds are far more than the loading time of interest that corresponded to single leg stance [16]. In order to predict the behaviour of cartilages under short-term condition, the properties is applicable and accurate enough since there are no significant changes in the contact behaviour of cartilages shortly after loading as demonstrated and proven by Donzelli et al. [17]. Surface layers at femur and tibia were extracted by following the anatomical shapes of cartilages at knee joint. The thickness of the region of interest from the bones that represents the cartilages were developed according to previous literature that ranges from 1.54 to 2.98 mm [18]. Figure 2 shows the modelling process of articular cartilages for femur and tibia at the knee.

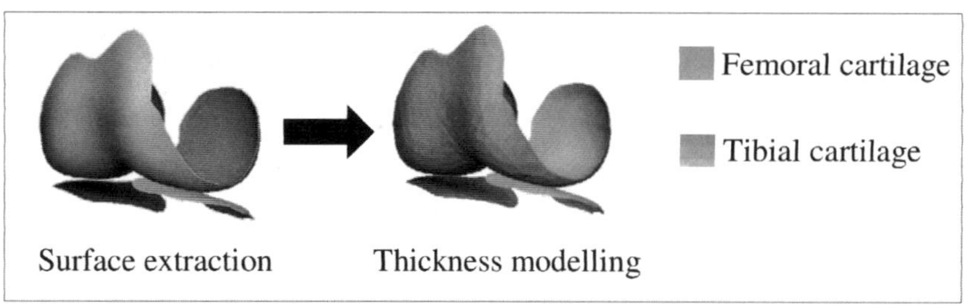

Figure 2. Modelling of articular cartilages at knee joint using surface extraction and extrude function.

2.3. Modelling of Ligaments

Ligaments are soft collagenous tissue bundles that connect one bone to another bone to form joints and enable movements [19]. Linear spring elements were used to model four ligaments at knee joint. Anterior Cruciate ligament (ACL), Posterior Cruciate ligament (PCL), Medial Collateral ligament (MCL) and Lateral Collateral ligament (LCL) were developed accordingly between femur and tibia bones by referring to anatomical attachment sites [20].

2.4. Finite Element Modelling

Finite element models of bones and cartilages were imported into 3-Matic (software version 6.1, Materialise, Leuven, Belgium) for mesh editing. Triangular surface meshes were generated on the surface layer of bones and cartilages. The bone geometry of the knee joint was checked to ensure that it is in a proper anatomical size and shape. Boolean operation was performed to ensure that there are no intersection bodies exist in the finite element models. The surface meshes for all models were converted into volume meshes to generate 3D solid models. The knee joint were set into two different degrees of knee flexion which are at 0° and 30° of flexion as illustrated in Figure 3. Both of the knee joint models with different knee flexion were exported into Patran files for biomechanical analysis in the simulation software.

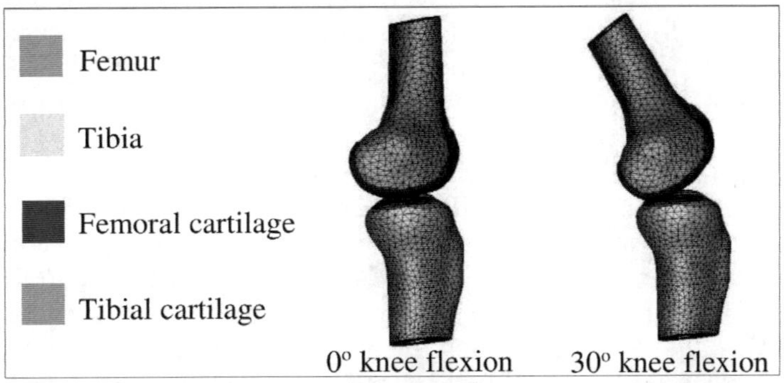

Figure 3. Finite element models of knee joint with 0° and 30° knee flexion.

2.5. Biomechanical Analysis

The finite element models were imported into MSC.Marc Mentat (software version 2005r2, MSC.Software Corporation, Germany) to be assigned with respective geometrical and mechanical properties [10]. Table 1 below shows the mechanical properties of bones, cartilages and ligaments that are used in the simulation. Young modulus (E) and Poisson's ratio (v) of the tissues are set based on previous literatures [20, 21]. Linear link elements that represents ACL, PCL, MCL and LCL were created by using spring elements between femur and tibia. The stiffness coefficient (K) value of each ligament bundles ranging from 20 to 75 N/mm as shown in the table [22]. Next, contact bodies were created between cortical bones and cancellous bone of femur and tibia together with the articular cartilages [24]. Compression loads that vary from 100 to 1000 N were applied axially on femur for full knee extension. Meanwhile, 35 N of compression load was assigned for both knee joint models with different knee flexion as shown in Figure 4.

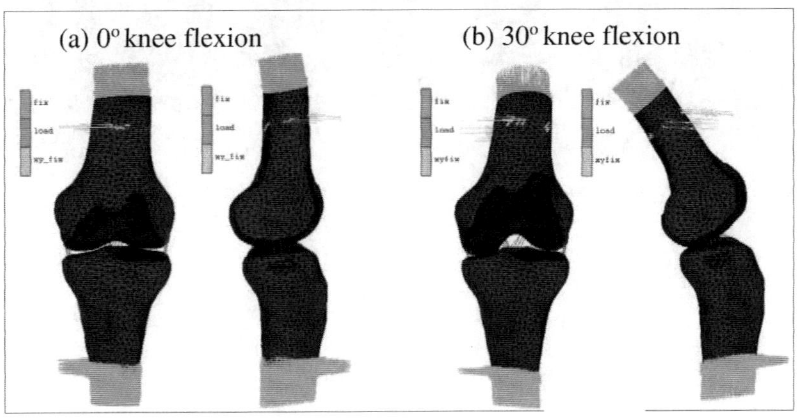

Figure 4. Finite element analysis of knee joint with boundary conditions of 35 N compression load and fixed displacement at two different knee flexion; (a) flexion of 0°, (b) flexion of 30°.

Table 1. Mechanical properties of bones, cartilages and stiffness coefficient value for ligaments.

Tissue	K (N/mm)	E (MPa)	ν
Cortical bone	-	16200	0.36
Cancellous bone	-	389	0.3
Articular cartilages	-	10	0.4
ACL	75	-	-
PCL	75	-	-
LCL	20	-	-
MCL	70	-	-

3. Result and Discussion

3.1. Finite Element Analysis

Results from the analysis are presented in the following figures. Figure 5 shows the contour plot of peak Von Mises Stress (VMS) at articular cartilages of knee joint and the variation of the predicted peak VMS results using finite element analysis are explained using graph in Figure 6. From the presented figures, the value of peak VMS at knee joint increases with increment of compression loading force. Minimum value of peak VMS for both analyses were during 100 N loading, with 1.69 MPa from literature and 1.012 MPa from simulation. Meanwhile, the maximum value of peak VMS were at 1000 N compression load, with 3.25 MPa and 2.928 MPa respectively [21].

Based on Figure 6, the results from the literature were observed to have higher magnitude as compared to the simulation results obtained in this study. This might be due to the variables exist in the finite element models that include value of mechanical properties assigned to the models and their geometrical sizes. Generally, a larger Young modulus would increase VMS while increase in Poisson's ratio would result in lower VMS [21]. Nonetheless, the mechanical behaviours of the cartilages at the knee joint in terms of peak VMS between the literature and the finite element analysis from this study exhibit similar patterns.

On the other hand, Figure 7 illustrated the peak VMS for knee joint with two different flexion. It could be observed that the value of peak VMS was higher in 30° knee flexion as compared to knee with 0° knee flexion. The contact area between the cartilages at knee joint were noticed to be reduced in value when the knee flexion increases from 0° to 30°. In general, the value of pressure increases when the surface area decreases. This is also proven by previous literature that reported increases in knee flexion will decrease the contact area and increase the contact pressure at knee joint [23]. Thus, results from this simulation that proved the contact pressure increases when the knee flexion angle increases were in coincident agreement with the experimental works.

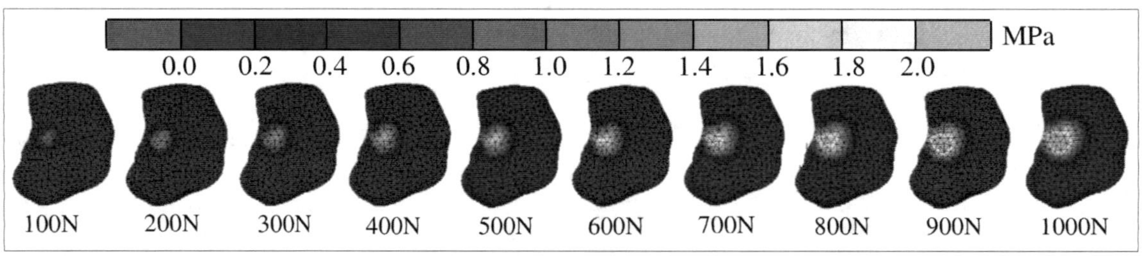

Figure 5. Contour plot of Peak Von Mises Stress at articular cartilages from FEA.

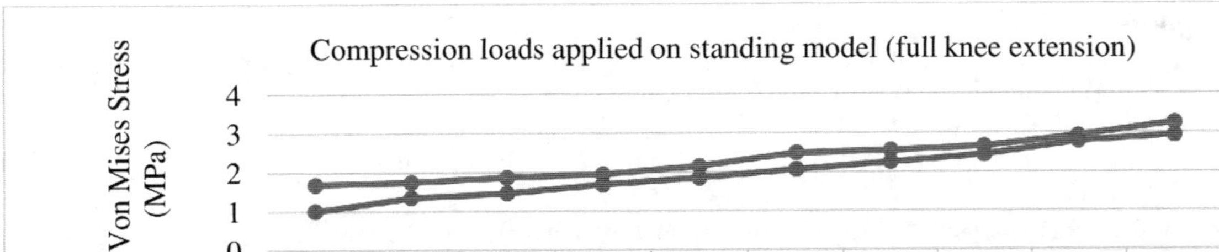

	100N	200N	300N	400N	500N	600N	700N	800N	900N	1000 N
Wang et. al	1.69	1.75	1.87	1.95	2.15	2.49	2.55	2.65	2.92	3.25
Simulation	1.012	1.356	1.466	1.681	1.854	2.062	2.251	2.434	2.778	2.928

Figure 6. Peak Von Mises Stress at articular cartilages on knee joint from previous literature work and finite element simulation.

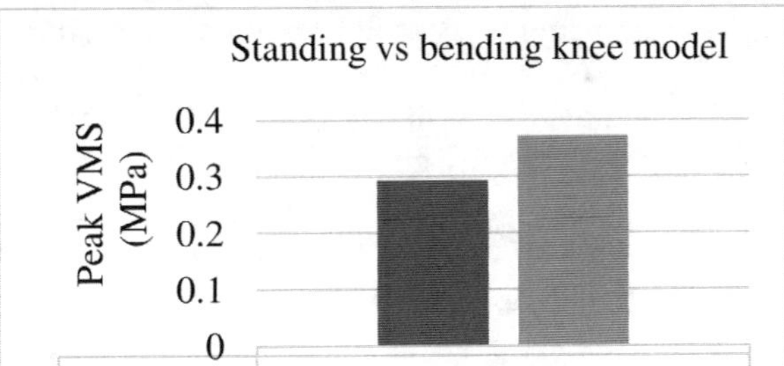

0 degree	0.2934
30 degree	0.3712

(a)

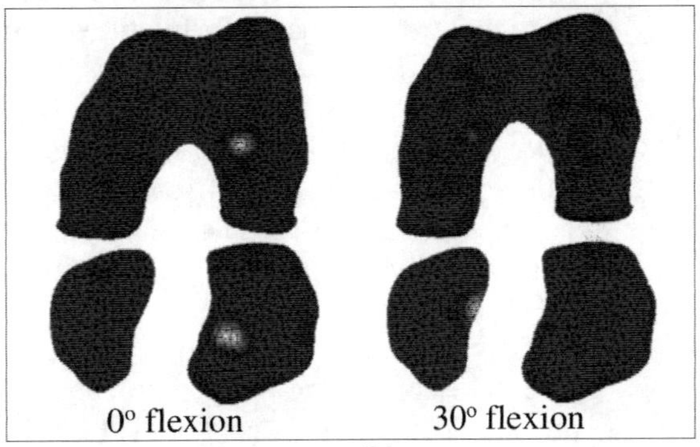

(b)

Figure 7. Peak Von Mises Stress at articular cartilages on knee joint at 0° and 30° knee flexion; (a) graph with table, (b) contour plot.

4. Conclusion

The 3D finite element models of human knee joint based on CT data were developed and analysed successfully by using Mimics, 3-Matic and MSC.Marc Mentat softwares. The cancellous bones, cortical bones, cartilages and ligaments were modelled to complete the knee joint. From the results and discussions on the models and simulation, it can be concluded that the finite element analyses of the 3D knee joint were in an agreement with previous literature works in terms of maximum value of peak VMS at contact point on articular cartilages with value of 3.25 MPa and 2.928 MPa respectively. In addition to that, the increase in contact pressure when the degree angle of knee flexion increase that was observed from the simulation was in concurrent with reported experimental investigation.

Acknowledgments

The authors would like thank to the Medical Devices and Technology Centre, School of Biomedical Engineering and Health Sciences, Faculty of Engineering and Universiti Teknologi Malaysia for the research facilities and laboratories. This work was financially supported by the Universiti Teknologi Malaysia under the Research University Grant (20H20, 4J358, 15J84 and 20H26), Ministry of Energy, Science, Technology, Environment and Climate Change (EF0618I1166 and 4S144) Fundamental Research Grant Scheme (FRGS) Ministry of Education Malaysia (278784-294883) and Zamalah UTM Scholarship.

Journal of Physics: Conference Series **1372** (2019) 012068 doi:10.1088/1742-6596/1372/1/012068

References

[1] Fregly, B.J., Besier, T.F., Lloyd, D.G., Delp, S.L., Banks, S.A., Pandy, S.A., D'Lima, D.D.: Grand challenge competition to predict in vivo knee loads. Journal of orthopaedic research **30**(4) (2012) 503-513

[2] Ganvir, S.D., Ganvir, S., Nagarale, A.V.: Biomechanics of joints vol 1 ed. Jaypeedigital (2013) 260

[3] Woo, S.L., Debski, R.E., Withrow, J.D., Janaushek, M.A.: Biomechanics of knee ligaments. Am J sports med **27**(4) (1999) 533-543

[4] Mündermann, A., Dyrby, C.O., D'Lima, D.D., Colwell, C.W.J., Andriacchi, T.P.: In vivo knee loading characteristics during activities of daily living as measured by an instrumented total knee replacement. Journal of orthopaedic research **26**(9) (2008) 1167-1172

[5] Dai, B., Mao, D., Garrett, W.E., Yu, B.: Anterior cruciate ligament injuries in soccer: Loading mechanisms, risk factors, and prevention programs. Journal of sport and health science **3**(4) (2014) 299-306

[6] Haufe, T., Förch, S., Müller, P., Plath, J., Mayr, E.: The role of a primary arthroplasty in the treatment of proximal tibia fractures in orthogeriatric patients. BioMed research international (2016) 5

[7] Peña, E., Calvo, B., Martínez, M.A., Doblaré, M.: A three-dimensional finite element analysis of the combined behavior of ligaments and menisci in the healthy human knee joint. Journal of biomechanics **39**(9) (2006) 1686-1701

[8] Totoribe, K., Chosa, E., Yamako, G., Hamada, H., Ouchi, K., Yamashita, S., Deng, G.: Finite element analysis of the tibial bone graft in cementless total knee arthroplasty. J orthop surg res **13**(1) (2018) 113

[9] Radke, H., Aron, D.N., Applewhite, A., Zhang, G.: Biomechanical analysis of unilateral external skeletal fixators combined with im-pin and without im-pin using finite-element method. Veterinary surgery **35**(1) (2006) 15-23

[10] Guo, Y., Zhang, X., Chen, W.: Three-dimensional finite element simulation of total knee joint in gait cycle. Acta mechanica solida sinica **22**(4) (2009) 347-351

[11] Shriram, D., Kumar, G., Cui, F., Dave, L., Subburaj, K.: Evaluating the effects of material properties of artificial meniscal implant in the human knee joint using finite element analysis. Sci rep **7**(1) (2017) 6011

[12] Kazemi, M., Dabiri, Y., Li, L.: Recent advances in computational mechanics of the human knee joint. Computational and mathematical methods in medicine **5** (2013) 718423

[13] Bae, J.Y., Kim, G.H., Seon, J.K., Jeon, I.: Finite element study on the anatomic transtibial technique for single-bundle anterior cruciate ligament reconstruction. Medical & biological engineering & computing **54**(5) (2016) 811-820

[14] Aspar, M.A.S.M., Razak, N.A.A., Kadir, M.R.A., Manap, H.: Development of 3-dimensional model of femur bone considering cortical and cancellous structures. International journal of engineering technology and sciences (IJETS) **7**(1) (2017) 8

[15] Fox, A.J.S, Bedi, A., Rodeo, S.A.: The basic science of articular cartilage: structure, composition, and function. Sports health **1**(6) (2009) 461-468

[16] Armstrong, C.G., Lai, W.M., Mow, V.C.: An analysis of the unconfined compression of articular cartilage. J biomech eng **106**(2) (1984) 165-173

[17] Donzelli, P.S., Spilker, R.L., Ateshian, G.A., Mow, V.C.: Contact analysis of biphasic transversely isotropic cartilage layers and correlations with tissue failure. J biomech **32**(10) (1999) 1037-1047

[18] Faisal, A., Ng, S.C., Goh, S.L., Lai, K.W.: Knee cartilage segmentation and thickness computation from ultrasound images. Medical & biological engineering & computing **56**(4) (2018) 657-669

[19] Gerard, B.D., Tortora, J.: Principles of anatomy and physiology vol 12 ed. John wiley & sons, (2009) 1264

[20] Wan, C., Hao, Z., Li, Z., Lin, J.: Finite element simulations of different hamstring tendon graft lengths and related fixations in anterior cruciate ligament reconstruction. Medical & biological engineering & computing **55**(12) (2017) 2097-2106

[21] Wang, Y., Fan, Y., Zhang, M.: Comparison of stress on knee cartilage during kneeling and standing using finite element models. Med eng phys **36**(4) (2014) 439-447

[22] Germain, F., Rohan, P.Y., Rochcongar, G., Rouch, P., Thoreux, P., Pillet, H., Skalli, W.: Role of ligaments in the knee joint kinematic behavior: development and validation of a finite element model. Computational biomechanics for medicine (2016) 15-26

[23] Marchetti, D.C., Phelps, B.M., Dahl, K.D., Slette, E.L., Mikula, J.D., Dornan, G.J., Bucci, G., Turnbull, T.L., Singleton, S.B.: A contact pressure analysis comparing an all-inside and inside-out surgical repair technique for bucket-handle medial meniscus tears. Arthroscopy **33**(10) (2017) 1840-1848

[24] Ramlee, M.H., Sulong, M.A., Garcia-Nieto, E., Penaranda, D.A., Felip, A.R., Kadir, M.R.A.: Biomechanical features of six design of the delta external fixator for treating Pilon fracture: a finite element study. Medical & biological engineering & computing **56**(10) (2018) 1925-1938

Fibre based bioplastic film from *Morus sp.* (mulberry) leaves for medical purpose

Izzat Aiman Aziz[1], Che Wan Sharifah Robiah Mohamad[1*] , Robiyanti Adollah[1]

[1]Biomedical Electronic Engineering Program, School of Mechatronic Engineering, Universiti Malaysia Perlis, 02060 Arau, Perlis.

*robiah@unimap.edu.my

Abstract. New trend was shown the application of biodegradable plastic was used in the medicinal field especially as a template for the 3rd and 4th degree of burning is extensively studied. Currently, a bioplastic in medicinal applications for burning treatment comes from various materials based on fossil oil and synthetic chemical. *Morus sp.* among abundant plant and normally used for skin and inflammation treatment in traditional practices. The aim of this research is to synthesis bioplastic by using Polyvinyl alcohol (PVA) incorporates with *Morus sp.* leaves as the alternative of the skin template for a burning patient. With the phytochemical in the leaves such as phenolic, flavonoid and high content of antioxidant, hopefully, it will able to enhance the cell growth on the burning site without any contamination from the bacteria. 2 g Morus sp. leaves was blend with different concentration of the PVA (20%, 40%, 60%, 80%, and 100%) to produce plastic films. This mixture was stirred under constant magnetic stirring at 95°C for 1 hour. The mixtures are then pouring into a glass petri dish and dried in an oven at 50°C for 24 hours. The characterization of bioplastic is based on an antimicrobial test, tensile test, microstructure analysis using SEM and chemical composition analysed using FTIR. According to the results, a bioplastic made up from mixes *Morus sp.* leaves with 80% PVA has the best mechanical and antimicrobial properties. Therefore, the result showed a possible application of the film as an alternative for the medical purpose.

1. Introduction

Nowadays, antimicrobial and biodegradable films have been used in many fields due to their function and safety aspect. The antimicrobial film is a type of films or coatings with antimicrobial properties that can prevent the growth of microorganisms. Some studies have shown that the purpose of antimicrobial film produced was to ensure the safety, quality and shelf life of a product by preventing microbial growth [1]. Antimicrobial film inhibits the growth of microorganisms and pathogenic and increase the safety and quality of a product [2] compared to other films that can only increase in term of the safety aspect. By improving the concept of artificial skin or xenograft tissue which is a transplanted tissue from donor to a recipient of a different species, the films may able to reduce the rejection rate of the implants because most of the tissue implants will activate the host immune response which results in rejection of the implants [3]. But the films only cannot enhance cell growth and reduce the infection of the bacteria.

1

Morus sp. contains high antioxidant properties [4], and the plastic films able to reduce and prevent the inflammation and growth of microorganisms such as *Escherichia coli* and *Staphylococcus aureus*, and degraded harmlessly after completing their function which made it as a user-friendly template. *Morus sp.* is a tropical tree with fast-growing ability that possesses great potential in the medical field [4]. *Morus sp.* tree has been used for various health benefits and proved to be rich in mineral, dietary fiber, vitamins as well as it is considered as a good source of nutrients. The *Morus sp.* leaves are been choose because it possesses the highest active component such as antioxidants [4] that can lower the risk of infection. The phytochemical compounds in the *Morus sp.* leave which is flavonoids and phenolic compound. The *Morus sp.* leaves extract such phenolic compound will act as an active antimicrobial agent that can help to prevent the growth of bacteria and chronic inflammation [5]. By the incorporation between *Morus sp.* and poly-vinyl alcohol (PVA), the biodegradable plastic that contains antimicrobial properties and able to degrade harmlessly can to be produced.

PVA is a type of synthetic polymer which is soluble in water and crystalline in structure. This type of polymer has been used in industrial, commercial and one of the studies stated that PVA also has been used in pharmaceutical and biomedical areas due to its excellent film-forming [6]. PVA has a great potential to be used as a material for biofilm due to its good film-forming capability, odorless, non-toxic, high elasticity and tensile strength, transparency and resistance to the chemical [7]. By incorporation of *Morus sp.* in the production of antimicrobial and biodegradable plastic which partially replacing the plastic's polymer, it can reduce the production cost and increase the degradation process. Therefore, this project was carried out to develop a film based on blends of *Morus sp.* with PVA and to evaluate the characterization and the effectiveness of the film against the microbial growth

2. Materials and Methods

2.1. Chemical and Reagents

Fresh *Morus sp.* leaves were collected in Perlis, Malaysia. The *Morus sp* leaves were collected 2 hours earlier before the experiment is started to maintain the freshness of the leaves. Poly-vinyl alcohol (PVA) 96% hydrolysis, glycerol, Nutrient Agar (NA) and Tryptic soy broth (TSB) were purchased from Merck-Millipore (Selangor, Malaysia).

2.2. Preparation of samples

After collection, *Morus sp* leaves were washed with running tap water, followed by rinsing with distilled water [4] and left in room temperature for 5 minutes to dried water. After that, the *Moris sp* leaves were cut into small pieces [4] by using scissors and then placed in an incubator at 50°C for 24 hours [8] until it was completely dry. The dried leaves were grounded by using the electrical blender and then sieved through a plastic filter to get a fine powder. Finally, the *Moris sp.* leave powders were kept in a sealed container with silica to prevent the contamination of fungus.

2.3. Blend PVA/Morus sp.

The *Moris sp* leaves powders were blended with different ratios of PVA (20%/40%/60%/80%) and were processed into films by a casting method.

2.3.1. Preparation of Morus sp. solution.

2g of *Morus sp* leaves powders were mixed with 100ml of distilled water. The mixture was heated and stirred at 75°C for at least 1 hour on a magnetic stirrer until completely dissolved. This mixture was continue stirred at 85°C and 15ml of glycerol was added into the solution as the plasticizer which can improve the permeability of the films [9]. Heat and stir the solution for 1 hour until completely mixed.

2.3.2. Preparation of PVA solution.

20%, 40%, 60% and 80% of PVA were mixed with 100ml distilled water respectively. The mixture was heated and stirred at 85°C for 1 h until completely dissolved.

2.3.3. Preparation of bioplastic film.
A *Morus sp.* solution was added into the PVA solution. Then, the mixture was heated and slowly stirred at 85°C for 1 hour until completely mixed to avoid the formation of bubbles. The aliquot was spread over a circular glass petri dish with a diameter 10cm and placed into an incubator at 50°C for 24 hours. After 24 hours, bioplastic films peeled off from glass petri dish, labeled and placed into the desiccator for future analysis.

2.4. Antimicrobial test

2.4.1. Bacterial preparation.
Escherichia coli (E. coli) and *Staphylococcus aureus (Staphy)* were cultured on nutrient agar (NA) for 24 hours at 37°C to activated a strain and to get a single colony. Then, from the single colony, the bacteria were prepared for analysis by adjusted an inoculum to 0.5 McFarland standards ($5 \times 10^5 CFU/ml$).

2.4.2. Broth dilution method.
1g of each bioplastic film (20%/40%/60%/80%) was dissolved into 10ml of Tryptic Soy Broth (TSB) using a double-boiling method for the plastic film solution preparation. Then, 0.1g/ml of plastic film solution was added into 1ml of inoculum then tested in triplicate, respectively. The preparation of positive control using Doxycycline 100mg was dilute in 10ml of distilled water. Then, 10mg/ml of Doxycycline were added into 1ml of inoculum. For the negative control, only added 1ml of TSB into 1ml of inoculum. All test tubes then incubated at 37°C for 12 h to 24 h. After that, the determination of turbidity and growth of bacteria using OD_{600} was observed.

2.5. Characterization

2.5.1. Scanning electron microscope (SEM).
The surface and cross-section of synthesis bioplastic films were studied using the TM3000 Tabletop Microscope. The film samples were cut with a dimension of 10mm x 10 mm. The samples were coated using platinum at the lower surface and the side surface to make the films conductive. The magnification used to observe the surface morphology was 100X and 300X while for cross-section was 300X. SEM is used to identify the polymerization structure of the sample.

2.5.2. Fourier transform infrared spectroscopy (FTIR).
The FTIR spectra of synthesis bioplastic films were studied using a Perkin-Elmer Spectrum One Spectrometer, United States. The film samples were cut into (10mm x 10 mm) and the spectra were collected from 4000 cm^{-1} to 600cm^{-1} frequency range. All data analyzed using FTIR Spectrum Software with a 4 cm^{-1} spectral resolution.

2.5.3. Mechanical properties testing.
The tensile properties were tested using an Instron 5848 Micro Tester testing machine. The initial grip was set to 6cm and the load used was 1000N. The tensile strength (TS) and the elongation at break (EB) of the film samples (50mm x 10mm) were evaluated from the force-displacement data.

3. Results and discussions

3.1. Film production.
The structure and the ability to peel off of the films produced is increased as the ratio of PVA increase while the color of the films becomes lighter from dark green to light yellowish. Results indicated that the PVA was able to increase the structure of the films to become stronger due to the strong interaction between *Morus sp.* and PVA.

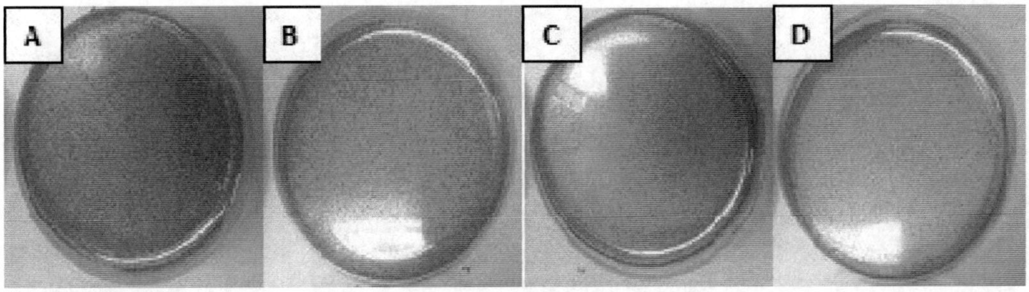

Figure 1. Morphology of (A) 20% PVA bioplastic film, (B) 40% PVA bioplastic film, (C) 60% PVA bioplastic film, and (D) 80% PVA bioplastic film

3.2. Antimicrobial test.

Figure 2 (A-B) shows the result of the antibacterial test of 20%, 40%, 60% and 80% synthesis plastic film respectively. Antimicrobial property is the important criteria for biodegradable films to have. It shows the ability of the films to inhibit bacteria and can safely be used for medical purposes. Based on the results, the ability to inhibit bacteria is increased as the PVA ratio is increased. 80% PVA film shows the lowest absorbance density when tested with *E. coli* and *Staph. aureus* indicating that the bacteria were inhibited and lysis in the medium.

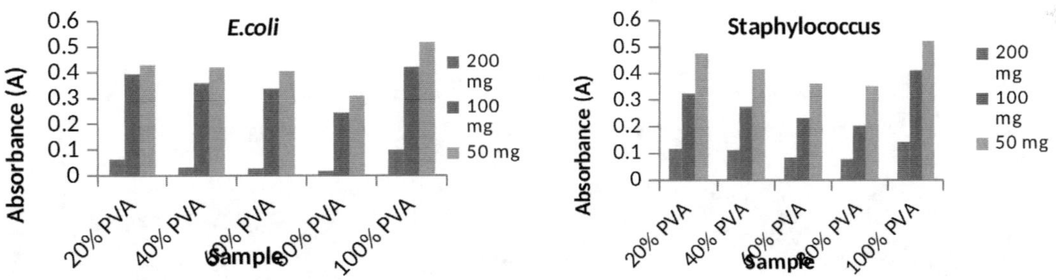

Figure 2. Graph show the (A) *E. coli* and (B) *Staph. aureus* of antimicrobial test with different concentration of PVA.

3.3. The microstructure of film.

Fig 2 (A-D) shows the result of microstructural analysis of the cross-section of 20%, 40%, 60% and 80% of PVA, respectively. The purpose of the test was to observe the microstructure of the synthesis plastic film based on cross-section using 300X of magnification. Figure 2 (D) show very good compatible morphologies with a smooth surface without a cavity. Thus, a smooth surface of the films resulted that the film becomes stronger and the bonding and arrangement of the particles are strong and close to each other. It shows by an increasing of PVA, the structure of the films becomes relatively smooth and homogeneous surfaces, without pores and crack. This condition was shown the compatibility of the PVA and *Morus sp.* as a biofilm [12]. SEM analysis on 20%, 40%, 60% PVA bioplastic film (Fig. 2A-2C) shown a roughness surface with undissolved particle and more voids. Presence of voids and undissolved particle may be due to the degree of the plasticizer in the synthesis plastic film [13].

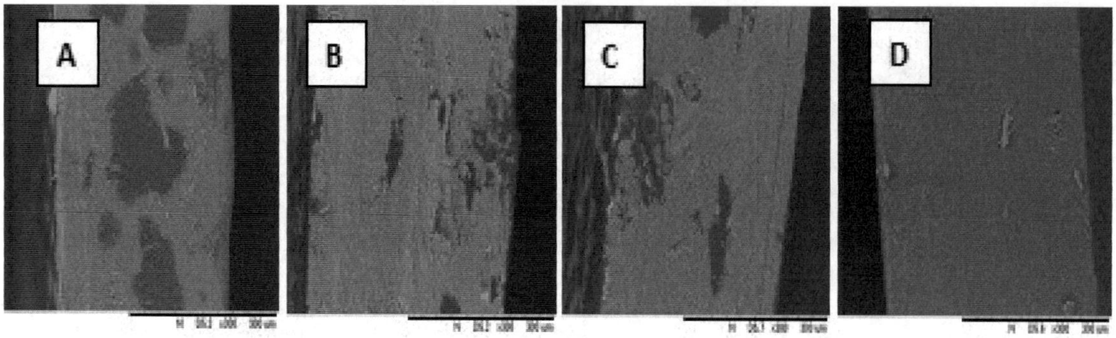

Figure 3. SEM show the (A) 20% PVA bioplastic film, (B) 40% PVA bioplastic film, (C) 60% PVA bioplastic film, and (D) 80% PVA bioplastic film.

3.4. FTIR.

The purpose of the test is to observe the interaction and chemical bonding between PVA and *Morus sp.* which is an important parameter to evaluate the strength and structure of the films. Fig. 4 shows that the –OH groups on PVA and starch formed hydrogen bonding interaction, which could improve the compatibility of *Morus sp.* and PVA. So, as the PVA ratio is increased, the interaction and the boding between *Morus sp.* and PVA becomes strong. Strong interaction could also indicate that the blend films are strong enough to use as the skin template.

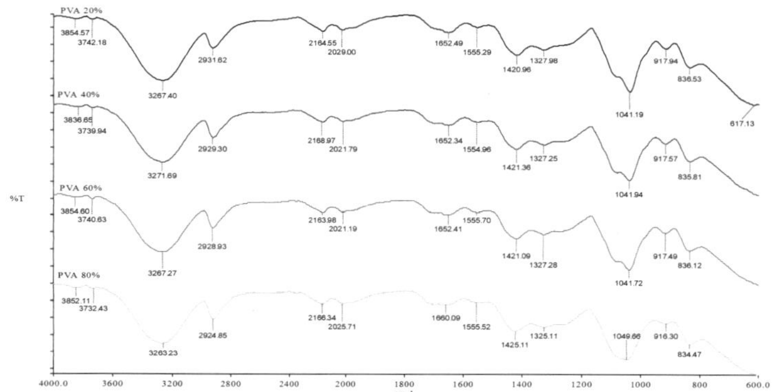

Figure 4. FTIR spectra are given for *Moris sp*. Leaves and bioplastic made with different percentage of PVA.

3.5. Mechanical properties.

One of the most important criteria required for biodegradable plastic is the mechanical stability of films during handling. Tensile strength (TS) and elongation at break (EB) are the most important parameters used to evaluate the mechanical properties of biodegradable films. Based on the results, the TS is increased as the PVA ratio increased while the EB is decreased [10]. The TS and EB indicate that the films become stronger and less elastic as the PVA ratio increased. This is because, intermolecular interaction and hydrogen bonds between polymer chains, make a rigid and brittle matrix with low elongation [11].

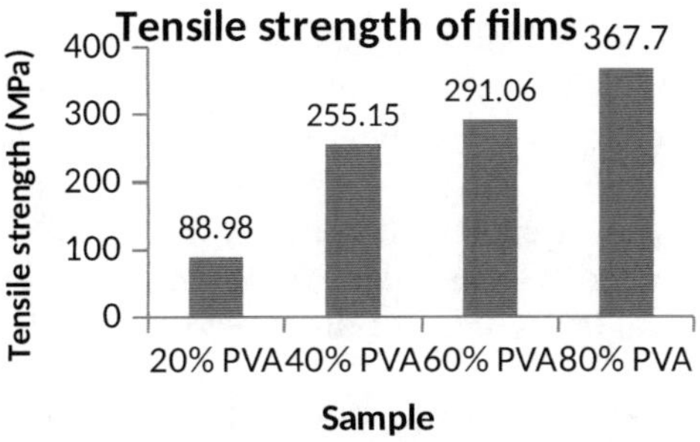

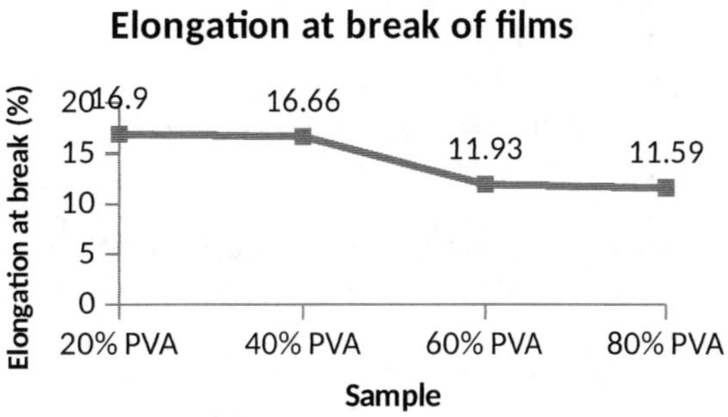

Figure 5. Tensile strength (A) and elongation (B) of *Moris sp.* Leaves and bioplastic film.

4. Conclusion

The incorporation of PVA with *Morus sp.* formed a smooth and strong structure of films due to the high compatibility of both materials. Increasing the ratio of PVA improved the structure and the physical properties of the resulted blend films, while it increases the color and transparency of the films. Also, results from the antimicrobial test demonstrated the ability of *Morus sp.* with the increasing ratio of PVA able to inhibit the growth of bacteria, *E. coli* and *Staphy*. According to the result, blending 80% PVA and *Morus sp.* leaves synthesis plastic film had the best structure, antimicrobial properties, mechanical and physical properties. These results suggest a high potential of this blend film be used as the template for skin burning.

From the SEM cross-section, it has shown this synthesis plastic film has good compatible morphologies between PVA and *Moris sp.* leaves. The FTIR spectra indicate that the –OH groups on PVA and starch formed hydrogen bonding interaction, which could improve the compatibility of *Morus sp.* and PVA. Strong interaction could also indicate that the blend films are strong enough to use as the skin template. A clear surface of the blend films exhibited a smooth surface. With the increasing of PVA, the structure of the films become smoother and less fracture can be found which was resulted that the film becomes stronger. The tensile strength (TS) increased with the increase of the PVA. However, the elongation at break (EB) decreased with the increase of PVA. Results show that the film becomes stronger and less elasticity as the PVA increased. Therefore, considering the possible application of biodegradable plastic for the medical purpose especially to the 3rd and 4th degree of skin burning.

Acknowledgments

The author would like to express the highest gratitude to Ministry of Higher Education, Malaysia for funding this research under the FRGS grant (FRGS/1/2015/SG05/UNIMAP/02/3) scheme and School of Mechatronic Engineering, Universiti Malaysia Perlis (UniMAP), Arau, Perlis, Malaysia.

5. References

[1] Lavoine N, Givord C, Tabary N, Desloges I, Martel B, and Bras J, 2014 Elaboration of a new antibacterial bio-nano-material for food-packaging by synergistic action of cyclodextrin and microfibrillated cellulose *Innov. Food Sci. Emerg. Technol.* **26** 330–340

[2] Alkan D and Yemenicioğlu A 2016 Potential application of natural phenolic antimicrobials and edible film technology against bacterial plant pathogens *Food Hydrocoll.* **55** 1–10

[3] Crupi A, Costa A, Tarnok A, Melzer S and Teodori L 2015 Inflammation in tissue engineering: The Janus between engraftment and rejection *Eur. J. Immunol.* **45** 3222–3236

[4] Mena E H, Liu T, Liao X Olajide, T M, and Huang J 2018 Effect of Superfine Grinding on the

Phytochemicals and Antioxidant Activities of Mulberry Leaves

[5] Sánchez-Salcedo E M, Mena P, García-Viguera C, Hernández F, and Martínez J J 2015 (Poly)phenolic compounds and antioxidant activity of white (Morus alba) and black (Morus nigra) mulberry leaves: Their potential for new products rich in phytochemicals *J. Funct. Foods.* **18** 1039–1046

[6] Silva G G D, Sobral P J A, Carvalho R A, Bergo P V A, Mendieta-Taboada O, and Habitante A M Q B 2008 Biodegradable Films Based on Blends of Gelatin and Poly (Vinyl Alcohol): Effect of PVA Type or Concentration on Some Physical Properties of Films *J. Polym. Environ.* **16** 276–285

[7] Bonilla J, Fortunati E, Atarés L, Chiralt A, and Kenny J M 2014 Physical, structural and antimicrobial properties of poly vinyl alcohol–chitosan biodegradable films *Food Hydrocoll.* **35** 463–470

[8] Chang Y C, Yang M Y, Chen S C., and Wang C J 2016 Mulberry leaf polyphenol extract improves obesity by inducing adipocyte apoptosis and inhibiting preadipocyte differentiation and hepatic lipogenesis *J. Funct. Foods.* **21** 249–262

[9] Dominguez-Martinez B M, Martínez-Flores H E, Berrios J D J, Otoni C G, Wood D F and Velazquez G 2017 Physical Characterization of Biodegradable Films Based on Chitosan, Polyvinyl Alcohol and Opuntia Mucilage *J. Polym. Environ.* **25** 683–691

[10] Bano I, Arshad M, Yasin T and Ghauri M A 2019 Preparation, characterization and evaluation of glycerol plasticized chitosan/PVA blends for burn wounds *Int. J. Biol. Macromol.* **124** 155–162

[11] Monjazeb Marvdashti L, Koocheki A and Yavarmanesh M 2017 Alyssum homolocarpum seed gum-polyvinyl alcohol biodegradable composite film: Physicochemical, mechanical, thermal and barrier properties," *Carbohydr. Polym.* **155** 280–293

[12] Ghaderi J, Hosseini S F, Keyvani N and Gómez-Guillén M C 2019 Polymer blending effects on the physicochemical and structural features of the chitosan/poly(vinyl alcohol)/fish gelatin ternary biodegradable films *Food Hydrocoll.* **95** 122–132

[13] Ramakrishnan N, Sharma S, Gupta A and Alashwal Y 2018 Keratin based bioplastic film from chicken feathers and its characterization", *International Journal of Biological Macromolecules*. **111** 352-358

Development of a Biomechanical Model to Study the Effects of Lifting Load on Sciatic Pain

Shehla Inam[1*], Zainab Ali[1], Haleema Wahab[1], Hina Shaheen[1], Faisal Amin[1], Muhammad Zia Ur Rehman[1], Hamza Toor[1] and Muhammad Shafique[1]

[1]Department of Biomedical Engineering, Riphah International University, Islamabad Pakistan

*shehlainam10@gmail.com

Abstract. Sciatica is a frequently encountered leg pain that originates in the human lower back and travels down the buttocks through the large sciatic nerve present in the back of the leg. Due to lifting heavy loads at different angles, the humans could acquire sciatica at different levels. Thus, the reason of using biomechanical model was to mimic the human spine in such a way that the increased force while lifting the load could easily be measured. This could help to beware the humans to not lifting load at these angles. This paper presents the findings of an experiment conducted to measure the increased forces at 6 different angles i.e. 95°, 105°, 115°, 125°, 135° and 145°. Experiments were carried out by connecting the model and a load cell with PASCO CAPSTONE. Results were concluded using both statistical and graphical analysis. The resultant force increased as the angle of bending was increased due to increased load on the back model. Stair climbing is one of the unique exercises which is frequently used and helps in maintaining the leg muscles active and healthy.

1. Introduction

Sciatica is a cause of extreme leg pain and lack of adequate power, strength or physical or mental ability. The pain originates in the lower region of the back and goes through the entire sciatic nerve in the leg as shown in figure 1. Some primary sciatic patients are treated with normal medication, but others require surgery. About 90% cases are caused by a herniated disc in which nerve root is compressed [1]. The relationship between the body mass index and low back pain was evaluated with the lumbar and thoracic curves for patients with low back pain which is caused by several factors including increased body mass index and changes in lumbar kyphosis and lordosis [2]. Sciatica treated with surgery is comparatively more convenient and effective than prolonged medication [3].The medicines discovered for sciatic disease so far is not really efficient [4]. Sciatica could be treated easily with surgery in the beginning but after one year, the body loses its immunity and a person starts facing the same problem again [5]. Usually disc herniation and low back pain result from excess physical work, lifting load, acquiring bad posture and height, factors like obesity, body stress, smoking and poor health [6]. Low back pain in initial stage is usually treated within weeks but extreme pain or other disabilities can take long time for recovery [7]. It has been observed that bio psychosocial rehabilitation with functional restoration can help reduce the pain and chronic low back pain [8]. Lumbar disc herniation usually occurs at L4/L5 and L5/S1 levels in more than 90% cases [9]. The normal lumbosacral angle in upright position of about 319 males was observed to be 41.1° with 7.7° of standard deviation [10].Sciatic pain is usually observed during the summer or autumn due to increased physical activities and drying up of joint fluid [11]. The electrical stimulation significantly treated the lumbar disc herniation-induced sciatica, reduced pain, lessened the clinical symptoms and signs of patients, regulated the peripheral ROS level, and prevented the oxidative damage of myocardial tissues [12]. Study of effects of acupuncture therapy in the patients suffering with Postpartum sciatica was carried out for 111 women patients. The acupuncture therapy had relatively effective results as compared to bed rest [13]. In recent years, transplantation of stem cells into transacted sciatic nerve in animal models demonstrates clinical improvement, inducing vigorous nerve regeneration accompanied by myelin synthesis [14]. Sciatica can also be well treated with the help of epidural steroid injection via transformational route [15].

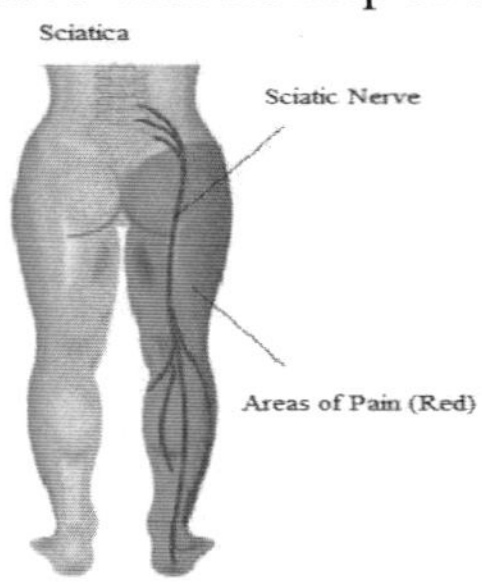

Figure 1. Sciatic nerve and areas of sciatic pain. [16]

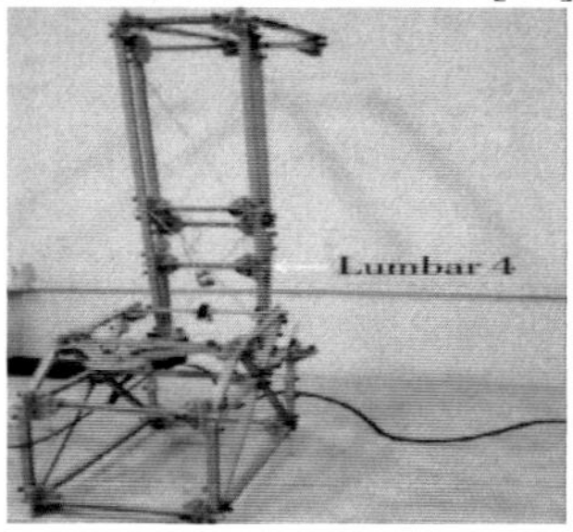

Figure 2. Location of lumbar 4 of human back model.

2. Methodology

2.1. Designing the Model
The human back model was designed manually from human structure model PASCO ME-7001 as shown in figure 2. For locating lumbar 4 vertebra, the total length of human skeleton was measured in inches. Similarly, the total length of human back model designed was also measured in inches. The average result of the two was assumed to be lumbar 4 region of the model where basically, the sciatic nerve initiates. The human back model was connected to PASCO Capstone software. The muscles of back model were mimicked using tension cords. The resultant forces on different muscle positions using different loads at different angles were calculated with the help of load cell attached to model. The angles were set using goniometer. The length of skeleton measured was 13 inches and 2 cm and that of human back model was measured 26 inches. The average of the two calculated was 19.5 inches which was assumed to be lumbar 4 of the human back models as shown in table 1.

2.2. The Experiment
The experiment was conducted by connecting human back model to PASCO Capstone, where different forces were calculated. Four different muscle positions with reference to lumbar 4 of the models were taken and two loads i.e. 100 and 150 grams were attached to each muscle position at different angles i.e. 95°, 105°, 115°, 125°, 135° and 145°. The cords mimicked as muscle positioned are given in table 2. Initially, a 100 grams load was attached to position 1 at an angle of 95° as revealed in figure 3 (a). The forces acting on position 1 could be observed through load cell on PASCO. The average of these forces was then calculated for this angle. Similarly, the forces on the same position for the other angles i.e. 105°, 115°, 125°, 135° and 145° were calculated as shown in figure 3 (b), (c), (d), (e) and (f) respectively. The same load i.e. 100 grams was attached to the other positions and corresponding average resultant forces at desired angles were observed. Then, the average resultant forces were measured by attaching 150 grams load to each muscle position at each angle.

Table 1. Location of L4 on human back model.

Length of human skeleton	Length of human back model	Average
13-inch 2 cm	26 inches	19.5 inch

Table 2. Muscles and their locations.

Positions	Mimicked muscles of model	Location in inches
1	Iliocostalis Lumborum	26
2	Latissimus dorsi	15.4
3	Iliocostalis	12
4	Semispinalis	10.2

3. Results and Discussion

The resultant forces for each muscle position at different angles measured by using the two loads are discussed below:

3.1. 100 grams load

Initially, a 100 grams load was hanged to back model at the four different muscle positions by changing different angles.

3.1.1. At Position 1

The position 1 refers to the muscle Iliocostalis Lumborum. Figure 4 shows the average resultant forces at different angles for position 1 that were recorded with 100 grams load hanged. From the graph, it is observed that with increasing angles, the force at position 1 increases. There are maximum chances of rise of sciatic pain with the maximum angle i.e. 145°. This is because the distance between load and lower back increases.

3.1.2. At Position 2

Figure 5 also that with increasing angles, the force at position 2 increases. Here also, there are maximum chances of rise of sciatic pain with the maximum angle i.e. 145° because of the distance that increased between load and lower back.

3.1.3. At Position 3

Figure 6 shows that with increasing angles, the force at position 3 increases. There are maximum chances of rise of sciatic pain with the maximum angle i.e. 145°. This is because of the increase in distance between load and lower back.

3.1.4. At Position 4

Applying the same protocol, a 100 grams load was attached to position 4 of the model. The average of the resultant force readings at different angles are given in the figure 7 where the average resultant force increases for the lower angles i.e. 95°, 105° and 115°. However, the trend is different for the next two angles i.e. 125° and 135° where the resultant forces decrease respectively perhaps due to acting of less resultant forces on these angles while for angle 145°, the value of resultant force again increases to 11.60 N.

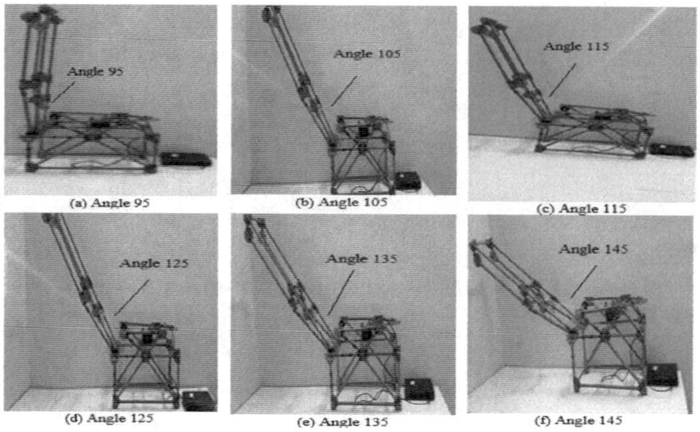

Figure 3. Six bending angles of human back model.

3.2. 150 grams load

Similarly, 150 grams load was hanged to the 4 positions and average resultant forces were observed on PASCO at different angles.

3.2.1. At Position 1

From figure 8, it is observed that with increasing angles, the force at position 1 increases. There are maximum chances of rise of sciatic pain with the maximum angle i.e. 145° due to increase of distance between load and back.

3.2.2. At Position 2

Figure 9 shows that as the angle increases, the average resultant force increases correspondingly with 150 grams load. There are more chances of rise of sciatic pain with the maximum angle i.e. 145° which shows that as the lifted load is away from the lumbar region, the stress increases at the back.

3.2.3. At Position 3

Figure 9 aslo shows that as the angle increases, the average resultant force increases correspondingly with 150 grams load. There are more chances of rise of sciatic pain with the maximum angle i.e. 145° which shows that as the lifted load is away from the lumbar region, the stress increases at the back.

3.2.4. At Position 4

The same protocol for position 4 was followed using 150 grams load. The average resultant force can be observed in figure 11. It shows that the average resultant forces increase with increasing angle from 95 degrees to 145 degrees due to increased distance from load.

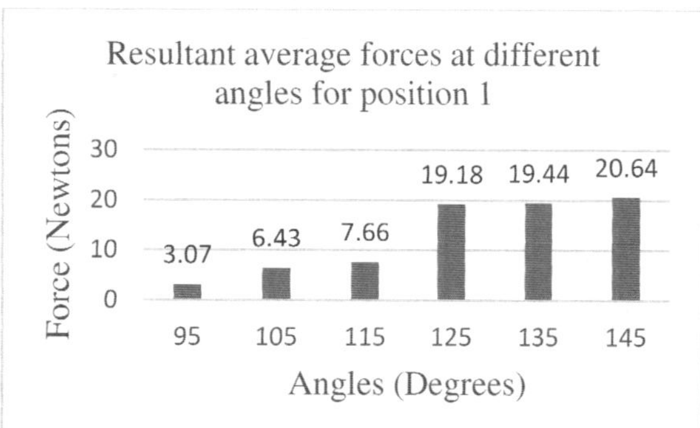

Figure 4. Resultant forces at position 1 for 100g

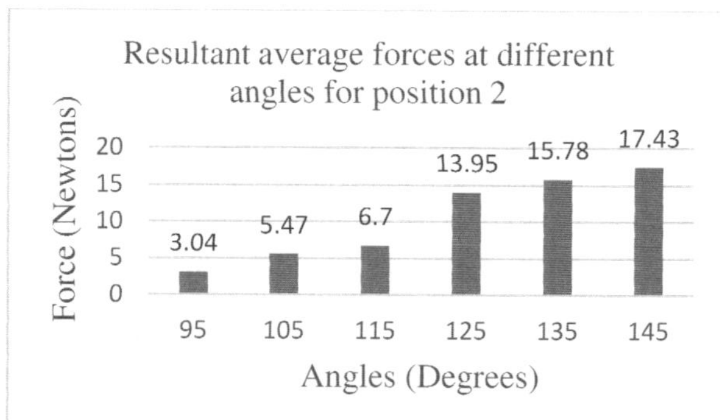

Figure 5. Resultant forces at position 2 for 100g

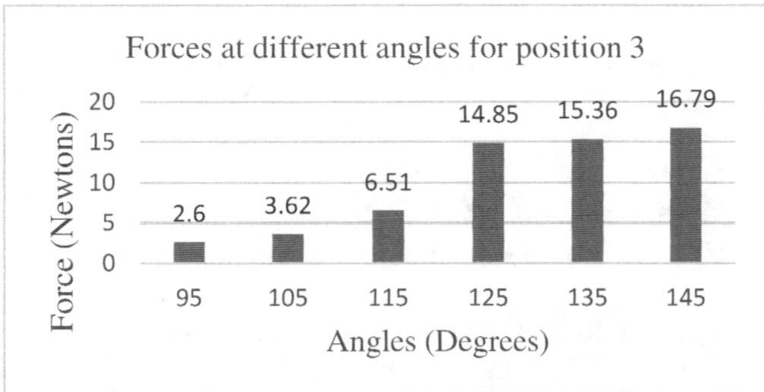

Figure 6. Resultant forces at position 3 for 100g

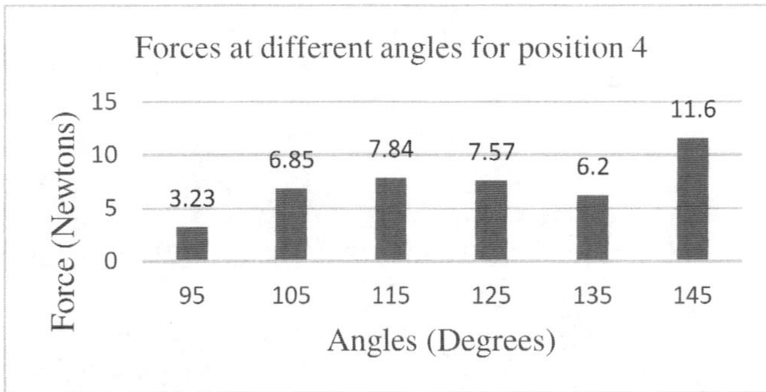

Figure 7. Resultant forces at position 4 for 100g

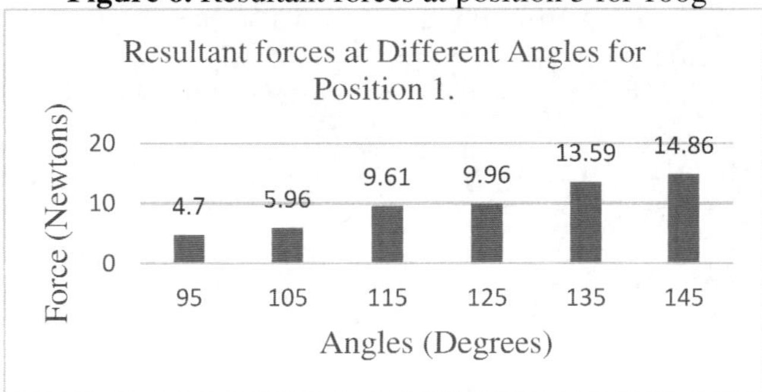

Figure 8. Resultant forces at position 1 for 150g

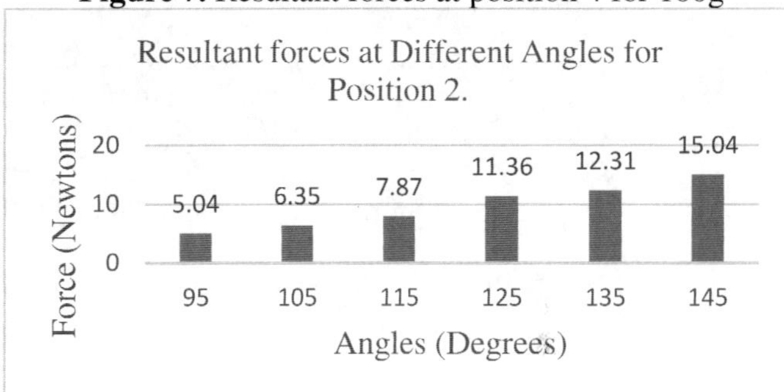

Figure 9. Resultant forces at position 2 for 150g

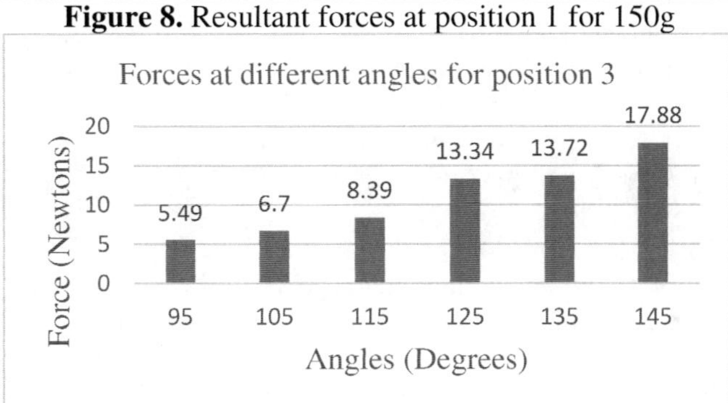

Figure 10. Resultant forces at position 3 for 150g

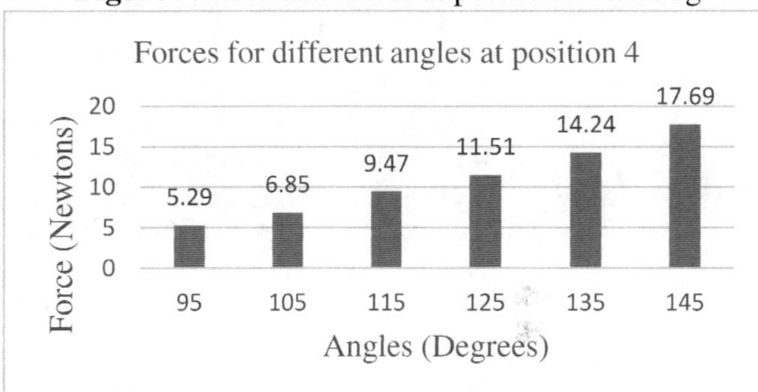

Figure 11. Resultant forces at position 4 for 150g

4. Conclusion

In cases where full restoration is not possible or in degenerative conditions, the maintenance approach is used within the biomechanical approach to enable preservation of the client's physical performance capabilities and slow declines in impairments and task abilities. Therefore, development of sciatic biomechanical model would help in understanding the effects of lifting different loads in order to decrease the pain among patients. The force acting along the back muscles of sciatic patient increases with increasing angle of bending and by adding more load forces. By increasing the load force from 100 to 150 grams in accordance with six different angles and four positions, the resultant force increased accordingly. However, for position 4 of 100 grams load, the trend became slightly different. The values of resultant force decreased for angle 125° and 135° which might be because of mechanical or human error. This shows that as sciatica itself is painful, the added weight of the lifting an object causes increased stress on the back muscles and lifting heavy loads or bending over too much strains the back muscles causing pain in the lower back region.

Acknowledgments

The authors are grateful to the Biomedical Engineering Department of Riphah International University, Islamabad for providing them the necessary equipment in a laboratory for performing the experiment.

5. References

[1] Koes BW, Van Tulder MW and Peul WC 2007 Diagnosis and treatment of sciatica *BMJ.* **334** 1313–1317

[2] Mohammad H, Fateme G and Asghar A 2007 The Relationship between Lumbar and Thoracic Curves with Body Mass Index and Low Back Pain in Students of Zahedan University of Medical Sciences *J Med Sci.* **7** 984-990

[3] Van D H W, Peul WC, Koes BW, Brand R, Keivit J, Thomeer RT and Leiden The HagueSpine Intervention Prognostic Study Group 2008 Prolonged conservative care versus early surgery in patients with sciatica from lumbar disc herniation: cost utility analysis alongside a randomised controlled trial. *BMJ.* **336** 1351-4

[4] Rafael ZP, Chris GM, Manuela LF, Paulo HF, Mark H, Vinicius CO, Andrew JM, Bart K2012 Drugs for relief of pain in patients with sciatica: systematic review and meta-analysis *BMJl.* **344**

[5] Wilco CP, Wilbert BH, health economist, Ronald B, statistician, Ralph T W M T and BartW 2008 Prolonged conservative care versus early surgery in patients with sciatic caused by lumbar
disc herniation: two year results of a randomised controlled trial *BMJ.* 1355-8

[6] Marku H 1989 Risk factors for low back pain and sciatica *Ann Med.* **21** 257-264

[7] Pengel LH, Herbert RD, Maher CG and Refshauge KM 2003 Acute low back painsystematic review of its prognosis *BMJ.* **327** 323

[8] Guzmán J, Esmail , Karjalainen K, Malmivaara A, Irvin E, Bombardier C 2001 Multidisciplinary rehabilitation for chronic low back pain: systematic review *BMJ.* **322** 1511-6

[9] Chou R, Qaseem A, Snow V, Casey D, Cross JT Jr, Shekelle P and Owens DK 2007 Diagnosis and treatment of low back pain: a joint clinical practice guideline from the American College of Physicians and the American Pain Society *Ann Intern Med.* **147** 47891

[10] Theodore EK and H.Keith H 1971 Measurement of the normal lumbosacral angle *AJR.***113** 642-645

[11] Robert WB 2018 What Causes Sciatica? Sciatica and Chronic Pain 2018 *SpringerLink.* 19-32

[12] Lulu W, Weiqiang F, Caihong Y, Minglei L and Guisen Sun 2018 Clinical effects of electrical stimulation therapy on lumbar disc herniation-induced sciatica and its influence on peripheral ROS level *JMNI.* **18** 393-398

[13] Bing SH, Yang L and Tong G 2018 Preliminary Clinical Evaluation of Acupuncture Therapy in Patients With Postpartum Sciatica *JMWH.*

[14] Dadon-Nachum M, Melamed E and Offen D Stem cells treatment for sciatic nerve injury *Expert Opin Biol Ther.* **11** 1591-7

[15] Ritesh P 2016 Efficacy of Epidural Steroid Injection in Management of Lumbar Prolapsed Intervertebral Disc: A Comparison of Caudal, Transforaminal and Interlaminar Routes *J Clin Diagn Res.* **10**

[16] Sciatica *Ravenswood Chiropractic & Wellness Center*

Feature detection of mitral valve based on image processing technique: A review

Lina Farhana Mahadi[1], Nabilah Ibrahim[2*], Mohd Thariq Zaluwi[3], Muhammad Haniff S.M. Johan[4]

[1,2]Department of Electrical and Electronic Engineering, Faculty of Electrical and Electronic Engineering, University Tun Hussein Onn Malaysia (UTHM), 86400, Parit Raja, Batu Pahat, Johor, Malaysia
[3]Department of Cardiology (Invasive and Non-invasive Cardiac Lab), Pantai Hospital Klang, A Branch of Pantai Medical Centre Sdn. Bhd., Selangor, Malaysia
[4]Department of Engineering, Pantai Hospital Klang, A Branch of Pantai Medical Centre Sdn. Bhd., Selangor, Malaysia
*nabilah@uthm.edu.my

Abstract. Mitral valve prolapse is a standout amongst the most common heart conditions, which may influence up to five to 10% of typical populace and a standout amongst the most questionable ones. The advent of computer aided technologies has contributed widely in medical application. This paper discussed on a methodological study on medical image processing to evaluate accurate diagnosis in the field of computer vision. The fundamental of image processing is to enhance the raw images obtain from machine so the useful information can be extracted from it. This paper deals with the application of the image processing on mitral valve images for further research purpose. Ultrasound data taken with subject diagnosed by mitral valve disease is analyse to determine the diameter of the valve. Based on the experiment, the average of mitral valve diameter is 21.82 mm which represent the abnormal length of mitral valve for adult.

1. Introduction

The two chambers (atrium and ventricle) of the left side of the heart were separated by mitral valve. As the heart contract, the leaflets of the mitral valve bulge (prolapse) into the left atrium in mitral valve prolapse [1]. Mitral valve prolapse causes the leaking of blood from the ventricle into the atrium which known as Mitral valve regurgitation [2]. In this condition, the blood may flow in two directions during the contraction. Supposedly, the blood will flow from the ventricle through the aortic valve. However, in this case some of the blood will flow back into the atrium. Therefore, the blood volume and pressure in the area will increase due to the leakage. Ultrasound imaging has been widely used as one of diagnostic method. The information of the diagnosis of disease can be extracted by the image produce with high frequency sound wave. Ultrasound machines are non-invasive, low cost and save time, but it depends mostly on the operator's skills and experiences to capture clear images. Image processing is important in order to absorb value information within the image which including the operation of converting the image properties and perform some operation to enhance the image. Despite the measurement of blood flow velocity, researchers found that the study of the factor that trigger the velocity in heart is more reliable, which focus on valve patency behaviour [3]. The normal diameter of the mitral annulus is 2.7 to 3.5 centimeters, and the circumference is 8 to 9 centimeters [4]. The purpose of this article is to deal with echocardiographic assessment by the image processing on the mitral valve.

2. Related work

1.1. Image denoising

Noise is an undesired data present in an image which can be removed with filters. In digital image processing, filters can be connected to an image in two different ways, which incorporate spatial and frequency domain. Based on its nature, noise can be modelled as either an additive or a multiplicative process.

$$g(x,y) = f(x,y) + n(x,y)$$
$$g(x,y) = f(x,y) * n(x,y) \quad (1)$$

In image processing, filters can be categorized into convolution based, hybrid and order statistics filters [4]. Apparently, Wiener filters shows promising result in removing Gaussian as well as Poisson noise [5]. Moreover, Mean and Gaussian filters also perform well in removing Gaussian noise compared to median filters [3, 6]. In some cases, by using a frequency domain method, the wiener filter has been applied on noisy image which removed the Speckle, Poisson and Gaussian noise present in the image [8]. Other than that, WB-Filter for Medical Image denoising which is a combination of median filtering and bilateral Filtering is implemented in [7,9] to determine the best performance and comparability to the median, wiener and bilateral filters. In this filter each pixel in an image is replaced by a weighted average of intensity values from nearby pixels and second filter used to minimize the mean square error between estimated and the desired. Different remedies of median filters have been proposed, in the literature like, adaptive median filter [10], the multi-state median filter [11], median filters based on homogeneity information [12,13]. Due to its good denoising power [12] and computational efficiency, the median filter was the most popular non-linear filter for removing impulse noise. However, some details of the original image might be smeared by the filter during high noise levels. A new approach to identify pixels corrupted with salt and pepper noise has been proposed where the noisy pixels are replaced with median value in their vicinity [13]. The algorithm developed by [14, 16] was used to remove the noise and adjusting the contrast and image enhancement [15]. The result of applying these algorithms on a sample of echocardiography image is shown in the figure 1.

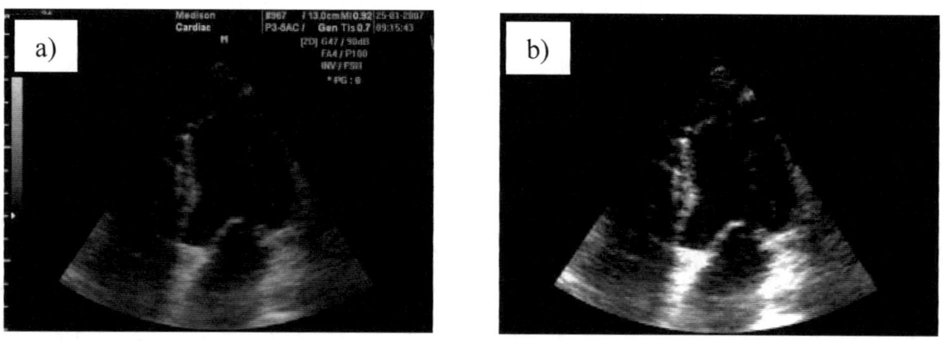

Figure 1. (a) Main input image (b) The result of using image enhancement

2.2. Image segmentation

Segmentation is the dividing of a picture into significant areas, most as often as possible to recognize items or locales of intrigue ("forefront") from everything else ("foundation"). In reference [22], the creator proposed the denoising strategy for medicinal pictures utilizing thresholding and streamlining utilizing a stochastic and randomized method of Genetic Algorithm. Thresholding is as indicated by the intensity/brightness is a basic strategy for pictures which contain strong articles on a foundation of various, however uniform, brightness. Every pixel is contrasted with the edge: if it worth is higher than the limit, the pixel is viewed as "frontal area" and is set to white, and on the off chance that it is not exactly or equivalent to the edge it is considered "foundation" and set to dark. The accomplishment of thresholding depends fundamentally on the choice of a fitting edge [23-25].

3. Application in mitral valve detection

3.2. Sampling and image acquisition

Ultrasound images were obtained by the S5-1 sector probe, which was connected to ultrasonic diagnosis equipment Philips iE33, located in Medical Imaging Laboratory Pantai Hospital Klang. A subject who suffers from mitral valve disease was selected in this study. The images were scanned from the left parasternal window. The parasternal long axis (PLAX) appeared in figure 2 is acquired by putting the transducer in the three to four left intercostal spaces near the sternum with the beam arranged toward the patient's right shoulder. This direction cut through the heart on a long axis from base to apex. After collecting the video file from the ultrasound imaging system, the video file was extracted into frame image by using MATLAB and continue with image processing.

3.2. Image processing

In this study the region of interest is on the mitral valve. Therefore, we create fixed coordinate pixel to crop the original image to only view selected region. The colour image is then converted into grayscale to reduce the complexity from a 3D pixel value (RGB) to a 2D value which ranges between 0 to 255 (8-bit unsigned integers). The concept of unsharp masking is to remove a scaled blur version of the image from the original. The result appears to be a better and sharp image; crisper and clearly defined edges. The images then undergo more edge detection, skeletons to reduce all objects in an image to lines, without changing the essential structure of the image and finally finding the endpoint of the skeleton for further research purpose. The 3x3 images produce when undergoes overall image processing can be seen in figure 3.

3.3. Mitral valve diameter measurement

The diameter of the mitral valve is then being measured using 'imtool' in MATLAB software. The instrument indicates the separation in information units controlled by the XData and YData properties where the worth is gotten in pixel esteem, as a matter of course. To obtain the measurement in proper units, the pixel value has been manually calculated. The calculation to measure distance in millimeter (mm) is

$$(N\ /\ M) * 10 = A\ mm \tag{2}$$

since 1 centimeter (cm) = 37.795275591 pixel (X), where N is a measured pixel value from image, M is a pixel value for 1 cm in image and A is a measurement in distance (mm). In this section, the endpoint produced in the last step of image processing is being used for measurement of the mitral valve diameter. Note that the measurements are initially in pixel value before being converted into millimeters unit. The example of measurement in pixel on the endpoint of the skeleton image by using 'imtool' in MATLAB software can be shown in figure 4. The pixel region of the endpoint skeleton images is in binary.

The skeletonization is based on the thinning morphological operation. In binarized picture, pixels that are set to 1 distinguish local maxima; every other pixel are set to 0. Local maxima are related pieces of pixels with a reliable intensity value, and whose external cutoff pixels all have a lower value. Meanwhile, Table 1 shows the overall measurement result of the mitral valve in pixel and millimeter unit per frame and then converted into per time as shown in figure 5. The video taken was acquired in complete four beats of QRS complex wave which relates to the depolarization of the right and left ventricles of the human heart and compression of the huge ventricular muscles. However, the measurement of the mitral valve diameter in millimeter was calculated from frame image 4 until 36 which consisting 3 complete pulse waves considering the errors that might be perform before and after the data acquisition. Based on the graph in figure 5 it can be seen that the value of the diameter of mitral valve were fluctuated over frames due to systolic and diastolic of the heart. Originally, the images have been extracted from the video to 25 image frames per second. Therefore, the video produced about 49 frame that is 1.96s. The diameter of mitral valve per time unit for this patient can be observed as shown in figure 5.

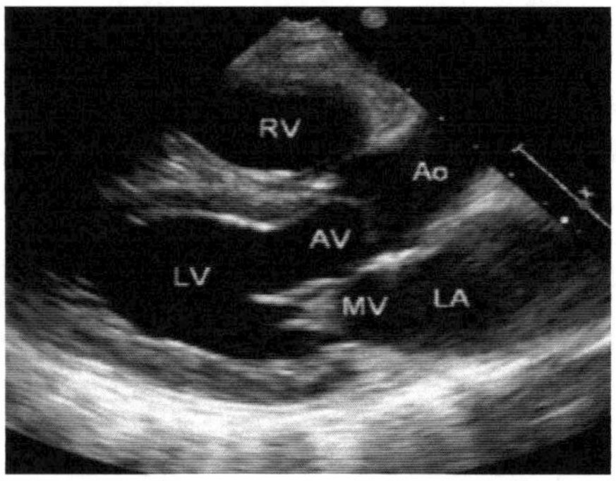

Figure 2. The parasternal long axis (PLAX) view. LA, RA, LV, and RV represent left/right atrium, and left/right ventricle, particularly, while MV and Ao is mitral valve and aorta.

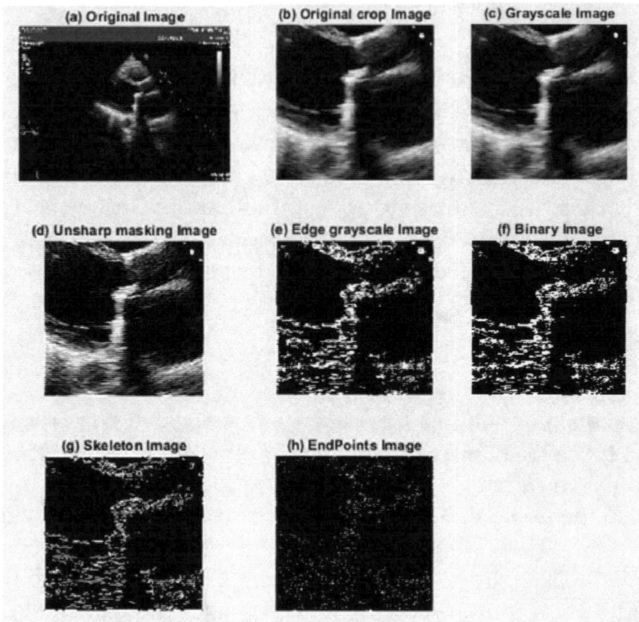

Figure 3. The images 3x3 produce in MATLAB software when undergoes image processing: (a) Original input image, (b) Region of interest, (c) Grayscale image, (d) Unsharp masking image, (e) Edge detection image, (f) Binary image, (g) Skeleton image, (h) Endpoints image.

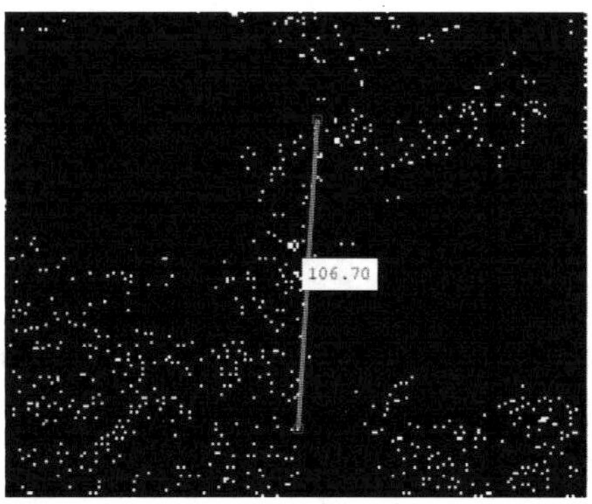

Figure 4. The mitral valve diameter in pixel unit distance using 'imtool' on endpoint of skeleton image.

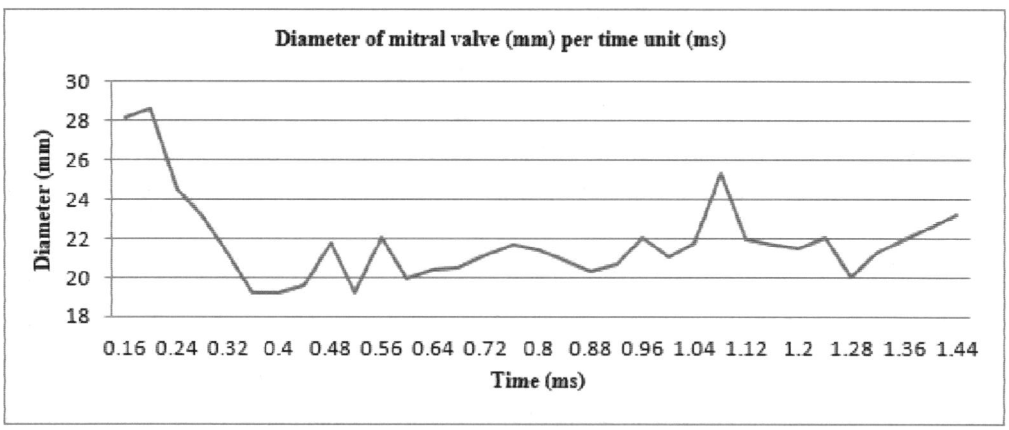

Figure 5. The mitral valve diameter per time unit

Table 1. The mitral valve diameter measurement in pixel unit

Frame images (n)	Pixel distance (X)	Frame images (n)	Pixel distance (X)
4	106.7	20	81.01
5	108.06	21	78.66
…	…	…	…
		36	87.89

Based on the previous research it is found that image processing is vital to ensure all the value information is being recorded. In this study, the mitral valve is being measured based on the video captured and divided into consecutive of image frames. The range of frame utilized is from four until 36. The result of the mitral valve diameter measurement obtained will be used to determine the size of the template for template matching process. Since the standard diameter of mitral valve annulus is 27 to 35 millimeters, it proves the mitral valve in the subject is abnormal since the average value obtained is 21.82 mm. This is one of the first few steps in order to create the optimum size of the template matching for tracking the mitral valve and hence monitoring the blood flow. This study may serve as guidelines purposes for future tracking method.

4. Conclusion

Cardiac diseases are a major health concern worldwide. In particular, useful information about the cardiac function can be extracted by analyzing of echocardiography sequences. However, echocardiographic images are difficult to analyze. In this paper, a relative study is done as a review of various methods of image processing for mitral valve disease detection. The testing on the application of several techniques in image processing has been by shown by the early study of mitral valve detection mentioned above. The diameter of the mitral valve is being calculated and the measurements are fluctuated over image frame. Based on the experiment, the average of mitral valve diameter is 21.82 mm which represent the abnormal length of mitral valve for adult. Hence this review will highlight new panoramas for developing more robust technique.

Acknowledgments

This work is supported by Graduate research Grant (GPPS) and Fundamental Research Grant (FRGS: K048). The authors would like to express their gratefulness for the support and generosity from Universiti Tun Hussein Onn Malaysia (UTHM) and Pantai Hospital Klang, Medical Imaging Laboratory for having the space doing the experiment and analysis also data collecting for this research.

References

[1] Shah PM 2010 Current concepts in mitral valve prolapse-diagnosis and management *J. Cardiol* Vol. **56**, **2** 125−133
[2] Tang Z. et al. 2019 Mitral Annular and Left Ventricular Dynamics in Atrial Functional Mitral Regurgitation: A Three Dimensional and Speckle-Tracking Echocardiography Study *J. American Soc. Echo.* Vol. **32 (4)** 503−513
[3] Moral S. et al. 2016 Differential Diagnosis and Clinical Implications of Remnants of the Right Valve of the Sinus Venosus *J. American Soc. Echo* Vol. **29 (3)**, 183−194
[4] Catherine M. Otto and Robert Bonow 2009 Valvular Heart Disease: A Companion to Braunwald's Heart Disease *Elsevier Health Sci*
[5] R. Swaminathan 2016 Application of Spatial Domain Filters on Noisy Images using MATLAB *Int. J. Comp. Appl.* Vol.**134**, **2** 0975 – 8887
[6] S. Sridhar, *Digital Image Processing.:* Oxford University Press.

[7] VandanaVikas Thakare Pratap Singh Chahar 2015 Performance Comparison of Various Filters for Removing Gaussian and Poisson Noises *Int. Res J. Eng. Tech. (IRJET)* e-ISSN: Vol. **2**, **5** 2395–0056
[8] Nayan Patel Abhishek Shah Mayur Mistry Kruti Dangarwala 2014 A Study of Digital Image Filtering Techniques in Spatial Image Processing," *Int. Conf. on Convergence of Technology IEEE*
[9] Manoj Gupta, Sumit Srivastava, Ashok Kumar Nagawat Pawan Patidar 2010 Image Denoising by Various Filters for Different Noise," *Int. J. of Comp Appl*. Vol. **9**, **4** 0975 – 8887
[10] Dr. P. Suresh, Dr. M. Suganthi S. Senthilraja 2014 Noise Reduction in Computed Tomography Image Using WB Filter *Int.J. Sci. Eng. Res*.
[11] Nilamani Bhoi 2009 Development of Some Novel Spatial-Domain and Transform Domain Digital Image Filters *Nat. Inst Tech, Rourkela, India*
[12] Hwang H and Haddad R A 1995 Adaptive median filters: new algorithms & results *IEEE Trans. Image. Process.* 499–502
[13] Chen T and Wu H R 2001 Space variant median filters for the restoration of impulse noise corrupted images *IEEE Trans. Circ. Syst. II*, **48**, 784–789
[14] Eng H Land Ma K K 2010 Noise adaptive softswitching median filter *IEEE Trans. Image Process.* **10** 242–251
[15] Liu J C and Nair A S Pok G 2003 Selective removal of impulse noise based on homogeneity level informatiion," *IEEE Trans. Image Process. V*ol **12** 85–92
[16] Alireza Ahmadian, Amrollah Gorgian Mohammadi, Parastoo Farnia Meysam Siyah Mansoory 2012 Mitral Valve Prolapse Detection Using Landmark Extraction from Echocardiography Sequencesn *34th Ann. Int. Conf. IEEE EMBS*
[17] Dr. B L Deekshatulu, Dr. K Lal Kishore, Y RakeshKumar S Suryanarayana 2009 Novel Impulse Detection Technique for Image Denoising *J. Theoretical Appl. Info. Tech.*
[18] Mihai M. Croitoru, akhil Bidani, Joseph B. Zwischenberger, and Jr John W. Clark. Yong Yue 2006 Nonlinear multiscale wave-let diffusion for speckle suppression and edge enhancement in ultrasound images. *IEEE Trans. Med. Imaging* **25(3)** 297–311
[19] Yasser Kadah Banazier A. Abrahim 2011 Speckle Noise Reduction Method Combining Total Variation and Wavelet Shrink-age for Clinical Ultrasound Imaging *IEEE*
[20] P Soille 1999, *Springer-Verlag* 170–171.
[21] Geoff Dougherty 2009 Digital Image Processing for Medical Applications. California State University, Channel Islands: Cam-bridge University Press
[22] S C Prasanna Kumar WilliamThomas H M 2015 A review of segmentation and edge detection methods for real time image processing used to detect brain tumors *IEEE Int. Conf. Computational Intelligence and Computing Research*
[23] Annie Delouche, Alain Herment, Frédérique Frouin and Be-noît Nadjia Kachenoura 2007 Automatic Detection of End Systole within a Sequence of Left Ventricular Echocardiographic Im-ages using Autocorrelation and Mitral Valve Motion Detection in *Proc. of the 29th Annual Int. Conf. of the IEEE EMBS*, Cité Internationale, Lyon, France, 4504–4507.

Journal of Physics: Conference Series **1372** (2019) 012072 doi:10.1088/1742-6596/1372/1/012072

Hardened Glass Particle and Carbon Black Using Resin for Potential Electromagnetic Shielding in Biomedical Electronic Equipments

Cheng Ee Meng[1*], Mohd Shukry Abdul Majid[1], Shahriman Abu Bakar[1,5], Nashrul Fazli Mohd Nasir[1], Khor Shing Fhan[2], You Kok Yeow[3], Lee Kim Yee[4], Mohd Ridzuan Mohd Jamir[1], Tan Wei Hong[1], Shahrir Ridha Shaharuddin[1] and Beh Chong You[1]

[1]School of Mechatronic Engineering, Universiti Malaysia Perlis (UniMAP), Pauh Putra Campus, 02600 Arau, Perlis, Malaysia
[2]School of Electrical Systems Engineering, Universiti Malaysia Perlis (UniMAP), Pauh Putra Campus, Arau, Perlis 02600, Malaysia
[3]School of Electrical Engineering, Faculty of Engineering, Universiti Teknologi Malaysia Skudai, Johor 81310, Malaysia
[4]Lee Kong Chian Faculty of Engineering & Science, Tunku Abdul Rahman University Sungai Long Campus, Jalan Sungai Long, Sungai Long City, Cheras, 43000 Kajang, Selangor, Malaysia
[5]UniMAP Motor Sports Technology Unit (MoTECH), Universiti Malaysia Perlis (UniMAP) Pauh Putra Campus, 02600 Arau, Perlis, Malaysia.

*emcheng@unimap.edu.my

Abstract. This study aims to investigate the electromagnetic properties of a composite of hardened glass particle and carbon using resin in determining the absorbability of the composite for electromagnetic interference (EMI). The electromagnetic absorption measurement was conducted using T-R method (transmission-reflection) by measuring the dielectric constant, loss factor, real and imaginary part of permeability, reflection coefficient and transmission coefficient. The measurement is taken for X band frequency and Ku band frequency within the frequency range from 8.2GHz to 12.4GHz and 12.4GHz to 18GHz, respectively. Dielectric constant of sample decrease when frequency increases. The high dielectric constant of material implies that the material has high absorbability of field energy, resulting in high attenuation of the applied field.

1. Introduction

Recently, the implementation of wireless technology has increased rapidly. Many wireless devices are ubiquitous in our daily life. The wireless technology brings massive convenience to all walk of life nowadays. There are also several EMI issues have been arisen. As a result, various electromagnetic absorbers were proposed to overcome this problem. The early research on electromagnetic-wave absorber was conducted during the second war by Germans [1]. They introduced a kind of paint that can act as the absorbing material. It was the earliest absorbing material that has been introduced. Nowadays, radar absorbing material for enhancement in national security is also triggered among the countries [1].

Electromagnetic interference in medical device occurs when another external source of electromagnetic energy interferes the electrical medical devices. As the popularity of wireless device increase, there will also be a rise in electromagnetic radiation within the segment of spectrum [2]. Various metallic materials have been used as EMI shielding materials such as steel, copper, aluminium, and nickel. Although these metallic materials are effective in shielding the electromagnetic wave, limitations are still available. Generally, metallic materials are heavy, susceptible to oxidation (rust) and corrosion which can lead to difficulty in electromagnetic shielding electronic device [3].
Glass is a material that can be implemented in many applications because it has many potential values. Generally, this material can be recycled limitlessly without losing its quality and also the purity of this material [4]. As a result, glass particle has been selected to be backbone material which can act as a matrix material in the fabrication of absorbing material incorporating with carbon black using epoxy/resin. A composite with glass particle and carbon black are proposed in this work.

2. Methodology

2.1 Sample Preparation

Samples were prepared by integrating carbon black powder with glass powder using epoxy/resin. Two dimensions of sample need to be prepared for measurement using waveguide WR90 and WR62. The prepared samples are in ratio of 20:80, 30:70, 60:40, 70:30, 80:20 and 90:10 of carbon black to glass particles.

The resin/epoxy and carbon black are commercially available. Glass particle is acquired by crushing the domestic glassware. The glass particle needs to be compressed and ground into powder form using a pestle and a mortar. Lastly, resin/epoxy is applied to harden and integrate glass powder and carbon black into a cuboid shape which needs to be fitted into sample holder for waveguide WR62 and WR90. The dimension of both sample holders are tabulated in Table 1.

Table 1. Operation frequency range and dimension of sample holders

Sample holder	Frequency Range (GHz)	Internal Dimensions (mm)
WR90	8.2 – 12.4	22.860 × 10.160
WR62	12.4 – 18.0	15.799 × 7.899

2.2 Measurements

Calibration on P-series network analyzer (PNA) is conducted before measurement. The systematic error can be avoided through the calibration using standard loads, i.e. air, water and shorting block. The required measurement parameters, i.e. frequency range, the dimension of the waveguides and sample under test need to be defined. The operation frequency range in this work is X band (8.2GHz to 12.4GHz) and Ku band (12.4GHz to 18GHz).

The calibration needs to be conducted before measurement. The electrical standards used during the calibration process are short, line and thru. Before the measurement is conducted, the sample must be placed in sample holder. Then, the sample holder with the sample placed between waveguides without a gap. The measured data are dielectric constant (ε'), loss factor (ε''), reflection coefficient (S11), transmission coefficient (S21) and absorption coefficient (A).

3. Experimental Results

3.1 Dielectric measurements

In Figure 1, the sample with 30% carbon black varies insignificantly with frequency compared to samples with carbon black 60% to 90%. ε' for sample with 30% present at the highest level. The percentage of carbon black exceeds the percolation threshold is the main factor. When carbon exceeds the percolation threshold agglomeration of carbon occurs. The agglomeration causes the increment of conductivity. Hence, ε' decreases. Sample with 20% is exempted because the carbon content is too low and its dielectric behaviour is dominated by glass particles and resin. Overall, ε' of all samples decreases when frequency increases [5] at X band. It is due to inconsistency between relaxation frequency of composites and the operating frequencies of the applied field. The similar observation can be noticed through Figure 5 for Ku Band. The variation of ε' at Ku band as shown in Figure 2 seem more dispersive. It is attributed to inhomogeneity of the prepared samples. The drastic is conducted within frequency range from 14.5 GHz to 16 GHz are most probably due to inactivation of a particular type of polarisation.

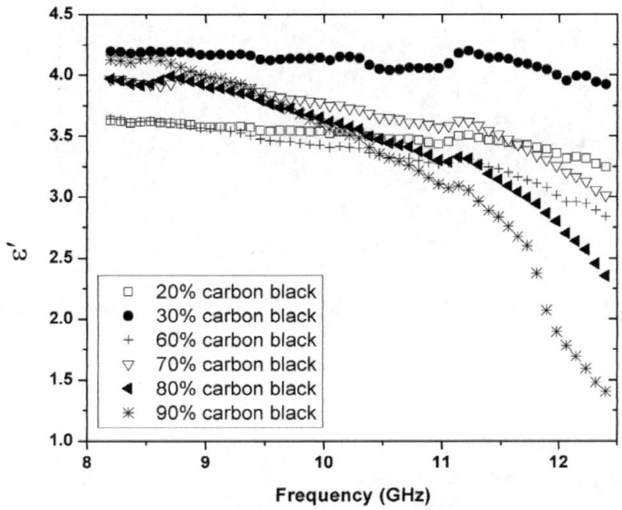

Figure 1. Dielectric constant (ε') versus frequency at X band

Dielectric loss factor, ε" for all samples at X band and low percentage of carbon content change insignificantly. When carbon content ≥ 80%, ε" increases remarkably at high frequencies, especially 90% carbon black. It shows highest loss might be due to conduction loss when conductivity of sample of 90% carbon black increases due to presence of carbon black in substantial amount. This insignificant variation of ε' as exhibited by samples with low carbon black as shown in Figure 1 explain the behavior of ε" in Figure 3. The polarisation direction is able to remain align with the applied field. Hence, ε" is low due to low energy dissipation. However, the inactivation of polarization at Ku Band that exhibited by the drastic decrement of ε' as shown in Figure 2 implies the asynchronous of polarization between applied field and sample. It increases the friction loss where the energy dissipated in heat loss. As a result, it shows peak for ε" in Figure 4.

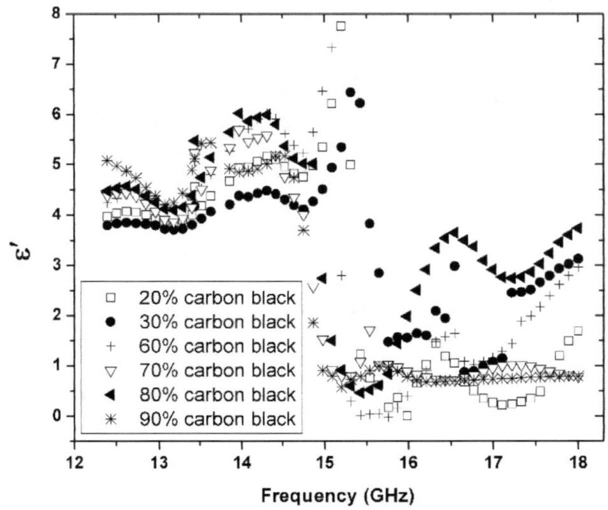

Figure 2. Dielectric constant (ε') versus frequency at Ku band

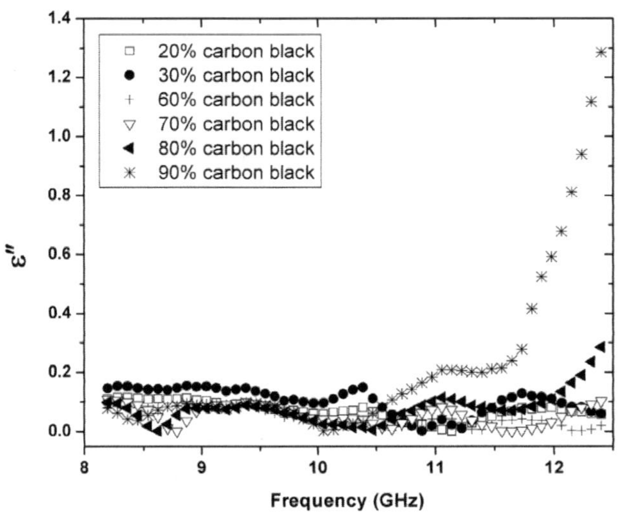

Figure 3. Loss factor (ε") versus frequency at X band

3.2 Reflection, Transmission and Absorption Measurement

In Figure 5, reflection coefficient, S11 for all samples decrease when frequency increases. It implies that mismatch impedance between sample decline when frequency increases. S11 increases when an impedance of sample largely deviates from the characteristic impedance of waveguide, i.e. 50 Ω. The consistent decrement of ε' in Figure 3 lead to trendline in Figure 5 where ε' contributes to capacitive reactance. Subsequently, this capacitive reactance is used to determine the impedance. On the other hand, the sinusoid trendline as exhibit in Figure 4 leads to the variation from a negative gradient to a positive gradient at higher frequency as shown in Figure 6.

In Figure 7 and Figure 8, it can be noticed that the gradient of their trendline is in contrary with Figure 5 and Figure 6. When the reflection energy decline, the transmission energy and absorbed energy (absorption coefficient) increases. It is consistent with literature [6] where absorption coefficient increases over frequency. The absorption coefficient in percentage as shown in Figure 9 and Figure 10. These percentage is acquired through $A(\%) = 1 - S11^2 - S21^2$. Sample with 90% carbon black exhibit the highest absorption among other samples at Ku band where it can up to 30% of absorption. It shows the highest absorption too when compared with X band.

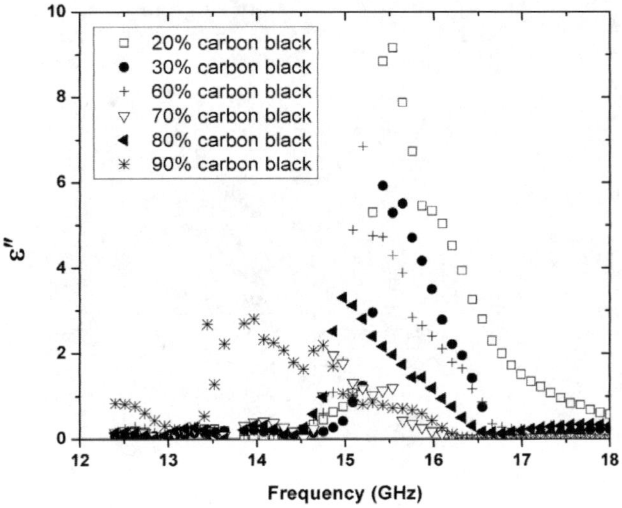

Figure 4. Loss factor (ε") versus frequency at Ku band

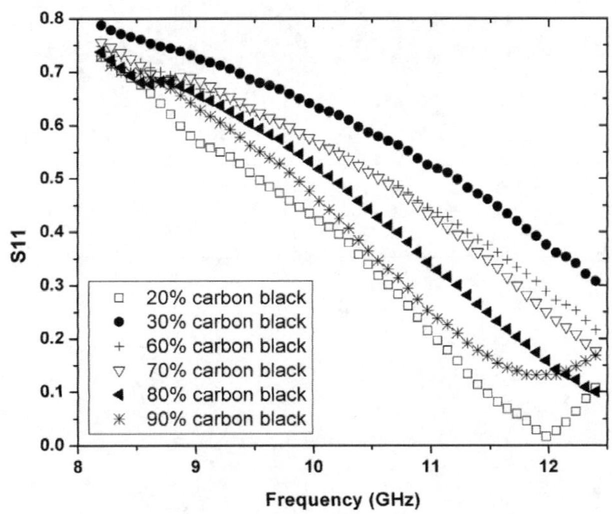

Figure 5. Reflection coefficient response to frequency at X band

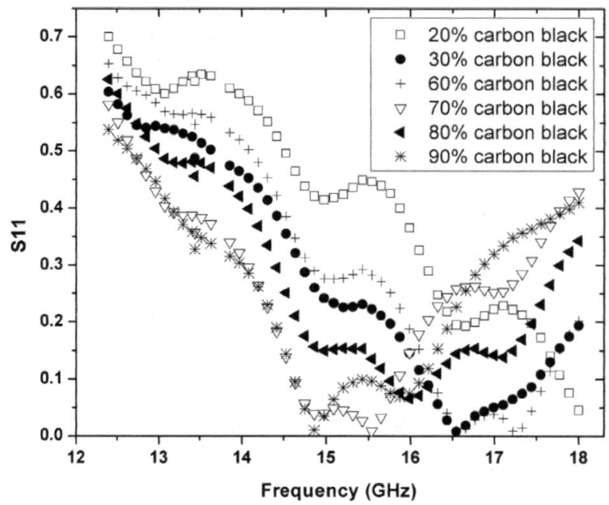

Figure 6. Reflection coefficient response to frequency at Ku band

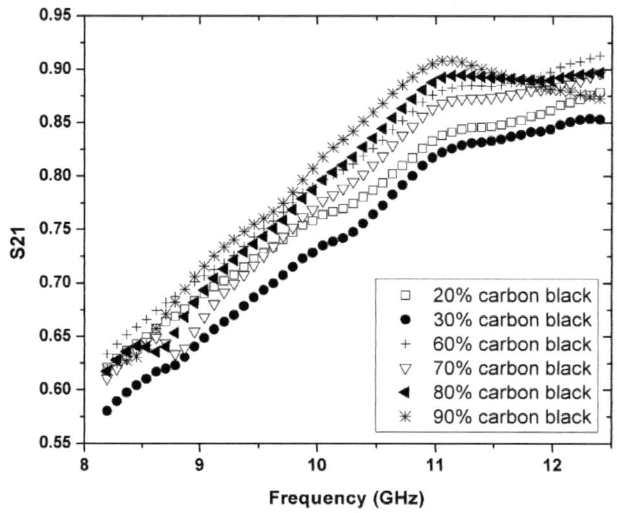

Figure 7. Transmission coefficient response to frequency at X band

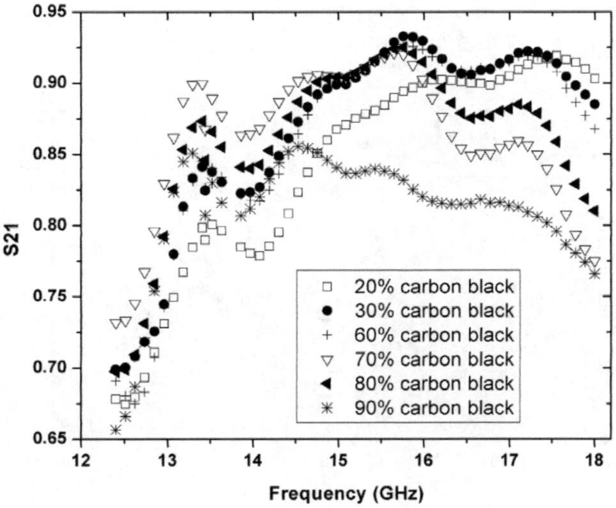

Figure 8. Transmission coefficient response to frequency at Ku band

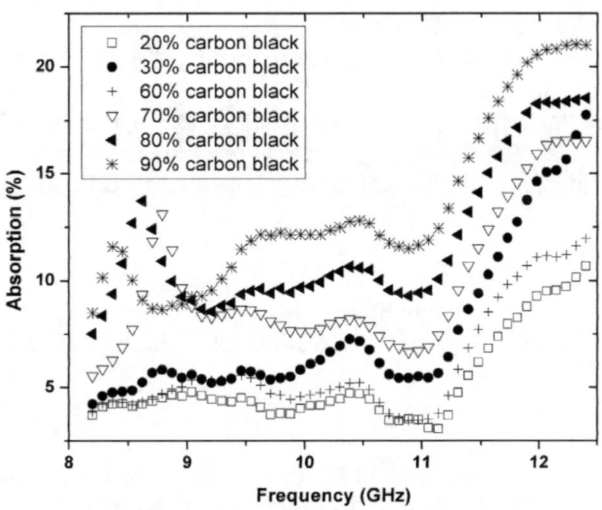

Figure 9. Absorption coefficient at X band

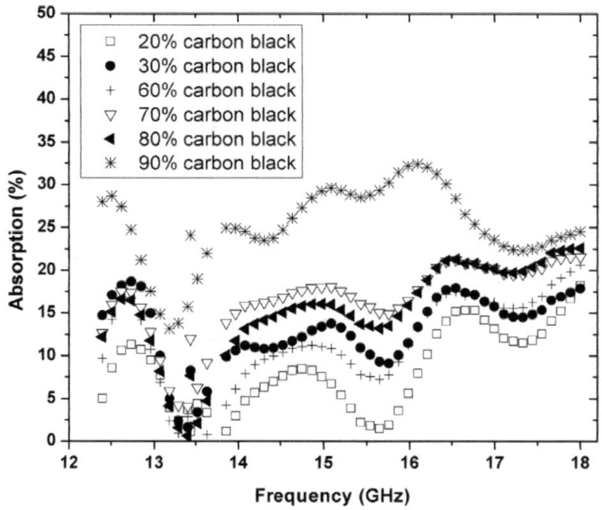

Figure 10. Absorption coefficient at Ku band

4. Conclusions

This study is conducted to investigate the dielectric properties and propagation mechanism in the hardened glass particle and carbon black using resin. The dielectric, reflection and transmission measurement are conducted using rectangular waveguide in conjuction with P-series network analyser. It can be observed that sample with high percentage of exhibit low dielectric constant, ε' at X band due to higher conductivity, whereas ε' shows drastic decrement at Ku band because there is polarization mechanism become inactive. It leads to a peak of ε" at Ku Band. This dielectric behaviors cause decrement and increment of reflection coefficient and transmission coefficient, respectively at both bands. This results justify the highest absorption up to 30 % at Ku Band.

Acknowledgement

The authors wish to express the highest gratitude to Ministry of Higher Education, Malaysia for funding this research under Research Acculturation Grant Scheme (RAGS) (Project no.: 9018-00058) and Universiti Malaysia Perlis for the provision of technical assistance and supports to accomplish this research project.

References

[1] Micheli D, Vricella A, Pastore R, and Marchetti M 2014 Synthesis and electromagnetic characterization of frequency selective radar absorbing materials using carbon nanopowders *Carbon N. Y.* **77** 756–774.

[2] Kaur M, Kakar S, and Mandal D Electromagnetic interference *2011 3rd Int. Conf. Electron. Comput. Technol.* pp. 1–5.

[3] Nam W, Lee H K, and Jang J H Electromagnetic interference shielding/absorbing characteristics of CNT-embedded epoxy composites 2011 *Compos. Part A Appl. Sci. Manuf.* **42** pp. 1110–1118.

[4] "Recycling," *Glass Packaging Institute*. [Online]. Available: http://www.gpi.org/recycling/glass-recycling-facts. [Accessed: 12-May-2015].

[5] Cheng E M, Abdul Malek M F, Khor S F, You K Y, Lee K Y, Rojan M A, Abu Bakar S, Mohd Nasir N F, Zakaria Z, and Tan W H Reflection and dielectric measurement for salinity of water using microstrip loop antenna and dielectric probe 2016 *Int. J. GEOMATE* **11** pp. 2335-2340.

[6] Che Mansor M F, Cheng E M, Abd. Malek M F, Beh H G, Chik A, Khor S F, Rojan M A, Mohd Nasir N F, Ong S A, and Abu Bakar S Study on 10 GHz electromagnetic wave absorption of cullet doped with Iron (II, III) Oxide *2015 2nd International Conference on Biomedical Engineering (ICoBE)* DOI: 10.1109/ICoBE.2015.7235922.

Effect of pressure and infill density parameter setting on morphological and mechanical properties of polycaprolactone printed scaffold using desktop 3D bioprinter

Khalida Fakhruddin[1], Saiful Izwan Abd Razak[1,2,3,*], Muhammad 'Ammar Murad[1], Nadirul Hasraf Mat Nayan[3,4] and Shafizah Sa'adon[1]

[1] School of Biomedical Engineering and Health Sciences, Faculty of Engineering, Universiti Teknologi Malaysia, 81300, Skudai, Johor, Malaysia
[2] Centre for Advanced Composite Materials, Universiti Teknologi Malaysia, 81300, Skudai, Johor, Malaysia
[3] Microelectric and Nanotechnology-Shamsuddin Research Centre (MiNT-SRC), Universiti Tun Hussein Onn Malaysia, 86400, Parit Raja, Johor, Malaysia.
[4] Faculty of Engineering Technology, Universiti Tun Hussein Onn Malaysia, KM1 Jalan Panchor, 84600, Pagoh, Johor, Malaysia

* saifulizwan@utm.my

Abstract. Three dimensional (3D) bioprinting is the process of building a 3D construct containing biological cells. There are varies set of printing parameter involved in 3D bioprinting system that will affect the structure and the mechanical properties of the bioprinted construct. This study was done to investigated the effect of extrusion pressure and infill density on the polycaprolactone (PCL) 3D printed construct's morphology, printing accuracy and compressive strength. Based on morphological evaluations using scanning electron microscope, the printed scaffold was suggested to be at the best condition at pressure 60psi using 80% infill setting. Besides, the mechanical testing revealed the printed scaffold was successfully fabricated without losing its mechanical integrity. This is a good indication for PCL as structural host material to obtain functionalized 3D scaffold using bioprinting method.

1. Introduction

Bioprinting is the process of dispensing cell-laden biomaterials for the construction of 3D functional living tissues or artificial organs [1]. On other hand, 3D printing technique also has been primarily used to create acellular 3D scaffolds and molds which then can be seeded with cells post-fabrication [2]. Material used to print in 3D bioprinting process such as combination of living biological cells, polymers, chemical factors, and biomolecules are known as bioinks [3]. There are three major methods of 3D bioprinting technologies that commonly used which are inkjet printing, laser-assisted printing, and extrusion printing [1].

Extrusion based bioprinting basically used a high air pressure (pneumatic) or mechanical force to dispense bioinks, by applying a non-stop force so that the extrusion printing can be printed uninterrupted rather than a single bioink droplet [1, 4]. Through this method, the resolution of 200 μm can be achieved, but compared to laser-based or inkjet-based this is considerably low [4]. In this system, the bioinks are required to be compatible where it has suitable rheological properties and crosslinking mechanisms to enable accurate and precise deposition. The material should be sufficiently viscous to be dispensed as a free standing filament and it must have sufficient strength and stiffness to maintain structural integrity after printing.

Polycaprolactone (PCL) is a semi crystalline polymer with about -60 °C for its glass transition temperature [5]. With just at point 58 to 60°C, PCL can be melt down, making it suitable with other polymers [6, 7]. Compared to polylactic acid (PLA), the degradation rate for PCL is much lower, thus make it valuable base polymers for a long term drug delivery devices. Besides, PCL possessed a good rheological and viscoelestic property among resorbable polymers which make it suitable candidate to be studied.

In bioprinting process, there are several parameter that need to be consider such as pressure, shear stress, temperature, feed rate, dispensing speed, and nozzle diameter [4]. All these parameter will affect the 3D printed construct structure and also their mechanical properties. Currently, there are limited studies focus on the accuracy of printing technique using this PCL. Here, this study aim to test the printability that is the ability of material to be accurately and precisely deposited with desired degree of spatial and temporal control. The morphology and mechanical properties of printed scaffold will be investigated as indication for best parameter printing. This study is important as it can be served as guideline to print 3D construct using PCL with high accuracy.

2. Methodology

2.1. Printing process

PCL (BioBots) in solid beads form, white in color and odorless was used in the printing process to generate 3D scaffold. A dual layered 9.9 mm x 9.9 mm x 1.8 mm grid was designed by using CAD software (SolidWorks). Then the .stl file was converted to .gcode file by loaded into repetier host (Slic3r) to adjust the preferred printing parameters. 5 mg of PCL were filled into metal syringe capped with 30 gauge metal needle. The design 3D construct was printed by using commercial desktop 3D bioprinter (Biobots, USA) at 100°C to allow the PCL to melt. The extrusion occurs at layer height of 0.05 mm and printing speed was set to 1 mm/s. The extrusion pressure was set at 60 psi, 80 psi and 100 psi with infill density percentage of 60%, 80% and 100%. The printed scaffolds then were analyzed using scanning electron microscope and compression test.

2.2. Scanning electron microscope

The post-print morphology and the geometry structure of the printed scaffold were observed using tabletop scanning electron microscope (SEM) (Hitachi TM3000) at 50X magnification and further analyzed by using imageJ software.

2.3. Accuracy

The SEM micrographs were analyzed using ImageJ to find dimensions of the printed area. The printed area, A_i (mm^2) was compared to the design area, A (mm^2). The printing accuracy for each sample was calculated using following equation, with average of three grids used.

$$\text{Printing accuracy (\%)} = \left[1 - \frac{(A_i - A)}{A}\right] \times 100 \qquad (1)$$

2.4. Compression test

Uniaxial compression tests were performed using universal tensile machine (INSTRON 5566) at a loading rate 0.1 mm/s with a load 10N load cell. The upper and lower sample surfaces were fixed to the platens of the tester to ensure there was no slippage during tests. For each test, three samples were used and their average values calculated.

3. Results and discussion

Microscopic examinations were carried out on the printed scaffolds at different pressure and infill density percentage. From the SEM image of the upper view of the printed construct, the construct shows the porous structure which indicates it is suitable for cell attachment (Figure 1a) [8]. Apart from that, the strand thickness also were observed and measured from SEM image using imageJ. The strand thicknesses of extruded PCL scaffold through 30G needle are shown in Table 1. The printed construct print with 60 psi and 80 % infill density produced the thinnest strand (1.25 mm) where printed construct print with 100 psi and 100 % infill density produced the thickest strand (1.53mm). It can be seen that at same infill density percentage, the increasing extrusion pressure will produced thicker strand width construct resulting from high extrusion pressure forces the PCL through the needle gauge. Action of gravity also might have an impact to the PCL ink as it will caused an increase in width, making pooling inevitable [9]. Therefore, there is a limit to the minimum achievable strand thickness.

The printing accuracy was calculated using previous mentioned formula (1). Results show that increasing the extrusion pressure with fixed infill density percentage results in less accurate printing (Figure 2). However, printing with increasing infill density with fixed extrusion pressure does not produce any specific pattern for the printing accuracy. Printing with 60 psi of extrusion pressure and 80 % infill density will produce highest accuracy (88.78%) of 3D printed construct. According to Di *et al.*, the final geometric accuracy is a combination of many factors, including the effects of depositing materials in successive layers [10]. Here, this study shows the effect of extrusion pressure on multilayered grid dimension and found that accuracy reduces with printing in increasing extrusion pressure. The compressive modulus was found highest (17.8 MPa) when printing at low pressure of 60 psi with 80% infill density percentage (Table 1).

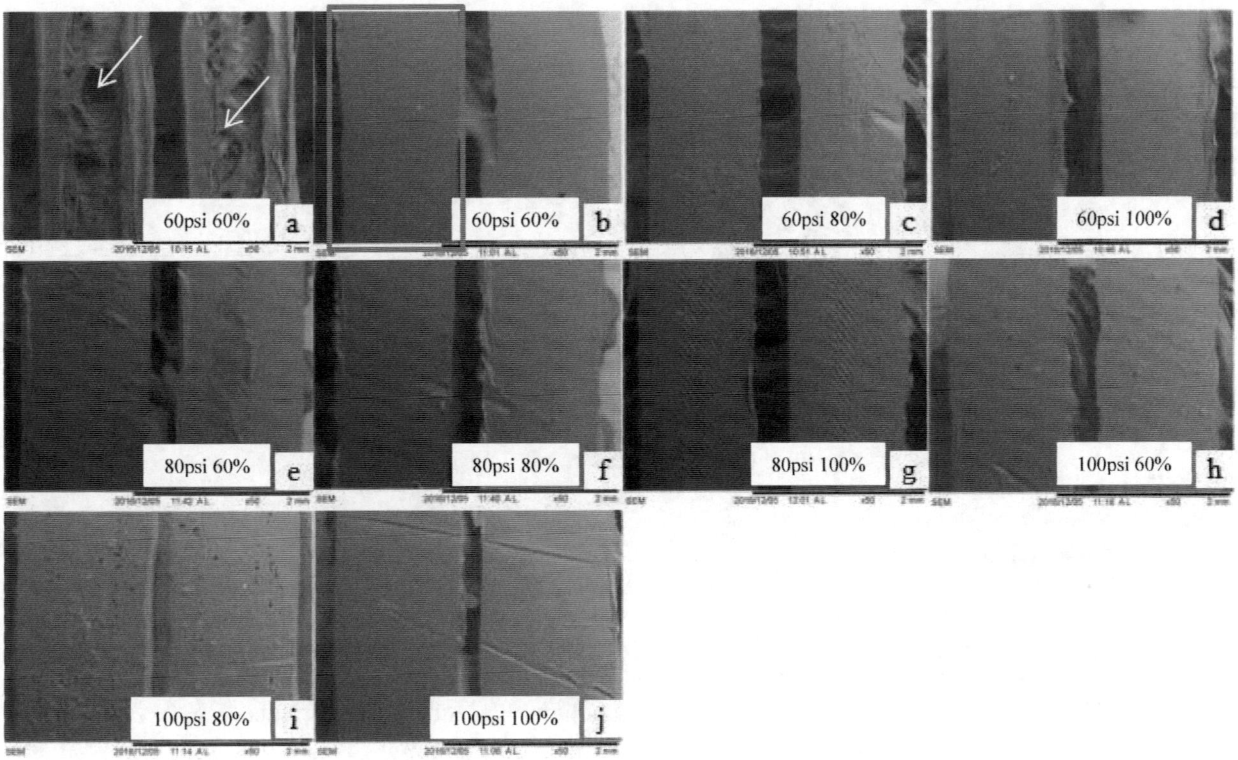

Figure 1 SEM image of the 3D printed construct. An upper view of the printed construct with porous structure showed with arrow (a), bottom view of printed construct, different strand (red box) thickness produced (b-j).

Table 1 Strand width, printing accuracy and compressive modulus of the 3D printed construct

Pressure (psi)	Infill Density (%)	Strand Width (mm)	Printing Accuracy (%)	Compressive Modulus (MPa)
60	60	1.31	87.12	17.1
	80	1.25	88.78	17.8
	100	1.35	88.26	16.3
80	60	1.43	86.74	16.1
	80	1.39	87.04	16.7
	100	1.39	87.64	16.3
100	60	1.41	86.72	16.3
	80	1.49	85.88	15.9
	100	1.53	85.99	16.5

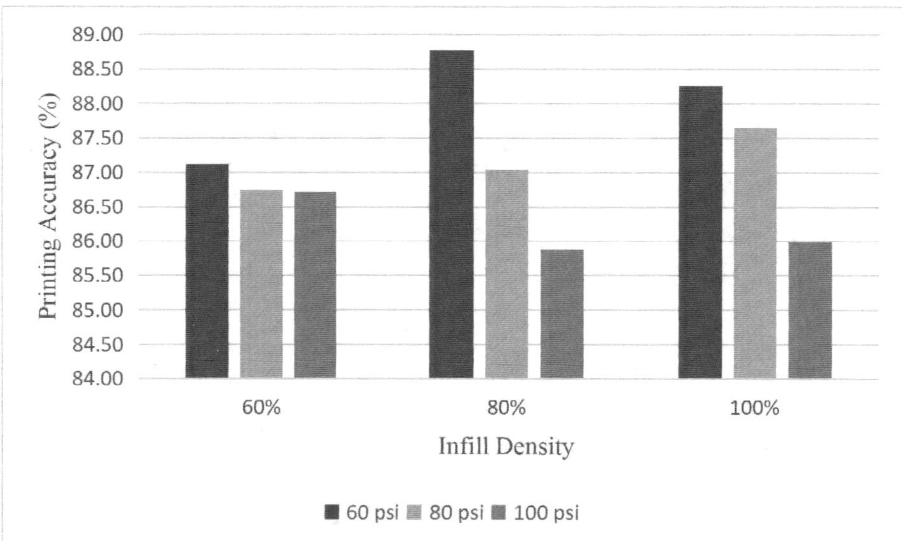

Figure 2 Printing accuracy of 3D printed construct

4. Conclusion

Herein, this study have presented and described the effect of extrusion pressure and infill density on the structure, print accuracy and mechanical properties of the printed construct. Based on the result, the best parameters achieve to fabricate dual layer PCL scaffold were at pressure 60psi and by using infill density of 80% where it produce construct with highest printing accuracy and compressive strength. From the value obtained, it can be concluded that the parameter successfully fabricate the scaffold construct with a good flow and without losing its mechanical strength.

Acknowledgment
The authors would like to acknowledge Ministry of Higher Education Malaysia for funding through Fundamental Research Grant Scheme.

References

[1] Mandrycky C., Wang Z., Kim K., Kim D.H..: 3D bioprinting for engineering complex tissues. Biotechnology advances (1996) 422-434
[2] Kolesky D.B., Truby R.L., Gladman A., Busbee T.A., Homan K.A., Lewis J.A.: 3D bioprinting of vascularized, heterogeneous cell-laden tissue constructs. Advanced materials (2014) 3124-3130
[3] Zhang L.G., Fisher J.P., Leong K.: 3D Bioprinting and nanotechnology in tissue engineering and regenerative medicine *Academic Press (*2015)
[4] Atala A. and Yoo J. J.: Essentials of 3D biofabrication and translation. Academic Press (2015)
[5] Gunatillake P.A.,Adhikari R.: Biodegradable synthetic polymers for tissue engineering. Eur Cell Mater. (2003) 1-16
[6] Chua C.K., Yeong W.Y.: Bioprinting: principles and applications (Vol. 1). World Scientific Publishing Co Inc. (2014)
[7] Woodruff M.A., Hutmacher D.W.: The return of a forgotten polymer—polycaprolactone in the 21st century. Progress in polymer science (2010) 1217-1256
[8] Hollister S.J.: Porous scaffold design for tissue engineering. Nature materials (2005) 518
[9] Chang C.C., Boland E.D., Williams S.K., Hoying J.B.: Direct-write bioprinting three-dimensional biohybrid systems for future regenerative therapies. Journal of Biomedical Materials Research Part B: Applied Biomaterials (2011) 160-170
[10] Di Guiseppe M., Law N., Webb B., Macrae R., Liew L.J., Sercombe T.B., Dilley R.J., Doyle B.J.: Mechanical Behaviour of Alginate-Gelatin Hydrogels for 3D Bioprinting. Journal of the mechanical behavior of biomedical materials (2017) 150-157

Predictive Systems: Role of Feature Selection in Prediction of Heart Disease

[1]Debjani Panda, [2] Ratula Ray, [3]Azian Azamimi Abdullah, [4]Satya Ranjan Dash*
[1]School of Computer Engineering, KIIT Deemed to be University, Patia, Bhubaneswar, Odisha, India, E-mail: pandad25@gmail.com
[2]School of Biotechnology, KIIT Deemed to be University, Patia, Bhubaneswar, Odisha, India, E-mail: ratularay@gmail.com
[3]School of Mechatronic Engineering, Universiti Malaysia Perlis, Malaysia
[4]School of Computer Applications, KIIT Deemed to be University, Patia, Bhubaneswar, Odisha, India, E-mail: sdashfca@kiit.ac.in

Abstract: As per recent trends heart disease has become the major factor for untimely deaths. There are huge amounts of clinical data available from biomedical devices and various applications used by hospitals. Artificial Intelligence is rigorously being used in predicting conditions of heart patients. This is mainly achieved by machine learning where a model is trained with sample cases and is then used for prediction of the ailment as per data available from clinical tests of the patient. This paper focuses in analyzing the accuracy of various classification algorithms, when they are supervised by set of features. Feature selection plays an important role in eliminating redundant and irrelevant features and reduces the training cost and time of the predictive models. The classification algorithms, which have been analyzed include Naive Bayes, Random Forest, Extra Trees and Logistic regression which have been provided with selected features using least absolute shrinkage and selection operator (LASSO) and Ridge regression. The accuracy of the classifiers shows remarkable improvement after using feature selection. The prediction has improved on an average by 33.3% using Lasso regression as compared to 30.73% using ridge regression.

Key words: Naive Bayes, Random forest, Ensemble Trees, Logistic regression, Lasso, Ridge

1. Introduction

Heart Disease has been identified as one of the most alarming factors causing untimely deaths worldwide [1]. The ailment is associated with several symptoms, which are due to reduced blood flow to the organs of the human body. A patient mostly suffers from edema, arrhythmia, and high blood pressure and has difficulty in breathing [2]. With onset of one or more symptoms, several clinical tests are performed which are mostly invasive in nature and also takes a considerable amount of time before the disease is predicted by the doctors. Early diagnostics with Electrocardiogram (ECG) signals [3, 4] and other data from blood tests can prevent fatal occurrences.

Considerable amount of research has already been done for predicting the heart ailment much faster based on some rudimentary information available with patient's history. These involve mostly ECG signals, cholesterol level, blood pressure, sugar levels in the patient's blood, exercise induced pain, etc. The patient's information is a resultant of data mining wherein the tests results are stored and retrieved from Hospital database when there is a need. The major challenge with the vast amount of information is the selection of the desired features, which will result in accurate prediction by the predictive algorithms.

1.1. Data Mining

Data Mining plays a very crucial role in storing data, retrieving useful data and making use of this data to retrieve desired results. In health sector, data mining has become all the more important due to its effectiveness in providing the classification algorithms, the desired set of features which directly impacts their performance. Huge sets of data are generated everyday from the biomedical devices and clinical tests of patients, which are stored in the database. The prediction models are built by training classification algorithms with sample data containing useful features/attributes .The selected features play a pivotal role in obtaining the desired accuracy from any prediction system.

1.2. Features Selection

Feature selection is one of the important aspects in data mining [5]. Its necessity is felt due to very high dimensionality of data sets and growing computational methodologies of the target problems. Data mining aids in storing huge data and these data is full of noise i.e. redundant and irrelevant features. Feature selection is the pre -processing step where the noise is filtered, resulting in reducing the dimensionality of the data set and aids in creating computationally effective models with less time and cost.

1.2.1. Objectives of feature selection

- Reducing the dimensionality without loss of originality.
- Improve accuracy of prediction of the model.
- Create a subset of M features, where M < N, so that value of criterion function is optimized over all subsets of M [6].

1.2.2. *Types of Feature Selection.* The basic steps involved in feature selection are **generation of subset**, **evaluation of subset**, **stopping criterion** and **validating the subset** [7]. Usually, three types of methods are used in feature selection viz. **Filter Method**, **Wrapper Method** and **Hybrid (Embedded) Method**.

1.2.2.1. *Filter Method.* These methods are applied before the processing of data as a pre-step before training models. This method is used where the classification algorithms are independent of the selected features. On the other hand the attributes are chosen on the basis of their output results/scores from different statistical methods used for their validation. These set of features are checked for their inter-dependency with output attribute so that the correct subset is chosen. Some examples of Filter method are Linear discriminate analysis (LDA), Pearson's correlation, Chi-Square, etc.

1.2.2.2. *Wrapper Method:* Wrapper Methods are involved in generating a subset of features and training the model. Then after measuring the accuracy of the model, the features are added or removed from the subset. This iteration goes until, the best subset is found. These methods are very resource hungry and also consume more time than filter methods. Examples of these types of methods are forward feature selection, backward feature elimination, recursive feature elimination, etc.

1.2.2.3. *Embedded Method:* This is a combination of both Filter and wrapper methods. Here algorithms have their own built in feature selection criteria. These help in generating the best subset and provide the same to the training model. This type of method normally gives much more accurate prediction. Examples of embedded (hybrid) method include LASSO regression, Ridge Regression, Memetic Algorithm, etc.

In this paper features selection has been done using Lasso and Ridge Regression. The accuracy of different classification algorithms are measured both before and after implementing feature selection.

- *LASSO Regression:* Lasso stands for Least Absolute Selection and Shrinkage Operator. The 'lasso' minimizes the residual sum of squares subject to the sum of the absolute value of the coefficients being less than a constant [8].
- *Ridge Regression:* This type of regression helps in dealing with variance caused due to multi-colinearity of variables. It helps in reducing the variance, which is a resultant of non-linear relationships between two independent variables. It performs L2 regularization by adding a penalty to square of magnitude of coefficients.

2. Literature survey
In our literature survey, the papers that we have studied have been summarized in Table 1 below.

Table 1. Detailed Analysis of Feature Selection Methods and Characteristics.

S. No.	Feature Selection Type	Classifiers Used	Data Set	Methods	Characteristics and Results
1.	Hybrid	Random forest	Statlog Heart Dataset	Rough Set Theory	Finding unique objects based on their attribute values. Calculates Lower approximation and Upper Approximation for each object. The lower value determines its belongingness to a class. [9]
2.	Hybrid	Regression Analysis models, SVM with different kernels, Decision Trees, MLP, Naive Bayes	Statlog Heart Dataset	Regression models, Fisher filtering (FF), and relief algorithm	SVM accuracy improved by 2.2% after feature selection and KNN gave better results. SVM with Relief gave 84.81% accuracy. [10]
3.	Hybrid	SVM, KNN, Levenberg-Marquardt neural network (LM NN), scalar conjugate gradient neural network (SCG NN).	ECG signals with 200 features	Improved version of bat algorithm for measuring accuracy	Best 20 attributes have been selected for study. LM NN was the best classifier with 98.9% accuracy. [11]
4.	Hybrid	SVM	UCI	Infinite Latent Feature selection technique.	When attributes are more (like 55 to 57 attributes) then performance of CFS (correlation Feature Search) and LLCFS (Local Learning based clustering Feature Search) is better. However, with less no. of attributes, CFS is much better than ILFS [12].
5.	Hybrid	Logistic regression, KNN, ANN, SVM, DT, NB and Random forest	Cleveland Dataset from UCI	Relief, minimal-Redundancy-Maximal-Relevance (mRMR), Shrinkage and Selection Operator (LASSO).	6 attributes selected from 13 after feature selection. Logistic regression shows improvement in accuracy from 84% to 89 % and SVM shows improvement in accuracy from 86 % to 88%. [13]
6.	Hybrid	Naive Bayes, SVM, C 4,5 Tree and Multi Layer Perceptron feed forward networks	UCI	Genetic Algorithm wrapped with Bayes-Naive classifier	Genetic algorithm with Bayesian Fitness function is iterated to get best set of features. GA with BN has given highest accuracy of 85.50%. [14]

3. Implementation/Methods

3.1. Flowchart

The workflow diagram of our experiment is attached in Fig.1 in which the performance of the data set is measured before and after applying feature selection techniques of Lasso and Ridge.

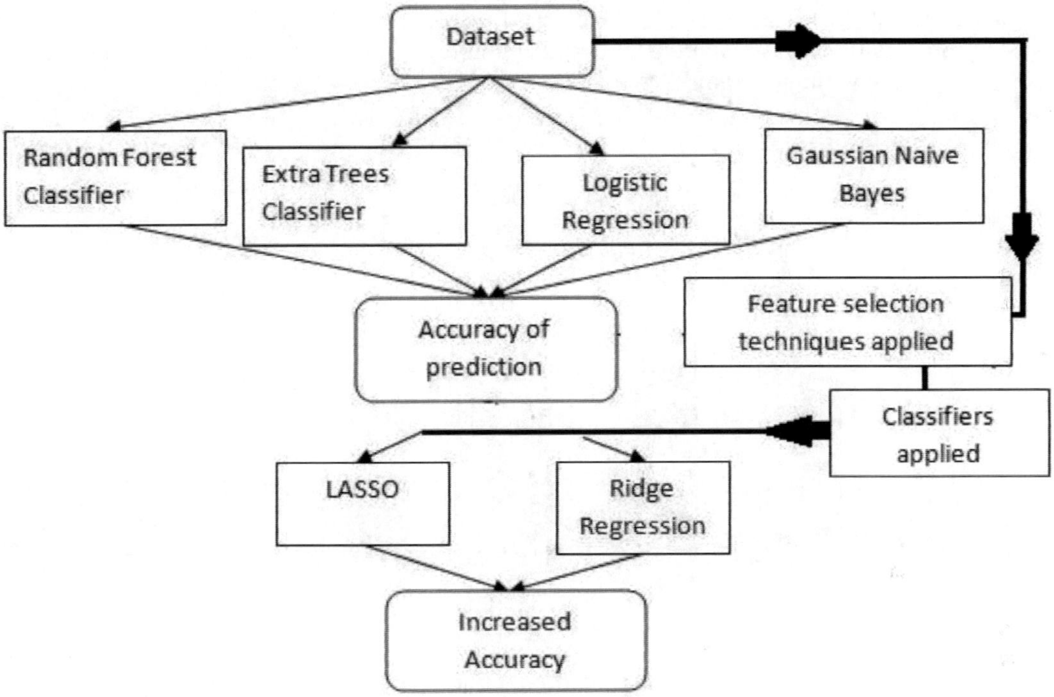

Figure 1. Workflow Diagram of Measurement of Prediction Models with and without Feature Selection.

3.2. Dataset

The Cleveland heart dataset has been used from UCI database [15]. This database contains 76 attributes, one of which is predicted. The attributes for patient identification have been dropped (attributes 1, 2, 75 & 76) and 72 attributes have been considered for the study. The dataset also contains missing values, which have been dropped. After removing the missing values, 115 instances have been considered.

3.3. Results and discussion

We have used the Accuracy as the metrics for statistical analysis of the datasets. Table 2 gives the result for the numeric data used for analysis of occurrence of heart disease in patients.

Table 2. Statistical results determining the performance of the classifiers for predicting heart disease.

Classifiers	Accuracy (Without Filters)	Accuracy (With Lasso)	Accuracy (With Ridge)
Random Forest Classifier	47.02%	84.98%	85.31%
Extra Trees Classifier	55.83%	90.32%	84.77%
Gaussian Naive Bayes Classifier	57.17%	94.92%	94.92%
Logistic Regression	40.73%	63.73%	59.12%

Amongst the four different classifiers used, we find Gaussian Naive Bayes Classifier shows remarkable results with an accuracy score of 94.92% after using feature selection methods like Lasso and Ridge regression. Dimensionality reduction helps in getting much more accurate predictions with the same data set. Also, we have observed that Lasso regression with classifiers has given better results in most of the cases. Figure 2 shows the graphical representation of the comparison of the classifiers with and without feature selection.

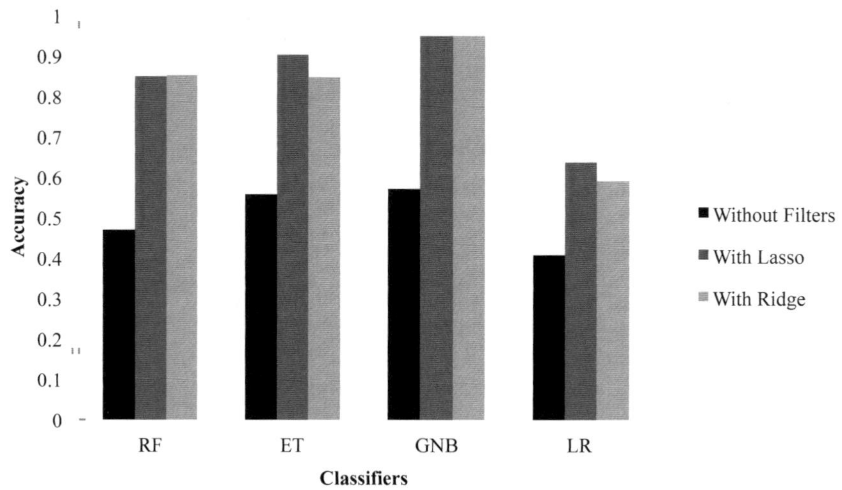

Figure 2. Graphical representation of the Random Forest (RF), Extra Trees (ET), Gaussian NB (GNB) and Logistic Regression (LR) classifiers with and without feature selection.

4. Conclusion and Future Prospects

The embedded feature selection methods using Ridge and Lasso regression involves using two types of penalty functions. L1 and L2 are loss functions, which help in minimizing the error in regression. Ridge regression uses L2 wherein the penalty is sum of squares of coefficients of the variables. In Lasso, the sum of the absolute values of the coefficients is considered as the penalty that is L1. Lasso basic objective is shrinkage of an absolute value (L1 penalty) towards zero rather than a using sum of squares (L2 penalty).

The limitation of ridge regression is that it cannot reduce the coefficients to zero. Here, either all or none of the coefficients are selected. On the other hand, LASSO performs both parameter shrinkage and variable selection, owing to its capability to reduce the coefficients of collinear variables to zero. It helps to select the required variable(s) out of the given n variables while performing lasso regression.

Future work can be carried out in combining optimization techniques with feature selection to study the impact on various classifiers and providing a cost and time effective approach for prediction of heart disease using non-invasive methods.

References

[1] Bui AL, Horwich TB, Fonarow GC. Epidemiology and risk profile of heart failure. Nature Reviews Cardiology. 2011 Jan;8(1):30.
[2] Durairaj M, Ramasamy N. A comparison of the perceptive approaches for preprocessing the data set for predicting fertility success rate. Int. J. Control Theory Appl.. 2016;9(27):255-60.
[3] Tripoliti EE, Papadopoulos TG, Karanasiou GS, Naka KK, Fotiadis DI. Heart failure: diagnosis, severity estimation and prediction of adverse events through machine learning techniques. Computational and structural biotechnology journal. 2017 Jan 1;15:26-47.
[4] Dash SR, Sheeraz AS, Samantaray A. Filtration and Classification of ECG Signals. In Handbook of Research on Information Security in Biomedical Signal Processing 2018 (pp. 72-94). IGI Global.
[5] Urbanowicz RJ, Meeker M, La Cava W, Olson RS, Moore JH. Relief-based feature selection: introduction and review. Journal of biomedical informatics. 2018 Jul 18.
[6] Narendra PM, Fukunaga K. A branch and bound algorithm for feature subset selection. IEEE Transactions on computers. 1977 Sep 1(9):917-22.
[7] Dash M, Liu H. Feature selection for classification. Intelligent data analysis. 1997 Jan 1;1(1-4):131-56.
[8] Tibshirani R. Regression shrinkage and selection via the lasso. Journal of the Royal Statistical Society: Series B (Methodological). 1996 Jan;58(1):267-88.
[9] Yekkala I, Dixit S. Prediction of Heart Disease Using Random Forest and Rough Set Based Feature Selection. International Journal of Big Data and Analytics in Healthcare (IJBDAH). 2018 Jan 1;3(1):1-2.
[10] Takci H. Improvement of heart attack prediction by the feature selection methods. Turkish Journal of Electrical Engineering & Computer Sciences. 2018 Jan 27;26(1):1-0.
[11] Kora P, Kalva SR. Improved Bat algorithm for the detection of myocardial infarction. SpringerPlus. 2015 Dec 1;4(1):666.
[12] Le HM, Tran TD, Van Tran La. Automatic heart disease prediction using feature selection and data mining technique. Journal of Computer Science and Cybernetics. 2018 Aug 8;34(1):33-48.
[13] Haq AU, Li JP, Memon MH, Nazir S, Sun R. A Hybrid Intelligent System Framework for the Prediction of Heart Disease Using Machine Learning Algorithms. Mobile Information Systems. 2018;2018.
[14] Mokeddem S, Atmani B, Mokaddem M. Supervised feature selection for diagnosis of coronary artery disease based on genetic algorithm. arXiv preprint arXiv:1305.6046. 2013 May 26.
[15] Frank A, Asuncion A. UCI Machine Learning Repository [http://archive.ics.uci.edu/ml]. Irvine, CA: University of California. School of information and computer science. 2010;213:2-.

Journal of Physics: Conference Series **1372** (2019) 012075 doi:10.1088/1742-6596/1372/1/012075

Gait Asymmetry Assessment using Muscle Activity Signal: A Review of Current Methods

Husnir Nasyuha Abdul Halim [1], Aizreena Azaman [1,2*], Haidzir Manaf [3], Syafiqah Saidin[1,4], Izwyn Zulkapri[1,2], Azli Yahya[1]

[1]School of Biomedical Engineering and Health Sciences, Universiti Teknologi Malaysia, Johor Bahru, Johor, MALAYSIA
[2]Sport Innovation Technology Center, Universiti Teknologi Malaysia, Johor Bahru, Johor, MALAYSIA
[3]Centre of Physiotherapy, Faculty of Health Sciences, Universiti Teknologi MARA Puncak Alam Campus, 43200 Puncak Alam, Selangor, MALAYSIA
[4]IJN-UTM Cardiovascular Engineering Centre, Institute of Human Centered Engineering, Universiti Teknologi Malaysia, Johor Bahru, Johor, MALAYSIA

*aizreena@biomedical.utm.my

Abstract. Gait asymmetry is a type of gait characteristics when there is difference in gait parameters statistically, measured bilaterally between left and right limbs. Gait asymmetry assessment is used to observe changes or deviation in gait due to pathological condition, effect of rehabilitation program or to give insight on effect of gait on stability and fall-risk. The assessments of gait asymmetry could be measured by using spatiotemporal, kinetics, kinematics parameters or by analysis of muscle activity signals obtained from surface electromyography (EMG). However, EMG-based assessment for gait asymmetry is not well explored compared to assessment using other gait parameters. This review aims to compare research designs, methods and procedure of previous studies that utilized EMG for gait asymmetry analysis. Therefore, any research in the future that involved gait asymmetry measurement could take note on and produce more reliable findings.

1. Introduction

Gait is defined as patterns or manner of locomotion from a point to a point such as walking, running, hopping, and skipping as the result of limb movement across the body [1]. It encompasses high interaction of central nervous system and musculoskeletal system to ensure the body stay upright in stable manner during movement [2].

The purpose of quantitative gait analysis is to study the biomechanics of walking and consequently assist in treatment or rehabilitation of patients with abnormal gait [3]. There are various kinds of data that are able to describe gait pattern such as spatiotemporal, kinetics, kinematics and also muscle activity [4]. Application of gait analysis is primarily for medical purpose however; it is extensively applied for other purpose such as sport science [5].

Gait asymmetry characteristics is one of the assessments that can be analysed to measure deviation in gait between right and left limbs. Gait asymmetry assessment could give insight on effect of pathological condition such as stroke, cerebral palsy, Parkinson's disease, osteoarthritis and other medical condition concerning on gait such as injury or orthoses usage [6]. It also gives information on stability or risk of fall of able-bodied gait. Electromyography (EMG); the recording of muscle electrical activity also had been used in gait asymmetry measurement although it is not extensively used compared to other gait parameters [7]. By using EMG, extended information on walking pattern could be provided as it describes muscle activation, the process that initiate gait cycle.

2. Methodology

In this review, method for data acquisition and processing of surface EMG for gait analysis will be overview from previous studies. Aspect of study design such as subjects, selected assessed muscles, gait procedure, and asymmetry measures methods will also be discussed thoroughly. The search will be performed through electronic databases (ScienceDirect, IEEE Xplore and Web of Science) to find research conducted from 1995 to 2018 with the keyword gait, symmetry, asymmetry combined with electromyography (EMG).

There are total of 18 journal articles and conference proceedings [8-25] selected from the literature search of research done from year 1995 to 2018. All literatures were collected from databases of ScienceDirect, IEEE Xplore and Web of Science. The details of journal selection procedure are presented in figure 1.

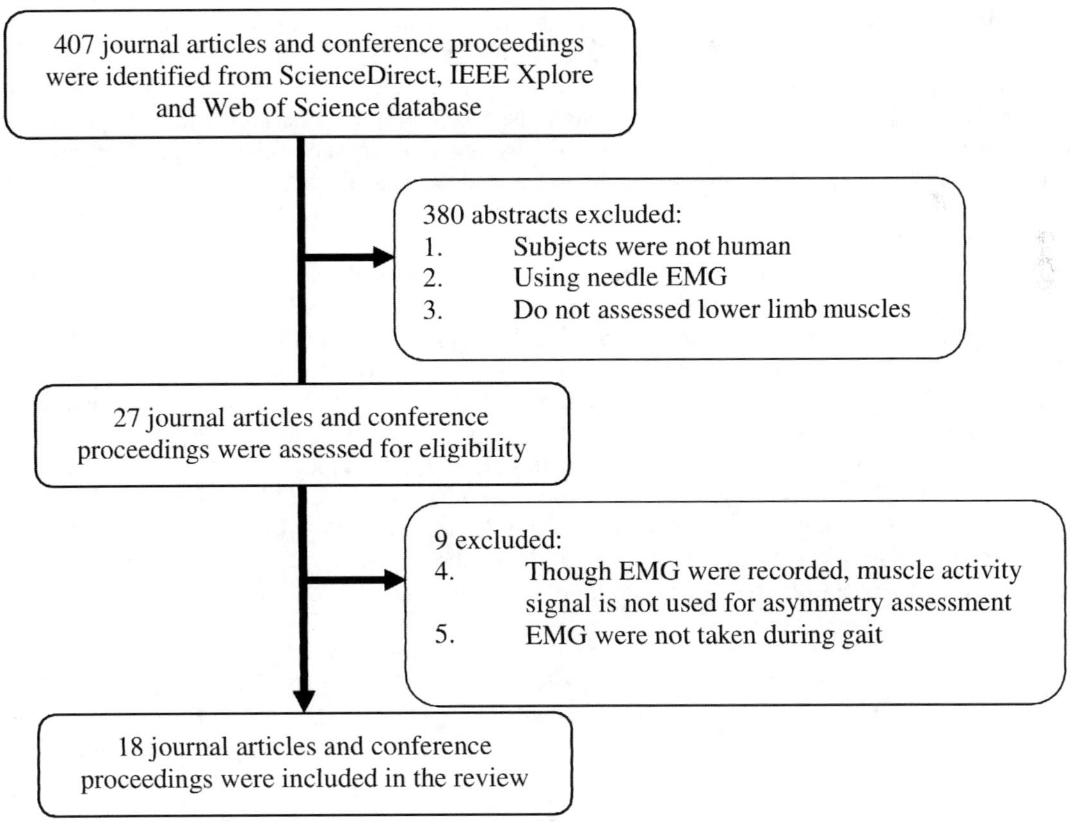

Figure 1. Flowchart of selecting journal articles and conference proceedings.

Journal of Physics: Conference Series **1372** (2019) 012075 doi:10.1088/1742-6596/1372/1/012075

3. Results and Discussion

The studies included mostly had observational designs to observe gait asymmetry because of certain condition while other studies were experimental studies to evaluate the impact of gait training program or rehabilitation [8, 14], assistive device during gait [11, 16, 25] and after-effect of surgery [12, 19]. The studies reviewed here involved subjects in various kind of condition that causes asymmetric gait for example neuromusculoskeletal diseases such as stroke [10, 13, 20, 22, 24, 25], osteoarthritis [21], multiple sclerosis [18], cerebral palsy [18], cerebral hypoxia [18], and Parkinson's disease [8], hemiparesis [11, 18], patients with abnormal gait because of surgery [12], traumatic brain injury [18], amputees [9, 16], spinal cord injury [14], ligament injury [19], and genetic pathology [18]. Normal subjects were involved in majority of the studies for comparison purpose but there are three studies that involved only normal subjects to analyse dominant vs non-dominant limb effect on gait [15], balance in normal people [23] and another study that modulated limping gait on healthy people using split-belt treadmill [17].

3.1. Study Protocol and Muscles Assessed

Most studies record data of walking gait at preferred speed of the subjects except [16, 17, 23, 25] but all with different length of walkway ranging from 3 to 15 m. Some studies used treadmill [11-13, 17, 23, 25] and allow or purposely used assistive device for experimental purposes such as orthosis [14, 18, 25], prostheses [16] and other walking aid [18, 25].

Only one study stated the reason of choosing muscles assessed which is [18] that referred to physical medicine doctors and rehabilitation physicians to give recommendation on which muscles that best represent gait. The two muscles that were assessed in majority of the studies are tibialis anterior (TA) [13, 14, 18, 20, 21, 23, 25] and gastrocnemius muscles [13, 14, 18, 19, 21, 22, 25]. Both muscles are the prime mover of actions at the ankle joint. Tibialis anterior is the muscle that responsible for dorsiflexion and gastrocnemius playing the role in plantar flexion movement [26].

Other muscles that were assessed are also prime mover that are responsible for other two joints of lower limbs; hip and knee, for example vastus lateralis, biceps femoris and rectus femoris [26]. Study by Burnett et al. [15] also assessed the muscles at the trunk and abdomen because the study protocol is not only involved walking but also involve sit-to-stand, and stand-to-sit tasks.

3.2. EMG Signal Processing Technique

The EMG data were sampled at frequency ranging from 500-2160Hz but the majority of the studies recorded with sampling rate of 1000Hz [9]. The Nyquist rate state that sampling rate must be at least double the highest frequency components of the signal to ensure the signals are not distorted [27]. For surface EMG, the frequency components are between 400-500Hz so the studies reviewed here follows the recommended sampling rate which is in the range of 800-1000Hz [27]. For the signal conditioning of EMG, all studies vary on the type and cut-off frequency of the filters based on the needs of the raw signals. Based on all studies included, only four studies [15, 17, 22, 23] did not mention rectification of the signals. A study [25] done the signal processing of EMG based on SENIAM recommendation where firstly, the data should be high-pass filtered using 4th order Butterworth with a cut-off frequency of 10 Hz in order to remove motion artefacts. Then, the data should be full-wave rectified and low-pass filtered, also with 10Hz cut-off frequency of 4th order Butterworth.

3.3. EMG Asymmetry Assessment Approaches

Although EMG is often used in gait analysis, it rarely used for gait asymmetry quantification compared to other gait data; spatiotemporal, kinetics or kinematics data [7]. Several studies quantify asymmetry of gait using EMG signals in order to especially analyse muscle activation pattern in pathologic gait [8, 10-14, 18, 20, 21], however on the remaining studies EMG asymmetry were compared with other gait parameters. According to S. Viteckova et al. [7], gait asymmetry assessment approaches can be classified into four types; discrete, complete gait cycle, statistically-based and nonlinear methods. Every approach has pros and cons depending on the type of gait data used for the assessment, the need or purpose of the study.

Majority of the studies assessed asymmetry characteristics of EMG signals using statistical approaches such as cross-correlation [8], auto-correlation [14], t-tests [9, 19], ANOVA [12, 16, 25], or mean and standard deviation comparison [11, 15, 22]. In the other hand, other studies measure asymmetry using discrete approaches [15, 17, 20, 21, 23] with different kind of formulas to quantify asymmetry index of selected EMG data. The remaining studies observe qualitatively the asymmetry of the EMG profile based on its shape and timing. The details of the asymmetry assessment of EMG are as shown in table 1 and table 2 below.

Table 1. EMG asymmetry using statistical approaches

Method		**EMG feature**
Statistical	Cross-correlation	Latency corrected ensemble average (LCEA) of each muscle over the strides in a trial [8]
	Auto-correlation	EMG profile [14]
	t-tests	-Mean of integrated EMG and mean of ratios during first and second half of stance time and swing time [9] -Mean and SD of knee contact force estimated using EMG-driven musculoskeletal model of the knee
	ANOVA	-Peak activity of the gastrocnemius muscle [12], integrated EMG [16] -Ensemble averaged EMG profiles [25]
	Mean and SD	-EMG linear envelope [11] -Root mean square error [22] -Muscle synergies based on EMG envelope [24]

Table 2. EMG asymmetry assessment using discrete calculation

Method	**Reference**	**EMG feature**	**Formula**
Discrete measurement	[15]	RMS during stance phase	Symmetry Index $SI = \frac{RMS_{ND,stance}}{RMS_{D,stance}}$ ND: non-dominant limb D: dominant limb
	[17]	Fast and slow leg muscle contributions C_m during tied-belt and split-belt walking on a treadmill [17]	$SI(\%) = \frac{C_{m,fast} - C_{m,slow}}{\left(\frac{1}{2}\right)(C_{m,fast} + C_{m,slow})} \times 100$
	[20]	Maximum root mean square (RMS) during stance phase	$Symmetry\ ratio = \frac{paretic\ side\ parameter}{non-paretic\ side\ parameter}$

	[21]	Mean muscle activity (mV) during the complete gait cycle	Asymmetry Index (ASI) $ASI(\%) = \lvert \frac{2 \times (MA_L - MA_R)}{MA_L + MA_R} \rvert \times 100\%$ MA_L and MA_R representing individual mean muscle activities of the left and right limb, respectively
	[23]	Mean EMG integral	Symmetry index $SI = \frac{(X_R - X_L) \times 2}{(X_R - X_L)} \times 100$ Where X_R is signal from right leg and $X_L$is signal from left leg

4. Conclusion

This paper intended to review methodologies used for gait asymmetry assessment using muscle activity signal measured by surface EMG. Compared to other gait features, muscle activity signals using EMG is least used to measure gait pattern especially for asymmetric characteristics. It was not possible to recommend the best methods considering various factors such as subjects involved, joint of interest, and the purpose of study. However, this review highlighted the previous studies methodologies to observe the trend so that future research could refer and acquire more reliable findings. Future research will be able to compare and find the relationship of gait asymmetry measures from EMG with other method or technology in gait studies.

Acknowledgements

The authors would like to express the highest gratitude to our colleagues from Universiti Teknologi Malaysia and UiTM Puncak Alam how provided insight and expertise that greatly assisted the manuscript.

References

[1] Kirtley, C.: Introduction in Clinical Gait Analysis., Churchill Livingstone: Edinburgh. (2006) 5-14.
[2] Sadeghi, H., et al.: Symmetry and limb dominance in able-bodied gait: a review. Gait & Posture, (2000) **12**(1): 34-45.
[3] Whittle, M.W.: Chapter 4 - Methods of gait analysis, in Gait Analysis (Fourth Edition)., Butterworth-Heinemann: Edinburgh. (2007)137-175.
[4] Introduction: Theory and practice in gait analysis A2 - Kirtley, Chris, in Clinical Gait Analysis. (2006) Churchill Livingstone: Edinburgh. ix-xii.
[5] Muro-de-la-Herran, A., Garcia-Zapirain, B., and Mendez-Zorrilla, A.: Gait Analysis Methods: An Overview of Wearable and Non-Wearable Systems, Highlighting Clinical Applications. Sensors (Basel, Switzerland), (2014) **14**(2): 3362-3394.
[6] Cabral, S., et al.: Inter-session agreement and reliability of the Global Gait Asymmetry index in healthy adults. Gait & Posture. (2017) **51**: 20-24.
[7] Viteckova, S., et al.: Gait symmetry measures: A review of current and prospective methods. Biomedical Signal Processing and Control.(2018). **42**: 89-100.
[8] Miller, R.A., et al.: Components of EMG symmetry and variability in parkinsonian and healthy elderly gait. Electroencephalography and Clinical Neurophysiology/Electromyography and Motor Control. (1996). **101**(1): 1-7.
[9] Isakov, E., Keren, O., and Benjuya, N.: Trans-tibial amputee gait: Time-distance parameters and EMG activity. Prosthetics and Orthotics International. (2000) **24**(3): 216-220.

[10] Roetenberg, D., et al.: Surface electromyography analysis for variable gait. Gait & Posture. (2003) **18**(2): 109-117.

[11] Hwang, I.S., et al.: Electromyographic analysis of locomotion for healthy and hemiparetic subjects—study of performance variability and rail effect on treadmill. Gait & Posture.(2003) **18**(1): 1-12.

[12] De Visser, E., et al.: Altered phase-transitions in tibialis anterior and medial gastrocnemius during walking after limbsaving surgery. Clinical Neurophysiology. (2005) **116**(12): 2741- 2747.

[13] Higginson, J.S., et al.: Muscle contributions to support during gait in an individual with post-stroke hemiparesis. Journal of Biomechanics. (2006) **39**(10): 1769-1777.

[14] Wang, P. and K.H. Low: Qualitative evaluations of gait rehabilitation via EMG muscle activation pattern: Repetition, symmetry, and smoothness. in 2009 IEEE International Conference on Robotics and Biomimetics (ROBIO). (2009).

[15] Burnett, D.R., et al.: Symmetry of ground reaction forces and muscle activity in asymptomatic subjects during walking, sit-to-stand, and stand-to-sit tasks. Journal of Electromyography and Kinesiology. (2011) **21**(4): 610-615.

[16] Fey, N.P., G.K. Klute, and R.R. Neptune: The influence of energy storage and return foot stiffness on walking mechanics and muscle activity in below-knee amputees. Clinical Biomechanics. (2011)**26**(10): 1025-1032.

[17] Jansen, K., et al.: Muscle contributions to center of mass acceleration adapt to asymmetric walking in healthy subjects. Gait & Posture (2013) **38**(4): 739-744.

[18] Bervet, K., et al.: KeR-EGI, a new index of gait quantification based on electromyography. Journal of Electromyography and Kinesiology. (2013) **23**(4): 930-937.

[19] Gardinier, E.S., et al.: Knee Contact Force Asymmetries in Patients Who Failed Return-to-Sport Readiness Criteria 6 Months After Anterior Cruciate Ligament Reconstruction. The American Journal of Sports Medicine. (2014) **42**(12): 2917-2925.

[20] Jeong, M.-G. and J.-H. Kim: Comparison of Both Legs EMG Symmetry during Over-Ground Walking and Stair Walking in Stroke Patients. The Journal of Korean Physical Therapy, (2015) **27**(4): 228-233.

[21] Schmidt, A., et al.: Unilateral hip osteoarthritis: Its effects on preoperative lower limb muscle activation and intramuscular coordination patterns. Gait & Posture. (2016) **45**: 187-192.

[22] Wang, W., et al.:Post-stroke asymmetry of muscle contractions for knee flexion and extension during walking. in 2016 9th International Congress on Image and Signal Processing, BioMedical Engineering and Informatics (CISP-BMEI). (2016).

[23] Park, Y., et al.: A gait imbalance evaluation system using multi-sensors. in 2016 IEEE Conference on Computational Intelligence in Bioinformatics and Computational Biology (CIBCB). (2016).

[24] Barroso, F.O., et al.: Combining muscle synergies and biomechanical analysis to assess gait in stroke patients. Journal of Biomechanics. (2017) **63**: 98-103.

[25] van Kammen, K., et al.: Differences in muscle activity and temporal step parameters between Lokomat guided walking and treadmill walking in post-stroke hemiparetic patients and healthy walkers. Journal of NeuroEngineering and Rehabilitation. (2017) **14**: 32.

[26] Marieb, E.N. and K. Hoehn: Human Anatomy & Physiology. 2013: Pearson.

[27] Ives, J.C. and J.K. Wigglesworth: Sampling rate effects on surface EMG timing and amplitude measures. Clinical Biomechanics. (2003) **18**(6): 543-552.

Investigation on Green Properties of Magnesium Alloy AZ31 Feedstocks Molded via Metal Injection Molding Process

Nooraizedfiza Zainon[1*], **Mohd Afian Omar**[2], **Nur Liyana Mohd Rosli** [1], **Marina Marzuki** [1]

[1]School of Manufacturing Engineering, Universiti Malaysia Perlis, Kampus Tetap Pauh Putra, 02600 Arau, Perlis, Malaysia.
[2]Advance Materials Research Center (AMREC), SIRIM Bhd., Kulim Hi-Tech Park, Kedah, Malaysia.

*nooraizedfiza@unimap.edu.my

Abstract. Powder loading is an important parameter because of its role in part shrinkage, dimensional stability, and final density. Thus, this research investigated the properties of the molded magnesium feedstocks in determining the best powder loading for producing magnesium part through metal injection molding process. In this research, the feedstocks with three magnesium alloy AZ31 powder loading of (63, 65, 67) vol%, was prepared comprising the binder component of paraffin wax, high density polyethylene, waste plastic, and stearic acid (PW/HDPE/WP/SA) with weight fraction 55/21/14/10, respectively. After injection molding, the density, diamensional accuracy, and strength of the green part was investigated. The findings reveal that 65 vol% of powder loading resulting to the best green properties.

1. Introduction

The typical Metal Injection Molding (MIM) process comprises four stages which named mixing, injection molding, debinding and sintering. The mixture of powder and the binder system is called the feedstock which produced by the first step of MIM process. The feedstock is expected to be homogenous where each powder particles has been enveloped by a very thin film of binder and has a tight contact to each other. The feedstock desired to provide the rheological properties for injection molding [1-3]. The second stage in MIM is injection molding. The feedstock is injected into a desired shape by heating in the molding machine and forcing it under pressure into the tool cavity. This is accomplished by heating the feedstock to a temperature which allows it to flow. The temperature should be under the degradation temperature of the binder. Pressure is maintained on the feedstock during cooling in the mold to minimize the void formation [4-5]. Once sufficient cooling achieved, the green molding is ejected from the mold that has an adequate strength to be handled. Conventional plastic injection molding equipment is used to mold the feedstock into desired shapes. Although the flow behavior of MIM is quite similar to the plastic injection molding, typical MIM feedstocks flow less easily and freeze faster than the plastics. This is due to their high viscosity and high thermal conductivity which results in incomplete mold filling, especially in thin sections [6-8].

The solid loading is an important parameter in metal injection molding because of its role in part shrinkage, dimensional stability, and final density. Low solids loading causes part dimensional variations during debinding and sintering, as well as product with low density. On the other hand, if the solid concentration is very high, the feedstock will be inhomogeneous and will not flow steadily in the injection molding machine generating dimensional inconsistencies in molded parts, with each pat having different dimensions [9-12]. In recent years there have been many investigations regarding powder loading of metal injection molding feedstock. Poh et al., [13] uses the thermogravimetric analysis and differential scanning calorimetry were used to determine the volume fraction of powder. The result reveals the highest powder loading was 46.96%. They also proved that TGA suitable for quantifying the point to point powder fraction in green compacts for ceramic materials as well as quality control of the feedstock. Aggarwal et al. [14] determined the optimal and critical solid loading based on simulation and rheological behavior (such as flowability, activation energy, and moldability index). Sotomayor et al. [15] studied on powder injection molding of premixed ferritic and austenitic stainless steel powders. They found that the optimum powder loading was established at 68 vol.%. Hidalgo et al. [16] reported the optimal loading of Invar 36 feedstocks resulted to be 57.5 vol.%. Schaper et al., [17] has successfully introduced AZ81 into the MIM process. However, the evaluation on green properties was not detailed discussed.

Thus, the aim of this study is to investigate the green properties of magnesium feedstocks henceforth the best feedstock formulation are determined. The density, dimensional accuracy, and green strength of the molded part have been studied.

2. Methodology

The Mg AZ31 feedstocks which containing certain fraction of metal powder, wetting agent, backbone polymer, and additives were developed to produce the green molded part. The magnesium powder with an average particle size of 22 μm supplied by MGH Industrial Co. Ltd, China with density of 1.734 g/cm^3 was used as a metal powder. The high density polyethylene (HDPE) with melting point of 134.5 °C which supplied by Titan Polyethylene (M) Sdn. Bhd, Johor Bharu and waste plastic (WP) with melting point 135 °C which supplied by SLT plastic Sdn. Bhd. Penang were used as a backbone polymer for the green molded parts with weight fraction of 21% and 14 %, respectively. Paraffin wax (PW) with a molecular weight of 380 g/mol and melting point of 59.4 °C used as a wetting agent. While, stearic acid (SA) with molecular weight 284.48 g/mol and melting point of 62.2 °C was used as a additives. Firstly, the feedstocks was prepared using Brabender at 160 °C at 50 rpm for 90 minutes. Three batches of the feedstocks were prepared with different powder/binder volume percentage which are (63/37), (65/35), and (67/33) as studied previously [18]. Then, the Mg feedstocks were molded on the vertical MCP 100KSA injection molding machine in tensile bar shape. According to [9] temperature used for the injection molding process was 140°C.

After injection molding, the green molded part were visually inspected to be free from any molding defect such as cracking, surface voids or distortion. The dimensions of the parts were measured and recorded. The recorded dimensions were very useful for further investigation, so that the changes in dimensions in subsequent processing could be followed. Each specimen was measured along its length and in three places, roughly equidistant from one another, across its thickness and width. The binder distributions in green mold parts were also examined on both the free and fracture surface using Hitachi TM3000 Tabletop Scanning Electron Microscope (SEM).

After injection molded, the molded part was then being measure for its density. The density of the molded part was then measured by applying Archemedes principles (ASTM B 962-14) [19] using electronic densimeter MD 300S. The densimeter machine provides a precision analytical balance which will permit readings within 0,01% of the test specimen and a beaker or vessel that is suitable for holding water of sufficient adaptability to immerse the specimens of various size. The test specimens may be supported in a wire basket. The samples were weighted in air and then in water. The density and the porosity of the molded part is given by Equation 1 and 2, respectively.

$$\rho_g = \frac{\rho_w W_a}{W_a W_b} \tag{1}$$

$$P_a = \left(\frac{W_c - W_a}{W_c - W_b} \right) \times 100 \quad (2)$$

where, ρ_w is a density of water, W_a is a weight of the sample on the air, W_b is a weight of the sample in water (g), and W_c is a weight after immersed in water.

The sample was then being measured for the green strength. The Autograph Shimadzu Precision Universal Tensile Machine used to perform the three point bending test (ASTM B 925-15) [20] on the molded part. 30 mm of support rollers with 50 mm span length used in the test and the peak load at fracture was recorded. The strength (S_g) of the green molded part was then calculated using Equation 3.

$$S_g = \frac{3PL}{2t^2 w} \quad (3)$$

Where P is a force required re to rupture, L is a length of the specimen span of fixture, t is a thickness of the sample, w is a width of the sample.

3. Results and discussions

Figure 1 shows the tensile shape of the injection molded parts (green molded parts) with a single gate which located at the end of the sample. At the beginning of the molding process, the molded parts facing several normal defects observed such as flow mark and powder/binder separation as shown in the Figure 2 (a) and (b), respectively. The flow mark may occurs due to insufficient heat on the Mg feedstock in the mold. Besides, the flow mark defects also may occurs due to the unstable flow at the flow front. Therefore, lowering injection speed and increasing the Mg feedstock temperature was done to troubleshoot this defect. Another defect observed was poor appearance around the gate which called powder/binder separation defect.

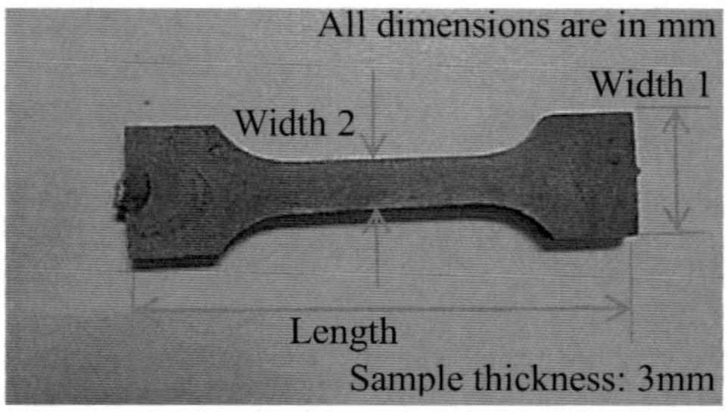

Figure 1. The dimensions of the green molded part.

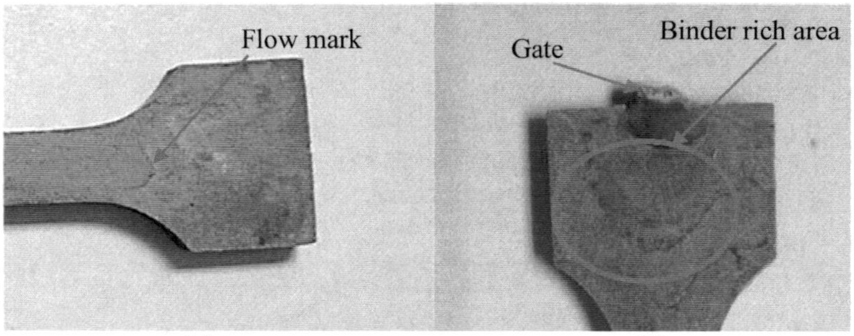

(a) **(b)**

Figure 2. The defects occurs on the molded parts (a) Flow mark (b) powder/binder seperation.

As the Mg feedstock passes through the narrow gate into the mold cavity, binder separation can occur due to powder migration from high shear rate region to a low shear rate region. The powder/binder separation become more serious at the end of filling stage and during the pressure holding stage wherein the Mg feedstock flows slowly and the pressure builds up rapidly. To fill in the clearance in the die cavity resulting from thermal shrinkage, more feedstock is pushed into these openings by the holding pressure. Since the movement of the metal powders in the feedstock is slow, due to the high interparticle friction, the binder which has low viscosity was forced to flow through interparticle intersections. Thus, the binder content at the gate areas is increased, causing the molding defects, particularly at the gate, where the shear rate is the highest. The separation will occur easily and form binder rich areas at the surface of the part near the gate.

According to [21] the gate redesign helps to alleviate this powder/binder separation problem. If the original gate has a narrow opening and is located at a thin section, it can be enlarged and relocated to a thick section. As a result. The level of powder/binder separation will be reduced and the flow marks will be minimized. However in this study, no redesign were made on the mold since thoroughly investigation are required on redesigning the mold which are out of the scope for this study.

3.1. Dimensional accuracy

The Mg feedstock with 67 vol% of powder loading exhibit the nearest values to the mold dimensions at all measured dimensions. Figure 3 shows the effect of powder loading on the percentage of molded shrinkage for Mg feedstocks containing 63, 65, and 67 vol% powder loading. The dimensions measured were length, which is parallel to melt flow, width perpendicular to the melt flow, and thickness which parallel to the mold opening direction. It expressed a decrease in dimensional variation as the powder loading increases. The increase of powder loading will produce greater number of point contacts between powder particles and restrict the movement of the particles. It is also translated into a smaller volume of binder and therefore smaller shrinkage occurred during solidification due to mold-melt interaction effect [22]. The percentage of the shrinkage measured also indicated that the shrinkage observed for the injected specimens are not uniform and was not according to the melted feedstock flow direction.

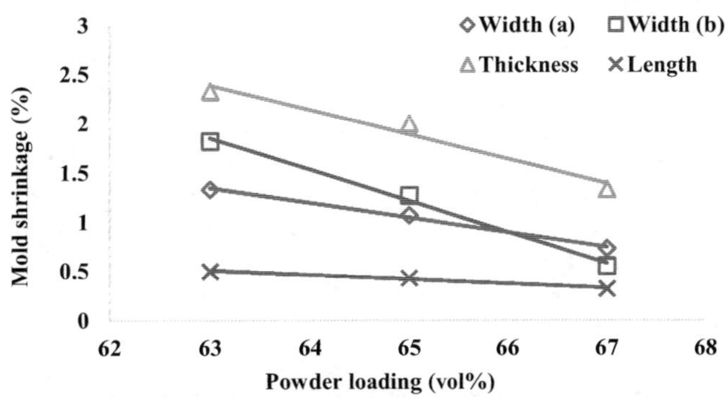

Figure 3. Shrinkage on the green molded parts at different dimensions for various powder loading.

3.2. Density of Molded Parts

Increasing the powder loadings result in greater packing density. The theoretical green density for (63, 65, and 67) vol% powder are summarized in Table 1. The densities are slightly lower than the theoretical which indicates the presence of a small fraction of voids in the molding. High green density of 67 vol% shows 1.79% reduction from the theoretical. Figure 4.55 shows the experimental and theoretical value for various powder loading. It shows that green molded parts with 63 vol% exhibits green density of 1.20-1.28 g/cm^3 while 65 vol% exhibits green density of 1.30 – 1.32 g/cm^3 and 67 vol% exhibits green density of 1.30-1.34 g/cm^3. The theoretical density for injected parts containing 63, 65, and 67 vol% are

1.31, 1.33, and 1.34 g/cm^3, respectively. Increasing the powder loading results in the greater packing density. The experimental density were slightly lower than the theoretical density which indicates the presence of small fraction of voids in the molding [23]. The parts containing 65 and 67 vol% powder loading both showing 98.70% of theoretical density as compared to 63 vol% which shows 94.98% of theoretical density. The better transmitted of molding pressure during injection, resulted in better packing of the Mg feedstock into the mold.

Table 1: The density, porosity and density reduction for the green molded parts.

Feedstock (vol%)	Theoretical density (g/cm^3)	Green density (g/cm^3)	Apparent porosity (%)	Percentage of density reduction (%)
63	1.3051	1.2396	5.1724	5.0190
65	1.3300	1.3122	0.0550	1.3380
67	1.3413	1.3234	1.2350	1.7900

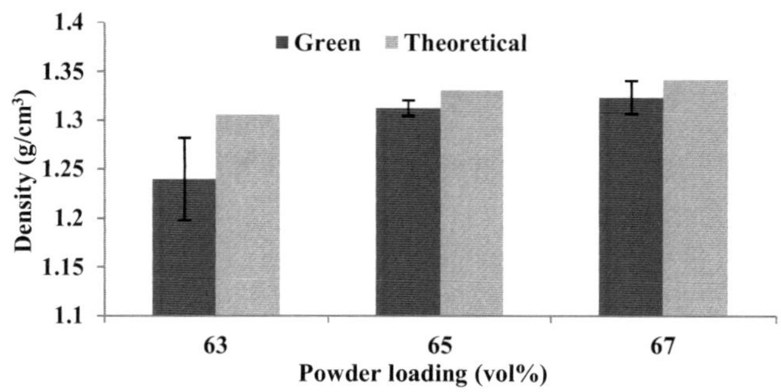

Figure 4: Graph of density of green molded parts for various powder loading.

3.3. Strength of Molded Parts

Table 2 shows the result of the 3 point bending test conducted on the green molded part to verify the strength provided by the binder content in the samples. The result reveals that the feedstock prepared with 37 vol.% of binder exhibits the highest strength after injection molded. The results proved that the binder plays an important role to strengthen the sample. The green strength decreased from 11.87 N/mm^2 to 6.13 N/mm^2 as the powder loading increased from 65 vol% to 67 vol%. The green strength decreased substantially with the increase of powder loading which translated into higher content of backbone polymer and therefore increase the strength of the green parts. This is because HDPE has the maximum strength as compared to the other components in the Mg feedstock. The increment of powder loading not only causes the percentage of backbone polymer (HDPE) decrease, but also decreases the forces between molecules by dispersing in HDPE molecule chains [8,24].

Table 2: The 3 point bending test results of molded parts with various powder loading.

Powder loading (vol%)	Maximum force (N)	Maximum Strength (N/mm^2)	Maximum Displacement (mm)	Maximum Strain (%)
63	7.8281	11.8358	0.6583	0.6077
65	5.5234	8.3512	0.5241	0.3711
67	4.0531	6.1282	0.2735	0.1936

4. Conclusions

For green molded part produce by magnesium feedstock, the densities for all formulation are slightly lower than the theoretical. The highest percentage of powder loading (67 vol.%) resulted to the highest density of molded parts which has only 1.79% of density reduction from the theoretical. However, the powder loading of 65 vol.% showing the lowest range of density measured for each samples which is indicated the best homogeneity among the feedstocks.

Acknowledgement

This research is supported by Fundamental Research Grant Scheme (FRGS) under a grant number FRGS/1/2016/TK03/UNIMAP/03/7 from Ministry of Education (MOE). Special thanks to Advanced Materials Process and Design Research Group, School of Manufacturing Engineering, UniMAP for the knowledgeable views during completing this research.

References

[1] Li, N., and Zheng, Y.: Novel magnesium alloys developed for biomedical application: A review. Journal of material science and technology **29**(6) (2013) 489-502

[2] Wolff, M., Schaper, J., Dahms, M., Ebel, T., Kainer, K. T.: Magnesium powder injection moulding for biomedical application. Powder metallurgy **57** (2014) 331-340

[3] Wolff, M., Schaper, J.G., Dahms, M., Ebel, T., Kainer, K.U., Klassen, T.: Magnesium powder injection moulding for biomedical application. In: Proceeding of the Euro powder metallurgy: Biomaterials, Salzburg, Brussels (2014)

[4] Santos, P., Niinomi, M., Liu, H., Cho, K., Nakai, M., Itoh, Y., Ikeda, M.: Fabrication of low-cost beta-type Ti-Mn alloys for biomedical applications by metal injection molding process and their mechanical properties. Journal of mechanical behavior of biomedical materials **59** (2016) 497-507

[5] Wolff, M., Schaper, J.G., Dahms, M., Ebel, T., Willumeit-Römer, R., and Klassen, T.: Metal injection molding (MIM) of Mg-alloys. In: Proceeding of 147th Annual Meeting and Exhibition, Volume 1., Pheonix, Switzerland (2018) 239–251

[6] Wolff, M., Schaper, J., Suckert, M., Dahms, M., F., F., Ebel, T., Klassen, T.: Metal injection molding (MIM) of magnesium and its alloys. Metals **6**(5) (2016) 118

[7] Wolff, M., Schaper, J., Sucket, M., Dhams, M., Ebel, T., Willumeit-Romer, R., Klassen, T.: Magnesium powder injection molding (MIM) of orthopedic implants for biomedical applications. The journal of the minerals, metals & materials society **68** (2016) 1191-1197

[8] Moballegh, L., Morshedian, J., Esfandeh, M.: Copper injection molding using a thermoplastic binder based on paraffin wax. Materials letters **59** (2005) 2832-2837

[9] Nooraizedfiza, Z., Salmah, H., Omar, M.A.: The effect of shear rate and temperature on rheology properties of magneisum metal injection molding feedstock. International review of mechanical engineering **7**(7) (2013) 1378-1383

[10] Romero, A., Herranz, G.: Development of feedstocks based on steel matrix composites for metal injection molding. Powder technology **308** (2017) 472-478

[11] Shivashankar, T., Enneti, R., Park, S. G., Atre, S.: The effects of material attributes on powder-binder separation phenomena in powder injection molding. Powder Technology **243** (2013) 79-84

[12] Haghshenas, M.: Mechanical characteristics of biodegredable magnesium matrix composites: A review. Journal of magnesium and alloys, **5** (2017) 189-201

[13] Poh, L., Della, C., Ying, S., Goh, S., Li, Y.: Powder distribution on powder injection moulding of ceramic green compacts using thermogravimetric analysis and differential scanning calorimetry. Powder technology **328** (2018) 256-263

[14] Aggarwal, G., Smid, I., Park, S. J., Randall, M. G.: Development of niobium powder injection molding Part II: Debinding and sintering. Refractory metals and hard materials **25** (2007) 226-236

[15] Sotomayor, M., Levenfeld, B., Varez, A.: Powder injection molding of premixed ferritic and austenitic stainless steel powders. Materials science and engineering A **528** (2011) 3480-3488
[16] Hidalgo, J., Morales, A. J., Barriere, T., Gelin, J. C., Torralba, J. M.: Water soluble Invar 36 feedstock development for μPIM. Journal of materials processing technology **214** (2014) 436-444
[17] Schaper, J.G., Wolff, M., Wiese, B., Ebel, T., Regine W.R.: Powder metal injection moulding and heat treatment of AZ81 Mg alloy. Journal of materiasl processing technology **267** (2019) 241-246
[18] Nooraizedfiza, Z., Omar, M.A., Rosliza, S., Marina, M.: Elucidating the influences of torque and energy in determining optimal loading. Materials today: proceedings **16** (2019) 2144–2152
[19] ASTM B962-14.: Standard test methods for density of compacted or sintered powder metallurgy products using Archimedes' Principle. In: ASTM International. West Conshohocken, PA (2014)
[20] ASTM B925-15.: Standard practices for production and preparation of powder metallurgy test specimens. In: ASTM International. West Conshohocken, PA (2015)
[21] Hwang, K.S. 2012 *Handbook of Metal Injection Molding*, ed D Heaney (Cambridge: Woodhead Publisher Limited) p. 109-131
[22] Hwang, K.S.: Common defects in metal injection molding (MIM). In: Handbook of Metal Injection Molding. (2012) 109-131
[23] Gershon, A.L., Gyger Jr., L.J., Bruck, H.A., Gupta, S.K.: Thermoplastic Polymer Shrinkage in Emerging Molding Processes. Experimental mechanics **48**(6) (2008) 789-798
[24] Raza, M., Faiz, A., Muhamad, N., Sulong, A., Omar, M., Akhtar, M., Aslam, M.: Effects of solid loading and cooling rate on the mechanical properties and corrosion behavior of powder injection molded 316L stainless steel. Powder Technology **289** (2016) 135-142.

The Optimization of the Hydroxyapatite (HA) Material Characteristics Produced From Corbiculacea (Etok) Shells

Mohd Riza Roslan[1], Nashrul Fazli Mohd Nasir[1, 2,*], Mohd Farid Abdul Khalid[2], Nur Farahiyah Mohammad[1], Cheng Ee Meng [1, 2], Nor Nabielah Nazihah Hashim[1], Beh Chong You [1], Mohd Shukry Abd. Majid[3] and Nasrul Amri Mohd Amin[3]

[1]School of Mechatronic Engineering, Universiti Malaysia Perlis, Kampus Pauh Putra, 02600 Arau, Perlis, Malaysia
[2]Microwave Research Institute (MRI), Universiti Teknologi MARA, 40450 Shah Alam, Selangor
[3]Mechanical Engineering Program, School of Mechatronic Engineering, Universiti Malaysia Perlis, Perlis, Malaysia

*nashrul@unimap.edu.my
Abstract. Hydroxyapatite commonly used in medical field for tissue disease, bone disease, drug delivery and also for non-medical field purpose such as wastewater treatment. Due to excellent biocompatibility and bioactive properties, hydroxyapatite (HA) which is a group of calcium phosphate (CaP) and similar to the natural bone composition has been broadly ventured in medical area. Synthesising the HA was done via precipitation method by implementing the *Corbiculacea* (Etok) shells as the resource of calcium precursors. Calcination of Corbiculacea (Etok) shells took place to convert calcium carbonate ($CaCO_3$) to calcium oxide(CaO). CaO were then reacted with water to form calcium hydroxide ($Ca(OH)_2$) and later, reacted with phosphoric acid(H_3PO_4). This experiment manipulated the CaP ratio to study the properties HA formed. CaP ratios used are 1.50, 1.67 and 2.00. The characterization were performed using Energy Dispersive X-Ray Spectroscopy (EDS), Scanning Electron Microscopy (SEM) and Fourier Transformation Infrared (FTIR).

1. Introduction

Bone tissue defects due to injury are a major concern in orthopedic surgery [1]. Technology on biomaterials product had produced autologous bone grafts and allografts to overcome the defect extensively in orthopedic field. The chemical formula of hydroxyapatite (HA) is ($Ca_{10}(PO_4)_6(OH)_2$) and it is widely used in orthopedic and biomedical application [1]. HA's chemical composition is similar to the natural bone, thus encourage the production of synthetic bone which is predominantly based on HA material [2-5]. HA can be synthesized using various methods and the most practical method is by the precipitation method. In this study, precipitation was chosen due to its simplicity and convenience. Recently, the interest of utilizing waste materials for various applications had increased. Hence, the potential biomaterials product from the shell wastes especially due to the high content of $CaCO_3$ could be further ventured [6]. To date, the study of HA based on Malaysian seashells apart from common cockle (*Anadara Granosa*) can be further explored. Therefore, the objective of this study is to determine the characteristics of HA from *Corbiculacea* (Etok) shell and manipulating Ca/P ratio in synthesizing HA.

2. Material and methods

2.1 Materials

Corbiculacea (Etok) shells were purchased from local market in Kelantan. Phosphoric acid with 95% and ammonia with 25% were purchased from Merck. The chemical is used without further purification.

2.2 Synthesis of Hydroxyapatite

Corbiculacea (Etok) shells were cleaned from debris before further process. Ammonia and phosphoric acid were purchased from Merck and directly used without any purification. The first step was to crush and sieve the shells into micrometre size to fasten the process of calcination. High temperature was used for calcination which is about 1000°C in 90 minutes to produce calcium oxide (CaO). CaO was then reacted with distilled water to produced calcium hydroxide$(Ca(OH)_2)$. This was followed by the addition of phosphoric acid(H_3PO_4) via titration. The phosphoric acid as supplied was added in different amounts; 1.80ml, 1.64 ml and 1.35ml in respective to the 1.50, 1.67 and 2.00 CaP ratio. While titration of phosphoric acid(H_3PO_4) took place, the ammonium solution was added drop wise at rate 5.5ml/minutes to maintain the solution at pH 11. The solution was let to age over night to produce HA precipitation. After 24 hours, the solution was filtered and the precipitation was washed with distilled water followed by drying. After that, the dried HA was powdered by using mortar and pestle.

2.3 Characterization of Hydroxyapatite powder

There are few test performed for the characterization of HA nanocrystals made from *Corbiculacea* (Etok) shells. The first test was Scanning Electron Microscopy (SEM) to study the morphology of the HA via HITACHI TM 3000 Tabletop Microscope. The same testing modality was used for Energy Dispersive X-ray Spectrometry (EDS) which determine the chemical elements existed in the synthesized HA. The third test was Fourier transform infrared spectroscopy (FTIR) for determining the functional group of a pure HA using Perkin Elmer Spectrum 65 through KBr method. EDS was carried out to determine the element or chemical composition of the hydroxyapatite. Three varieties Ca/P ratio used in this study which is 1.50, 1.67 and 2.00. Three variety of parameter used in this studies showed that compositional analysis of powders reveals the presence of Ca, P and O elements. The larger peaks for hydroxyapatite were O elements and comparatively smaller peaks for Ca and P elements.

3. Result and Discussion

3.1 Scanning Electron Microscopy (SEM)

The purpose of SEM is to determine the morphology of the HA powdered. Figure 1, 2 and 3 shows the image of HA morphology with different Ca/P ratio. The morphology of the HA images are shown below with different Ca/P molar ratio. HA with Ca/P ratio 1.67 shows less agglomeration meanwhile for HA with Ca/P ratio 1.50 and 2.00 show a more agglomerated HA powder. Agglomeration occurs due to the van der Walls attraction between HAP during calcination process [7]. All samples show uniform and spherical shape, although irregular. Basically, the temperature used during calcination process will affect the morphology of the particles [6]. Another factor that will affect the morphology of the HA is the phosphate-containing solutions. This had consequently affected the morphology of the HA obtained, for example platelet-like, needle like, or rod-like microstructure [8]. Thus, the morphology of HA can be optimized by the phosphate solution.

Figure 2. SEM image for Ca/P 1.67

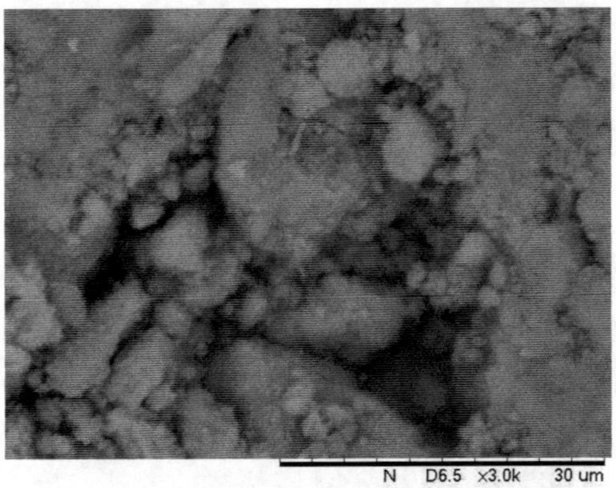

Figure 1. SEM image for Ca/P 1.50

Figure 3. SEM image for Ca/P 2.00

3.2 Energy Dispersive X-ray Spectrometry (EDS)

Three varieties of Ca/P ratio was used in this study which is 1.50, 1.67 and 2.00. The result EDS analysis from Table 1 showed that the actual result HA Ca/P ratio is different from calculation as much as 0.03± to 0.08±. Besides, EDS showed that compositional powders consist of Calcium (Ca), Phosphorus (P) and Oxygen (O) elements. The larger peaks for HA is an O element and comparatively smaller peaks for Ca and P elements. Increasing the Ca/P ratio showed an increment of O element which might contribute from oxide mass during calcination.

Table 1. Result of EDS

HA Ca/P molar ratio	Element O Atomic %	Element Ca Atomic %	Element P Atomic %	Result HA Ca/P ratio
1.50	72.21	17.04	10.76	1.58
1.67	73.07	16.73	10.20	1.64
2.00	77.15	15.38	7.48	2.05

Journal of Physics: Conference Series **1372** (2019) 012077 doi:10.1088/1742-6596/1372/1/012077

3.3 Fourier Transform Infrared Spectroscopy (FTIR)

FT-IR wave spectrum presented in Figure 4.0, 5.0 and 6.0 for Ca/P ratio 1.50, 1.67 and 2.0 respectively. There were eight clear peaks exist in the spectrum for both HA sample pH 6 and pH 12. Whereas, there were nine distinct peaks in HA sample pH 7. Phosphate, hydroxyl and carbonate bond/group were present among those peaks.

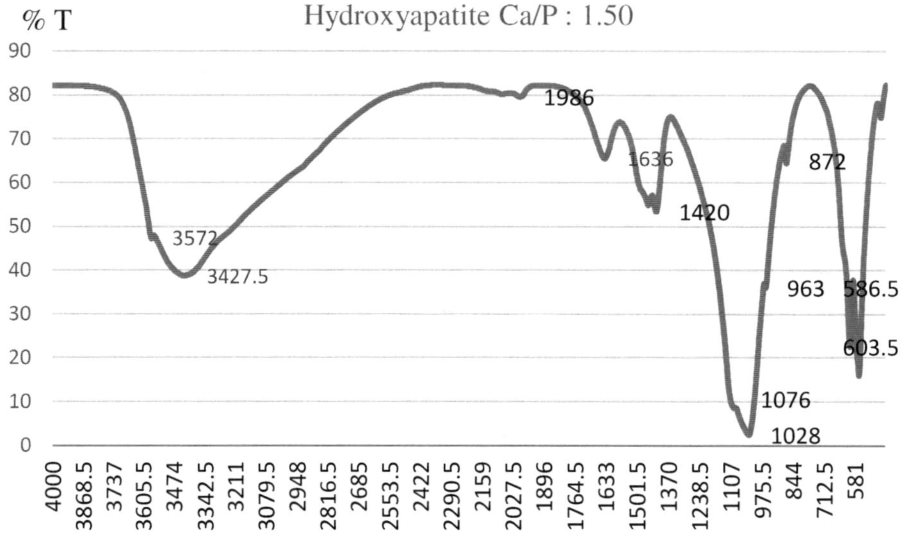

Figure 4. HA spectra with Ca/P ratio 1.50

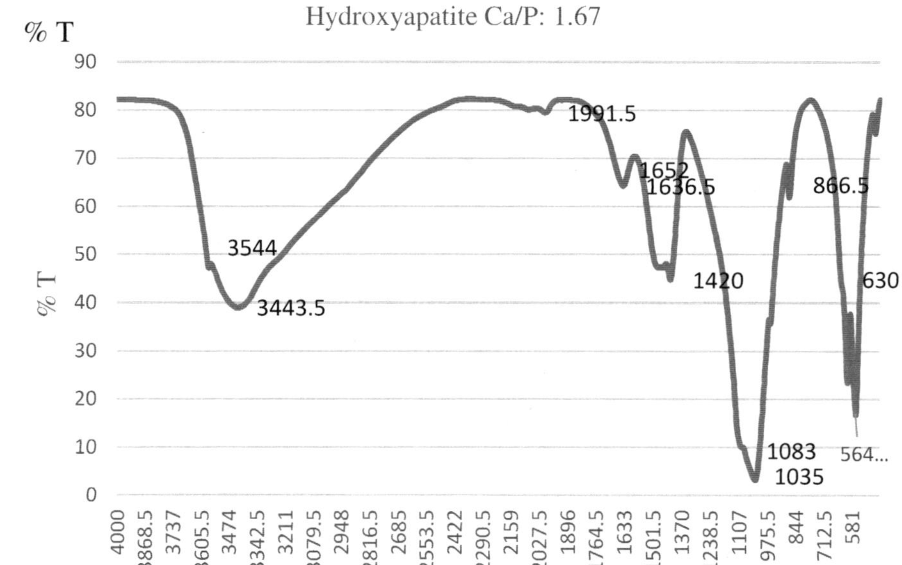

Figure 5. HA spectra with Ca/P ratio 1.67

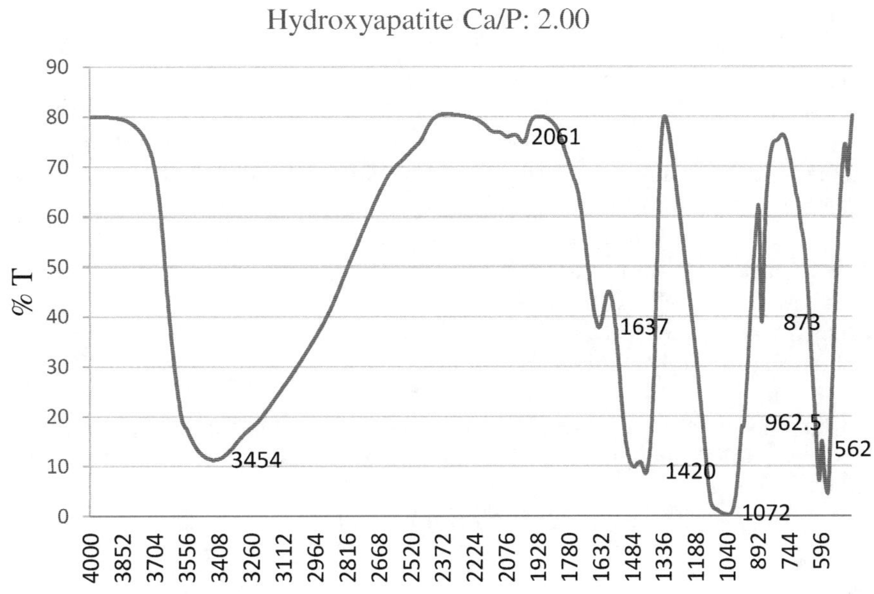

Figure 6. HA spectra with Ca/P ratio 2.00

Figure 4 above showed the FTIR spectrum for hydroxyapatite with Ca/P 1.50. At bands 3572cm^{-1}, stretching vibration of free hydroxyl group happened. Phosphate group absorption in HA at range 1028cm^{-1} to 1076cm^{-1}. The PO4^{-3} v4 mode can be found at 586.5cm^{-1} and 603.5cm^{-1}. The band at 872cm^{-1} indicates the presence of CO3^{-2} group appeared [8]. Meanwhile, Figure 5 shows the FTIR spectra of HA with Ca/P 1.67. Band at 1636.5cm^{-1} corresponds to the stretching mode of HO- and stretching vibration of phosphate group took place at 1083cm^{-1} and 1035cm^{-1}. Band at 564.5cm^{-1}corresponds to PO4^{-3} which is present at calcined sample [9]. The peak of band at 3443.5cm^{-1} is a typical assignment of stretching mode OH^{-}ion [10]. A context of pure HA structure should not present any elements of carbonate vibration mode at 1420cm^{-1}. The presence may due to atmospheric adsorption after the synthesis process which is unavoidable in chemical synthesis technique [10]. Last is Figure 6; HA with Ca/P molar ratio 2.00. The phosphate ions, PO4^{-3} are a principal molecular component in HA given in IR absorbance in range of 550cm^{-1} to 1200cm^{-1} regions. At bands 873cm^{-1} and 1080cm^{-1} correspond to the stretching vibration of PO4^{-3} and 562cm^{-1} due to bending of PO4^{-3}. Stretching mode of OH- ions occurs at peak 3453cm^{-1}. One of the distinct bands was observed at 1637cm^{-1} which rise from stretching mode and bending mode molecules of OH^{-}[11].

4. Conclusion

In conclusion, HA can be synthesized from Corbiculacea (Etok) shells and the Ca/P ratio affects the HA characteristic. It was clear that each of the samples exhibited distinct features proved by the SEM and the EDS analysis. All HA made with Corbiculacea (Etok) shells exhibited three main chemical compositions present in the commercialized HA such as Calcium, Phosphorous and Oxygen. As for FTIR analysis, the presence of HPO_4^{2-} and OH^{-1} groups in all three samples confirmed the formation of HA. The presences of CO_3^{2-} band in all samples were because of the air contamination after the calcination of Corbiculacea (Etok) shells and during HA synthesizing process.

References

[1] Bhatt T D, R. A., and Rozental," Hand Clin., vol. 28, no. 4, pp. 457–468, 2012.
[2] Popa G T et al. 2018 Materials (Basel) **11**. 2081.
[3] Roslan M R, Nasir N F M, Cheng E M and Mamat N 2016 International Conference on Electrical, Electronics, and Optimization Techniques (ICEEOT) (Chennai: IEEE) p 1560
[4] Razali K R, Nasir N M, Cheng E M, Tan M K, Zakaria A and Mamat N 2016 J. Eng. Appl. Sci. (Asian Res. Publ. Netw.) **11**, 4987
[5] Nasir N F M, Khan S F, Baling J S, Roslan M R, Cheng E M, Amin N A M, Abdullah M N 2018 J. Eng. Res.& Edu. 10, 83
[6] Saharudin S H, Shariffuddin J H, Ismail A, and Mah J H 2018, Bull. Mater. Sci.
[7] D. Xidaki et al. 2018 Materials (Basel) **11**, no. 4
[8] Hosseini M. R. K, Mirhadi B., Mehrazin S M,, Yazdanian, M and Motamedi M 2017 Trauma Mon. **22**
[9] Mahmud M M, et al. 2015 Int. J. Innov. Res. Sci. Eng. Technol, **4**, 5039.
[10] Supelano D P, Estupinan N M,, Garcia H A, & Ballesteros S J 2018 Chem. Eng. Trans. 64, pp. 307.
[11] Ragu P, Senthilarasan A,, & Sakthivel K 2014 Int. J. Eng. Res. Appl, **4**, 50.

Optimization of Dual Pathway AV Nodal Conduction Model

Maxim Ryzhii[1*], Elena Ryzhii[2]

[1]Department of Computer Science, University of Aizu, Aizu-Wakamatsu, Japan
[2]Department of Anatomy and Histology, Fukushima Medical University, Fukushima, Japan

*m-ryzhii @ u-aizu.ac.jp

Abstract. In this work we tested our previously proposed heterogeneous oscillator model of atrioventricular node with respect to dual pathway physiology in order to optimize the model parameters. For this purpose we compared three model variants with different number of pathway node elements represented by nonlinear Aliev-Panfilov equations. The parameters controlling coupling between the pathway elements and the refractoriness of the fast pathway were selected for each model variant to obtain qualitative correspondence to the all range of functional atrioventricular node behavior. The results demonstrate that all considered model variants are able to reproduce normal cardiac signals conduction, AV node reentry, automaticity (pacemaking) and filtering function. However, the model variant with smallest number of the pathway elements better correspond to human electrophysiology, in particular to filtering high atrial rates by AV node. In addition, it is preferred for use in low-power systems with real-time simulations, such as ICD and pacemaker validation and testing hardware, as well as for educational and research purposes.

1. Introduction

Atrioventricular (AV) node is located between the upper (atria) and lower (ventricles) chambers of the heart. It realizes three important functions: 1) to filter (slow) excessively fast impulses from atria; 2) to be a potential substrate for arrhythmias (AV node reentry); 3) to function as an independent pacemaker (automaticity). According to recent experimental studies, the AV junction pacemaker in human heart is located between fast (FP) and slow (SP) pathways and His bundle (see, for example [1]). Currently there are very few AV models of dual conduction pathway [2–8]. The detailed multi-variable cardiac transmembrane multicellular models like [2, 3] are too complex and do not have the ability to effectively simulate cardiac electrical activity during arrhythmia in real time that is crucial for close-loop ICD and pacemaker testing and educational purposes. The nonlinear model [4] is based on Lienard-transformed van der Pol-Duffing and FitzHugh-Nagumo models and also contains over hundred elements. The models [5, 6] are emulating models that cannot be used for the arrhythmia research. In the works [7, 8] the authors utilized simple functional mathematical model of AV node with inclusion of dual pathway physiology. This model consists of exponential mathematical equations and is based on rabbit heart preparation data. However the model does not include other parts of the heart and is unable to reproduce the reentry phenomena.

The model considered in this paper can provide a simplified solution to some drawbacks of the above mentioned models. In our previous studies [9–11], we developed and implemented a nonlinear heterogeneous oscillator model for cardiac conduction system that includes natural pacemakers

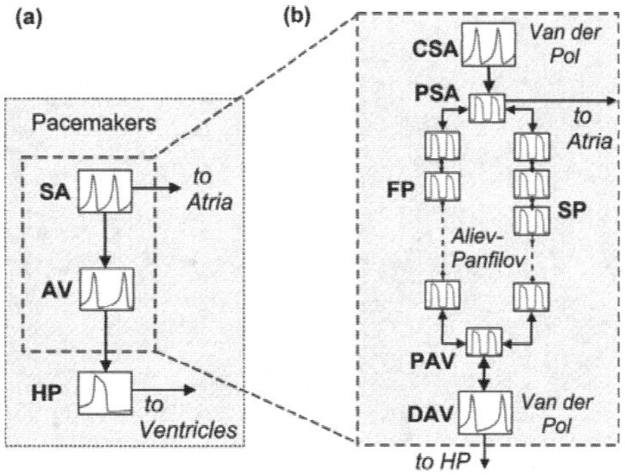

Figure 1. Schematic view of the original simplified SA-AV part of cardiac conduction system model (a) and AV node dual pathway model block under consideration (b).

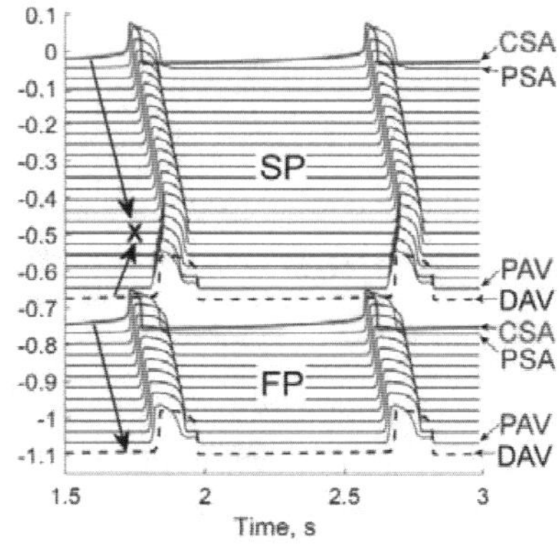

Figure 2. Normal pacing at 70 bpm for model B. Bold arrows indicate propagation of excitation from CSA to DAV through SP and FP.

[sinoatrial (SA) and AV nodes, and His-Purkinje system (HP), see Fig. 1(a)] and atrial and ventricular muscles (not shown). The model reproduces action potentials for each pacemaker and calculates synthetic ECG waveforms in a variety of typical normal and pathological electrical heart activity. Recently we presented the AV node dual pathway model block [12] which is expected to be a part of our heterogeneous oscillator model of the whole cardiac conduction system [9–11].

Our aim is to develop a simplified dual pathway model of AV node which should be fast enough to allow the realistically reproduce of cardiac conduction phenomena in real-time simulations with the smallest possible number of elements without changing their quantity and properties. This task requires optimization of the dual pathway model structure to provide acceptable balance between the realistic physiology, structural simplicity, and efficient real-time computation, especially taking into account future hardware model implementation with less powerful microcontrollers. The considered optimized model block, being incorporated into our heterogeneous oscillator model can be used for pacemaker and ICD testing software, education, and arrhythmia research.

2. Model of AV Dual Pathway

Schematic view of the AV node dual pathway model is shown in Fig. 1. The model block [Fig. 1(b)] was developed in MATLAB/Simulink environment and will replace original SA-AV part [Fig. 1(a)] of our previously developed heterogeneous cardiac conduction system model [9–11]. In the proposed AV node block, central sinoatrial node (CSA) and distal atrioventricular node (DAV) are represented by modified Van der Pol equations [9, 13]. We placed the atrioventricular pacemaker in DAV node according to the findings on human heart anatomy [1]. The peripheral sinoatrial node (PSA), fast (FP) and slow (SP) pathway node elements, and proximal AV node (PAV) are represented by Aliev-Panfilov (A-P) model [14]. In order to examine the model block ability to reproduce the whole range of AV conduction phenomena, we considered three model variants with different number of FP and SP units: (A) 5 and 11, (B) 9 and 13, and (C) 13 and 27, respectively [see Fig. 1(b)]. The A-P model is described by the following set of nonlinear differential equations:

$$\frac{du_i}{dt} = -k_1 u_i(u_i - a_1)(1 - u_i) - u_i v_i + f(u_{i-1} - 2u_i + u_{i+1}) \tag{1}$$

$$\frac{dv_i}{dt} = \varepsilon(u_i, v_i)\big(-v_i - k_2 u_i(u_i - a_2)\big) \tag{2}$$

$$\varepsilon(u, v) = \varepsilon_0 + \frac{\mu_1 v}{u + \mu_2}$$

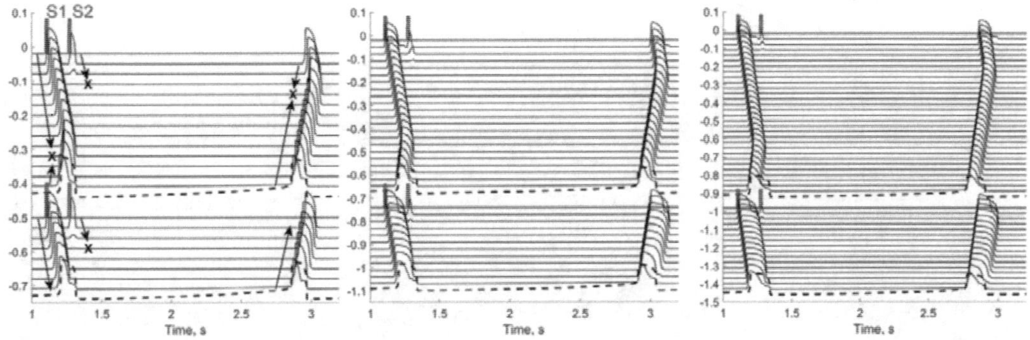

Figure 3. S1-S2 protocol stimulation with 165 ms interval for models A (left), B (center), and C (right), respectively. DAV pacemaking activity is seen in the right side of each panel. Arrows indicate propagation of action potentials.

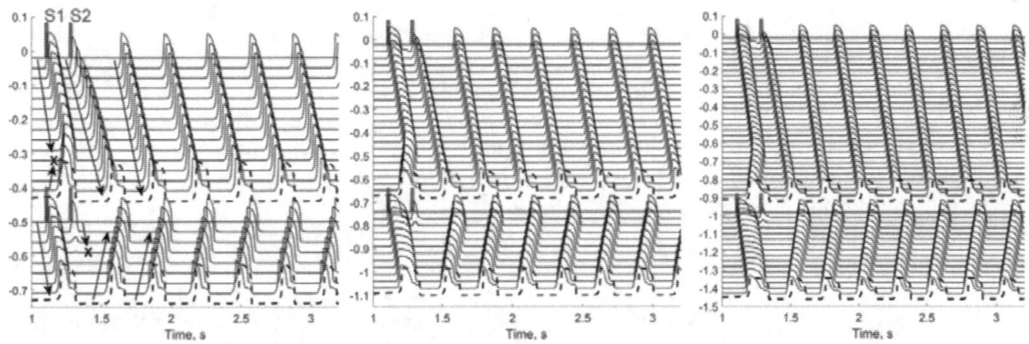

Figure 4. S1-S2 protocol stimulation with 175 ms interval for models A (left), B (center), and C (right), respectively.

Here u is the action potential (in mV), and v is recovery variable that controls repolarization. Last term in the right side of (1) represents the diffusive coupling of each element with its neighbors with the coupling coefficient f. Since the signal propagation time through a single A-P element (and, thus, the total propagation time through SP and FP) depends on this parameter, we selected values of f in order to obtain total propagation time through FP 75 ms in all three model variants. This time is approximately 1.5 times higher than 50 ms for rabbit heart used in [2]. Thus, the values of $f = f_{SP} = f_{FP} = 19, 42, 78$ were chosen for the variants A, B, and C, respectively.

The parameter ε_0 controls the refractoriness and action potential duration of each A-P unit [14, 15]. We set $\varepsilon_0 = 0.05$ for the SP units in all the model variants considered, and $\varepsilon_0 = 0.035$, 0.006 and 0.003 for the FP unis of models A, B, and C, respectively, to make the action potential durations longer in FP [1]. Other parameters were chosen to set the amplitude and action potential duration at 90% repolarization of A-P units about ~100 mV (with -80 mV resting potential) and APD_0 ~100–160 ms, respectively. All the parameters were selected to obtain AV nodal reentry similar to that demonstrated in [2], taking into account 1.5 times increase in S1-S2 intervals with respect to human heart.

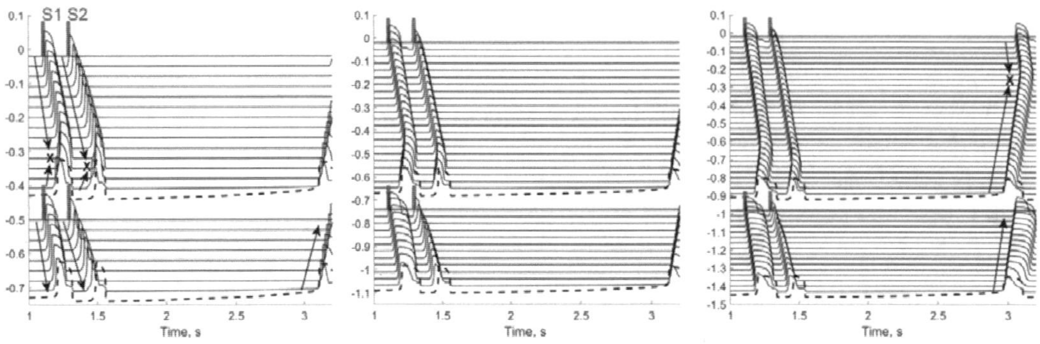

Figure 5. S1-S2 protocol stimulation with 185 ms interval for models A (left), B (center), and C (right), respectively.

3. Results

First we simulated normal pacing conditions with 70 bpm sinus rhythm set for CSA. In all three model variants the excitation propagated to PAV and further to DAV through FP, while conduction through SP was terminated due to retrograde propagation from PAV. This effect is indicated for variant B in Fig. 2 by thick arrows. In this and subsequent figures top part corresponds to action potentials propagation in SP, and bottom part – in FP.

Next, we replaced the CSA with a unit producing two consequent rectangular pulses with amplitude of 100 mV, resting potential -80 mV, and 15 ms width, and applied this S1-S2 stimulation to all model variants. Pertinent simulation results with obtained three characteristic responses – normal, AV node reentry, and AV automaticity pacing in the absence of stimulation from atria are shown in Figs. 3–5. In the Figs. 3-5 left, center, and right panels correspond to the variants A, B, and C, respectively. In Fig. 3, upon S1 stimulation the excitation propagated normally through FP, while the response to the S2 stimulation applied after 165 ms interval decayed in both pathways due to their refractoriness. In the absence of stimulation from atria (CSA), the DAV become leading pacemaker and caused retrograde propagation through FP [1] after approximately 1.5 seconds (right side of each panel in Fig. 3), which corresponds to its internal rate of 37 bpm.

With increase of the S1-S2 interval to 175 ms we observed AV nodal reentry (Fig. 4) in all model variants. In this case, in response to S2 stimulation, action potentials managed to pass through SP, but failed to propagate through FP due to its refractory state. Upon the impulse in the SP reaches PAV, the FP units become re-excitable, and impulses propagate retrogradely through the FP to PSA, resulting in another PSA excitation and anterograde propagation of impulses through the SP.

When the S1-S2 interval was longer (185 ms) the AV node reentry did not take place, as seen in Fig. 5. In this case the S1-S2 interval was longer than SP and FP refractory periods. The automaticity behavior of DAV with retrograde propagation through FP was also observed this case similar to that in the case of short S1-S2 intervals (Fig. 3).

The AV node filtering function at high rates produced by CSA is shown in Figs. 6 and 7. In contrast to method used in [2], here we considered regular fast atria rhythms, so called atrial flutter, since it better illustrates relation between atrial and ventricular rates when AV node limits dangerous surplus impulses from CSA. Figure 6 demonstrates the transition of CSA-DAV ratio from 1:1 to 1:2 as AV node blocks excessive action potentials coming to ventricles from atria with only each second impulse reaching DAV. This transition took place at 305 bpm for model variant A, and at 340 bpm for both variants B and C. Further increase of CSA rates leads to transition from 1:2 to 1:3 CSA-DAV ratio (see Fig. 7) appearing in the model A at 405 bpm, in the model B at 460 bpm, and in the model C at 462 bpm.

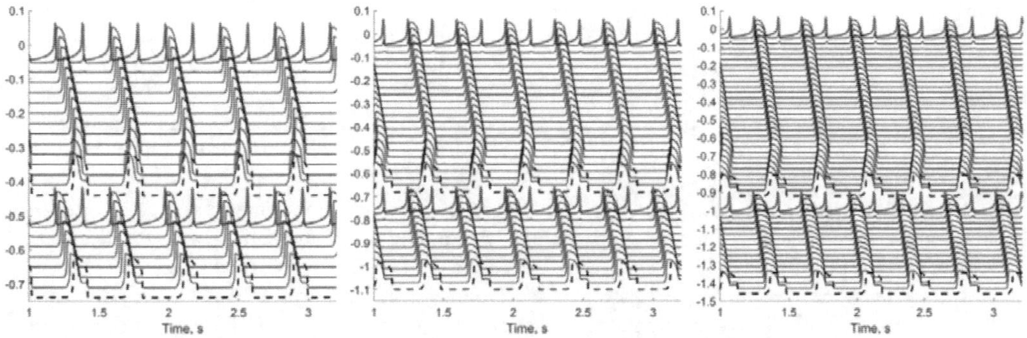

Figure 6. Filtering properties of the models: transition from 1:1 to 1:2 CSA-DAV ratio - model A, 305 bpm (left), model B, 340 bpm (center), and model C, 340 bpm (right).

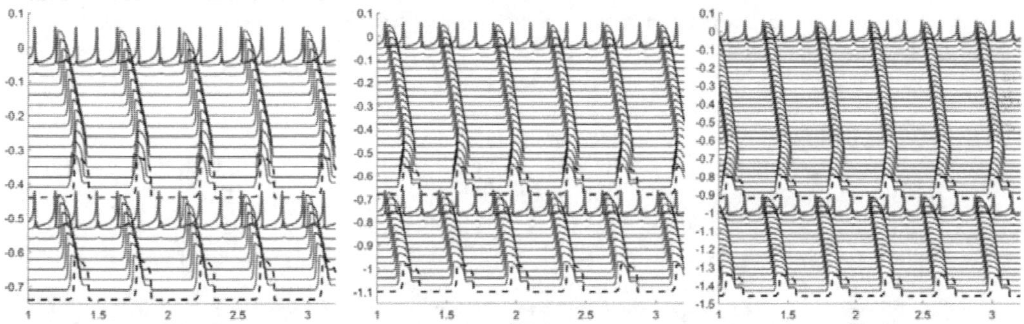

Figure 7. Filtering properties of the models: transition from 1:2 to 1:3 CSA-DAV ratio – model A, 405 bpm (left), model B, 460 bpm (center), and model C, 462 bpm (right).

The above results demonstrate that the model with smaller number of elements is preferable since it can reproduce all AV node physiological phenomena. That is, the observed in the model A AV node filtering at lower heart rates better corresponds to clinical findings [1]. Moreover, smaller number of pathway units is desirable for possible hardware implementation with less powerful microprocessor systems to provide real time simulations [5].

4. Conclusion

In this work we tested our previously proposed dual pathway AV nodal conduction model in MATLAB/Simulink environment in order to find optimal model parameters. We considered and compared three model variants with different number of pathway node elements. All tested models were able to reproduce normal cardiac signals conduction, AV node automaticity, reentry, and filtering function. We selected the parameter controlling the refractoriness of the fast pathway and coupling coefficient between the pathway elements in order to obtain qualitative correspondence to the all range of functional AV node behavior in each model variant.

We can conclude that the model variant with reduced number of pathway elements better corresponds to human AV nodal electrophysiology and is preferable for utilization in the low power hardware systems with real-time response simulations, such as ICD and artificial pacemaker validation and testing hardware, as well as for educational and research purposes.

Acknowledgement
The work is supported by JSPS KAKENHI Grant Number 17K00411.

References

[1] Kurian, T., Ambrosi, Ch., Hucker, W., Fedorov, V.V., Efimov, I.R.: Anatomy and electrophysiology of the human AV node. Pacing and clinical electrophysiology **33**(6) (2010) 754–762

[2] Inada, S., Hancox, J.C., Zhang, H., Boyett, M.R.: One-dimensional mathematical model of the atrioventricular node including atrio-nodal, nodal, and nodal-his cells. Biophysical journal **97** (2009) 2117–2127

[3] Li, J., Inada, S., Schneider, J.E., Zhang, H., Dobrzynski, H., Boyett, M.R.: Three-dimensional computer model of the right atrium including the sinoatrial and atrioventricular nodes predicts classical nodal behaviours. PLoS ONE **9**(11) (2014) e112547

[4] Podziemski, P., Zebrowski, J.J.: Liénard-type models for the simulation of the action potential of cardiac nodal cells. Physica D **261** (2013) 52–61

[5] Yip, E., Andalam, S., Roop, P.S., Malik, A., Trew, M.L., Ai, W., Patel, N.: Towards the emulation of the cardiac conduction system for pacemaker validation. ACM transactions on cyber-physical systems **2**(4) (2018) article 32

[6] Ai, W., Patel, N.D., Roop, P.S., Malik, A., Andalam, S., Yip, E., Allen, N., Trew, M.L.: A parametric computational model of the action potential of pacemaker cells. IEEE transactions on biomedical engneering **65**(1) (2018) 123–130

[7] Climent, A.M., Guillem, M.S., Zhang, Y., Millet, J., Mazgalev, T.N.: Functional mathematical model of dual pathway AV nodal conduction. American journal of physiology Heart and circulatory physiology **300**(4) (2011) H1393–H1401

[8] Masè, M., Glass, L., Disertori, M., Ravelli, F.: Nodal recovery, dual pathway physiology, and concealed conduction determine complex AV dynamics in human atrial tachyarrhythmias. American journal of physiology Heart and circulatory physiology **303**(10) (2012) H1219–1228

[9] Ryzhii, E., Ryzhii, M.: A heterogeneous coupled oscillator model for simulation of ECG signals. Computer methods and programs in biomedicine **117**(1) (2014) 40–49

[10] Quiroz-Juárez, M.A., Jiménez-Ramírez, O., Vázquez-Medina, R., Ryzhii, E., Ryzhii, M., Aragón, J.L.: Cardiac conduction model for generating 12 lead ECG signals with realistic heart rate dynamics. IEEE transactions on nanobioscience **17**(4) (2018) 525–532

[11] Quiroz-Juárez, M.A., Jiménez-Ramírez, O., Vázquez-Medina, R., Ryzhii, E., Ryzhii, M., Aragón, J.L.: ECGHOSM (ECG heterogeneous oscillator simulation model) [Source code] https://codeocean.com/capsule/2723361/tree/v1 (2018)

[12] Ryzhii, M., Ryzhii, E.: Development of simplified model of atrioventricular node with dual pathway. In: Proceedings of the 2018 IEEE International Conference on Bioinformatics and Biomedicine. (2019) 2800

[13] Postnov, D., Han, S.K., Kook, H.: Synchronization of diffusively coupled oscillators near the homoclinic bifurcation. Physical review E **60(3)** (1999) 2799–2807

[14] Aliev, R.R., Panfilov, A.V.: A simple two-variable model of cardiac excitation. Chaos, solitons and fractals **7**(3) (1996) 293–301

[15] Hurtado, D.E., Kuhl, E.: Computational modeling of electrocardiograms: Repolarization and T-wave polarity in the human heart. Computational methods biomechanics and biomedical engineering **17**(9) (2014) 986–996

An Automated Intelligent Identification and Counting System Procedure for Tuberculosis

Fatin Atiqah Rosli[1], Mohd Yusoff Mashor[*1], Siti Suraya Md Noor[2]

[1]School of Mechatronic Engineering, Universiti Malaysia Perlis, 02600, Perlis, Malaysia.

[2]Department of Microbiology, School of Medical Science, Universiti Sains Malaysia Kubang Kerian, Kelantan, Malaysia

*yusoff@unimap.edu.my

Abstract. Tuberculosis (TB) is an infectious disease caused by Mycobacterium Tuberculosis or TB Bacilli. Currently, the classification of TB bacilli is carried out by microbiologist by using Ziehl-Nielsen (ZN) stained smear sputum slide under a light microscopy. However, the manual evaluation is time-consuming and lead to slow decision. Furthermore, the sensitivity is less due to incline of human error which lead to inaccurate conclusion. Therefore, this study proposes an intelligence identification and counting system to detect the presence of TB bacilli in the ZN-stained smear sputum image. This system is designed to identify the presence of TB bacilli and count the number of TB bacilli by applying digital image processing and artificial intelligence techniques. In image acquisition, there are 70 samples images of ZN-stained smear sputum image were collected from Hospital Universiti Sains Malaysia (HUSM) Kubang Kerian, Kota Bharu, Kelantan, Malaysia. The image processing technique consists of contrast enhancement, segmentation, and feature extraction. The contrast of original image was enhanced by the combination of global enhancement, local enhancement and Contrast Limited Adaptive Histogram Equalization (CLAHE). Then, the enhanced image was segmented using color thresholding and the features were extracted consists of on 18 colour features, 15 shape features and 5 texture features. Afterward, the features underwent feature selection to select the relevant features by using Neighborhood Component Analysis (NCA) and ReliefF Analysis. The study showed that there are relevant features were chosen by ReliefF at feature weight more than 0.004 including (8 colour features, 11 shape feature and 3 texture features) for improving the performance and accuracy of Multilayer Perceptron (MLP) trained by Scaled Conjugate Gradient (SCG). For classification process, MLP, k-Nearest Neighborhood (k-NN) and Support Vector Machine (SVM) are used with 6 folds cross-validation. It was found that MLP has the highest of accuracy, sensitivity and specificity with 93.8%, 93.4% and 94.1% respectively.

1. Introduction

Tuberculosis (TB) is an infectious disease transmitted from a person to another person by coughing or breathing in the air containing bacteria called Mycobacterium Tuberculosis or TB Bacilli. It can be divided into Pulmonary TB (PTB) and Extra-Pulmonary TB (EPTB). TB mainly target the lungs, but also can affect any part of the body such as bone and joints, kidneys, brain, and lymph node.

According to The World Health Organization (WHO) [1], TB is the top 10 leading causes of death worldwide in 2017. Estimated 10 million people were infected by TB and 1.6 million of people die from the disease. There were 5.8 million of men, 3.2 million of women and 1 million of children are suspected in the overall 10 million that were infected with TB. It shows that more than half of the overall cases are represented by men.

Nowadays, manual diagnostic procedure is commonly used to detect the presence of TB by microbiologist. In many countries, sputum smear microscopy is the main tool for laboratory diagnosis of TB [2]. Although it requires simple laboratory, this procedure reveal low detection rate and less sensitivity for determining a positive TB [3]. Moreover, the diagnosis under the microscope can lead to eye pain and poor posture during screening because it takes 15 minutes per slides and the microbiologist require to confirm hundreds of slides may per person. The smear microscopy method is categorised as manual diagnostic procedure. Currently, the manual diagnostic procedure is quite outdated compared to other current technology worldwide, include in every field. Thus, artificial intelligence (AI) is explored to develop an autonomous system in detection of TB bacilli

Samuel *et al.* [4] proposed a system using Inception V3-SVM model in transfer learning and fine-tuning for classification of infected and non-infected by TB Bacilli. The system had the limitation in low availability of the dataset and decreased the sensitivity because unable to support adaptive learning. The average accuracy of the classifier is 95.05%.

Panicker *et al.* [5] has proposed a system to detect TB Bacilli by using pixel classification for detection TB Bacilli. Pixel classification specifically used the convolution neural network (CNN) in the architecture of the network which extracted a patch in images and probability value was generated to determine the presence of bacilli. The architecture of this classifier consisted of 2 convolution layers contained 32 filter maps, 3 convolution layer contained 64 filters and 1 convolution layer with 128 filters.

Mithra *et al.* [2] claimed that Gaussian-Fuzzy-Neural network (GFNN) classifier had achieved better performance than another model for diagnosising of TB Bacilli specifically in determining images into the bacilli and few bacilli. Besides that, the counting bacilli in the overlapping bacilli were computed by GFNN. The GFNN model was used as a classifier to classify bacilli into three categories which were few bacilli, overlapping bacilli and non-bacilli. Moreover, the model was also used for counting the bacilli by computing few bacilli and overlapping bacilli.

The current research consists of two parts which is image processing and classification. Image processing consists of contrast enhancement, segmentation and filtering. Then, the segmented images undergo feature extraction and feature selection. After that, the selected features become inputs for classifiers to determine the best classifier by evaluating accuracy, sensitivity and specificity.

2. Methodology

Figure 1 shows the flowchart for the methodology. The process involved image pre-processing, contrast enhancement, segmentation, filtering, feature extraction and classification. The ZN-stained smear sputum slides were acquired from Hospital Universiti Sains Malaysia (HUSM) Kubang Kerian, KotaBharu,Kelantan,Malaysia.

2.1. Image Pre-Processing

In pre-processing method, the background and foreground were adjusted in order to get a clear view of TB Bacilli and partially eliminate the undesired object in the images such as a bluish colour of sputum and other artefacts. The foreground is enhanced so that the purple reddish TB Bacilli become clearer. This method has three parts which are global contrast enhancement, local contrast enhancement and contrast limited adaptive histogram equalization (CLAHE).

2.2. Segmentation

The color thresholding is implemented in order to segment the non-reddish colour and reddish colour of TB bacilli. The colour model implements the *LAB* colour space to adjust the pixels so that the reddish colour was segmented. This method uses Color Thresholding Apps provided by MATLAB R2017a. The outputs are the RGB image and binary image. This is due to regionprops() requires binary image to proceed. The RGB image of colour thresholding is used only for display, however the binary image is used to further filtering the unwanted objects within the image. The unwanted objects areas as known as artefacts. Then, the original colour of the image was retrieved based on the resulted segmented binary image.

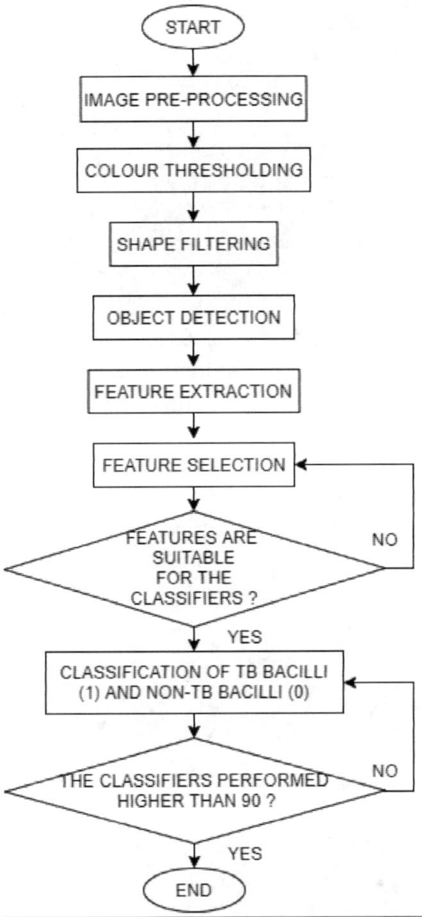

Figure 1 The flowchart of methodology

2.3. Feature Extraction and Feature Selection

2.3.1. Feature Extraction

A total of the 38 features have been extracted, namely 18 colour, 15 shape and 5 texture. The colour features were extracted from the mean, variance and standard deviation of RGB colour and HIS colour space. The shape features were extracted from *regionprop()* function which are perimeter, area, solidity, diameter and eccentricity. Other shape features are compactness, circularity, the ratio of major over minor axis and 7 Hu's Moment. Lastly, the texture features were extracted by using *grayprops()*where the RGB segmented image converted into grayscale image. These features are correlation, energy, contrast, homogeneity, and entropy.

2.3.2. Feature Selection

There are 38 extracted features from the previous process have undertaken feature selection to avoid overfitting. There are two types of feature selection methods that were used in this study, namely Neighborhood Component Analysis (NCA) and Relieff Analysis. In MATLAB, ReliefF and NCA are the built-in function. For ReliefF, the function is *relief()* that give feature weight as the output. Usually, the feature weight which assigns as zero, that indicates the irrelevant feature and whereas NCA uses function *fscnca()* that will give rank and feature weight as outputs. In NCA, the feature weight assigns either zero or negative value is the irrelevant feature.

2.4. Classification

After extracting and selecting the features, the subsets of data were used for classification. This classification used 3 models of classifier for detecting TB bacilli in the segmented image. These 3 models are Multilayer Perceptron (MLP), K-Nearest Neighbour (K-NN) and Support Vector Machine (SVM). Each classifier has different approach in order to classify different objects existent within the image. All classifiers will implement k-fold cross validation. The analysis is further when the data split into training and testing by using k-fold cross validation approach. The number of k will be varied in order to train the data and evaluated the accuracy, sensitivity and specificity of the classifier. The best classifier has the highest of these three evaluation parameters.

3. Experimental Result and Discussion

The procedure was evaluated using 70 images of ZN-stained smear sputum that were collected from Hospital Universiti Sains Malaysia (HUSM) Kubang Kerian, Kota Bharu, Kelantan, Malaysia. Figure 2 shows images to illustrate the process of image enhancement, filtering and segmentation where the image was segmented until only TB bacilli and other reddish colour of objects were remained in the image. The images of the process are shown in Figure 2(a) until Figure 2(g)

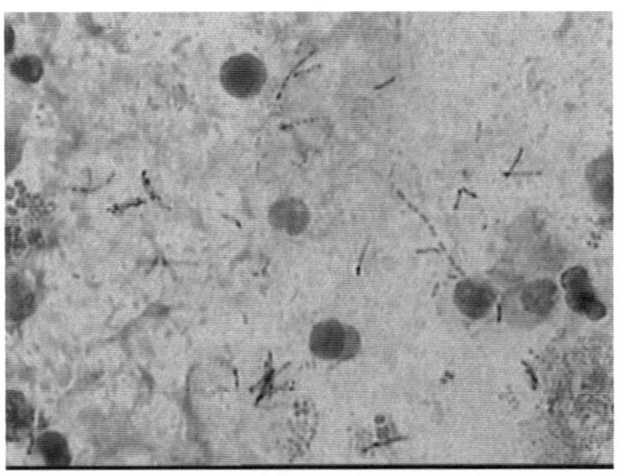

(a) Original Image

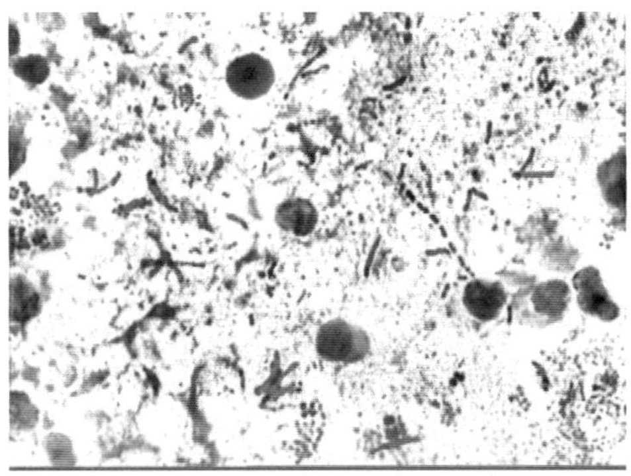

(b) Enhanced Image

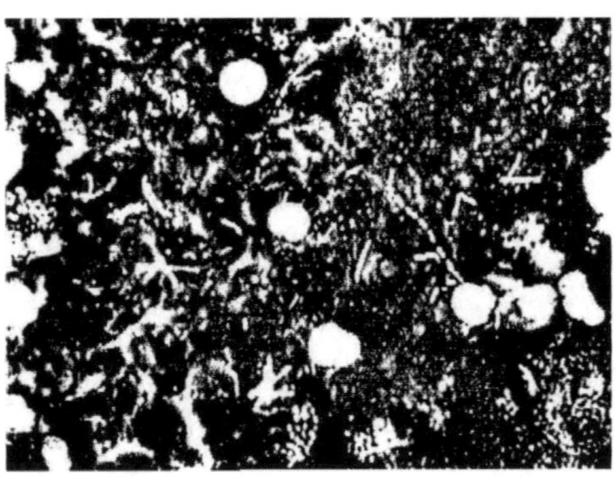

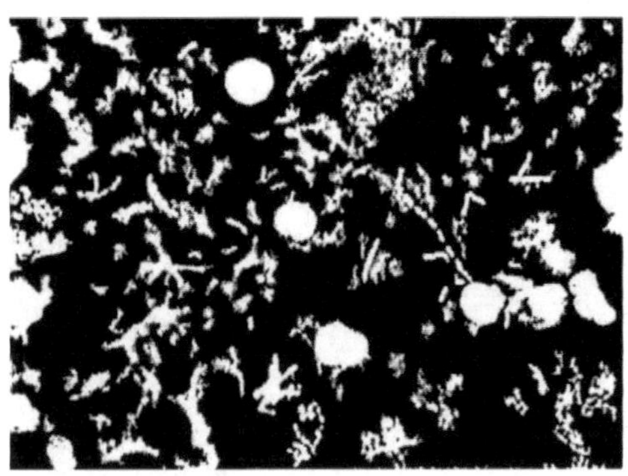

(c) Conversion from RGB to binary

(d) After noise filtering

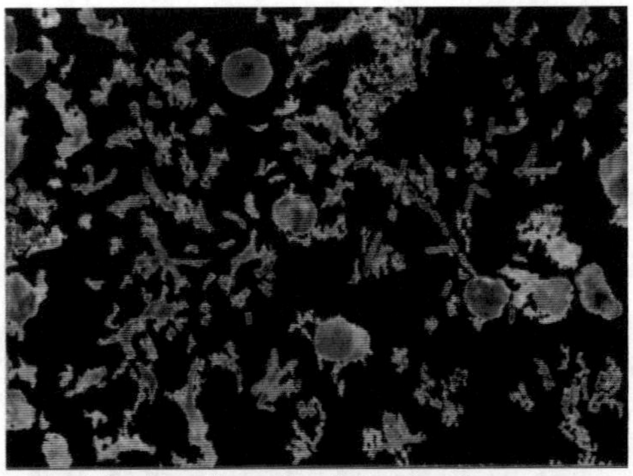

(e) Retrieved original colour

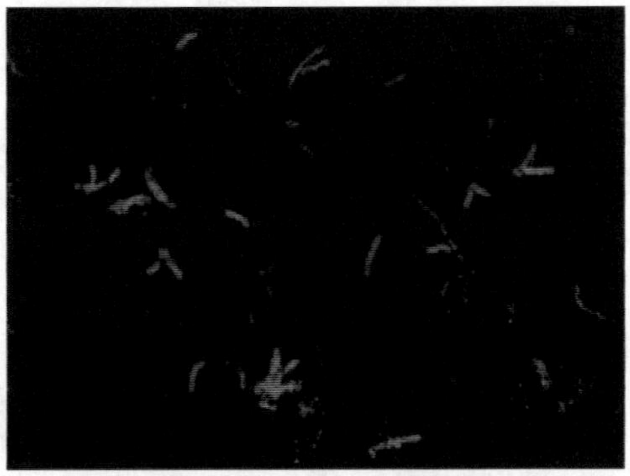

(f) Filtering artefacts

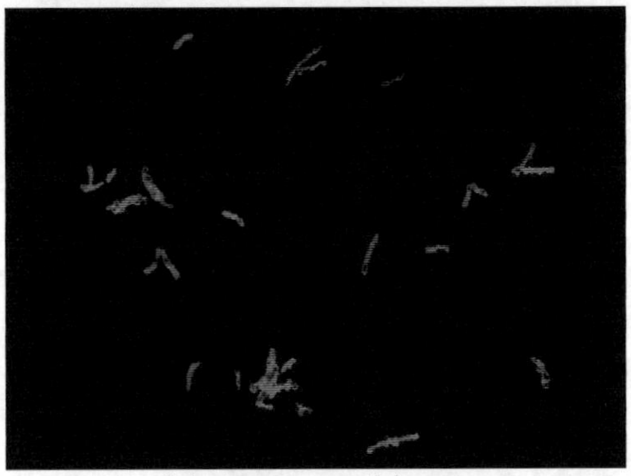

(g) segmented image

Figure 2 The TB bacilli image segmented process

There are 38 features that have been extracted in this study. However, the features have to be selected in order to improve the performance of classifier. Neighbour Component Analysis (NCA) and ReliefF were used for this task. As a result, 22 features have been selected with the lowest MSE by ReliefF as shown in Figure 3. Table 1 shows the performance of the three classifiers, evaluated using 6 folds cross-validation. The best performance id MLP 15 hidden layer that yielded the accuracy, sensitivity and specificity of 93.8%, 93.4% and 94.1% respectively.

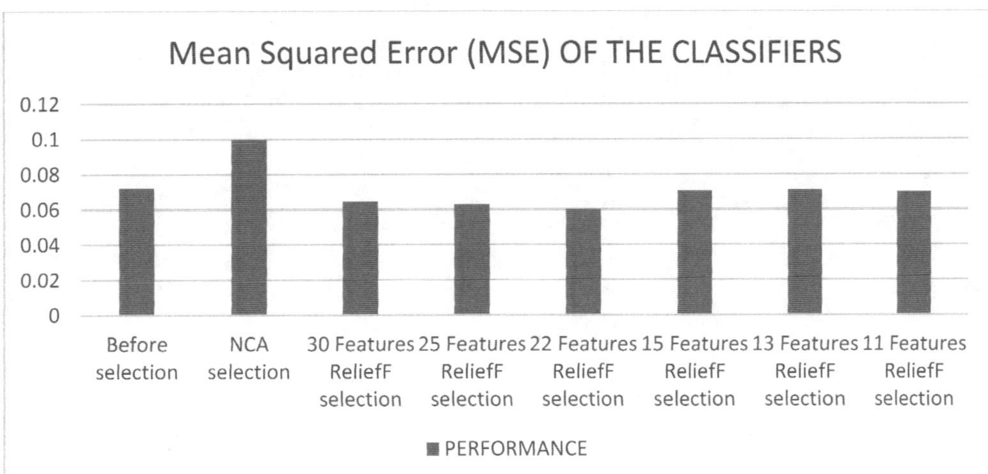

Figure 3. The performance of the classifiers

Table 1 Performance od classifiers

Classifier	Accuracy (%)	Sensitivity (%)	Specificity (%)
MLP trained by SCG	93.8	93.4	94
k-NN (k =5)	85.0	82.1	87.9
SVM (polynomial kernel function)	91.1	91.2	90.9

4. Conclusion

The proposed image segmentation procedure has successfully segmented TB and non-TB bacilli from the background sputum. After the image processing was performed, 528 objects have been detected in the images that containing TB bacilli and non-TB bacilli. There were 38 features that have been extracted in this study, NCA and ReliefF algortihms were then used to select the most suitable features. The best accuracy was produced by ReliefF with 22 selected features. The selected features are 8 colour features. MLP, k-NN and SVM were used for classification. All the classifier yield good classification performance with the accuracy of higher than 80%. MLP trained by yielded the highest accuracy, sensitivity and specificity with 93.8%, 93.4% and 94.1% respectively.

Acknowledgements

The authors would like to express the highest gratitude to Universiti Malaysia Perlis and Prof. Dr. Mohd. Yusoff Mashor as my final year project's supervisor which he had giving me completely support, guidance and motivation through all the procedure of this study. I am genuinely thankful to have support and consolation from my family and friends who teach and share their knowledge with me. Lastly, I might want to express my earnest thanks toward other people who had helped me in the advancement of the project.

References

[1] Global tuberculosis report 2018. Geneva: World Health Organization; 2018. Licence: CC BY-NC-SA 3.0 IGO.
[2] Organización Mundial de la Salud, *Laboratory diagnosis of tuberculosis by sputum microscopy*. 2013.
[3] J. C. S. Ngabonziza *et al.*, "Diagnostic performance of smear microscopy and incremental yield of Xpert in detection of pulmonary tuberculosis in Rwanda," *BMC Infect. Dis.*, vol. 16, no. 1, pp. 1–7, 2016.
[4] R. Dinesh Jackson Samuel and B. Rajesh Kanna, "Tuberculosis (TB) detection system using deep neural networks," *Neural Comput. Appl.*, vol. 4, pp. 1–13, 2018.
[5] R. O. Panicker, K. S. Kalmady, J. Rajan, and M. K. Sabu, "Automatic detection of tuberculosis bacilli from microscopic sputum smear images using deep learning methods," *Biocybern. Biomed. Eng.*, vol. 38, no. 3, pp. 691–699, 2018.

Journal of Physics: Conference Series **1372** (2019) 012080 doi:10.1088/1742-6596/1372/1/012080

Image Enhancement Using Modified Partial Contrast Technique in Ziehl-Neelsen Sputum Slide Images

R. A. A. Raof[1], M. Y. Mashor[1], S. S. M. Noor[2]

[1]Universiti Malaysia Perlis, Pauh Putra, Perlis, Malaysia
[2]Universiti Sains Malaysia, Kubang Kerian, Kelantan, Malaysia
Corresponding Author E-mail: rafikha@unimap.edu.my

Abstract. Image contrast and brightness for sputum slide images are among important factors that can affect the accuracy of image analysis process. Microscopic images usually have the problem of low contrast and blurry, which are caused by a couple of factors such as uneven thickness of the slides and manual preparation of the slide in the first place. In this research, image enhancement procedure has been considered for either increasing the image quality, increasing the visibility of tubercle bacilli or to ease the segmentation process. Partial contrast technique is considered for the contrast enhancement method. The results show that the modified technique helps to improve tubercle bacilli visibility. Hence the resulted images would become useful for the microbiologists and technologists for further analysis of Ziehl-Neelsen (ZN) sputum slide.

1. Introduction

Several computer-aided image diagnosis methods for tuberculosis (TB) detection have been described in the literature [1 – 5]. These methods can be divided into two approaches: using fluorescence and light microscope. Early works focused on using the fluorescence microscope [1 - 3]. Images viewed using a fluorescence microscope is more sensitive to TB bacilli and the screening process can be conducted quickly under lower magnification, compared to the light microscope [4]. However, it is expensive and difficult to maintain, thus limiting the use of fluorescence microscope in low and medium income countries. Therefore, recent works in detecting pulmonary tuberculosis (PTB) used images acquired from both fluorescence and light microscope [5 - 7]. The specimens are stained using Ziehl-Neelsen (ZN) staining procedure to visualize the bacilli. The automated technique can save the time and cost involved, with reduced human error.

Image contrast and brightness for sputum slide images are among important factors that can affect the accuracy of image analysis process. Microscopic images usually have the problem of low contrast and blurry, which are caused by a couple of factors which have been mentioned earlier. In this paper, image enhancement procedure has been considered for either increasing the image quality, increasing the visibility of TB bacilli or to ease the segmentation process. Modified approaches to perform image enhancement using partial contrast algorithm is presented. The performances of the method are being presented through the resulted image of Ziehl-Neelsen slide after the image enhancement process has been done.

2. Methodology

The aim of partial contrast is to generate an image with higher contrast than the original image. Stretching method is carried out by modifying each pixel value to a new value using pre-specified function. Linear stretching will generally improve the overall contrast of an image. Contrast stretching is a process that applies auto-scaling method, which is a linear mapping function. It is usually used to enhance the brightness as well contrast level of the image. The general mapping function is shown in Equation 1 [8].

$$p_k = \frac{(max - min)}{(f_{max} - f_{min})}(q_k - f_{min}) + min \tag{1}$$

Referring to Equation 1, f_{max} and f_{min} are the maximum and minimum colour level in an input image. max and min are the desired maximum and minimum colour level in the output image. q_k is the colour level of the input pixel while p_k is the colour level of the output pixel. The combination of stretching and compressing process is called partial contrast. A part of the intensity level is being stretched to a new range, while other intensity levels left is being compressed to a different new range as well. The stretching and compressing processes are illustrated by Fig. 1.

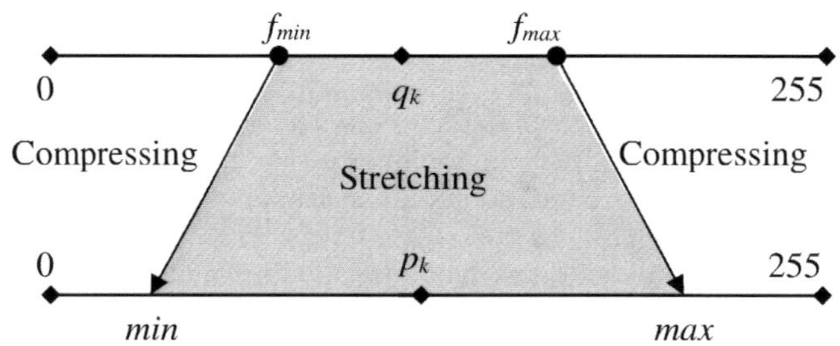

Figure 1: Partial contrast process

The process illustrated by Fig. 1 can be put into a mathematical function such as in Equation 2.

$$p_k = \begin{cases} \dfrac{min}{f_{min}}(q_k) & ; \text{for } q_k < f_{min} \\ \dfrac{(max - min)}{(f_{max} - f_{min})}(q_k - f_{min}) + min & ; \text{for } f_{min} \le q_k \le f_{max} \\ \dfrac{(255 - max)}{(255 - f_{max})}(q_k - f_{max}) + max & ; \text{for } q_k \ge f_{max} \end{cases} \tag{2}$$

By applying this technique, the value of pixels within the range of f_{min} and f_{max} will be stretched to a wider new range of min and max. On the other hand, the pixels with values that are outside of this range will be compressed.

In the original algorithm, the technique is meant to be applied to monochrome image that only have one intensity level, which is its greyscale intensity ranging from 0 to 255. However, since the sputum slide images are colour images, modifications need to be made to adopt the original algorithm for implementation on colour images. Colour imaging may be considered as a layered intensity image processing scheme. This is based on the assumption that each of the colour layers can be processed separately and finally the results are combined. In this study, the process of stretching and compressing for partial contrast technique is implemented as illustrated in Fig. 1.

Each pixel in colour images are represented by three components; red (R), green (G) and blue (B). Therefore, in order to accommodate the algorithm of partial contrast, each parameter in Equation 2 will need to consider for the values from these RGB components. Therefore, the technique is done separately for all three components.

Some modifications have been made to the original partial contrast algorithm to be applied to the sputum slide images. The procedure for implementing partial contrast technique in this study is mentioned as follows:

i) Read the intensity value of RGB component for each pixel of the original image.
ii) Develop the RGB histograms for original image using the intensity values obtained.
iii) Find the intensity value of $f_{min(R)}$, $f_{min(G)}$, $f_{min(B)}$, $f_{max(R)}$, $f_{max(G)}$, $f_{max(B)}$, $min(R)$, $min(G)$, $min(B)$, $max(R)$, $max(G)$ and $max(B)$ for each parameter.
iv) Perform partial contrast algorithm for each of the RGB component based on Equation 2.
v) Construct the output image using the combination of RGB components obtained in step (iv).

$f_{min(R)}$, $f_{min(G)}$, $f_{min(B)}$, $f_{max(R)}$, $f_{max(G)}$ and $f_{max(B)}$ are the original images' minimum and maximum values for R, G and B component, respectively. These values are defined by percentage and the real intensity value for each of the parameters will be calculated from the RGB histograms. Meanwhile values of $min(R)$, $min(G)$, $min(B)$, $max(R)$, $max(G)$ and $max(B)$ are the minimum and maximum values for desired intensity level of the output images for R, G and B component respectively.

The real intensity values for parameters $f_{min(R)}$, $f_{min(G)}$, $f_{min(B)}$, $f_{max(R)}$, $f_{max(G)}$ and $f_{max(B)}$ defined in step (iii) of previous procedure are calculated using the steps as follows:

i) Choose a specific percentage value for each of the specified parameter.
ii) Initialize the current intensity value to 0.
iii) Find out the total number of pixels (from the whole image) for each of the specified parameter which is having the current intensity value.
iv) Sum up the number of pixels obtained in (iii) with the previous summation value (summation value starts with 0).
v) Calculate the current percentage of the summation value by using Equation 3.

$$\text{current percentage} = \left(\frac{\text{summation}}{\text{totalnumber of pixel}}\right) * 100 \qquad (3)$$

vi) If the result of current percentage is less than the percentage value chosen in (i), increase the current intensity value by 1 and go back to step (iii).
vii) If the result of current percentage is equal or more than the percentage value chosen in (i), then take the current intensity value as the real intensity value for the specified parameter.

The reason of using percentage value for the parameters is because each of the images has different minimum and maximum intensity level. Therefore, by using percentage value it can ensure that the values to be stretched or compressed are within the range that is suitable to that particular image. On the other hand, if the values to be stretched are fixed, for example 30 for minimum (f_{min}) and 200 for maximum (f_{max}) values, the images that are having a minimum intensity of 60 (min) and maximum intensity of 150 (max) would not experience a complete partial contrast process on its pixels.

In this research, the percentage value for step (iii) that has been chosen for parameters $f_{min(R)}$, $f_{min(G)}$ and $f_{min(B)}$ is 10% while percentage value for $f_{max(R)}$, $f_{max(G)}$ and $f_{max(B)}$ is 90%. These values are chosen because in the RGB histograms of ZN sputum slide images, the values in between these percentage often contains relevant information of TB bacilli, which is the object of interest. Meanwhile the darkest 10% and the brightest 10% of the histograms are usually pixels representing unnecessary objects in the image.

On the other hand, the real intensity value that has been chosen for parameter $min(R)$, $min(G)$ and $min(B)$ is 10 while intensity value for $max(R)$, $max(G)$ and $max(B)$ is 245. After partial contrast, the intensity level for RGB components will be spread out to values between 10 and 245. Fig. 2 illustrates the process of compressing and stretching the pixels to new intensity value.

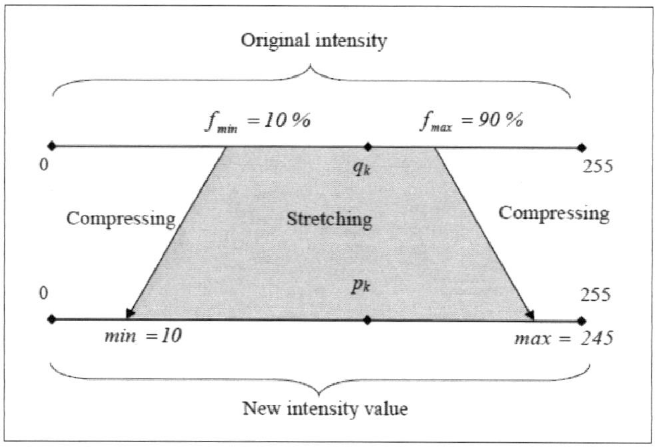

Figure 2: Partial Contrast Technique applied to sputum slide image

The usage of partial contrast technique helps to increase the contrast for images with low contrast. Images can appear too bright if the sputum smear on the specimen slide is too thin at that particular field that is being captured. On the other hand, images can appear darker if the sputum smear on the specimen slide is too thick.

3. Results

Fig. 3 shows the resulted image for partial contrast process together with their histograms.

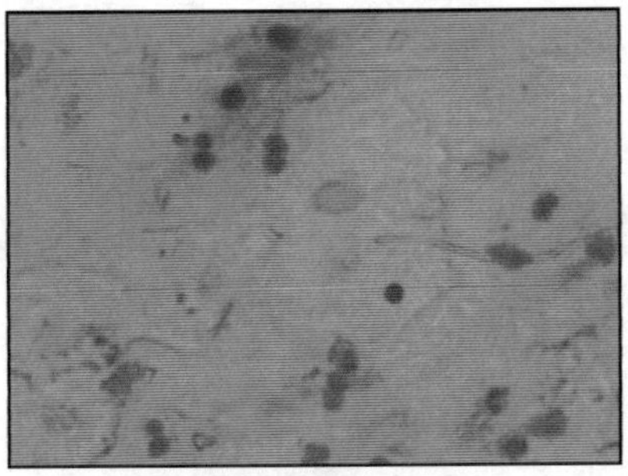

(a) Original image

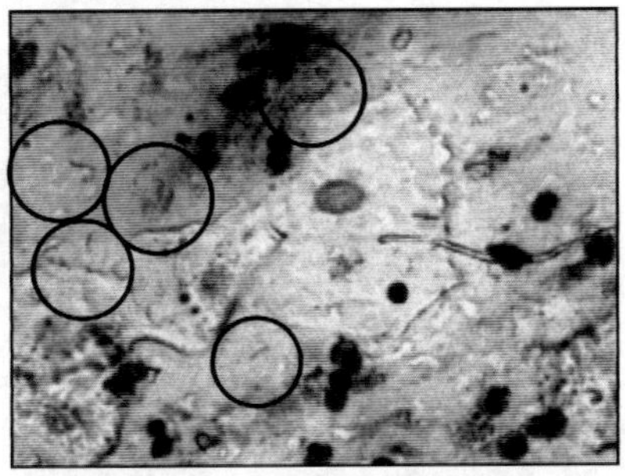

(b) Partial contrast image

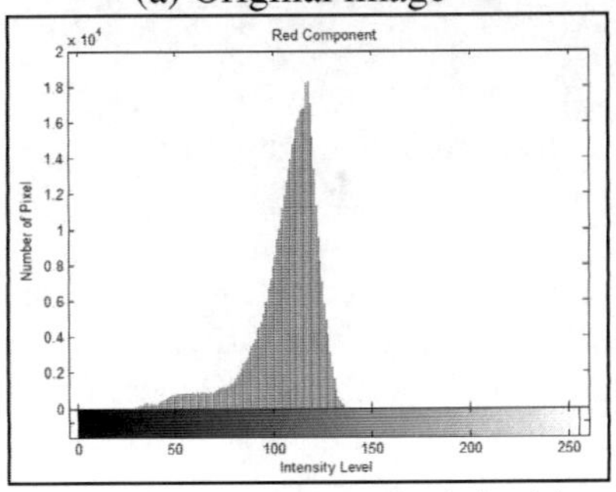

(c) Red component of original image

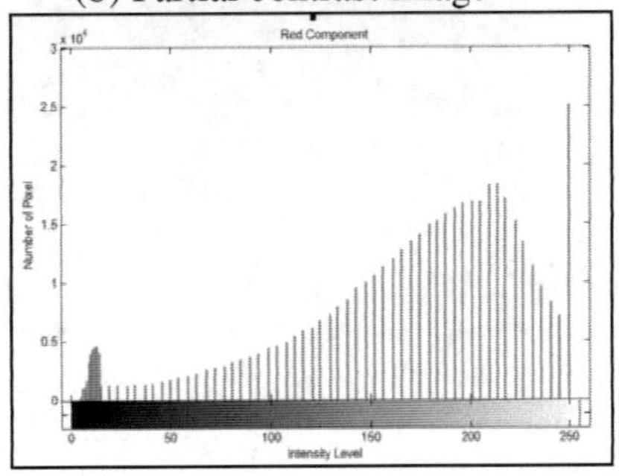

(d) Red component of output image

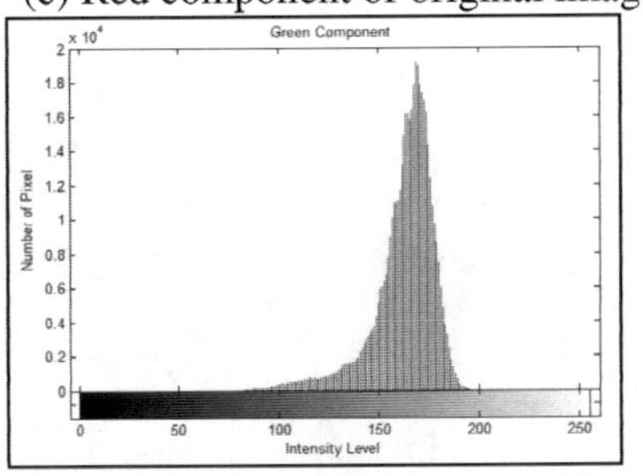

(e) Green component of original image

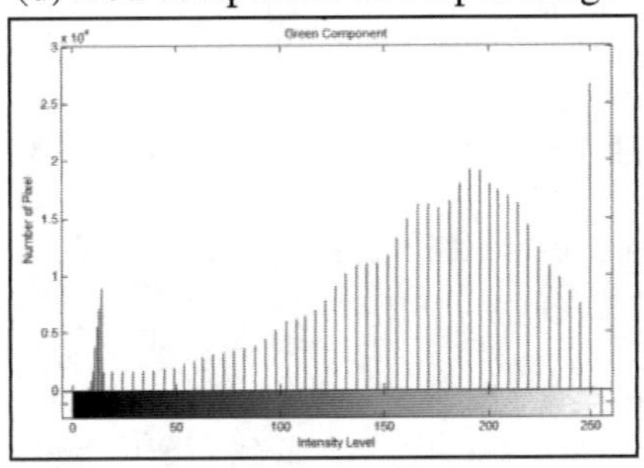

(f) Green component of output image

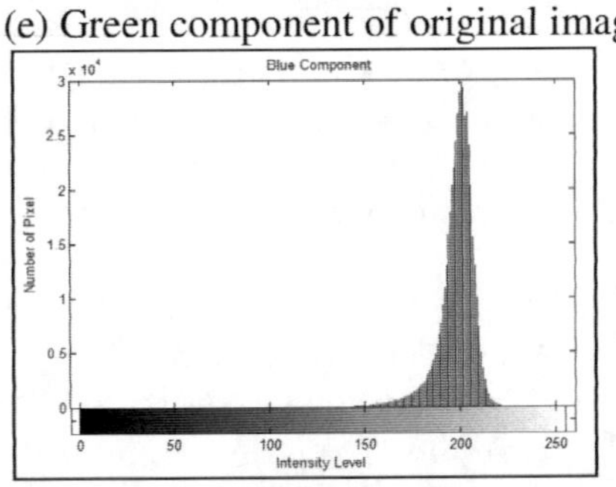

(g) Blue component of original image

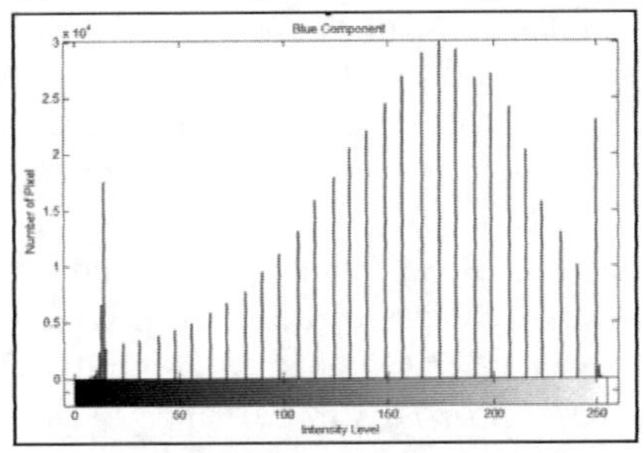

(h) Blue component of output image

Figure 3: Original Low Contrast Image and Its Partial Contrast Image

Partial contrast has successfully enhanced the contrast of the original image. TB bacilli can be seen clearly in the resulted images compared to the original image. The parameters' value for the particular image shown in Fig. 3(a) are $f_{min(R)} = 85$, $f_{max(R)} = 123$, $f_{min(G)} = 143$, $f_{max(G)} = 178$, $f_{min(B)} = 187$ and $f_{max(B)} = 207$. These are real values that are previously obtained from the percentage values chosen for each parameter. The values have spread out the intensity level of the image and hence significantly increase the contrast of the image especially the pixels that represents the TB bacilli. The result of partial contrast performed on other low contrast image is shown in Fig. 4.

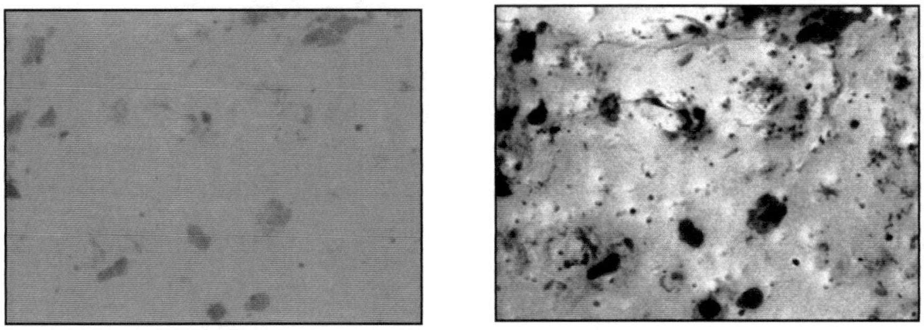

(a) Original image (b) Partial contrast image
Figure 4: Result of Partial Contrast on Other Image

The resulted pixels from the procedure for each RGB components are combined after the partial contrast process. It can be seen that the contrast in dark pixels increases and thus provides better visual experience. It should be noted that because of this, the colour ratio will change. Hence, the resulted image appearance also will look artificial. However, the object of interest in the image which is the TB bacilli will appear clearer than the original image. This will help the microbiologists and technologists to identify the TB bacilli easily.

4. Conclusion

The modified partial contrast algorithm is proven to be able to increase the contrast range for images with average brightness while lighting up darker images. It also helps to improve TB bacilli visibility. Hence the resulted images would become useful for the microbiologists and technologists for further analysis of ZN sputum slide.

References

[1] K. Veropoulos, C. Campbell, and G. Learmonth (1998), 'Image Processing and Neural Computing Used in the Diagnosis of Tuberculosis'. Proc. IEE Colloquium on Intelligent Methods in Healthcare and Medical Applications (Digest No. 1998/514), , York, UK.

[2] K. Veropoulos, C. Campbell, Learmonth, G., B. Knight, and J.Simpson (1998), 'The Automated Identification of Tubercle Bacilli using Image Processing and Neural Computing Techniques'. Proc. 8th International Conference on Artificial Neural Networks, Skövde, Sweden, pp. 797-802.

[3] Khutlang, R., Krishnan, S., Dendere, R., Whitelaw, A., Veropoulos, K., & Learmonth, G. (2010). Classification of mycobacterium tuberculosis in images of ZN-stained sputum smears. IEEE Transactions on Information Technology in Biomedicine, 14(4), 949-957.

[4] Santiago-Mozos, R., Fernando, P., Michael, G. M., & Antonio, A. (2014). An Automated Screening System for Tuberculosis. IEEE Journal of Biomedical and Health Informatics, 18(3), 855-862.

[5] Cicero F. F. C. F., & Marly, G. F. C. (2012). Sputum Smear Microscopy for Tuberculosis: Evaluation of Autofocus Functions and Automatic Identification of Tuberculosis Mycobacterium. Understanding Tuberculosis - Global Experiences and Innovative Approaches to the Diagnosis, 277-292.

[6] R. Singhal and V. P. Myneedu (2015). Microscopy as a diagnostic tool in pulmonary tuberculosis. International Journal of Mycobacteriology, vol. 4, no. 1, pp. 1–6.

[7] Chang, E. W., Page, A. L. & Bonnet M. (2016). Light-emitting diode fluorescence microscopy for tuberculosis diagnosis: a metaanalysis. Eur Respir J 47, 929–937.

[8] Weeks, J. R. (1996). Fundamentals of Electronic Image Processing. Bellingham: SPIE Press.

Antibacterial Ability of Mesoporous Carbonated Hydroxyapatite

Nur Farahiyah Mohammad[1,*], Farah Syazana Ahmad Fadzli[1], Siti Shuhadah Md Saleh[2], Che Wan Sharifah Robiah Mohamad[1], and Mustafa Ali Azhar Taib[3]

[1]Biomedical Engineering Research Group (BERG), School of Mechatronic Engineering, Pauh Putra Campus, Universiti Malaysia Perlis, 02600 Arau, Perlis, Malaysia

[2]School of Materials Engineering, Universiti Malaysia Perlis, 02600 Arau, Perlis, Malaysia

[3]Advanced Technology Training Centre (ADTEC) Taiping, PT 15643, Kamunting Raya, Mukim Asam Kumbang, 34600 Kamunting, Perak, Malaysia

* farahiyah@unimap.edu.my

Abstract. Mesoporous carbonated hydroxyapatite (Meso-CHA) is a bioceramic materials that offer good biocompatibility and bioactivity properties that suitable for bone defects or diseases treatment and therapy. Antibacterial study is very important to ensure that materials is biocompatible enough to be used on or inside the human body. The aim of this study is to investigate the antibacterial ability of the Meso-CHA (without addition of antibacterial agents) towards Escherichia coli (*E. coli*) bacteria. Meso-CHA samples were synthesized using chemical precipitation method where surfactant P-123 was used to introduce pores within nanoparticles structure. X-ray Diffraction (XRD) and Fourier Transform Infrared (FTIR) analysis proved that pure phase of Meso-CHA was obtained. Synthesized Meso-CHA demonstrated higher pore properties (surface area = 146.92 m^2g^{-1}, pore size = 2.35 nm, and pore volume = 0.2437 cm^3g^{-1}) compared to commercial HA (surface area = 7.84 m^2 g^{-1}, pore size = 2.05 nm, and pore volume = 0.0085 cm^3 g^{-1}). The antibacterial study demonstrated that Meso-CHA has low antibacterial properties with Minimum Inhibitory Concentration (MIC) of 200 $mg.ml^{-1}$ compared to HA. Carbonate ion addition into the Meso-CHA structure does not improve the antibacterial ability of the materials.

1. Introduction

Bioceramic materials are one of the biomaterials that are widely used in various type of biomedical applications. Some of the examples are calcium phosphates, alumina, zirconia, and bioactive glasses [1]. Hydroxyapatite ($Ca_{10}(PO_4)_6(OH)_2$, HA) is one of the most studied materials from the calcium phosphates family for application in biomedical field because of its good biocompatibility property. Moreover, HA possesses chemical similarity to bones and hard tissues components where it can form bonds with surrounding tissues which is suitable for bone implantation, bone coating, dental application, and drug delivery application [2-6].

Previously, several studies had demonstrated that porous bioceramics like porous HA has better potential in bone implants application rather than using dense HA. Materials with porous structure allows better bone ingrowth due to its bioactive property [2] and porous structure. Porous materials also offer better efficiency in drug delivery application with suitable pore characteristics [7, 8]. Porous material is an amorphous solids or crystalline that consists of empty pores, cavities, or channels within the particles where according to the International Union of Physical and Applied Chemistry (IUPAC), pore sizes can be divided into three categories which are micropore (<2 nm), mesopore (2-5 nm), and macropore (>50 nm) [9]. Materials with mesopore structure is said to has a good potential for drug delivery application where it offers smaller pore size, larger surface area, and higher porosity compared to macroporous material which has lower porosity and exhibit burst release profile [10].

The main target of the mesoporous bioceramic is to be used in the bone defects treatment and therapy. Thus, it is crucial to choose the materials that mimic closely with the bone tissue mineral component. The major inorganic composite of our natural bones and teeth is a biological apatite, the impure and non-stoichiometric version of HA [11] that contains several other ions such as CO_3^{2-}, Mg^{2+}, Na^+, Fe^{2+}, HPO_4^{2-}, F^- and Cl^-. Therefore, carbonated hydroxyapatite that is nonstoichiometric, nanosized with low crystallinity and contain carbonate ion is a good alternative for any application, treatment and therapy related to bones.

Biocompatibility is the important aspect for biomaterials; this includes the cytotoxicity and antibacterial properties of the materials. It is important that the materials did not facilitate any bacterial infection and could also act as an antibacterial agent to prevent bacterial colonization on prostheses and implants [12]. Thus, this study aims to investigate the antibacterial properties of the in house synthesised mesoporous carbonated hydroxyapatite using Escherichia coli (*E. coli*) bacteria.

2. Methodology

2.1. Materials

Calcium nitrate tetrahydrate ($Ca(NO_3)_2.4H_2O$) and diammonium hydrogen phosphate ($(NH_4)_2.HPO_4$) were used as the calcium and phosphate precursor respectively. Carbonate ion was introduced into the structure with the addition of ammonium hydrogen carbonate (NH_4HCO_3). Non-ionic triblock co-polymer Pluronics® P123, ($PEO_{19}PPO_{69}PEO_{19}$) donated by BASF, USA was used as the surfactant for pores template. Sodium hydroxide (NaOH) was used to set pH of the mixture at pH 11 throughout the mixing process. Commercial HA, (HAP-200) obtained from Taihei, Japan was used as a control sample.

2.2. Preparation of Mesoporous Carbonated Hydroxyapatite (Meso-CHA)

Meso-CHA was synthesized using precipitation method. The Ca/P ratio was set to be 1.67. Surfactant-calcium ion mixed solution was prepared by dissolving 3 g of P123 in 100 ml of distilled water followed by the addition of 9.45 g (0.04 mol) of $Ca(NO_3)_2.4H_2O$. The solution was then stirred for 30 minutes until all solids were dissolved. Then, 3.17 g (0.024 mol) of $(NH_4)_2.HPO_4$ was dissolved in 60 ml of distilled water followed by the addition of 3.795 g of ammonium hydrogen carbonate (NH_4HCO_3); the mixture was called as phosphate-carbonate mixed solution. Subsequently, phosphate-carbonate mixed solution was dripped slowly into the surfactant-calcium mixed solution under continuous stirring, yielded a solution contain white precipitate. The alkanity of the white solution was maintained at pH 11 by NaOH drops. The white solution then was aged in room temperature for 24 hours. Then, the white solution was centrifuged at 3000 rpm for 20 minutes to obtain the white precipitate. The precipitate washed and centrifuged repeatedly for five times with distilled water. Next, the white precipitate was dried in an oven at 100°C for 24 hours. The dried white precipitate was then ground into fine powder using a mortar and pestle and further calcined in a furnace at 550°C for 6 hours which hence completes the synthesis process of mesoporous carbonated hydroxyapatite powder.

2.3. Characterization

The phase and crystallographic structures of synthesized sample were identified by Rikagu Ultima IV, X-ray Diffraction (XRD) diffractometer using CuKα radiation ($\lambda = 0.15406$) over the range of $10° \leq 2\theta \leq 90$. The Fourier Transform Infrared (FTIR) analysis was performed using PIKE Technologies spectrophotometer in the range of wave numbers of 4000-400 cm^{-1} for the identification of functional group that represent carbonated hydroxyapatite. Morphology of the Meso-CHA nanoparticles were observed using JSM-6490LV, JEOL Scanning Electron Microscope (SEM). Nitrogen adsorption-desorption analysis was conducted using Micromeritrics TriStar to evaluate the surface area and pore characteristics of the samples. Specific surface area of the samples was calculated using Brunauer-Emmett-Teller (BET) while pore size distribution (PSD) of the samples was measured from the desorption data of the isotherms using Barrett-Joyner-Halenda (BJH) model.

2.4. Antibacterial Activity of Mesoporous Carbonated Hydroxyapatite

The antibacterial activity of mesoporous carbonated hydroxyapatite was conducted using the broth dilution method. Fresh overnight grown Escherichia coli (*E. coli*) bacteria was prepared and were transferred aseptically into distilled water before adjusting its turbidity to 0.5 McFarland consisting 1-2 $\times 10^8$ CFU/ml bacteria. Next, Meso-CHA powder were extracted in Tryptic Soy Broth (TSB) before continuing with two-fold serial dilution consisting of 10 different concentrations of sample solutions where TSB were also used as the broth medium for the test. After sample dilution, bacterial inoculum was diluted into TSB until it gives the concentration of 5×10^5 CFU/ml bacteria before adding into the diluted sample solutions. Then, the solutions were incubated for 18 hours at 35°C. Minimum inhibitory concentration (MIC) of the solutions were measured using Genesys 20, UV-VIS spectrophotometer.

3. Results and discussions

3.1. Characterization of Meso-CHA Nanoparticles

The XRD pattern of all sample in Figure 1 are attributed to CHA reflections and no characteristic peaks of impurities are observed which are well consistent with PDF Number 01-089-7834 with lattice constant of a = b = 9.3892 Å and c = 6.9019 Å with space group of hexagonal $P6_3/m$. The three main characteristic peaks of CHA at $2\theta = 25.8°$, 31.8°, and 33.0° that corresponding to diffraction plane of (002), (211), and (300) are clearly observed in the XRD pattern of both samples. The XRD peaks of Meso-CHA is broader compared to HA that is narrower and sharper especially at diffraction peak of plane (211), (112), (300), and (202). This result indicate that Meso-CHA is less crystalline compared to HA sample. Material with lower crystallinity will degrade faster than high crystalline materials. This characteristic is appropriate for Meso-CHA since it is targeted to be applied as bone scaffold or drug delivery carrier which need to be degrade at certain period of time.

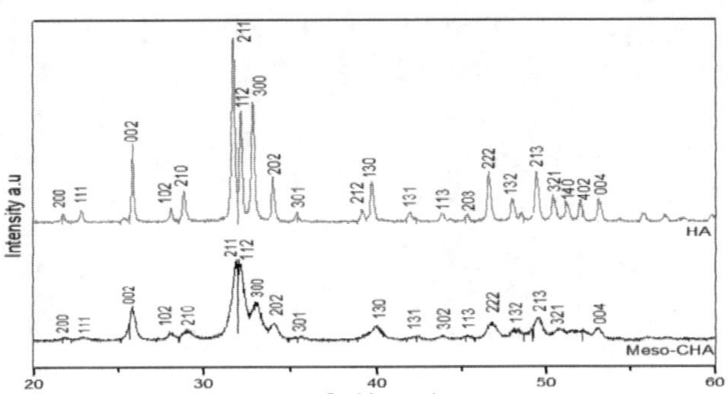

Figure 1. XRD patterns of Meso-CHA and commercial HA

Figure 2 shows the FTIR spectrums of Meso-CHA and HA. The absorption peaks at 1041.74 cm^{-1} and 1044.71 cm^{-1} attributed to the stretching vibration (v_3) of the phosphate (PO_4^{3-}) groups for Meso-CHA and HA, respectively. While, the absorption peaks at 572.23 cm^{-1} and 573.90 cm^{-1} are corresponding to bending vibration (v_4) of the phosphate (PO_4^{3-}) groups Meso-CHA and HA, respectively. The characteristic bands of B-type CO_3^{-2} substitution are observed at 880.6 cm^{-1} and 1474.66 cm^{-1} for Meso-CHA while commercial HA at 882.2 cm^{-1} and 1449.10 cm^{-} [13]. Even though CO_3^{-2} ions are present in both samples, Meso-CHA has more intense carbonate bend compare to HA. This result is due to the addition of carbonate ion during the synthesis process. The broad bend at 3773.38 cm^{-1} (Meso-CHA) and 3465.18 cm^{-1} (HA) corresponding to OH group [14, 15].

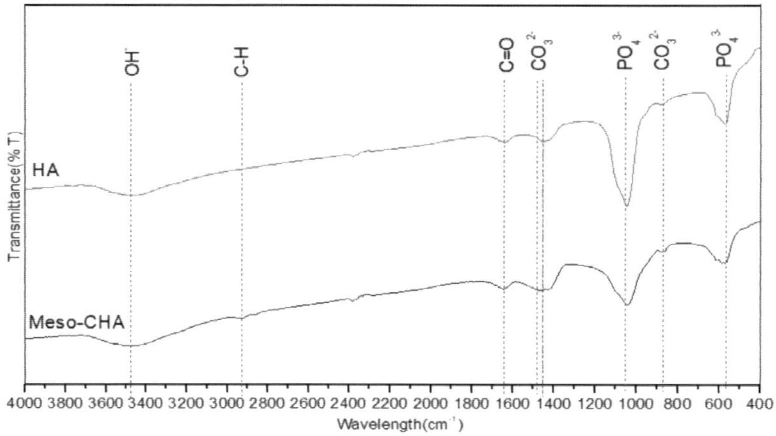

Figure 2. FTIR spectra of meso-CHA and commercial HA

3.2. *Morphology of Meso-CHA Nanoparticles*

Figure 3 (a) and (b) shows the SEM images of Meso-CHA and HA samples at 20,000× magnification. SEM images revealed that Meso-CHA consists of fine spherical nanoparticle and highly agglomerated (Figure 3(a)). Unfortunately, due to highly agglomerated nanoparticles, the spherical-shaped of Meso-CHA was not able to be seen clearly in the SEM image. Similarly, HA nanoparticles are highly agglomerated, however HA has rod-like structure that look like rice grains (Figure 3(b)). The agglomeration in both sample is mostly caused by the Van der Waals force between the nanoparticles [28].

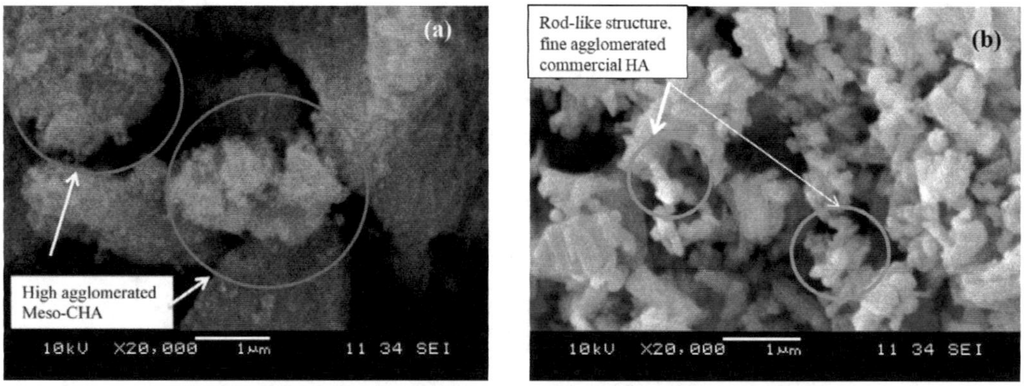

Figure 3. SEM images of (a) Meso-CHA and (b) HA

3.3. Pore Characterisation

Figure 4 (a) shows the nitrogen adsorption-desorption isotherms of the samples. Meso-CHA exhibit Type IV curve with H1 hysteresis loop after the calcination, and a well-defined step at approximately P/P0 0.89 to 0.99, implying a small pore size distribution which is often observed at agglomerates or compacts of approximately uniform spheres in a fairly regular array [16]. Meso-CHA surface area (S_{BET}) is 146.92 m^2g^{-1}. Contrarywise, CHA sample showed a Type II character indicates that non-porous particles were produced after the calcination with a low specific surface areas (S_{BET}), of 7.84 m^2g^{-1}.

Table 1 shows that surface areas (S_{BET}), pore size and pore volume of Meso-CHA is higher than HA. Pore size distributions of the samples are shown in Figure 4 (b). The pore size of Meso-CHA is mainly distributed at 2.32 nm which is falls within the mesopore range (2-50 nm). Meanwhile, HA exhibited pore size of 2.05 nm with a very low pore volume which is 0.0085 cm^3g^{-1}; that most probably due to inter-particulate pores (pores between the particles). This result strongly supported the isotherm result earlier that commercial HA is consisting of non-porous nanoparticles, while Meso-CHA consists of mesoporous nanoparticles.

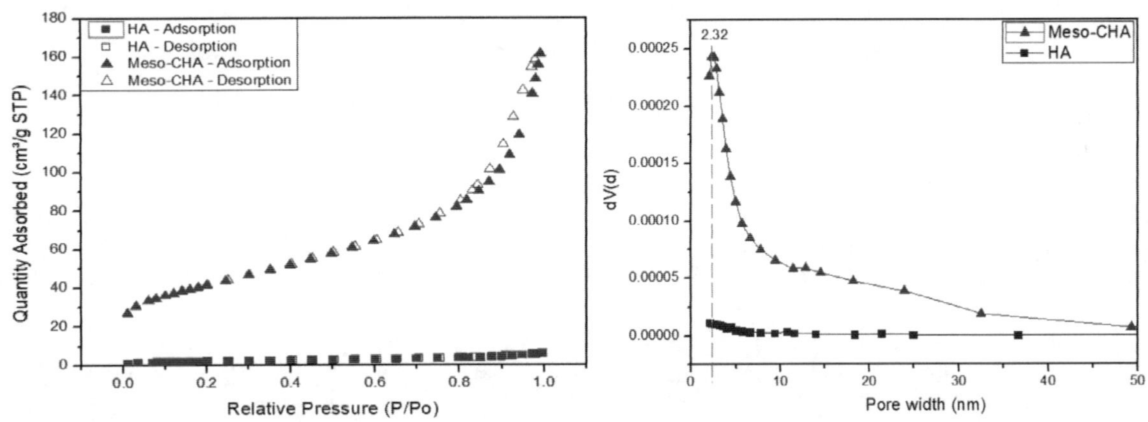

Figure 4. (a) N_2 adsorption-desorption isotherms and (b) BJH pore size distribution of Meso-CHA and HA

Table 1. Surface areas (S_{BET}), pore size and pore volume of Meso-CHA and HA

Sample	BET surface area (m^2g^{-1})	BJH pore size (nm)	BJH pore volume (cm^3g^{-1})
Meso-CHA	146.92	2.32	0.2437
HA	7.84	2.05	0.0085

3.4. Antibacterial Study

Figure 5 shows the antibacterial ability of Meso-CHA and HA. Based on the graph, Meso-CHA shows higher absorbance percentage value than the control (*E.coli* in broth) at most solution concentrations compared to HA. However, the highest Meso-CHA concentration which is 200 $mg.ml^{-1}$, the absorbance percentage of Meso-CHA sample solution was at its lowest which is at 77% with slight difference to HA which is 76%. Then, at sample concentration of 100 $mg.ml^{-1}$, the percentage of Meso-CHA sample solution increases abruptly to 101% compared to commercial HA which is 80%.

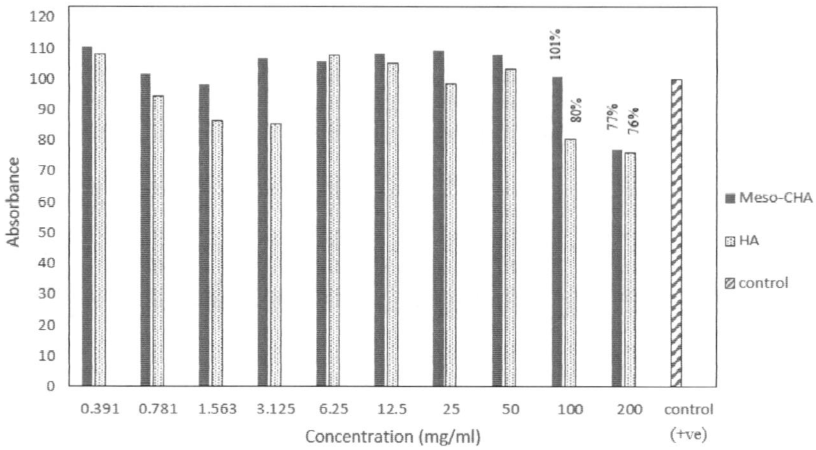

Figure 5. Result of antibacterial ability of meso-CHA and commercial HA

A higher absorbance percentage value indicates that the solutions have higher bacteria growth which will also cause the solution to be cloudy. So, the antibacterial ability of the sample will be low due to a larger number or bacteria grown in the solution and the Meso-CHA has low ability to inhibit the bacteria growth in the solution. On the other hand, a clearer solution will produce lower absorbance percentage value indicate that a smaller number of bacteria is present and it shows that the sample is able to inhibit bacteria growth.

In this study, the antibacterial ability of Meso-CHA was found to be low compared to commercial HA. This may be due antibacterial agent such as silver or copper ion was not introduced into the Meso-CHA structure despite of knowing that HA is a weak antibacterial material [17]. The main aim of this study is to investigate the antibacterial ability of Meso-CHA without the additional of metal ions to see whether it has a better antibacterial ability compared to commercial HA or not. Therefore, it can be concluded that Meso-CHA apparently has low antibacterial ability than commercial HA with the minimum inhibitory concentration (MIC) of 200 $mg.ml^{-1}$.

4. Conclusion

In conclusion, Meso-CHA with high surface area (146.92 m^2g^{-1}) was successfully synthesized through precipitation method using P-123 surfactant as pore template. The pore size of Meso-CHA is 2.32 nm which is within the range of mesopore sizes. Meso-CHA exhibit low antibacterial ability compared to HA. For future work, antibacterial agents such as silver or copper ion should be doped into the Meso-CHA to enhance antibacterial properties of the material.

Acknowledgement

Authors would like to acknowledge School of Mechatronic Engineering, Universiti Malaysia Perlis (UniMAP), School of Material Engineering, UniMAP and School of Bioprocess Engineering, UniMAP and Advanced Technology Training Centre (ADTEC) Taiping for providing the equipment and facilities to conduct this study.

References

[1] Hench L L and Wilson J 1993 *An Introduction to Bioceramics* (Singapore: World Scientific)
[2] Mohammad N F, Yeoh F Y, Ling N P and Othman R 2012 Synthesis & Characterization of Mesoporous Hydroxyapatite by Co-Precipitation Method. In: *Asian International Conference on Materials, Minerals and Polymer,* ed A Z A Azhar, *et al.* (Penang: School of Materials and Mineral Resources Engineering, Universiti Sains Malaysia) pp 626-33
[3] Mohammad N F, Othman R and Yee-Yeoh F 2014 Nanoporous hydroxyapatite preparation methods for drug delivery applications *Reviews on Advanced Materials Science* **38** 138-47
[4] Shepherd J H, Shepherd D V and Best S M 2012 Substituted hydroxyapatites for bone repair *J. Mater. Sci.: Mater. Med.* **23** 2335-47
[5] Šupová M 2015 Substituted hydroxyapatites for biomedical applications: A review *Ceram. Int.* **41** 9203-31
[6] Iconaru S L, Chapon P, Le Coustumer P and Predoi D 2014 Antimicrobial activity of thin solid films of silver doped hydroxyapatite prepared by sol-gel method *ScientificWorldJournal* **2014** 165351-
[7] Guo Y-J, Wang Y-Y, Chen T, Wei Y-T, Chu L-F and Guo Y-P 2013 Hollow carbonated hydroxyapatite microspheres with mesoporous structure: Hydrothermal fabrication and drug delivery property *Mater. Sci. Eng., C* **33** 3166-72
[8] Cheah W K, Ooi C W, Othman R and Yeoh F Y 2012 Synthesis of Mesoporous Hydroxyapatite Through a Soft Template Route Using Non-ionic Surfactant *Asean Engineering Journal Part B* **1** 57-69
[9] Rouquerol J, Avnir D, Fairbridge C W, Everett D H, Haynes J H, Pernicone N, Ramsay J D F, Sing K S W and Unger K K 1994 Recommendations for the Characterization of Porous Solid *Pure Appl. Chem.* **66** 1739-58
[10] Mohammad N F, Othman R and Yeoh F Y 2015 Controlling the pore characteristics of mesoporous apatite materials: Hydroxyapatite and carbonate apatite *Ceram. Int.* **41** 10624-33
[11] Zhao Q, Wang T, Wang J, Zheng L, Jiang T, Cheng G and Wang S 2012 Fabrication of mesoporous hydroxycarbonate apatite for oral delivery of poorly water-soluble drug carvedilol *J. Non-Cryst. Solids* **358** 229-35
[12] Riaz M, Zia R, Ijaz A, Hussain T, Mohsin M and Malik A 2018 Synthesis of monophasic Ag doped hydroxyapatite and evaluation of antibacterial activity *Materials Science and Engineering C* **90** 308-13
[13] Guo Y-P, Yao Y-B, Guo Y-J and Ning C-Q 2012 Hydrothermal fabrication of mesoporous carbonated hydroxyapatite microspheres for a drug delivery system *Microporous Mesoporous Mater.* **155** 245-51
[14] Ślósarczyk A, Paszkiewicz Z and Paluszkiewicz C 2005 FTIR and XRD evaluation of carbonated hydroxyapatite powders synthesized by wet methods *Journal of Molecular Structure* 657-61
[15] Guo Y J, Long T, Chen W, Ning C Q, Zhu Z A and Guo Y P 2013 Bactericidal property and biocompatibility of gentamicin-loaded mesoporous carbonated hydroxyapatite microspheres *Materials science & engineering. C, Materials for biological applications* **33** 3583-91
[16] Sing K S W, Everett D H, Haul R A W, Moscou L, Pierotti R A, Rouquerol J and Siemieniewska T 1985 Reporting Physisorption Data For Gas/Solid Systems with Special Reference to the Determination of Surface Area and Porosity *Pure Appl. Chem.* **57** 603-19
[17] Yingguang L, Zhuoru Y and Jiang C 2007 Preparation, Characterization and Antibacterial Property of Cerium Substituted Hydroxyapatite Nanoparticles *Journal of Rare Earths* **25** 3-7

AUTHOR INDEX

AUTHOR INDEX

AUTHOR INDEX

Institute of Physics
Dirac House, Temple Back
Bristol BS1 6BE UK

ISSN: 1742-6588
ISBN 978-1-7138-1129-9

CURRAN ASSOCIATES INC.
proceedings.com